W9-CMP-673

WORLD *of* GENETICS

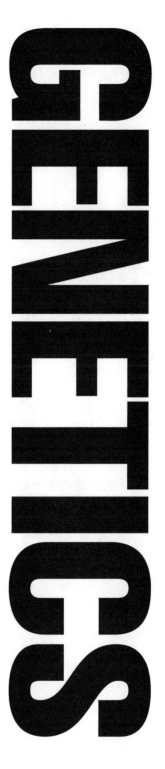

WORLD *of*

K. Lee Lerner and Brenda Wilmoth Lerner, *Editors*

Volume 2

M-Z
General Index

GALE GROUP

THOMSON LEARNING

Detroit • New York • San Diego • San Francisco
Boston • New Haven, Conn. • Waterville, Maine
London • Munich

K. Lee Lerner and
Brenda Wilmoth Lerner, *Editors*

GALE GROUP STAFF

Kimberley A. McGrath, *Senior Editor*

Maria Franklin, *Permissions Manager*

Margaret A. Chamberlain, *Permissions Specialist*
Shalice Shah-Caldwell, *Permissions Associate*

Mary Beth Trimper, *Manager, Composition and Electronic Prepress*
Evi Seoud, *Assistant Manager, Composition and Electronic Prepress*
Dorothy Maki, *Manufacturing Manager*
Wendy Blurton, *Buyer*

Michelle DiMercurio, *Senior Art Director*
Michael Logusz, *Art Director*

Barbara J. Yarrow, *Manager, Imaging and Multimedia Content*
Robyn V. Young, *Project Manager, Imaging and Multimedia Content*
Leitha Etheridge-Sims, Mary K. Grimes, and David G. Oblender, *Image Catalogers*
Pam A. Reed, *Imaging Coordinator*
Randy Bassett, *Imaging Supervisor*
Robert Duncan, *Senior Imaging Specialist*
Dan Newell, *Imaging Specialist*

ISBN 0-7876-4958-9 (set)
0-7876-5068-4 (vol. 1)
0-7876-5069-2 (vol. 2)

Printed in Canada
10 9 8 7 6 5 4 3 2 1

Contents

Introduction vii

How to Use This Book x

Entries

Volume 1: A-L 1

Volume 2: M-Z 441

Sources Consulted 747

Historical Chronology 763

General Index 785

INTRODUCTION

The *World of Genetics* is devoted to the study of the most intimate and fundamental mechanisms of life. Genetics is the study of heredity—of new and vigorous scientific explanations of how life arose and continues on Earth.

The *World of Genetics* is a collection of 800 entries on topics covering a range of interests—from biographies of the pioneers of genetics to explanations of the most current research. Despite the complexities of genetics, along with the fast pace of research and innovation, every effort has been made to set forth entries in everyday language and to provide generous explanations of the most important terms used by professional scientists.

World of Genetics articles are designed to instruct, challenge, and excite less-experienced students, while providing a solid foundation and reference for more advanced students. Genetics is a dynamic field. Almost daily, new discoveries extend our understanding of genetics and new technologies arise to expand applications of those discoveries. The pace of change and innovation is challenging to all publications regarding genetics. For example, during the writing of *World of Genetics*, researchers announced the completion of the sequencing of the human genome—and that the human genome contained far fewer genes that originally thought. Accordingly, although *World of Genetics* has attempted to incorporate the latest findings and applications, the emphasis is on providing students and readers with the basic information and insights into critical thinking that will enable a greater understanding of the news and current events related to genetics.

The "central dogma" of molecular biology (i.e., that the genetic information in DNA is transcribed into RNA, and that RNA conveys that genetic information to where it can be translated into protein structure) provides the common thread with which the tapestry of *World of Genetics* articles are woven. Whenever possible, we have attempted to explain the relationship of the central dogma to the specific topics included in *World of Genetics*.

The term genetics was coined at the beginning of the twentieth century to separate new forms of scientific inquiry from previous studies of generation, inheritance, or heredity. Classical genetics was originally included in a broad area of science known as generation, a term that once encompassed the study of reproduction, embryology, development and differentiation, regeneration of parts, and genetics.

The Greek philosopher Aristotle (384–322 B.C.) raised fundamental questions about reproduction, development, and heredity. According to Aristotle, the female parent contributed only unorganized matter to the new individual, while the male provided the form. Aristotle described two models of development known as preformation and epigenesis. Preformationist theories hold that a miniature individual preexists in either the egg or the semen and begins to grow into its adult form when properly stimulated. The theory of epigenesis, which Aristotle favored, asserts that the new organism is gradually produced from an undifferentiated mass by the addition of parts. Artistotle's theories formed the core of biological thought for hundreds of years. Indeed, ideas about generation changed very little change from the time of Aristotle to the publication of *On the Generation of Animals* (1651) by William Harvey (1578–1657), who was the first naturalist to argue that all living beings arose from eggs.

A new way of thinking about heredity, fertilization, and development was made possible by the establishment of cell theory in the 1830s by Matthias Jacob Schleiden (1804–1881) and Theodor Schwann (1810–1882), who realized the importance of Robert Brown's (1773–1858) discovery of the cell nucleus. August Weismann (1834–1914) proposed the theory of the continuity of the germplasm and predicted the reduction division of the chromosomes during the formation of the germ cells. In *Cell-Formation and Cell-Division* (1875) Eduard Strasburger (1844–1912) described the division of plant cells. Walter Flemming's (1843–1905) *Cell Substance, Nucleus, and Cell Division* (1882) established a basic framework for the stages of cell division. Flemming used the term chromatin for the nuclear substance and gave the name mitosis to cell division. Heinrich W. G. Waldeyer (1836–1921) introduced the term chromosome in 1888.

Many scientists were interested in theories of heredity in connection with Charles Darwin's (1809–1882) theory of evolution. But the work most closely associated with the development of modern genetics was the study of plant hybridization. Joseph Gottlieb Koelreuter (1733–1806) was one of the first botanists to systematically make and test hybrids. Koelreuter's work was extended by Carl Friedrich von Gaertner (1772–1850).

In the 1860s, Gregor Mendel (1822–1884) carried out a remarkable series of hybridization experiments and systematically analyzed the results of his tests. Although Mendel is generally regarded as the founder of modern genetics and the basic laws of genetics (segregation and independent assortment are known as Mendel's laws), his work was ignored for almost 40 years. Mendel's laws were rediscovered at the beginning of the nineteenth century by Hugo de Vries (1848–1935), Carl Correns (1864–1935), and Erik von Tschermak (1871–1962).

By 1910, some biologists thought that the behavior of the chromosomes during gamete formation and fertilization indicated that the hereditary factors could be on the chromosomes. The work of work of Walter S. Sutton (1877–1916), Theodor Boveri (1862–1915), Nettie M. Stevens (1861–1912), Edmund B. Wilson (1856–1939), and others suggested that the chromosomes were individuals in the morphological and the functional sense.

The chromosome theory was established by Thomas Hunt Morgan (1866–1945). Working with mutants of the fruit fly, *Drosophila melanogaster,* Morgan and his coworkers proved that genes are located on the chromosomes in a specific linear sequence. Morgan and his associates constructed the first "chromosome map" and by 1915 they had established four groups of linked factors that corresponded to the four pairs of *Drosophila* chromosomes. F. A. Janssens's (1863–1924) "chiasmatype hypothesis" provided a cytological basis for the recombination of genetic factors.

In the 1920s, Hermann Joseph Muller (1890–1967) demonstrated that it was possible to induce mutations in *Drosophila* using x rays. Most of these induced mutations were stable over many generations and behaved like typical Mendelian factors when subjected to breeding tests.

Modern genetics can be seen as the result of the integration of three lines of investigation: classical breeding tests, cytology, and biochemistry. Classical genetics could answer questions about how genes were transmitted, but it could not answer questions about the chemical nature of the gene or the mechanism of gene action. Cytological studies became very sophisticated during the 1930s with the development of ultraviolet microspectrophotometry and special staining techniques. Ingenious use of such techniques provided some insights into the chemical nature of the chromosomes. Nevertheless, some scientists thought it was impossible to explain genes in terms of a specific chemical substance. Others thought that the gene might be a complex chemical entity, probably a protein. Until the 1950s, nucleic acids, which had been discovered in the 1860s by Johann Friedrich Miescher (1844–1895), were thought to be too simple to serve

as the hereditary material. Phoebus A. Levene's (1869–1940) work suggested that DNA was basically a simple and repetitious polymer. Two forms of nucleic acids were identified: deoxyribonucleic acid (DNA) and ribonucleic acid (RNA).

In 1944, Oswald T. Avery (1877–1955) and his associates discovered that the transforming principle of certain bacteria seemed to be DNA. However, Avery could not reconcile the apparent biological activity of DNA with prevailing ideas about its chemical structure. After reading Avery's paper, Erwin Chargaff (1905–) proved that nucleic acids were actually complex chemicals. Chargaff also realized that there were certain regularities about the base composition of DNA.

Research on bacteria and bacteriophage provided another important approach to understanding the nature of the gene. In 1946, Joshua Lederberg (1925–) demonstrated that bacteria undergo genetic recombination. Max Delbrück (1906–1981) demonstrated the value of bacterial viruses as model systems for geneticists. In 1952, Alfred Hershey (1908–1997) and Martha Chase (1927–) conducted experiments that seemed to prove that DNA was the genetic material. This convinced James Watson (1928–) and Francis Crick (1916–) that understanding the chemical structure of DNA would reveal the nature of the gene. In 1953, Watson and Crick discovered the DNA double helix. Watson and Crick based their model on a combination of model building and x-ray data collected by Rosalind Franklin (1920–1958), Maurice Wilkens (1916–), and others.

The Watson-Crick double helix explained Chargaff's rules. The two chains of the double helix are held together by hydrogen bonds between the purine and pyrimidine bases; adenine always paired with thymine, and guanine paired with cytosine. Any sequence of bases could occur on one chain, but the rules of base pairing automatically determined the sequence of bases on the other chain. The double helix also suggested a copying mechanism for the genetic material. Scientists proved that each original DNA strand became associated with a newly synthesized strand of DNA. Enzymes that catalyze the synthesis of DNA and enzymes that use DNA as a template to synthesize RNA were discovered.

Once the chemical nature of the gene was understood, scientists discovered many aspects of how genes work. Watson summarized the fundamental principles of molecular genetics in what was called the central dogma of molecular biology: DNA is transcribed into RNA that is, in turn, translated into protein. The process of transferring information from DNA to RNA is called transcription. Transferring sequence information from nucleic acids to proteins is known as translation. The process of protein synthesis involves the use of different forms of RNA known as messenger RNA, and transfer RNA, and the cytoplasmic entities known as ribosomes. Eventually, Howard Temin (1934–1994) and David Baltimore (1938–) proved that information could also flow from RNA to DNA by means of a viral enzyme called reverse transcriptase. The retroviruses have RNA rather than DNA as their genetic material. HIV, the virus associated with AIDS, is a retrovirus.

In 1965 François Jacob (1920–), Jacques Monod (1910–1976), and André Lwoff (1902–1994) shared the Nobel

Prize for their discoveries about the way genes regulate the synthesis of proteins. Their theoretical framework for genetic regulation was called the operon theory. Although the operon model worked well for bacteria, further research indicated that genetic regulation in higher organisms was considerably more complicated. Moreover, in higher organisms, some traits did not show Mendelian segregation and seemed to be inherited only through the maternal line. In the 1960s, scientists discovered that certain cytoplasmic entities, such as mitochondria and chloroplasts, contain their own DNA.

Studies of inherited metabolic diseases provided another approach to the biochemical mechanism of gene action. George W. Beadle (1903–1989) and Edward Tatum (1909–1975) established the one gene-one enzyme concept through their studies of the red bread mold *Neurospora crassa*. The relationship between genes and enzymes had been suggested in 1908 by Archibald Edward Garrod (1857–1936) on the basis of his studies of alkaptonuria, cystinuria, albinism, and pentosuria, which Garrod called "inborn errors of metabolism." In 1949, Linus Pauling (1901–1994) and his associates established the molecular basis of a genetic disease by analyzing the abnormal hemoglobin produced by patients with sickle-cell anemia.

Since the 1950s, studies of the molecular biology of the gene have provided new answers to the fundamental questions about the mechanism of inheritance and the relationship between genes and gene products, especially in bacteria and viruses. By the 1970s, studies of reverse transcriptase, ribozymes, transposable genetic elements, recombinant DNA, cell growth and differentiation, had provided many unexpected insights into the nature of gene regulation in higher organisms.

During the last ten years, geneticists have developed tools that allow them to recreate steps in the evolution of organisms within a laboratory environment. These tools provide the means to carry out experiments that nature alone is incapable of performing. With the techniques of recombinant DNA technology, geneticists have learned how to transplant genes from one organism to another, thus reshuffling genetic material in ways never experienced in the evolution of life on Earth. Such knowledge and our ability to apply it to new purposes have profound implications for all living organisms on this planet, and to "life as we know it" can now be added, to a small but significant degree, "life as we make it." More than ever before, human actions affect the world ecosystem in unprecedented ways and the new biology requires not just an understanding of scientific concepts, but also a strong ethical basis and responsible actions.

One of the most powerful tools developed in recent years was the polymerase chain reaction (PCR). It allows for the amplification of minute samples of DNA and has quickly advanced molecular analysis, gene cloning, genome sequencing and many other fields of biology, including such socially important disciplines as forensic science. In 1993, Kary Mullis received the Nobel Prize for developing the PCR technique. In the same year a man was arrested for a murder committed eight years earlier. The suspect was convicted on the basis of DNA obtained from a few cells remaining on cigarettes which

he had smoked—a dramatic demonstration of how PCR can be applied in unprecedented ways.

Barely a week passes without the appearance of some news item about a novel genetic discovery or application. A list of recent media items could easily fill a book. In 1997, we witnessed the cloning of the first mammal, a sheep called Dolly, and subsequently, several other farm animals have been successfully cloned using similar methods. Dramatic discoveries have been made in agriculture and the impact of genetic research can be seen even at the corner grocery store. To eat or not to eat genetically engineered products is a question asked by many people after the Flavr Savr™ tomato hit the supermarkets in 1993. This new variety of tomato contains a gene allowing it to remain firm longer during the ripening process, thus extending its shelf life. Many people, however, fear that genetically manipulated food may have unexpected side effects for those who eat them. Data presented in 1998 by Arpad Pusztai, showing the effects of genetically manipulated (GM) potatoes on rat development, did nothing to assuage public fears. Pusztai's experiments involved feeding rats on GM potatoes for ten days, after which it was found that some of the rats' immune systems were weakened and also that some of their organs atrophied (shrank) or did not develop properly. Although Pusztai's work is regarded as controversial, it nevertheless suggests that much research into the safety of GM food remains to be done.

In the field of medical genetics, developments have been such that the metabolic basis of several hundred inherited disorders is now known and many defective genes resulting in such disorders have already been isolated. Significantly, in 1993, the specific genetic defect responsible for Huntington's disease was discovered. The mutation is located on human chromosome 4 and consists of an abnormally expanded sequence of trinucleotide CAG repeats in the genetic code. This discovery lead to the rapid development of a blood test, which is now generally available and can determine if the abnormal gene is present. Thus, it is possible to make a definitive diagnosis of Huntington's disease in individuals who have symptoms suggestive of this disorder and also to test for the presence of the abnormal gene in family members who are at risk of inheriting it. Here is one prime example of a disease whose diagnosis necessarily goes hand in hand with genetic counseling. Patients who are positive for the defective gene may be advised about their own futures, and also that of any children that they may have.

In 1996, scientists identified the site of a gene associated with some forms of Parkinson's disease and in the same year, gained significant progress in understanding the cancer-associated gene BRCA1. BRCA1 results in an increased susceptibility to breast and ovarian cancer in certain families. Later, a second breast cancer gene, BRCA2, was also identified. Apart from studying the genetic basis of specific diseases, researchers are also approaching the fascinating question of which aspects of our personality and behavior are controlled by the genetic constitution. Ongoing studies (e.g., at the University of Minnesota) using identical twins raised apart show that many aspects of human behavior and personality appear to be influenced by genetic components. There appears to be evidence emerging suggesting that genes play a significant role in determining some human behaviors such as alcoholism or sexuality.

Yet, in spite of the huge advances in genetic medicine, there is still a wide gulf between the isolation of a gene and the identification and characterization of its product. The product of the Huntington's disease gene, for example, is a protein called huntingtin, yet nine years after its discovery the function of this protein is still not completely clear. It is also necessary to be careful when interpreting genetic data as quite often, as in the case of the Parkinson's disease and breast cancer, the genetic defects only account for some cases and there appear to be non-genetic factors playing an equally important role in establishing these diseases. It is still believed, with some justification, that the identification of a genetic cause for a disease can lead to the development of treatments by, for example, gene therapy. Recent debates on this subject have included questions around performing somatic and germline gene therapy, and many ethical issues still remain to be resolved.

Genome sequencing has been one of the scientific highlights of the last ten years, with complete sequences now existing for a number of microbes, as well as for eukaryotic organisms such as the yeast (*Saccharomyces cerevisiae*), the nematode worm (*Coenorhabditis elegans*), the fruit fly (*Drosophila melanogaster*), the thale cress (*Arabidopsis thaliana*) and in very recent months, the mouse (*Mus musculus*) and rice (*Oryza sativa*) must be added to the list. Naturally, the genome sequence that has received the most publicity is our own, that of *Homo sapiens*. Since the announcement of the working draft of the entire human genome (HUGO) sequence in the summer of 2000, scientists have been working hard to make sense of the information. HUGO was the first "big science" project in biology, and formed the focus for a mass collective effort by thousands of researchers worldwide. Interestingly, one of the results emerging from the data is that an analysis of the complete human genome sequence is far from adequate to explain the various complicated intricacies of the human organism. So now, in the post-genomic era, the focus is changing to the analysis of the complete human protein complement—a project known as the human proteome (HUPO) project. This will be many times more complicated than the sequencing of the human genome. It will rely on a combination of genetic techniques, bioinformatics, and the powerful new high-throughput chip technologies. Gene chips can be used to quickly compare the levels of expression of many thousands of genes. They already have an application in medicine as they can be used to rapidly diagnose inherited diseases as well as to detect the presence of viruses, or changes in patterns of gene expression in various tissues. With the enormous amount of biological data that is now being generated it is clear that the analytical tools of the future will be based on a combination of molecular biology, microprocessor technology, and biocomputing.

We hope that *World of Genetics* inspires a new generation of scientists who will join in the exciting quest to unlock the remaining secrets of life, and provides valuable information to anyone interested in a world of genetics that plays an increasingly important role in our everyday lives.

K. Lee Lerner & Brenda Wilmoth Lerner, editors
Florence, Italy
June, 2001

K. Lee Lerner is a physicist, lecturer, and former fellow of the Science Research & Policy Institute. Brenda Wilmoth Lerner is a widely published author and science correspondent who has written extensively on the history and applications of medical science.

How to Use the Book

The articles in the book are meant to be understandable by anyone with a curiosity about genetics. Cross-references to related articles, definitions, and biographies in this collection are indicated by **bold-faced type**, and these cross-references will help explain and expand the individual entries. Although far from containing a comprehensive collection of genetic disorders, *World of Genetics* carries specifically selected entries designed to represent a broader class of genetic disorders with similar genetic and molecular mechanisms. For those readers interested in genetic disorders, the editors highly recommend the *Gale Encyclopedia of Genetic Disorders* as an accompanying reference to *World of Genetics*.

World of Genetics contains many articles addressing the impact of genetics on history, ethics, and society.

This first edition of *World of Genetics* has been designed with ready reference in mind:

- **Entries are arranged alphabetically**, rather than by chronology or scientific field.

- **Bold-faced terms** direct reader to related entries.

- **"See also" references** at the end of entries alert the reader to related entries not specifically mentioned in the body of the text.

- A **Sources Consulted** section lists the most worthwhile material we encountered in the compilation of this volume. It is there for the inspired reader who wants more information on the people and discoveries covered in this volume.

- The **Historical Chronology** includes many of the most important events in the biological sciences spanning the period from 900 B.C. through June 2001.

- A **comprehensive General Index** guides the reader to all topics and persons mentioned in the book. Bolded page references refer the reader to the term's full entry.

Although there is an important and fundamental link between the composition and shape of biological molecules and their functions in biological systems, a detailed understanding of biochemistry is neither assumed or required for *World of Genetics*. Accordingly, students and other readers should not be intimidated or deterred by the complex names of biochemical molecules (especially the names for particular proteins, enzymes, etc.). Where absolutely necessary, sufficient information regarding chemical structure is provided. If desired, more information can easily be obtained from any basic chemistry or biochemistry reference.

Advisory Board

In compiling this edition, we have been fortunate in being able to rely upon the expertise and contributions of the following scholars who served as academic and contributing advisors for *World of Genetics*, and to them we would like to express our sincere appreciation for their efforts to ensure that *World of Genetics* contains the most accurate and timely information possible:

- **Marcelo Amar, M.D.**, *Senior Fellow, Molecular Disease Branch*, National Institutes of Health (NIH) Bethesda, Maryland
- **Robert G. Best, Ph.D.**, *Director, Division of Genetics, Department of Obstetrics and Gynecology*, University of South Carolina School of Medicine Columbia, South Carolina
- **Wei Chin, Ph.D.**, *Department of Molecular & Cellular Biology*, Harvard University Cambridge, Massachusetts
- **Antonio Farina, M.D., Ph.D.**, *Department of Embryology, Obstetrics, and Gynecology*, University of Bologna Bologna, Italy
- **Edward J. Hollox, Ph.D**, *Institute of Genetics, Queen's Medical Centre*, University of Nottingham, Nottingham, England
- **Eric v.d. Luft, Ph.D., M.L.S.**, *Curator of Historical Collections*, SUNY Upstate Medical University Syracuse, New York
- **Lois N. Magner, Ph.D.**, *Professor Emerita of History*, Purdue University
- **Abdel Hakim Ben Nasr, Ph.D.**, *Department of Genetics, Molecular Oncology and Development Program at the Boyer Center for Molecular Medicine*, Yale University School of Medicine. New Haven, Connecticut
- **Judyth Sassoon, Ph.D., ARCS**, *Department of Chemistry and Biochemistry*, University of Bern Bern, Switzerland
- **Michael J Sullivan, M.D., Ph.D., FRACP**, *Cancer Genetics Laboratory*, University of Otago New Zealand
- **Constance K. Stein, Ph.D.**, *Director of Cytogenetics, Assistant Director of Molecular Diagnostics*, SUNY Upstate Medical University Syracuse, New York
- **Diego F. Wyszynski, M.D., Ph.D.**, *Department of Medicine, Epidemiology & Biostatistics* Boston University School of Medicine Boston, Massachusetts

Acknowledgments

It has been our privilege and honor to work with the following contributing writers and scientists who represent scholarship in genetics spanning five continents: Mary Brown; Janet Buchanan, Ph.D.; Joshua Buhs; Sherri Chasin Calvo; Bryan Cobb, M.S., *Ph.D.; Thomas Drucker, graduate student, University of Wisconsin; Ellen Elghobashi; David Evans, Ph.D.; Sandra Galeotti, M.S.; James Hoffmann; Brian Hoyle, Ph.D.; Evelyn Kelly, Ph.D.; Lynn Lauerman; Nicole LeBrasseur, M.S, graduate student, University of Michigan; Adrienne Wilmoth Lerner, graduate student, Vanderbilt University; Lee W. Lerner; Ricki Lewis, Ph.D.; Agnieszka Lichanska, Ph.D.; Jill Liske, M.Ed.; C.C. Love; Ann Marsden; Kyla Maslaniec; Leslie Mertz, Ph.D.; Stacy Murray; Kelli Miller; Shawn Oset; Jennifer A. Owens; Barry A. Palevitz, Ph.D.; Borut Peterlin, M.D.; Sudip Kumar Rakshit, Ph.D.; Michelle Rose; Lissa Rotundo; Jennifer R. Samuels; Elizabeth D. Schafer, Ph.D.; Diego Sebiastiani; Sharonda L. Shula; Tabitha Sparks, Ph.D.; Susan Thorpe-Vargas, Ph.D.; David Tulloch, Ph.D.; Meagan R. Washington; Stephanie Watson, James Welch; Richard Weikart, Ph.D.; Audra Wolfe, graduate student, University of Pennsylvania; and Xiaomei Zhu, Ph.D. (*) Anticipated by date of publication

Special advisor for graphics:

Evelyn B. Kelly, Ph.D.
Professor of Education Saint Leo University, Florida

Many of the academic advisors for *World of Genetics*, along with others, authored specially commissioned articles within their field of expertise:

Marcelo Amar, M.D.
Genetic dyslipidemias

Robert G. Best, Ph.D.
Molecular cytogenetics

Janet Buchanan, Ph.D.
Predictive genetic testing

Antonio Farina, M.D., Ph.D.
Statistical analysis of genetic syndromes

Edward J. Hollox, Ph.D.
Ancient DNA

Ricki Lewis, Ph.D.
Microchimerism

Adrienne Wilmoth Lerner (graduate student, Vanderbilt University)
Archaeogenetics: Genetic studies of ancient peoples

Eric v.d. Luft, Ph.D., M.L.S.
Twin studies

Abdel Hakim Ben Nasr, Ph.D.
Locus control regions of gene expression

Michael J. Sullivan, M.D., Ph.D., FRACP
Cancer genetics

Constance K. Stein, Ph.D.
Medical genetics

Diego F. Wyszynski, M.D., Ph.D.
Cleft lip/palate

The editors would like to add a special thank you to Lois Magner, Ph.D. and Dr. Judyth Sassoon, Ph.D. for their contributions to—and review of—contents contained in the introduction to *World of Genetics*. In addition, the editors especially thank Dr. Constance Stein and Kristina Amy Miller for their generous contributions of drawings and diagrams included in *World of Genetics*.

The editors gratefully acknowledge the assistance of Stacey Blachford in the compilation of articles related to genetic disorders and diseases. The editors are also indebted to Robyn Young for her patient guidance through the complexities and difficulties related to graphics. Last, but certainly not least, the editors thank Ms. Kimberley McGrath, without whose wisdom, wit, and guidance in all things, *World of Genetics* would have remained but a heap of 100% recycled electrons.

Also befitting a book on genetics, the editors lovingly dedicate this book to their parents: James Richard Lerner, Shirley LaWave Emminger Lerner, Allwyn Nathan Wilmoth Jr., and Karen LaVerne Glover Wilmoth whose dedication and love passed on the legacy of the generations and nurtured our bodies, souls, and love of science.

Cover

The image on the cover depicts an example of Pedigree analysis.

M

MacLeod, Colin Munro (1909-1972)
Canadian American microbiologist

Colin Munro MacLeod is recognized as one of the founders of molecular biology for his research concerning the role of **deoxyribonucleic acid (DNA)** in **bacteria**. Along with his colleagues Oswald Avery and Maclyn McCarty, MacLeod conducted experiments on bacterial **transformation** which indicated that DNA was the active agent in the genetic transformation of bacterial cells. His earlier research focused on the causes of pneumonia and the development of serums to treat it. MacLeod later became chairman of the department of microbiology at New York University; he also worked with a number of government agencies and served as White House science advisor to President John F. Kennedy.

MacLeod, the fourth of eight children, was born in Port Hastings, in the Canadian province of Nova Scotia. He was the son of John Charles MacLeod, a Scottish Presbyterian minister, and Lillian Munro MacLeod, a schoolteacher. During his childhood, MacLeod moved with his family first to Saskatchewan and then to Quebec. A bright youth, he skipped several grades in elementary school and graduated from St. Francis College, a secondary school in Richmond, Quebec, at the age of fifteen. MacLeod was granted a scholarship to McGill University in Montreal but was required to wait a year for admission because of his age; during that time he taught elementary school. After two years of undergraduate work in McGill's premedical program, during which he became managing editor of the student newspaper and a member of the varsity ice hockey team, MacLeod entered the McGill University Medical School, receiving his medical degree in 1932.

Following a two-year internship at the Montreal General Hospital, MacLeod moved to New York City and became a research assistant at the Rockefeller Institute for Medical Research. His research there, under the direction of Oswald Avery, focused on pneumonia and the pneumococcal infections which cause it. He examined the use of animal antiserums (liquid substances that contain proteins that guard against antigens) in the treatment of the disease. MacLeod also studied the use of sulfa drugs, synthetic substances that counteract bacteria, in treating pneumonia, as well as how pneumococci develop a resistance to sulfa drugs. He also worked on a mysterious substance then known as "C-reactive protein," which appeared in the blood of patients with acute infections.

MacLeod's principal research interest at the Rockefeller Institute was the phenomenon known as bacterial transformation. First discovered by **Frederick Griffith** in 1928, this was a phenomenon in which live bacteria assumed some of the characteristics of dead bacteria. Avery had been fascinated with transformation for many years and believed that the phenomenon had broad implications for the science of biology. Thus, he and his associates, including MacLeod, conducted studies to determine how the bacterial transformation worked in pneumococcal cells.

The researchers' primary problem was determining the exact nature of the substance which would bring about a transformation. Previously, the transformation had been achieved only sporadically in the laboratory, and scientists were not able to collect enough of the transforming substance to determine its exact chemical nature. MacLeod made two essential contributions to this project: he isolated a strain of *Pneumococcus* which could be consistently reproduced, and he developed an improved nutrient culture in which adequate quantities of the transforming substance could be collected for study.

By the time MacLeod left the Rockefeller Institute in 1941, he and Avery suspected that the vital substance in these transformations was DNA. A third scientist, **Maclyn McCarty**, confirmed their hypothesis. In 1944, MacLeod, Avery, and McCarty published "Studies of the Chemical Nature of the Substance Inducing Transformation of Pneumococcal Types: Induction of Transformation by a Deoxyribonucleic Acid Fraction Isolated from *Pneumococcus* Type III" in the *Journal of Experimental Medicine*. The article proposed that DNA was the material which brought about genetic transformation. Though the scientific community was slow to recognize the article's significance, it was later hailed as the beginning of a revolution that led to the formation of molecular biology as a scientific discipline.

MacLeod married Elizabeth Randol in 1938; they eventually had one daughter. In 1941, MacLeod became a citizen of the United States, and was appointed professor and chairman of the department of microbiology at the New York University School of Medicine, a position he held until 1956. At New York University he was instrumental in creating a combined program in which research-oriented students could acquire both an M.D. and a Ph.D. In 1956, he became professor of research medicine at the Medical School of the University of Pennsylvania. MacLeod returned to New York University in 1960 as professor of medicine and remained in that position until 1966.

From the time the United States entered World War II until the end of his life, MacLeod was a scientific advisor to the federal government. In 1941, he became director of the Commission on Pneumonia of the United States Army Epidemiological Board. Following the unification of the military services in 1949, he became president of the Armed Forces Epidemiological Board and served in that post until 1955. In the late 1950s, MacLeod helped establish the Health Research Council for the City of New York and served as its chairman from 1960 to 1970. In 1963, President John F. Kennedy appointed him deputy director of the Office of Science and Technology in the Executive Office of the President; from this position he was responsible for many program and policy initiatives, most notably the United States/Japan Cooperative Program in the Medical Sciences.

In 1966, MacLeod became vice-president for Medical Affairs of the Commonwealth Fund, a philanthropic organization. He was honored by election to the National Academy of Sciences, the American Philosophical Society, and the American Academy of Arts and Sciences. MacLeod was en route from the United States to Dacca, Bangladesh, to visit a cholera laboratory when he died in his sleep in a hotel at the London airport in 1972. In the *Yearbook of the American Philosophical Society,* Maclyn McCarty wrote of MacLeod's influence on younger scientists, "His insistence on rigorous principles in scientific research was not enforced by stern discipline but was conveyed with such good nature and patience that it was simply part of the spirit of investigation in his laboratory."

MAJOR HISTOCOMPATIBILITY COMPLEX (MHC)

In humans, the proteins coded by the genes of the major histocompatibility complex (MHC) include human leukocyte antigens (HLA), as well as other proteins. HLA proteins are present on the surface of most of the body's cells and are important in helping the immune system distinguish "self" from "non-self" molecules, cells, and other objects.

The function and importance of MHC is best understood in the context of a basic understanding of the function of the immune system. The immune system is responsible for distinguishing foreign proteins and other antigens, primarily with the goal of eliminating foreign organisms and other invaders that can result in disease. There are several levels of defense characterized by the various stages and types of immune response.

Present on **chromosome** 6, the major histocompatibility complex consists of more than 70 genes, classified into class I, II, and III MHC. There are multiple alleles, or forms, of each HLA **gene**. These alleles are expressed as proteins on the surface of various cells in a co-dominant manner. This diversity is important in maintaining an effective system of specific immunity. Altogether, the MHC genes span a region that is four million base pairs in length. Although this is a large region, 99% of the time these closely-linked genes are transmitted to the next generation as a unit of MHC alleles on each chromosome 6. This unit is called a haplotype.

Class I MHC genes include HLA-A, HLA-B, and HLA-C. Class I MHC are expressed on the surface of almost all cells. They are important for displaying antigen from viruses or parasites to killer T-cells in cellular immunity. Class I MHC is also particularly important in organ and tissue rejection following transplantation. In addition to the portion of class I MHC coded by the genes on chromosome 6, each class I MHC protein also contains a small, non-variable protein component called beta2-microglobulin coded by a gene on chromosome 15. Class I HLA genes are highly polymorphic, meaning there are multiple forms, or alleles, of each gene. There are at least 57 HLA-A alleles, 111 HLA-B alleles, and 34 HLA-C alleles.

Class II MHC genes include HLA-DP, HLA-DQ, and HLA-DR. Class II MHC are particularly important in humoral immunity. They present foreign antigen to helper T-cells, which stimulate B-cells to elicit an antibody response. Class II MHC is only present on antigen presenting cells, including phagocytes and B-cells. Like Class I MHC, there are hundreds of alleles that make up the class II HLA **gene pool**.

Class III MHC genes include the complement system (i.e. C2, C4a, C4b, Bf). Complement proteins help to activate and maintain the inflammatory process of an immune response.

When a foreign organism enters the body, it is encountered by the components of the body's natural immunity. Natural immunity is the non-specific first-line of defense carried out by phagocytes, natural killer cells, and components of the complement system. Phagocytes are specialized white blood cells that are capable of engulfing and killing an organism. Natural killer cells are also specialized white blood cells that respond to **cancer** cells and certain viral infections. The complement system is a group of proteins called the class III MHC that attack antigens. Antigens consist of any molecule capable of triggering an immune response. Although this list is not exhaustive, antigens can be derived from toxins, protein, carbohydrates, **DNA**, or other molecules from viruses, **bacteria**, cellular parasites, or cancer cells.

The natural immune response will hold an infection at bay as the next line of defense mobilizes through acquired, or specific, immunity. This specialized type of immunity is usually what is needed to eliminate an infection and is dependent on the role of the proteins of the major histocompatibility complex. There are two types of acquired immunity. Humoral immunity is important in fighting infections outside the body's cells, such as those caused by bacteria and certain viruses. Other types of viruses and parasites that invade the cells are better fought by cellular immunity. The major players in acquired immunity are the antigen-presenting cells (APCs), B-cells, their secreted antibodies, and the T-cells. Their functions are described in detail below.

In humoral immunity, antigen-presenting cells, including some B-cells, engulf and break down foreign organisms. Antigens from these foreign organisms are then brought to the outside surface of the antigen-presenting cells and presented in conjunction with class II MHC proteins. The helper T-cells recognize the antigen presented in this way and release cytokines, proteins that signal B-cells to take further action. B-cells are specialized white blood cells that mature in the bone marrow. Through the process of maturation, each B-cell develops the ability to recognize and respond to a specific antigen. Helper T-cells aid in stimulating the few B-cells that can recognize a particular foreign antigen. B-cells that are stimulated in this way develop into plasma cells, which secrete antibodies specific to the recognized antigen. Antibodies are proteins that are present in the circulation, as well as being bound to the surface of B-cells. They can destroy the foreign organism from which the antigen came. Destruction occurs either directly, or by tagging the organism, which will then be more easily recognized and targeted by phagocytes and complement proteins. Some of the stimulated B-cells go on to become memory cells, which are able to mount an even faster response if the antigen is encountered a second time.

Another type of acquired immunity involves killer T-cells and is termed celluar immunity. T-cells go through a process of maturation in the organ called the thymus, in which T-cells that recognized self antigens are eliminated. Each remaining T-cell has the ability to recognize a single, specific, non-self antigen that the body may encounter. Although the names are similar, killer T-cells are unlike the non-specific natural killer cells in that they are specific in their action. Some viruses and parasites quickly invade the body's cells, where they are hidden from antibodies. Small pieces of proteins from these invading viruses or parasites are presented on the surface of infected cells in conjunction with class I MHC proteins, which are present on the surface of most all of the body's cells. Killer T-cells can recognize antigen bound to class I MHC in this way, and they are prompted to release chemicals that act directly to kill the infected **cell**. There is also a role for helper T-cells and antigen-presenting cells in cellular immunity. Helper T-cells release cytokines, as in the humoral response, and the cytokines stimulate killer T-cells to multiply. Antigen-presenting cells carry foreign antigen to places in the body where additional killer T-cells can be alerted and recruited.

The major histocompatibility complex clearly performs an important role in functioning of the immune system. Related to this role in disease immunity, MHC is also impor-tant in organ and tissue transplantation, as well as playing a role in susceptibility to certain diseases. HLA typing can also provide important information in parentage, forensic, and anthropologic studies.

There is significant variability of the frequencies of HLA alleles among ethnic groups. This is reflected in anthropologic studies attempting to use HLA-types to determine patterns of migration and evolutionary relationships of peoples of various ethnicity. Ethnic variation is also reflected in studies of HLA-associated diseases. Generally speaking, populations that have been subject to significant patterns of migration and assimilation with other populations tend to have a more diverse HLA gene pool. For example, it is unlikely that two unrelated individuals of African ancestry would have matched HLA types. Conversely, populations that have been isolated due to geography, cultural practices, and other historical influences may display a less diverse pool of HLA types, making it more likely for two unrelated individuals to be HLA-matched.

There is a role for HLA typing of individuals in various settings. Most commonly, HLA typing is used to establish if an organ or tissue donor is appropriately matched to the recipient for key HLA types, so as not to elicit a rejection reaction in which the recipient's immune system attacks the donor tissue. In the special case of bone marrow transplantation, the risk is for graft-versus-host disease (GVHD), as opposed to tissue rejection. Because the bone marrow contains the cells of the immune system, the recipient effectively receives the donor's immune system. If the donor immune system recognizes the recipient's tissues as foreign, it may begin to attack, causing the inflammatory and other complications of GVHD. As advances occur in transplantation medicine, HLA typing for transplantation occurs with increasing frequency and in various settings.

There is an established relationship between the **inheritance** of certain HLA types and susceptibility to specific diseases. Most commonly, these are diseases that are thought to be autoimmune in nature. **Autoimmune diseases** are those characterized by inflammatory reactions that occur as a result of the immune system mistakenly attacking self tissues. The basis of the HLA association is not well understood, although there are some hypotheses. Most autoimmune diseases are characterized by the expression of class II MHC on cells of the body that do not normally express these proteins. This may confuse the killer T-cells, which respond inappropriately by attacking these cells. Molecular **mimicry** is another hypothesis. Certain HLA types may look like antigens from foreign organisms. If an individual is infected by such a foreign **virus** or bacteria, the immune system mounts a response against the invader. However, there may be a cross-reaction with cells displaying the HLA type that is mistaken for foreign antigen. Whatever the underlying mechanism, certain HLA-types are known factors that increase the relative risk for developing specific autoimmune diseases. For example, individuals who carry the HLA B-27 allele have a relative risk of 150 for developing ankylosing spondylitis—meaning such an individual has a 150-fold chance of developing this form of spinal and pelvic arthritis, as compared to someone in the general population. Selected associations are listed below, together with the approximate corresponding relative risk of disease.

HLA-disease associations

(Disease, MHC allele, Approximate relative risk):

- Type 1 diabetes, DR3, 5
- Type 1 diabetes, DR4, 5
- Type 1 diabetes, DR3 + DR4, 20-40
- Narcolepsy, DR2, 260-360
- Ankylosing spondylitis, B27, 80-150
- Reiter's disease, B27, 37
- Rheumatoid arthritis, DR4, 3-6
- Myasthenia gravis, B8, 4
- Lupus, DR3, 2
- Graves disease, DR3, 5
- Multiple sclerosis, DR2, 3
- Celiac disease, DR3 and DR7, 5-10
- Psoriasis vulgaris, Cw6, 8

In addition to autoimmune disease, HLA-type less commonly plays a role in susceptibility to other diseases, including cancer, certain infectious diseases, and metabolic diseases. Conversely, some HLA-types confer a protective advantage for certain types of infectious disease. Also, there are rare immune deficiency diseases that result from inherited mutations of the genes of components of the major histocompatibility complex.

Among other tests, HLA typing can sometimes be used to determine parentage, most commonly paternity, of a child. This type of testing is not generally done for medical reasons, but rather for social or legal reasons.

HLA-typing can provide valuable DNA-based evidence contributing to the determination of identity in criminal cases. This technology has been used in domestic criminal trials. Additionally, it is a technology that has been applied internationally in the human-rights arena. For example, HLA-typing had an application in Argentina following a military dictatorship that ended in 1983. The period under the dictatorship was marked by the murder and disappearance of thousands who were known or suspected of opposing the regime's practices. Children of the disappeared were often adopted by military officials and others. HLA-typing was one tool used to determine non-parentage and return children of the disappeared to their biological families.

HLA-typing has proved to be an invaluable tool in the study of the evolutionary origins of human populations. This information, in turn, contributes to an understanding of cultural and linguistic relationships and practices among and within various ethnic groups.

See also Antibody and antigen; Immunogenetics; Immunological analysis techniques

MALARIA • *see* ADAPTATION AND FITNESS

MALPIGHI, MARCELLO (1628-1694)
Italian physician

In the second half of the seventeenth century, Marcello Malpighi used the newly invented microscope to make a number of important discoveries about living tissues and structures, earning himself enduring recognition as a founder of scientific microscopy, histology (the study of tissues), **embryology**, and the science of plant anatomy.

Malpighi was born at Crevalcore, just outside Bologna, Italy. The son of small landowners, Malpighi studied medicine and philosophy at the University of Bologna. While at Bologna, Malpighi was part of a small anatomical society headed by the teacher Bartolomeo Massari, in whose home the group met to conduct dissections and vivisections. Malpighi later married Massari's sister.

In 1655, Malpighi became a lecturer in logic at the University of Bologna; in 1656, he assumed the chair of theoretical medicine at the University of Pisa; in 1659, he returned to Bologna as lecturer in theoretical, then practical, medicine; from 1662 to 1666, he held the principal chair in medicine at the University of Messina; finally in 1666, he returned again to Bologna, where he remained for the rest of his teaching and research career. In 1691, at the age of sixty-three, Malpighi was called by his friend Pope Innocent XII to serve as the pontiff's personal physician. Reluctantly, Malpighi agreed and moved to Rome, where he died on November 29, 1694, in his room in the Quirinal Palace.

Early in his medical career, Malpighi became absorbed in using the microscope to study a wide range of living tissue—animal, insect, and plant. At the time, this was an entirely new field of scientific investigation. Malpighi soon made a profoundly important discovery. Microscopically examining a frog's lungs, he was able for the first time to describe the lung's structure accurately—thin air sacs surrounded by a network of tiny blood vessels. This explained how air (oxygen) is able to diffuse into the blood vessels, a key to understanding the process of respiration. It also provided the one missing piece of evidence to confirm William Harvey's revolutionary theory of the blood circulation: Malpighi had discovered the capillaries, the microscopic connecting link between the veins and arteries that Harvey—with no microscope available—had only been able to postulate. Malpighi published his findings about the lungs in 1661.

Malpighi used the microscope to make an impressive number of other important observations, all "firsts." He observed a "host of red atoms" in the blood—the red blood corpuscles. He described the papillae of the tongue and skin—the receptors of the senses of taste and touch. He identified the rete mucosum, the Malpighian layer, of the skin. He found that the nerves and spinal column both consisted of bundles of fibers. He clearly described the structure of the kidney and suggested its function as a urine producer. He identified the spleen as an organ, not a gland; structures in both the kidney and spleen are named after him. He demonstrated that bile is secreted in the liver, not the gall bladder. In showing bile to be a uniform color, he disproved a 2,000-year-old idea that the bile was yellow and black. He described glandular adenopathy, a syndrome rediscovered by Thomas Hodgkin (1798–1866) and given that man's name 200 years later.

Malpighi also conducted groundbreaking research in plant and insect microscopy. His extensive studies of the silkworm were the first full examination of insect structure. His detailed observations of chick embryos laid the foundation for microscopic embryology. His botanical investigations established the science of plant anatomy. The variety of Malpighi's microscopic discoveries piqued the interest of countless other researchers and firmly established microscopy as a science.

MALTHUS, THOMAS ROBERT (1766-1834)

English economist

Thomas Robert Malthus had an enormous impact on scholars of biology, human populations and economics. Malthus was born southwest of London in Surrey. He entered Jesus College at Cambridge in 1784 and graduated four years later, at which time he was ordained in the Church of England. Malthus served for a brief time as a curate in a parish not far from where he was born. He later became·a Fellow of Jesus College and also a Fellow of the Royal Society. In 1805, Malthus became a professor of history and political economy at the East India Company's college at Haileybury in Hertfordshire.

Malthus is remembered not because of the holy orders that he took. Rather, his essays on population brought recognition to him worldwide. Malthus was not optimistic with regard to the lot of working humans at the time. He believed that the means of subsistence would always be in short supply because of human reproduction. Malthus thought that human reproduction would always outstrip the capacity of the land to produce food. Ultimately, it was the availability of food that would set population limits. Famine, disease, and infant mortality would take their toll because of inadequate food supplies. The views of Malthus were articulated at a time when it was hoped that the poor laws would encourage population growth and the increased population would insure national wealth. The dire views of Malthus were exactly opposite those of most other scholars at the time. He urged that a nation balance production with consumption. He suggested that public works would provide relief for the working poor. Malthus published some of his ideas in his *Essay on the Principle of Population* in 1798. The essay was revised and enlarged for a total of six editions all of which carried his name except the first which was anonymous.

Charles Darwin (1809–1882) acknowledged that he read Malthus. Darwin recognized that uncontrolled reproduction coupled with limited means of subsistence would create a struggle for existance with survival of only those best suited. Animals, as humans, vary in capabilities related to survival. The struggle, according to Darwin, selected those best fit for survival. Darwin had just returned from the voyage of the *Beagle* and he recognized that the Malthusian population concept applies to animals as well as humans. Thus, Malthus indirectly contributed to Darwin's theory of **natural selection**.

Alfred Russell Wallace (1823–1913), whose views on natural **selection** were similar to Darwin, also acknowledged reading Malthus.

Malthus, who believed that public works were a means of minimizing economic distress of the poor, was said to have influenced the English economist John Maynard Keynes (1883–1946) who advocated governmental aid for full employment as a means of digging out of economic depression. And, certainly the administration of Franklin Delano Roosevelt (1882-1945) did in fact institute vast public projects to rescue the poor and to stimulate production.

Public concern for overpopulation and inadequate food supply is greater now than it was in the time of Malthus. Malthus did not provide easy answers for this problem. He was against contraception and against abortion. In lieu of disease and famine, he offered delayed marriage as a means of population control. The American Nobel Laureate Norman E. Borlaug (1914–), the father of the green revolution, has averted famine for much of the world, at least for the time being, by vastly increasing the food supply with modern agriculture. Here is an instance where the means of subsistence has increased at an incredible pace not foreseen by Malthus. Also, modern refrigeration has decreased food spoilage and waste. Hence, there has been a period when food supplies have been relatively abundant. Modern birth control methods offer options not previously available for population control. In some countries, laws are being changed in an effort to discourage large families. As of 2000, the world population surpassed six billion people. The concerns of Thomas Robert Malthus, how human society will handle the problem of limited means of subsistence and the pressing burden of population growth, remain relevant over two centuries after his death.

See also Darwinism; Evolutionary mechanisms; Population genetics; Sociobiology; Survival of the fittest

MANGOLD, HILDE PROESCHOLDT (1898-1924)

German biologist

German biologist Hilde Mangold studied the process by which part of an embryo—sometimes called an organizer—causes other parts of the embryo to form specific types of tissues and organs. This process is called embryonic induction. Working under the direction of German biologist **Hans Spemann**, (1869–1941), Mangold helped discover the location of the organizer in amphibians.

Mangold had attended the University of Frankfurt, where she had once been to a lecture by Hans Spemann. Spemann was studying the processes that control the development of amphibian embryos, and Mangold was intrigued by his experiments. After she graduated, she moved to Freiburg, Germany, where Spemann was head of the Zoological Institute. Mangold began working on an advanced degree, and Spemann suggested a series of experiments for her doctoral thesis.

Spemann proposed that Mangold should work with early embryos of two **species** of newts. Her experiments would involve transplanting a portion of one embryo to a different part of an embryo of the other species. Because one newt species had light cells, and one had dark cells, she would be able to determine exactly what happened to the transplant as the embryo developed.

Earlier experiments (using a single species of newt) had shown that the fate of most transplanted cells was determined by the recipient embryo rather than by the donor. For instance, if cells that would normally become belly skin were transplanted to an area that would normally become back skin, the transplanted cells would become back skin. This result indicated that the fate of the cells was not yet determined when they were transplanted.

This was not the case for one region of the early embryo—the region that would eventually form the neural tube. (The neural tube, in turn, eventually forms the newt's brain and spinal cord.) When these cells were transplanted, they continued to form a neural tube regardless of their location. In other words, the fate of these cells was already determined in the early embryo.

It was these cells that would eventually form a neural tube, that Mangold transferred between the two species. To her and Spemann's surprise, Mangold found that in such a transplant, most of the cells making up the resulting neural tube came from the recipient of the transplant rather than the donor. Therefore, the transplant cells were somehow able to control not only their own fate, but also the fate of the cells surrounding them. They induced these cells to develop into tissues and organs that they would not normally form. In fact, not only would a second spinal cord and brain be produced, but other internal organs as well. The eventual result was a secondary embryo attached to the main embryo (similar to a conjoined twin).

Spemann named this transplanted region the organizer because it was capable of reorganizing the cells in the recipient embryo. Mangold and Spemann wrote a paper on their results. (The paper was also Mangold's thesis for her doctorate degree.) Just as the paper was being published, Mangold was killed (at the age of 26) when a heater in her kitchen exploded. Spemann, however, went on to win a Nobel Prize in 1935 for the discovery of the organizer. (Mangold's death made her ineligible.) This Nobel Prize was one of the few that have ever been awarded for work based on a doctoral thesis, in this case, the thesis of Hilde Mangold.

See also Cell differentiation; Developmental genetics; Embryology; Embryology, the history of developmental and generational theory

MARGULIS, LYNN (1938-)
American biologist

Lynn Margulis is a theoretical biologist and professor of botany at the University of Massachusetts at Amherst. Her research on the evolutionary links between cells containing nuclei (**eukaryotes**) and cells without nuclei (prokaryotes) led her to formulate a symbiotic theory of **evolution** that was initially spurned in the scientific community but has become more widely accepted.

Margulis, the eldest of four daughters, was born in Chicago. Her father, Morris Alexander, was a lawyer who owned a company that developed and marketed a long-lasting thermoplastic material used to mark streets and highways. He also served as an assistant state's attorney for the state of Illinois. Her mother, Leone, operated a travel agency. When Margulis was fifteen, she completed her second year at Hyde Park High School and was accepted into an early entrant program at the University of Chicago.

Margulis was particularly inspired by her science courses, in large part because reading assignments consisted not of textbooks but of the original works of the world's great scientists. A course in natural science made an immediate impression and would influence her life, raising questions that she has pursued throughout her career: What is **heredity**? How do genetic components influence the development of offspring? What are the common bonds between generations? While at the University of Chicago she met Carl Sagan, then a graduate student in physics. At the age of nineteen, she married Sagan, received a B.A. in liberal arts, and moved to Madison, Wisconsin, to pursue a joint master's degree in zoology and **genetics** at the University of Wisconsin under the guidance of noted **cell** biologist Hans Ris. In 1960, Margulis and Sagan moved to the University of California at Berkeley, where she conducted genetic research for her doctoral dissertation.

The marriage to Sagan ended before she received her doctorate. She moved to Waltham, Massachusetts, with her two sons, Dorion and Jeremy, to accept a position as lecturer in the department of biology at Brandeis University. She was awarded her Ph.D. in 1965. The following year, Margulis became an adjunct assistant of biology at Boston University, leaving 22 years later as full professor. In 1967, Margulis married crystallographer Thomas N. Margulis. The couple had two children before they divorced in 1980. Since 1988, Margulis has been a distinguished university professor with the Department of Botany at the University of Massachusetts at Amherst.

Margulis' interest in genetics and the development of cells can be traced to her earliest days as a University of Chicago undergraduate. She always questioned the commonly accepted theories of genetics, but also challenged the traditionalists by presenting hypotheses that contradicted current beliefs. Margulis has been called the most gifted theoretical biologist of her generation by numerous colleagues. A profile of Margulis by Jeanne McDermott in the *Smithsonian* quotes Peter Raven, director of the Missouri Botanical Garden and a MacArthur fellow: "Her mind keeps shooting off sparks. Some critics say she's off in left field. To me she's one of the most exciting, original thinkers in the whole field of biology." Although few know more about cellular biology, Margulis considers herself a "microbial evolutionist," mapping out a field of study that doesn't in fact exist.

As a graduate student, Margulis became interested in cases of **non-Mendelian inheritance**, occurring when the genetic make-up of a cell's descendants cannot be traced solely to the genes in a cell's **nucleus**. For several years, she concentrated her research on a search for genes in the cytoplasm of cells, the area outside of the cell's nucleus. In the early 1960s, Margulis presented evidence for the existence of extranuclear genes. She and other researchers had found **DNA** in the cytoplasm of plant cells, indicating that heredity in higher organisms is not solely determined by genetic information carried in the cell nucleus. Her continued work in this field led her to formulate the serial endosymbiotic theory, or SET, which offered a new approach to evolution as well as an account of the origin of cells with nuclei.

Prokaryotes—bacteria and blue-green algae, now commonly referred to as cyanobacteria—are single-celled organisms that carry genetic material in the cytoplasm. Margulis proposes that eukaryotes (cells with nuclei) evolved when different kinds of prokaryotes formed symbiotic systems to enhance their chances for survival. The first such symbiotic fusion would have taken place between fermenting **bacteria** and oxygen-using bacteria. All cells with nuclei, Margulis contends, are derived from bacteria that formed symbiotic relationships with other primordial bacteria some two billion years ago. It has now become widely accepted that mitochondria—those components of eukaryotic cells that process oxygen—are remnants of oxygen-using bacteria. Margulis's hypothesis that cell hairs, found in a vast array of eukaryotic cells, descend from another group of primordial bacteria much like the modern spirochaete still encounters resistance, however.

The resistance to Margulis's work in microbiology may perhaps be explained by its implications for the more theoretical aspects of evolutionary theory. Evolutionary theorists, particularly in the English-speaking countries, have always put a particular emphasis on the notion that **competition** for scarce resources leads to the survival of the most well-adapted representatives of a **species** by **natural selection**, favoring adaptive genetic mutations. According to Margulis, natural **selection** as traditionally defined cannot account for the "creative novelty" to be found in evolutionary history. She argues instead that the primary mechanism driving biological change is symbiosis, while competition plays a secondary role.

Margulis doesn't limit her concept of symbiosis to the origin of plant and animal cells. She subscribes to the **Gaia hypothesis** first formulated by James E. Lovelock, British inventor and chemist. The Gaia theory (named for the Greek goddess of Earth) essentially states that all life, as well as the oceans, the atmosphere, and Earth itself are parts of a single, all—encompassing symbiosis and may fruitfully be considered as elements of a single organism.

Margulis has authored more than one hundred and thirty scientific articles and ten books, several of which are written with her son Dorion. She has also served on more than two dozen committees, including the American Association for the

Advancement of Science, the MacArthur Foundation Fellowship Nominating Committee, and the editorial boards of several scientific journals. Margulis is co—director of NASA's Planetary Biology Internship Program and, in 1983, was elected to the National Academy of Sciences.

See also Evolutionary mechanisms; Gaia hypothesis

MARKER · *see* GENETIC MARKER

MASTER GENE · *see* GENE NAMES AND FUNCTIONS

MATERNAL INHERITANCE

Maternal **inheritance** is a type of uniparental inheritance in which all the progeny of a mating inherit the **genotype and phenotype** of the female parent. It is an extreme form of **non-Mendelian inheritance** and is seen as the failure of the progeny to display Mendelian segregation for certain characters. This bias in maternal genotype is established at, or soon after, the production of the **zygote** following mating, when the female sex **cell** (or **gamete**), containing the maternal genetic information in the **nucleus**, unites with the male gamete, containing information from the male. The nuclei of the male and female gametes fuse and the genetic information from mother and father is mixed. In the vast majority of **species**, the female gamete is physically larger than the male gamete and provides the cytoplasm for the developing embryo. This cytoplasm also contributes the organelles that are normally located there, as well as a number of proteins and messenger RNAs (mRNAs), encoded by the maternal nucleus prior to fusion. The organelles contain their own genetic information and the phenotypes controlled by their genes are said to exhibit maternal inheritance. In contrast, those phenotypes that are controlled by factors found in the zygote cytoplasm, but which are under the control of the maternal nucleus (such as the proteins and mRNAs) are said to express a maternal effect. An example of a maternal effect is the direction of coiling in the shells of pond snails or the patterning of the embryo in fruit flies.

True maternal inheritance results from the presence of **DNA** in organelles, such as **mitochondria** (in all species) and chloroplasts (in plants). This DNA is inherited independently of nuclear genes. Organelle DNA has been physically sequestered from nuclear DNA and its **transcription, translation**, and protein expression occurs within the same organelle compartment in which it resides. By contrast, nuclear DNA is expressed in the cell cytoplasm. Conditions in the organelle are different from those in the nucleus and organelle DNA, therefore, evolves at its own distinct rate. If inheritance is uniparental, there can be no **recombination** between parental genomes, and even in those cases where organelle genomes are inherited from both parents, recombination does not occur. Since organelle DNA has a different **replication** system from that of the nucleus, the **mutation** rate can be different. For example, **mitochondrial DNA** in mammals accumulates mutations more rapidly than nuclear DNA.

The classic study of maternal inheritance was performed in 1909 by the German plant geneticist Carl Correns (1864–1933). Correns was working on the four o'clock plant *(Mirabilis jalapa),* a variegated plant showing patches of green and white tissue. Some branches carry only green leaves while others carry only white leaves. Flowers can develop at different locations on the plant and can be intercrossed in a variety of different combinations by transferring the pollen from one flower to another. The leaves that emerge always corresponded to the color of the leaf of the female. For example, regardless of whether the male pollen is from a green, variegated, or white leaf, if the female flower comes from a region where the leaves are green, all the progeny leaves will be green. This is because differences in leaf color are due to the presence of either green or colorless chloroplasts that are never transmitted through the pollen of the male parent. All the organelle DNA that is found in the embryo after **fertilization** is from the female partner and determines the color of the leaves.

Knowledge of maternal inheritance has been applied in studies on human **evolution**. One consequence of maternal inheritance is that the sequence of mitochondrial DNA is more sensitive than nuclear DNA to reductions in the size of the breeding population. Comparisons of mitochondrial DNA sequences in a range of human populations allow evolutionary trees to be constructed. The divergence among human mitochondrial DNA spans 0.57%, and a tree can be constructed in which the mitochondrial variants diverged from a common (African) ancestor. The rate at which mammalian mitochondrial DNA accumulates mutations is 2-4% per million years. Such a rate would generate the observed divergence over an evolutionary period of 140,000-280,000 years. This implies that the human race is descended from a single female (the "mitochondrial Eve"), who lived in Africa around 200,000 years ago.

See also Chromosome; Extranuclear inheritance; Plant genetics

MATHEMATICAL THEORIES AND MODELS OF GENETICS

Mathematical theory and mathematical techniques are often critical tools in genetic research. Careful studies of data allow researchers to probe selected regions of molecules, or to characterize the distribution of genes in large populations. Incorporated into such studies are mathematical models for the genetic code and for molecular reactions (e.g., **transcription** and **translation** processes). In addition, mathematical models are used to study both the frequency and the mechanisms of genetic processes (e.g., **mutation** frequency or the number of possible errors in translation that can affect **protein synthesis**.

In addition to extensive use of statistical analysis, a number of other mathematical theories play an increasingly important role in **genetics**.

Group theory, for example, is one of several mathematical tools utilized to provide an accurate description of genetic molecules and processes. Of special interest to mathematicians and geneticists are mathematical descriptions of the structure of **DNA**, **RNA**, and of mutational processes that change the structure and stability of these fundamental molecules. Using some of the same mathematical techniques used in the study of subatomic physics, mathematicians and geneticists use advanced group theory based techniques to develop increasing accurate mathematical descriptions of genetic processes. On a much larger scale, the nature of quantifiable changes in the **gene pool** of a population provide insights to the **evolutionary mechanisms** that produce genetic processes that resist mutational change (especially mutations that are lethal). Scientists and mathematicians are able to use group theory based models to study this stability.

Number theories play an increasing important part unraveling the mathematical complexities of genetics. DNA is a double stranded polymer consisting of **nucleotide** subunits coiled around each other into a double helical shape. The genetic code for all process is derived from the **sequencing** and combination of just four nitrogenous **bases** (adenine (A), thymine (T), cytosine (C), and guanine (g) one of which is a part of each nucleotide in the DNA chain. In RNA molecules, the base uracil (U) takes the place of thymine. Because almost all the of the genetic complexities of an organism are contained in subtle variations in the sequencing or reading of these four bases, mathematical models often prove helpful in predicting outcomes. One of the earliest uses of mathematics in genetics research was based upon the observation the number of possible code combinations is vastly more than the 20 different **amino acids** that are the building blocks of protein. Based simple on mathematics, it was clear that the genetic code must have a large number of redundancies and therefore is a **degenerate code**. In contrast, however, to this mathematical degeneracy, the redundant code and 20 amino acids produce an incredible array of protein structures.

Despite the wide diversity of **gene** products the code is fundamentally so simple that, especially considering the mathematical complexities, there is a high degree of fidelity to the genetic process (i.e., a relative lack of errors in **replication**, transcription, and translation. Some researchers use a group theory concept called a crystal basis to devise mutation models that use mathematical operators (plus and minus signs) to simulate nitrogenous base additions and **deletions** from DNA sequences.

Another theory being explored for usefulness toward explaining DNA reactions is a mathematical theory termed Knot theory. Because the double stranded DNA molecule must continually wind and unwind into its condensed and active states, Knot theory, a variant of topological theory, provides a mathematical means to describe such changes. One question addressed by knot theory is whether the winding process is seemingly as random as once thought—or whether DNA coiling takes place to facilitate the most rapid uncoiling possible to make available the DNA for replication or transcription.

Knot theory can also be used to quantitatively describe coiling in protein—especially in the important enzyme subclass of proteins.

Combinatorics represents yet another mathematical approach to the description of biological systems. Combinatorics is the study of combining objects by various rules to create new arrangements of objects. The objects can be anything from points and numbers to apples and oranges or genes. Use of combinatorics has grown rapidly in the last two decades, making critical contributions to genetics, computer science, operations research, finite probability theory, and cryptology. Computers and computer networks operate with finite data structures and algorithms that makes them perfect for enumeration and graph theory applications. In addition to genetics, research areas such as neural networking rely on the contribution made by combinatorics.

See also Bacterial genetics; Biochemistry; Chromatin; Chromosome; DNA replication; DNA structure; Double helix; Gene mapping; Molecular biology and molecular genetics

MATING · *see* GENETIC IMPLICATIONS OF MATING AND MARRIAGE CUSTOMS

MATTHAEI, JOHANN HEINRICH (-)
German biochemist

Heinrich Matthaei conducted experiments in 1961 that revealed the first "word" of the genetic code. In **protein synthesis**, groups of three **DNA bases**, called **codons**, specify which amino acids will be added to a growing popypeptide. A complementary strand of **RNA** conveys the genetic information to the **enzymes** of protein synthesis. Working with American biochemist Marshall Nirenberg, Matthaei showed that the polypeptide poly-phenylalanine is produced when a poly-U RNA molecule is added to a cell-free protein synthesis system. In other words, the RNA triplet UUU codes for a protein molecule consisting solely of the **amino acid** phenylalanine. They identified the polypeptide product by conducting twenty different experiments, each containing a single radiolabeled amino acid; the one that had contained radiolabled phenylalanine now contained a radiolabeled molecule of higher molecular weight. The technique was quickly adopted by other researchers eager to uncover the relationship between other codons and amino acids. Within seven years of the discovery of Matthaei and Nirenberg, the genetic code was complete.

Matthaei had come to the United States to complete a one-year NATO postdoctoral research fellowship. Having completed a degree in plant physiology at the University of Bonn in 1956, he arrived at the Cornell University laboratory of British-born American botanist Frederick Steward (1904–1993) hoping to conduct research on cell-free protein synthesis. Steward was not encouraging, however, and Matthaei inquired whether he could work with Nirenberg—only recently a postdoctoral fellow himself—at the National Institutes of Health (NIH). Matthaei, a technical wizard, solved many of the methodological problems that had slowed Nirenberg in his attempt to break the code. Nirenberg's and Matthaei's biochemical approach to the coding problem contrasted dramatically with the biophysical, mathematical, and cryptoanalytical approaches that had stymied a peculiar group of leading scientists, including British biophysicst **Francis Crick**, Russian-born American theoretical physicist **George Gamow**, and Hungarian-born American physicist Leo Szilard (1898–1964).

Unhappy with working conditions and the distribution of credit at the NIH, Matthaie returned to Germany in 1962. He obtained a position at the Max Plank Institute for Experimental Medicine at Göttingen, where he is now Professor Emeritus. Although he continued to conduct research on protein synthesis, Matthaei soon fell behind in the race to solve the code. Nirenberg, along with the American chemist **Robert Holley** and Indian-born American biochemist **Har Gobind Khorana**, received the Nobel Prize in 1968 for the interpretation of the genetic code.

MAYER, ADOLF (1843-1942)
German microbiologist

German microbiologist Adolf Mayer was one of the first scientists to study tobacco **mosaic** disease. Although he incorrectly concluded that it was caused by **bacteria**, other scientists would draw upon his work as they eventually discovered the disease's true cause—the **tobacco mosaic virus** (also known as TMV). TMV was the first **virus** ever identified.

Mayer attended college at the Universities of Heidelberg, Ghent, and Halle. In 1876, he moved to the Netherlands, where he became head of the Agricultural School at Wageningen. He was soon approached by a group of Dutch tobacco farmers whose crops were being ravaged by disease. They hoped Mayer could find a way to prevent the disease from spreading. (Tobacco had been brought to Europe from the New World, and by the middle of the 1800s, it was a major Dutch crop.)

Because it caused a mottled pattern to form on the plants' leaves, Mayer named it the tobacco mosaic disease. He began a series of experiments to determine its cause. He found that when he took sap from a diseased plant and applied it to a healthy plant, the healthy plant became diseased. Therefore, he concluded that the disease was infectious rather than hereditary. (Mayer may have been the first person to infect plants with a virus as part of an experiment.) Next, he took infected sap and passed it through a single layer of filter paper. The filtrate (the liquid that passed through the filter) remained infectious. From this evidence, Mayer concluded that the cause of the disease was not a fungus because any fungi would have been too large to pass through the filter.

However, when Mayer passed infected sap through two layers of filter paper, he found that the filtrate was no longer infectious. This evidence suggested to him that the cause of the disease was bacterial, because bacteria would normally pass through a single filter, but not two. However, when Mayer looked at infected sap under a light microscope, he could see no bacteria. In addition, he was unable to grow the infectious agent in artificial cultures like he could with other types of bacteria. In an attempt to account for these seemingly conflicting results, Mayer concluded that the cause of tobacco mosaic disease was an unusually small **species** of bacteria.

In 1892, however, Russian botanist Dmitri Ivanovski (1864–1920) found that sap infected with tobacco mosaic disease did in fact remain infectious even after passing through two layers of filter paper. Six years later, Dutch microbiologist Martinus Beijerinck (1851–1931) confirmed Ivanovski's results. He went on to conclude that a previously undiscovered type of infectious agent (rather than bacteria) caused the disease. As a result, these two men, rather than Mayer, are generally credited with the discovery of viruses.

Scientists were eventually able to offer explanations for the results of some of Mayer's experiments. For instance, Mayer was unable to see an infectious agent under a light microscope because viruses are much too small to be seen with such an instrument. In addition, he was incapable of growing the infectious agent artificially because viruses can only reproduce inside living cells.

See also Agricultural genetics; Plant breeding and crop improvement; Viral genetics

MAYR, ERNST (1904-)
German-American biologist

Considered one of the century's most important evolutionary biologists, Ernst Mayr made major contributions to ornithology, evolutionary theory, and the history and philosophy of biology. He is best known for his work on speciation—how one **species** arises from another. In his more than sixty years in the United States, however, he has published hundreds of articles and more than a dozen books. Through these writings, Mayr not only clarified certain aspects of earlier scientific theories but proposed new theories which have changed the course of biological research.

Mayr was born in Kempten, Germany, near the borders of Austria and Switzerland. He was one of three sons of Helene Pusinelli Mayr and Otto Mayr, who was a judge. As a boy, Mayr enjoyed bird watching. He received a broad education, including Latin and Greek. In 1923, he followed in the footsteps of several physicians in his family and began studying for a medical degree at the University of Greifswald. Within two years, however, he had become so enthralled with the evolutionary theories and the exploratory voyages of nineteenth-century British naturalist **Charles Darwin** that he switched from medicine to zoology. He moved to the University of Berlin, where he had once worked at the zoo-

logical museum during his summer vacations. In 1926, he received his Ph.D. in zoology from the university. Soon afterward, he became the zoological museum's assistant curator.

While still working at the museum, Mayr went to Budapest in 1928 to attend a zoological conference. There he met Lionel Walter Rothschild, a British baron and well-known zoologist. Impressed by the young man, Lord Rothschild asked him to lead an ornithological expedition to Dutch New Guinea, in the southwest Pacific. Mayr jumped at the chance. New Guinea at that time was extremely inaccessible, but Mayr was eager to investigate the birds of several remote mountain ranges. The trip was not easy, and Mayr's party suffered a variety of illnesses and injuries. However, Mayr was undaunted and he decided to remain in the region, making a second expedition sponsored by the University of Berlin to mountain ranges in the Mandated Territory of New Guinea. Then in 1929 and 1930, Mayr participated in a third expedition, the American Museum of Natural History's Whitney South Sea Expedition to the Solomon Islands. The experiences and insights crowded into these few years in the south Pacific were to stimulate Mayr's thinking about biology and the development of species for decades to come.

When the expedition to the Solomons was over, Mayr was invited to be a Whitney research associate in ornithology at the American Museum of Natural History in New York City. In 1932, he was named associate curator of the museum's bird collection and he decided to stay in the United States, eventually becoming an American citizen. Over the next decade, Mayr worked at identifying and classifying bird species, studying their geographical distribution and relationships. These years resulted in two of his most influential books.

Mayr published the first of these books in 1941, the *List of New Guinea Birds*. This book, more complex than its title implies, explores the ways closely related species can be distinguished from one another and how variations can arise within a species. In a short article in *Science,* **Stephen Jay Gould** calls each species Mayr discusses in this book "a separate puzzle, a little exemplar of scientific methodology." In writing the *List of New Guinea Birds,* Gould observes, Mayr "sharpened his notion of species as fundamental units in nature and deepened his understanding of evolution."

While most biologists in the 1930s and 1940s accepted the broad premise of Darwin's theory about **evolution**—that species change and evolve through a process called **natural selection**, sometimes loosely called "the **survival of the fittest**"—there was little understanding of how the process worked. If the fittest members of an animal population were the ones that survived, where did those especially well adapted creatures come from initially? These questions were complicated during this period by the fact that there was no clear understanding of exactly what constituted a plant or animal species. Instead, there were two conflicting approaches: one school of thought tried to classify species by their shape and appearance, and another tried to identify species through their genes.

In December 1939, Mayr attended a lecture series at Yale University given by a well-known geneticist named Richard Goldschmidt, who argued that new species can arise through sudden genetic **mutation**. Goldschmidt believed that these changes could take place within a single generation, and Mayr was appalled by what he heard. As Fred Hapgood explained in *Science 84,* "What Mayr heard in those lectures seemed so wrong that he decided, in his own words, to 'eliminate' those ideas 'from the panorama of evolutionary controversies.'" Mayr was convinced that extremely long periods of time were required for the development of a new species, and he set out to demolish Goldschmidt's argument.

The result was *Systematics and the Origin of Species,* which Mayr published in 1942. Based partly on what he had learned in the South Pacific, Mayr argued that geographic speciation is the basic process behind the formation of new species. This theory had been advanced more than a hundred years earlier—even before Darwin—but it had never taken hold. Mayr showed how the process works: when a few animals become separated from their original population and breed among themselves over a great many generations, they eventually change so much that they can no longer breed with their original group. For example, birds from the mainland that once settled on an island may look like their ancestors and will have many similar genetic traits, yet the two groups will not be able to interbreed. The island birds have then become a new species. This concept, which Mayr gradually elaborated, was to form the core of his thinking.

Mayr's ideas about speciation not only found general acceptance, but also won him great respect in scientific circles. *Systematics and the Origin of Species* has been called the "bible" of a generation of biologists. E. O. Wilson remembers how in this book Mayr offered him the "theoretical framework on which to hang facts and plan enterprises" as reported in *Science 84*. Wilson continued, "He gave taxonomy an evolutionary perspective. He got the show on the road."

Over the next several years, Mayr continued to expand and refine his ideas about speciation. In 1946, he founded the Society for the Study of Evolution, becoming its first secretary and later its president. In 1947, he founded the Society's journal, *Evolution,* and served as its first editor. In 1953, at the age of forty-nine, he was named **Alexander Agassiz** Professor of Zoology at Harvard. From 1961 to 1970, he was director of Harvard's Museum of Comparative Zoology, and during those years, he brought about an important expansion of the museum, which is now a major center of biological research. By this time, Mayr was recognized as a leader of what has been called "the modern synthetic theory of evolution." In this context, he is often mentioned with three other eminent researchers in the field: George Gaylord Simpson, **Theodosius Dobzhansky**, and Julian Huxley. In 1963, Mayr published *Animal Species and Evolution,* in which he wrote of man's place in the ecosystem.

In 1975, Mayr retired from Harvard as emeritus professor of zoology. However, he continued to work intensely and his interests continued to expand. He changed careers, as Stephen Jay Gould has observed; from a scientist he became a historian of science. He undertook to write not only about evolution, but also about the entire history of biology. In 1982, he published *The Growth of Biological Thought,* intended to be

Ernst Mayr. *AP/Wide World Photos. Reproduced by permission.*

the first of two volumes. Then in 1991, at the age of eighty-seven, he published yet another carefully wrought discussion of evolution, *One Long Argument.* In that book, he wrote, "The basic theory of evolution has been confirmed so completely that modern biologists consider evolution simply a fact.... Where evolutionists today differ from Darwin is almost entirely on matters of emphasis. While Darwin was fully aware of the probabilistic nature of **selection**, the modern evolutionist emphasizes this even more. The modern evolutionist realizes how great a role chance plays in evolution."

Mayr married Margarete Simon on May 4, 1935, and they had two daughters. For many years, Mayr was a familiar figure around Harvard, a lean man with brown eyes and white hair. Many people noted that Mayr's natural assertiveness and strong writing style added to the weight of his arguments, helping to win over others. During an exceptionally long career, Mayr was awarded ten honorary degrees, including Doctor of Science degrees from both Oxford and Cambridge Universities and a Doctor of Philosophy degree from the University of Paris. In 1954, he was elected to the National Academy of Sciences. His numerous prizes include the Darwin-Wallace Medal in 1958, the Linnean Medal in 1977, the Gregor Mendel medal in 1980, and the Darwin Medal of the Royal Society in 1987. In 1983, he received the Balzan Prize, which has been called the equivalent of the Nobel Prize in the biological sciences.

See also Ecological and environmental genetics; Evolution, evidence of; Evolutionary mechanisms; Natural selection

McCarty, Maclyn (1911-)

American bacteriologist

Maclyn McCarty is a distinguished bacteriologist who has done important work on the biology of streptococci and the origins of rheumatic fever, but he is best known for his involvement in early experiments which established the function of **DNA**. In collaboration with **Oswald Avery** and **Colin Munro MacLeod**, McCarty identified DNA as the substance which controls **heredity** in living cells. The three men published an article describing their experiment in 1944, and their work opened the way for further studies in bacteriological physiology, the most important of which was the demonstration of the chemical structure of DNA by **James Watson** and **Francis Crick** in 1953.

McCarty was born in South Bend, Indiana. His father worked for the Studebaker Corporation and the family moved often, with McCarty attending five schools in three different cities by the time he reached the sixth grade. In his autobiographical book, *The Transforming Principle,* McCarty recalled the experience as positive, believing that moving so often made him an inquisitive and alert child. He spent a year at Culver Academy in Indiana from 1925 to 1926, and he finished high school in Kenosha, Wisconsin. His family moved to Portland, Oregon, and McCarty attended Stanford University in California. He majored in biochemistry under James Murray Luck, who was then launching the *Annual Review of Biochemistry.* McCarty presented public seminars on topics derived from articles submitted to this publication, and he graduated with a B.A. in 1933.

Although Luck asked him to remain at Stanford, McCarty entered medical school at Johns Hopkins in Baltimore in 1933. He was married during medical school days, and he spent a summer of research at the Mayo Clinic in Minnesota. After graduation, McCarty spent three years working in pediatric medicine at the Johns Hopkins Hospital. Even in the decade before penicillin, new chemotherapeutic agents had begun to change infectious disease therapy. McCarty treated children suffering from pneumococcal pneumonia, and he was able to save a child suffering from a streptococcal infection, then almost uniformly fatal, by the use of the newly available sulfonamide antibacterials. Both of these groups of **bacteria**, *Streptococcus* and the *Pneumococcus*, would play important roles throughout the remainder of McCarty's career.

McCarty spent his first full year of medical research at New York University in 1940, in the laboratory of W. S. Tillett. In 1941, McCarty was awarded a National Research Council grant, and Tillett recommended him for a position with Oswald Avery at the Rockefeller Institute, which was one of the most important centers of biomedical research in the United States. For many years, Avery had been working with Colin Munro MacLeod on pneumococci. In 1928, the British microbiologist **Frederick Griffith** had discovered what he called a "transforming principle" in pneumococci. In a series of experiments now considered a turning point in the history of **genetics**, Griffith had established that living individuals of one strain or variety of pneumococci could be changed into another, with different characteristics, by the application of material taken from dead individuals of a second strain. When McCarty joined Avery and MacLeod, the chemical nature of this transforming material was not known, and this was what their experiments were designed to discover.

In an effort to determine the chemical nature of Griffith's transforming principle, McCarty began as more of a lab assistant than an equal partner. Avery and MacLeod had decided that the material belonged to one of two classes of organic compounds: it was either a protein or a nucleic acid. They were predisposed to think it was a protein, or possibly **RNA**, and their experimental work was based on efforts to selectively disable the ability of this material to transform strains of pneumococci. Evidence that came to light during 1942 indicated that the material was not a protein but a nucleic acid, and it began to seem increasingly possible that DNA was the molecule for which they were searching. McCarty's most important contribution was the preparation of a deoxyribonuclease which disabled the transforming power of the material and established that it was DNA. They achieved these results by May of 1943, but Avery remained cautious, and their work was not published until 1944.

In 1946, McCarty was named head of a laboratory at the Rockefeller Institute which was dedicated to the study of the streptococci. A relative of pneumococci, streptococci is a cause of rheumatic fever. McCarty's research established the important role played by the outer cellular covering of this bacteria. Using some of the same techniques he had used in his work on DNA, McCarty was able to isolate the **cell** wall of the streptococcus and analyze its structure.

McCarty became a member of the Rockefeller Institute in 1950; he served as vice president of the institution from 1965 to 1978, and as physician in chief from 1965 to 1974. For his work as co-discoverer of the nature of the transforming principle, he won the Eli Lilly Award in Microbiology and Immunology in 1946 and was elected to the National Academy of Sciences in 1963. He won the first Waterford Biomedical Science Award of the Scripps Clinic and Research Foundation in 1977 and received honorary doctorates from Columbia University in 1976 and the University of Florida in 1977.

See also Transformation

McClintock, Barbara (1902-1992)

American geneticist

Barbara McClintock was a pioneering American geneticist whose discovery of transposable or jumping genes in the 1940s baffled most of her contemporaries for nearly three decades. McClintock spent nearly fifty years working apart from the mainstream of the scientific community. Yet her colleagues had such a high regard for her as an adherent to rigid scientific principles that they accepted her discovery of transposable genes decades before others could confirm her observations. McClintock was eventually awarded the Nobel Prize in medicine or physiology in 1983 for this prescient discovery.

McClintock was born on June 16, 1902, in Hartford, Connecticut. After graduating from Erasmus High School in Brooklyn, McClintock enrolled at Cornell University in 1919. McClintock became interested in the study of the cells, known as cytology, under the tutelage of Lester Sharp, a professor who gave her private lessons on Saturdays. She was invited to take graduate-level **genetics** courses while still in her junior year. She received her B.S. in 1923 and entered graduate school, where she majored in cytology and minored in genetics and zoology. At that time, geneticists favored studies of the fruit fly *Drosophila* that produces a new offspring every ten days. This rapid production of successive generations offered geneticists the opportunity to see quickly the results of genetic traits passed on through crossbreeding. It was studies of *Drosophila* that produced much of the early evidence of the relationship between genes and chromosomes. Chromosomes are strands of **DNA (deoxyribonucleic acid)** that contain the genes that pass hereditary traits from one generation to the next. At Cornell, the main focus of genetic research was corn, or maize, whose varicolored kernels, relatively long life spans, and larger chromosomes (which could be more easily viewed under the microscope) offered geneticists the opportunity to identify specific genetic processes.

While still in graduate school, McClintock had refined and simplified a technique originally developed by John Belling to prepare slides for the study of **chromosome** structures under a microscope. McClintock made modifications to this technique that enabled her to apply it to detailed chromosomal studies of maize. She obtained her M.A. degree in 1925 and her Ph.D. two years later and then was appointed an instructor in Cornell's botany department. McClintock's research at that time focused on linkage groups, the inherited sets or groups of genes that appear on a chromosome. Geneticists had already discovered these linkage groups in *Drosophila* and McClintock set out to relate specific linkage groups to specific chromosomes in maize.

In 1931, McClintock and Harriet Creighton published a landmark study proving a theory geneticists had previously believed without proof: that a correlation existed between genetic and chromosomal crossover. Their study revealed that genetic information was exchanged during the early stages of **meiosis**, the process of **cell division**. They found that this exchange occurred when parts of **homologous chromosomes** (chromosomes on which particular genes are identically located) were exchanged in the same division that produced sex cells. This groundbreaking work and McClintock's successive studies further establishing this relationship eventually led to her election to the National Academy of Sciences in 1944 and presidency of the Genetics Society of America in 1945.

In 1936, she received an appointment to the University of Missouri as an assistant professor of botany, her first faculty appointment. During this time, she performed further experiments delineating the cellular processes of chromosomal interactions and their effects on large-scale mutations. Although this appointment provided a base for McClintock to continue

Barbara McClintock. *UPI/Bettmann. Reproduced by permission.*

her work, the university decided that after five years McClintock did not fit into its future plans.

In the summer before the bombing of Pearl Harbor in December 1941, Marcus Rhoades obtained an invitation for his out-of-work friend and colleague to spend the summer at the Cold Springs Harbor Laboratory in New York. Run by the Carnegie Institute of Washington, the laboratory was a self-contained facility that had its own summerhouses for researchers. On the first of December 1941, McClintock was offered a one-year position at Cold Springs, where she would spend the remainder of her career. By the summer of 1944, McClintock had initiated the studies that would lead to her discovery of genetic transposition. She had noticed different colored spots that did not belong on the green or yellow leaves of a particular plant. She surmised that the larger the discoloration patch, the earlier the **mutation** had occurred, believing that many large patches on the leaves meant the mutation had occurred early in the plant's development. From this observation, McClintock determined that mutations occurred at a constant rate that did not change within a plant's life cycle, which led her to the concept of regulation and control in the passing on of genetic information. Investigating how this passing on of genetic information could be regulated, McClintock next noticed that in addition to these regular mutations there also occurred exceptions, in which there were different types of mutations not normally associated with the plant. Convinced that something must occur at the early stages of meiosis to cause these irregular mutations, McClintock worked with maize to identify what it might be.

McClintock discovered kernels on a self-pollinated ear of corn that had distinctive pigmentation but should have been clear, suggesting a loss of some genetic information that normally would have been passed on to inhibit color. Finally, after two years, she found what she called a controlled breakage in the chromosome and in 1948, she coined the term **transposition** to describe how an element is released from its original position on the chromosome and inserted into a new position. As a result of this jumping **gene**, plant offspring could have an unexpected pattern of **heredity** due to a specific genetic code that other offspring did not have. In fact, two transposable genes were involved in the process: one, which she called a dissociator gene, allowed the release of the activator gene that could then be transposed to a different site.

In 1950, McClintock published her research on transposition, but her work was not well received. Her discovery went against the genetic theory then current that genes were stable components of chromosomes. It wasn't until the 1970s, after technology had been developed that enabled geneticists to study genes on the molecular, rather than cellular, level—that McClintock's ideas were truly understood by the scientific community. Her discovery presaged many later discoveries, such as genetic imprinting or the presetting of genetic activity. Her work also would eventually be used to explain **inheritance** patterns that seemed to lie outside the strict Mendelian law based on simple ratios of dominant and recessive genes. In 1983, McClintock was awarded the Nobel Prize for physiology or medicine for her discovery of mobile genetic systems.

McClintock spent the remainder of her life at Cold Spring Harbor, studying transposition. She was acknowledged as a true pioneer in every sense of the word, and many of her fellow scientists believe that her accomplishments came from an intense desire for knowledge and the commitment to keep working on a problem.

McClung, Clarence Erwin (1870-1946)

American zoologist

Clarence McClung, co-discoverer of the **chromosome** responsible for **sex determination**, was born in Clayton, California. After public high school in Columbus, Kansas, he received a degree from the University of Kansas School of Pharmacy in 1892, taught chemistry and pharmacy for a year, then enrolled in the University of Kansas as an undergraduate. As the protégé of histologist Samuel Wendell Williston (1851–1918), McClung received his B.A. in 1896, M.A. in 1898, and Ph.D. in 1902. While a graduate student at Kansas, he also studied under **Edmund Beecher Wilson** (1856–1939) at Columbia University and William Morton Wheeler (1865–1937) at the University of Chicago. The University of Kansas named him assistant professor of zoology in 1898, associate professor in 1901, and full professor in 1906. Among his students was **Walter Stanborough Sutton**. He also served the University of Kansas from 1902 to 1912 as curator of paleontogical collections and from 1902 to 1906 as dean of the medical school.

From 1912 until he retired in 1940, McClung directed the zoological laboratories at the University of Pennsylvania. He died in Swarthmore, Pennsylvania.

Hermann Henking (1858–1942) noticed in 1891 that a certain chromosome did not pair with the others during **meiosis** in the wasp *Pyrrhocoris*. Building upon Henking's work, McClung began investigating sex determination in the late 1890s. He observed similar uneven pairings in the grasshopper *Xiphidium* in 1899, dismissed inferences that the leftover body was a nucleolus, and named it the "accessory chromosome." Henking and McClung each saw that in any given sperm sample, roughly half had the accessory chromosome, and that females typically had an even number of chromosomes, while males had an odd number, one less. Both suspected that sex determination might be linked to this fact.

McClung's landmark article, "The Accessory Chromosome: Sex Determinant?" in the *Biological Bulletin* reported his discovery that sex is determined during **fertilization** by the presence or absence of the accessory chromosome. McClung mistakenly believed that the egg, responsive to the environment of the mother, could influence or even "select" which sperm fertilized it, and thereby influence the determination of the sex of the **zygote**. He also erred in thinking that the **X chromosome**, as the accessory chromosome soon came to be called, determined maleness. He thought that only males had an accessory chromosome. Rather, as Wilson and Nettie Stevens (1861–1912) showed in 1905, males have one accessory chromosome, while females have two.

Henking's and McClung's work set the stage for Stevens to discover the **Y chromosome** in 1905. Subsequent to her work, the female and male genotypes were respectively designated XX and XY.

See also Heredity; Sex chromosomes; Sex determination

McKusick, Victor A. (1921-   )

American physician

Victor A. McKusick has been called the "father of medical genetics" due his major contributions to the field that have spanned the second half of the twentieth century. From cardiologist, to geneticist, to educator, administrator, and author, McKusick has had an impact on nearly every major milestone in modern **genetics**. It has been suggested that through his writing and teaching he has probably influenced all geneticists in the world today.

McKusick was born in Maine, one of a pair of identical **twins**. He was an undergraduate at Tufts College in the early 1940s, and then attended medical school at Johns Hopkins University School of Medicine where he received his M.D. in cardiology in 1946. Between 1946 and 1954, McKusick progressed from resident to instructor in Medicine at Johns Hopkins University, and in 1954, he accepted a faculty appointment at that institution.

Although trained as a cardiologist, McKusick soon became interested in genetics by working with Marfan syndrome patients. He noticed that in addition to heart defects, Marfan individuals had skeletal abnormalities and dislocated lenses. Because these features occurred together repeatedly, he suggested that they were common characters in an inherited connective tissue disorder, observations that led to his first book, *Heritable Disorders of the Connective Tissue* published in 1956. Thus began a lifelong search to identify other previously undetected genetic disorders in his patients. His success in this endeavor was noticed by the chairman of medicine who suggested that McKusick should consider establishing a specialty clinic. This clinic soon became one of the first Divisions of **Medical Genetics** in the United States.

McKusick's new clinic afforded many opportunities to investigate new genetic diseases. He worked with members of the Amish population and described a variety of inherited maladies, including a hemolytic anemia and several liver disorders. His association with patients with metabolic diseases allowed him to describe two new forms of inherited short stature. In addition to identifying the diseases clinically, McKusick became interested in looking for the genes that caused the diseases. Studies of patients with X-linked disorders revealed a linkage between the genes for **colorblindness** and the enzyme glucose-6-phosphate dehydrogenase (G6PD). Several years later, McKusick succeeded in mapping the Duffy blood group **gene** to **chromosome** 1. This was the first time a gene had been specifically localized to an autosome.

To make genetics more available and understandable to students, teachers, and researchers, McKusick cofounded the Short Course in Medical and Experimental Mammalian Genetics. he course is held for two weeks each summer at the Jackson Laboratory in Bar Harbor, Maine. Top researchers and teachers from around the world are invited as instructors, and the course covers the most up-to-date information on mammalian genetics. The course remains one of the premier genetics series, and applications always exceed the capacity of the meeting. Admission is kept purposely low to allow one on one interaction between the participants and the faculty.

In 1966, McKusick published the first version of *Mendelian Inheritance in Man*, a compilation of all genetic disorders that had been described to that point. This book has been added to regularly over the years, growing to three thick volumes, and has come to be one of the primary references used by geneticists. The extensive listings of genetic diseases including all known associated genes, gene mutations, clinical findings, historical background, and literature citations provide an invaluable resource for anyone trying to diagnose a patient or better understand a genetic abnormality. The advent of the Internet provided an opportunity to make the contents even more widely available, so the entries were converted to computer format and allowed the creation of the "Online Mendelian **Inheritance** in Man" or OMIM web database. The rapid advances in medical genetics provide a continuous stream of suggested additions to this work resulting in daily reviews and updates of the material. The website can be easily reached by anyone with Internet access making it of the most useful tools in genetics.

As early as 1969, McKusick suggested the possibility of **sequencing** the human **genome** and over the years, he continued to promote the idea at workshops and seminars. In 1988, his vision resulted in the founding of HUGO, the Human Genome Organization, a group dedicated to the advancement of human genome sequencing. One year later, he was instrumental in obtaining the funding necessary to start the **Human Genome Project**.

McKusick's work has garnered many important awards and honors that recognize the importance and excellence of his contributions to medicine and genetics. A short list includes membership in the National Academy of Science and the International Pediatrics Hall of Fame, founding co-editor of *Genomics*, and receipt of the Lasker Award, William Allan Award (American Society of Human Genetics), and the John Phillips Award (American College of Physicians). Seventeen academic institutions have awarded him honorary M.D., Ph.D., or Sc. D. degrees. He has served as president for a number of different organizations, and has been editor or sat on the editorial boards of many journals.

McKusick was part of the birth of medical genetics in this country and has actively participated in and encouraged its growth through the years. In addition, he has served as its self appointed historian, collecting images and memories of people, places and events that have shaped medical genetics, and he shares these with colleagues and students at the Bar Harbor Short Course and other scientific gatherings. His pioneering efforts have built a firm foundation for upon which the science of modern genetics continues to grow.

See also Genetic mapping; Inborn errors of metabolism

MEDICAL GENETICS

Medical **genetics** is the study of human **genetic variation** of medical significance. It is a multi-dimensional discipline that encompasses elements from molecular, cellular, organismal, and population biologies. Although once thought to be a relatively esoteric science, genetics now plays a key role in essentially all subspecialties of medicine. It has been suggested that, with the exception of trauma, genetics may be the underlying cause of all medical complaints. Therefore, when a **gene** that causes a particular disease is identified, it becomes important to understand the **inheritance** and expression of that gene.

The era of medical genetics was born when Archibold Garrod identified the first genetic diseases, the **inborn errors of metabolism**. Over the years, other similar disorders have been identified and were shown to be caused by defects, or mutations, in particular biochemical pathways. Examples include **phenylketonuria** (PKU), **Tay Sachs disease**, hypercholesterolemia, **cystic fibrosis**, and galactosemia. These are often progressive diseases with severe mental and physical complications that may result in the death of the patient. In some cases, the negative consequences may be avoided if the disease is treated early. Because of this, most states now mandate a newborn screening program to evaluate all babies for several

Medical students receive instruction on the increasingly important role of genetics in diagnosis and treatment of patients. *Custom Medical Stock Photo. Reproduced by permission.*

of the most severe disorders. Another recent discovery is that inherited genes play a role in drug metabolism and the efficacy of drug treatment. The new field of **pharmacogenetics** is beginning to elucidate to complex role of genes as they relate to this aspect of biochemical genetics.

Cytogenetics is the oldest of the medical genetic sciences, but it did not become important clinically until 1959, when the presence of an extra copy of **chromosome** 21 (Trisomy 21) was first implicated in **Down syndrome**. Since then, many other chromosome abnormalities have been directly associated with different diseases. By viewing metaphase cells with the light microscope, trained personnel can detect a broad range of numerical and structural chromosome anomalies that can then be correlated to specific diagnoses ranging from Down syndrome and Turner syndrome to various types of **cancer**. Recent collaboration with molecular geneticists has provided a new tool, fluorescence *in situ* hybridization, or **FISH**, that allows visualization of smaller abnormalities using fluorescently labeled molecular probes.

Molecular diagnostics is the newest specialty field in genetics and allows a better understanding of **heredity** and disease at the molecular level. Disorders due to mutations as small as a single base change in the **DNA** can now be

explained. Once the **mutation** is known, clinical laboratory testing can usually be set up. For diseases such as cystic fibrosis, **Huntington disease**, and **fragile X syndrome**, molecular assays are now the standard of care. The next step is to look for ways to reverse or repair the defect or to completely replace the mutation with a fully functional gene. Advances in these areas provide the foundations for **gene therapy**.

For the patient, the key to medical genetics is the physician, or clinical geneticist, who evaluates the problem and makes the diagnosis. The medical history, pedigree, and physical examination of the patient are reviewed to narrow the possibilities, and then laboratory testing is performed. The final diagnosis is based on all of the information obtained. The clinical interpretation, potential treatment, prognosis, and recurrence risks are then provided to the patient and the patient's family with the aid of a genetic counselor. Genetic counselors are very important in this process since they are specially trained to turn complex medical jargon into terms patients can easily understand. Once these steps are completed, much of the care and treatment of the patient can be carried out by other medical specialists.

Medical genetics, therefore, reflects a team approach to patient care with input from each of the relevant specialty areas (clinical genetics, cytogenetics, molecular genetics, bio-chemical genetics, and genetic counseling) as well as clinicians with expertise in other areas of medicine. Genetics is, however, a continually evolving science, so in addition to clinical endeavors, medical geneticists are active in research. Spearheaded by the **Human Genome Project**, genes that cause disease are continually being identified. In time, better laboratory tests, more effective treatments, and potentially cures for serious diseases will be available. This represents a major advance over the time when medicine was content to describe a disease and utilize treatments that were merely palliative (i.e., although a patient felt better, he or she still had the underlying disease). Medical genetics represents the promise of medicine for the future—the ability to provide targeted cures for diseases.

See also Cancer genetics; Cytogenetics; Fragile X syndrome; Gene therapy; Genetic Counseling; Molecular diagnostics; Pharmacogenetics

MEIOSIS

Meiosis, also known as reduction division, consists of two successive **cell** divisions in **diploid** cells. The two cell divisions are similar to **mitosis**, but differ in that the chromosomes are duplicated only once, not twice. The end result of meiosis is four daughter cells, each of them **haploid**. Since meiosis only occurs in the sex organs (gonads), the daughter cells are the gametes (spermatozoa or ova), which contain hereditary material. By halving the number of chromosomes in the sex cells, meiosis assures that the fusion of maternal and paternal gametes at **fertilization** will result in offspring with the same **chromosome** number as the parents. In other words, meiosis compensates for chromosomes doubling at fertilization. The two successive nuclear divisions are termed as meiosis I and meiosis II. Each is further divided into four phases (prophase, metaphase, anaphase, and telophase) with an intermediate phase (interphase) preceding each nuclear division.

The events that take place during meiosis are similar in many ways to the process of mitosis, in which one cell divides to form two clones (exact copies) of itself. It is important to note that the purpose and final products of mitosis and meiosis are very different.

Meiosis I is preceded by an interphase period in which the **DNA** replicates (makes an exact duplicate of itself), resulting in two exact copies of each chromosome that are firmly attached at one point, the centromere. Each copy is a sister chromatid, and the pair are still considered as only one chromosome. The first phase of meiosis I, prophase I, begins as the chromosomes come together in homologous pairs in a process known as **synapsis**. **Homologous chromosomes**, or homologues, consist of two chromosomes that carry genetic information for the same traits, although that information may hold different messages (e.g., when two chromosomes carry a message for eye color, but one codes for blue eyes while the other codes for brown). The fertilized eggs (zygotes) of all sexually

reproducing organisms receive their chromosomes in pairs, one from the mother and one from the father. During synapsis, adjacent chromatids from homologous chromosomes "cross over" one another at random points and join at spots called chiasmata. These connections hold the pair together as a tetrad (a set of four chromatids, two from each homologue). At the chiasmata, the connected chromatids randomly exchange bits of genetic information so that each contains a mixture of maternal and paternal genes. This "shuffling" of the DNA produces a tetrad, in which each of the chromatids is different from the others, and a **gamete** that is different from others produced by the same parent. **Crossing over** does, in fact, explain why each person is a unique individual, different even from those in the immediate family.

Prophase I is also marked by the appearance of spindle fibers (strands of microtubules) extending from the poles or ends of the cell as the nuclear membrane disappears. These spindle fibers attach to the chromosomes during metaphase I as the tetrads line up along the middle or equator of the cell. A spindle fiber from one pole attaches to one chromosome while a fiber from the opposite pole attaches to its homologue.

Anaphase I is characterized by the separation of the homologues, as chromosomes are drawn to the opposite poles. The sister chromatids are still intact, but the homologous chromosomes are pulled apart at the chiasmata. Telophase I begins as the chromosomes reach the poles and a nuclear membrane forms around each set. Cytokinesis occurs as the cytoplasm and organelles are divided in half and the one parent cell is split into two new daughter cells. Each daughter cell is now haploid (n), meaning it has half the number of chromosomes of the original parent cell (which is diploid-2n). These chromosomes in the daughter cells still exist as sister chromatids, but there is only one chromosome from each original homologous pair.

The phases of meiosis II are similar to those of meiosis I, but there are some important differences. The time between the two nuclear divisions (interphase II) lacks **replication** of DNA (as in interphase I). As the two daughter cells produced in meiosis I enter meiosis II, their chromosomes are in the form of sister chromatids. No crossing over occurs in prophase II because there are no homologues to synapse. During metaphase II, the spindle fibers from the opposite poles attach to the sister chromatids (instead of the homologues as before). The chromatids are then pulled apart during anaphase II. As the centromeres separate, the two single chromosomes are drawn to the opposite poles. The end result of meiosis II is that by the end of telophase II, there are four haploid daughter cells (in the sperm or ova) with each chromosome now represented by a single copy. The distribution of chromatids during meiosis is a matter of chance, which results in the concept of the **law of independent assortment** in **genetics**.

The events of meiosis are controlled by a protein enzyme complex known collectively as maturation promoting factor (MPF). These **enzymes** interact with one another and with cell organelles to cause the breakdown and reconstruction of the nuclear membrane, the formation of the spindle fibers, and the final division of the cell itself. MPF appears to work in a cycle, with the proteins slowly accumulating during interphase, and then rapidly degrading during the later stages of meiosis. In effect, the rate of synthesis of these proteins controls the frequency and rate of meiosis in all sexually reproducing organisms from the simplest to the most complex.

Meiosis occurs in humans, giving rise to the haploid gametes, the sperm and egg cells. In males, the process of gamete production is known as spermatogenesis, where each dividing cell in the testes produces four functional **sperm cells**, all approximately the same size. Each is propelled by a primitive but highly efficient flagellum (tail). In contrast, in females, oogenesis produces only one surviving egg cell from each original parent cell. During cytokinesis, the cytoplasm and organelles are concentrated into only one of the four daughter cells-the one which will eventually become the female **ovum** or egg. The other three smaller cells, called polar bodies, die and are reabsorbed shortly after formation. The process of oogenesis may seem inefficient, but by donating all the cytoplasm and organelles to only one of the four gametes, the female increases the egg's chance for survival, should it become fertilized.

The process of meiosis does not work perfectly every time, and mistakes in the formation of gametes are a major cause of genetic disease in humans. Under normal conditions, the four chromatids of a tetrad will separate completely, with one chromatid going into each of the four daughter cells. In a disorder known as nondisjunction, chromatids do not separate and one of the resulting gametes receives an extra copy of the same chromosome.

The most common example of this mistake in meiosis is the genetic defect known as Down syndrome, in which a person receives an extra copy of chromosome 21 from one of the parents. Another fairly common form of nondisjunction occurs when the sex chromosomes (XX, XY) do not divide properly, resulting in individuals with **Klinefelter syndrome** or Turner syndrome. Other mistakes that can occur during meiosis include **translocation**, in which part of one chromosome becomes attached to another, and deletion, in which part of one chromosome is lost entirely. The severity of the effects of these disorders depends entirely on the size of the chromosome fragment involved and the genetic information contained in it. Modern technology can detect these genetic abnormalities early in the development of the fetus, but at present, little can be done to correct or even treat the diseases resulting from them.

See also Cell proliferation; Mitosis

MELANOMAS · *see* CANCER GENETICS AND ONCOGENES

MEMES · *see* DEME AND MEME GROUPS

MENDEL, JOHANN GREGOR (1822-1884)
Austrian biologist

The science of **genetics** can trace its origins to biologist Gregor Mendel. In meticulous studies with pea plants, Mendel acquired the experimental data necessary to formulate the laws of **heredity**.

Born in Heinzendorf, Austria (now the Czech Republic), Mendel was the son of a peasant farmer and the grandson of a gardener. As a child, Mendel benefited from the progressive education provided by the local vicar, and he eventually enrolled at the Philosophical Institute in Olmutz (now Olomouc). However, Mendel's worsening financial condition repeatedly forced him to suspend his studies, and in 1843, he entered the Augustinian monastery at Brünn (now Brno).

Although Mendel felt no personal vocation at the time, he believed that the monastery would provide him the best opportunity to pursue his education without the financial worries. He took the name Gregor and eventually was placed in charge of the monastery's experimental garden. In 1847, he was ordained as a priest. Four years later, he was sent to the University of Vienna to study zoology, botany, chemistry, and physics. Following his studies, he returned in 1854 to the monastery and also began teaching the natural sciences at the Brno Technical School.

From then until 1868, in his limited spare time, Mendel performed most of his now-famous heredity experiments. No one had yet been able to make any statistical analysis in breeding experiments, but Mendel's strong background in the natural sciences and his coursework in principles of combinatorial operations prompted him to try. Mendel worked mostly with pea plants, carefully selecting pure varieties that had been cultivated for several years under strictly controlled conditions. He crossed different plants until he produced seven easily distinguishable seed and plant variations (yellow vs. green seeds, wrinkled vs. smooth seeds, tall vs. short plant stems, etc.).

Mendel discovered that, while short plants produced only short offspring, tall plants produced both tall and short offspring. Since only about one-third of the tall plants produced other tall plants, Mendel concluded that there must be two types of tall plants: those that bred true and those that did not.

Mendel continued experimenting, attempting to find intermediate varieties of the offspring by crossing these different plants. In other words, if a tall plant was crossed with a short plant, Mendel expected the result would be a medium-sized plant. Mendel soon found that this was not the case. Mendel crossed short plants with tall plants, planted the seeds from that union, then self-pollinated the plants from this second generation. He followed the results by counting and recording each generation. All of the offspring that sprouted from the short-tall cross were tall, but the offspring from the self-pollination of those tall plants gave him half tall plants (non-pure), one-quarter pure tall, and one-quarter pure short. Tallness, the more powerful characteristic (the one that shows up the most), was dubbed the dominant trait. Shortness, the weaker characteristic (the one that is frequently masked), was called therecessive trait. It did not seem to matter whether

Gregor Mendel. *U.S. National Library of Science. Reproduced by permission.*

Mendel used male or female plants, the results were always the same. Mendel's quiet, methodical investigation took over eight years to complete and involved more than 30,000 plants.

The results of Mendel's initial plant breeding experiments formed the basis of his first law of heredity: the law of segregation. This law states that hereditary units (genes) are always in pairs, that genes in a pair separate during division of a **cell** (the sperm and egg each receive one member of the pair), and that each **gene** in a pair will be present in half the sperm or eggs.

Mendel's further experiments established a second law: the **law of independent assortment**. This law states that each pair of genes is inherited independently of all other pairs. However, it holds true only if the characteristics are located on different chromosomes. By sheer coincidence, Mendel had indeed selected such characteristics. But genes located on the same **chromosome**, as was **Thomas Hunt Morgan** later discovered, are usually inherited together.

In all, Mendel uncovered the following basic laws of heredity: 1) heredity factors must exist; 2) two factors exist for each characteristic; 3) at the time of sex cell formation, heredity factors of a pair separate equally into the gametes (the sperm or the egg); 4) gametes bear only one factor for each characteristic; 5) gametes join randomly no matter what factors they carry; 6) different hereditary factors sort independently when gametes are formed.

Mendel, however, never received acknowledgment during his lifetime for the important contribution he had made to the study of heredity. Although Mendel carefully documented his experiments, presented his findings to the Brünn Society for the Study of Natural Science in 1865, and published *Experiments with Plant Hybrids* the following year, the scientific community was indifferent. Botanists, including Karl Wilhelm von Nägeli, to whom Mendel sent his work, were unaccustomed to statistical analysis. Also, scientists as a whole were hesitant to give credence to such novel theories regarding heredity from such an obscure man.

Mendel died in Brünn on January 6, 1884. Ironically, because of Mendel's refutation of the intermediacy theory that he himself had once posited, **Charles Darwin**'s evolutionary theory was greatly bolstered, for prior to Mendel **natural selection** was believed to be counteracted or compromised by repeated blending of gene characteristics throughout the hereditary cycle. Not until 1900, when Mendel's pioneering work was rediscovered by **Hugo de Vries** and others, did Mendel begin to receive scientific recognition.

See also Dominant genes and traits; Mendelian genetics and Mendel's laws of heredity; Recessive genes and traits

MENDELIAN GENETICS AND MENDELIAN LAWS OF HEREDITY

The foundations of the modern science of **genetics** were laid by **Gregor Mendel**, an Austrian Monk, who carried out experiments on the **inheritance** of characters between generations. Mendel worked on inheritance in sweet-peas, and selected characters that bred true; that is, the characters did not blend into one another in the next generation. Characters chosen for study by Mendel included flower color (such as red versus white), plant height (tall versus dwarf), seed coat (smooth-coated seeds verses wrinkled seeds), pod length (long pods versus short pods), and so on. Mendel eventually formulated the three laws of genetics, known today as the Mendelian laws of inheritance. These are the law of segregation, the **law of independent assortment**, and the law of dominance. Mendel's work went unnoticed for nearly two decades after his death in 1887, but was eventually recognized widely by the scientific community.

In order to understand Mendel's three laws of inheritance, it is necessary to review the plant breeding experiments which inspired the laws. Two of the three laws involve dihybrid crosses where two different sets of traits are studied together. For example, if tall plants (controlled by two dominant alleles, TT), red-flowering (controlled by two dominant alleles, RR), sweet-peas are pollinated by dwarf (controlled by two recessive alleles, tt), white-flowering (controlled by two recessive alleles), plants, the pollen and ova will contain either TR or tr, which represent only a single allele of each set of genes. When **fertilization** takes place, the resulting first filial generation (F1) will all have progeny of the same outward appearance (phenotype); they will all be tall, red-flowering (TtRr) sweet-peas. The progeny of the first filial generation (the F2 generation) are also all tall, red-flowering

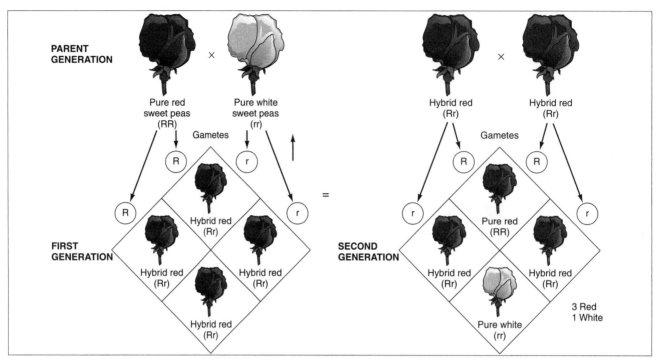

Mendel's law of segregation.

plants, because R, the allele for red and T, the allele for tall are dominant to r and t, which are known as recessive alleles. Since sweet-peas are self-pollinating and each plant now produces gametes of four different genotypes, namely TR, Tr, tR, and tr. When the seeds resulting from the random combination of gametes with one of these genotypes, the second filial generation (F2) will have four different phenotypes: tall with red flowers, tall with white flowers, dwarf with red flowers, and dwarf with white flowers, in the proportion 9:3:3:1.

Mendel found that sweet peas with the same phenotype (purple flowers) had different genotypes which were only expressed in subsequent generations. Purple flowers were produced when the alleles contained at least one allele for purple flowers, which is dominant over the allele for white flowers. White flowers were produced when either both alleles (coded for white flowers the condition) which is the homozygous recessive allele.

During Mendel's time neither the existence of genes nor their structure and function were understood. Indeed, chromosomes remained unknown for several years after Mendel's death. Mendel's experimental plants had factors that occurred in pairs. These factors have only one member of a pair in the gametes (pollen and ova). Mendel described the three laws of inheritance that described the passage of genes from one generation to the next.

Mendel's first law of inheritance is the law of segregation. This states that genes segregate during **gamete** formation into their different alleles. The two members of a pair of alleles separate (segregate) into two different gametes, and exert their influence in the offspring as one of a new pair of alleles.

Segregation is the result of the separate carriage of genes on chromosomes, which are not altered or blended by forming pairs. A **gene** for red flowers in the sweet-pea does not become diluted from having been paired with the gene for white flowers, and is passed to subsequent generations unaltered in the gametes.

The second Mendelian law of inheritance, the law of independent assortment, describes the chance distribution of alleles to the gametes (ova or spermatozoa). If an individual has two pairs of alleles, Aa and Bb, it's gametes will contain equal numbers of the four possible combinations (AB, Ab, aB, ab), with one member from each pair. Independent assortment applies only to genes lying on different chromosomes, and does not-apply to linked genes on the same chromosomes. The F2 generation shows a ratio of 9 Tall red-flowers 3 Dwarf Red-flowering, 3 Tall White-flowering, and 1 Dwarf, White-flowering. A monohybrid cross, which involves a single character, such as plant height produces 12 Tall plants, and 4 Dwarf plants, in a typical 3:1 ratio. Each pair of alleles making up a gene, whether controlling plant height or flower color behaves though each pair were the only gene present for each pair segregates independently of other characters.

The third Mendelian law of inheritance, the law of dominance, states that **heredity** factors (genes) work together as sets, usually as pairs of alleles. The total number of different alleles represents some of the variation available to **species**. Frequently, in heterozygotes (with one dominant and one recessive allele) only the dominant allele of the gene is expressed in the phenotype, the recessive allele being concealed and is not expressed.

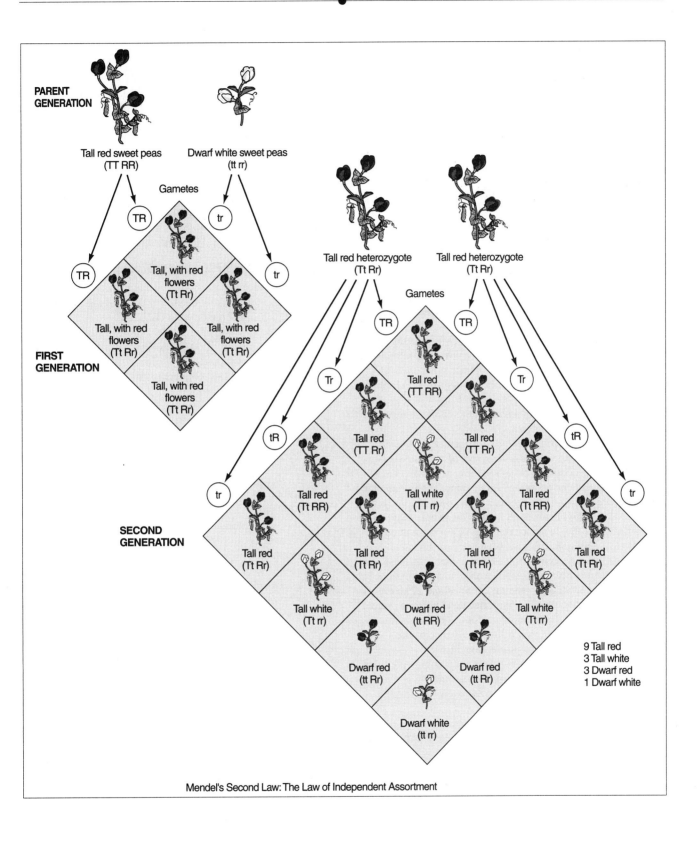

PARENT
GENERATION

Tall red sweet peas
(TT RR)

Dwarf white sweet peas
(tt rr)

Gametes

TR

tr

TR

tr

FIRST
GENERATION

Tall, with red
flowers
(Tt Rr)

Tall, with red
flowers
(Tt Rr)

Tall, with red
flowers
(Tt Rr)

Tall, with red
flowers
(Tt Rr)

Tall red heterozygote
(Tt Rr)

Tall red heterozygote
(Tt Rr)

Gametes

TR

TR

Tr

Tr

tR

tR

tr

tr

SECOND
GENERATION

Tall red
(TT RR)

Tall red
(TT Rr)

Tall red
(TT Rr)

Tall red
(Tt RR)

Tall white
(TT rr)

Tall red
(Tt RR)

Tall red
(Tt Rr)

Tall red
(Tt Rr)

Tall red
(Tt Rr)

Tall red
(Tt Rr)

Tall white
(Tt rr)

Dwarf red
(tt RR)

Tall white
(Tt rr)

Dwarf red
(tt Rr)

Dwarf red
(tt Rr)

Dwarf white
(tt rr)

9 Tall red
3 Tall white
3 Dwarf red
1 Dwarf white

Mendel's Second Law: The Law of Independent Assortment

MENTAL ILLNESS, GENETIC FACTORS

Mental illnesses are all, at root, disorders of thought or emotion, stemming from physical problems. Over the past decade, evidence is accumulating concerning the genetic basis or contribution to these physical aspects of mental illness—the chemistry at nerve **cell** junctions, the metabolism in different regions of the brain, and anatomical anomalies. For example, nerve cell development may become impaired during fetal development. In other cases, chemical malfunctions, at least partially attributable to genetic malfunction, may impair normal nerve cell communication.

Most mental disorders fall into three broad categories: schizophrenia and related disorders, mood disorders (major depression, bipolar disorder) and anxiety disorders (such as compulsive disorder and panic disorder).

The genetic contributions to schizophrenia are subtle. Brain scans of identical **twins** have shown that the brain of the affected individual is differs slightly in the smaller size of certain cavities and in the smaller size of the thalamus, which acts to route sensory signals in the brain. In other mental tasks, involving the recognition of sounds, the involved regions of the brain are underactive in schizophrenia patients. Other developmental differences have been identified. All these point to a genetic involvement. Since 1986, a study at the City of Hope has been ongoing in an attempt to delineate a genetic predisposition to schizophrenia.

In the past decade, a number of mental complications have been implicated as being the result of genetic miscues. Compulsive and addictive behaviors have a genetic basis—heroin addicts have an abnormally long **gene** on **chromosome** 11. The gene functions to produce a receptor for the brain chemical dopamine. Elsewhere, studies on people diagnosed as neurotic have shown defective forms of a protein that functions to clear serotonin from between neurons. Elevated serotonin levels have been associated with anxiety. At least five genes have so far been associated with manic depression. In 1997, researchers linked a gene on chromosome 22 to obsessive-compulsive disorder. The purported gene makes a protein that functions to rid the brain of neurotransmitter chemicals. The malfunctioning gene may allow persistence of the chemical, resulting, perhaps, in recurrent thoughts.

Research over the past two decades has indicated that there is likely not a single gene whose malfunction is responsible for a mental disorder. Rather, there is probably a genetic complexity, in which multiple genes act in concert with non-genetic factors to produce a risk of mental disorder. Part of this risk probably involves the timing of the expression of risk genes during brain development.

See also Behavioral genetics

MESELSON, MATTHEW STANLEY (1930-)

American molecular biologist

Matthew Meselson, in collaboration with biologist **Franklin W. Stahl**, showed experimentally that the **replication** of **deoxyri-**bonucleic acid (**DNA**) in **bacteria** is semiconservative. Semiconservative replication occurs in a double stranded DNA molecule when the two strands are separated and a new strand is copied from the parental strand to produce two new double stranded DNA molecules. The new double stranded DNA molecule is semiconservative because only one strand is conserved from the parent; the other strand is a new copy. (Conservative replication occurs when one offspring of a molecule contains both parent strands and the other molecule offspring contains newly replicated strands) The classical experiment revealing semiconservative replication in bacteria was central to the understanding of the living **cell** and to modern molecular biology.

Matthew Stanley Meselson was born May 24, 1930, in Denver, Colorado. After graduating in 1951 with a Ph.D. in liberal arts from the University of Chicago, he continued his education with graduate studies at the California Institute of Technology in the field of chemistry. Meselson graduated with a Ph.D. in 1957, and remained at Cal Tech as a research fellow. He acquired the position of assistant professor of chemistry at Cal Tech in 1958. In 1960, Meselson moved to Cambridge, Massachusetts to fill the position of associate professor of natural sciences at Harvard University. In 1964, he was awarded professor of biology, which he held until 1976. He was appointed the title of Thomas Dudley Cabot professor of natural sciences in 1976. From that time on, Meselson held a concurrent appointment on the council of the Smithsonian Institute in Washington, D.C.

After graduating from the University of Chicago, Meselson continued his education in chemistry at the California Institute of Technology. It was during his final year at Cal Tech that Meselson collaborated with Franklin Stahl on the classical experiment of semiconservative replication of DNA. Meselson and Stahl wanted to design and perform an experiment that would show the nature of **DNA replication** from parent to offspring using the **bacteriophage** T4 (a **virus** that destroys other cells, also called a phage). The idea was to use an isotope to mark the cells and centrifuge to separate particles that could be identified by their DNA and measure changes in the new generations of DNA. Meselson, Stahl, and Jerome Vinograd originally designed this technique of isolating phage samples. The phage samples isolated would contain various amounts of the isotope based on the rate of DNA replication. The amount of isotope incorporated in the new DNA strands, they hoped, would be large enough to determine quantitatively. The experiments, however, were not successful. After further contemplation, Meselson and Stahl decided to abandon the use of bacteriophage T4 and the isotope and use instead the bacteria *Echerichia coli* (*E. coli*) and the heavy nitrogen isotope 15N as the labeling substance. This time when the same experimental steps were repeated, the analysis showed three distinct types of bacterial DNA, two from the original parent strands and one from the offspring. Analysis of this offspring showed each strand of DNA came from a different parent. Thus the theory of semiconservative replication of DNA had been proven. With this notable start to his scientific career Meselson embarked on another collaboration, this time with biologists **Sydney Brenner**, from the Medical Research Council's Division of Molecular Biology in Cambridge, England, and **François Jacob** from the Pasteur Institute Laboratories in Paris, France. Together, Meselson, Brenner, and Jacob performed a series of experiments in which they showed that when the bacteriophage T4 enters a bacterial cell, the phage DNA incorporates into the cellular DNA and causes the release of **messenger RNA**. Messenger **RNA** instructs the cell to manufacture phage proteins instead of the bacterial cell proteins that are normally produced. These experiments led to the discovery of the role of messenger RNA as the instructions that the bacterial cell reads to produce the desired protein products. These experiments also showed that the bacterial cell could produce proteins from messenger RNA that are not native to the cell in which it occurs.

In his own laboratory at Harvard University, Meselson and a postdoctoral fellow, Robert Yuan, were developing and purifying one of the first of many known **restriction enzymes** commonly used in molecular biological analyses. Restriction **enzymes** are developed by cultivating bacterial strains with phages. Bacterial strains that have the ability to restrict foreign DNA produce a protein called an enzyme that actually chews up or degrades the foreign DNA. This enzyme is able to break up the foreign DNA sequences into a number of small segments by breaking the double stranded DNA at particular locations. Purification of these enzymes allowed mapping of various DNA sequences to be accomplished. The use of purified restriction enzymes became a common practice in the field of molecular biology to map and determine contents of many DNA sequences.

After many years working with the bacteria *E. coli*, Meselson decided to investigate the fundamentals of DNA replication and repair in other organisms. He chose to work on the fruit fly called *Drosophila melanogaster*. Meselson discovered that the fruit fly contained particular DNA sequences that would be transcribed only when induced by heat shock or stress conditions. These particular heat shock genes required a specific setup of DNA **bases** upstream of the initiation site in order for **transcription** to occur. If the number of bases was increased or reduced from what was required, the **gene** would not be transcribed. Meselson also found that there were particular DNA sequences that could be recombined or moved around within the entire **chromosome** of DNA. These moveable segments are termed transposons. Transposons, when inserted into particular sites within the sequence, can either turn on or turn off expression of the gene that is near it, causing mutations within the fly. These studies contributed to the identity of particular regulatory and structural features of the fruit fly as well as to the overall understanding of the properties of DNA.

Throughout his career as a scientist, Meselson has written over 50 papers published in major scientific journals and received many honors and awards for his contributions to the field of molecular biology. In 1963, Meselson received the National Academy of Science Prize for Molecular Biology, followed by the Eli Lilly Award for Microbiology and Immunology in 1964. He was awarded the Lehman Award in

1975 and the Presidential award in 1983, both from the New York Academy of Sciences. In 1990, Meselson received the Science Freedom and Responsibility Award from the American Association for the Advancement of Science. Meselson has also delved into political issues, particularly on government proposals for worldwide chemical and biological weapon disarmament.

See also Viral genetics; Restriction enzymes; Transposition

MESSENGER RNA (MRNA)

Genes transcribe their encoded sequences as a **RNA** template that plays the role of precursor for messenger RNA (mRNA), being thus termed pre-mRNA. Messenger RNA is formed through the splicing of exons from pre-mRNA into a sequence of **codons**, ready for protein **translation**. Therefore, mRNA is also termed mature mRNA, because it can be transported to the cytoplasm, where protein translation will take place in the ribosomal complex.

Transcription occurs in the **nucleus**, through the following sequence of the events. The process of **gene** transcription into mRNA in the nucleus begins with the orginal **DNA** nitrogenous base sequence represented in the direction of transcription (e.g. from the 5' (five prime) end to the 3' (three prime end) as DNA 5'... AGG TCC TAG TAA...3' to the formation of pre-mRNA (for the exemplar DNA cited) with a sequence of 3'... TCC AGG ATC ATT...5' (exons transcribed to pre-mRNA template) then into a mRNA sequence of 5'... AGG UCC UAG UAA...3' (codons spliced into mature mRNA).

Messenger RNA is first synthesized by genes as nuclear heterogeneous RNA (hnRNA), being so called because hnRNAs varies enormously in their molecular weight as well as in their **nucleotide** sequences and lengths, which reflects the different proteins they are destined to code for translation. Most hnRNAs of eukaryotic cells are very big, up to 50,000 nucleotides, and display a poly-A tail that confers stability to the molecule. These molecules have a brief life, being processed during transcription into pre-mRNA and then in mRNA through splicing.

The molecular weight of mRNAs also varies in accordance with the protein size they encode for translation and they necessarily are much bigger than the protein itself, because three nucleotides are needed for the translation of each **amino acid** that will constitute the polypeptide chain during **protein synthesis**. Therefore, the molecular weight of mRNAs varies from some hundreds to thousands of Daltons. In prokaryotic cells, mRNA molecules can even be longer. **Bacteria**, for instance, have a long mRNA molecule that can be translated from different regions, thus resulting in more than one different protein, depending upon the site where the translation starts. Prokaryotic mRNA molecules usually have a short existence of about 2-3 minutes, but the fast bacterial mRNA turnover allows for a quick response to environmental changes by these unicellular organisms. In mammals, the average life span of mRNA goes from 10 minutes up to 2 days. Therefore, eukaryotic **cell** in mammals have different molecules of mRNA that show a wide range of different degradation rates. For instance, mRNA of regulatory proteins, involved either in cell metabolism or in the cell cycle control, generally has a short life of a few minutes, whereas mRNA for globin has a half-life of 10 hours.

The enzyme RNA-polymerase II is the transcriptional element in eukaryotic cells that synthesizes messenger RNA. The general chemical structure of most eukaryotic mRNA molecules contain a 7-methylguanosine group linked through a triphosphate to the 5' extremity, forming a cap. At the other end (i.e., 3' end), there is usually a tail of up to 150 adenylils or poly-A. One exception is the histone mRNA that does not have a poly-A tail. It was also observed the existence of a correlation between the length of the poly-A tail and the half-life of a given mRNA molecule.

The RNA molecules are linear polymers that share a common basic structure comprised of a backbone formed by an alternating polymer of phosphate groups and ribose (a sugar containing five carbon atoms). Organic nitrogenous **bases** i.e., the purines adenine (A) and guanine (A), and the pyrimidines cytosine (C) and uracil (U) are linked together through phosphodiester bridges. These four nitrogenous bases are also termed heterocyclic bases and each of them combines with one of the riboses of the backbone to form a nucleoside, such as adenosine, guanosine, cytidine, and uridine. The combination of a ribose, a phosphate, and a given nitrogenous base by its turn results in a nucleotide, such as adenylate, guanylate, cytidylate, uridylate. Each phosphodiester bridge links the 3' carbon at the ribose of one nucleotide to the 5' carbon at the ribose of the subsequent nucleotide, and so on. RNA molecules fold on themselves and form structures termed hairpin loops, because they have extensive regions of complementary guanine-cytosine (G-C) or adenine-uracil (A-U) pairs. Nevertheless, they are single polynucleotide chains.

The mRNA molecules contain at the 5' end a leader sequence that is not translated, known as UTR (untranslated region) and an initiation codon (AUG), that precedes the coding region formed by the spliced exons, which are termed codons in the mature mRNA. At the end of the coding region, three termination codons (UAG, UAA, UGA) are present, being followed by a trailer sequence that constitutes another UTR, which is by its turn followed by the poly-A tail. The stability of the mRNA molecule is crucial to the proper translation of the transcript into protein. The poly-A tail is responsible by such stability because it prevents the precocious degradation of mRNA by a 3'to 5' exonuclease (a cytoplasmatic enzyme that digests mRNA starting from the extremity 3' when the molecule leaves the cell nucleus). The mRNA of histones, the nuclear proteins that form the **nucleosomes**, do not have poly-A tails, thus constituting an exception to this rule. The poly-A tail also protects the other extremity of the mRNA molecule by looping around and touching the 7-methylguanosine cap attached to the 5' extremity. This prevents the decapping of the mRNA molecule by another exonuclease. The removal of the 7-methylguanosine

exposes the 5' end of the mRNA to digestion by the 5'to 3' exonuclease exonuclease (a cytoplasmatic enzyme that digests mRNA starting from the 5' end). When the translation of the protein is completed, the enzymatic process of deadenylation (i.e., enzymatic digestion of the poly-A tail) is activated, thus allowing the subsequent mRNA degradation by the two above mentioned exonucleases, each working at one of the ends of the molecule.

See also Bacteria; Bacterial genetics; Codons; Exons and introns; Gene splicing; Histones and histone conservation; Nuclease; Nucleic acid; Nucleotide; Polymerase (DNA and RNA); RNA function

METACENTRIC CHROMOSOME · *see*
CHROMOSOME STRUCTURE AND MORPHOLOGY

METAPHASE · *see* CELL DIVISION

METHYLATION

Addition of a methyl group (-CH3) to the **DNA** base, cytosine, creates a modified base called 5-methylcytosine, found in both prokaryotes and **eukaryotes**. This process is catalyzed by **enzymes** called methylases. In **bacteria**, methylation is part of a protective mechanism against foreign DNA. In mammalian cells, methylation of cytosine takes place specifically when it is located immediately 5' to a guanine, a structure called a CpG site; dense clusters of these sites are called CpG islands.

DNA, once methylated, tends to stay methylated through subsequent **cell** divisions, due to work of the maintenance methylase, DNA methyltransferase 1 (Dnmt 1). As DNA is replicated, it becomes transiently hemi-methylated, with methyl groups on the template strands but not the daughter strands. Dnmt 1 recognizes these structures and methylates CpG sites on daughter strands, thereby maintaining the methylation patterns.

Methylation is generally a silencing mechanism in mammalian cells, inhibiting transcriptional activity from the modified genes, including transposons and retroviruses that have accumulated in the mammalian **genome**. It is important for specific tissue and developmental patterns of **gene expression** and plays essential roles in **X chromosome inactivation**, **genomic imprinting**, genome stability and **chromosome** segregation. Normal methylation depends upon the functioning of a wide metabolic network in the cell that involves enzymes, vitamins, and nutrients such as folates.

One way that methylated DNA can affect **transcription** is if binding proteins recognize methylated sites, bind to the DNA and also bind to enzymes (deacetylases) that then alter nearby histones. This changes the local **chromatin** structure and renders methylated DNA inaccessible to the transcriptional machinery. Sometimes, however, methylation is associated with active alleles, in which case it likely interferes with recruitment of transcriptional repressors.

Structural genes are generally hypomethylated in tissues where they are actively expressed and hypermethylated in tissues where they are inactive. As part of the normal regulatory mechanism, methylation frequently acts to suppress the tissue-specific genes whose activity is not required in the particular cell type.

In mammals, X-chromosome inactivation in females is a dosage compensation mechanism, since females have two X chromosomes whereas males have one. Methylation of the inactive X, particularly at **gene** promoters, seems to function as a maintenance mechanism for X inactivation.

Genetic imprinting is another normal process, whereby the two parental alleles are differentiated for selective expression. The importance of methylation in at least the maintenance of imprints is clear, and the nature of the de novo imprint is a subject of active research. Imprinted genes are rich in CpG islands near to clustered direct repeats and most show differences in DNA methylation status between maternal and paternal alleles. To re-set the imprint during transmission between generations, the previous imprint is erased with genome-wide demethylation in germ cells. A *de novo* imprint is then reestablished using specific methylases, though how the sites for methylation are recognized is not yet clear. Once methylated in the parental germ cells, the patterns are maintained after **fertilization** by Dnmt 1.

Failure of the normal methylation mechanisms of the cell can lead to a variety of consequences. In most **tumor** cells, methylation patterns are disrupted, typically showing both genome-wide hypomethylation and region-specific hypermethylation. Each may contribute to tumor progression. Hypomethylation is likely involved with the widespread chromosomal alterations, whereas hypermethylation of specific promoters, notably in **tumor suppressor genes** can functionally inactivate the genes, providing growth advantage to the cell. Methylator phenotype describes a pathway that results in instability through the simultaneous silencing of multiple genes, and is involved with tumors with sporadic mismatch repair deficiency. Because of the involvement of hypermethylation in some cancers, demethylating drugs such as 5-azacytidine are being tested in clinical trials.

The loss of normal methylation control is responsible for a number of non-cancer genetic syndromes. **Fragile X syndrome** is caused by an expanded trinucleotide (CGG) repeat region in the X-linked FMR1 gene. The expanded region is hypermethylated, thereby repressing the gene's expression. Patients with ICF Syndrome (immunodeficiency, centromeric instability, and facial anomalies) show selective undermethylation of pericentromeric satellite DNA, with related chromosome abnormalities, due to mutations in a de novo methyltransferase gene (Dnmt3b). Rett syndrome involves mutations in the gene for a protein that binds to methylated DNA, and which is involved in the chain of regulatory events. Genes that are normally imprinted express from only one parental allele, and are susceptible to the effects of genetic alterations of the complimentary allele. Example of such

abnormalities include Prader-Willi, Angelman and Beckwith-Wiedemann syndromes.

he importance of DNA methylation as a regulatory mechanism in many aspects of cell function has become increasingly apparent with research in the past few years. The major question still puzzling scientists regards the exact nature of the basic signal that alerts the cell to first methylate or demethylate a particular site.

See also Angelman syndrome; Bases; Cancer genetics; Cell differentiation; Control sequences; DNA binding proteins; DNA repair; Familial polyposis; Fragile sites; Gene expression; Hot spots; Locus control regions of gene expression; Prader-Willi syndrome; Repeated sequences; Trinucleotide expansion

MICROBIAL GENETICS

Microbial **genetics** is a branch of genetics concerned with the transmission of hereditary characters in microorganisms. Within the usual definition, microorganisms include prokaryotes like **bacteria**, unicellular or mycelial **eukaryotes** e.g., yeasts and other fungi, and viruses, notably bacterial viruses (bacteriophages). Microbial genetics has played a unique role in developing the fields of molecular and **cell** biology and also has found applications in medicine, agriculture, and the food and pharmaceutical industries.

Because of their relative simplicity, microbes are ideally suited for combined biochemical and genetic studies, and have been successful in providing information on the genetic code and the regulation of **gene** activity. The **operon** model formulated by French biologists **François Jacob** (1920–) and **Jacques Monod** (1910–1976) in 1961, is one well known example. Based on studies on the induction of **enzymes** of lactose catabolism in the bacterium *Escherichia coli,* the operon has provided the groundwork for studies on **gene expression** and regulation, even up to the present day. The many applications of microbial genetics in medicine and the pharmaceutical industry emerge from the fact that microbes are both the causes of disease and the producers of **antibiotics**. Genetic studies have been used to understand variation in pathogenic microbes and also to increase the yield of antibiotics from other microbes.

Hereditary processes in microorganisms are analogous to those in multicellular organisms. In both prokaryotic and eukaryotic microbes, the genetic material is **DNA**; the only known exceptions to this rule are the **RNA** viruses. Mutations, heritable changes in the DNA, occur spontaneously and the rate of **mutation** can be increased by mutagenic agents. In practice, the susceptibility of bacteria to mutagenic agents has been used to identify potentially hazardous chemicals in the environment. For example, the Ames test was developed to evaluate the mutagenicity of a chemical in the following way. Plates containing a medium lacking in, for example, the nutrient histidine are inolculated with a histidine requiring strain of the bacterium *Salmonella typhimurium.* Thus, only cells that revert back to the **wild type** can grow on the medium. If plates are exposed to a mutagenic agent, the increase in the number of mutants compared with unexposed plates can be observed and a large number of revertants would indicate a strong mutagenic agent. For such studies, microorganisms offer the advantage that they have short mean generation times, are easily cultured in a small space under controlled conditions and have a relatively uncomplicated structure.

Microorganisms, and particularly bacteria, were generally ignored by the early geneticists because of their small in size and apparent lack of easily identifiable variable traits. Therefore, a method of identifying variation and mutation in microbes was fundamental for progress in microbial genetics. As many of the mutations manifest themselves as metabolic abnormalities, methods were developed by which microbial mutants could be detected by selecting or testing for altered phenotypes. Positive **selection** is defined as the detection of mutant cells and the rejection of unmutated cells. An example of this is the selection of penicillin resistant mutants, achieved by growing organisms in media containing penicillin such that only resistant colonies grow. In contrast, negative selection detects cells that cannot perform a certain function and is used to select mutants that require one or more extra growth factors. Replica plating is used for negative selection and involves two identical prints of colony distributions being made on plates with and without the required nutrients. Those microbes that do not grow on the plate lacking the nutrient can then be selected from the identical plate, which does contain the nutrient.

The first attempts to use microbes for genetic studies were made in the United States shortly before World War II, when George W. Beadle (1903–1989) and Edward L. Tatum (1909–1975) employed the fungus, *Neurospora,* to investigate the genetics of tryptophan metabolism and nicotinic acid synthesis. This work led to the development of the "one gene one enzyme" hypothesis. Work with bacterial genetics, however, was not really begun until the late 1940s. For a long time, bacteria were thought to lack **sexual reproduction**, which was believed to be necessary for mixing genes from different individual organisms—a process fundamental for useful genetic studies. However, in 1947, **Joshua Lederberg** (1925–) working with Edward Tatum demonstrated the exchange of genetic factors in the bacterium, *Eschereichia coli.* This process of DNA transfer was termed conjugation and requires cell-to-cell contact between two bacteria. It is controlled by genes carried by **plasmids**, such as the fertility (F) factor, and typically involves the transfer of the plasmid from donor torecipient cell. Other genetic elements, however, including the donor cell **chromosome**, can sometimes also be mobilized and transferred. Transfer to the host chromosome is rarely complete, but can be used to map the order of genes on a bacterial **genome**.

Other means by which foreign genes can enter a bacterial cell include **transformation**, **transfection**, and **transduction**. Of the three processes, transformation is probably the most significant. Evidence of transformation in bacteria was first obtained by the British scientist, Fred Griffith (1881–1941) in the late 1920s working with *Streptococcus pneumoniae* and the process was later explained in the 1930s by Oswald Avery (1877–1955) and his associates at the Rockefeller Institute in New York. It was discovered that certain bacteria exhibit competence, a state in which cells are able to take up free DNA released by other bacteria. This is the process known as transformation, however, relatively few microorganisms can be naturally transformed. Certain laboratory procedures were later developed that make it possible to introduce DNA into bacteria, for example electroporation, which modifies the bacterial membrane by treatment with an electric field to facilitate DNA uptake. The latter two processes, transfection and transduction, involve the participation of viruses for nucleic acid transfer. Transfection occurs when bacteria are transformed with DNA extracted from a bacterial **virus** rather than from another bacterium. Transduction involves the transfer of host genes from one bacterium to another by means of viruses. In generalized transduction, defective virus particles randomly incorporate fragments of the cell DNA; virtually any gene of the donor can be transferred, although the efficiency is low. In specialized transduction, the DNA of a temperate virus excises incorrectly and brings adjacent host genes along with it. Only genes close to the integration point of the virus are transduced, and the efficiency may be high.

After the discovery of DNA transfer in bacteria, bacteria became objects of great interest to geneticists because their rate of reproduction and mutation is higher than in larger organisms; i.e., a mutation occurs in a gene about one time in ten million gene duplications, and one bacterium may produce ten billion offspring in 48 hours. Conjugation, transformation, and transduction have been important methods for mapping the genes on the chromosomes of bacteria. These techniques, coupled with restriction enzyme analysis, **cloning** DNA **sequencing**, have allowed for the detailed studies of the bacterial chromosome. Although there are few rules governing gene location, the genes encoding enzymes for many biochemical pathways are often found tightly linked in operons in prokaryotes. Large scale sequencing projects revealed the complete DNA sequence of the genomes of several prokaryotes, even before eukaryotic genomes were considered.

See also Cloning vector; Fertility factor; Fungal genetics; Viral genetics

MICROCHIMERISM

In Greek mythology, a chimera is a fire-breathing monster with the head of a lion, the body of a goat, and the tail of a serpent. In biology, a chimera is an organism consisting of tissues whose cells derive from two or more genetic sources. An apple tree, for example, can include grafts that enable it to produce a variety of apple types. In humans, such genetic mosaics reflect specific, intimate circumstances. The phenomenon is termed microchimerism because a one-cell population is typically much smaller than the other.

An organ transplant is an unnatural procedure that suddenly deposits many cells from one individual into another whose own organ has failed. It is similar to grafting in plants, and in fact, if the donor tissue reacts against the recipient's body, the condition is termed "graft-versus-host disease" (GVHD). When a transplant works, a peaceful co-existence evolves as recipient cells infiltrate graft tissue, and perhaps also as the transplanted cells alter their **gene expression** to modify their **cell** surfaces so that they are less likely to provoke the recipient's immune system. Microchimerism in transplant medicine does not refer to the transplanted material, but to small numbers of donor cells found outside the region of the transplant. For example, following transplantation of a liver, lymphoid and dendritic cells of the immune system often turn up outside of the liver, but with cell surfaces that match those of the donor, not the recipient.

Seeding of an organ recipient's body appears to be rather common. In one short-term study of nine women who had received lungs, upon autopsy, donor white blood cells were found in the recipients' own lungs, as well as in their skin, liver, kidneys and lymph nodes. In a longer-term study, 5 of 15 people who had lived at least 20 years with a transplanted kidney displayed microchimerism. More than half of heart transplant recipients show microchimerism after three years with the new organ.

Microchimerism is most easily detected in females who harbor male cells, because the presence of a **Y chromosome** highlights the unusual cells. A female who has received bone marrow from a male, for example, will produce white blood cells whose nuclei can be stained to reveal Y chromosomes—she essentially produces "male" blood.

A second situation that sets the stage for microchimerism is the intimate relationship of pregnancy, when one or two individuals develop within the body of another. **Twins** can swap cells while in the uterus. When this occurs between dizygotic (fraternal) twins during the time that the immune system is forming, each can develop tolerance against tissues of the other, so that after birth, they can donate blood to each other, even if their blood types are not compatible.

Cells also travel from fetus to pregnant woman, and in the other direction as well. As long ago as 1893, doctors noted on autopsies that trophoblasts, which are cells that form at the edge of the placenta, were in the lungs of pregnant women. In 1969, researchers detected Y chromosomes in the nuclei of white blood cells in the circulation of pregnant women, and a 1977 study identified fetal blood cells in 40–70% of pregnant women by the third trimester. Fetal microchimerism can be long-lasting. In 1996, Diana Bianchi's group at the New England Medical Center detected **fetal cells** in a woman's blood 27 years after that fetus occupied her body.

The discovery of microchimerism has spurred development of prenatal tests based on sampling **DNA** from fetal cells in maternal blood. The telltale cells are rare—one fetal cell per 100,000–1,000,000 maternal cells. Parallel to this research has been the emerging hypothesis that microchimerism gone awry might cause autoimmune disease, specifically systemic sclerosis, which is more commonly known as systemic scleroderma ("hard skin"). This condition may instead arise from persisting fetal cells that provoke a woman's immune system to produce

autoantibodies. Symptoms of scleroderma include the hallmark skin hardening, fatigue, swollen joins, stiff fingers, and damage to internal organs.

In 1998, J. Lee Nelson, of the Fred Hutchinson Cancer Research Center in Seattle and colleagues a the University of Chicago Medical Center and New England Medical Center identified more persistent microchimerism among women who have scleroderma than among controls. They looked at 40 women who had borne sons many years earlier—17 women with scleroderma, seven healthy patients who were sisters of affected women, and 16 healthy controls—using the **polymerase chain reaction** to amplify Y-specific DNA sequences in T cells in skin. The levels were elevated in the affected women, but most telling was the fact that these women had immune systems that were remarkably similar to those of the sons—specifically, in the nature of the major histocompatibility class I and II antigens, as well as minor histocompatibility antigens. Additional evidence that scleroderma arises from a malfunctioning immune system is that the symptoms resemble those of GVHD, and antigens on the cells of scleroderma lesions match those of transplanted cells that cause GVHD.

From these clues has emerged a compelling hypothesis. Because healthy mothers of sons also often have traces of Y **chromosome** sequences, such fetal seeding of maternal tissues may be normal, and perhaps part of a mechanism to dampen a pregnant woman's immune response so that she does not reject a fetus. But, if the fetus and the pregnant woman are too alike immunogenetically, the process somehow goes awry, and errant fetal cells colonize unusual tissues, such as the skin. Many years later, the woman's immune system recognizes the cells as foreign and initiates an immune response—which appears to be autoimmune ("against self").

Researchers at first thought that this new explanation for an autoimmune disease might apply to the other autoimmune conditions that affect mothers more than others. Candidates include multiple sclerosis, systemic lupus erythromatosis, and primary biliary cirrhosis (PBC). However, results of a similar search for telltale Y sequences in mothers of sons with PBC did not yield the same outcome as the scleroderma study. In this second disorder, healthy controls had the same levels of Y-specific DNA in liver cells as affected women. Therefore, microchimerism as a cause of an autoimmune disorder may depend upon the nature of the affected tissue.

See also Immunogenetics; Transplantation genetics

MIESCHER, II, JOHANN FRIEDREICH (1844-1895)
Swiss biochemist

DNA is now a household word, but its existence was unknown until 1869. During that year, Johann Friedreich Miescher II first isolated DNA from cells. Born in Basel, Switzerland in 1844, he was the son and nephew of distinguished anatomists at the University of Basel. Miescher himself studied medicine at the University, but decided to study chemistry instead because he was concerned that his partial deafness, caused by an earlier bout with typhus, would impair his ability as a physician.

Miescher became a student of Felix Hoppe-Seyler, whose laboratory was located in the castle of Tubingen, Germany. In 1869, Miescher collected the dressings of wounded soldiers and scraped the pus off the dressings. From the white blood cells in the pus he isolated the nuclei from the cytoplasm; and from the nuclei, he extracted a substance that contained phosphorus in addition to the carbon, hydrogen, oxygen, and nitrogen typically found in organic molecules. Miwscher determined that this material was not protein because it was unaffected by the action of pepsin. He called the material "nuclein." Only after 1889 was nuclein called nucleic acid.

The discovery of nuclein astonished Hoppe-Seyler, who had previously discovered the substance lecithin. Up to that point, lecithin was the only natural substance known to contain both phosphorus and nitrogen. Hoppe-Seyler wouldn't allow Miescher to publish his work for two years, because he wanted to investigate nuclein himself. He was able to extract nuclein from **yeast** cells.

In 1855, Miescher founded Switzerland's first physiological institute. Apart from his discovery of nucleic acids, he is also credited with the discovery that the rate of breathing is determined, using a feedback mechanism, by the concentration of carbon dioxide in the blood.

Miescher also continued his work on nuclein extracted from salmon sperm. He was interested in the chemical agent of **fertilization**, and said in 1874, "If one wants to assume that a single substance...is the specific cause of fertilization than one should undoubtedly first of all think of nuclein." Eighty years later, Watson and Crick would prove his hunch to be correct, but Miescher did not follow it up himself. Albrecht Kossle, a later student of Hoppe-Seyler, analyzed the chemical content of nuclein in detail, and in 1910, was awarded the Nobel Prize in medicine or physiology for his work on proteins and nucleic acids.

When Miescher became ill from tuberculosis, he required treatment in a sanatorium. Ever the scientist, he took advantage of the opportunity to study the effects of high altitude on the makeup of blood. He observed that the count of erythrocytes increased with the altitude. Miescher died of tuberculosis in Davis, Switzerland, in 1895. Miescher is honored today by the Fried Reich Miescher Institute in Basel, and the Fried Reich Miescher Laboratory of the Max Planck Society in Tubing.

See also DNA (deoxyribonucleic acid)

MILLER, STANLEY L. (1930-)
American chemist

Stanley Lloyd Miller is most noted for his experiments that attempted to replicate the chemical conditions that may have first given rise to life on earth. In the early 1950s he demon-

strated that amino acids could have been created under primordial conditions. Amino acids are the fundamental units of life; they join together to form proteins, and as they grow more complex they eventually become nucleic acids, which are capable of replicating. Miller has hypothesized that the oceans of primitive earth were a mass of molecules, a prebiological "soup," which over the course of a billion years became a living system.

Miller was born in Oakland, California, the youngest of two children. His father, Nathan Harry Miller, was an attorney and his mother, Edith Levy Miller, was a homemaker. Miller attended the University of California at Berkeley and received his B.S. degree in 1951. He began his graduate studies at the University of Chicago in 1951.

In an autobiographical sketch entitled "The First Laboratory Synthesis of Organic Compounds under Primitive Earth Conditions," Miller recalled the events that led to his famous experiment. Soon after arriving at the University of Chicago, he attended a seminar given by **Harold Urey** on the origin of the solar system. Urey postulated that the earth was reducing when it was first formed—in other words, there was an excess of molecular hydrogen. Strong mixtures of methane and ammonia were also present, and the conditions in the atmosphere favored the synthesis of organic compounds. Miller wrote that when he heard Urey's explanation, he knew it made sense: "For the nonchemist the justification for this might be explained as follows: it is easier to synthesize an organic compound of biological interest from the reducing atmosphere constituents because less chemical bonds need to be broken and put together than is the case with the constituents of an oxidizing atmosphere."

After abandoning a different project for his doctoral thesis, Miller told Urey that he was willing to design an experiment to test his hypothesis. However, Urey expressed reluctance at the idea because he considered it too time consuming and risky for a doctoral candidate. But Miller persisted, and Urey gave him a year to get results; if he failed he would have to choose another thesis topic. With this strict deadline Miller set to work on his attempt to synthesize organic compounds under conditions simulating those of primitive earth.

Miller and Urey decided that ultraviolet light and electrical discharges would have been the most available sources of energy on earth billions of years ago. Having done some reading into amino acids, Miller hypothesized that if he applied an electrical discharge to his primordial environment, he would probably get a deposit of hydrocarbons, organic compounds containing carbon and hydrogen. As he remembered in "The First Laboratory Synthesis of Organic Compounds": "We decided that amino acids were the best group of compounds to look for first, since they were the building blocks of proteins and since the analytical methods were at that time relatively well developed." Miller designed an apparatus in which he could simulate the conditions of prebiotic earth and then measure what happened. A glass unit was made to represent a model ocean, atmosphere, and rain. For the first experiment, he filled the unit with the requisite "primitive atmosphere"—methane, hydrogen, water, and ammonia—and then submitted the mixture to a low-voltage spark over night. There was a layer of hydrocarbons the next morning, but no amino acids.

Miller then repeated the experiment with a spark at a higher voltage for a period of two days. This time he found no visible hydrocarbons, but his examination indicated that glycine, an **amino acid**, was present. Next, he let the spark run for a week and found what looked to him like seven spots. Three of these spots were easily identified as glycine, alpha-alanine, and beta-alanine. Two more corresponded to a-amino-n-butyric acid and aspartic acid, and the remaining pair he labeled A and B.

At Urey's suggestion, Miller published "A Production of Amino Acids under Possible Primitive Earth Conditions" in May of 1953 after only three-and-a-half months of research. Reactions to Miller's work were quick and startling. Articles evaluating his experiment appeared in major newspapers; when a Gallup poll asked people whether they thought it was possible to create life in a test tube; seventy-nine percent of the respondents said no.

After Miller finished his experiments at the University of Chicago, he continued his research as an F. B. Jewett Fellow at the California Institute of Technology from 1954 to 1955. Miller established the accuracy of his findings by performing further tests to identify specific amino acids. He also ruled out the possibility that **bacteria** might have produced the spots by heating the apparatus in an autoclave for eighteen hours (fifteen minutes is usually long enough to kill any bacteria). Subsequent tests conclusively identified four spots that had previously puzzled him. Although he correctly identified the a-amino-n-butyric acid, what he had thought was aspartic acid (commonly found in plants) was really iminodiacetic acid. Furthermore, the compound he had called A turned out to be sarcosine (N-methyl glycine), and compound B was N-methyl alanine. Other amino acids were present but not in quantities large enough to be evaluated.

Although other scientists repeated Miller's experiment, one major question remained: was Miller's apparatus a true representation of the primitive atmosphere? This question was finally answered by a study conducted on a meteorite which landed in Murchison, Australia, in September of 1969. The amino acids found in the meteorite were analyzed and the data compared to Miller's findings. Most of the amino acids Miller had found were also found in the meteorite. On the state of scientific knowledge about the origins of human life, Miller wrote in "The First Laboratory Synthesis of Organic Compounds" that "the synthesis of organic compounds under primitive earth conditions is not, of course, the synthesis of a living organism. We are just beginning to understand how the simple organic compounds were converted to polymers on the primitive earth...nevertheless we are confident that the basic process is correct."

Miller's later research has continued to build on his famous experiment. He is looking for precursors to **ribonucleic acid (RNA)**. "It is a problem not much discussed because there is nothing to get your hands on," he told Marianne P. Fedunkiw in an interview. He is also examining the natural occurrence of clathrate hydrates, compounds of ice and gases that form under high pressures, on the earth and other parts of the solar system.

Miller has spent most of his career in California. After finishing his doctoral work in Chicago, he spent five years in the department of biochemistry at the College of Physicians and Surgeons at Columbia University. He then returned to California as an assistant professor in 1960 at the University of California, San Diego. He became an associate professor in 1962 and eventually full professor in the department of chemistry.

Miller served as president of the International Society for the Study of the Origin of Life (ISSOL) from 1986 to 1989. The organization awarded him the Oparin Medal in 1983 for his work in the field. Outside of the United States, he was recognized as an Honorary Councilor of the Higher Council for Scientific Research of Spain in 1973. In addition, Miller was elected to the National Academy of Sciences. Among Miller's other memberships are the American Chemical Society, the American Association for the Advancement of Science, and the American Society of Biological Chemists.

See also Evolutionary mechanisms; Origin of life

MILLER-UREY EXPERIMENT

A classic experiment in molecular biology and **genetics**, the Miller-Urey experiment, established that the conditions that existed in Earth's primitive atmosphere were sufficient to produce amino acids, the subunits of proteins comprising and required by living organisms. In essence, the Miller-Urey experiment fundamentally established that Earth's primitive atmosphere was capable of producing the building blocks of life from inorganic materials.

In 1953, University of Chicago researchers **Stanley L. Miller** and Harold C. Urey set up an experimental investigation into the molecular origins of life. Their innovative experimental design consisted of the introduction of the molecules thought to exist in early Earth's primitive atmosphere into a closed chamber. Methane (CH_4), hydrogen (H_2), and ammonia (NH_3) gases were introduced into a moist environment above a water-containing flask. To simulate primitive lightning discharges, Miller supplied the system with electrical current.

After a few days, Miller observed that the flask contained organic compounds and that some of these compounds were the amino acids that serve as the essential building blocks of protein. Using chromatological analysis, Miller continued his experimental observations and confirmed the ready formation of amino acids, hydroxy acids, and other organic compounds.

Although the discovery of **amino acid** formation was of tremendous significance in establishing that the raw materials of proteins were easily to obtain in a primitive Earth environment, there remained a larger question as to the nature of the origin of genetic materialsmdash;in particular the origin of **DNA** and **RNA** molecules.

Continuing on the seminal work of Miller and Urey, in the early 1960s Juan Oro discovered that the **nucleotide** base adenine could also be synthesized under primitive Earth conditions. Oro used a mixture of ammonia and hydrogen cyanide (HCN) in a closed aqueous enviroment.

Oro's findings of adenine, one of the four nitrogenous **bases** that combine with a phosphate and a sugar (deoxyribose for DNA and ribose for RNA) to form the nucleotides represented by the genetic code: (adenine (A), thymine (T), guanine (G), and cytosine (C). In RNA molecules, the nitrogenous base uracil (U) substitutes for thymine. Adenine is also a fundamental component of **adenosine triphosphate** (**ATP**), a molecule important in many genetic and cellular functions.

Subsequent research provided evidence of the formation of the other essential nitrogenous bases needed to construct DNA and RNA.

The Miller-Urey experiment remains the subject of scientific debate. Scientist continue to explore the nature and composition of Earth's primitive atmosphere and thus, continue to debate the relative closeness of the conditions of the Miller-Urey experiment (e.g., whether or not Miller's application of electrical current supplied relatively more electrical energy than did lightning in the primitive atmosphere). Subsequent experiments using alternative stimuli (e.g., ultraviolet light) also confirm the formation of amino acids from the gases present in the Miller-Urey experiment. During the 1970s and 1980s, astrobiologists and astrophyicists, including American physicist Carl Sagan, asserted that ultraviolet light bombarding the primitive atmosphere was far more energetic that even continual lightning discharges. Amino acid formation is greatly enhanced by the presence of an absorber of ultraviolet radiation such as the hydrogen sulfide molecules (H_2S) also thought to exist in the early Earth atmosphere.

Although the establishment of the availability of the fundamental units of DNA, RNA, and proteins was a critical component to the investigation of the origin of biological molecules and life on Earth, the simple presence of these molecules is a long step from functioning cells. Scientists and evolutionary biologists propose a number of methods by which these molecules could concentrate into a crude **cell** surrounded by a primitive membrane.

See also Alternate genetic codes; Ancient DNA; Archae; Archaeogenetics: Genetic studies of ancient peoples; Base pairing (bp); Biochemistry; Chargaff's rules; Chromosome; Codons; Comparative genomics; DNA structure; DNA synthesis; Double helix; Mitochondrial inheritance; Origin of life; RNA function

MILSTEIN, CÉSAR (1927-)

English biochemist

César Milstein conducted one of the most important late-twentieth-century studies on antibodies. In 1984, Milstein

received the Nobel Prize in physiology or medicine, shared with **Niels K. Jerne** and Georges Kohler, for his outstanding contributions to immunology and immunogenetics. Milstein's research on the structure of antibodies and their genes, through the investigation of **deoxyribonucleic acid** (**DNA**) and **ribonucleic acid** (**RNA**), has been fundamental for a better understanding of how the human immune system works.

Milstein was born on October 8, 1927, in the eastern Argentine city of Bahía Blanca, one of three sons of Lázaro and Máxima Milstein. He studied biochemistry at the National University of Buenos Aires from 1945 to 1952, graduating with a degree in chemistry. Heavily involved in opposing the policies of President Juan Peron and working part-time as a chemical analyst for a laboratory, Milstein barely managed to pass with poor grades. Nonetheless, he pursued graduate studies at the Instituto de Biología Química of the University of Buenos Aires and completed his doctoral dissertation on the chemistry of aldehyde dehydrogenase, an alcohol enzyme used as a catalyst, in 1957.

With a British Council scholarship, he continued his studies at Cambridge University from 1958 to 1961 under the guidance of **Frederick Sanger**, a distinguished researcher in the field of **enzymes**. Sanger had determined that an enzyme's functions depend on the arrangement of amino acids inside it. In 1960 Milstein obtained a Ph.D. and joined the Department of Biochemistry at Cambridge, but in 1961, he decided to return to his native country to continue his investigations as head of a newly-created Department of Molecular Biology at the National Institute of Microbiology in Buenos Aires.

A military coup in 1962 had a profound impact on the state of research and on academic life in Argentina. Milstein resigned his position in protest of the government's dismissal of the Institute's director, Ignacio Pirosky. In 1963 he returned to work with Sanger in Great Britain. During the 1960s and much of the 1970s, Milstein concentrated on the study of antibodies, the protein organisms generated by the immune system to combat and deactivate antigens. Milstein's efforts were aimed at analyzing myeloma proteins, and then DNA and RNA. Myeloma, which are tumors in cells that produce antibodies, had been the subject of previous studies by Rodney R. Porter, MacFarlane Burnet, and Gerald M. Edelman, among others.

Milstein's investigations in this field were fundamental for understanding how antibodies work. He searched for mutations in laboratory cells of myeloma but faced innumerable difficulties trying to find antigens to combine with their antibodies. He and Köhler produced a **hybrid** myeloma called hybridoma in 1974. This **cell** had the capacity to produce antibodies but kept growing like the cancerous cell from which it had originated. The production of monoclonal antibodies from these cells was one of the most relevant conclusions from Milstein and his colleague's research. The Milstein-Köhler paper was first published in 1975 and indicated the possibility of using monoclonal antibodies for testing antigens. The two scientists predicted that since it was possible to hybridize antibody-producing cells from different origins, such cells could be produced in massive cultures. They were, and the technique

César Milstein. *Photo Researchers, Inc. Reproduced by permission.*

consisted of a fusion of antibodies with cells of the myeloma to produce cells that could perpetuate themselves, generating uniform and pure antibodies.

In 1983 Milstein assumed leadership of the Protein and Nucleic Acid Chemistry Division at the Medical Research Council's laboratory. In 1984 he shared the Nobel Prize with Köhler and Jerne for developing the technique that had revolutionized many diagnostic procedures by producing exceptionally pure antibodies. Upon receiving the prize, Milstein heralded the beginning of what he called "a new era of immunobiochemistry," which included production of molecules based on antibodies. He stated that his method was a by-product of basic research and a clear example of how an investment in research that was not initially considered commercially viable had "an enormous practical impact." By 1984 a thriving business was being done with monoclonal antibodies for diagnosis, and works on vaccines and **cancer** based on Milstein's breakthrough research were being rapidly developed.

In the early 1980s Milstein received a number of other scientific awards, including the Wolf Prize in Medicine from the Karl Wolf Foundation of Israel in 1980, the Royal Medal from the Royal Society of London in 1982, and the Dale

Medal from the Society for Endocrinology in London in 1984. He is a member of numerous international scientific organizations, among them the U.S. National Academy of Sciences and the Royal College of Physicians in London.

See also Antibody and antigen; Antibody, monoclonal; Autoimmune diseases; Immunogenetics; Transplantation genetics

MIMICRY

Mimicry may broadly be defined as imitation or copying of an action or image. In biological systems, mimicry specifically refers to the fascinating resemblance of an organism, called the "mimic," to another somewhat distantly related organism, called the "model." The set of mimic and model **species** involved is often referred to as a mimicry complex. Usually through escape from predation, the mimicry of a trait or traits helps the mimic to survive. This, coupled with the fact that the resemblance traits are genetically based, implies that mimicry complexes have been shaped by **natural selection**. There are two major types of mimicry, Batesian and Müllerian, named after the naturalists that first theorized them upon their observations of butterflies. There are a few other types that are not as prevalent, such as aggressive mimicry.

In 1862, H.W. Bates presented an hypothesis explaining the similar color patterns in several species sets of tropical butterflies in different families. His hypothesis was one of the early applications of **Charles Darwin**'s theory of natural **selection**. Bates reasoned that an edible butterfly species that was susceptible to predation would evolve, due to selection by a bird predator, to look like an unpalatable, or distasteful model species. If the mimic was rarer than the model, then birds would encounter the distasteful model more frequently, and would learn to avoid all butterflies that looked like the distasteful ones. In fact, the relative rarity of the model was to Bates a prerequisite for such a phenomenon to evolve. As mimicry theory has progressed, mathematical models show that relative abundances of models and mimics, as well as relative palatability of the two species, will determine the outcome.

In the 1870s, Fritz Müller theorized a different type of mimicry. His idea, also based on sets of butterfly species, was that several species, all somewhat distasteful, would evolve to look like each other. Such an evolutionary strategy would, in effect, reduce predation on any of the species because the predator would learn to avoid a single color pattern, but since all of them had the same pattern, they would all be safe from predation. The rarer form, say species 1, would eventually converge to look like the more common form, species 2, as the individuals that looked too different from species 2 would be rapidly selected out by predators. Since species 2 was more common, the predator would have had more experience with it and would have had more opportunity to learn to avoid it than with species 1, the rare species. Individuals of species 1 that resembled species 2 would benefit from the predator's learned

For protecton, this Indian leaf butterfly is able to mimic the look of its surroundings. *Photograph by F. Poelking. Stock Market/Zefa Germany. Reproduced by permission.*

avoidance of species 2, and thus would proliferate. The species would evolve to share a similar pattern as relative frequencies shifted. If the two species were equal in abundance, Müller reasoned, it would not be possible to distinguish mimic from model, as both had converged on a common phenotype, or appearance.

A less common but equally fascinating type of mimicry involves not only a model and a mimic, but a "dupe" species that is tricked by the mimicry. In the previously noted types of mimicry, the dupe is the predator who is tricked out of a potential food source, but in aggressive mimicry, the word is especially appropriate as being duped as lethal. In aggressive mimicry, the mimic is a predator who imitates, usually in behavior, a model species in order to draw in a dupe, who then becomes prey. An example of this occurs in spiders of the family Mimetidae (mimic), who attempting to draw in spiders of other species (dupe) as prey items, produce vibrations on the webs of the dupe that mimic the prey items (model) of the dupe. When the dupe is tricked, and approaches what it thinks is food, the mimic attacks it and eats it. Bolus spiders are another type of aggressive mimic. They produce chemicals that mimic the sex pheromones of particular moth species. When male moths approach what they perceive to be a female

in order to mate with her, they are caught by the bolus spider and become prey.

See also Adaptation and fitness

MISSENSE MUTATIONS

A missense **mutation** is a change in the nucleic acid base sequence (e.g., the polynucleotide sequence found in **DNA**) that results in a change in the order of amino acids in the protein chain resulting from the **translation** of that sequence.

As with all mutations, a missense mutation is a change in a **gene** or **chromosome**. There are many types of mutations, including those resulting in a substitution of **bases** in the DNA sequence (i.e., one **nucleotide** is exchanged for another in the polynucleotide sequence. The effects of these mutations depend on the type, extent, and placement of the mutation in the DNA sequence. The effects of mutations range from silent mutations that do not result in a change in the translated protein sequence to mutations that are lethal and prevent embryonic development.

Changes in DNA **sequencing** are natural. Changes that occur during **replication** are usually repaired by specialized

DNA repair mechanisms. When these mechanisms fail to correct a change, (revert the sequence of bases back to their normal state) a mutation results. As a result of a missense mutation, instead of the insertion of the proper **amino acid** into the chain of amino acids bonded together to form a protein, another amino acid is inserted. Unless there is another mutation in a gene that codes for a **transfer RNA** (tRNA) that allows the suppression of a missense mutation, the structure and function of the protein can be dramatically altered by the improper substitution of an amino acid.

A well-known example of a missense mutation occurs with the sickle **cell** gene. A substitution of thymine for adenine at the nucleotide 17 of the gene that contains the instructions to construct the beta chain of the hemoglobin molecule results in a substitution of the amino acid valine for the amino acid glutamic acid in the beta chain. This subtle alteration of genetic sequence results in profound changes in the shape and physiological responses of the hemoglobin molecule that, in turn, can lead to sickle cell disease. At the molecular level, the substitution of thymine for adenine results in a change in the three base (triplet) sequence that determines the codon sequence. Instead of a glutamine-adenine-glutamine (GAG) triplet that instructs the insertion of glutamic acid into the growing protein molecule during translation, a codon is constructed from a glutamine-thymine-glutamine (GTG) triplet that codes for valine.

For example, if the base sequence of the coding stand of DNA is GAG, because of restricted base paring (e.g., A-T and C-G) the anticoding DNA strand used as a **transcription** template contains a CTC sequence. Because uracil (U) replaces thymine in **mRNA**, the DNA coding strand sequence (the strand that has the same sequence as does the mRNA transcript) results in a mRNA codon sequence of GAG. This codon sequence is ultimately translated to insert glutamic acid into the translated protein's sequence of amino acids. A missense mutation resulting in the substitution of thymine (T) for adenine (A) in the coding strand sequence results in DNA sequence of GTG that results in a GUG mRNA codon. As a result of the missense mutation, and the changes in the **codons** transcribed. Instead of the insertion of the amino acid glutamic acid into the lengthening amino acid chain during the translation process, the amino acid valine is inserted in its place.

Missense mutation suppressors are similar to nonsense mutation suppressor in that are mutant genes that result in the production of transfer ribonucleic acids (tRNA) that have anticodons altered so that they have the ability to correct the errant instructions produced by missense mutations. In addition to DNA repair mechanisms, extragenic suppressor mutations termed missense mutation suppressors can also allow the proper synthesis of proteins despite the presence of a missense mutation.

See also Base pairing (bp); Deletions; Frame shifts; Point mutation

MITOCHONDRIA

Mitochondria are double membrane-bound cytoplasmic organelles (they are enclosed by two independent lipoprotein membranes) found in almost all higher plant and animal cells. Although they vary considerably in size and shape even within a single **cell** over time, mitochondria are usually bean-shaped or thread-like with an average diameter of about 0.5 nm and a length ranging from one to hundreds of meters. Mitochondria often move rapidly around the cell interior, constantly changing size or shape. There might be a few giant interconnected mitochondrial networks in a cell at one time, or they may break up into hundreds or thousands of individual mitochondria under different conditions.

Mitochondria serve as the principle energy source for all animal cells and for plant cells when light is absent. A complex set of **enzymes** and cofactors located both in the interior of the mitochondrion and in its inner membrane carry out a series of reactions that break down energy-rich organic compounds. The energy released by these reactions is captured by a process called oxidative phosphorylation and used to generate **adenosine triphosphate** (ATP), a stable chemical compound that serves as the currency for macromolecular synthesis, mechanical motion, transmembrane transport, and almost all other metabolic and physiological activities at the cellular level.

It is widely accepted by scientists that mitochondria probably arose sometime far back in history as a result of a symbiotic relationship between a prokaryotic (bacterial-like) cell and a nucleus-containing proto-eukaryotic cell. Mitochondria have many similarities to free-living **bacteria** including their own genetic material and a bacterial-like protein synthesizing system. There are also many similarities between the enzymes that carry out oxidative phosphorylation in mitochondria and those that are responsible for photosynthesis in chloroplasts, suggesting that one of these systems might be evolutionary descendants of the other.

See also Evolution; Organelles and subcellular genetics

MITOCHONDRIAL DNA

Mitochondria are cellular organelles that generate energy in the form of **ATP** through oxidative phosphorylation. Each **cell** contains hundreds of these important organelles. Mitochondria are inherited at conception from the mother through the cytoplasm of the egg. The mitochondria, present in all of the cells of the body, are copies of the ones present in at conception in the egg. When cells divide, the mitochondria that are present are randomly distributed to the daughter cells, and the mitochondria themselves then replicate as the cells grow.

Although many of the mitochondrial genes necessary for ATP production and other genes needed by the mitochondria are encoded in the **DNA** of the chromosomes in the **nucleus** of the cell, some of the genes expressed in mitochondria are encoded in a small circular **chromosome** which is contained within the mitochondrion itself. This includes 13 polypeptides, which are components of oxidative phosphorylation **enzymes**, 22 **transfer RNA** (t-RNA) genes, and two genes for **ribosomal RNA** (r-RNA). Several copies of the mitochondrial chromosome are found in each mitochondrion. These chromosomes are far smaller than the chromosomes found in the nucleus, contain far fewer genes than any of the **autosomes**, replicate without going through a mitotic cycle, and their morphological structure is more like a bacterial chromosome than it is like the chromosomes found in the nucleus of **eukaryotes**.

Genes which are transmitted through the mitochondrial DNA are inherited exclusively from the mother, since few if any mitochondria are passed along from the sperm. Genetic diseases involving these genes show a distinctive pattern of **inheritance** in which the trait is passed from an affected female to all of her children. Her daughters will likewise pass the trait on to all of her children, but her sons do not transmit the trait at all.

The types of disorders which are inherited through mutations of the mitochondrial DNA tend to involve disorders of nerve function, as neurons require large amounts of energy to function properly. The best known of the mitochondrial disorders is Leber hereditary optic neuropathy (LHON), which involves bilateral central vision loss, which quickly worsens as a result of the death of the optic nerves in early adulthood. Other mitochondrial diseases include Kearns-Sayre syndrome, myoclonus epilepsy with ragged red fibers (MERFF), and mitochondrial encephalomyopathy, lactic acidosis, and stroke-like episodes (MELAS).

See also Maternal inheritance; RNA (ribonucleic acid)

MITOCHONDRIAL INHERITANCE

Mitochondrial **inheritance** is the study of how mitochondrial genes are inherited. **Mitochondria** are cellular organelles that contain their own **DNA** and **RNA**, allowing them to grow and replicate independent of the **cell**. Each cell has 10,000 mitochondria each containing two to ten copies of its **genome**. Because mitochondria are organelles that contain their own genome, they follow an inheritance pattern different from simple Mendelian inheritance, known as extranuclear or **cytoplasmic inheritance**. Although they posses their own genetic material, mitochondria are semi-autonomous organelles because the nuclear genome of cells still codes for some components of mitochondria.

Mitochondria are double membrane-bound organelles that function as the energy source of eukaryotic cells. Within the inner membrane of mitochondria are folds called cristae that enclose the matrix of the organelle. The DNA of mitochondria, located within the matrix, is organized into circular duplex chromosomes that lack histones and code for proteins, rRNA, and tRNA. A nucleoid, rather than a nuclear envelope, surrounds the genetic material of the organelle. Unlike the DNA of nuclear genes, the genetic material of mitochondria does not contain introns or repetitive

sequences resulting in a relatively simple structure. Because the chromosomes of mitochondria are similar to those of prokaryotic cells, scientists hold that mitochondria evolved from free-living, aerobic **bacteria** more than a billion years ago. It is hypothesized that mitochondria were engulfed by eukaryotic cells to establish a symbiotic relationship providing metabolic advantages to each.

Mitochondria are able to divide independently without the aid of the cell. The chromosomes of mitochondria are replicated continuously by the enzyme DNA **polymerase**, with each strand of DNA having different points of origin. Initially, one of the parental strands of DNA is displaced while the other parental strand is being replicated. When the copying of the first strand of DNA is complete, the second strand is replicated in the opposite direction. **Mutation** rates of mitochondria are much greater than that of nuclear DNA allowing mitochondria to evolve more rapidly than nuclear genes. The resulting phenotype (cell death, inability to generate energy, or a **silent mutation** that has no phenotypic effect) is dependent on the number and severity of mutations within tissues.

During **fertilization**, mitochondria within the sperm are excluded from the **zygote**, resulting in mitochondria that come only from the egg. Thus, **mitochondrial DNA** is inherited through the maternal lineage exclusively without any **recombination** of genetic material. Therefore, any trait coded for by mitochondrial genes will be inherited from mother to all of her offspring. From an evolutionary standpoint, Mitochondrial Eve represents a single female ancestor from who our mitochondrial genes, not our nuclear genes, were inherited 200,000 years ago. Other women living at that time did not succeed in passing on their mitochondria because their offspring were only male. Although the living descendants of those other females were able to pass on their nuclear genes, only Mitochondrial Eve succeeded in passing on her mitochondrial genes to humans alive today.

See also Extranuclear inheritance

MITOSIS

Mitosis is the process during which two complete, identical sets of chromosomes are produced from one original set. This allows a **cell** to divide during another process called cytokinesis, thus creating two completely identical daughter cells.

In order for an organism to grow and develop, the organism's cells must be able to duplicate themselves. Three very basic events must take place to achieve this **duplication**: the **deoxyribonucleic acid DNA**, which makes up the individual chromosomes within the cell's **nucleus** must be duplicated; the two sets of DNA must be packaged up into two separate nuclei; and the cell's cytoplasm must divide itself to create two separate cells, each complete with its own nucleus. The two new cells, products of the single original cell, are known as daughter cells.

During much of a cell's life, the DNA within the nucleus is not actually organized into the discrete units known as chro-

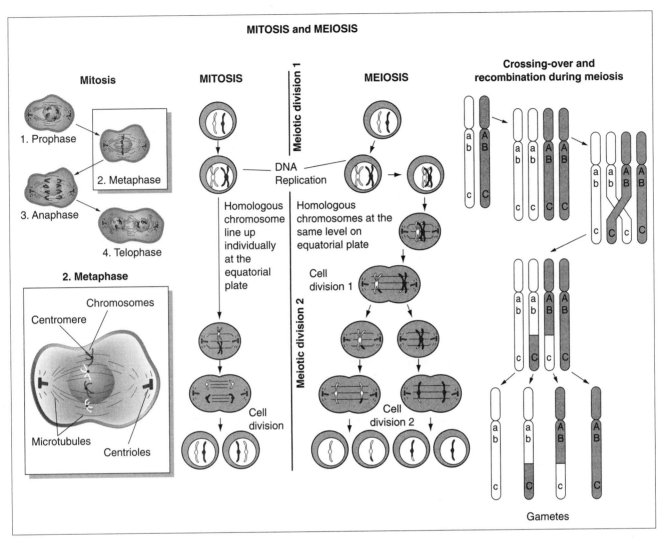

Comparison of mitosis and meiosis.

mosomes. Instead, the DNA exists loosely within the nucleus, in a form called **chromatin**. Prior to the major events of mitosis, the DNA must replicate itself, so that each cell has twice as much DNA as previously. Mitosis is then ready to begin.

The first stage of mitosis is called prophase. During prophase, the DNA organizes or condenses itself into the specific units known as chromosomes. Chromosomes appear as double-stranded structures. Each strand is a replica of the other and is called a chromatid. The two chromatids of a **chromosome** are joined at a special region, the centromere. Structures called centrioles position themselves across from each other, at either end of the cell. The nuclear membrane then disappears.

During the stage of mitosis called metaphase, the chromosomes line themselves up along the midline of the cell. Fibers called spindles attach themselves to the centromere of each chromosome.

During the third stage of mitosis, called anaphase, spindle fibers will pull the chromosomes apart at their centromere

(chromosomes have two complementary halves, similar to the two nonidentical but complementary halves of a zipper). One arm of each chromosome will migrate toward each centriole, pulled by the spindle fibers.

During the final stage of mitosis, telophase, the chromosomes decondense, becoming unorganized chromatin again. A nuclear membrane forms around each daughter set of chromosomes, and the spindle fibers disappear. Sometime during telophase, the cytoplasm and cytoplasmic membrane of the cell split into two (cytokinesis), each containing one set of chromosomes residing within its nucleus.

Mitosis always creates two completely identical cells from the original cell. In mitosis, the total amount of DNA doubles briefly, so that the subsequent daughter cells will ultimately have the exact amount of DNA initially present in the original cell. Mitosis is the process by which all of the cells of the body divide and therefore reproduce. The only cells of the body which do not duplicate through mitosis are the sex cells egg and **sperm cells**. These cells undergo a slightly different type of **cell division** called **meiosis**, which allows each sex cell produced to contain half of its original amount of DNA, in anticipation of doubling it again when an egg and a sperm unite during the course of conception.

MODEL ORGANISM RESEARCH

The rapid development genetic technology during the second half of the twentieth century was made possible by meticulous scientific studies on nonhuman organisms and populations. Ethical and practical considerations often prevent the use of human subjects in most basic science research. Accordingly, many of the technologies that have recently allowed the **sequencing** of the human **genome** and the mapping of genes owe a great debt to work done with **species** as phylogenically different as *Drosophilae melanogaster* (a fruit fly), *Escherichia coli* (a **bacteria**), *Saccharomyces cerevisiae* (a **yeast**), and *Mus musculus* (the common laboratory mouse). Each of these species present unique advantages in genetic studies (e.g., rapid generational turnover, etc) and each of these model organisms present workable genomes that are much more easily studied than the human genome. Studies involving such organisms fall under the broad classification of model organism research.

Model organism research is an integral part of modern molecular biology and allows researchers to more easily gather and evaluate data relating to a variety of genetic processes ranging from basic genetic structures and reactions involving **DNA** to the mechanisms of genetic regulation.

There exists a much greater variety in **gene expression** (phenotype) between species that there is difference in the fundamental workings of their genetic mechanisms, especially as related to the fundamental mechanisms of **replication, transcription**, and **translation**. Because to this proximity, at the genetic level is often easier to extrapolate and apply findings based upon model organism research to human **genetics**.

The biological similarities between mice and men make **laboratory mouse genetics** especially significant and applicable to human genetics research. Mice and men share many similar sequences of DNA, and many genes have been conserved (remained the same) throughout evolutionary history. Such evolutionary studies are made easier by the ongoing construction of physical maps for both species that will more clearly delineate the similarities and differences between them. Comparisons of the similarities and differences between the genomes will allow more exacting use of mouse data in the search for understanding of human diseases and of the mechanisms of **pharmacogenetics**.

See also Bacterial genetics; Bases; Chromosomal mutations and abnormalities; DNA (Deoxyribonucleic acid); Gene mutation; Plant genetics; Viral genetics

MODIFER GENE · *see* GENE NAMES AND FUNCTIONS

MOEWUS, FRANZ (1908-1959)
German geneticist

Franz Moewus is perhaps remembered more often for scrutiny of his scientific veracity, rather than his pioneering work on the **genetics** and reproductive cycle of the single-celled algae *Chlamydomonas. Chlamydomonas* has a complex life cycle, having both asexual and sexual phases. Moewus claimed that copulation only occurred between male and female types, and that **sexual reproduction** furthermore depended on the presence of certain chemical factors. This early work lead Moewus to path-breaking research in the late 1930s on the genetic control of biochemical pathways. This work was similar in concept to, and was published a year before, the one-gene, one-enzyme theory of the American geneticists **George Beadle** and **Edward Tatum**. By 1940, however, some scientists believed Moewus's data to be statistically improbable—too good to be true. Moewus was accused of selecting only the data that fit his theoretical models, or perhaps of making up his experiments altogether. The American zoologist and geneticist Tracy Sonneborn (1905–1981) defended Moewus until the mid-1950s, arguing that his theoretical contributions remained valid even if the quantitative data were in doubt. After unsuccessfully trying to replicate Moewus's experiments for over a year, the American geneticist Francis Ryan (1916–1963) declared Moewus's work unreliable in an article in *Science.* Though Moewus' scientific reputation was tarnished, other scientists, most notably the American geneticist Ruth Sager (1918–1997), continued to use *Chlamydomonas* with great success in their own research.

Moewus was born in Berlin, Germany, to a family of uniform tailors to the Prussian Royal Guard. Moewus studied biology in a home laboratory stocked with a microscope, science textbooks, and chemicals supplied by a relative while attending gymnasium in the early 1920s. Instead of studying towards the civil service exam, Moewus proceeded directly to doctoral studies in botany at the University of Berlin. It was here that Moewus met German protozoologist Max Hartmann (1867–1962) and began his work on sexuality in algae. Hartmann and the German Nobel Prize-winning biochemist Richard Kuhn (1900–1967) mentored Moewus during his years in Germany, helping him to locate temporary positions at the Kaiser Wilhelm Institute for Biology in Berlin-Dahlem and later, the Chemical Institute of the Kaiser Wilhelm Institute for Medical Research in Heidelberg. Moewus managed to continue his research throughout World War II. Moewus and his wife began a more transient existence after the controversy surrounding his work grew, spending first two years in Australia, then a year in Ryan's laboratory, and eventually finding work in laboratories associated with the University of Miami, Florida.

Much of the outstanding work in molecular biology in the 1950s through the 1970s depended on genetic analyses of single-celled organisms. Although Moewus pioneered techniques in **microbial genetics**, Beadle and Tatum are generally credited for establishing the field. Beadle and Tatum later received the Nobel Prize; Moewus faded into obscurity as a laboratory technician in Florida.

See also Asexual generation and reproduction; Experimental organisms; Microbial genetics; Sexual reproduction

MOLECULAR BIOLOGY AND MOLECULAR GENETICS

At its most fundamental level, molecular biology is the study of biological molecules and the molecular basis of structure and function in living organisms.

Molecular biology is an interdisciplinary approach to understanding biological functions and regulation at the level of molecules such as nucleic acids, proteins, and carbohydrates. Following the rapid advances in biological science brought about by the development and advancement of the Watson-Crick model of **DNA (deoxyribonucleic acid)** during the 1950s and 1960s, molecular biologists studied **gene** structure and function in increasing detail. In addition to advances in understanding genetic machinery and its regulation, molecular biologists continue to make fundamental and powerful discoveries regarding the structure and function of cells and of the mechanisms of genetic transmission. The continued study of these processes by molecular biologists and the advancement of molecular biological techniques requires integration of knowledge derived from physics, chemistry, mathematics, **genetics, biochemistry, cell** biology and other scientific fields.

Molecular biology also involves organic chemistry, physics, and biophysical chemistry as it deals with the physicochemical structure of macromolecules (nucleic acids, proteins, lipids, and carbohydrates) and their interactions. Genetic materials including DNA in most of the living forms or **RNA (ribonucleic acid)** in all plant viruses and in some animal viruses remain the subjects of intense study.

The complete set of genes containing the genetic instructions for making an organism is called its **genome**. It contains the master blueprint for all cellular structures and activities for the lifetime of the cell or organism. The human genome consists of tightly coiled threads of deoxyribonucleic acid (DNA) and associated protein molecules organized into structures called chromosomes. In humans, as in other higher organisms, a DNA molecule consists of two strands that wrap around each other to resemble a twisted ladder whose sides, made of sugar and phosphate molecules are connected by rungs of nitrogen-containing chemicals called **bases** (nitrogenous bases). Each strand is a linear arrangement of repeating similar units called nucleotides, which are each composed of one sugar, one phosphate, and a nitrogenous base. Four different bases are present in DNA adenine (A), thymine (T), cytosine (C), and guanine (G). The particular order of the bases arranged along the sugar-phosphate backbone is called the DNA sequence; the sequence specifies the exact genetic instructions required to create a particular organism with its own unique traits.

Each time a cell divides into two daughter cells, its full genome is duplicated; for humans and other complex organisms, this **duplication** occurs in the **nucleus**. During **cell division** the DNA molecule unwinds and the weak bonds between the base pairs break, allowing the strands to separate. Each strand directs the synthesis of a complementary new strand, with free nucleotides matching up with their complementary bases on each of the separated strands. Nucleotides match up according to strict base-pairing rules. Adenine will pair only with thymine (an A-T pair) and cytosine with guanine (a C-G pair). Each daughter cell receives one old and one new DNA strand. The cell's adherence to these base-pairing rules ensures that the new strand is an exact copy of the old one. This minimizes the incidence of errors (mutations) that may greatly affect the resulting organism or its offspring.

Each DNA molecule contains many genes, the basic physical and functional units of **heredity**. A gene is a specific sequence of **nucleotide** bases, whose sequences carry the information required for constructing proteins, which provide the structural components of cells and tissues as well as **enzymes** for essential biochemical reactions.

The central dogma of molecular biology is that DNA is copied to make **mRNA (messenger RNA)**, and mRNA is used as the template to make proteins. Formation of RNA is called **transcription** and formation of protein is called **translation**. Transcription and translation processes are regulated at various stages and the regulation steps are unique to prokaryotes and **eukaryotes**. DNA regulation determines what type and amount of mRNA should be transcribed, and this subsequently determines the type and amount of protein. This process is the fundamental control mechanism for growth and development (morphogenesis).

All living organisms are composed largely of proteins, the end product of genes. Proteins are large, complex molecules made up of long chains of subunits called amino acids. The protein-coding instructions from the genes are transmitted indirectly through messenger ribonucleic acid (mRNA), a transient intermediary molecule similar to a single strand of DNA. For the information within a gene to be expressed, a complementary RNA strand is produced (a process called transcription) from the DNA template in the nucleus. This messenger RNA (mRNA) moves from the nucleus to the cellular cytoplasm, where it serves as the template for **protein synthesis**. Humans synthesize more than 30,000 different types of proteins.

Twenty different kinds of amino acids are usually found in proteins. Within the gene, sequences of three DNA bases serve as the template for the construction of mRNA with sequence complimentary **codons** that serve as the language to direct the cell's protein-synthesizing machinery. Cordons specify the insertion of specific amino acids during the synthesis of protein. For example, the base sequence ATG codes for the **amino acid** methionine. Because more than one codon sequence can specify the same amino acid, the genetic code is termed a **degenerate code** (i.e., there is not a unique codon sequence for every amino acid).

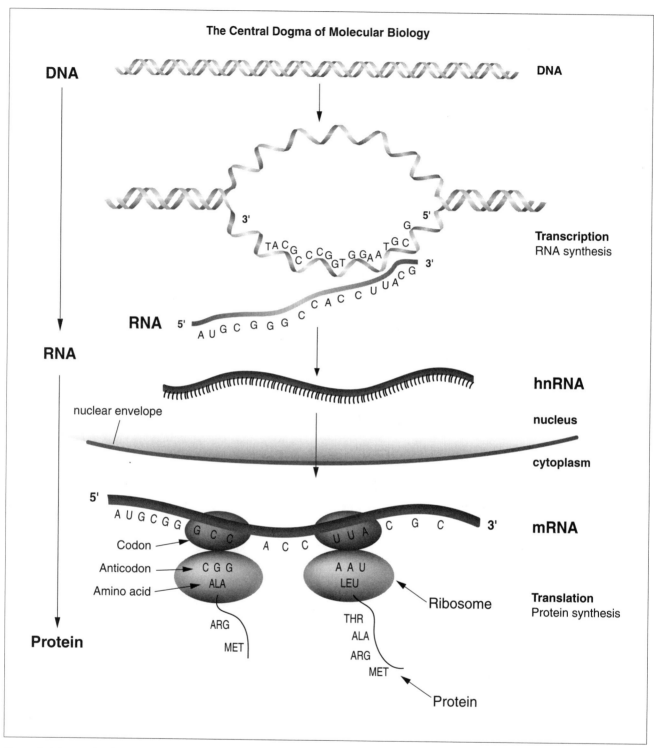

The central dogma of molecular biology, DNA to RNA to protein.

Areas of intense study by molecular biology include the processes of **DNA replication**, repair, and **mutation** (alterations in base sequence of DNA). Other areas of study include the identification of agents that cause mutations (e.g., ultra-violet rays, chemicals) and the mechanisms of rearrangement and exchange of genetic materials (e.g. the function and control of small segments of DNA such as **plasmids**, transposable elements, insertion sequences, and transposons to obtain recombinant DNA).

Recombinant DNA technologies and genetic engineering are an increasingly important part of molecular biology. Advances in **biotechnology** and molecular medicine also carry profound clinical and social significance. Advances in molecular biology have led to significant discoveries concerning the mechanisms of the embryonic development, disease, immunologic response, and **evolution**.

See also Cell differentiation; Chromosome structure and morphology; Gene therapy; Genetic engineering technology; Geneticists and careers in genetics

MOLECULAR CLOCKS

The molecular clock hypothesis, first advanced in 1965 by **Linus Pauling** (1901-1994) and Emile Zuckerkandl, suggests that the **mutation** driven changes (substitutions) in the **amino acid** sequences of proteins has a characteristic substitution rate that is related to time. The identification of such clocks would allow biologists to assign accurate dates to evolutionary history.

Upon comparing the amino acid sequences for hemoglobin molecules taken from different **species**, Pauling and Zuckerkandl noticed that the number of amino acid substitutions increased with the presumed evolutionary distance between species (i.e., how high up on the "evolutionary tree" a species appeared). One proposed mechanism that might account for such changes was an evolutionary clock in which the ticks of the clock are the number of amino acids substitutions that occur over time. Such molecular clocks are a predicted consequence of the neutral theory of molecular **evolution** (i.e., the theory that molecular evolution was primarily the result of genetic drift of neutral mutations rather than **natural selection**). Molecular clocks also allow scientists to determine the balance between the effects of natural **selection**, mutation, and genetic drift on evolutionary changes occurring in genes and proteins.

Evolution is defined as the process of biological change over time. These changes may be accomplished (mediated) by multiple, complex, and interdependent processes. Evidence bounds that species of organisms change, and that the rate of change depends upon the particular trait or characteristic under scrutiny. It is also known, however, that some characteristics are conservative and change slowly. Despite profound differences in appearance (morphological differences), the molecular structures fundamental to all life are essentially similar. The subtle changes of these fundamental molecules, in particular as measured by the substitution of nucleotides in nucleic acids (i.e., **DNA** and **RNA**) and amino acids in proteins, create the wide diversity of life on Earth.

An accurate molecular clock would make it possible to put a timepiece to evolutionary events. Instead of settling for relative relationships, a functional molecular clock could apply absolute dating to the development and divergence of life (**phylogenetic** history).

Scientists who study evolution continue to study the interactions of the mechanisms of evolutionary change.

According to evolutionary theory stressing natural selection, most changes (mutations) have a selective advantage or disadvantage and neutral mutations are rare. According to the neutral evolutionary theory, however, most changes at the molecular level do not carry a selective advantage or disadvantage and are thus, selectively neutral. To account for differences in the genetic composition of the wide variety of life forms, this would mean that the rate of neutral mutations is high. Although neither selection based theory or neutral based theory excludes the other, the exact extent of the influence of each process on evolution is debated among biologists.

Because the molecular clock hypothesis is related to the neutral theory of evolution, first argued by **Motoo Kimura**. Accordingly, evidence for or against the molecular clock hypothesis is often used to argue for or against the neutral theory. According to neutral theory, molecules such as DNA, RNA and proteins with fewer active sites (functional constraints) evolve faster because the number of effectively neutral mutants is correspondingly higher. According to this argument, mutations that are deleterious are quickly lost and, while they occur, advantageous mutants are rare. By definition, neutral mutants have no effect on **fitness**. Supporting the neutral theory, Kimura observed that silent sites and pseudogenes evolved many times faster than coding sequences.

Complicating the accuracy, reliability, and applicability of molecular clocks, however, is the fact that not all DNA variation (e.g., silent mutations, synonymous **codons**, pseudogenes and non-coding DNA) translates into protein variation. Further, not all protein variation translates into differences that can be observed (phenotypic variation) and not all phenotypic variation ultimately makes a difference in fitness. Changes in phenotype may, in some circumstances, be due more to differences in **gene regulation** than in sequence alterations. Moreover, although advances in **genetics** research are rapid, researchers do not yet fully understand the exact relationship between changes in molecules (biochemical evolution) and changes in genes (genetic **differentiation**).

There is substantial evidence that different molecular clocks tick (undergo changes) at different rates. The rate itself may also change during a given span of time (episodic change). Some interpretations of data may also be hindered because of inaccurate assumptions about the pattern of evolution and of the extent of the genetic differences between species. Recent estimates, made in 2001, of the number of genes carried by humans, reduced prior estimates of the number of genes carried from 80,000 to about 30,000 (less than twice the number of genes found in the fruit fly (*Drosophila melanogaster*). Such data may lead to revision in current concepts regarding the evolutionary distance between species, and the interpretation of molecular clocks. It is also possible that the mechanisms of selection, mutation, and drift may have different effects at the levels of molecules or species.

Scientists continue to study molecular clocks in an attempt make the molecular clock hypothesis more useful and accurate. Although evolutionary theory is a well-established and well-supported cornerstone of modern science, there are a few critics who argue that the data regarding molecular clocks actually argues against some parts of evolutionary theory. Such arguments are most often the result of an incorrect understanding of **evolutionary mechanisms**. For example, some arguments fail distinguish between the functional and non-functional parts of proteins, or that parts of proteins may not have had the same function in the past as they do now. In fact, the critical or active portions of molecules or proteins are usually limited to a very limited portion of the molecule (site specific) and are but a fraction of both the total mass and volume of the protein. Another fundamental error involves arguments that rely on the existence of existing (extant) primitive organisms. There are no extant primitive organisms. For example, all **eukaryotes** have in common that a primitive common ancestor was derived from prokaryotic **bacteria** and that, therefore, all eukaryotes have evolved for the same amount of time. Accordingly, no eukaryote group or species (e.g., shark or man) is more primitive with regard to the amount of time they have been subjected to evolutionary mechanisms. Seemingly primitive characteristics are simply highly efficient and conserved characteristics that have changed little over time.

The molecular clock hypothesis, as well as the interpretation of data related to genetic change, has become a part of controversial interpretations regarding the exact nature of the relationships between species (e.g., the relationships between man, chimps and apes). Molecular clocks are also used to develop often controversial interpretations regarding the origin, geographic arrival time, and differentiation, of aboriginal peoples.

Until more is known about the mechanisms of genetic change, and of the differences between species, the ability to use molecular clocks to fix absolute dates to evolutionary events will remain limited.

See also Adaptation and fitness; Evolutionary mechanisms; Natural selection; Selection; Transcription; Translation

MOLECULAR CYTOGENETICS

The introduction of **fluorescent dyes** into molecular biology techniques in the 1980s greatly improved techniques for visualizing **DNA probes**, and has allowed for the development of methods to visualize specific **DNA** sequences on human chromosomes using the microscope. The term "fluorescent *in situ* hybridization" (**FISH**) was adopted as a descriptive name for this methodology which combines techniques of molecular biology and cytogenetics (the study of chromosomes).

Using this technique, DNA probes are labeled with fluorescent dyes and hybridized to metaphase **chromosome** spreads on glass slides. DNA stains are used to visualize the unlabeled portions of the chromosomes. Thus, this method has become a means of identifying specific DNA sequences on metaphase chromosomes.

Initially, only single-color detection was possible, which limited most applications to the study of a single **gene** or region of the chromosome. The development of more sophisticated fluorescence filters and a variety of different dyes with their own unique fluorescence spectra, however, permit the simultaneous detection of multiple colored dyes, and thus, the analysis of multiple genes or chromosome regions in a single experiment. In addition, by using various combinations of five dyes, it is possible to discriminate 31 different dye combinations at a time.

There are a number of clever innovations in molecular cytogenetics, which have greatly increased the power of these methods beyond the simple applications of FISH. These include multiplex FISH (M-FISH) in which each chromosome is demarcated with a different color assigned by a computer software application based on the combination of fluorescent dyes present, spectral karyotyping (SKY) which produces a result similar to that in M-FISH using interferometry in place of standard fluorescence microscopy, cross-species color banding (RX-FISH) which uses probes produce from gibbons and other primates to produce a color banding pattern which is unique to each chromosome, padlock probes which interlock a circular fluorescent probe onto metaphase chromosomes potentially enabling detection of even subtle mutations which are not generally evident from chromosome studies, and primed *in situ* hybridization (PRINS) which uses the metaphase chromosome on a glass slide as a template for a PCR-type reaction to produce a result similar to that produced by conventional FISH methodology.

The introduction of each of these new tools to the molecular cytogenetic toolbox has increased the utility and flexibility of fluorescent microscopy in research and clinical applications in **genetics**. These techniques are used to identify genetic abnormalities in interphase or metaphase cells, to detect submicroscopic **deletions** and duplications associated with a growing variety of clinical syndromes, to accurately characterize complex chromosome rearrangements which occur frequently in **tumor** cells, and to compare the genomic organization of different **species** with each other.

See also Cancer genetics

MOLECULAR DIAGNOSTICS

Molecular diagnostics is a specialized division of clinical laboratory medicine or pathology that includes tests involving nucleic acids (i.e., **DNA** or **RNA**). Nucleic acid based tests allow physicians and genetic counselors more accurate diagnosis of many acquired infectious diseases, cancers, and drug sensitivities. In addition, molecular diagnostics is a powerful tool in determining inherited disorders, whether an individual is a **carrier** of a particular genetic disorder, or whether an individual has a genetic constitution that leads to higher levels of susceptibility or predisposition to genetic disease.

Study of the molecules DNA or RNA reveals information about an individual's genes through a search for **genetic variation** or abnormalities, providing an interface between molecular research and applied medicine. It may overlap with or complement cytogenetic or biochemical approaches used in other **genetics** laboratories, in which chromosomes or protein products of genes, respectively, are studied. Because of the similarity of laboratory procedures, the term might also refer to non-medical applications, such as for forensics or determination of family relationships (paternity or immigration matters, for example). It might also be considered to include methods to identify and track infectious diseases. With molecular diagnostics, one examines the genotype, rather than the phenotype of a family, an individual, or a **cell**.

Sometimes the molecular information is used to help diagnose a symptomatic patient. Other times, tests may be undertaken before there is evidence of disease, such as when a fetus is tested prenatally (prenatal diagnosis), when an adult has testing for a genetic disease with onset in later life (predictive testing) or when groups or populations are tested for susceptibilities (genetic screening).

Molecular diagnostics may be used to study someone's genetic constitution, meaning the genes inherited at conception and present in all cells, carried by chromosomes from the nuclei, or in **mitochondria**. In other situations, notably for the study of cancers, tests may be to look for changes that have occurred after conception in specific cells of the body, such as tumors. Other applications may look at genes of infectious organisms.

The most common source of genetic material for diagnostics has been the nucleus-containing white blood cells. With improved laboratory methods it has become possible to work with smaller samples, or with traces of cells from other tissues. For prenatal diagnosis, **fetal cells** are usually extracted from the surrounding amniotic fluid or from a part of the placenta, or, in recent experiments, from the pregnant mother's blood. Some tissues such as cells from a scraping of the inside of the cheek are easier to obtain than blood is, and will do for certain purposes. Sometimes blood from a finger prick onto a piece of filter paper, or tissue from an old pathology slide is enough.

The prototype for molecular diagnostics was perhaps the prenatal diagnosis of **sickle-cell anemia** in the late 1970s. Because the globin **gene** was already well characterized and the disease-causing **mutation** known, a direct test could be developed. Many of the tests used over the subsequent 20 years, however, were done for genes that had not yet been identified but whose approximate location was known. Such indirect tests make use of markers near enough to the disease gene to be considered linked to that gene. The markers, often restriction fragment length polymorphisms (RFLPs), can be tracked to determine a linkage pattern specific for a family. That information may then allow testing of other family members whose status is unknown. Since the indirect tests involve more work and sampling of family members, and are usually less diagnostic, many have been replaced by direct tests once the genes and relevant mutations have been identified. When the **cystic fibrosis** gene was mapped to **chromosome** seven in 1985, nearby markers were used to analyze families at risk; once the gene was cloned in 1989, it became possible to search

directly for the (now almost 1,000) known disease-causing mutations.

Numerous techniques have evolved to search DNA for polymorphisms, point mutations, **deletions**, insertions, imprinting, etc. Methods such as Southern blots require relatively large DNA samples, hours of work, and long waits for results. Invention of the **polymerase chain reaction** (PCR) in 1985 greatly advanced molecular diagnostics, since it allowed efficient use of samples, materials and time. DNA **sequencing**, once a very inefficient way to look for diagnostic changes, is now applied widely, largely due to automation and other technological improvements. Robots now take over monotonous laboratory tasks involved with large numbers of analyses. Microarrays of test fragments on tiny plates create vehicles to assay hundreds or thousands of DNA variations simultaneously in a single sample.

The purpose of molecular diagnostics is to provide information to people about their own genetic makeup, and assist them to make choices about how to optimize their health, well-being and reproductive options in accordance with their genotype. As research advances, treatment and prevention strategies will be increasingly available and more specific to the individual.

See also Amniocentesis; Blotting analysis; Cancer genetics; Chromosome mapping and sequencing; Ethical issues of modern genetics; Gene linkage; Gene mutation; Genetic disorders and diseases; Genetic marker; Genetic testing and screening; Molecular biology and molecular genetics; Prenatal diagnostic techniques

MONACO, ANTHONY (-)
British geneticist

Anthony Monaco is a well-respected researcher in the field of neurogenetics. He is currently the principal scientist and director of the Wellcome Trust Centre for Human Genetics (Oxford, United Kingdom). Under his direction, the center has become a leading institute for the study of complex genetic diseases. Monaco is responsible for elegant work on the **cloning** of Duchenne muscular dystrophy, as well as classic and insightful work during the early days of **gene** cloning.

From the beginning of his career, Monaco was intrigued by the brain and behavior. He therefore majored in neuroscience as an undergraduate at Princeton University. His early research as an M.D./Ph.D student at Harvard Medical School in the mid-1980s led to the isolation of candidate genes that cause Duchenne muscular dystrophy, results that were published in the highly influential scientific journal, *Nature*. In 1988, after receiving his graduate degrees, Monaco decided to forego his medical internship training and instead moved to London to work at the Imperial Cancer Research Fund, where he became involved in the **human genome project** in its early stages. In 1990, he moved to Oxford. Eight years later, he was appointed director of the human genetics center, whose focus lies in uncovering the genetic basis behind complex but common diseases.

His current research interests lie in the genetic basis of neurodevelopmental disorders in children, particularly, autism, dyslexia, and specific language impairment. His research in the area of language impairment involves finding the genes that are behind the difficulties that 2-5% of children have in acquiring language skills, despite normal intelligence and adequate opportunity. In particular, his research group is interested in the pathways that lead to the disorders. His group has mapped the gene involved in a rare speech and language disorder to a small region on **chromosome** 7, work that should lead to the isolation of the causal gene itself. Other diseases that his group studies include chorea acanthocytosis, X-linked Dystonia-Parkinsons, Menkes disease, and Charcot-Marie-Tooth peripheral neuropathy. In December 1998, Monaco received the Medical Research Council Award for his work on the genetics of autism.

See also Behavioral genetics; Dystrophinopathies; Genetic disorders and diseases; Neurological disorders, inherited

MONOD, JACQUES LUCIEN (1910-1976)

French biologist

French biologist Jacques Lucien Monod and his colleagues demonstrated the process by which messenger **ribonucleic acid** (**mRNA**) carries instructions for **protein synthesis** from **deoxyribonucleic acid** (**DNA**) in the **cell nucleus** out to the **ribosomes** in the cytoplasm, where the instructions are carried out.

Jacques Monod was born in Paris, on February 9, 1910. In 1928, Monod began his study of the natural sciences at the University of Paris, Sorbonne where he went on to receive a B.S. from the *Faculte des Sciences* in 1931. Although he stayed on at the university for further studies, Monod developed further scientific grounding during excursions to the nearby Roscoff marine biology station.

While working at the Roscoff station, Monod met André Lwoff, who introduced him to the potentials of microbiology and microbial nutrition that became the focus of Monod's early research. Two other scientists working at Roscoff station, **Boris Ephrussi** and Louis Rapkine, taught Monod the importance of physiological and biochemical **genetics** and the relevance of learning the chemical and molecular aspects of living organisms, respectively.

During the autumn of 1931, Monod took up a fellowship at the University of Strasbourg in the laboratory of Edouard Chatton, France's leading protistologist. In October 1932, he won a Commercy Scholarship that called him back to Paris to work at the Sorbonne once again. This time he was an assistant in the Laboratory of the Evolution of Organic Life, which was directed by the French biologist Maurice Caullery. Moving to the zoology department in 1934, Monod became an assistant professor of zoology in less than a year. That summer, Monod also embarked on a natural history expedition to Greenland aboard the *Pourquoi pas?* In 1936,

Monod left for the United States with Ephrussi, where he spent time at the California Institute of Technology on a Rockefeller grant. His research centered on studying the fruit fly (*Drosophila melanogaster*) under the direction of **Thomas Hunt Morgan**, an American geneticist. Here Monod not only encountered new opinions, but he also had his first look at a new way of studying science, a research style based on collective effort and a free passage of critical discussion. Returning to France, Monod completed his studies at the Institute of Physiochemical Biology. In this time he also worked with Georges Teissier, a scientist at the Roscoff station who influenced Monod's interest in the study of bacterial growth. This later became the subject of Monod's doctoral thesis at the Sorbonne where he obtained his Ph.D. in 1941.

Monod's work comprised four separate but interrelated phases beginning with his practical education at the Sorbonne. In the early years of his education, he concentrated on the kinetic aspects of biological systems, discovering that the growth rate of **bacteria** could be described in a simple, quantitative way. The size of the colony was solely dependent on the food supply; the more sugar Monod gave the bacteria to feed on, the more they grew. Although there was a direct correlation between the amount of food Monod fed the bacteria and their rate of growth, he also observed that in some colonies of bacteria, growth spread over two phases, sometimes with a period of slow or no growth in between. Monod termed this phenomenon diauxy (double growth), and guessed that the bacteria had to employ different **enzymes** to metabolize different kinds of sugars.

When Monod brought the finding to Lwoff's attention in the winter of 1940, Lwoff suggested that Monod investigate the possibility that he had discovered a form of enzyme adaptation, in that the latency period represents a hiatus during which the colony is switching between enzymes. In the previous decade, the Finnish scientist, Henning Karstroem, while working with protein synthesis had recorded a similar phenomenon. Although the outbreak of war and a conflict with his director took Monod away from his lab at the Sorbonne, Lwoff offered him a position in his laboratory at the Pasteur Institute where Monod would remain until 1976. Here he began working with Alice Audureau to investigate the genetic consequences of his kinetic findings, thus beginning the second phase of his work.

To explain his findings with bacteria, Monod shifted his focus to the study of enzyme induction. He theorized that certain colonies of bacteria spent time adapting and producing enzymes capable of processing new kinds of sugars. Although this slowed down the growth of the colony, Monod realized that it was a necessary process because the bacteria needed to adapt to varying environments and foods to survive. Therefore, in devising a mechanism that could be used to sense a change in the environment, and thereby enable the colony to take advantage of the new food, a valuable evolutionary step was taking place. In Darwinian terms, this colony of bacteria would now have a very good chance of surviving, by passing these changes on to future generations. Monod summarized his research and views on relationship between the roles of random chance and adaptation in evolution in his 1970 book *Chance and Necessity.*

Between 1943 and 1945, working with Melvin Cohn, a specialist in immunology, Monod hit upon the theory that an inducer acted as an internal signal of the need to produce the required digestive enzyme. This hypothesis challenged the German biochemist Rudolf Schoenheimer's theory of the dynamic state of protein production that stated it was the mix of proteins that resulted in a large number of random combinations. Monod's theory, in contrast, projected a fairly stable and efficient process of protein production that seemed to be controlled by a master plan. In 1953, Monod and Cohn published their findings on the generalized theory of induction.

That year Monod also became the director of the department of cellular biology at the Pasteur Institute and began his collaboration with **François Jacob**. In 1955, working with Jacob, he began the third phase of his work by investigating the relationship between the roles of **heredity** and environment in enzyme synthesis, that is, how the organism creates these vital elements in its metabolic pathway and how it knows when to create them.

It was this research that led Monod and Jacob to formulate their model of protein synthesis. They identified a **gene** cluster they called the **operon**, at the beginning of a strand of bacterial DNA. These genes, they postulated, send out messages signaling the beginning and end of the production of a specific protein in the cell, depending on what proteins are needed by the cell in its current environment. Within the operons, Monod and Jacob discovered two key genes, which they named the **operator** and structural genes. The scientists discovered that during protein synthesis, the operator gene sends the signal to begin building the protein. A large molecule then attaches itself to the **structural gene** to form a strand of mRNA. In addition to the operon, the regulator gene codes for a repressor protein. The repressor protein either attaches to the operator gene and inactivates it, in turn, halting structural gene activity and protein synthesis; or the repressor protein binds to the regulator gene instead of the operator gene, thereby freeing the operator and permitting protein synthesis to occur. As a result of this process, the mRNA, when complete, acts as a template for the creation of a specific protein encoded by the DNA, carrying instructions for protein synthesis from the DNA in the cell's nucleus, to the ribosomes outside the nucleus, where proteins are manufactured. With such a system, a cell can adapt to changing environmental conditions, and produce the proteins it needs when it needs them.

Word of the importance of Monod's work began to spread, and in 1958 he was invited to become professor of biochemistry at the Sorbonne, a position he accepted conditional to his retaining his post at the Pasteur Institute. At the Sorbonne, Monod was the chair of chemistry of metabolism, but in April 1966, his position was renamed the chair of molecular biology in recognition of his research in creating the new science. Monod, Jacob, Lwoff won the 1965 Nobel Prize

Jacques Lucien Monod.

in physiology or medicine for their discovery of how genes regulate cell metabolism.

See also Bacterial genetics; Operon; Proteins and enzymes

MONOZYGOTIC TWINS • *see* TWINS

MOORE, STANFORD (1913-1982)
American biochemist

Stanford Moore's work in protein chemistry greatly advanced understanding of the composition of **enzymes**, the complex proteins that serve as catalysts for countless biochemical processes. Moore's research focused on the relationship between the chemical structure of proteins, which are made up of strings of amino acids, and their biological action. In 1972 he was awarded the Nobel Prize in chemistry with longtime collaborator **William Howard Stein** for providing the first complete decoding of the chemical composition of an enzyme, **ribonuclease** (RNase). This discovery provided scientists with insight into **cell** activity and function, which has had important implications for medical research.

Moore was born on September 4, 1913, in Chicago, Illinois, but spent most of his childhood in Nashville, Tennessee, where his father was a professor at Vanderbilt University's School of Law. In 1935, Moore earned a B.A. in chemistry from Vanderbilt. Moore continued his education in organic chemistry at the graduate school of the University of Wisconsin and completed the Ph.D. in 1938. Though Moore considered attending medical school, he accepted a position in 1939 as a research assistant in the laboratory of German chemist Max Bergmann at the Rockefeller Institute for Medical Research (RIMR), later renamed the Rockefeller University.

Bergmann's research group focused on the structural chemistry of proteins. During his early years at RIMR, Moore questioned whether proteins actually had specific structures. The direction of research in Bergmann's laboratory was greatly influenced by the arrival of William Howard Stein. At Bergmann's suggestion, Moore and Stein began a long-lived and successful collaboration. With the exception of his wartime service, the years abroad, and a year at Vanderbilt University, Moore remained at RIMR all of his professional life. In 1952 he became a member of the institute, and his association with RIMR/Rockefeller University continued until his death in 1982.

From 1942 to 1945, Moore worked as a technical aide in the National Defense Research Committee of the Office of Scientific Research and Development (OSRD). As an administrative officer for university and industrial projects, Moore studied the action and effects of mustard gas and other chemical warfare agents. As a technical aide, Moore also coordinated academic and industrial studies on the actions of chemical agents. In 1944 he was appointed to the project coordination staff of the Chemical Warfare Service and continued to contribute to this research until the end of the war in 1945.

Though his initial investigation of chromatography— the process of separating the components of a solution— began in the late 1930s, the war interrupted this work. After the war and following the death of Max Bergmann in 1944, Moore returned to RIMR to resume work with William Howard Stein. This marked a productive period for the two men, leading to their work on chromatographic methods for separating amino acids, peptides (compounds of two or more amino acids), and proteins. Moore's work in chromatography was influenced by the methods of paper chromatography developed by A. J. P. Martin and Richard L. M. Synge of England. However, limitations in these earlier methods prohibited the study of protein chemistry; new techniques in chromatography had to be developed so that amino acids could be separated. Moore and Stein utilized column chromatography, in which a column or tube is filled with material that separates the components of a solution. In 1948 they successfully separated amino acids by passing the solution through a column filled with potato starch. The process was time consuming, however, and presented inadequate separations for **amino acid** analysis. To facilitate the procedure, Moore and Stein replaced the filler material with a synthetic ion exchange resin, which separated components of a solution by electrical charge and size. In 1949 they successfully separated amino acids from blood and urine.

Further interruptions in the work at RIMR occurred in 1950, when Moore held the Francqui Chair at the University of Brussels (where he established a laboratory for amino acid analysis), and in 1951, when he was a visiting professor at the University of Cambridge (England) working with **Frederick Sanger** on the amino acid sequence of insulin.

In 1958 Moore and Stein contributed to the development of the automated amino acid analyzer. This instrument facilitated the complete amino acid analysis of a protein in twenty-four hours; previously employed procedures, such as chromatography, had required up to one week. The automated technique afforded researchers a tool with which to separate and study the large chemical sequences in protein molecules. This instrument is used worldwide for the study of proteins, enzymes, and hormones as well as the analysis of food. While Moore and Stein endeavored to determine the chemical composition of proteins, scientists concurrently worked to determine their three-dimensional structure.

By 1959 Moore and Stein had determined the amino acid sequence of pancreatic ribonuclease (RNase), a digestive enzyme that breaks down **ribonucleic acid** (**RNA**) so that its components can be reused. They discovered that ribonuclease is made up of a chain of 124 amino acids, which they identified and sequenced. This marked the first complete description of the chemical structure of an enzyme, a discovery that earned Moore and Stein the 1972 Nobel Prize in chemistry (an award shared with **Christian Boehmer Anfinsen** of the National Institutes of Health). Understanding protein structure is essential to understanding biological function, which opens the door to the treatment of disease. Moore's findings influenced research in neurochemistry and the study of such diseases as sickle-cell anemia. Scientists later discovered that related ribonucleases are present in nearly all human cells, which prompted studies in the fields of **cancer** and malaria research.

In 1969, Moore returned to Rockefeller University following one year spent at Vanderbilt University School of Medicine as a visiting professor of health sciences. Upon his return, the team of Moore and Stein resumed their studies of protein chemistry with an investigation of deoxyribonuclease, the enzyme that breaks down **deoxyribonucleic acid** (**DNA**).

In addition to scientific work, Moore also served as an editor of *The Journal of Biological Chemistry,* treasurer and president of the American Society of Biological Chemistry, and president of the Federation of American Societies for Experimental Biology. Moore received many honors, including membership in the National Academy of Sciences and receipt of the Richards Medal of the American Chemical Society and the Linderstrom-Lang Medal. He was also awarded honorary degrees by the University of Brussels, the University of Paris, and the University of Wisconsin.

Moore was diagnosed with amyotrophic lateral sclerosis (Lou Gehrig's disease) some time in the early 1980s. After Moore's death in 1982, his estate passed to Rockefeller University.

See also Amino acid disorders and screening; Biochemical analysis techniques; Blotting analysis; DNA fingerprinting and forensics; DNA structure; Enzymes; Eukaryotes; Eukaryotic genetics; Nuclease; Proteins and enzymes; Protein electrophoresis

MORGAN, LILIAN VAUGHAN SAMPSON (1870-1952)

American cytologist

Lilian Morgan was the first person to detect the attached X and closed X chromosomes. Morgan made these discoveries while investigating the fruit fly *Drosophila melanogaster's* **X chromosome**. This information impacted **genetics** by offering evidence for the **chromosome** theory of **heredity**, which her husband, **Thomas Hunt Morgan**, and his colleagues were developing. Researching independently without the benefit of an institutional affiliation, Morgan was the sole author of ten scientific papers about her fruit fly discoveries. Geneticists such as **Barbara McClintock** benefitted from Morgan's research regarding attached X **inheritance** patterns. Morgan's work established a fundamental concept of genetics upon which other researchers elaborated, and aided geneticists in their efforts to confirm how organisms' characteristics are physically transmitted. Many basic genetics courses require students to demonstrate the principles outlined by Morgan.

Born in Hallowell, Maine, Morgan studied with **cell** biologist Edmund B. Wilson at Bryn Mawr College. Wilson considered Morgan one of his most capable students, and she researched with him at the Marine Biological Laboratory in Woods Hole, Massachusetts, after she graduated in 1891. As the most outstanding graduate in her class, she received the European Fellowship to attend the University of Zurich for one year to study with Arnold Lange, a comparative anatomist. Morgan returned to Bryn Mawr to earn an advanced biology degree. Wilson, who had begun teaching at Columbia University, introduced her to Thomas Hunt Morgan, his replacement at Bryn Mawr, who directed her graduate studies and for whom she worked as a research assistant. Morgan received a master's degree in 1894. For the next seven summers, she was an investigator at the Marine Biological Laboratory, studying amphibian reproduction and development.

She married Thomas Hunt Morgan in 1904, and moved to New York City when he accepted a professorship in experimental zoology at Columbia University. During the first summer they were married, she investigated marine regeneration at the Stanford Marine Laboratory in California while her husband was a visiting scholar. Morgan ceased researching while she raised the couple's four children. Spending summers with her family at the Morgans' Woods Hole home, Morgan directed scientific educational activities for the researchers' children.

When she resumed her investigations, the scientifically versatile Morgan, who had previously focused on **embryology**, began studying *Drosophila* genetics at Columbia and utilized her husband's laboratories and fruit fly stock. She was not her husband's assistant, and was not affiliated with the research he conducted. On February 12, 1921, Morgan found a unique female **mosaic** fruit fly among her specimens that was a gynandromorph, or **hermaphrodite**, differing from other flies by displaying both male and female traits. Seeking a cytological explanation, she realized that the fly had two connected X chromosomes which it had inherited as a unit. Morgan developed the concept of the attached X chromosome by crossing the female with normal male flies to determine how the attached X chromosome appeared in each generation. The mother's X chromosome sex-linked recessive genes were transmitted to the daughters instead of sons as occurred in the reproduction of normal fruit flies. The father's dominant **gene** passed to sons instead of daughters. The father's X chromosome consistently passed solely to male offspring.

Her first paper discussing the *Drosophilia* X chromosome appeared in 1922. Morgan's discovery was significant to contemporary geneticists because it reinforced the hypothesis that chromosomes played the key role in organisms inheriting traits. Morgan's work also validated Calvin Bridges's balance theory concerning the determination of gender according to the ratio of sex chromosomes and **autosomes**. Morgan bred an attached X strain of *Drosophila* which were used to advance crossing-over investigations of chromosome behavior during **meiosis** She applied her techniques to additional fruit fly X chromosome experiments addressing such subjects as polyploidy and chromosome shape and behavior. Her methodology provided a fundamental technique for other geneticists to appropriate.

In 1928, Morgan moved to Pasadena, California, when her husband began directing the biology program at the California Institute of Technology (Caltech). Provided laboratory access, she continued her *Drosophila* genetics research. At Caltech in 1933, she announced that she had discovered a closed X chromosome. In some daughters of attached X female fruit flies, a maternal X chromosome formed a ring which could act unstable during **mitosis** and development. The next year, Morgan traveled to Europe when her husband accepted the Nobel Prize that he had won the previous autumn. She collaborated with her husband to write two articles concerning maintaining a fruit fly stock center. He died in 1945. Appointed a research associate at Caltech the next year, she continued her investigations until her death.

See also Chromosomal mutations and abnormalities; Experimental Organisms

MORGAN, THOMAS HUNT (1866-1945)

American geneticist

Thomas Hunt Morgan, along with **William Bateson** was the co-founder of modern **genetics**. Morgan was the first to show that **genetic variation** occurs through numerous small mutations.

As child growing up in rural Kentucky, Morgan was surrounded by nature and wildlife. Perhaps that environment contributed to his intense interest in biology, for Morgan later majored in zoology at State College of Kentucky. After his graduation in 1886, Morgan investigated chemistry and morphology (the study of organism development to better understand evolutionary relationships) at Johns Hopkins University, completing his doctorate in 1890. From his graduate days on, Morgan believed that **heredity** was in some way central to

Thomas Hunt Morgan.

understanding all biological phenomena—especially development and **evolution**. His persistence in trying to prove and develop heredity theories led to his winning the Nobel Prize for physiology or medicine in 1933.

In 1903, there were several attempts to explain variations in plants and animal **species**. One was **Charles Darwin**'s theory of **natural selection**, a process by which organisms best adapted to local environments leave more offspring that survive to spread their favorable traits throughout a population. Morgan wondered how complex organisms such as humans could have evolved from such a process. To him, the theory seemed incomplete. Morgan viewed natural **selection** as a process that sorted out variations in an organism, not as one that created the variations. So what was it that determined whether a baby would be a boy or a girl, or whether it would have blue eyes or green eyes? The three widely known heredity theories of the time offered competing explanations: the Mendelian (or **gene**) theory, the **chromosome** theory, and the **mutation** theory.

Gregor Mendel, by cross-breeding pea plants, had first determined some of the rules of inheritable traits—those of **sex determination**, **gene linkage** (**inheritance** of characteristics together), and **mimicry**. Advocates of the chromosome theory maintained that genes located on chromosomes were responsible for specific inherited traits. Morgan was skeptical of the Mendelian and chromosome theories because the conclusions

were speculative, based on nothing more than observation, inference, and analogy. Morgan wanted to be able to draw firm, rigorous, testable conclusions based on quantitative and analytical data. His strong belief in experimental analysis attracted Morgan to **Hugo de Vries**'s mutation theory. De Vries, a Dutch botanist, had physical evidence that large-scale variations in one generation could produce offspring that were of a different species than their parent plants. Morgan set out to test de Vries's theory in animals and also to disprove the other heredity theories.

Morgan's first experiments using the fruit fly (*Drosophila melanogaster*) were unsuccessful. Morgan was not able to duplicate the magnitude of mutations that de Vries had claimed for plants. Then in 1910, Morgan noticed a natural mutation in one of the male fruit flies: it had white eyes instead of red. He began breeding the white-eyed male to its red-eyed sisters and found that all of the offspring had red eyes. When Morgan bred those offspring, he found that they produced a second generation of both red- and white-eyed fruit flies. Morgan was fascinated to find that all of the white-eyed flies were male. He traced the unusual finding to a difference between male and female chromosomes. The white-eye gene of the fruit fly was located on the male **sex chromosome**. By studying future generations of fruit flies, Morgan found that genes were linearly arranged on chromosomes. His work with the fruit fly strongly backed Mendel's gene concept and, moreover, established that chromosomes definitely carried genetic traits. For the first time, the association of one or more hereditary characteristics with specific chromosomes was clear, thereby unifying Mendelian "trait" theory and chromosome theory.

Morgan, working with students Hermann Muller, Alfred H. Sturtevant, and Calvin Bridges, later went on to develop and perfect these concepts of linkage by explaining why, for instance, occasionally white-eyed female flies were found in his studies. Morgan concluded that traits found on the same chromosome were not always inherited together. This genetic "mistake" was called **crossing over**, because one chromosome actually exchanged material with (or crossed over to) another chromosome. This process was an important source of genetic diversity. In 1915, Morgan, along with his students, published the culmination of his work, *The Mechanism of Mendelian Heredity*. These results provided the key to all further work in the area of genetics and laid the groundwork for all genetically-based research.

In 1904, Morgan married Lilian Vaughan Sampson, who assisted in his research. They had one son and three daughters. Morgan died in 1945, at age 79.

See also Mendelian genetics and Mendel's laws of heredity

Mosaic

Each multicellular organism is the result of repeated divisions of the original fertilized egg. Before each division, a cell's **DNA** is copied so that each new **cell** receives a complete copy of its **genome**. Copying errors (mutations) occasionally happen. If an error goes uncorrected, then all its progeny cells will carry the change, as well, and will be genetically different from the original cell. The resulting organism is called a genetic mosaic. Patches of skin different in color from surrounding skin are an example of a mosaic effect.

Another cause of mosaicism is the X-chromosome inactivation that occurs in all female mammals. In humans, for example, X inactivation is thought to happen at or before the 32-cell stage. In this process, one of the X chromosomes in each cell is randomly inactivated. For a given cell, all its progeny cells have the same **X chromosome** inactivated. Therefore, a female mammal is in effect a genetic mosaic because some of its cells express the alleles on the maternal X **chromosome**, and some the alleles on the paternal X chromosome. A visually striking example of this phenomenon occurs in the calico cat. The X chromosome of the cat contains a coat-color **gene** that has two alleles—orange and black. If the cat is heterozygous for this gene, then some of its cells and their progeny will express the orange allele and others the black allele, resulting in the familiar colored patches of fur.

It is also possible to be mosaic for number of chromosomes. Occasionally during **mitosis**, a chromosome is lost from one daughter cell, causing it to have one chromosome too few. One of the more common examples of this is the loss of the **Y chromosome** from a cell in a male, resulting in a person having some cells that have the normal 46, XY complement and others that have the 45,X (Turner syndrome) complement. Another example is the loss of an X chromsome from a cell of a normal XX female, resulting in a mosaic containing 45,X and 46,XX cells.

In all cases of mosaicism, the phenotypic effect on the organism is determined by the location of the changed cells and by their number, which is in turn dependent upon how early in development the change occurred. There is one extreme condition in *Drosophila* known as bilateral gynandromorph, in which an X chromosome is lost from one cell during the first mitotic division. The half of the fly missing an X chromosome expresses the recessive characteristics, white eye and miniature wing; whereas the other half, which is heterozygous for both genes, expresses the dominant characteristics, red eye and normal wing.

See also Cell differentiation; Genotype and phenotype; Mitosis

MULLER, HERMANN JOSEPH (1890-1967)

American geneticist

Hermann Joseph Muller was the first to show that genetic mutations can be induced by exposing chromosomes to x rays. For this demonstration, he was awarded the 1946 Nobel Prize in physiology or medicine. He also took up a crusade to improve the condition of the human **gene pool** by calling for a cessation of the unnecessary use of x rays in medicine and a

halt to nuclear bomb testing in order to prevent further damage to the genetic makeup of the human population.

Hermann Joseph Muller was born on December 21, 1890, in New York City. Hermann attended Morris High School in the borough of the Bronx in New York City. On a scholarship, Muller enrolled at Columbia University in 1907 and majored in genetics. He received his bachelors degree in 1910 and then continued at both Cornell Medical School and Columbia for his master's degree, studying the transmission of nerve impulses.

In 1912, Muller began working with A. H. Sturtevant, Calvin B. Bridges, and **Thomas Hunt Morgan**, a zoologist who was performing ground-breaking work in genetics. The group researched *Drosophila melanogaster*, a fruit fly with a brief three-week breeding cycle that makes them ideal for genetic study. *Drosophila* also have only four pairs of chromosomes, the dark-staining microscopic structures within the **nucleus** of each **cell**. Muller and other designed experiments to study mutations resulting in abnormal traits that seem to arise spontaneously in the fruit fly population. The mutations were tracked in order to infer which part of each **chromosome** contained the **gene** responsible for a particular trait, such as eye color or wing shape.

Muller's doctoral thesis in 1916 was on **crossing over**, a phenomenon discovered in 1909 by a Belgian scientist, F. A. Janssens, when he noticed that during the **duplication** and separation of like chromosomes, sometimes part of a chromosome would break off and reattach at a comparable place on the other chromosome. If two genes were far apart on a chromosome, then it would be more likely that a break could occur between them. Thus, a high frequency of crossing over observed between any two traits would mean a long distance between the genes, while a low frequency of crossing over between two traits would mean the genes were close together. The team used this information to map each chromosome in order to show how genes for each trait might be arranged along its length. The findings of the group were published in *The Mechanism of Mendelian Heredity*, written in 1915.

In 1916, Muller took a teaching position at Rice Institute in Texas, where he did further research in genetics, especially mapping modifier genes, which seem to control the expression of other genes. Upon his return to Columbia two years later, Muller did some of his most important theoretical work. Realizing that genes on the chromosomes are self-replicating and are responsible for synthesizing the other components of cells, he theorized that all life must have started out with molecules that were able to self-replicate, which he likened to naked genes. These molecules, he suggested, must have been something like viruses, a very astute hypothesis given the little that was known about viruses at the time.

In 1921, Muller began work as an associate professor teaching genetics and **evolution** and doing research on **mutation** at the University of Texas in Austin, where he remained until 1932. Muller had grown impatient with waiting for mutations to happen by chance, so he began seeking methods of hastening rates of mutation. In 1919, he had discovered that higher temperatures increase the number of mutations, but not always in both chromosomes in a chromosome pair. He deduced that mutations must involve changes at the molecular or sub-molecular level. He struck on the idea of using x rays instead of heat to induce mutations, and by 1926 he was able to confirm that x rays greatly increased the mutation rate in *Drosophila*. He also concluded that most mutations are harmful to the organism, but are not passed on to future generations since the individual affected is unlikely to reproduce; nonetheless, he suggested, if the rate of harmful mutations were to become too high, a **species** might die out.

Muller reported his success in inducing mutations in a 1927 article in *Science* entitled "Artificial Transmutation of the Gene." The article gained him international status as an innovator and introduced other scientists to a technique for studying a large number of mutations at once. This led to the realization that mutations are actually chemical changes that can be artificially induced with any number of other chemicals. It also helped spawn the infant study of radiation genetics.

Muller left Texas in 1932 and moved to Berlin to work at the Kaiser Wilhelm Institute. There he spent a year as a Guggenheim fellow doing research on mutations and exploring the structure of the gene. Muller's next stop was the Soviet Union, where he stayed from 1933 to 1937. At the Academy of Sciences in both Leningrad and Moscow he studied radiation genetics, cytogenetics, and gene structure.

The next year Muller got a job at the Institute of Animal Genetics in Edinburgh, Scotland, again working on radiation genetics upon his return to the United Stated in 1940. He continued his research at Amherst College and, starting in 1945, at Indiana University, where he was appointed professor of zoology and where he stayed until his death on April 5, 1967.

In 1946, Muller was awarded the Nobel Prize for medicine or physiology for his important work on mutations. Muller was also a member of the National Academy of Sciences and a fellow of the Royal Society. He used the opportunity of his world fame to campaign for many social concerns sparked by his interest in the genetic health of the human population. Muller spoke out against needless x rays in medicine and for safety in protecting people regularly exposed to x rays. In the 1950s he campaigned to outlaw nuclear bomb tests because nuclear fallout would cause mutations in future generations. Toward the end of his life, Muller argued that the human race should take action in order to keep healthy genes in the population. His idea came out of the belief that modern culture and technology suspend the process of **natural selection** and thus increase the number of mutations in human genes. Further, Muller argued that there should be programs to promote **eugenics**, literally good genes. Muller the idea of establishing sperm banks in which the sperm of exceptionally healthy and gifted men would be frozen as an endowment to be used for future generations. The concept of such massive intervention in the human gene pool and in the private lives of individuals was and remains highly controversial.

See also Mutation; Radiation mutagenesis

MÜLLER, JOHANN FRIEDRICH THEODOR (FRITZ) (1822-1897)
German naturalist

Fritz Müller was among the earliest German scientists to support **Charles Darwin**'s theory of **evolution**. Born the son of a Protestant minister in Windischholzhausen, near Erfurt, Germany, Müller was always predisposed toward science. His maternal grandfather was the prominent apothecary and chemist, Johann Bartholomäus Trommsdorff (1770–1837). The whole family shared a lively interest in natural history.

After attending the Erfurt Gymnasium from 1835 to 1840, where he became fluent in six languages, Müller studied pharmacy in Naumburg. In 1841 he switched to zoology, first at the University of Berlin, then at the University of Greifswald, then again at Berlin, where he received his Ph.D. in 1844 with a dissertation on leeches directed by the famous physiologist and comparative anatomist, Johannes Peter Müller (1801–1858), who was not related to him. He taught unsuccessfully at the Erfurt Gymnasium in 1845, then decided to study medicine. Having satisfied all the requirements for the M.D. at Greifswald by 1849, he was nevertheless refused the degree because of his support for leftist rebels in the 1848 Revolutions.

Discouraged by German culture and politics, and attracted by the tremendous opportunities for naturalist field work in South America, Müller moved in 1852 to Blumenau, Brazil. He farmed until 1856, taught mathematics in Desterro until 1867, held several civil service posts until 1876, then received a stimulating appointment as traveling naturalist for the National Museum of Brazil in Rio de Janeiro. All the while he conducted zoological research and published scientific papers, especially in botany and entomology. His book, *Für Darwin* (1864) provided cogent empirical support for Darwin's theories, and prompted Darwin to initiate a fruitful scientific correspondence with Müller that lasted until Darwin's death.

Müller's most important scientific contribution was his theory of **mimicry**, still known as "Müllerian mimicry," which superseded "Batesian mimicry" shortly after he published it in a series of articles from 1878 to 1883. According to the theory of **Henry Walter Bates**, the model/mimic relationship is one-to-one, that is, to ensure its own protection, one harmless, weak, or edible **species** imitates the appearance of one dangerous, strong, or inedible species. Müller accepted this, but added that there also exists a multilateral, mutually protective relationship among several unrelated mimics of the same model, so that, when a predator kills a mimic, all other species mimicking that model are subsequently protected.

A modest, amiable gentleman, Müller was still in trouble nearly his entire life because he held fast to his liberal political and social views in the face of conservative forces such as the Prussian Ministry of Education, the Society of Jesus, the Brazilian government, and majority public opinion. His last years were not happy. The National Museum fired him and denied his pension in 1891. He was imprisoned and court-martialed by Brazilian insurgents in 1894. Although the deaths of his wife and daughter nearly deprived him of his will to live, Müller continued to work until he died in Blumenau.

See also Evolutionary mechanisms; Selection

MULLIS, KARY BANKS (1944-)
American biochemist

American biochemist Kary Mullis designed **polymerase chain reaction** (PCR), a fast and effective technique for reproducing specific genes or **DNA** fragments that is able to create billions of copies in a few hours. Mullis invented the technique in 1983 while working for Cetus, a California **biotechnology** firm. After convincing his colleagues of the importance of his idea, they eventually joined him in creating a method to apply it. They developed a machine that automated the process, controlling the chain reaction by varying the temperature. Widely available because it is now relatively inexpensive, PCR has revolutionized not only the biotechnology industry, but also many other scientific fields and it has important applications in law enforcement, as well as history. Mullis shared the 1993 Nobel Prize in chemistry with **Michael Smith** of the University of British Columbia, who also developed a method for manipulating genetic material.

Kary Banks Mullis was born in Lenoir, North Carolina, on December 28, 1944. As a high school student, Mullis designed a rocket that carried a frog some 7,000 ft (2,133.6 m) in the air before splitting open and allowing the live cargo to parachute safely back to earth. Mullis entered Georgia Institute of Technology in 1962 and studied chemistry. As an undergraduate, he created a laboratory for manufacturing poisons and explosives. He also invented an electronic device stimulated by brain waves that could control a light switch.

Upon graduation from Georgia Tech in 1966 with a B.S. degree in chemistry, Mullis entered the doctoral program in biochemistry at the University of California, Berkeley. In Berkeley, at that time, was growing interest in hallucinogenic drugs; Mullis taught a controversial neurochemistry class on the subject. At the age of twenty-four, he wrote a paper on the structure of the universe that was published by *Nature* magazine. He was awarded his Ph.D. in 1973 and he accepted a teaching position at the University of Kansas Medical School in Kansas City, where he stayed for four years. In 1977, he assumed a postdoctoral fellowship at the University of California, San Francisco. In 1979, he accepted a position as a research scientist with a growing biotech firm, Cetus Corporation, in Emeryville, California that was in the business of synthesizing chemicals used by other scientists in genetic **cloning**.

Reproducing **deoxyribonucleic acid** (DNA) had long been an obstacle to anyone working in molecular biology. The most effective way to reproduce DNA was by cloning. Although cloning technology represented a significant scientific advance, it was still a cumbersome process in certain respects. DNA strands are long and complicated, composed of many different chromosomes; the problem was that most

Kary Mullis (seated). *Archive Photos. Reproduced by permission.*

genetic engineering projects were tasks that involved tiny fragments of the DNA molecule, almost infinitesimal sections of a single strand. Cloning works by inserting the DNA into **bacteria** and waiting while the reproducing bacteria create copies of the DNA. The cloning process is not only time-consuming, but it replicates the whole strand, increasing the complexity. The revolutionary advantage of PCR is its selectivity; it is a process that reproduces specific genes on the DNA strand millions or billions of times, effectively allowing scientists to amplify or enlarge parts of the DNA molecule for further study.

Scientists at Cetus developed a commercial version of PCR and a machine called the Thermal Cycler; with the addition of the chemical building blocks of DNA, called nucleotides, and a biochemical catalyst called **polymerase**, the machine would perform the process automatically on a target piece of DNA. The machine is so economical that even a small laboratory can afford it.

The selectivity of the PCR process, as well as the fact that it is simple and economical, have profoundly changed the course of research in many fields. In the field of **genetics**, the process has been particularly important to the Human **Genome** Project—the massive effort to map human DNA. **Nucleotide** sequences that have already been mapped can now be filed in a computer, and PCR enables scientists to use these codes to rebuild the sequences, reproducing them in a Thermal Cycler. The ability of this process to reproduce specific genes, thus effectively enlarging them for easier study, has made it possible for virologists to develop extremely sensitive tests for acquired immunodeficiency syndrome (**AIDS**), capable of detecting the **virus** at early stages of infection. There are many other medical applications for PCR and it has been particularly useful for diagnosing genetic predispositions to diseases such as **sickle cell anemia** and **cystic fibrosis**.

PCR has also revolutionized evolutionary biology, making it possible to examine the DNA of woolly mammoths and the remains of ancient humans found in bogs. PCR can also answer questions about more recent history. For example, it has been used to identify the bones of Czar Nicholas II of Russia who was executed during the Bolshevik revolution. Scientists at the National Museum of Health and Medicine in Washington, D.C., are preparing to use PCR to amplify DNA from the hair of Abraham Lincoln, as well blood stains and bone fragments, in an effort to determine whether he suffered from a disease called Marfan's syndrome. In law enforcement, PCR has made genetic fingerprinting more accurate and effective; it has been used to identify murder victims, and to overturn the sentences of men wrongly convicted of rape. Some have suggested that PCR can be used to create tags or markers for industrial and biotechnological products, including oil and other hazardous chemicals, to insure that they are used and disposed of in a safe manner.

In 1986, Mullis left Cetus to work for Xytronyx, a San Diego research firm, where he became director of molecular biology. Two years later, he left to become a private biochemical research consultant. In 1993, Mullis won the Nobel Prize in Chemistry.

See also Polymerase chain reaction (PCR); Clones and cloning

MULTIFACTORIAL OR MULTIGENIC DISORDER • *see* POLYGENIC DISORDER

MULTIFACTORIAL TRANSMISSION

Diseases and health traits that involve multiple genes and complex patterns of **inheritance** are generally described as exhibiting multifactorial transmission. The term multifactorial means many factors. By contrast, simple genetic traits like **cystic fibrosis**, which are caused by a **mutation** in a single **gene**, tend to be inherited in predictable patterns that are either autosomal or sex-linked, and they are either dominant or recessive. Multifactorial traits, however, may involve genes on both **autosomes** and sex-chromosomes at the same time. Some of the genes may be dominant acting, and others recessive. There may be significant interaction with nongenetic factors like nutrition or other environmental exposures. Traits influenced by multiple genes that do not involve any significant interactions environmental factors are called polygenic traits. When environmental and genetic factors both are involved, the degree of influence attributable to **genetics** is called heritability. Careful study of identical and fraternal twin pairs is one method of measuring heritability of multifactorial traits.

Multifactorial traits which involve features that can be measured on a continuous numerical scale like weight, blood pressure, or IQ, are called metric traits. Metric traits tend to fit onto a bell-shaped distribution in which there are as many individuals who are above the average as there are below the average. For example, in a population of individuals with an average IQ of 100, there will be approximately the same number of individuals with IQ over 120 as there are individuals with IQ below 80. On average, an individual's measured value will be midway between their values of their biological parents. Naturally, there is considerable variation between different individuals and the actual value may be well above or well below either parent.

Many multifactorial traits do not involve measurable quantifiable variables. An example of this includes structural **birth defects** such as congenital heart defects. Such traits are called discontinuous or threshold traits. One way to understand these traits is to imagine that there is some measurable variable called liability. How to measure liability is unknown, and like the metric traits, it is distributed on a bell-shaped curve. Everyone has some amount of liability for each of these traits, but a baby will only have the trait or disease if the amount of liability is so high that it exceeds some threshold. Parents of a baby who has the trait typically have higher than average liability, and therefore the chance that their other children will also have the trait is increased.

Considering structural birth defects which have multifactorial transmission, there is some clustering in families, but the numbers of offspring showing the trait are significantly lower than what would be predicted from a single gene model. This is what would be expected from the liability model. Because liability cannot be measured and many different factors may be involved, most cases of these birth defects occur without warning in families without any history of the disorder. Examples of multifactorial traits that cause birth defects include **spina bifida**, anencephaly, cleft lip, cleft palate, pyloric stenosis, along with many others.

See also Cleft lip/palate; Polygenic inheritance and disorders; Twin studies

MULTIMERIC ENZYMES

All **enzymes** are proteins. An enzyme is very specific in the reaction it catalyzes. The reaction requires the binding of the substrate to an **active site** in the enzyme. In some cases, association of more than one polypeptide molecule is required for the formation of the active site. The resulting quaternary structure, involving the interaction between polypeptides, is called a multimeric enzyme.

Multimeric enzymes have various functions, constituent molecules and shapes. A few examples will suffice to indicate the diversity of these enzymes. Lactose dehydrogenase and glyceraldehyde 3-phosphate dehydrogenase are tetrameric—they have four subunits—and are donut like in shape. Triose phosphate isomerase consists of two subunits arranged as a barrel. Finally, glutamine synthetase is a dodecmer, consisting of two hexagonal rings, with a water-filled channel through the middle, stacked against each other.

Multimeric enzymes can affect the accuracy of a complementation analysis. Complementation is a test of **gene** function. A complementation analysis asks if two versions of the same region of the **chromosome**, located in the same **cell** are mutated, acting independently can supply all the functions necessary for a wild-type phenotype. For example, when the two mutations affect copies of the same gene such that neither is capable of generating a wild-type product then the resulting strain will have a mutant phenotype. On the other hand, if the two mutations affect different genes, so that each copy of the region of **DNA** is able to generate some of the gene products required, then between the two regions of DNA all the gene products necessary for a wild-type phenotype might result. A **complementation test** looks for restoration of the wild-type phenotype in pairings between two mutant organisms.

The intereference in complementation analysis by multimeric enzymes, termed negative complementation, is rare, but real. The interference occurs in the product of a cross between two mutants where the mutations are in genes coding for components of the enzyme. A multimeric enzyme can be generated whose subunits come from both the mutant and wild-type genes. Thus, even though a normal complement of genes for the enzyme subunits exists, random chance creates an enzyme in which one or more of the subunits are coded for by mutant copies of the genes. The activity of the enzyme, if not completely abolished, can be diminished. Because of the negative complementation phenomenon, it can be difficult to gauge the number of genes in the region of DNA under study.

See also Gene regulation

MULTIPLEXING

Multiplexing is an approach to the **sequencing** of **DNA** (**deoxyribonucleic acid**) that uses several pooled samples simultaneously. This greatly increases the speed at which sequencing can be accomplished. The method was described in 1998.

For multiplexing, the different DNA molecules to be analyzed are linked to a set of identifier tags before the analysis begins. The tagged DNA molecules are then pooled together, their numbers are chemically increased, and then they are chemically fragmented. The resulting reaction products are separated on the basis of size on electrophoretic gels, and then are transferred to another support. This support can then be probed, as many times as there are tags in the original pooled sample, to extract information on the identity of the DNA fragments.

Another application of multiplexing involves the use of fluorescent compounds. The multiplex multifluor technique employs laser light to scan DNA-containing material as the material flows through a thin tube. The DNA tag contains a molecule that can fluoresce when excited by the laser light.

Analysis of the patterns of fluorescence enables the chemical content of the sample.

The ability to rapidly sequence different areas of a **genome** at the same time greatly reduced the time required to sequence the genetic material. This was one of the reasons that the human genome could be sequenced in little over a decade.

Multiplexing has also been instrumental in elucidating the genomc sequence of other organisms, such as *Arabidopsis thaliana,* a model plant sequenced by the international *Arabidopsis* Genome Initiative (AGI) group in 2000.

MURRAY, ANDREW (-)
British-born American molecular biologist

British born American molecular biologist Andrew W. Murray is a professor of biology at Harvard University and serves as the director of the Center for Genomic Research. In 1983, Murry and Jack W. Szostak created the first artificial **chromosome**.

Murray's current research interests include studies regarding the structure and **evolution** of cells. Special areas of research under Murray's direction include studies regarding transmission of genetic information during **cell division**. In particular, as director of the Center for Genomic Research, Murray oversees research into the mechanisms of chromosomal sorting and segregation during **meiosis**. Research into genetic regulation and mechanisms of transmission during **cell** division may play a critical role in furthering medical understandings of the nature of cancerous cells. One aspect of many cancers is an unchecked growth or breakdown in self-regulation of the cell division process.

Using organism as diverse as **yeast** and amphibian eggs, scientists at CGR study the mechanisms of chromatid formation and separation.

Although his parents were American, Murray was born in England. In addition to his work at Harvard Murray holds the position of Associate Professor of Physiology at the University of California at San Francisco.

Murray found early inspiration to become a scientist while visiting America. In particular, Murray received early encouragement form MIT researcher Nancy Hopkins. Murray earned degree in biochemistry from Cambridge in 1978 and then undertook his graduate work at Harvard. While a graduate student, Murray conducted his thesis research under the direction of Szostak. At the time, Murray's work utilized innovative techniques to study yeast plasmid function and control. During their work with yeast, Murray and Szostak developed the methodology to create stable yeast artificial chromosomes. Yeast artificial chromosomes (**YAC**) are **cloning vectors** that can carry large **DNA** fragments. This large carrying capacity allows scientists the ability to study large regions of DNA while it remains intact. YACs also allow the **cloning** of large genes or **gene** complexes.

See also Bacterial artificial chromosome (BAC); Cancer; Cell cycle (Eukaryotic), genetic regulation of; Cell cycle (Pro-karyotic), genetic regulation of; Cell division; Chromosome; Gene; Gene regulation; Genomic library

MUTAGEN

Mutagens are chemicals or physical factors (such as radiation) that increase the rate of **mutation** in the cells of **bacteria**, plants, and animals (including humans). In the living **cell**, **DNA** undergoes frequent chemical change, especially during **replication**. Most of these changes are quickly repaired. Those that are not, result in a mutation. Accordingly, one form of mutation results from a failure of **DNA repair**. Most mutagens are of natural origin. Mutagenesis is a **DNA replication** failure that results in a mutation. Mutagenesis may occur as a result of a mutatgen, or occur spontaneously.

Mutagens can be found throughout nature. Very small doses of a mutagen usually have little effect while large doses of a mutagen could be lethal. DNA in the nuclei of all cells encodes proteins, which play important structural and functional (metabolic) roles in the cell. Mutagens typically disrupt the DNA of cells, causing changes in the proteins that the cell produces, which can lead to abnormally fast growth (**cancer**), or even cell death. In rare incidences, mutagens may cause protein changes that are beneficial to the cell.

Early physicians detected tumors in patients more than 2,000 years before the discovery of chromosomes and DNA. In 500 B.C., the Greek Hippocrates named crab-shaped tumors cancer (meaning crab). The first mutagens to be identified were carcinogens, or cancer-causing substances.

In England in 1775, Dr. Percivall Pott wrote a paper on the high incidence of scrotal cancer in chimney sweeps who were typically boys small enough to fit inside chimneys and clean out the soot. Pott suggested that chimney soot contained carcinogens that could cause the growth of the warts seen in scrotal cancer. More than 150 years later, chimney soot was found to contain hydrocarbons capable of mutating DNA.

In France in the 1890s, Bordeaux wine workers showed an unusually high incidence of skin cancer on the back of the neck. These workers spend their days bending over in the fields picking grapes, exposing the back of their necks to the sun. The ultraviolet (UV) radiation in natural sunlight was later identified as a mutagen.

Mutagens can be found in foods, beverages, and drugs. Sometimes a substance is mutagenic because it is converted in the body into something harmful. In many industrialized countries, regulatory agencies are responsible for testing food and drugs to insure that the public is not unknowingly exposed to mutagens. However, some mutagen-containing substances are not tightly controlled. One such substance suspectd of being a mutagen is found in the tobacco of cigars and cigarettes.

Some mutagens occur naturally, and some are synthetic. Cosmic rays from space are natural, but they can be mutagenic. Some naturally occurring viruses are considered mutagenic since they can insert themselves into host DNA. Hydrogen and atomic bombs are manmade, and they emit harmful radiation. Radiation from nuclear bombs and gaseous particles from nitrogen mustard and acridine orange have been used destructively in war. On the other hand, some mutagens are used constructively to kill bacteria that could grow in human foods, such as the small doses of nitrites used to preserve some meat. Even though nitrites can be mutagenic, they are still used because without the nitrites these meats could cause botulism.

Mutagens affect DNA in different ways. Some mutagens, such as nitrogen mustard, bind to a base and cause it to make a different **amino acid**. These mutagens cause point mutations, because they change the genetic code at one specific location or base in such a way that the instructions for a protein's amino acid sequence are also altered.

Mutagens such as acridine orange work by deleting or inserting one or more **bases** into the DNA molecule, shifting the frame of the triplet code for an amino acid. Deletion and insertion mutations causing frame-shift mutations can change a long string of amino acids, which can severely alter the structure and function of a protein product.

Normal cells recognize cues from their environment and respond with specific reactions, but cells impaired by a mutation do not behave or appear normal, and are said to be transformed.

The significance of mutations is profoundly influenced by the distinction between germline and soma. Mutations in somatic (body) cells are not transferred to offspring. Mutations that occur in a somatic cell, in the bone marrow or liver for example, may damage the cell, make the cell cancerous or even kill the cell. Whatever the effect, the ultimate fate of that somatic mutation is to disappear when the cell in which it occurred, or its owner, dies. However, mutated DNA can only be passed to the next generation if it is present in a germ cell such as spermatozoa and ova (eggs), each of which contribute half of the DNA of the new organism. Germline mutations will be found in every cell descended from the **zygote** to which that mutant **gamete** contributed. If an adult is successfully produced, every one of its cells will contain the mutation. Included among these will be the next generation of gametes, so if the owner is able to become a parent, that mutation will pass down to yet another generation.

Chemical mutagens are classified as alkylating agents, cross-linking agents, and polycyclic aromatic hydrocarbons (PAHs). Alkylating agents act by adding molecular components to DNA bases, which alters the protein product. Cross-linking agents create covalent bonds with DNA bases, while PAHs are metabolized by the human body into other potentially mutagenic molecules.

Radiation is another potent mutagen. For biologists, the most significant forms of radiation are light, heat, and ionizing radiation. Ionizing radiation can penetrate cells and create ions in the cell contents. These, in turn, can cause permanent alterations in DNA; that is, mutations. Ionizing radiation includes: x rays, gamma rays, and the sub-atomic particles—neutrons, electrons ("beta" particles), and alpha particles (helium nuclei). High energy radiation passes through cells and tissues, and can induce breaks in chromosomes that result in rearrangements of entire sections of the chromosomes. UV radiation causes covalent bonds to form between neighboring thymine bases in the DNA, so altering the DNA product at that location.

Mutagens are often associated with specific cancers in humans. Aromatic amines are mutagens that can cause bladder cancer. Tobacco taken in the form of snuff contains mutagens that can cause nose tumors. Tobacco smoke contains mutagens such as PAHs and nitrosamine (a type of alkylating agent), as well as toxins such as carbon monoxide, cyanide, ammonia, arsenic, and radioactive polonium. Although tobacco products are legal and widely available, many physicians and government agencies warn about the health risks linking smoking with several types of cancer and heart disease.

In 1973, Bruce Ames, a professor of biochemistry and molecular biology at the University of California at Berkeley, introduced the now widely-used Ames test to identify potential mutagens. Suspected mutagens are mixed with a defective strain of the bacteria *Salmonella*, which only grows if it is mutated. Substances that allow the *Salmonella* to grow are considered mutagenic.

In addition to mutagen-induced DNA changes, spontaneous mutations occur in the dividing cells of the human body every day. The nuclei of the cells have repair **enzymes**, which remove mutations and restore mutated DNA to its original form. If these natural DNA repair mechanisms fail to keep up with the rate of mutation or the repair mechanisms themselves are defective, disease can result. This latter case is one of the suspected mechanisms in lung cancer due to cigarette smoking, where the nicotine in the smoke is thought to block an important repair process in the lungs.

Some aging mechanisms appear to be related to mutagenic oxidants produced as by-products of normal metabolism. These oxidants, such as hydrogen peroxide, are similar to the same mutagens produced by high energy radiation, and cause damage to DNA, proteins, and lipids. Decay of **mitochondria** with time, due to oxidative damage, also appears to play a major role in aging. The degenerative diseases of aging, such as cancer, cardiovascular disease, cataracts, and brain dysfunction, are also suspected to be related to mutagenic oxidative changes.

See also Cancer genetics; Gene mutation; Missense mutations; Neutral mutation; Radiation mutagenesis; Samesense mutation; Silent mutation; Spontaneous mutations and reversions

MUTANTS: ENHANCED TOLERANCE OR SENSITIVITY TO TEMPERATURE, PH, ALKALINITY

Microorganisms have optimal environmental conditions under which they grow best. Classification of microorganisms in terms of growth rate dependence on temperature includes the thermopiles, the mesophiles and psychrophiles. Similarly, while most organisms grow well in neutral pH conditions, some organisms grow well under acidic conditions, while others can grow under alkaline conditions. The mechanism by which such control exists is being studied in detail. This will overcome the need to obtain mutants by a slow and unsure process of acclimatization.

When some organisms are subjected to high temperatures, they respond by synthesizing a group of proteins that help to stabilize the internal cellular environment. These, called heat shock proteins, are present in both prokaryotes and **eukaryotes**. Heat stress specifically induces the **transcription** of genes encoding these proteins. Comparisons of **amino acid** sequences of these proteins from the **bacteria** *Escherichia coli* and the fruit fly *Drosophila* show that they are 40–50% identical. This is remarkable considering the length of evolutionary time separating the two organisms.

Fungi are able to sense extracellular pH and alter the expression of genes. Some fungi secrete acids during growth making their environment particularly acidic. A strain of *Asperigillus nidulans pac* encodes a regulatory protein that activates transcription of genes during growth under alkaline conditions and prevents transcription of genes expressed in acidic conditions. A number of other genes originally found by analysis of mutants have been identified as mediating pH regulation, and some of these have been cloned. Improved understanding of pH sensing and regulation of **gene expression** will play an important role in **gene** manipulation for **biotechnology**.

The pH of the external growth medium has been shown to regulate gene expression in several enteric bacteria like *Vibrio cholerae*. Some of the acid-shock genes in *Salmonella* may turn out to assist its growth, possibly by preventing lysosomal acidification. Interestingly, acid also induces virulence in the plant pathogen (harmful microorganism) *Agrobacterium tumefaciens*.

Study of pH-regulated genes is slowly leading to knowledge about pH homeostasis, an important capability of many enteric bacteria by which they maintain intracellular pH. Furthermore, it is felt that pH interacts in important ways with other environmental and metabolic pathways involving anaerobiosis, sodium (Na+) and potassium (K+) levels, **DNA repair**, and amino acid degradation. Two different kinds of inducible pH homeostasis mechanisms that have been demonstrated are acid tolerance and the sodium-proton antiporter NhaA. Both cases are complex, involving several different stimuli and gene loci.

Salmonella typhimurium (the bacteria responsible for typhoid fever) that grows in moderately acid medium (pH 5.5-6.0) induces genes whose products enable cells to retain **viability** (ability to live) under more extreme acid conditions (below pH 4) where growth is not possible. Close to 100% of acid-tolerant (or acid-adapted) cells can recover from extreme-acid exposure and grow at neutral pH. The inducible survival mechanism is called acid tolerance response. The retention of viability by acid-tolerant cells correlates with improved pH homeostasis at low external pH represents inducible pH homeostasis.

Cells detect external alkalization with the help of a mechanism known as the alkaline signal **transduction** system. Under such environmental conditions, an inducible system for internal pH homeostasis works in *E. coli*. The so-called sodium-proton antiporter gene NhaA is induced at high external pH in the presence of high sodium. The NhaA antiporter acts to acidify the cytoplasm through proton/sodium exchange. This allows the microorganism to survive above its normal pH range. As *B. alkalophilus* may have as many as three sodium—proton antiporters, it is felt that the number of antiporters may relate to the alkalophilicity of a **species**.

The search for extremophiles has intensified recently. Standard **enzymes** stop working when exposed to heat or other extreme conditions, so manufacturers that rely on them must often take special steps to protect (stabilize) the proteins during reactions or storage. By remaining active when other enzymes would fail, enzymes from extremophiles (extremozymes) can potentially eliminate the need for those added steps, thereby increasing efficiency and reducing costs in many applications.

Many routes are being followed to use the capacity that such extremophiles possess. First, the direct use of these natural mutants to grow and produce the useful products. Also, it is possible with recombinant **DNA** technology to isolate genes from such organisms that grow under unusual conditions and clone them on to a fast growing organism. For example, an enzyme alpha-amylase is required to function at high temperature for the hydrolysis of starch to glucose. The gene for the enzyme was isolated from *Bacillus stearothermophilus,* an organism that is grows naturally at 194°F (90°C), and cloned into another suitable organism. Finally, attempts are being made to stabilize the proteins themselves by adding some groups (e.g. disulfide bonds) that prevent its easy denaturation. This process is called protein engineering.

Conventional mutagenesis and **selection** schemes can be used in an attempt to create and perpetuate a mutant form of a gene that encodes a protein with the desired properties. However, the number of mutant proteins that are possible that are possible after alteration of individual nucleotides within a **structural gene** by this method is extremely large. This type of mutagenesis also could lead to significant decrease in the activity of the enzyme. By using a set techniques that specifically change amino-acids encoded by a cloned gene, proteins with properties that are better than that obtained from the naturally occurring strain can be obtained. Unfortunately, it is not possible to know in advance which particular amino acid or short sequence of amino acids will contribute to particular changes in physical, chemical or kinetic properties. A particular property of a protein, for example, will be influenced by amino acids quite far apart in the linear chain as a consequence of the folding of the protein, which may bring them into close proximity. The amino acid sequences that would bring about change in physical properties of the protein can be obtained after characterization of the three dimensional structure of purified and crystallized protein using **x-ray crystallography** and other analytical procedures. Many approaches are being tried to bring about this type of "directed mutagenesis" once the specific **nucleotide** that needs to be altered is known.

See also Adaptation and fitness; Chemical mutagenesis; Ecological and environmental genetics; Enzymes, genetic manipulation of; Mutations and mutagenesis

MUTATION

A mutation is a sudden change in **DNA** (**deoxyribonucleic acid**), the genetic material of life. Mutation is a major evolutionary force that results from wide range of factors and that carries a wide range of results. Mutations may carry no discernable effects or act to significantly increase or decrease **genetic variation** within a population.

Although some mutations have no visible effects, there are mutations that can cause dramatic changes in an organism's appearance, behavior, or health. Organisms born with mutations can look very different from their parents. **Albinism**, for example, results from mutations that eliminate skin pigment. Dwarfism can be the result of mutations that affect growth hormones. Mutations are usually harmful, but some may help an organism survive, by proving to be beneficial to the **species**. Regardless, mutations are a major driving force behind biological change (**evolution**).

Every **cell** contains DNA on threadlike structures called chromosomes. Stretches of DNA that code for specific proteins are known as genes. If the DNA of a particular **gene** is altered, that gene is considered a mutant gene. Such mutations may have little or no impact on **protein synthesis**. If, however, the mutation adversely impacts protein synthesis (i.e., if the mutation causes the protein for which it codes to be missing or defective) it is possible that the loss or alteration of the protein will result in either the death of the organism (usually early in embryonic development) or a significant phenotypic change (a visible difference in the organism). Albinism, for instance, is the result of one missing protein. This understanding of genetic **inheritance** is based upon experiments conducted by Thomas H. Morgan in 1910 with fruit fly mutations, and experiments conducted by **George W. Beadle** and **Edward L. Tatum** in the 1940s on bread mold mutations.

Errors in DNA take several forms. DNA itself is made up of subunits known as **nucleotide bases**. There are four kinds of bases: adenine, cytosine, guanine, and thymine. They are referred to by their initials: A, C, G, and T. DNA can be thought of as a code (the genetic code) written with these four letters. Mutations, in the strictest sense, are changes in the genetic sequence or code.

The term mutation is also more broadly used in **genetics**. For example, errors in all or part of a **chromosome** are another form of mutation termed chromosomal abnormalities. Humans normally have 23 pairs of chromosomes. (Each pair of chromosomes is distinct under the microscope and scientists have numbered them for ease of identification.) An extra chromosome can have an enormous effect. Three copies of chromosome 21, for instance, results in **Down syndrome**. People with Down syndrome have a unique physical appearance and are developmentally disabled. If parts of non-homologous chromosomes swap pieces, the result is a **translocation**. Such translocations can also carry a significant impact. A translocation between chromosome 9 and chromosome 22, for example, is associated with a certain type of leukemia.

Mutations that occur in an organism's egg or **sperm cells** are known as germinal mutations. Germinal mutations can be passed on to the organism's offspring. Mutations that occur in cells other than the sex cells are known as somatic mutations and can not be passed on. Accordingly, some causes of mutation will affect only the somatic cells of the organism exposed. Other causes will affect the germ cells and may be passed on to many succeeding generations. In this manner, a mutation can become common in certain populations.

Mutations are a normal occurrence. Mutation rates vary depending on the gene in question. The opportunity for mutation exists every time a cell replicates. Almost always, DNA reproduces itself correctly. Yet if the genetic code is somehow altered—if part of it is deleted, duplicated, or switched —the result is a mutation. Generally, cells that divide many, many times in a lifetime are more at risk for errors than cells that divide less frequently.

Uncontrolled cell growth, known as **cancer**, is also a kind of mutation. Some types of cancers are associated with environmental factors such as smoking. Many researchers assert that repeated exposure to cigarette smoke may cause a somatic mutation in the lung cells that leads to lung cancer. Other environmental factors that are known to cause mutations include exposure to radiation, pesticides, asbestos, and some (now banned) food additives. Factors that cause mutations are termed mutagens. Those that cause cancer are known as carcinogens.

Scientists have found that carcinogens work by engineering mutations in important genes. A gene known as p53, for instance, helps prevent the growth of tumors. Yet, exposure to ultraviolet light and other environmental stimuli may cause that gene to mutate. In its mutated form, the gene no longer prevents tumors. People with two copies of the mutated gene are at greater risk develop some forms of cancer. If mutated genes responsible for cancer are present in the egg or sperm cells, then a susceptibility to cancer may be passed on to future generations. This mechanism may account for cancers that have a hereditary association in families, and points to the existence of a germinal mutation.

Developing embryos and fetuses are especially at risk for mutation. Their cells divide very rapidly and become increasingly specialized for specific tasks. Accordingly, physicians often warn pregnant women to avoid excessive x rays and some medications in order to protect developing **fetal cells.**

For every human trait, there was once a person or small group in which the genetic mutation first appeared. Certain mutations occurred long ago and are now part of what is considered the normal genetic constitution shared by the majority of humans. Other traits occur only in certain populations of people. **Cystic fibrosis**, for instance, is most common in people of northern European descent. **Sickle cell anemia**, a serious blood disease, occurs frequently in people of African and Mediterranean ancestry. **Tay-Sachs disease**, a fatal disorder, is

Genetic mutations occur in humans and in nature. *JLM Visuals. Reproduced by permission.*

found primarily in ethnic Jews with eastern European ancestors.

Whether a mutation is detrimental (deleterious) to useful to a population may depend on external or environmental circumstances. For example, while two copies of the mutant sickle cell anemia gene may result in serious illness, one copy confers a resistance to malaria. Such selective advantage proves useful to people living in the tropics where malaria is common and, as a group, outweighs the deleterious effects of sickle cell. For these reasons, sickle cell genes are preserved in some populations. Scientists have hypothesized that some advantage must be conferred upon people with single copies of the cystic fibrosis gene or the Tay-Sachs disease gene.

Over millions of years, advantageous mutations have allowed life to develop and diversify from primitive cells into the multitude of species on Earth. Evolution is led by mutation. Mutations that allow an organism to survive and reproduce better than other members of its species are valuable. Mutations become especially important when an organism's environment is changing.

Animal and plant breeders use mutations to produce new or improved species of crops and livestock. Manipulation of mutations can result in crops that are resistant to drought or insects and that have a high yield per acre.

See also Adaptation and fitness; Agricultural genetics; Alleles and allotype; Biodiversity; Cancer genetics; Comparative genomics; Evolutionary mechanisms; Gametogenesis; Mutations and mutagenesis; Population genetics; Selection

MUTATIONS AND MUTAGENESIS

A **mutation** is any change in genetic material that is passed on to the next generation. The process of acquiring change in genetic material forms the fundamental underpinning of **evolution**. Mutation is the source of **genetic variation** in humans and other life forms. Depending on the organism or the source of the mutation, the genetic alteration may be an alteration in the organized collection of genetic material or a change in the composition of an individual **gene**.

Mutations may have little impact, or they may produce a significant positive or negative impact, on the health, competitiveness, or function of an individual, family, or population. A recently isolated series of gene mutations has been shown to alter the regulation of electrical activity in nerve cells, which may be the cause of one or more types of inherited epilepsy.

Mutations arise in different ways. An alteration in the sequence, but not in the number of nucleotides in a gene is a **nucleotide** substitution. Two types of nucleotide substitution mutations are missense and **nonsense mutations**. **Missense mutations** are single base changes that result in the substitution of one **amino acid** for another in the protein product of the gene. Nonsense mutations are also single base changes, but create a termination codon that stops the **transcription** of the gene. The result is a shortened, dysfunctional protein product.

Another mutation involves the alteration in the number of **bases** in a gene. This is an insertion or deletion mutation. The impact of an insertion or deletion is a frameshift, in which the normal sequence with which the genetic material is interpreted is altered. The alteration causes the gene to code for a different sequence of amino acids in the protein product than would normally be produced. The result is a protein that functions differently or not all, as compared to the normally encoded version.

Genomes naturally contain areas in which a nucleotide repeats in a triplet. Trinucleotide repeat mutations, an increased number of triplets, are now known to be the cause of at least eight genetic disorders affecting the nervous or neuromuscular systems.

Mutations arise from a number of processes collectively termed mutagenesis. Frameshift mutations, specifically insertions, result from mutagenic events where **DNA** is inserted into the normally functioning gene. The genetic technique of insertional mutagenesis relies upon this behavior to locate target genes, to study **gene expression**, and to study protein structure-function relationships.

DNA mutagenesis also occurs because of breakage or base modification due to the application of radiation, chemicals, ultraviolet light, and random **replication** errors. Such mutagenic events occur frequently, and the **cell** has evolved repair mechanisms to deal with them. High exposure to DNA damaging agents, however, can overwhelm the repair machinery.

Genetic research relies upon the ability to induce mutations in the lab. Using purified DNA of a known restriction map, site-specific mutagenesis can be performed in a number of ways. Some **restriction enzymes** produce staggered nicks at the site of action in the target DNA. Short pieces of DNA (linkers) can subsequently be introduced at the staggered cut site, to alter the sequence of the DNA following its repair. Cassette mutagenesis can be used to introduce selectable genes at the specific site in the DNA. Typically, these are drug-resistance genes. The activity of the insert can then be monitored by the development of resistance in the transformed cell. In deletion formation, DNA can be cut at more than one restriction site and the cut regions can be induced to join, eliminating the region of intervening DNA. Thus, **deletions** of defined length and sequence can be created, generating tailor-made deletions. With site-directed mutagenesis, DNA of known sequence that differs from the target sequence of the original DNA, can be chemically synthesized and introduced at the target site. The insertion causes the production of a mutation of pre-determined sequence. Site-directed mutagenesis is an especially useful research tool in inducing changes in the shape of proteins, permitting precise structure-function relationships to be probed. Localized mutagenesis, also known as heavy mutagenesis, induces mutations in a small portion of DNA. Mutations are identified by the classical technique of phenotypic identification— looking for an alteration in appearance or behavior of the mutant.

Mutagenesis is exploited in **biotechnology** to create new **enzymes** with new specificity. Simple mutations will likely not have as drastic an effect as the simultaneous alteration of multiple amino acids. The combination of mutations that produce the desired three-dimensional change, and so change in enyzme specificity, is difficult to predict. The best progress is often made by creating all the different mutational combinations of DNA using different **plasmids**, and then using these plasmids as a mixture to transform *Escherichia coli* **bacteria**. The expression of the different proteins can be monitored and the desired protein resolved and used for further manipulations.

See also Chemical mutagenesis; Chromosomal mutations and abnormalities; Gene mutation

N

NATHANS, DANIEL (1928-1999)

American molecular biologist

Daniel Nathans is best known for his work with **restriction enzymes**, which are used to cut or break **DNA (deoxyribonucleic acid)** molecules. This technique, first applied to **gene** study by Nathans, led the way in studying the structure of viruses and opened the door for recombinant DNA research and **genetic mapping**. His work was recognized in 1978, when he shared the Nobel Prize in physiology or medicine with **Werner Arber** and **Hamilton O. Smith**.

Nathans was born in Wilmington, Delaware, the last of nine children born to Samuel and Sarah Nathans, Russian Jewish immigrants. Nathans received his B.A. from the University of Delaware in 1950 and his M.D. from Washington University in St. Louis in 1954. It was during the summer after his first year of medical school that Nathans had his initial exposure to laboratory work.

After medical school, Nathans completed a one-year internship at Columbia-Presbyterian Medical Center. After this, he spent two years (1955–57) at the National Cancer Institute as a clinical associate studying **protein synthesis**. In 1956, Nathans married Joanne Gomberg, with whom he had three sons. Returning to Columbia-Presbyterian, Nathans completed his residency in 1959. That same year Nathans won a United States Public Health Service grant to do biochemical research at Rockefeller University in New York with Fritz Lipmann and Norton Zinder. It was at this point that Nathans fully committed to work in the laboratory rather than in a clinical practice. In New York, Nathans continued his work on protein synthesis and began viral research, mostly related to host-controlled variations in viruses.

In 1962, Nathans began his long relationship with Johns Hopkins University as assistant professor of microbiology and director of genetics. He was elevated to associate professor in 1965 and full professor in 1967. He was named director of the molecular biology and genetics department in 1972 and Boury Professor of Molecular Biology and Genetics in 1976, positions he retained for many years.

In 1962, when Nathans first arrived at Johns Hopkins, Werner Arber, at Basel University in Switzerland, predicted the existence of an enzyme capable of cutting DNA at specific sites. Deoxyribonucleic acid (DNA) is assumed to be the source of autoreproduction in many viruses. An ability to cut or cleave the DNA into specific and predictable fragments was important to greatly improving our capabilities for researching and understanding viruses. The necessity of specific and predictable fragments relates to the need of the scientist to know the fragment he or she is studying is identical to the fragment any other scientist would get following the same laboratory procedure.

In 1968, Arber was halfway to his goal, finding an enzyme (type I) capable of cleaving DNA, but in seemingly random patterns. In 1969, Hamilton O. Smith, a colleague of Nathans at Johns Hopkins, wrote to Nathans (who was in Israel at the time) to tell him he had developed a type II enzyme. This enzyme, named Hind II, was capable of cleaving DNA into specific and predictable fragments.

At this time, Nathans was working on a simian **virus** (SV40) which causes tumors in monkeys. SV40 was particularly impervious to then-current methods of study, so Nathans immediately saw an application of Smith's tool. Nathans, with Kathleen Danna, used Hind II to cut SV40 into eleven pieces and show its method of **replication**.

The combined efforts of Arber, Smith, and Nathans over a period of more than a decade led to their receipt of the Nobel Prize in physiology or medicine in 1978. Their inter-laboratory cooperation greatly advanced the potential for consistent DNA and gene research.

Nathans continued his work with Hind II and cleared the path for much of the work that has been done since in research on DNA function and structure (such as restrictions maps, used to define **DNA structure**). This early work has also led to the

Daniel Nathans. *UPI/Corbis-Bettmann. Reproduced by permission.*

area of recombinant DNA research, which involves the process of joining two DNA fragments from separate sources into one molecule. Since this field of research was uncharted territory and carried some risks, including the creation of new pathogens, Nathans was among an early group of scientists who, in 1974, encouraged the publication of research guidelines and some self-imposed limits on DNA research. Despite the risks, recombinant DNA research has been put to good use in creating supplies of heretofore scarce **enzymes** and hormones, including insulin.

In the 1980s, Nathans's research continued to be linked closely to DNA and genetics. A good portion of his scientific work during this time related to the effect of growth factors on genes and **gene regulation**. Nathans was a member of the American Academy of Arts and Sciences. He was a senior investigator at the Howard Hughes Medical Institute and a member of the National Academy of Sciences. He has served on the editorial board of *Proceedings of the National Academy of Sciences* and has been a regular contributor. He maintained his association with the National Cancer Institute and authored dozens of articles for several scientific journals. Nathans, a major player in scientific research and education in the mid to late part of the twentieth century, died in Baltimore at age 71.

See also Restriction enzymes

NATURAL SELECTION

Although not the only agent of evolutionary change, natural **selection** is certainly the most important mechanism of adaptive evolutionary change in populations of organisms. Through the process of natural selection, individual organisms become better adapted to their local environment and thus acquire greater **fitness**, defined as an increase in reproductive success. If individual organisms vary in their ability to survive and reproduce, and those variations are inherited from parent to offspring, traits that are favored in the current biological conditions will spread in the population.

Natural selection as a means for evolutionary change in living creatures was first proposed, simultaneously, by two British naturalists, **Charles Robert Darwin** (1809–1882) and **Alfred Russel Wallace** (1823–1913), in a joint paper presented to the Linnean Society in London in July 1858. It was a case of a simultaneous discovery. Since the early 1840s, Darwin had been at work compiling his book *On the Origin of Species by Means of Natural Selection*, laboriously presenting his theory and its ramifications; but he was stunned when in 1858 he received a letter from Wallace and a manuscript in which the young scientist detailed a strikingly similar theory. With some anxiety, Darwin contacted his friend, naturalist Sir **Charles Lyell** (1797–1875), who responded by arranging the joint announcement of Darwin's and Wallace's theory before the audience of the Linnean Society. The following year, Darwin published a condensed version of his book, ultimately establishing himself as the principal creator of the theory of natural selection.

A great deal of the material in the *Origin of Species* was inspired by Darwin's earlier travels aboard the *HMS Beagle*, on which he served as ship's naturalist during the years 1831–1836. The vessel circumnavigated the globe. Although the voyage ultimately focused on surveying the coast of South America, including the Galapagos Islands, Darwin's observations spanned the globe. The notebooks he produced during this time contained his detailed documentation of the great variety of living creatures he saw and his efforts to understand their sometimes curious features. He observed that marsupials, mammals who carry their young in a pouch, are almost entirely restricted to the Australian continent (although he could think of no environmental factor unique to Australia that could explain the need for pouches). Specimens of mockingbirds and finches collected in different islands in the Galapagos were so distinct from one another that Darwin apparently came to doubt the fixity of **species**; after the *Beagle's* return, he set about gathering evidence for evolutionary change and a mechanism to account for it.

Darwin did not develop his theory of natural selection, however, until some time after his return to London. Then, in the late 1830s, he engaged in the study of the writings of numerous philosophers and statisticians, including the economist T. R. Malthus. Malthus's *Essay on the Principle of Population* (1798) contained a critical idea that Darwin used in developing his argument, namely that uncontrolled growth of the human population must lead to famine and the ultimate elimination of a significant portion of the human race. Herein lies the foundation for Darwin's notion of the "struggle for existence" in a world in which all organisms are observed to produce many more offspring than are needed to maintain their numbers, and where superior variations would be preserved at the expense of less favorable alternatives.

Darwin was not the first to propose a theory suggesting that change in the forms of living things might have an environmental cause. One earlier proposal, put forward by the French naturalist Jean-Baptiste Lamarck (1744–1829), argued that organisms could improve themselves by their own efforts, and pass these improvements on to their offspring. This "inheritance of acquired characters" could explain, for example, the giraffe's long neck, which it would presumably develop by trying to reach leaves in high tree branches, stretching its neck to do so. Other early evolutionists expounded on the internal perfecting tendencies of living things, as if each somehow aspired toward an evolutionary goal. Darwin's contribution was different in that he proposed a testable theory that was based on the interactions between individuals, each engaged in the struggle to survive and reproduce in the local environment. Because there would never be enough food, nest sites, or mates for all progeny of all individuals produced, only the best competitors would survive and reproduce. The natural outcome of this struggle is that the features of the superior competitors become better represented in the population—not because of an internal "perfecting tendency," but because they provide greater fitness.

The modern understand of **genetics**, base upon the transmission of genetic information via genes comprised of **DNA**, completely disproves the errant Lamarckian theory. Although there are types of selection pressures that can be subsets of natural selection, there is abundant laboratory confirmation of the role of natural selection in evolutionary theory.

The differential survival and reproduction of living organisms is measured as fitness, which is technically defined as the number of offspring produced that survive to maturity. Fitness is a relative concept; measuring the relative rate of increase of one genotype, the genetic constitution of an individual, over alternative genotypes. Evolutionary biologists investigating the process of natural selection generally measure either the relative survival of individuals possessing some trait of interest, or the change in **gene** frequencies (the rate at which a gene increases in a population), between successive generations. Both measures are difficult to obtain, since they require documenting the survival of individuals and their offspring in the lab or in the field, and establishing a pattern of **inheritance** for the trait of interest. The process is further complicated since most observable traits—that is, most aspects of an organism's phenotype—are controlled by more than one gene. Individual genes may assort in different ways between generations. Furthermore, they frequently interact to produce complicated effects. Sorting out the connection between phenotype and genotype is not usually an easy task, but it has been done for some traits in some organisms.

The traits that confer the greatest fitness will depend on the local environment, and what proves advantageous in one

environment may be quite unsuitable in another. An example is **sickle-cell anemia**, a disease that, if left untreated, kills the affected individuals, who carry two sickle-cell genes. However, in populations where the disease malaria occurs, the sickle-cell trait will increase because heterozygotes—those with one sickle-cell and one normal gene—have increased resistance to malaria, and only mild anemia. Thus, natural selection produces genetic change in populations that may fluctuate or even reverse themselves, in accordance with local environmental conditions.

The use of the insecticide DDT provides a well-known example of natural selection in operation. When DDT is first sprayed on an area populated by insects, the population declines abruptly; the poison can remain effective on such populations for up to ten years or more. The initial effectiveness then begins to diminish, so that increasingly greater applications are required to have the same impact. The reason is that DDT, a potent nerve poison, places a strong selective force on the insects; any that remain after the application of the pesticide are likely have some resistance to the poison, sometimes due to a single gene enabling the insects to detoxify it. Such a gene would quickly spread under a selective regime featuring repeated applications of the poison. What is more, once insects have evolved resistance to one insecticide, the time required to evolve resistance to others is reduced. It appears that once selection has honed the mechanisms of detoxification, those mechanisms are fairly easily appropriated for handling new toxins, allowing resistance to appear in one or two years instead of ten.

See also Acquired character; Alternation of generation; Altruism; Animal husbandry; Antibiotics; Biodiversity; Character displacement; Characters and traits; Comparative genomics; Conservation genetics and biodiversity; Darwinism; Directional selection; Disruptive selection; Embryology, the history of developmental and generational theory; Epidemiology and genetics; Ethical issues of modern genetics; Evolution, evidence of; Evolutionary mechanisms; Fertility factor; Fitness; Founder effect; Gradualism vs. Catastrophism; Heterozygous advantage; Lamarckism; Mendelian genetics and Mendel's laws of heredity; Mutations and mutagenesis; Origin of life; Philosophy of genetics: Neo-Darwinism and the modern synthesis; Population genetics; Quantum speciation; Rare genotype advantage; Selection; Social Darwinism; Sociobiology; Species; Spontaneous generation theory; Survival of the fittest

NAUDIN, CHARLES (1815-1899)
French biologist

Charles Naudin performed experiments on plant hybridization and theorized about the nature of **heredity**. A contemporary of Gregor Mendel, Naudin pursued a similar experimental direction. Due to Naudin's lack of statistical analysis, however, along with some unfortunate accidents, Naudin did not present the scientific community with strong evidence of his conclusions, leading to a delay in the realization of the benefits of his experimentation.

The son of a struggling entrepreneur, Naudin graduated from university at Montpellier, France in 1837. He then moved to Paris, finding work as a bookkeeper, tutor, private secretary, and gardener in order to support further study. After obtaining his doctorate from Paris in 1842, he became a teacher. In 1846, Naudin was given an opportunity, joining the herbarium at the Museum of Natural History, and then becoming professor of Zoology at the College Chaptal.

Naudin was forced to resign from these hard won posts almost as soon as he obtained them, due to a severe nervous disorder, which left him deaf, and in constant pain. His academic career seemingly over, Naudin entered an unsettled period, finally settling at Collioure, in 1869, and establishing a small, private experimental garden. He sold seeds and specimens to make a living, while pursuing his interest in experimental horticulture. Naudin's botanical efforts gained reward in 1878, when he was made director of the experimental garden at Antibes, finally finding financial security. However, further tragedies challenged his scientific progress, including the loss of his eyesight and the death of his children.

Despite these setbacks, Naudin managed to perform a number of innovative experiments in acclimatization, economic botany, and most importantly, plant hybridization. Earlier work had suggested that hybrids—crosses between two races, breeds, strains, or varieties—might be a means for the creation of new **species**. This theory gained new popularity with the work of Charles Darwin. Naudin became interested in the evolutionary nature of hybrids, but altered his views when he observed that most hybrids were infertile, and that such crosses did not seem to survive many generations.

Naudin's experiments led him to conclude that first generation hybrids were uniform in nature, most having a mixture of characteristics from both parents. In subsequent generations from **hybrid** plants, however, there was great diversity and a strong tendency to revert to one of the parent types. Naudin concluded that nature abhors hybrids, and that within such plants occurred a kind of battle to return to a natural, pure form between the "specific essences" from the parent plants. He argued that hybrids could not result in new species, as the new, mixed forms that appeared in early generations would disappear in later generations, eventually reverting completely to a parental type.

Naudin combined his results into a law of segregation, which bears a strong resemblance to **Gregor Mendel**'s own law of the same name. However, Mendel simplified the problem, focusing on specific traits rather than the plant as a whole, and supported his work with statistical analysis. Naudin's conclusions were based on scientific intuition, and this meant that while he often came to correct assumptions, he could not support them with strong evidence or theory.

Bad weather, poor choice of plants, lack of space and money, and other misfortunes also plagued Naudin's experiments. A **virus**, which was not identified until the following century infected many of his plants, and gave them unusual characteristics. Accidents such as this led Naudin to follow a number of fruitless paths in his research, and reach false conclusions.

Naudin's ideas were dealt a serious blow when **Charles Darwin** heavily criticized them. Darwin supported a theory of blended **inheritance**, in which each parent contributed equally, and where each offspring was still, in some small part, influenced by all previous generations. Eventually, Darwin did incorporate Naudin's ideas into his own work, as the weight of experimental evidence for non-blended inheritance grew. As a result, many of the late nineteenth century theories for heredity became confused and vague, using a hodge-podge of possible methods. It was not until the rediscovery of Mendel's work in the early twentieth century that a more coherent explanation became available, and the early hybridization work of Naudin and others was re-examined gained scientific appreciation.

See also Hybridization of plants

NEEDHAM, JOHN TURBERVILLE (1713-1781)

English priest and naturalist

John Turberville Needham dedicated himself to the Roman Catholic religion and was ordained in 1738. Much of his life thereafter, however, was devoted to the natural sciences. Stimulated by his readings of 'animacules,' the term for microbes at that time, he studied in London and Paris from 1746 to 1749. Needham became a vocal proponent of the theories of **spontaneous generation** (life arising from inorganic matter) and vitalism (life operating outside the laws of chemistry and physics).

Needham performed experiments on spontaneous generation in mutton broth and hay infusions. To see if organisms came from the outside or were generated from within the fluid, he heated flasks of broth and then corked them tightly. Needham's observations of the appearance of microorganisms shortly after the fluid cooled were taken as evidence supportive of spontaneous generation. He published his observations and interpretations in 1748 in a work entitled *Observations upon the Generation, Composition, and Decomposition of Animal and Vegetable Substances*. However, performance of more careful experiments by the Italian biologist Lazzaro Spallanzani (1729–1799) and French biologist, astronomer, and mathematician Pierre-Louis Moreau de Maupertuis (1698-1759), helped refute the theory of spontaneous generation. Maupertuis and Spallanzani ensured in similar experiments that the boiling process was complete, thereby eliminating all the microorganisms initially present in the flask. Maupertuis' observations were published in 1751, as *Systéme de la Nature*.

In 1767, Needham retired to the English seminary at Paris to pursue his scientific passions. He continued to serve as the director of the Imperial Academy in Brussels until the year before his death in 1781.

See also History of genetics: Ancient and classical views of heredity; Spontaneous generation theory

NEEDHAM, JOSEPH (1900-1995)

English physician and biochemist

Joseph Needham is best known for his authorship of the seventeen volumes of *Science and Civilization in China*. This series married his lifelong interests in science and religious philosophy, which began during his medical and biochemistry training at Cambridge.

The son of a physician, Needham earned a doctoral degree in 1924 from the University of Cambridge. He then joined the newly formed Dunn Institute of Biochemistry, his principle affiliation for the remainder of his life. In 1931, Needham published his three-volume *Chemical Embryology*. Containing a comprehensive history of **embryology** in the introduction, *Chemical Embryology* was one of the first historical accounts of developmental biology, and Needham's first contribution among many to the history of natural sciences.

In the late 1930s, his collaboration with Chinese biochemists sparked his interest in China's language and civilization. This interest prompted his relocation to China from 1942 to 1946, as director of the Sino-British Science Co-operation Office, a British scientific mission.

His experiences in China led to his spearheading the inclusion of science in the then formative UNESCO. He became the first head of UNESCO's science division. After serving as director from 1946 to 1948, he returned to Cambridge and began work on *Science and Civilization*.

This series surveys the history of Chinese chemistry, mechanics, navigation, medicine, and other disciplines. The work examined the relationship between the Confucian and Taoist traditions and Chinese scientific innovation and explored the differences between Chinese and Western philosophies of scientific inquiry.

See also History of genetics: ancient and classical views of heredity; Philosophy of genetics: Neo-Darwinism and the modern synthesis

NEEL, JAMES V. (1915-2000)

American physician

James V. Neel was a scientist, clinician, and teacher in the field of Genetics. His accomplishments span over half a decade and include such diverse issues as the genetic basis of **sickle cell anemia** and the effects of atomic radiation on humans.

James Neel was born in Ohio in 1915, and attended the College of Wooster, Wooster, Ohio, where he received his Bachelor's of Arts degree in 1935. Under the tutelage of Dr. **Curt Stern**, Neel studied *Drosophila* genetics and was awarded a Ph.D. from the University of Rochester, Rochester, NY, in 1939. After teaching briefly at Dartmouth College, Hanover, New Hampshire, he moved to Columbia University where he had a National Research Council Fellowship in Zoology. He then returned to the University of Rochester, where he had been accepted into medical school and completed his M.D. in 1944, followed by a one-year internal medicine internship at

Strong Memorial Hospital, also in Rochester, N.Y. Military service in the Army Medical Corps briefly interrupted his career, but in 1946, Neel accepted a faculty appointment at the University of Michigan, Ann Arbor, Michigan, where he remained until his retirement in 1985.

Neel's wartime service with the Atomic Bomb Casualty Commission stimulated his interest in the effects of radiation on the human body. Over the years, he spent a significant amount of time in Japan, working with survivors of the Nagasaki and Hiroshima atomic bombings. To gain more insight into the effects of radiation, Neel also traveled to the Bikini Atoll where the testing of the atomic bomb had resulted in radioactive fallout and a potential risk for the native inhabitants. The areas he investigated included the consequences of radiation exposure on the blood system, risks to pregnant women and the potential for abortions, stillbirths, and malformations, changes in birth weight and sex ratio in newborns, and the potential for brain damage in fetuses and young children. His publications on these findings are seminal reports in the field and have stimulated others to investigate the issues further.

Population genetics was a favorite subject for Neel, and he headed several projects on this topic. One investigation dealt with consanguineous marriages and the affects of **inbreeding** on populations. He also spent time in the Amazon rain forest studying the **inheritance** of various traits in isolated tribes. Evaluations of different populations lead him to propose that many of the mutations currently present in the human **gene pool** and considered deleterious may have a greater significance. This gave rise to his "thrifty gene" hypothesis that suggests some disease causing mutations may be remnants of earlier, more difficult times when certain traits may have been beneficial to the human **species**. Obesity, for example, may have been a way for the body to collect and store life-sustaining materials as a hedge against lean times when food might be less available.

The data he collected and his thoughts on genetics were published in over 400 articles. Neel also authored several books, including *Physician to the Gene Pool*.

Neel contributed a great deal to early genetic studies, but perhaps his most enduring legacy is at the University of Michigan Medical School. Realizing the potential importance of genetics research to medicine, Neel established the first Department of Human Genetics at a university in the United States. Genetics had only recently begun to find favor in the world of medicine, so the founding of an independent department was considered a radical move at the time. Neel shepherded it through its early years, serving as chairman from 1956 through 1981. The Department is now considered one of the premier graduate genetics programs in the United States. Many renown practicing geneticists of today were trained or directly influenced by James Neel.

As an adjunct to the academic department, Neel was also involved in establishing one of the first genetics clinics in the United States at the University of Michigan Hospital. Genetic counseling became a key feature of the services offered, and a new training program for master's degree level personnel was created.

For his considerable contributions to the fields of genetics and medicine, Neel received numerous awards and citations including the 1960 Lasker Award, the William Allen Award from the American Society of Human Genetics, membership in the American Academy of Arts and Sciences and the National Academy of Science, the 1975 National Medal of Science, and the 1995 James D. Bruce Award from the American College of Physicians. Upon his retirement in 1985, he was named the Lee R. Dice Distinguished University Professor Emeritus of Human Genetics. During his lifetime, he sat on several editorial boards and many committees from a variety of organizations such as the Salk Institute, the United States Department of Energy, and the World Health Organization.

Neel died at the age of 84. As a pioneer in the field of human genetics, his research forms a foundation that will continue to be built on for years to come.

See also Radiation mutagenesis

NERVOUS SYSTEM · *see* NEUROLOGICAL AND OPTHALMOLOGICAL GENETICS

NEURAL TUBE DEFECTS (NTDs)

Neural tube defects are a group of severe **birth defects** in which the brain and spinal cord are malformed and lack the protective skeletal and soft tissue encasement.

Incomplete formation and protection of the brain or spinal cord with bony and soft tissue coverings during the fourth week of embryo formation are known collectively as neural tube defects. Lesions may occur anywhere in the midline of the head or spine. These defects are among the most common serious birth defects, but they vary considerably in their severity. In some cases, the brain or spinal cord is completely exposed, in some cases protected by a tough membrane (meninges), and in other cases covered by skin.

Most neural tube defects (80-90%) occur as isolated defects. Neural tube defects of this variety are believed to arise through the combined influence of genetic and environmental forces. This multifactorial causation presumes that one or more predisposing genes collaborate with one or more environmental influences to lead to the birth defect. Poor nutrition is correlated as an environmental risk factor and hereditary defects in the absorption and utilization of folic acid are presumptive genetic predisposing factors. After a couple has one infant with a neural tube defect, the recurrence risk is 3–5%. After the birth of two NTD-affected infants, the risk increases to 8–10%.

When neural tube defects occur concurrently with other malformations there is a greater likelihood of an underlying specific genetic or environmental cause. Genetic causes include **chromosome** aberrations and single **gene** mutations.

Environmental causes include maternal diabetes mellitus, exposure to prolonged hyperthermia, and seizure medications during the early months of pregnancy.

Spina bifida accounts for about two-thirds of all neural tube defects. The spine defect may appear anywhere from the neck to the buttocks. In its most severe form, termed spinal rachischisis, the entire spinal canal is open exposing the spinal cord and nerves. More commonly, the defect appears as a localized mass on the back that is covered by skin or by the meninges.

Anencephaly, the second most common neural tube defect, accounts for about one-third of cases. Two major subtypes occur. In the most severe form, all of the skull bones are missing and the brain is exposed in its entirety. The second form in which only a part of the skull is missing and a portion of the brain exposed is termed meroacrania.

Encephaloceles are the least common form of neural tube defects, comprising less than ten percent of birth defects. With encephaloceles, a portion of the skull bones is missing leaving a bony hole through which the brain and its coverings herniate. Encephaloceles occur in the midline from the base of the nose, to the junction of the skull and neck. As with spina bifida, the severity varies greatly. At the mildest end of the spectrum, encephalocele may appear as only a small area of faulty skin development with or without any underlying skull defect. At the severe end of the spectrum, most of the brain may be herniated outside of the skull into a skin-covered sac.

Neural tube defects occur worldwide. It appears that the highest prevalence (about one in 100 pregnancies) exists in certain northern provinces in China; an intermediate prevalence (about one in 300-500 pregnancies) has been found in Ireland and in Central and South America; and the lowest prevalence (less than one in 2,000 pregnancies) has been found in the Scandinavian countries. In the United States, the highest prevalence has occurred in the Southeast. Worldwide there has been a steady downward trend in prevalence rates over the past 50-70 years.

Because of the faulty development of the spinal cord and nerves, a number of consequences are commonly seen in spina bifida. As a rule, the nerves below the level of the defect develop in a faulty manner and fail to function, resulting in paralysis and loss of sensation below the level of the spinal lesion. Since most defects occur in the lumbar region, the lower limbs are paralyzed and lack normal sensation. Furthermore, the bowel and bladder have inadequate nerve connections, causing inability to control bladder and bowel function. Sexual function is likewise impaired. Hydrocephaly develops in most infants either before or after surgical repair of the spine defect.

In anencephaly, the brain is destroyed by exposure during intrauterine life. Most infants with anencephaly are stillborn, or die within the initial days or weeks after birth.

Infants with encephaloceles have variable neurologic impairments depending on the extent of brain involvement. When only the brain covering is involved, the individual may escape any adverse effect. However, when the brain is involved in the defect, impairments of the special senses such as sight and hearing and cognitive impairments commonly result.

At birth, the diagnosis of a neural tude defect is usually obvious based on external findings. Prenatal diagnosis may be made with ultrasound examination after 12-14 weeks of pregnancy. Screening of pregnancies can be carried out at 16 weeks by testing the mother's blood for the level of alpha-fetoprotein. Open neural tube defects leak this fetal chemical into the surrounding amniotic fluid, a small portion of which is absorbed into the mother's blood.

No treatment is available for anencephaly. Aggressive surgical and medical management has improved survival and function of infants with spina bifida. Surgery closes the defect, providing protection against injury and infection. Ambulation may be achieved with orthopedic devices. A common complication that may occur before or after surgical correction is the accumulation of excessive cerebral spinal fluid (hydrocephaly) in the major cavities (ventricles) within the brain. Hydrocephaly is usually treated with the placement of a mechanical shunt, which allows the cerebral spinal fluid from the ventricles to drain into the circulation or another body cavity. A number of medical and surgical procedures have been used to protect the urinary system. Encephaloceles are usually repaired by surgery soon after birth. The success of surgery often depends on the amount of brain tissue involved in the encephalocele.

It has been found that 400 micrograms of folic acid taken during the periconceptional period (2–3 months prior to conception, and 2–3 months following conceptions) protects against most neural tube defects. While there are a number of foods (green leafy vegetables, legumes, liver, orange juice) that are good sources of natural folic acid, synthetic folic acid is available in over-the-counter multivitamins and a number of fully fortified breakfast cereals.

Additionally, a population-wide increase in folic acid intake has been achieved through the fortification of enriched cereal grain flours since January 1998, a measure authorized by the United States Food and Drug Administration. The increased blood folic acid levels achieved in recent years has likely resulted from the synergy of dietary, supplementation, and fortification sources of folic acid.

Infants with anencephaly are usually stillborn or die within the initial days of life. Eighty to ninety percent of infants with spina bifida survive with surgery. Paralysis below the level of the defect, including an inability to control bowel and bladder function, and hydrocephaly are complications experienced by most survivors. Intellectual function is normal in most cases.

The prognosis for infants with encephaloceles varies considerably. Small encephaloceles may cause no disability whether surgical correction is performed or not. Infants with larger encephaloceles may have residual impairment of vision, hearing, nerve function, and intellectual capacity.

See also Chromosomal mutations and abnormalities; Genetic disorders and diseases; Hereditary diseases; Inherited cancers; Mutations and mutagenesis

NEUROLOGICAL DISORDERS, INHERITED

The enormous complexity of the brain's structure and function allows many possibilities for genetic errors to occur. The same is true for the eye, an extension of the brain.

Some of these errors are known as structural defects. One such defect, holoprosencephaly, results in a single brain ventricle where a pair usually occurs, and malformation of facial features, including cyclopia and the lack of a nose. Although such cases are often sporadic, they sometimes follow a Mendelian pattern. Hydrocephalus, which results from too much cerebrospinal fluid for the size of the brain, enlarges the ventricles, causing intracranial pressure. It can be X-linked or multifactorial in origin. In some 2-3 per 10,000 live births, the neural tube of the embryo fails to close completely. This can cause varying sensory and motor defects, depending on which part of the spinal cord and nerves are affected. The condition may be autosomal dominant, and can be detected in pregnancy by measuring the level of maternal serum alpha-fetoprotein.

Up to 3% of the general population are affected by some type of mental (psychomotor) retardation. **Down Syndrome**, one of the more common types, is associated with characteristic facial features, congenital heart disease, and problems with vision and hearing. It is caused by a trisomy of the 21st **chromosome**, specifically, the genetic material in the region from q22.1–q.22.2. Fragile-X Syndrome is an X-linked disorder caused by high numbers of CCG trinucleotide repeats. In addition to retardation, it causes large testes, unusual facial features, connective tissue abnormalities, and often attention-deficit disorder, poor coordination, and seizures. **Rett syndrome** is X-linked, but lethal in males, so only females inherit it. These girls seem normal at birth, but at about one year of age, head growth stops, and developmental milestones are lost.

Seizure disorders tend to be more common in some families, but most have polygenic or multifactorial causes. A few, such as La Fora body disease, which is autosomal recessive, are associated with a specific **gene**.

Neurocutaneous syndromes result in developmental anomalies of the skin and nervous system. One of the better known examples is neurofibromatosis, of which there are two types. Both are autosomal dominant, and cause cafe-au-lait spots on the skin and neurofibromas, which are soft tumors that develop from peripheral nerve sheaths. Also associated with the disease are demineralization of the bone and resulting fractures, scoliosis, and an increased risk of tumors in the nervous system. Neurofibromatosis 1 is often associated with learning disability or attention-deficit disorder. The neurofibromin gene has significant homology to a known tumor-suppressor gene.

Diseases of the motor neurons are genetically heterogeneous with different ages of onset. While all cause decreases in movement, some cause death in infancy while others allow a normal life span. Degeneration of motor neurons causes amyotropic lateral sclerosis, which results in progressive

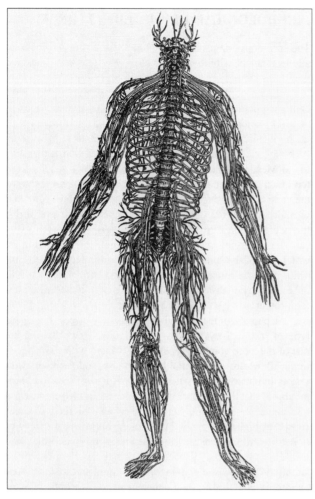

The human nervous system. *Bettmann Archive/Newsphotos, Inc. Reproduced by permission*

weakness and eventually death from weakness of the respiratory muscles.

There is a host of diseases of the sensory neurons. Hereditary radicular neuropathy (autosomal dominant) causes progressive loss of pain and temperature perception. Congenital sensory neuropathy (autosomal recessive) results in total absence of pain, and absence of fungiform taste buds, gag reflexes, and deep tendon reflexes.

Huntington disease is caused by decreased neurotransmitters in the basal ganglia. It causes lack of coordination, dementia, personality changes, and eventual death. An autosomal dominant disorder, it is the result of an increased number of CAG trinucleotide repeats in the Huntington gene. The higher the number of repeats, the earlier the onset of the disease. Another disease of the basal ganglia is Parkinson syndrome. It causes a progressive resting tremor, muscle rigidity, and slowness of movement. About ten percent of cases are familial, inherited in autosomal dominant fashion with variable penetrance.

Among inherited ophthalmological disorders, retinitis pigmentosa is one of the most common. It is associated with abnormal retinal vessels and changes of pigment within the retina. It causes night blindness, followed by loss of central vision. Mutations in any one of three different genes can cause this condition. Another hereditary vision disorder is stationary night blindness. Caused by a dysfunction of the rods, it has an early onset but does not progress. Color blindness is an X-linked recessive disorder that occurs in 8% of Caucasian males. It is caused by mutations in the genes for red, green, and blue photopigments. Other disorders include congenital cataracts and aniridia, the absence of an iris.

See also Multifactorial transmission

NEUROLOGICAL AND OPHTHALMOLOGI-CAL GENETICS

Neurological **genetics** is the study of the genetic causes of neurological diseases, which can affect either central nervous system (brain and spinal cord) or the peripheral nervous system. Ophthalmological genetics is the study of the genes responsible for ocular genetic disorders. Because the neural part of the retina is an extension of the brain, many neurological disorders have an ocular component. To develop new ways of treating both diseases, scientists study the developmental, metabolic and environmental factors regulating the expression of the genes involved in the diseases.

The mammalian eye is a complex organ that develops from mesoderm, ectoderm, and endoderm, and contains multiple **cell** types including neuronal cells. Inherited eye diseases are the main cause of blindness in the developed countries and although they can affect all parts of the eye (including the lens, cornea and optic nerve), retinal disorders are the most common. The genetic defects can cause dysfunction of structural proteins, the phototransduction, color perception or metabolic pathways and can also result in full or partial absence of anatomical structures e.g., aniridia (absence of iris caused by a **mutation** in the Pax6 **gene**).

The rod photoreceptor cells must maintain the discs in their outer segments in a precisely ordered structural arrangement to enable efficient light detection. The integrity of this structure is maintained by tetramers formed by RDS/peripherin and ROM1 proteins. Mutations in ROM1 gene cause retinitis pigmentosa (RP), while changes in RDS protein cause a number of clinically different diseases (e.g., macular dystrophy, rod -cone dystrophy, and RP).

The main component of the rod outer segment discs is rhodopsin (Rho) that plays a central role in phototransduction by capturing photons and initiating a signal **transduction** pathway in the photoreceptor cells. Mutations in rhodopsin cause retinitis pigmentosa.

The phototransduction in the photoreceptor cells is dependent on a sequential activation of GTP interacting proteins such as: transducin, guanylate cyclase, phosphodiesterase, and GTPase regulator. Mutations in these genes cause Leber's congenital amaurosis (LCA1), cone-rod dystrophy (CORD6), retinitis pigmentosa, and congenital stationary

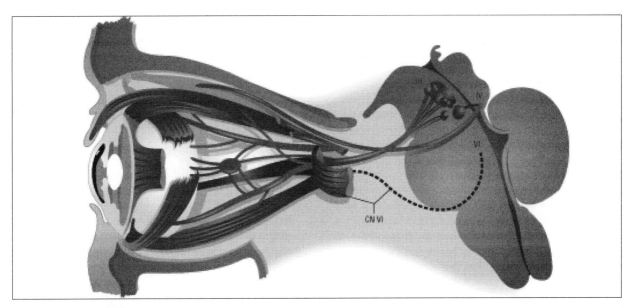

Duane syndrome pathology, shown here, results in an inability of the eye to turn outwards beyond the midline. Duane syndrome is a congenital disorder occurring in the left eye more often than the right, and is more often in females than males. *Bettmann Archive/Newsphotos, Inc. Reproduced by permission*

night blindness (CSNB3). Desensitization of G-protein coupled receptors is carried out by arrestin which also functions in endocytosis of the receptors. Although no mutations have been demonstrated in arrestin to date, it can form stable complexes with rhodopsin that cause retinal degeneration and can be an important factor in pathology of Rho-linked RP (approx. 25% of RP cases). Signal transduction from the receptor to the brain depends on the neurotransmission. The transmission of signal from the photoreceptors to retinal neurons is affected by mutations in the NYX gene causing X-linked congenital stationary blindness (CSNB1).

Rod outer segments are constantly being renewed and the cell components are being recycled. Therefore, genes whose products are involved in membrane transport and phagocytosis, such as the ABCR gene (mutations cause Stargard's macular dystrophy) or intracellular transport, such as myosin (MYO7) (interacts with actin filaments and mutations cause Usher syndrome) are important for the function of the photoreceptors. Mutations in ubiquitously expressed genes not always affect all of the tissues in the body, for example changes in tissue inhibitor of metalloproteinase 3 (TIMP3). Haploinsufficiency of TIMP3 causes Sorby's fundus dystrophy by inducing thickening of the Bruch's membrane by an unknown mechanism and does not affect any other tissues. Metabolic abnormalities in the retina can also be caused by excessive accumulation of the metabolites in the photoreceptor cells. Retina and choroid are affected by mutations in the mitochondrial ornithine aminotransferase that cause Gyrate atrophy.

Mutations in some genes cause both neurological and ophthalmic disorders, e.g. Leber's hereditary optic neuropathy (LHON) a disease with mitochondrial **inheritance**. There is, however, a significant group of neurological defects with no ophthalmic component (e.g., Huntington's disease).

Neurological defects can result from mutations in genes encoding components of ion transport channels or intra- and inter-cellular signalling, extracellular matrix or cytoskeleton, or result from haploinsufficiency during embryonic development and from mutations leading to neuronal cell death.

Ion channels are important for signal transduction in muscles and along the neurons. The main types of channels involved are sodium, potassium and calcium. Mutations in the genes encoding the proteins forming the channels can cause lack of the normal channel inactivation (e.g. sodium ion channel in paramyotonia congenita), increase in ion (chloride) conductance leading to repetitive firing (myotonia congenita - CLCN1) and altered kinetics and/or interference in ion (potassium) channel assembly (Episodic ataxia - KCN1).

Neuronal dystrophies can also be caused by mutations in cytoskeleton proteins, e.g. dystrophin in Duchenne muscular dystrophy or in structural proteins such as collagens or laminin in congenital muscular dystrophies, leading to abnormalities of sarcolemma and basement membrane structure in muscle cells.

A number of inherited disorders (Huntington's disease, Kennedy's disease or spinocerebral ataxia 6) are caused by expansion of polyglutamine repeats. Such changes in proteins lead to a progressive neurodegeneration with symptoms specific for a particular disease. This process is attributable to accumulation of inclusion bodies containing abnormal protein and subsequent neuronal cell death. Although the genetic problem has been identified in these diseases, the actual pathway leading to cell death is not clear.

Both neurological and ophthalmological genetics will benefit from the completion of the human **genome sequencing**, as the identification of the candidate genes can be done in a relatively short time after the initial linkage analysis.

See also Colorblindness; Cytogenetics; Genetic anticipation; Medical genetics

NEUTRAL MUTATION

A neutral **mutation**, which can also be called a **silent mutation**, is, like all genetic mutations, a change in the **deoxyribonucleic acid (DNA)** that comprises the **heredity** material in all living organisms. Specifically, a neutral mutation is a mutation that does not affect the particular phenotype whose expression is under the genetic control of the mutated **gene**. An example of a neutral mutation would be the replacement of one base with another in the deoxyribonucleic acid comprising the gene, with the replacement having no effect on the protein encoded by the gene.

Neutral mutations can be divided into two kinds. First, the mutation can occur in a region of DNA called junk DNA—a region of DNA that is nonfunctional in terms of the phenotype of the organism. Only mutations in the coding regions of DNA contribute to the mutational change in organisms over time. The second type of neutral mutation may leave the phenotype unchanged, but create the possibility of a future mutation having an effect. For example, a neutral mutation in a dominant form of a gene may allow another mutation to change the dominant phenotype of the gene to a recessive phenotype when that gene is passed to the next generation.

The neutral theory of molecular **evolution** proposes that most of the molecular **genetic variation** that occurs—that is, changes in the **bases** of the DNA—is selectively neutral. Genetic drift, rather than **natural selection**, is the driver of evolutionary change. Mutations that adversely affect an organism will be selected out in evolution. Conversely, mutations that confer a benefit will be retained. But, because neutral mutations are not harmful and so are allowable in an evolutionary sense. The extent to which the latter drive evolution is not clear as yet.

According to Sir **Francis Crick**, not all DNA variation translates into protein variation (e.g., silent mutations, synonymous **codons**, pseudogenes and non-coding DNA). Further, not all protein variation translates into phenotypic variation and not all phenotypic variation is subsequently translated into differences in **fitness**. Accordingly, phenotypic changes may be due more to differences in **gene regulation** than in sequence alterations.

In contrast to the long held agreement that the rate of point mutations is uniform over the entire **genome** of an organism and that, therefore, **nucleotide** substitution rate variations among DNA regions reflected differential selective constraints, some researchers argue that evidence for variations in mutation rates among regions of the mammalian genome were due to the differences in the timing of chromosomal **replication** within germline cells. Such observations also provide a mechanism to explain why some selectively neutral silent nucleotide substitution sites do not appear to "tick" to the same molecular clock.

There is also a rational argument to be made regarding the plausibility that **selection** is simply weaker at the at the molecular level and that, while selection can operate at the molecular level—the mechanisms of mutation and drift predominate.

See also Adaptation and fitness; Evolutionary mechanisms; Gene pool; Heterozygote; Homozygous

NEUTRAL SELECTION

Neutral **selection** describes changes in the **gene pool** of a **species** that are a result of accumulated random neutral changes that do not convey any particular advantage to a species. Accordingly, neutral selection does not depend upon adaptation, **fitness**, and **natural selection**.

Evolution describes biological change over time. Scientists who study evolution argue whether natural selection or random changes have more profoundly influenced the course of evolution. According to evolutionary theory based upon natural selection, changes in the **gene** pool are a result of differing fitness among individuals with the population. These differences in fitness manifest result in a selective advantage or disadvantage; neutral mutations are rare. According to the neutral theory, however, most changes at the molecular level do not carry a selective advantage or disadvantage and are thus, selectively neutral. To account for wide differences in genetic make-up this would mean that the rate of neutral mutations is very high. Although neither theory necessarily excludes the other, the importance of each process on evolution is remains an important question for biologists.

The concept of neutral selection led to the advancement of neutral theory by **Motoo Kimura** in the 1960s as a reaction to classic "cost of selection" arguments that for significant genetic changes to occur there must be an expresssed difference within individuals that causes genetic deaths. A genetic death occurs whenever individuals within a population fail to reproduce, or fail to produce fertile offspring.

The higher the selective pressures (measured by the selective coefficient) the greater percentage of a population suffers genetic death. If, for example, the pressures of selection are too high, a given population might face extinction because not enough individuals will survive to successfully reproduce. Kimura and other evolutionary theorists argued that the types of selection pressures needed to produce the genetic differences between known species were so great that these pressures would have led to extinction rather than the manifest diversity of life found on Earth.

Kimura and other proponents of the neutral theory explained **genetic variation** between different species (or difference between individuals with a species) as a cumulative result of neutral mutations and genetic drift. Kimura based his neutral hypothesis on the fact that most genetic changes (mutations) are harmful (deleterious) and as thus quickly acted upon by natural selection. Because such deleterious mutations are quickly removed with a population, these mutations rarely get the chance to have any profound effect on the gene pool. Kimura argued that such selection was conservative because it actually opposed change by quickly removing mutations.

Kimura further argued that, if selection pressures were the dominant force in evolution, significant genetic changes could only occur if mutations were favorable, and such favorable mutations are rare.

Based on the established mechanisms of **mutation** and genetic drift, Kimura based the neutral selection theory on four premises. First, Kimura argued that the rate of evolution of proteins is too high to be a function of selection. Second, the rate of molecular evolution is more constant than the rate of morphological (shape or form) evolution and is constant in different lineages. This also allowed for the development of the concept of **molecular clocks** that relate mutations to time.

Kimura's third argument was that proteins, and parts of the same protein, may evolve at different rates. Those parts of proteins that are not important to the function of the protein may change faster than the more functionally important parts of the protein. According to neutral theory this is explained by fewer neutral mutations in important areas. Correspondingly, selection theory states that mutations in the active sites of proteins are more likely to have an advantageous or disadvantageous effect upon which selection can act.

Kimura's final argument asserted that **directional selection** would never have allowed the amount of heterozygosity and **polymorphism** observed in populations.

According to neutral theory, molecules such as **DNA**, **RNA** and proteins with fewer functional constraints (e.g., active sites) evolve faster because the number of effectively neutral mutants is correspondingly higher. Not all DNA variation, however, translates into protein variation (e.g., silent mutations, synonymous **codons**, pseudogenes and non-coding DNA). Further, not all protein variation translates into phenotypic variation, and not all phenotypic variation is subsequently translated into differences in fitness.

See also Adaptation and fitness; Evolutionary mechanisms; Homozygous

NEWBORN GENETIC SCREENING

Although most genetic conditions are relatively rare, certain genetic conditions occur with such high frequency in a population that they are of concern to most everyone. Such traits may be worthy of population-based screening provided that certain basic requirements are met. There are only a few checkpoints through which virtually everyone in a society passes, one of which is the newborn nursery. In addition, the newborn period is the earliest point following birth, and therefore any damage that is done to the body as a result of the disease has had minimal time to progress.

A disease much meet certain criteria if it is to be considered a valid candidate for newborn screening. First, the disease must be relatively common in the target population. It must have a serious health consequence, and the condition must be either treatable or preventable. There would be little benefit to the general population to screen for a disease that is extremely rare, bears little consequence to the individual, or for which nothing can be done.

There are also certain requirements of the proposed laboratory test to detect such diseases if screening is to be worthwhile. The test must be relatively accurate and reliable, inexpensive to carry out, and there must be very little risk of causing harm in collecting samples or conducting the test. A positive test is one that predicts the individual is at increased risk for the disease, and a negative test is generally regarded as normal. Reliability and accuracy are determined by how often the test detects the disease (sensitivity), how often the test comes back normal in people without the disease (specificity), and what proportion of the positive tests turn out to be true positives. A false negative result is one that comes back negative even though the person actually did have the disease. A false positive result is one that is regarded as positive even though the individual is healthy. The per test cost for screening is minimized by doing large volume testing in a centralized laboratory, and may amount to just a few cents per patient. The vast majority of testing is performed on blood spots collected from a heel prick of the newborn.

The screening program itself must also meet several criteria if a test is to prove worthwhile for a certain population. Participation must be voluntary, it must be made available to the vast majority of the babies, it must be considered acceptable to the population whom is being served by the screening program, and there must be a comprehensive system in place for education, counseling, and follow up of all screen positive cases.

Of the roughly 6,000 diseases and traits which appear to be caused by mutations involving a single **gene**, only a handful have been found to be suitable for screening in the United States. This includes **phenylketonuria** (PKU), congenital adrenal hyperplasia (CAH), galactosemia, **sickle cell anemia**, biotinidase deficiency, and various amino acidopathies. Congenital hypothyroidism, although not genetic, is readily treatable and is generally included in the newborn screening panel. The actual testing performed varies from state to state within the U.S. Newer technologies such as tandem mass spectrometry offer the prospect of simultaneous inexpensive screening for a wider array of biochemical abnormalities, although there is currently much discussion as to how such testing could best be implemented.

NICK TRANSLATION REACTION

Molecular biologists use a specialized technique termed nick **translation** to label double stranded **DNA** molecules. The technique prepares the DNA molecules for further **blotting analysis** studies (e.g., Southern and Northern blotting analysis).

The nick translation reaction is mediated by two critical **enzymes** that work, one after another, along the DNA molecule. In the first of the two reactions, one strand of DNA is nicked or cut by a Dnase (deoxyribonuclease I). The second enzyme critical to the reaction, *E. coli* **polymerase** I, aids in the addition of deoxynucleotides to the DNA molecule at the 3' (three prime) end of the nick or cut generated during the first part of the nick reaction. At the same time that the deoxyribonucleotides are added to the DNA strand at the 3' end of the nick, an exonuclease cuts deoxyribonucleotides from the 5' end of the nick.

The net effect of this cutting and removal of deoxyribonucleotides is a generalized rolling movement of the nick along the DNA strand in the direction of the 5' end of the nick from the 3' end of the nick. When scientists add labeled deoxynucleotide triphosphates to the medium, these easily identifiable molecules become incorporated into the molecules (**nucleotide**) residues being added to the original DNA molecule.

Following the labeling of the DNA molecule with radionucleotides (e.g., 32P deoxynucleoside triphosphates) the molecules are then ready for blotting analysis during which scientists transfer use the techniques of electrophoresis to separate component molecules. Once the molecules are separated they may be probed for specific proteins, nucleic acids, or particular sequences of amino acids or nucleotides.

See also Antibody and antigen; DNA probes; Protein electrophoresis; Restriction enzymes

NIRENBERG, MARSHALL WARREN
(1927-)
American biochemist

Marshall Nirenberg is best known for deciphering the portion of **DNA** (**deoxyribonucleic acid**) that is responsible for the synthesis of the numerous protein molecules which form the basis of living cells. His research has helped to unravel the DNA genetic code, aiding, for example, in the determination of which genes code for certain hereditary traits. For his contribution to the sciences of **genetics** and **cell biochemistry**, Nirenberg was awarded the 1968 Nobel Prize in physiology or medicine with **Robert W. Holley** and **Har Gobind Khorana**.

Nirenberg was born in New York City, and moved to Florida with his parents, Harry Edward and Minerva (Bykowsky) Nirenberg, when he was ten years old. He earned his B.S. in 1948 and his M.Sc. in biology in 1952 from the University of Florida. Nirenberg's interest in science extended beyond his formal studies. For two of his undergraduate years he worked as a teaching assistant in biology, and he also spent a brief period as a research assistant in the nutrition laboratory. In 1952, Nirenberg continued his graduate studies at the University of Michigan, this time in the field of biochemistry. Obtaining his Ph.D. in 1957, he wrote his dissertation on the uptake of hexose, a sugar molecule, by ascites **tumor** cells.

Shortly after earning his Ph.D., Nirenberg began his investigation into the inner workings of the genetic code as an American Cancer Society (ACS) fellow at the National Institutes of Health (NIH) in Bethesda, Maryland. Nirenberg continued his research at the NIH after the ACS fellowship

Marshall Nirenberg.

ended in 1959, under another fellowship from the Public Health Service (PHS). In 1960, when the PHS fellowship ended, he joined the NIH staff permanently as a research scientist in biochemistry.

After only a brief time conducting research at the NIH, Nirenberg made his mark in genetic research with the most important scientific breakthrough since **James D. Watson** and **Francis Crick** discovered the structure of DNA in 1953. Specifically, he discovered the process for unraveling the code of DNA. This process allows scientists to determine the genetic basis of particular hereditary traits. In August of 1961, Nirenberg announced his discovery during a routine presentation of a research paper at a meeting of the International Congress of Biochemistry in Moscow.

Nirenberg's research involved the genetic code sequences for **amino acids**. Amino acids are the building blocks of protein. They link together to form the numerous protein molecules present in the human body. Nirenberg discovered how to determine which sequences patterns code for which amino acids (there are about 20 known amino acids).

Nirenberg's discovery has led to a better understanding of genetically determined diseases and, more controversially, to further research into the controlling of hereditary traits, or genetic engineering. For his research, Nirenberg was awarded the 1968 Nobel Prize for physiology or medicine. He shared the honor with scientists Har Gobind Khorana and Robert W. Holley. After receiving the Nobel Prize, Nirenberg switched his research focus to other areas of biochemistry, including cellular control mechanisms and the cell **differentiation** process.

Since first being hired by the NIH in 1960, Nirenberg has served in different capacities. From 1962 until 1966 he was Head of the Section for Biochemical Genetics, National Heart Institute. Since 1966 he has been serving as the Chief of the Laboratory of Biochemical Genetics, National Heart, Lung and Blood Institute. Other honors bestowed upon Nirenberg, in addition to the Nobel Prize, include honorary membership in the Harvey Society, the Molecular Biology Award from the National Academy of Sciences (1962), National Medal of Science presented by President Lyndon B. Johnson (1965), and the Louisa Gross Horwitz Prize for Biochemistry (1968). Nirenberg also received numerous honorary degrees from distinguished universities, including the University of Michigan (1965), University of Chicago (1965), Yale University (1965), University of Windsor (1966), George Washington University (1972), and the Weizmann Institute in Israel (1978). Nirenberg is a member of several professional societies, including the National Academy of Sciences, the Pontifical Academy of Sciences, the American Chemical Society, the Biophysical Society, and the Society for Developmental Biology.

Nirenberg married biochemist Perola Zaltzman in 1961. While described as being a reserved man who engages in little else besides scientific research, Nirenberg has been a strong advocate of government support for scientific research, believing this to be an important factor for the advancement of science.

See also DNA structure; Sequencing

NITROGENOUS BASE · *see* BASES

NON-CODING DNA · *see* SELFISH DNA

NONDISJUNCTION · *see* CHROMOSOME SEGREGATION AND REARRANGEMENT

NONDISJUNCTION MOSAIC · *see* CHROMOSOME SEGREGATION AND REARRANGEMENT

NON-MENDELIAN INHERITANCE

Gregor Mendel was fortunate to have chosen some of the most genetically simple of characters in the garden pea for his seminal experiments that laid the foundation for the science of **genetics**. Each was determined by a single **gene** on a different **chromosome**, and each trait behaved as clearly dominant or recessive in this experimental system. This allowed Mendel to recognize patterns of **heredity**, which are described as the law of segregation of alleles, and the **law of independent assortment**. In its broadest sense, non-Mendelian **inheritance** includes any hereditary phenomena that do not appear to conform to Mendel's laws or to be attributable to single autosomal genes. It is not a clearly-defined classification, and is used quite variably in the scientific literature.

Arguably, the first form of non-Mendelian inheritance to be recognized was sex-linked, since Mendel studied only autosomal characters; however, this is rarely cited as such. Also, genes that are closely linked on the chromosome will tend to assort together rather than independently, but this too can be understood as part of classical inheritance.

Complex or multifactorial traits are determined by multiple genes and environmental factors, and therefore do not conform to simple Mendelian patterns. The human single-gene Mendelian disorders (such as those catalogued in McKusick's *Mendelian Inheritance in Man*) typically are individually rare, whereas the complex non-Mendelian disorders include more common afflictions such as heart disease, **cancer**, diabetes, psychiatric disorders, and many others that are becoming more experimentally accessible with recent advances in genomics.

Even with single-gene traits, various factors can alter the regularity of transmission and expression. The same genetic trait may be expressed differently in different individuals (variable expressivity), or not at all as in the case of nonpenetrance.

Anticipation refers to the progressively earlier appearance and increased severity of a disease in successive generations. Many such examples can now be explained by the phenomenon of dynamic mutations caused by expanded trinucleotide repeat regions. Superimposed upon these are parent-of-origin effects that make expansion in gametes of either females or males more likely.

Mendelian traits are determined by nuclear genes, but organelles such as chloroplasts (in plants) and **mitochondria** also have **DNA** containing genes, with their own inheritance characteristics. This form of non-Mendelian inheritance is called extranuclear, cytoplasmic, or **maternal inheritance**. The latter is due to the fact that sperm do not contribute to the cytoplasm of the **zygote**, and therefore all mitochondrial genes are maternally derived. The inheritance of traits from mitochondrial genes is therefore always from mother to offspring.

Situations in which the parental nuclear alleles are not represented equivalently in the offspring defy Mendel's Law of Segregation, and are variously called transmission bias, segregation distortion, or meiotic drive. The t-locus in mouse, segregation distorter in *Drosophilae,* and **retinoblastoma** locus in humans provide examples, with various underlying biological explanations, in which mutant alleles are differentially represented in offspring, with parent-of-origin effects.

The extreme in transmission bias is uniparental inheritance, in which only one parental genotype (for a given locus) is represented in the offspring. This can be caused by errors in chromosome segregation that result in two alleles being derived from one parent and none from the other. Rare cases of **cystic fibrosis** in which only one parent is a **carrier** can be explained by this non-Mendelian phenomenon. It is also an anomaly that creates problems for genes that are normally imprinted.

Epigenetic inheritance involves genetic changes other than changes in DNA sequence. The best understood mechanism for such change is DNA **methylation**, and the relevant phenomenon is imprinting. Although certain genes themselves may be inherited in a Mendelian fashion, their expression is over-ridden by the epigenetic factors to influence expression of the genes. At the level of the traits being studied, then, there is a difference between two types of crosses with respect to the sex of the parent transmitting a particular gene, and the outcome in the progeny. Classic examples include the insulin-like growth factor 2 (Igf2) gene in mouse, and the Prader-Willi/Angelman region in humans.

These ever-increasing exceptions to the rules of Mendelian inheritance, once deemed heretical or inconsistent with genetic transmission of any kind, are now the basis for some of the most intriguing and informative of genetics research.

See also Alternation of generation; Angelman syndrome; Autosomal inheritance; Characters and traits; Extranuclear Inheritance; Gene penetrance; Genetic anticipation; Genetic instability; Mendelian genetics and Mendel's laws of heredity; Mitochondrial inheritance; Ovum; Plant genetics; Polygenic inheritance and disorders; Slow viruses

NONSENSE MUTATIONS

A nonsense **mutation** is a change in the polynucleotide sequence that results in the production of a protein chain terminating codon that prematurely stops **protein synthesis** during the process of **translation**. As a result of a nonsense mutation, instead of the insertion of the proper **amino acid** into the chain of amino acids bonded together to form a protein, the synthesis of the protein prematurely stops. Unless there is another mutation in a **gene** that codes for a **transfer RNA** (tRNA) that allows the suppression of a nonsense mutation, a premature termination of protein synthesis results in the manufacture and release of an incomplete (truncated) and usually nonfunctional protein fragment.

A mutation is a change in a gene or **chromosome**. There are many types of mutations, including those resulting in a substitution of **bases** in the **DNA** sequence (i.e., one **nucleotide** is exchanged for another in the polynucleotide sequence). The effects of these mutations depend on the type, extent, and placement of the mutation in the DNA sequence. The effects

of mutations range from silent mutations that do not result in a change in the translated protein sequence to mutations that are lethal and prevent embryonic development.

A change (e.g., a **point mutation**) in the polynucleotide strand of DNA can result in an alteration of the sequence of three bases in the messenger **ribonucleic acid** (**mRNA**). When the change results in a nonsense mutation, the mRNA codon that normally relays the instruction to insert a specific amino acid in the lengthening amino acid chain during protein synthesis, instead orders the termination of protein synthesis.

For example, a nonsense mutation in the beta chain of hemoglobin, can result in hemoglobin (Hb) **thalassemia** (a general clinical term for a range of conditions in which amount of hemoglobin produced is less than normal). One particular form of Hb thalassemia can result from a nonsense mutation that produces an amber codon (chain termination codon) at a place in the codon sequence that normally carries the instruction to insert the amino acid lysine in the beta hemoglobin chain synthesized during translation.

Amber, ocher and opal mutations are types of nonsense mutations. An amber mutation in DNA results in the creation of a UAG (uracil-adenine-guanine) sequence codon (amber codon). Amber mutations result from any change in a strand of DNA that results in the creation of a UAG sequence codon in mRNA. An ocher mutation results in the creation of a UAA (uracil-adenine-adenine) mRNA codon (ocher codon). An opal mutation results in the production of a UGA codon sequence (umber or opal codon).

For example, if the base sequence of the coding stand of DNA is CAG, because of restricted base paring (e.g., A-T and C-G) the anticoding DNA strand used as a **transcription** template contains a GTC sequence. Because uracil (U) replaces thymine in mRNA, the DNA coding strand sequence (the strand that has the same sequence as does the mRNA transcript) results in a mRNA codon sequence of CAG. This codon sequence is ultimately translated to insert the amino acid glutamine (Gln) into the translated protein's sequence of amino acids. A point mutation resulting in the substitution of thymine (T) for cytosine (C) at the start of the coding strand sequence results in an amber nonsense mutation (i.e., a DNA sequence of TAG that results in an UAG amber codon in the mRNA). As a result of this particular nonsense mutation, instead inserting the amino acid glutamine into the lengthening amino acid chain during the translation process, the instruction is interpreted as a stop command and the synthesis of the protein is terminated.

Nonsense mutation suppressors are mutant genes that result in the production of transfer ribonucleic acids (tRNA) that have anticodons altered so that they have the ability to read through **codons** produced by nonsense mutations. Nonsense suppressors allow the continued synthesis of the protein. Accordingly, although the amino acid sequence is altered and the substitution of a different amino acid in the protein chain may result in changes in the shape or functioning of the protein, the suppression of the nonsense mutation allows the process of protein synthesis to continue uninterrupted.

See also Amber, ocher, and opal mutations

NORTHERN BLOTTING ANALYSIS · *see*

BLOTTING ANALYSIS

NUCELIC ACIDS

Nucleic acids are long chain molecules such as **DNA (deoxyribonucleic acid)** and **RNA (ribonucleic acid)** that link together individual nucleotides that are composed of a pentose sugar, a nitrogenous base, and one or more phosphate groups.

The most common nucleotides, the building blocks of nucleic acids found in DNA are deoxyadenylic acid, deoxyguanylic acid, deoxythymidylic acid, and deoxycytidylic acid. Each of these carries a nitrogenous base: deoxyadenylic acid contains adenine (A), deoxyguanylic acid contains gutamine (G), deoxythymidylic acid contains thymine (T), and deoxycytidylic acid contains cytosine (C). The nucleotides subunits of ribonucleic acid are adenylic acid, cytidylic acid, guanylic acid, and uridylic acid. Each of the RNA subunit nucleotides carries a nitrogenous base: adenylic acid contains adenine (A), cytidylic acid contains cytosine (C), guanylic acid contains guanine (G), and uridylic acid contains uracil.

Nucleic acids are complex molecules that contain a cell's genetic information and the instructions for carrying out cellular processes. In eukaryotic cells, the two nucleic acids, ribonucleic acid (RNA) and deoxyribonucleic acid (DNA), work together to direct **protein synthesis**.

A DNA molecule is made of phosphate-base-sugar **nucleotide** chains, and its three-dimensional shape affects its genetic function. In humans and other higher organisms, DNA is shaped in a two-stranded spiral helix organized into structures called chromosomes. DNA or RNA in some **bacteria** is circular. Most RNA molecules are single-stranded and take various shapes.

Nucleic acids were first identified by the Swiss biochemist **Johann Miescher**. Miescher isolated a cellular substance containing nitrogen and phosphorus. Thinking it was a phosphorus-rich nuclear protein, Miescher named it nuclein.

The substance identified by Miescher was actually a protein plus nucleic acid, as the German biochemist **Albrecht Kossel** discovered in the 1880s. Kossel also isolated nucleic acids' two purines (adenine and guanine) and three pyrimidines (thymine, cytosine, and uracil), as well as carbohydrates.

The American biochemist Phoebus Levene, who had once studied with Kossel, identified two nucleic acid sugars. Levene identified ribose in 1909 and deoxyribose (a molecule with less oxygen than ribose) in 1929. Levene also defined a nucleic acid's main unit as a phosphate-base-sugar nucleotide. The nucleotides' exact connection into a linear polymer chain was discovered in the 1940s by the British organic chemist Alexander Todd.

In 1951 American molecular biologist James Watson and the British molecular biologists **Francis Crick** and Maurice Wilkins developed a model of DNA that proposed its now accepted two-stranded helical shape in which adenine is always paired with thymine and guanine is always paired with the cytosine. In RNA, uracil replaces thymine.

During the 1960s scientists discovered that three consecutive DNA or RNA **bases** (a codon) comprise the genetic code or instruction for production of a protein. A **gene** is transcribed into **messenger RNA (mRNA)**, which moves from the **nucleus** to structures in the cytoplasm called **ribosomes**. **Codons** on the mRNA order the insertion of a specific **amino acid** into the chain of amino acids that are part of every protein. Codons can also order the **translation** process to stop. **Transfer RNA (tRNA)** molecules already in the cytoplasm read the codon instructions and bring the required amino acids to a ribosome for assembly.

Some proteins carry out **cell** functions while others control the operation of other genes. DNA is replicated (completely copied) when a cell prepares to divide, one copy (strand) ultimately goes to each of the two new cells in a process called semiconservative **replication**. Until the 1970s cellular RNA was thought to be only a passive **carrier** of DNA instructions. It is now known to perform several enzymatic functions within cells, including transcribing DNA into messenger RNA and making protein. In certain viruses called retroviruses, RNA itself is the genetic information. This, and the increasing knowledge of RNA's dynamic role in DNA cells, has led some scientists to believe that RNA was the basis for Earth's earliest life forms, an environment called the RNA World.

Since the 1970s nucleic acids' cellular processes have become the basis for genetic engineering, in which scientists add or remove genes in order to alter the characteristics or behavior of cells. Such techniques are used in agriculture, pharmaceutical and other chemical manufacturing, and medical treatments for **cancer** and other diseases.

See also DNA structure; Nucleotide; Transcription

NUCLEASE

Nucleases are ubiquitous phosphodiesterase **enzymes** that cleave phosphodiester bonds within nucleic acid molecules. They are indispensable for the both cellular and viral development. Nucleases that are specific for **DNA** are called deoxyribonucleases (DNases), while nucleases that recognize **RNA** (RNases) are called ribonucleases. There are many distinct types, all of which function to recognize a wide range of different yet types of molecules. Some nucleases, however, cleave nonspecific nucleic acid substrates. They also provide a variety of protective mechanisms. For example, they are responsible for degrading host **cell** DNA following viral infection and they are essential to **DNA repair** mechanisms such as **nucleotide** excision repair. Other important functions within the cell that nucleases play a role include **DNA synthesis**, DNA **recombination**, maturation of RNAs or **RNA splicing**, and DNA packaging involving either chromosomes or viral compartments. While some nucleases serve to degrade exposed single-stranded DNA, others are specific for double-standed DNA.

Exonucleases can remove nucleotides starting from the free ends of DNA inward, while endonucleases target somewhere within a DNA. Exonucleases can also operate unidirectionally in either the 5' (five prime) to 3' (three prime) direction

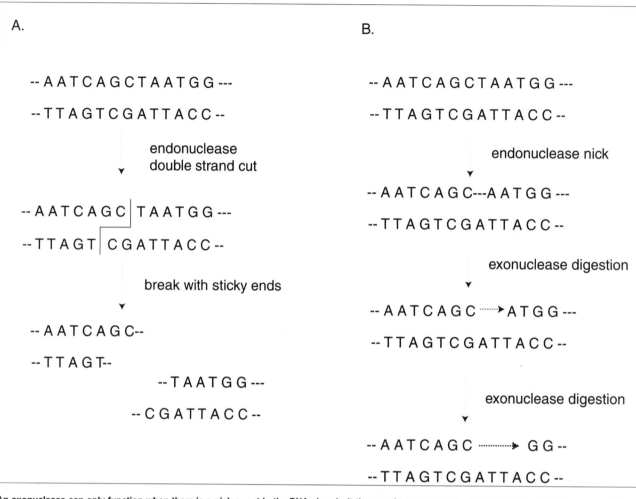

A.

-- A A T C A G C T A A T G G ---

-- T T A G T C G A T T A C C --

↓ endonuclease
double strand cut

-- A A T C A G C | T A A T G G ---

-- T T A G T | C G A T T A C C --

↓ break with sticky ends

-- A A T C A G C--

-- T T A G T--

-- T A A T G G ---

-- C G A T T A C C --

B.

-- A A T C A G C T A A T G G ---

-- T T A G T C G A T T A C C --

↓ endonuclease nick

-- A A T C A G C---A A T G G ---

-- T T A G T C G A T T A C C --

↓ exonuclease digestion

-- A A T C A G C ------→ A T G G ---

-- T T A G T C G A T T A C C --

↓ exonuclease digestion

-- A A T C A G C ------→ G G --

-- T T A G T C G A T T A C C --

An exonuclease can only function when there is a nick or cut in the DNA already. It then works to remove the bases from the DNA from one end to the other. An endonuclease cuts the DNA internally. Here, a double strand of DNA is represented by a sequence of paired bases (A,T C,G), A. An example of a two-strand internal cut by a restriction endonuclease. The enzyme severs the DNA strand leaving staggered or "sticky" ends, B. An internal single base nick by an endonuclease opens the DNA and allows digestion of one strand by an endonuclease (dashed arrow). *Courtesy of Dr. Constance Stein.*

or vice versa. Targeting of these enzymes to the appropriate site can be accomplished either with the help of specific proteins or by recognizing specific sequences. Endonucleases can affected by how the **chromosome** is packaged. For example, there are regions along the chromosome that are less tightly coiled around **nucleosomes**, making them vunerable to nuclease attack. These localized areas along the eukaryotic chromosomes are called nuclease hypersensitivity sites.

The field of molecular biology has greatly benefited from the identification and characterization of nucleases. Although many biologists aim to preserve DNA sequences, DNAases can be used in protocols that benefit from the removal or alteration of DNA. For example, DNases can be used for eliminating DNA during RNA isolations, identifying DNA-protein interactions (called DNase footprinting), or by creating nicks in the DNA prior to radiolabeling by nick **translation**. Rnase A is the most common endonuclease that degrades RNA and can be used to reduce RNA contamination in DNA isolation or plasmid

preparations. It can also be used in mapping mutations in RNA by cleaving RNA in RNA:DNA hybrids where there are single base mismatches. The most common type of nucleases used in the laboratory are **restriction enzymes**. Restriction enzymes are nucleases that recognize specific DNA sequences and provide a powerful tool in recombinant DNA technology. They can be used to digest DNA and clone the fragments for **sequencing** purposes. Additionally, they are used to identify differences in the sequences of two alleles that create restriction fragment length polymorphisms (RFLPs) when digested.

See also DNA binding proteins

NUCLEOSOMES

The **DNA** found inside the eukaryotic **nucleus** is usually associated with special classes of proteins that form a DNA-protein

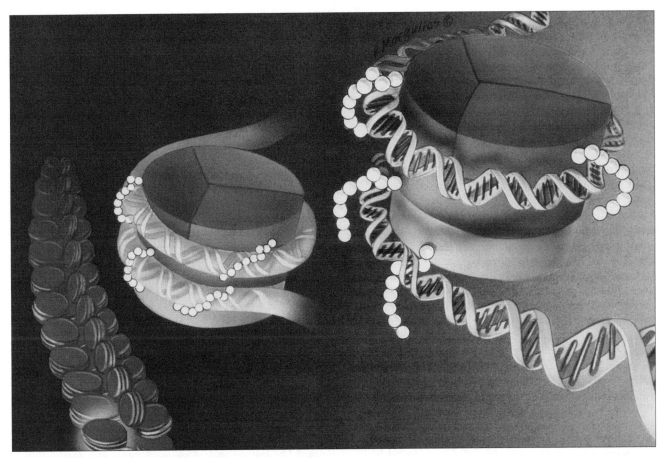

Illustration representing nucleosomes. © R. Margul. Reproduced by permission.

complex termed **chromatin**. In turn, chromatin clumps into tightly packed units termed nucleosomes. The structure of nucleosomes can restrict the accessibility of DNA and affect **gene regulation**.

The nuclei of most eukaryotic cells (cells with a nucleus defined by a nuclear membrane) contain nearly equal amounts of DNA and specialized proteins that bind to the DNA to form chromatin. Approximately half of the proteins associated with DNA are a form of very basic protein termed histones. There are five major classes or forms of histone proteins usually associated with DNA. The histones most commonly associated with DNA are designated H1, H2A, H2B, H3 and H4.

The packaging of DNA with histones and non-histone chromosomal proteins allows bonding to occur in the molecules surrounding the actual DNA nucleotides. This external or cooperative bonding allows chromatin to be packed into clumps and gives chromatin a "beads on a string" appearance under the **electron microscope**.

Histone H1 molecules play a critical role in the binding of nucleosome together into globular clumps. Although not all nucelosomes are packed the same way, H1 histones tend to bind to DNA in clusters of eight histones. Although bacterial DNA is not normally associated with proteins and histones, studies have demonstrated that histones are capable of binding bacterial DNA.

The nucleosomes (beads) are joined by a thin and tenuous looking short strand of DNA nucleotides that associate with far fewer protein molecules. This string-like strand of DNA is termed spacer DNA or linker DNA and may vary from 20 base pairs (bp) to 120 base pairs in length. The length of the spacer or linker DNA depends upon the organism, and the **cell type**. Nucleosome and chromatin structure also rapidly and dramatically change during **cell division**. Accordingly, the *in vivo* (in living cells) appearance of chromatin is dependant upon the cell's position in the cell cycle.

Scientists use a variety of equipment and a number of techniques to study the detailed structure of nucleosomes. These techniques include electron microscopy, DNA crystallization, neutron diffraction, protein assays and cross linking, x-ray diffraction studies, and variety of **immunological techniques**. Studies of the nucleosome core (nucleosome core particles or NCPs) establish that a nucleosome consists of DNA wrapped around a clump of histones. There are approximately 1.8 turns of DNA per nucleosome. The DNA is not wrapped smoothly, but is wound about the histones in a kinked manner to form superhelical twists that allow as much as eight times the amount of DNA to exist in the same volume.

Chromatin and nucleosome function, especially of subtle biochemical modifications that occur during specific stages of the cell cycle, comprise an active area of genetic research. These changes regulate the accessibility of DNA to the processes of **replication**, **transcription**, and repair. Nucleosomes can affect **gene expression** by limiting the access of DNA regulatory proteins. Some regulatory proteins can only bind to regions of DNA that contain low numbers or that are free of nucleosomes. Nucleosomes are not evenly distributed and a certain level of genetic regulation may be attributed to the non-random distribution of nucleosomes.

See also Amino acid; Bases; Chromosome structure and morphology; DNA structure; Double helix; Chromosomal mutations and abnormalities; Molecular biology and molecular genetics

NUCLEOTIDE

A nucleotide is a single chemical unit which, when bonded with other nucleotides, forms nucleic acids. Nucleic acids such as **deoxyribonucleic acid** (**DNA**) and **ribonucleic acid** (**RNA**) are the basis for all life on Earth.

Chemically, nucleotides are composed of three types of molecular groups including a sugar structure, a phosphate group, and a cyclic base. A sugar molecule is the primary structure for all nucleotides. In general, the sugars are composed of five carbon atoms with a number of hydroxyl (-OH) groups attached. The sugars differ depending on the type of nucleotide, and can be either D-ribose or 2-deoxy-D-ribose. When incorporated into a nucleotide, the sugar molecule exists in a closed ring structure.

A key part of a nucleotide is a heterocyclic base that is covalently bound to the sugar at its first carbon. These **bases** are either pyrimidine or purine groups, and they form the basis for the nucleic acid code. Two types of purine bases are found including adenine and guanine. In DNA, two types of pyrimidine bases are present, thymine, and cytosine. In RNA, the thymine base is absent and uracil is found instead.

A phosphate group makes up the final portion of a nucleotide. This group is derived from phosphoric acid and is covalently bonded to the sugar structure on the fifth carbon. Chemical linkages between nucleotides are made possible by the presence of the phosphate group. In a nucleic acid polymer, the phosphate group from one nucleotide is bonded to the third carbon on another nucleotide. Multiple bonds are made in this way creating a sequence of bases which become an organism's genetic code.

See also Double helix

NUCLEUS

The nucleus is a membrane-bounded organelle found in eukaryotic cells that contains the chromosomes and nucleolus. Intact eukaryotic cells are comprised of a nucleus and cyto-

plasm. A nuclear envelope encloses **chromatin**, the nucleolus, and a matrix which fills the nuclear space.

The chromatin consists primarily of the genetic material, **DNA**, and histone proteins. Chromatin is often arranged in fiber like nucleofilaments.

The nucleolus is a globular **cell** organelle important to ribosome function and **protein synthesis**. The nucleolus is a small structure within the nucleus that is rich in **ribosomal RNA** and proteins. The nucleolus disappears and reorganizes during specific phases of **cell division**. A nucleus may contain from one to several nucleoli. Nucleoli are associated with protein synthesis and enlarged nucleoli are observed in rapidly growing embryonic tissue (other than cleavage nuclei), cells actively engaged in protein synthesis, and malignant cells. The nuclear matrix itself is also protein rich.

The genetic instructions for an organism are encoded in nuclear DNA that is organized into chromosomes. Eukaryotic chromosomes are composed of proteins and nucleic acids (nucleoprotein). Accordingly, cell division and reproduction require a process by which the DNA (or in some prokaryotes, **RNA**) can be duplicated and passed to the next generation of cells (daughter cells).

It is possible to obtain genetic replicates through process termed nuclear transplantation. Genetic replicas are cloned by nuclear transplantation. The first **cloning** program using nuclear transplantation was able, as early a 1952, to produce frogs by nuclear transplantation. Since that time, research programs have produced an number of different **species** that can be cloned. More recently, sheep (Dolly) and other creatures have been produced by cloning nuclei from adult animal donors.

The cloning procedures for frogs or mammals are similar. Both procedures require the insertion of a nucleus into an egg that has been deprived of its own genetic material. The reconstituted egg, with a new nucleus, develops in accordance with the genetic instructions of the nuclear donor.

There are, of course, cells which do not contain the usual nuclear structures. Embryonic cleavage nuclei (cells forming a blastula) do not have a nucleolus. Because the cells retain the genetic competence to produce nucleoli, gastrula and all later cells contain nucleoli. Another example is found upon examination of mature red blood cells, erythrocytes, that in most mammals are without (devoid) of nuclei. The loss of nuclear material, however, does not preclude the competence to carry oxygen.

See also Cell; Cell division; Chromatin; Chromosome; Chromosome structure and morphology; DNA (deoxyribonucleic acid); DNA structure; Eukaryotes; Mitosis; Meiosis

NÜSSLEIN-VOLHARD, CHRISTIANE (1942-)
German geneticist

Christiane Nüsslein-Volhard, along with two American molecular biologists made important discoveries about how genes control the early development of embryos.

Christiane Nüsslein-Volhard was born in Magdeburg, Germany, the daughter of Rolf Volhard, an architect, and Brigitte (Hass) Volhard, a musician and painter. While few women of her generation chose scientific careers, Nüsslein-Volhard found that being female in a male-dominated field presented little in the way of an obstacle to her studies. She received degrees in biology, physics, and chemistry from Johann Wolfgang Goethe University in 1964 and a diploma in biochemistry from Eberhard-Karls University in 1968. In 1973, she earned a Ph.D. in biology and genetics from the University of Tübingen. Nüsslein-Volhard was married for a short time and decided to keep her husband's last name, as it was associated with her developing scientific career.

In the late 1970s, Nüsslein-Volhard finished post-doctoral fellowships in Basel, Switzerland, and Freiburg, Germany, and accepted her first independent research position at the European Molecular Biology Laboratory (EMBL) in Heidelberg, Germany. She was joined there by Eric F. Wieschaus who was also finishing his training. Because of their common interest in *Drosophila*, or fruit flies, Nüsslein-Volhard and Wieschaus decided to work together to find out how a newly fertilized fruit fly egg develops into a fully segmented embryo.

Nüsslein-Volhard and Wieschaus chose the fruit fly because of its fast embryonic development. They began to pursue a strategy for isolating genes responsible for the embryos' initial growth. This was a bold decision by two scientists just beginning their scientific careers; it wasn't certain that they would be able to actually isolate specific genes.

Their experiments involved feeding male fruit flies sugar water laced with **deoxyribonucleic acid** (DNA)—damaging chemicals. When the male fruit flies mated with females, the females often produced dead or mutated embryos. Nüsslein-Volhard and Wieschaus studied these embryos for over a year under a microscope which had two viewers, allowing them to examine an embryo at the same time. They were able to identify specific genes that basically told cells what they were going to be—part of the head or the tail, for example. Some of these genes, when mutated, resulted in damage to the formation of the embryo's body plan.

Nüsslein-Volhard and Wieschaus published the results of their research in the English scientific journal *Nature* in 1980. They received a great deal of attention because their studies showed that there were a limited number of genes that control development and that they could be identified. This was significant because similar genes existed in higher organisms and humans and, importantly, these genes performed similar functions during development. Nüsslein-Volhard and Wieschaus's breakthrough research could help other scientists find genes that could explain **birth defects** in humans. Their research could also help improve *in vitro* **fertilization** and lead to an understanding of what causes miscarriages.

In 1991, Nüsslein-Volhard and Wieschaus received the Albert Lasker Medical Research Award. During this time Nüsslein-Volhard had begun new research at the Max Planck Institute in Tübingen, Germany, similar to the work she did on the fruit flies. This time she wanted to understand the basic patterns of development of the zebra fish. She chose zebra fish as her subject because most of the developmental research on vertebrates in the past was on mice, frogs, or chickens, which have many technical difficulties, one of which was that one couldn't see the embryos developing. Zebra fish seemed like the perfect organism to study because they are small, they breed quickly, and the embryos develop outside of the mother's body. The most important consideration, however, was the fact that zebra fish embryos are transparent, which would allow Nüsslein-Volhard a clear view of development as it was happening. Despite her prize—winning research on fruit flies, Nüsslein-Volhard received skeptical feedback on her zebra fish work.

On October 9, 1995, in the midst of criticism about her new research, Nüsslein-Volhard (the first German woman to win in this category), Wieschaus, and **Edward B. Lewis** of the California Institute of Technology won the Nobel Prize in physiology or medicine for their work on genetic development in *Drosophila*. Lewis had been analyzing genetic mutations in fruit flies since the forties and had published his results independently from Nüsslein-Volhard and Wieschaus.

See also Embryology

Obesity, Genetic Factors

In the United States, more than 60 million adults are obese, meaning they are 20% or more above the ideal weight for a person's height. Complications from obesity can include heart disease, diabetes and high blood pressure. Obesity results from environmental and lifestyle factors, which can be controllable, and from physiological factors, for which there is no conscious control. The physiological factors are genetically regulated.

A genetic link to obesity was strongly inferred by the results of studies involving **twins** who were raised apart. The study allowed the influences of genetic make-up and environment to be distinguished. These studies determined that genetic factors contribute to about 40% of obesity variance in twins. Although genetic factors alone can not explain the large increase in the prevalence of childhood and adolescent obesity, interactions between **genetics** and environmental factors are likely. A person with a genetic disposition towards obesity, raised in an environment where consumption of high-fat food and little exercise are the norms, is likely to become obese. Susceptibility to obesity is largely a function of genetics, but the environment determines phenotypic expression.

Research into obesity has also utilized genetically engineered strains of mice. In 1997, researchers identified a **gene** in obesity-resisting mice called UCP2. The gene product may act to raise the body temperature, necessitating an increased use of calories. Conversely, malfunctioning UCP2 may increase the likelihood of fat accumulation, as the requirement for the expenditure of calories is less.

Five mouse strains currently exist for which obesity is a phenotypic trait. Research on these strains has identified a gene designated as Ob (for obese). The ob gene product is a protein called leptin. The gene has been designated LEP. Leptin acts on the hypothalamus, a portion of the brain that regulates body functions. If a large amount of leptin is produced, the hypothalamus reacts by reducing appetite and accelerating the body's metabolism. But if the Ob gene is

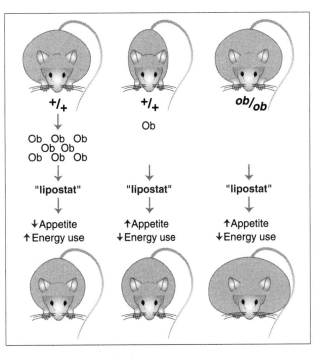

Results of mice being injected with varying levels of obese protein. *Reprinted with permission from Science, vol. 269, July 28, 1995. Illustration by K. Sutliff. Copyright 1995 American Association for the Advancement of Science.*

defective and the amount of leptin is reduced, the hypothalamus might be induced to continually signal a need for food.

LEP and four other genes have been implicated in human obesity: PCSK1, LEPR, POMC and MC4R. Other genes, which have been associated or linked with human obesity phenotypes now number above 200. The genetic link to obesity may be an evolutionary holdover. Primordial ancestors existed in a harsh environment in which subsistence depended on hunting and gathering. Physical activity was mandatory for survival and periodic famine was the norm. Today's environment is often one of plentiful, easily obtainable food. Because the **gene pool** has not changed substantially, the genes, which favored survival now, favor obesity.

See also Acquired character; Laboratory mouse genetics

OCHER MUTATION · *see* AMBER, OCHER, AND OPAL MUTATIONS

OCHOA, SEVERO (1905-1993)

Spanish-American biochemist

Severo Ochoa is best known for being the first to synthesize **ribonucleic acid** (**RNA**) outside the **cell**. He has also discovered several important metabolic processes. For his work with RNA, he received the 1959 Nobel Prize in physiology or medicine, along with his colleague, American biochemist **Arthur Kornberg**.

Ochoa was born in Luarca, Spain, where his father was a lawyer, and graduated from the University of Malaga in 1921. He received a medical degree in 1928 from the University of Madrid. After further studies in experimental biology, in 1940 he joined the Medical School faculty of Washington University in St. Louis. In 1942, he moved to New York University's College of Medicine, becoming chairman of the biochemistry department in 1954. Ochoa became an American citizen in 1956.

Ochoa's synthesis in 1955 of RNA was pure serendipity—an unexpected byproduct of his study of the way cells use glucose that is stored as **ATP** (**adenosine triphosphate**). Ochoa and a French associate, Marianne Grunberg—Manago, had purified an enzyme (now called polynucleotide phosphorylase) from the **bacteria** *Azotobacter vinelandii*. The two scientists were trying to study its reactions with ATP and other base-sugar combinations (called nucleosides) with one or three phosphate groups attached. No reaction occurred. However, when they added the enzyme and some magnesium to a nucleoside with two phosphate groups (diphosphate), over half of the nucleoside disappeared and some phosphorus was freed.

Ochoa traced the nucleoside to a new molecule that ultraviolet chromatography identified as a **nucleotide**. He then repeated the reaction with other nucleoside—diphosphates, in each case finding a nucleotide. Further analysis showed that the sugar was ribose, meaning that the reaction produced ribonucleic acid (RNA). Since the reaction was also reversible, Ochoa concluded that adding and removing phosphorus groups is a major mechanism in the synthesis and breakdown of nucleotide chains. Ochoa and other scientists used this method to decipher the genetic code. Later studies by others showed that RNA **polymerase**, instead of Ochoa's enzyme, is the main RNA synthesizing enzyme.

Ochoa is also known for his work on how the body uses carbon dioxide, and he helped identify a key compound in the metabolism of carbon dioxide. Ochoa also identified Krebs cycle reactions leading to energy storage in phosphate bonds.

OHNO, SUSUMO (1929-2000)

Japanese-American biologist

Susumo Ohno's long and productive scientific career, influential in many areas of biology and **genetics**, was begun because of his love of horses. Ohno received his D.V.M. from the Veterinary College of the Tokyo University of Agriculture and Technology in 1949. He also earned a Ph.D. in pathology and a D.Sc. in cytogenetics from Hokkaido University in Japan.

In 1952, Ohno took a research position at the City of Hope National Medical Center in Duarte, California, and there he stayed for almost half a century. One of his earliest contributions was the demonstration that the Barr body was actually the inactive **X chromosome** in females. This observation paved the way for the Lyon hypothesis.

His studies of the X **chromosome** also led him to formulate Ohno's law. Although genes are known to be juggled between different **autosomes** in the course of **evolution**, Ohno theorized that genes on the X chromosome could not be involved in such switches because they would result in different dosages of the genes since, once positioned on an autosome, the extra copy of a former X chromosome **gene** would not be inactivated. He reasoned that such dosage inequalities would be detrimental to the development of an organism, and would be eliminated by **natural selection**. Therefore, stated Ohno's law, genes that resided on the X chromosome in one mammalian **species** would reside on the X chromosome in all mammalian species.

During the fifteen years during which Ohno was chairman of the division of biology at City of Hope, his research focused on the genes and proteins that regulate **sex determination**. In 1981, The Ben Horowitz Chair in Genetics Research was established for him so that he could pursue his research ideas independently. It was the first time that the institution had so honored one of its scientists. Ohno used this academic freedom to study the relationship between musical composition and gene coding sequences, since both are based on repetition of specific themes. He established a set of rules for converting **DNA** sequences to music. Interestingly, using his rules, Ohno found a strong similarity between a particular **cancer** gene and a funeral march. In his work on cancer, Ohno also demonstrated that malignant cells tend to have an unstable number of chromosomes.

Ohno distinguished himself in yet another area by putting forth the idea that new genes arose in the course of evolution by **duplication** of existing genes. While the existing genes were necessary for their established function in the organism, the new copy, not necessary for survival, was free to mutate, occasionally resulting in a useful new protein. This idea, revolutionary when Ohno proposed it, is now well accepted. In his classic 1971 book, *Evolution by Gene Duplication,* Ohno coined the term "junk DNA."

In his almost 50-year career, Ohno was awarded membership in prestigious national scientific societies of many countries, including the United States, Japan, and Denmark.

See also Cancer genetics; Selfish DNA

OKEN, LORENZ (1779-1851)
German naturalist

Born as Lorenz Okenfuss on his father's farm near Bohlsbach bei Offenburg, Germany, Lorenz Oken shortened his rustic-sounding surname to prevent ridicule. Oken studied medicine at the University of Freiburg, but transferred to the University of Göttingen, where he took his medical degree in 1805. Oken taught medicine at Göttingen from 1805 to 1807 and at the University of Jena from 1807 to 1819.

While at Jena, Oken developed and correlated his interests in natural science and romantic philosophy. As he began to publish these ideas, Oken became a popular and controversial figure throughout the German-speaking world. In 1816, he founded the periodical *Isis* to showcase his views on science, philosophy, politics, and culture. Suppressed several times, it finally ceased during the 1848 Revolution. Oken was an early supporter of nationalist student movements and other anti-conservative causes. Because he played an active part in the notorious *Wartburgfest* of October 1817, *Isis* was banned in Jena in 1818 and a coalition of his ideological and personal enemies, including Johann Wolfgang von Goethe (1749–1832), eventually forced him to resign his Jena professorship.

Except for one semester at the University of Basel, Switzerland, in 1821–1822, Oken was out of academia until he became professor of physiology at the University of Munich in 1827. Again, his disputes caused trouble, and he had to resign that post in 1832. From 1833 until his death, Oken was professor of natural history at the University of Zurich. Among his students there were **Albert von Kölliker** (1817–1905) and **Carl Wilhelm von Nägeli** (1817–1891).

Besides hundreds of articles in *Isis*, Oken's thought is contained in two massive, multivolume works, *Lehrbuch der Naturphilosophie* (Textbook of Natural Philosophy) and *Allgemeine Naturgeschichte für alle Stände* (Universal Natural History for all Ranks). His romantic philosophy of nature consisted mainly in affirming the unity, interdependence, and equilibrium of all living things. Although he offered not much scientific evidence, he declared that higher animals were not created, but evolved from lower animals. More specifically, simple microscopic life was self-created when conditions favored it, and all other forms of life evolved from that time. He speculated on the origin of bone, especially the vertebrae and skull, and concluded that the skull evolved from specialized vertebrae.

The dominant philosopher of Oken's time, Georg Wilhelm Friedrich Hegel (1770–831), who wrote a dialectical, idealist, anti-romantic philosophy of nature, not only disagreed with Oken's reduction of human spirit, animal life, and plant growth to entirely natural or material states of affairs, but

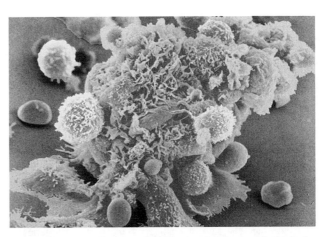

Scanning electron micrograph (SEM) image of a cancer cell attacked by tumor infiltrating lymphocytes. *Scanning electron micrograph by Jean Claude Revy. © Jean Claude Revy/Phototake. Reproduced by permission.*

also thought that Oken himself was misguided. Yet Oken's fearless enthusiasm for defying authority, engaging in scientific and philosophical polemics, and asking fundamental questions about the origin and **differentiation** of life set the stage for the fierce controversies over **evolution** that occurred in the generation after his death. In particular, many of Oken's ideas, such as that of the primeval slime, foreshadowed those of Ernst Haeckel (1834–1919). Oken also organized some of the first productive international scientific congresses.

See also Sociobiology

ONCOGENE

Oncogenes are a special type of **gene** that is capable of transforming host cells and triggering carcinogenesis. The name is derived from the Greek *onkos*, meaning bulk, or mass, because of the ability to cause **tumor** growth. Oncogenes were first discovered in retroviruses (viruses containing the enzyme **reverse transcriptase**, and **RNA**, rather than **DNA**) that were found to cause **cancer** in many animals (for example: feline leukemia **virus**, simian sarcoma virus). Although this is a relatively common mechanism of oncogenesis in animals, very few oncogene carrying viruses have been identified in man. The ones that are known include the papilloma virus HPV16 that is associated with cervical cancer, HTLV-1 and HTLV-2 associated with T-cell leukemia, and HIV-1 associated with Kaposi sarcoma.

Rather than viral infection, studies of humans led to the discovery of related genes called proto-oncogenes that exist naturally in the human **genome**. hese genes have DNA sequences that are similar to oncogenes, but under normal conditions, the proto-oncogenes do not cause cancer. However, specific mutations in these genes can transform them to an oncogenic form that may lead to carcinogenesis. So, in humans, there are two unique ways in which oncogenesis occurs, by true viral infection and by **mutation** of proto-oncogenes that already exist in human cells.

See also Cancer; Proto-oncogenes

ONCOGENE RESEARCH

Research into the structure and function of oncogenes has been a major endeavor for many years. The first **chromosome** rearrangement involving a **proto-oncogene** to be directly associated with **cancer** induction was identified in 1960. Since then, over 50 proto-oncogenes have been mapped in the human **genome**, and many cancer-related mutations have been detected. nce the role of oncogenes and proto-oncogenes in cancer was understood, the task of elucidating the exact mutations, specific breakpoints for translocations, and how protein products are altered in the disease process was undertaken.

Karyotype analysis has been used for many years to identify chromosome abnormalities that are specifically associated with particular types of leukemia and lymphoma aiding in diagnosis and the understanding of prognosis. ow that many of the genes involved in the chromosome rearrangements have been cloned, newer, more effective detection techniques, have been discovered. **FISH, fluorescence *in situ* hybridization**, uses molecular probes to detect chromosome rearrangements. Probes are developed to detect **deletions** or to flank the breakpoints of a **translocation**. or example, using a dual color system for chronic myelogenous leukemia (CML), a green probe hybridizes just distal to the c-*abl* locus on chromosome 9 and a red probe hybridizes just proximal to the locus on chromosome 22. In the absence of a rearrangement, independent colored signals (2 green and 2 red) are observed. When the rearrangement occurs, two of the fluorescent probes are moved adjacent to one another on one chromosome and their signals merge producing a new color (yellow) that can be easily detected (net result: 1 green, 1 red, and 1 yellow signal).

Other molecular techniques such as Southern blotting and PCR are also used for cancer detection and can identify point mutations as well as translocations. hese systems are set up such that one series of **DNA** fragments indicate no **mutation**, and a different size fragment or series of fragments will be seen if a mutation is present. All of the newer techniques are more sensitive than cytogenetic analysis and can pick up abnormal **cell** lines occurring at very low frequencies. Clinically, it may be useful to detect the disease in an early stage when there are fewer cancer cells present so that treatment may begin before severe symptoms are experienced. In addition, these techniques aid in detection of minimal residual disease (the presence of low levels of disease after treatment) and may give warning that the disease is returning.

A major breakthrough has come in treatment of diseases caused by oncogenes. The current standard of care for cancer patients has been chemotherapy and radiation therapy. This is successful in limiting or eradicating the disease, but, because the whole body is affected by these treatments, there are usually multiple side effects such as hair loss, nausea, fatigue, etc.

New drugs are designed to counteract the particular mutation associated with the patient's disease and thus are target specific. his is only possible if the mutation causing the disease is known and a treatment can be developed that inactivates the negative affect of that mutation. ecause only one cellular component is eaffected negative physical side effects may be reduced.

The most successful of these drugs to date is STI-571, or Gleevec, and was developed for use in patients with chronic myelogenous leukemia (CML). In CML, the proto-oncogene translocation results in overproduction of the enzyme tyrosine kinase. Gleevec is an inhibitor of tyrosine kinase and works at the cellular level to block excess enzyme activity. Although there are several different types of tyrosine kinase in humans, STI-571 is specific to the form produced by the CML mutation and does not affect other members of this enzyme family. The drug is therefore so specific, other cells and tissues in the body are not impacted, and there are few negative side effects resulting in a therapy that is much more tolerable to the patient. Early clinical trials showed such a high degree of success that the trails were terminated early and the drug was FDA approved and released for general use. There is now new evidence to suggest that this drug also may be effective for other diseases, including some types of solid tumors. This is clearly the way drug treatments will be designed in the future. By targeting only the defect and correcting that, a disease can be managed without impairing other aspects of a patient's health or quality of life.

Other types of ongoing research include further elucidation of normal proto-oncogene function and how the oncogenic mutations change cellular regulation. n particular, issues involving **oncogene** impact on **apoptosis**, programmed cell death, have become an important avenue of investigation. It has been shown that normal cells have a fixed life span but that cancer cells lose this characteristic and exhibit uncontrolled cell growth with aspects of immortality. A better understanding of the role oncogenes play in this process may give insight into additional ways to treat cancer.

See also Blotting analysis; Cancer genetics; Tumor suppressor genes

ONE GENE-ONE ENZYME

In 1941, American scientists **George Beadle** and **Edward Lawrie Tatum** proposed the one gene-one enzyme theory. The four main tenets of this theory (as modified by Tatum in 1959) were:

- All biochemical processes in all living organisms are under genetic control.
- All biochemical reactions in an organism are resolvable into separate steps.
- Each step or reaction is under the control of a single **gene**.
- **Mutation** of a single gene results in the loss of function of the appropriate enzyme.

In other words, each gene controls the reproduction, function, and specificity of a particular enzyme.

The theory was based on results originally obtained from *Neurospora crassa*, a fungus that was grown in a medium containing only the bare minimum of nutrients necessary (the fungus being capable of manufacturing the rest). After inducing mutations in the mold using radiation, some of the progeny were unable to grow on the medium. By testing with different supplements, it was found that the mutants had lost the ability to manufacture a single **amino acid**. By breeding the lab specimens with wild specimens, it was found that the mutation was transmitted in a simple Mendelian fashion. It was assumed that the ability to synthesize the appropriate amino acid was caused by the loss of a single enzyme. The work was supported by similar evidence found in humans, plants, and *Drosophila* (genus of fruit fly).

The hypothesis was further modified in 1962 by Vernon Ingram, and from it, the one gene-one polypeptide hypothesis was born. The modification arose from research conducted on **sickle cell anemia** and sickle **cell** trait. In 1949, it was proposed that sickling was caused by a single gene mutation, which was heterozygous in sickle cell trait individuals and homozygous in individuals with full sickle cell anemia. Simultaneously, it was also noted that the hemoglobin from normal individuals and that from sickle cell anemic individuals migrated differently on an electrophoresis plate, illustrating that there was a physical difference in the hemoglobin types and supporting the single gene mutation. A normal hemoglobin molecule is made of four different polypeptide chains—two identical alpha chains and two identical beta chains. All of the chains are approximately the same length, but they can be distinguished by their chemical and electrophoretic properties. Each of these chains contains approximately 140 amino acids, and Ingram analyzed them using a modified form of **Frederick Sanger's** protein analysis. This technique gave a fingerprint of the different hemoglobin types. The fingerprint showed that the differences between the two types of hemoglobin could be found in one peptide section of eight amino acids. When this section was isolated and analyzed, the only difference was in one amino acid (glutamic acid in normal and valine in sickle cell hemoglobin). The difference between these amino acids was one base in the triplet codon. Further analysis showed that amino acid changes in one chain were independent of changes in the other chain, suggesting that the genes determining the alpha and beta chains were located at different loci. The alpha and beta chains show independent assortment.

From this, it can be seen that hemoglobin is composed of two independent gene products, each of which is a separate polypeptide. The gene is a section of **DNA** that determines the amino acid sequence of a polypeptide. One gene codes for one polypeptide and several polypeptides may be required for a functional protein or enzyme.

See also Chromosome structure and morphology; Enzymes

ONTOGENY AND PHYLOGENY

Ontogeny refers to the development of an organism. With humans, ontogeny begins with the fertilized egg and continues through embryonic and fetal development, birth, maturation, and ultimately senescence. Clearly, many developmental phenomena in humans are found in other animals and seem to be related to events in the evolutionary history of the group. **Phylogeny** relates to the development of a group. Phylogeny is a history of a group of organisms from the beginning of life to the present time.

The terms ontogeny and phylogeny are grouped together frequently because of the fundamental **biogenetic law** espoused by the German scientist Ernst Heinrich Philip August Haeckel (1834–1919). Haeckel's law, also known as the theory of recapitulation, states that ontogeny is the short and rapid recapitulation of phylogeny. Haeckel, as had others before him, noted the similarity of embryonic forms within a group. Certainly a mouse, elephant, and human embryo appear remarkably similar in contrast to the vast differences in the adults of those forms. The similarity is thought to be due to common descent from a more primitive form. Some embryological similarities are ancient in origin. Consider the aortic arches of vertebrates. Aortic arches connect a ventral aorta to the dorsal aorta in the pharyngeal region of an early embryo. Blood is pumped from the beating heart via the ventral aorta to six arteries, which arch around the pharynx. Blood collects in the dorsal aorta and is distributed throughout the body and head of the embryo. During early embryonic development there are six aortic arches in all vertebrates. In fish and amphibian larvae, the posterior arches sprout a capillary bed and ultimately, the arches and their capillaries become the respiratory gills of the organism. Reptilian, avian, and mammalian embryos, being mostly terrestrial, have absolutely no need for functional gills. Nevertheless, as stated previously, they develop six aortic arches. That includes humans. The first and second arches of humans are mostly lost, portions of the third pair form the common carotid arteries, the left fourth contributes to the arch of the aorta, the right fourth forms part of the right subclavian artery, the fifth pair are lost, and the sixth contribute to the pulmonary arteries. One cannot but be impressed that humans, standing at the peak of organic **evolution**, have a vascular system with an early development greatly similar with the early development of lowly fishes and amphibians. Most would argue that this suggests descent from some common ancestor. Other human organ systems can be shown to have similarities in their development with corresponding organs in lower vertebrate forms.

Is this similarity in developmental pattern recapitulation? The answer to that rhetorical question is a firm negative. At no time does a human recapitulate the ancestral forms to which the human may be related. Humans never have the morphology of a mature fish nor do they have the anatomy of a mature amphibian. And, of course, they never recapitulate a mature mouse or mature elephant form. Thus, ontogeny does not recapitulate phylogeny. Clearly, the history of human evolution is not repeated in detail and thus, Haeckel was wrong.

However, the very existence of the six embryonic aortic arches in human embryos witnesses the close relationship of humans with all other vertebrate forms.

Phylogeny is a history. Not much historical information can be deduced from the study of embryos. Indeed, it may be stated that vertebrates are related and have similar early embryos. But, how are they related? Comparative anatomy, which involves seeking homologous structures in contemporary related forms, provides much better information on relationships, as does the paleontological record.

See also Developmental genetics; Embryology; Evolution; Evolutionary mechanisms; Phenotype

OOCYTE • *see* OVUM

OOGENESIS • *see* GAMETOGENESIS

OPAL MUTATION • *see* AMBER, OCHER, AND OPAL MUTATIONS

OPARIN, ALEKSANDR IVANOVICH (1894-1980)

Russian biochemist

Aleksandr Ivanovich Oparin was a prominent biochemist best known for his theory that life on earth originated from inorganic matter. Although a belief that life formed through **spontaneous generation** was prevalent up to the nineteenth century, that theory was disputed by the development of the microscope and the experiments of French scientist **Louis Pasteur**. Oparin's materialistic approach to the subject was responsible for a renewed interest in how life on earth originated. His book *The Origin of Life* outlined his basic theory that life originated as a result of **evolution** acting on molecules created in the primordial atmosphere through energy discharges. In addition to his work on the origin of life, he played a major role in the development of technical botanical **biochemistry** in the Soviet Union.

Oparin was born near Moscow, the youngest child of Ivan Dmitrievich Oparin and Aleksandra Aleksandrovna. He had a sister, Aleksandra, and a brother, Dmitrii. His secondary education was marked by high achievements in science. Oparin studied plant physiology at Moscow State University, graduating in 1917. He was a graduate student and teaching assistant at the university from 1921 until 1925. Oparin also studied at other institutes of higher learning in Germany, Austria, Italy, and France, but it is thought that he never earned a graduate degree (he was awarded a doctorate in biological sciences in 1934 by the U.S.S.R. Academy of Sciences).

Aleksei N. Bakh, Oparin's mentor during his years of graduate study, was to have great influence on Oparin's later role in the development of Soviet biochemistry. Bakh was well known internationally for his research in medical and industrial chemistry, and played an important role in the organization of the chemical industry in Russia. Bakh and Oparin cofounded the Institute of Biochemistry at the Academy of Sciences of the Soviet Union in Moscow in 1935. Oparin was appointed deputy director of the institute and held that position until 1946. After Bakh's death that same year, Oparin assumed the director's position, which he held throughout his life.

The practical aspects of Oparin's work during his association with Bakh in the early thirties involved biochemical research for increasing production in the food industry, work that was of extreme importance to the Soviet economy. Through his study of enzymatic activity in plants, he found that it was necessary for molecules and **enzymes** to combine in order to create starches, sugars, and other carbohydrates and proteins. He was able to show that this biocatalysis was the basis for producing many food products in nature. He held a post from 1927 through 1934 as assistant director and head of the laboratory at the Central Institute of the Sugar Industry in Moscow, where he conducted research on tea, sugar, flour, and grains. During this same period, he also taught technical biochemistry at the D. I. Mendeleev Institute of Chemical Technology. As professor at the Moscow Technical Institute of Food Production from 1937 to 1949, he continued his research of plant processes and began the study of nutrition and vitamins.

Oparin's biochemical research on plant enzymes and their role in plant metabolism, important for its practical application, would also be important for what was to be the focus of his career, the question of how life first appeared on earth. Oparin's first paper on this subject was presented to a meeting of the Moscow Botanical Society in 1922, and published in 1924 by the *Moscow Worker*. In it, Oparin discussed the problem of spontaneous generation, arguing that any differences between living and nonliving material could be attributed to physicochemical laws. This work went largely unnoticed, and Oparin did not seriously consider the topic again until the mid-thirties. In 1936, he published *The Origin of Life,* which modified and enlarged his earlier ideas. His ideas at this time were influenced not only by contemporary international thinking on astronomy, geochemistry, organic chemistry, and plant enzymology, but also by the dialectic philosophy espoused by Friedrich Engels, and the work of H. G. Bungenburg de Jong on colloidal coacervation. Translated into English in a 1938 edition, *The Origin of Life* was also revised and updated in 1941 and 1957. Although the later versions amended the original, the concept that life arose through a natural evolution of matter remained central, and he often described this concept metaphorically by comparing life to a constant flow of liquid in which elements within are constantly changed and renewed.

Oparin's theory that the origin of life had a biochemical basis was based on his suppositions concerning the condition of the atmosphere surrounding the primeval earth and how those conditions interacted with primitive organisms. It was his idea that the primeval atmosphere (consisting of ammonia, hydrogen, methane and water) in conjunction with energy (probably in the form of sunlight, volcanic eruptions, and lightning) gave this primitive matter its metabolic ability to grow and increase. He speculated that the first organisms had appeared in ancient seas between 4.7 and 3.2 billion years ago. These living organisms would have evolved from a nonliving coagulate, or gel-like, solution. Oparin argued that a separation process called coacervation occurred within the gel, causing nonliving matter at the multimolecular level to be chemically transformed into living matter. He further theorized that this chemical **transformation** was dependent upon protoenzymatic catalysts and promoters contained in the coacervates. From there, a process of **natural selection** began, which resulted in the formation of increasingly complex organisms and, eventually, primitive systems of life. Although others, such as de Jong and T. H. Huxley, would postulate that life arose from a kind of "sea jelly," Oparin's theory that nonliving material was a catalyst for the formation of living organisms is considered by many to be his special contribution to the issue.

Oparin's suppositions on life's origins were not merely theoretical. In laboratory experiments, he showed how molecules might combine to produce the needed protein structure for transformation. Experiments of other scientists, such as Stanley Lloyd Miller, **Harold Urey**, and **Cyril Ponnamperuma**, confirmed his initial experiments on the chemical structure necessary to produce life. Ponnamperuna took the work a step further when he altered Oparin's original experiments and was able to easily produce nucleotides, dinucleotides, and **adenosine triphosphate**, which also contribute to the formation of life. Building on Ponnamperuna's research, Oparin was able to produce droplets of gel that he called protobionts. He believed these protobionts were living organisms because of their ability to metabolize and reproduce. Although later research of scientists in both the Soviet Union and the West would develop independently of Oparin's biochemical experiments, he must be given credit for putting the question of the origin of life into the realm of modern science. It has been said that his work in this area opened the door, and scientists in the West walked through.

Oparin was a man of his time, and his thinking was greatly influenced by Charles Darwin's theory of natural **selection** and the ideological climate of dialectical materialism which pervaded Soviet society during the 1930s. Although Oparin was never a Communist Party member, both his writings and his research methods reflect a bias toward dialectical materialism. However, it has been suggested that his denigration of the science of **genetics** and his support of Trofim Denisovich Lysenko and the Marxist-Leninist ideology which permeated and controlled Soviet genetics at that time may have resulted from political pressure and a desire to protect his career, as much as philosophical and scientific belief. Whatever the reasons, Oparin used his influence and prestige as chief administrator of the U.S.S.R. Academy of Sciences from 1948 through 1955 to implement policies that advanced Lysenko's views at the expense of the advancement of Soviet

genetics. The influence of Lysenko and the Marxist-Leninist view of biology waned after Stalin's death in 1953. In 1956, as the result of a petition by 300 scientists calling for his resignation, Oparin was removed from his top position in the academy's biology division. He was replaced by Vladimir A. Engelhardt, a leading Soviet advocate of molecular biology. The 1950s saw an international explosion in the growth of molecular biology, but Oparin was severely critical of its principles. Although he considered the discoveries made by Watson and Crick concerning **DNA** to be important, he was skeptical of the idea of a genetic code, calling it "mechanistic reductionism." He did, however, support DNA research within his own Institute of Biochemistry during this time, and was a coauthor of papers discussing DNA and **RNA** in coacervate droplets.

Although Oparin's influence in Soviet science weakened in the early sixties, his international reputation, based on the origin of life theory, remained strong. This, coupled with his political reliability, led his government to send him abroad as a Soviet representative. Traveling by scientists in the Soviet Union was severely restricted in the 1950s, but Oparin was sent on official Soviet business not only to countries in the Eastern bloc and Asia, but to Europe and the United States as well. He also represented his country at international scientific and political conferences, such as the World Peace Council and the World Federation of Scientists.

His work brought him numerous honors. His awards from the Soviet Union include the A. N. Bakh Prize in 1950, the Elie Metchnikoff Gold Prize in 1960, the Lenin Prize in 1974, and the Lomonosov Gold Medal in 1979. The International Society for the Study of the Origin of Life elected him as its first president in 1970. He also was elected a member of scientific societies in Finland, Bulgaria, Czechoslovakia, East Germany, Cuba, Spain, and Italy.

Beginning in 1965 and continuing through 1980, the Soviet Union placed new emphasis on the science of genetics and molecular biology. However, Oparin's Institute of Biochemistry remained a stronghold of old-style biochemistry, and it eventually was bypassed by more progressive research institutions. Oparin died in Moscow at the age of 86.

See also Evolutionary mechanisms; History of genetics: ancient and classical views of heredity; Lysenkoism

OPERATOR

The operator is a regulatory region that controls the **transcription** of gene(s) located downstream.

The operator typically is located between the gene(s) it controls and the region of the **DNA** where the initiator of transcription, **RNA polymerase**, binds. When a repressor protein occupies the operator, the RNA polymerase is physically prevented from moving along the DNA and producing the **messenger RNA** (**mRNA**). When the repressor protein is detached from the operator, the polymerase is free to move and mRNA production, and ultimately, protein production, commences.

Typically, the operator is located just upstream of the genes it controls. This location allows RNA polymerase to bind to the DNA, but, if the operator is occupied, for the initiation of transcription to remain blocked. Once the operator is free, however, polymerase movement, and hence transcription, can commence immediately. The genetic system is thus primed for rapid activity.

In **bacteria**, several genes may be controlled simultaneously by the region to which the polymerase binds and one or more operators. When under such multiple control, the entire system is called an **operon**. The best-studied operon is the **lac operon** of *Escherichia coli.*

See also DNA (deoxyribonucleic acid); Gene regulation

OPERON

An operon is a single unit of physically adjacent genes that function together under the control of a single **operator gene**. The genes within an operon code for **enzymes** or proteins that are functionally related and are usually members of a single enzyme system. The operon is under the control of a single gene that is responsible for switching the entire operon "on" or "off." A repressor molecule that is capable of binding to the operator gene and switching it, and consequently the whole operon, off, controls the operator gene. A gene that is not part of the operon produces the repressor molecule. The repressor molecule is itself produced by a regulator gene. The repressor molecule is inactivated by a metabolite or signal substance (effector). In other words, the effector causes the operon to become active.

The **lac operon** in the bacterium *E. coli* was one of the first discovered and still remains one of the most studied and well known. The **deoxyribonucleic acid (DNA)** segment containing the lac operon is some 6,000 base pairs long. This length includes the operator gene and three structural genes (lac Z, lac Y, and lac A). The three structural genes and the operator are transcribed into a single piece of messenger **ribonucleic acid** (mRNA), which can then be translated. **Transcription** will not take place if a repressor protein is bound to the operator. The repressor protein is encoded by lac I, which is a gene located to the left of the lac promoter. The lac promoter is located immediately to the left of the lac operator gene and is outside the lac operon. The enzymes produced by this operon are responsible for the hydrolysis (a reaction that adds a water molecule to a reactant and splits the reactant into two molecules) of lactose into glucose and galactose. Once glucose and galactose have been produced, a side reaction occurs forming a compound called allolactose. Allolactose is the chemical responsible for switching on the lac operon by binding to the repressor and inactivating it.

Operons are generally encountered in lower organisms such as **bacteria**. They are commonly encountered for certain systems, suggesting that there is a strong evolutionary pressure for the genes to remain together as a unit. Operons have not yet been found in higher organisms, such as multicellular life forms.

A **mutation** in the operator gene which renders it nonfunctional would also render the whole operon inactive. As a direct result of inactivation, the coded pathway would no longer operate within the **cell**. Even though the genes are still separate individual units, they cannot function by themselves, without the control of the operator gene.

See also Structural gene

OPITZ, JOHN MARIUS (1935-)
German-American geneticist

John M. Opitz is renowned as one of the founders of clinical **genetics**. The Opitz G/BBB, Opitz-Kaveggia, and Smith-Lemli-Opitz syndromes all bear his name because of his research into the genetic roots of these pediatric abnormalities.

Opitz received his training in pediatrics and clinical genetics at the State University of Iowa College of Medicine and the University of Wisconsin. He held faculty positions with Montana State University and the University of Wisconsin, Madison, before joining the University of Utah as professor of pediatrics, human genetics, and obstetrics and gynecology.

Early research conducted by Opitz in the 1950s, along with Emil Witschi, focused on **embryology**, endocrinology, **sex determination** and sex **differentiation**. Other work involved the study of hereditary kidney disease and the role of endocrines in carbohydrate metabolism. Opitz was also part of the research group that completed the first successful immunologic analysis of the human growth hormone.

In his later work, Opitz sought to determine the genetic roots of **birth defects** and mental retardation. He discovered the genetic link to several pediatric abnormalities, describing three syndromes, which were consequently given his name: Opitz-G/BBB (otherwise known as Opitz syndrome or Opitz-Frias syndrome), Opitz-Kaveggia, and Smith-Lemli-Opitz. Opitz first described Opitz-G/BBB, an extremely rare syndrome characterized by a disproportionate skull, wide-set eyes, a cleft palate, as well as numerous other physical deformities, in 1969. It was originally thought to be two separate disorders, but since discovered to be one disorder with separate genetic origins—one caused by a **mutation** on the **x chromosome**, and the other by a lack of specific genetic material on **chromosome** 22.

Opitz and his team of researchers at the University of Utah also studied **pleiotropy**, or the ability of one genetic mutation to cause several malformations. An example of this can be found in Smith-Lemli-Opitz syndrome, a disorder caused by the body's inability to metabolize cholesterol. The syndrome can lead to severe mental retardation, cardiac defects, a cleft palate, and webbed toes.

Other research conducted by Opitz and his team included genetic influences on mental retardation, sex determination and differentiation, and skeletal abnormalities. Additional areas of study included morphology, embryology, and human development.

Opitz founded and served as editor-in-chief of the *Journal of Medical Genetics,* and was a founding member of the American Board of Medical Genetics. Opitz has authored some 400 textbooks and textbook chapters on the subject of clinical genetics.

See also Chromosome; Embryology; Gene mutation; Genetic defect; Genetic disorders and diseases

OPTHAMOLOGICAL GENETICS • *see*
NEUROLOGICAL AND OPTHALMOLOGICAL GENETICS

ORGANELLE AND SUBCELLULAR GENETICS

The mitochondrion of all **eukaryotes** and the chloroplasts of plant cells are the only organelles that have their distinct genomes. These genomes are made of a single, circular **DNA** molecule denoted mtDNA in mitochondrion and ctDNA in chloroplast. The **replication** and the mode of **inheritance** of organelle genomes are distinct from the nuclear genomes.

Mitochondrial genomes vary in size, among **species**, by up to one order of magnitude. Animal cells have a **genome** of approximately 16 kb (kilobases, 1,000 **bases**) and represent the smallest mitochondrial genomes in eukaryotes. **Yeast** possess a much larger genome that varies among the different strains but is about 80 kb, with the whole yeast **mitochondrial DNA** making up 18% of the total DNA of the yeast. Plant mitochondrial genomes are the largest and most complex. They show an extremely wide range of variation in DNA size. The smallest plant mitochondrial genome is around 100 kb, which make it very difficult to isolate in intact form. These genomes contain short homologous sequences that may undergo **recombination** thus generating small circular molecules that coexist with the intact ctDNA.

Eukaryotic cells may contain up to several hundred **mitochondria**. These mitochondria contain their own replication, **transcription** and **translation** systems. Nuclear genes, however, encode the majority of mitochondrial proteins which are synthesized in the cytosol and then targeted to the mitochondrion. Each mitochondrion can contain up to 10 copies of the circular genome. A special DNA **polymerase** replicates the mitochondrial genome. The complete **sequencing** and mapping of several mammalian mitochondrial genomes show extensive similarity in organization. The mammalian mitochondrial genome is extremely compact with many overlapping genes and no introns. This genome codes for 13 essential genes of biochemical pathways (e.g., oxidative phosphoryla-

tion), two rRNAs, and 22 tRNAs. The genetic code of mitochondria differs from the standard genetic code used by the cytosolic **ribosomes** and the bacterial ribosomes. There at least two **codons** for all the aminoacids and plus four termination codons. Yeast mitochondrial genome is much larger but code for only eight proteins. The mitochondrial products synthesized by this genome, both RNAs and proteins, are similar to those produced by the mammalian mitochondria. The most distinguishing feature of the yeast genome is the existence of interrupted loci. The introns in some are so large that their size is almost as large as the whole mammalian mitochondrial DNA.

The genomes of the chloroplast of different plant cells are relatively large but show considerable difference in overall length, between 100 kb and 200 kb. The complete sequence and mapping has been determined for some organisms. These sequences show a highly conserved overall **gene** number and organization in the different species. These genomes usually encode for 50-100 proteins as well as rRNAs and t RNAs. The majority of the characterized proteins are involved in **gene expression**, electron transfer or in photosynthesis. The latter form complexes located in the thylakoid membranes. Both protein-coding genes and those coding for tRNAs contain introns.

Both mitochondrial and chloroplast genomes are believed to have evolved through endosymbiosis. This model of organelle **evolution** proposes that eukaryotic cells captured **bacteria** that later provided the function of mitochondria and chloroplast. **Phylogenetic** studies based on DNA sequence analysis suggest that mitochondria and chloroplasts evolved separately from eubacterial lineages related to purple bacteria and cyanobacteria, respectively. Mitochondrion is presumed to have evolved from a species that is very similar to *Rickettsia*, an obligate the intracellular bacteria. Both mitochondria and chloroplasts follow a non-Mendelian mode of inheritance usually referred to as extranuclear or **cytoplasmic inheritance**. Most mtDNA is inherited from egg cells and thus show a maternal pattern of inheritance. This has implications both in studies of evolution between different populations of species as well as in the inheritance of some genetic diseases.

Because of the relative simplicity of the mitochondrial genome, genetic manipulation has advanced considerably. The first humans with half the nuclear genome from one mother and the mitochondrial genome from a donor mother have been delivered and are reported healthy. This genetic manipulation has now been carried out on about thirty children in the United States. Technically this amount to a kind of germline genetic modification and has therefore caused great reservation among the international biomedical community. Many scientists and physicians argue that there are serious ethical issues that should be addressed before such a practice should be allowed.

See also Cell division; Enzymes

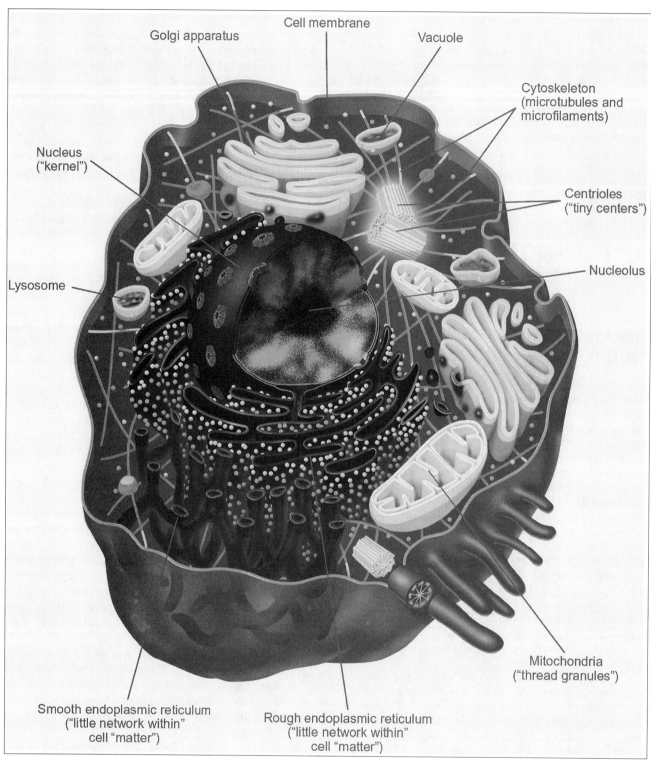

Cell organelles. *Articulate/Custom Medical Stock Photo. Reproduced by permission.*

As depicted by Michaelangelos's "The Creation, 1508-12", humans seek to understand the origin of life. Every generation adds new insights, whether artistic, philisophical, or scientific.

ORIGIN OF LIFE

The origin of life has been a subject of speculation in all known cultures and indeed, all have some sort of creation idea that rationalizes how life arose. In the modern era, this question has been considered in terms of a scientific framework, meaning that it is approached in a manner subject to experimental verification as far as that is possible. Radioactive dating suggests that Earth formed at least 4.6 billion years ago. Yet, the earliest known fossils of microorganisms, similar to modern **bacteria**, are present in rocks which are 3.5–3.8 billion years old. The earlier prebiotic era (i.e., before life began) left no direct record, and so it cannot be determined exactly how life arose. It is possible, however, to at least demonstrate the kinds of abiotic reactions that may have led to the formation of living systems through laboratory experimentation. It is generally accepted that the development of life occupied three stages: First, chemical **evolution**, in which simple geologically occurring molecules reacted to form complex organic polymers. Second, collections of these polymers self organized to form replicating entities. At some point in this process, the transition from a lifeless collection of reacting molecules to a living system probably occurred. The third process following organization into simple living systems was biological evolution, which ultimately produced the complex web of modern life.

The underlying biochemical and genetic unity of organisms suggests that life arose only once, or if it arose more than once, the other life forms must have become rapidly extinct. All organisms are made of chemicals rich in the same kinds of carbon-containing, organic compounds. The predominance of carbon in living matter is a result of its tremendous chemical versatility compared with all the other elements. Carbon has the unique ability to form a very large number of compounds as a result of its capacity to make as many as four highly stable covalent bonds (including single, double, triple bonds) combined with its ability to form covalently linked C—C chains of unlimited length. The same 20 carbon and nitrogen containing compounds called amino acids combine to make up the enormous diversity of proteins occurring in living things. Moreover, all organisms have their genetic blueprint encoded in nucleic acids, either **DNA** or **RNA**. Nucleic acids contain the information needed to synthesize specific proteins from their **amino acid** components. **Enzymes**, catalytic proteins, which increase the speed of specific chemical reactions, regulate the activity of nucleic acids and other biochemical functions essential to life, while other proteins provide the structural framework of cells. These two types of molecules, nucleic acids and proteins, are essential enough to all organisms that they, or closely related compounds, must also have been present in the first life forms.

Scientists suspect that the primordial Earth's atmosphere was very different from what it is today. The modern atmosphere with its 79% nitrogen, 20% oxygen, and trace quantities of other gases is an oxidizing atmosphere. The primordial atmosphere is generally believed not to have contained significant quantities of oxygen, having instead rather small amounts of gases such as carbon monoxide, methane, ammonia and sulphate in addition to the water, nitrogen and carbon dioxide which it still contains today. With these combinations of gases, the atmosphere at that time would have been a reducing atmosphere providing the hydrogen atoms for the synthesis of compounds needed to create life. In the 1920s, the Soviet scientist Aleksander Oparin (1894–1980) and the British scientist **J.B.S. Haldane** (1892–1964) independently suggested that ultraviolet (UV) light, which today is largely absorbed by the ozone layer in the higher atmosphere, or violent lightning discharges, caused molecules of the primordial reducing atmosphere to react and form simple organic compounds (e.g., amino acids, nucleic acids and sugars). The possibility of such a process was demonstrated in 1953 by Stanley Millar and **Harold Urey**, who simulated the effects of lightning storms in a primordial atmosphere by subjecting a refluxing mixture of water, methane, ammonia and hydrogen to an electric discharge for about a week. The resulting solution contained significant amounts of water-soluble organic compounds including amino acids.

The American scientist, Norman H. Horowitz proposed several criteria for living systems, saying that they all must exhibit **replication**, catalysis and mutability. One of the chief features of living organisms is their ability to replicate. The primordial self-replicating systems are widely believed to have been nucleic acids, like DNA and RNA, because they could direct the synthesis of molecules complementary to themselves. One hypothesis for the evolution of self-replicating systems is that they initially consisted entirely of RNA. This idea is based on the observation that certain **species** of **ribosomal RNA** exhibit enzyme-like catalytic properties and also all nucleic acids are prone to **mutation**. Thus RNA can demonstrate the three Horowitz criteria and the primordial world may well have been an "RNA world". A cooperative relationship between RNA and protein could have arisen when these self-replicating protoribosomes evolved the ability to influence the synthesis of proteins that increased the efficiency and accuracy of RNA synthesis. All these ideas suggest that RNA was the primary substance of life and the later participation of DNA and proteins were later refinements that increased the survival potential of an already self-replicating living system. Such a primordial pond where all these reactions were evolving eventually generated compartmentalization amongst its components. How such **cell** boundaries formed is not known, though one plausible theory holds that membranes first arose as empty vesicles whose exteriors served as attachment sites for enti-

ties such as enzymes and chromosomes in ways that facilitated their function.

See also Evolutionary mechanisms

ORTHOLOGY AND PARALOGY

Orthology and paralogy refer to comparisons between sequences of **deoxyribonucleic acid**. The sequences considered in both orthology and paralogy are identical to one another; also described as being homologous sequences.

The term orthology was coined in 1970 to describe an observed relationship between **gene** sequences. Orthologous genes are homologous genes present in organisms from two different **species** that originated from a common ancestor. In other words the genes are inherited through speciation. Despite this common origin, the genes need not necessarily have the same function.

Orthologous molecules provide useful information in taxonomic classification studies of organisms. The pattern of genetic divergence can be used to trace the relatedness of organisms. Two organisms that are very closely related can display very similar gene sequences. Conversely, an organism that is further removed evolutionarily from another organism can display a greater variation in the sequence of the gene in question.

Paralogous genes are homologous genes that occur in two organisms of the same species. But in one of the organisms the gene has been duplicated, so that there are two copies of the gene in one organism. Paralogous molecules provide useful information on the way genomes evolved. The **duplication** of genes and genomes have emerged s the important pathway to molecular innovation, including the **evolution** of developmental pathways.

The use of the terms orthology and paralogy has become somewhat controversial, as the terms have been misapplied. Orthology has also been used to signify genes that have the same function in different species, regardless of whether their sequences are identical, or even similar.

See also Gene divergence; Homologous chromosomes; Inheritance

OUTLAW GENE • *see* GENE NAMES AND FUNCTIONS

OVUM

Sex cells of most animals are either eggs in females (ova) or sperm in males. Eggs have the potential, when fertilized with the sperm of a male, to divide, form an embryo, and develop into an organism. In contrast to sperm, which are small, mobile, and produced in great numbers, eggs tend to be non-motile, larger, and produced in limited numbers.

Eggs are cells. That means that they contain a **nucleus** and cytoplasm and all of the other common **cell** structures. They differ from body cells, known as somatic cells, by the fact that they must eventually become **haploid** (having only one set of chromosomes) in preparation for fusion with the haploid sperm nucleus. The ovary is made up of **diploid** (having two sets of chromosomes) cells. The egg-forming ovarian cells are known as oogonia and are in fact **stem cells** that give rise to mature female gametes (sex cells). These diploid cells divide by ordinary **mitosis** to form more diploid oogonia. An egg forms when one of these cells doubles its **DNA** in preparation for meiotic maturation and forms a primary oocyte. The oocyte then begins a period of growth. During the growth period, the oocyte retains its double quantum of DNA. Ovulation, the release of the oocyte, is controlled by the pituitary follicle stimulating hormone (FSH) and luteinizing hormone (LH). The first meiotic maturation division is completed at ovulation forming a secondary oocyte and a polar body (a nonfunctioning cell containing a nucleus but very little cytoplasm), each with the diploid quantum of DNA. The second meiotic division occurs (in many animals, including humans) at the time of **fertilization** with the extrusion of the haploid second polar body which leaves the mature egg (ovum) with a haploid **chromosome** complement. Fertilization of the mature egg with a haploid sperm results in restoration of the diploid chromosome number and the beginning of embryonic development.

The larger size of eggs, compared with sperm, is attributed to the fact that eggs must contain all of the material necessary for the onset of growth and development. A fertilized egg cannot feed, and thus sufficient nutritional supplies must be stored to enable **cell division** to proceed. The built-in reserves of the egg must last until the developing organism becomes free-living or obtains parental nourishment. The care, or lack thereof, of parents correlates with egg size and abundance. For example, marine invertebrates tend to produce thousands of eggs which are, for the most part, tiny. Sea urchins do not nurture their young, and accordingly, they pro-

duce vast numbers of larvae that are highly vulnerable to predation. The sea urchin egg has a limited store of nutrients associated with its relatively small size and it rapidly develops into a free-living larva. The diameter of the sea urchin egg is about the same as for mammalian eggs which is approximately 0.004 in (0.1 mm). Despite its relatively small size, the sea urchin egg has a volume about 10,000 times larger than that of its sperm. Fish and amphibia have eggs much larger than sea urchins. The common leopard frog, *Rana pipiens*, has an egg about 0.04 in (1 mm) in diameter. Obviously, the roughly 3,000 eggs produced by the female frog are far fewer than those of the sea urchin. The frog ovum has sufficient nutrients to last the 10-14 days required to develop to a feeding larva (tadpole). Reptiles and birds have large eggs. The size of a chicken egg is well known. The largest egg known is that of the ostrich which may weigh 3 lb (1.4 kg). Fertilization of the chicken egg is internal. Sperm in the female reproductive tract encounter the huge egg (the yolk) just after ovulation. The newly fertilized egg continues down the reproductive tract gaining first an albumen layer and then a shell membrane and shell. The egg is then laid. No feeding is required of the developing chick because of the enormous stores of food. The egg at the time of laying already contains an embryo. The embryo, known as a blastodisc, develops after fertilization and is incubated during transit of the reproductive track. Twenty one days after laying, a fully developed chick emerges that can move and feed independently. In this instance, the large size of the ovum is associated with advanced development at hatching.

The human egg is extraordinarily tiny compared to the bird egg. The human egg is ovulated from the ovary, it is received by the upper end of the fallopian (uterine) tube, and it begins its transit to the uterus. Fertilization occurs in the upper end of the tube and development begins prior to the embryo entering the uterus (the womb). The embryo implants in the endometrium of the uterus shortly after its arrival in the womb and birth occurs approximately nine months later.

See also Gametogenesis; Sperm cells

P

PALEOPATHOLOGY: THE GENETICS OF PAST EPIDEMICS AND PLAGUES

Paleopathology is the study of the evidence of trauma, disease, and congenital defects in human remains. Archaeologists, geneticists, and physical-anthropologists, conduct paleopathology studies in order to evaluate the effects of disease upon ancient populations. Often, such research is conducted to garner more information about the biological and genetic characteristics of prehistoric or ancient populations, but sometimes paleopathology involves scientifically evaluating accounts of epidemics in historical records.

The earliest form of the science of paleopathology emerged in the 1600s. German naturalists, interested in newly discovered Egyptian mummies, carefully dissected and inspected the bones and tissues of specimen and attempted to compare the remains with contemporary cadavers. Early paleopathologists were able to identify striations and lesions on bone that indicated arthritis. In the late 1700s, scientists also cataloged distinctive marks on bone and tooth remains that were the result of deadly fevers.

Modern paleopathology is not limited to the study of mummified corpses. Various other types of remains, such as bone, teeth, blood, hair, fingerprints, and human waste, are now utilized in research. As medical technology has become more advanced, scientists have been able to conduct paleopathological analysis on smaller amounts of biological material. ith the aid of technology such as CAT scans and fiber optics, scientists are able to extract sample material without need of autopsy, thereby leaving remains relatively undisturbed and intact for future study.

The most common application of paleopathology is in studying patterns of disease in ancient individuals and populations. Osteologists (scientists who study bone remains) in the United States have devoted considerable effort to the study of the effects of European diseases on Native American population during the early colonial period (1492–1650). Analysis of remains not only shows the effect of disease upon individual specimen, but also adds to a larger understanding of the modes of transmission, virility, and mortality rates associated with epidemics. Paleopathology can also be used in combination with some forms of **population genetics**. For example, comparisons of analyzed remains from diverse geographic regions or kin associations assists in distinguishing possible genetic traits that aid in a population's resistance to certain diseases.

Paleopathology is one of the few means scientists have at their disposal to gain clues about the diet, health, pathology, and general genetic trends of ancient populations. Scientists also use paleopathology as a means of collecting "census material" or information regarding a specimen's age, sex, stature, and cause of death. From this data, anthropologists can estimate the general demographic composition of populations without the benefit of sometimes confused, scattered, fragmentary, or inaccurate written records. Even information about social customs and medicinal practices can sometimes be determined thorough paleopathological research on human materials. For example, skulls found in South America suggest that ancient peoples may have attempted to perform a primitive type of neurosurgery to relieve fevers or brain swelling. high levels of arsenic and mercury in hair samples from Medieval remains in France have raised questions about water contamination or possible medicinal use.

The context of paleopathological research is not limited to the ancient world. However, modern research of this type often falls under the label of forensic anthropology—the branch of science most popularly known for its applications in crime investigation.

See also Archaeogenetics: Genetic studies of ancient peoples

PALINDROME

A genetic palindrome refers to a sequence of nucleotides within a strand of **DNA (deoxyribonucleic acid)** or **RNA**

ABLE WAS I ERE I SAW ELBA

Palindrome, an inverted repeat, it reads the same forward and in reverse. *Courtesy of Dr. Constance Stein.*

(**ribonucleic acid**) that contains the same series of nitrogenous **bases** regardless from which direction the strand is analyzed. Akin to a language palindrome—wherein a word or phrase is spelled the same left-to-right as right-to-left (e.g., the word RADAR or the phrase "able was I ere I saw elba")—with genetic palindromes it does not matter whether the nucleic acid strand is read starting from the 3' (three prime) end or the 5' (five prime) end of the strand.

Related to the direction of **transcription** by RNA **polymerase**, DNA strands have upstream and downstream terminus defined by differing chemical groups at each end. The ends of each strand of DNA or RNA are termed the 5' (phosphate bound to the 5' position carbon) and 3' (phosphate bound to the 3' carbon) ends to indicate a polarity within the molecule. Using the letters A, T, C, G, to represent the nitrogenous bases adenine, thymine, cytosine, and guanine found in DNA, and the letters A, U, C, G to represent the nitrogenous bases adenine, uracil, cytosine, guanine found in RNA (Note that uracil in RNA replaces the thymine found in DNA), geneticists usually represent DNA by a series of base codes (e.g. 5' AATCGGATTGCA 3'). The base codes are usually arranged from the 5' end to the 3' end.

Because of specific **base pairing** in DNA (i.e., adenine (A) always bonds with (thymine (T) and cytosine (C) always bonds with guanine (G)) the complimentary stand to the sequence 5' AATCGGATTGCA 3' would be 3' TTAGCCTAACGT 5'.

With palindromes the sequences on the complimentary strands read the same in either direction. For example, a sequence of 5' GAATTC3' on one strand would be complimented by a 3' CTTAAG 5' strand. In either case, when either strand is read from the 5' prime end the sequence is GAATTC. Another example of a palindrome would be the sequence 5' CGAAGC 3' that, when reversed, still reads CGAAGC.

Palindromes are important sequences within nucleic acids. Often they are the site of binding for specific **enzymes** (e.g., restriction endobucleases) designed to cut the DNA strands at specific locations (i.e., at palindromes).

Palindromes may arise from breakage and chromosomal inversions that form inverted repeats that compliment each other. When a palindrome results from an **inversion**, it is often referred to as an inverted repeat. For example, the sequence 5' CGAAGC 3', if inverted (reversed 180°), still reads CGAAGC.

See also Mutations and mutagenesis; Intergenic regions

PALMER, JEFFREY DONALD (1955-)
American molecular biologist

Jeffrey Palmer has contributed to several research fronts in **plant genetics**, **evolution** and molecular systematics. His main, current interest is transfer of genes and introns between genetic compartments in cells and from organism to organism, a process called horizontal **gene flow**. In plants, those compartments consist of the **nucleus**, **mitochondria** and chloroplasts, each with their own **DNA genome**. Biologists believe that organisms have traded genes throughout the course of evolution. Palmer's research has particularly impacted current understanding of the origin of plastids from cyanobacterial endosymbionts, as well as the organization and evolution of the chloroplast genome. Palmer pioneered the field of chloroplast genomics starting with his graduate work in the early 1980s. By attracting numerous postdoctoral fellows and graduate students, who then established their own programs, Palmer jump-started research on the chloroplast genome and its application to evolutionary questions. According to Robert Jansen, Palmer's first postdoctoral fellow, "he's largely responsible for the field exploding into what it is now."

Palmer's expertise includes molecular biology, **genetics**, molecular evolution, plant systematics and bioinformatics. Palmer received a B.A. degree from Swarthmore College in 1977, and a Ph.D. from Stanford University in 1981. Following postdoctoral fellowships at the Carnegie Institution of Washington and Duke University, he joined the biology faculty of the University of Michigan in 1984. Palmer moved to Indiana University in 1989, where he is now Distinguished Professor and Chair of the Biology Department.

Palmer has also investigated the organization of mitochondrial genomes and transfer of mitochondrial genes to the nucleus. While such **gene** flow has been important in the history of all **eukaryotes** (organisms whose cells contain many discrete organelles), recent transfers have occurred primarily in plants. By **sequencing** the cytochrome oxidase 2 (*cox2*) and *rps10* genes of plant mitochondria, Palmer's laboratory has identified numerous cases of horizontal gene movement in flowering plants (angiosperms) such as legumes. Based on Palmer's research, it is known that mitochondrial genes moved frequently and recently into the nucleus, and much about how this process occurs. The same can be said for chloroplast genes.

Palmer and colleagues have also probed the origin and horizontal transfer of introns in plants. Introns are extra sequences within genes that are removed from the initial **mRNA** before **protein synthesis**. Multiple and frequent horizontal transfers of invasive, group 1 introns into the mitochondrial genome have occurred during angiosperm evolution. The Palmer laboratory has also discovered two separate lineages of plants whose mitochondrial genes evolve up to 1,000 times faster than in other plants, due in large part to differences in **mutation** rate. Palmer has parlayed much of his research on genomes into a clearer understanding of plant evolution. By combining sequence data from mitochondrial, chloroplast and nuclear genomes, his laboratory has dramatically increased the size of available data sets, thereby providing enough information to resolve early branches on the plant **phylogenetic** tree, including identification of the New Caledonian shrub *Amborella* as a remnant **species** of the most basal angiosperm group, or clade. His research team also helped establish the single origin, or monophyly, of the other major group of seed plants—gymnosperms like pines and cycads. It also placed a previously controversial group of plants called the *Gnetales* firmly within the gymnosperms. Lastly, by analyzing mitochondrial introns the Indiana researchers identified relatively simple species called liverworts as the earliest land plants. Palmer has made forays into several other research areas, including the evolution of ancient eukaryotes, the origin and evolution of a kind of intron (called a spliceosomal intron) that interrupts most nuclear genes, and the organization and function of the minimal plastids of apicomplexans—a group that includes non-photosynthetic, disease causing organisms such as *Plasmodium* and *Toxoplasma*.

Palmer has taught courses in introductory genetics, evolutionary biology, molecular systematics, and molecular biology laboratory. Palmer's research has been funded primarily by the National Institutes of Health, and the National Science Foundation (NSF). He has published more than 185 articles in a variety of scholarly books and journals. Palmer has also authored important reviews, including commentaries in *Nature* and *Science*. Palmer is in demand as a speaker at universities as well as national and international meetings. He has refereed papers for more than 45 journals; served on editorial boards of *The Plant Cell, Journal of Molecular Evolution,* and other publications; and participated in peer review panels for the United States Department of Agriculture. His research has been recognized with prestigious awards including a NSF Presidential Young Investigatorship, the David Starr Jordan Prize, the Arthur F. Thurneau professorship at Michigan, the Wilhelmine E. Key award from the American Genetics Association, and the Class of 1955 endowed professorship at Indiana. He is a member of the American Academy of Arts and Sciences and the National Academy of Sciences. Palmer credits his success to the many colleagues he has worked with, "I was fortunate to have a large number of exceptional people in the laboratory, especially early on. I want to pay tribute to them, and to the current people in my lab."

See also Evolution; Exons and introns; Genome; Molecular biology and molecular genetics; Phylogeny; Plant genetics

PARDEE, ARTHUR BECK (1921-)

American biochemist

Together with French molecular biologist **François Jacob** and French biochemist **Jacques Monod**, Arthur Pardee conducted pioneering research on protein regulation. Their teamwork is immortalized in the name of their most famous collaboration, the PaJaMo experiment. In the mid-1950s, Pardee, Monod, and other biochemists encountered several instances of feedback regulation, where the presence of an enzyme within a **cell**

inhibits its further production. Within the cell, some **enzymes** are produced continuously (constitutive production); others are synthesized only after the addition of a specific factor (inducible production). These regulatory characteristics are controlled by genetic factors.

In the 1957–1958 PaJaMo experiments, the scientists used the **bacteria** *Eschericia coli* K12 to demonstrate that instead of directly stimulating enzyme production, inducers work as anti-repressors. Later research demonstrated that the genes controlling synthesis, repression, and induction of a particular protein usually occur together in a regulatory unit, called an **operon**. Pardee and others began considering spatial (steric) mechanisms of protein regulation around 1960. In 1962, Monod proposed a general theory of **allosteric regulation**, generally accepted today, in which repressors and inducers change the spatial conformation (and, therefore, specificity) of proteins by binding at a second **active site**. These were crucial insights in understanding how genetic information is transferred between **DNA**, **RNA**, and protein.

Pardee was born in Chicago, Illinois, and completed his undergraduate studies at the University of California, Berkeley. After completing his Ph.D. at the California Institute of Technology, Pardee returned to Berkeley, first as an instructor, then as assistant, then associate professor. He joined Jacob and Monod at the Pasteur Institute in 1957–1958 as a National Science Foundation Senior Research Fellow. Upon his appointment to Princeton University in 1961, Pardee was instrumental in building the department of biochemistry. In 1975, he became Chief of the Division of Cell Growth Regulation of the Dana Farber Cancer Institute, Harvard Medical School, where he remains today.

Pardee's research at the Dana Farber Cancer Institute has established his reputation as a world expert on cell cycle control. He has received numerous honors and awards from the scientific community and has served as President of the American Society of Biological Chemists and the American Association for Cancer Research.

See also Gene induction; Gene regulation; Lac operon; Regulator gene; Regulatory region or sequence

PARKINSON'S DISEASE RESEARCH

Parkinson's disease (PD) is a progressive degenerative disease of the central nervous system, affecting individuals in adult life. It is a chronic condition that usually develops in people over the age of 50, and there are around 50,000 new cases each year in the United States. The pathology is a specific pattern of nerve **cell** (neuronal) degeneration in a region of the brain called the substantia nigra, as well as in other regions that control movement. It creates a shortage of the brain-signalling chemical (neurotransmitter) known as dopamine, resulting in impaired movement. Degenerating neurones in PD characteristically deposit intracellular structures known as Lewy bodies, thought to be altered cytoskeletal elements, which accumulate due to neuronal damage. The symptoms of PD include slowness of movements and reflexes (bradykinesia), muscular rigidity, resting tremor and difficulty with balance and walking. There is at present no specific biological test for the diagnosis of PD, and cases are frequently confirmed at autopsy. Despite its importance and severity and the many years of research, a cause has not yet been identified and there is, at present, no means of preventing PD, nor any proven permanent cure. Current treatment relies on symptom alleviation through drugs or surgical intervention.

One important aspect of current research into PD is an attempt to find a genetic link by means of population studies. A genetic link has been suspected for some time, but studies with **twins** have shown variable results, and suggest that the **genetics** of this disorder is quite complex. The disease is presently divided into sporadic PD and familial PD because of the apparent lack of any single, underlying genetic cause. Most cases of PD appear to be sporadic as are cases of several other neurodegenerative diseases, such as amyotrophic lateral sclerosis (ALS) and **Alzheimer's disease**. However, in many of these, multiple **gene** loci have been found which might give patients a predisposition for such diseases. Recently, familial PD was recognized to be more frequent than was previously thought, and some physicians say that PD is more prevalent in clinical practice than familial Alzheimer's disease or familial ALS.

Research into familial PD focuses on special populations of families. These can include families with many affected members, those that are geographically restricted, families in which the origin of the disorder can be traced back to a single individual, or those that have an early age of onset or unusual severity. Through linkage and allele sharing analysis, attempts are being made to locate genetic markers and also to determine the probability of penetrance, i.e., calculate the proportion of individuals who manifest such markers at the phenotype level. There are also efforts being made to identify the genes that increase the risk for the disorder. Aspects of interest include the determination of possible genetic heterogeneity and any evidence of multifactorial **inheritance** with environmental interaction. It is believed that if genetic factors for PD susceptibility are found, they will enhance epidemiological studies and might lead to the identification of susceptible groups or significant environmental **risk factors**. The difficulties in this work are that the signs and symptoms of familial PD do not differ from sporadic cases. They are also heterogeneous in onset and course, and wide variations of expression occur within families. The analysis of data from this research has so far shown some large families in which autosomal dominant inheritance with high penetrance is apparently demonstrated. In contrast, in small multicase families, the inheritance pattern is compatible with either autosomal dominance with reduced penetrance or multifactorial inheritance. The apparent paucity of parental consanguinity indicates no recessive inheritance, and X-linked inheritance is also thought not to be involved in PD.

There is also some evidence for possible environmental causes for PD. These have emerged from studies in humans and animals subjected to the neurochemical methylphenylte-trahydro-pyridine (MPTP). MPTP is broken down by the enzyme monoamine oxidase B into MPP+, which significantly interferes with the function of the mitochondrial respiratory chain and ultimately results in the death of neurones producing dopamine. From this, it appears that mitochondrial dysfunction may also play a role in the clinical presentation of PD.

Another important area of PD research is in improving methods of treating PD. Medical administration of the missing neurochemical dopamine is ineffective since it apparently does not enter the brain from the blood. Instead, a metabolic precursor of dopamine, L-DOPA, is used which can diffuse into the brain where it is converted to dopamine. PD is therefore most commonly treated with pharmaceutical products containing L-DOPA, but in more advanced cases neurosurgical procedures have to be used. With the advent of **gene therapy**, there is now hope that this will provide a feasible alternative for the long-term treatment of PD. Animal models have shown promising results with the long-term production of L-DOPA following just a single gene therapy treatment. Adeno-associated viral (AAV) vectors have been used to deliver two human genes to the specific area of the brain affected by the disease in PD rat models. Following gene transfer, the chemical synthesis of L-DOPA was demonstrated and the expression of the L-DOPA producing enzyme was stable for one year, throughout the duration of the study. There were no observable toxicities after treatment, and importantly, there were no other regions of the brain affected by the gene delivery. Thus, gene therapy may yet prove to be the most effective way of treating PD in the future.

See also Genetic disorders and diseases; Population genetics

PARTHENOGENESIS

Parthenogenesis is a type of **asexual reproduction** in which offspring (progeny) arise from unfertilized eggs. The phenomenon is known among many invertebrate and vertebrate groups, including some **species** of nematodes, gastropods, crustaceans, insects (especially honey bees and wasps), fishes, amphibians and reptiles. Few species reproduce solely by parthenogenesis and it is more common that episodes of parthenogenesis alternate with periods of **sexual reproduction**. Parthenogenesis may occur for several generations and then be followed at some point by sexual reproduction. During the sexual period males develop, produce sperm, and mate with females to fertilize their eggs. In some species, parthenogenesis actually appears to be an adaptation for survival in times of stress or serious population decline.

In situations where unfertilized eggs usually perish and only occasionally develop by parthenogenesis, the condition is known as exceptional parthenogenesis, while in those species where unfertilized eggs regularly develop to produce progeny, the condition is known as normal or physiological parthenogenesis. In facultatative parthenogenesis, the eggs develop whether they are fertilized or unfertilized. Several insect species demonstrate facultative parthenogenesis and their unfertilized eggs develop into only one sex, the other sex aris-ing from fertilized eggs. Thus, for example, in the honey bee (*Apis mellifica*), the males (drones) always develop from unfertilized eggs (arrhenotoky), whereas fertilized eggs develop either into sexually mature females (queens) or into female workers, depending on the nutrition of their respective larvae. A population of honey bees can become a degenerate "drone brood" if the available stock of sperm becomes exhausted, or if a queen is not fertilized. The population suf-fers a similar fate if a worker lays eggs in the absence of a queen. An alternative situation exists in the shield bug where only females hatch from unfertilized eggs (thelytoky), while in aphids, for example, unfertilized eggs develop into either males or females (amphitoky or deuterotoky).

Cyclic parthenogenesis or heterogony is a form of life cycle in which parthenogenic and sexual reproduction alter-nate. In rotifer worms and water fleas for example, certain eggs are capable of **fertilization** and if fertilized lie dormant through the winter. Such eggs are large, richly endowed with yolk and develop more slowly. In contrast, eggs that are laid parthenogenically in large numbers during the summer are small, have little yolk and develop rapidly. Especially compli-cated examples of heterogony have been discovered in aphids and gall wasps in which different generations of parthenogeni-cally produced females exhibit a pronounced **polymorphism** and also have different modes of existence. If sexual repro-duction is omitted from the cycle, for example because the cli-mate is favorable as may happen in the case of the vineyard pest, *Pylloxera vitifoliae*, then there is a continual succession of generations of parthenogenic females.

Parthenogenesis may be subdivided with regard to the **chromosome** complement of cells. In **haploid** parthenogene-sis, the progeny arise from haploid eggs. Thus honey bee drones are haploid, while the workers and queens are **diploid**. Haploid parthenogenesis has also been observed in some flow-ering plants e.g. thorn apple, tobacco, rice, maize and wheat. Diploid parthenogenic progeny can arise by various mecha-nisms, for example the haploid egg **nucleus** may fuse with the second polar body or two haploid cleavage nuclei may fuse. Alternatively, the egg may be diploid because **meiosis** does not occur completely during egg formation. Diploid partheno-genesis is much more common than haploid parthenogenesis and is known in insects such as the gall wasp and also in the roundworms and flukes as well as in various flowering plants such as the dandelion.

In addition to natural parthenogenesis, it is possible to induce artificial parthenogenesis. This has been achieved with representatives of practically all animal phyla, including amphibians and mammals. Nonfertilized eggs can be induced to develop by a variety of means, for example by exposure to hyper-or hypotonic solutions of salts, diluted acids, alkalis, toxins, narcotics, and also by physical methods such as tem-perature changes, ultraviolet or radium irradiation, brief elec-trical stimulation and puncture with a glass or platinum needle. In gynogenesis the sperm enters and activates the egg, but degenerates without its nucleus fusing with that of the egg. This can be performed experimentally in sea urchin or amphibian eggs, which are allowed to be penetrated by sperm from the same species, followed by destruction of the sperm nucleus by irradiation. Alternatively, egg development is induced by penetration (without fertilization) by sperm of dif-ferent species. The counterpart of gynogenesis is androgenesis (male parthenogenesis), in which progeny develop from a male **gamete**. Artificial androgenesis has been performed with sea urchin and amphibian sperm, by allowing them to pene-trate denucleated egg fragments of the same species, and also by damaging the egg nucleus of a normal sperm-penetrated and activated egg by radium irradiation. Artificial androgene-sis has also been performed in mouse sperm by allowing egg penetration by the sperm nucleus, then removing the female nucleus by suction.

PASTEUR, LOUIS (1822-1895)
French chemist

Louis Pasteur left a legacy of scientific contributions that include an understanding of how microorganisms carry on the biochemical process of fermentation, the establishment of the causal relationship between microorganisms and disease, and the concept of destroying microorganisms to halt the transmis-sion of communicable disease. These achievements led him to be called the founder of microbiology.

After his early education, Pasteur went to Paris to study at the Sorbonne, then began teaching chemistry while still a student. After being appointed chemistry professor at a new university in Lille, France, Pasteur began work on **yeast** cells and showed how they produce alcohol and carbon dioxide from sugar during the process of fermentation. Fermentation is a form of cellular respiration carried on by yeast cells, a way of getting energy for cells when there is no oxygen present. Pasteur found that fermentation would take place only when living yeast cells were present.

Establishing himself as a serious, hard-working chemist, Pasteur was called upon to tackle some of the problems plagu-ing the French beverage industry at the time. Of special con-cern was the spoiling of wine and beer, which caused great economic loss, and tarnished France's reputation for fine vin-tage wines. Vintners wanted to know the cause of l'amer, a condition that was destroying France's best burgundy wines. Pasteur looked at wine under the microscope and noticed that when aged properly, the liquid contained little spherical yeast cells. But when the wine turned sour, there was a proliferation of bacterial cells that produced lactic acid. Pasteur suggested that heating the wine gently at about 120°F (48.8°C) would kill the **bacteria** that produced lactic acid and let the wine age properly. Pasteur's book *Etudes sur le Vin*, published in 1866, was a testament to two of his great passions—the scientific method and his love of wine. It caused another French revolu-tion—one in winemaking, as Pasteur suggested that greater cleanliness was need to eliminate bacteria and that this could be accomplished using heat. Some wine-makers were initially reticent to heat their wines, but the practice eventually saved the wine industry in France.

The idea of heating to kill microorganisms was applied to other perishable fluids, including milk, and the idea of pasteurization was born. Several decades later in the United States, the pasteurization of milk was championed by American bacteriologist Alice Catherine Evans, who linked bacteria in milk with the disease brucellosis, a type of fever found in different variations in many countries.

In his work with yeast, Pasteur also found that air should be kept from fermenting wine, but was necessary for the production of vinegar. In the presence of oxygen, yeasts and bacteria break down alcohol into acetic acid, or vinegar. Pasteur also informed the vinegar industry that adding more microorganisms to the fermenting mixture could increase vinegar production. Pasteur carried on many experiments with yeast. He showed that fermentation can take place without oxygen (anaerobic conditions), but that the process still involved living things such as yeast. He did several experiments to show (as Lazzaro Spallanzani had a century earlier) that living things do not arise spontaneously but rather come from other living things. To disprove the idea of **spontaneous generation**, Pasteur boiled meat extract and left it exposed to air in a flask with a long S-shaped neck. There was no decay observed because microorganisms from the air did not reach the extract. On the way to performing his experiment Pasteur had also invented what has come to be known as sterile technique, boiling or heating of instruments and food to prevent the proliferation of microorganisms.

In 1862, Pasteur was called upon to help solve a crisis in another ailing French industry. The silkworms that produced silk fabric were dying of an unknown disease. Armed with his microscope, Pasteur traveled to the south of France in 1865. Here Pasteur found the tiny parasites that were killing the silkworms and affecting their food, mulberry leaves. His solution seemed drastic at the time. He suggested destroying all the unhealthy worms and starting with new cultures. The solution worked, and soon French silk scarves were back in the marketplace.

Pasteur then turned his attention to human and animal diseases. He supposed for some time that microscopic organisms cause disease and that these tiny microorganisms could travel from person to person spreading the disease. Other scientists had expressed this thought before, but Pasteur had more experience using the microscope and identifying different kinds of microorganisms such as bacteria and fungi.

In 1868, Pasteur suffered a stroke and much of his work thereafter was carried out by his wife Marie Laurent Pasteur. After seeing what military hospitals were like during the Franco-Prussian War, Pasteur impressed upon physicians that they should boil and sterilize their instruments. This was still not common practice in the nineteenth century.

Pasteur developed techniques for culturing and examining several disease-causing bacteria. He identified *Staphylococcus pyogenes* bacteria in boils and *Streptococcus pyogenes* in puerperal fever. He also cultured the bacteria that cause cholera. Once when injecting healthy chickens with cholera bacteria, he expected the chickens to become sick. Unknown to Pasteur, the bacteria were old and no longer virulent. The chickens failed to get the disease, but instead they received immunity against cholera. Thus, Pasteur discovered that weakened microbes make a good vaccine by imparting immunity without actually producing the disease.

Pasteur then began work on a vaccine for anthrax, a disease that killed many animals and infected people who contracted it from their sheep and thus was known as "wool sorters' disease." Anthrax causes sudden chills, high fever, pain, and can affect the brain. Pasteur experimented with weakening or attenuating the bacteria that cause anthrax, and in 1881 produced a vaccine that successfully prevented the deadly disease.

Pasteur's last great scientific achievement was developing a successful treatment for rabies, a deadly disease contracted from bites of an infected, rabid dog. Rabies, or hydrophobia, first causes pain in the throat that prevents swallowing, then brings on spasms, fever, and finally death. Pasteur knew that rabies took weeks or even months to become active. He hypothesized that if people were given an injection of a vaccine after being bitten, it could prevent the disease from manifesting. After methodically producing a rabies vaccine from the spinal fluid of infected rabbits, Pasteur sought to test it. In 1885, nine-year-old Joseph Meister, who had been bitten by a rabid dog, was brought to Pasteur, and after a series of shots of the new rabies vaccine, the boy did not develop any of the deadly symptoms of rabies.

To treat cases of rabies, the Pasteur Institute was established in 1888 with monetary donations coming from all over the world. It later became one of the most prestigious biological research institutions in the world. When Pasteur died in 1895, he was well recognized for his outstanding achievements in science.

See also Immunogenetics

PATAU SYNDROME

Patau syndrome, also called trisomy 13, occurs when a child is born with three copies of **chromosome** 13. Normally, two copies of the chromosome are inherited, one from each parent. The extra chromosome causes numerous physical and mental abnormalities. Owing mostly to heart defects, the lifespan of trisomy 13 baby is usually measured in days. Survivors have profound mental retardation.

Individuals normally inherit 23 chromosomes from each parent, for a total of 46 chromosomes. However, genetic errors can occur before or after conception. In the case of Patau syndrome, an embryo develops which has three copies of chromosome 13, rather than the normal two copies.

Trisomy 13 occurs in approximately 1 in 12,000 live births. In many cases, spontaneous abortion (miscarriage) occurs and the fetus does not survive. The risks of trisomy 13 seem to increase with the mother's age, particularly if she is older than her early thirties. Male and female children are equally affected, and the syndrome occurs in all races.

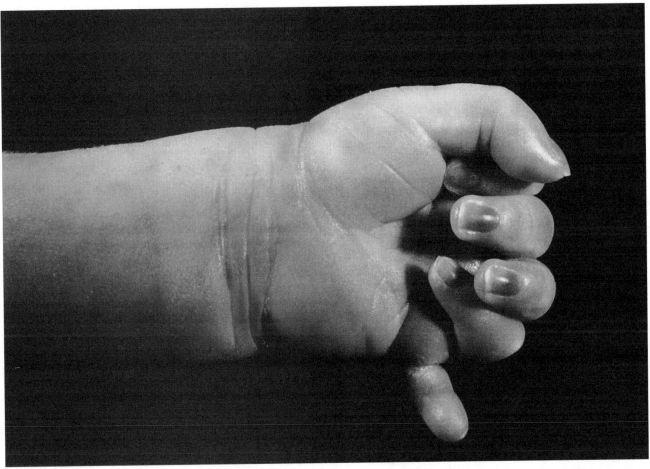

Postaxial polydactyly (extra digit), is often seen with cases of Patau syndrome. *Siebert/Custom Medical Stock Photo. Reproduced by permission.*

Patau syndrome is caused by the presence of three copies of chromosome 13. The presence of these three copies—rather then the normal two—is a random error and cannot be attributed to any action or lack of action on the part of the parents.

Newborns with trisomy 13 have numerous internal and external abnormalities. Commonly, the front of the brain fails to divide into lobes or hemispheres, and the entire brain is unusually small. Children who survive infancy have profound mental retardation.

Incomplete development of the optic (sight) and olfactory (smell) nerves often accompanies the brain defects, and the child may also be deaf. Frequently, a child with trisomy 13 has cleft lip, cleft palate, or both. Facial features are flattened and ears are malformed and lowset. Extra fingers or toes (polydactyly) may be present in addition to other hand and foot malformations.

In nearly all cases, trisomy 13 babies have respiratory difficulties and heart defects, including atrial and ventricular septal defects, patent ductus arteriosus, and defects of the pulmonary and aortic valves. Other organ systems may also be affected. The organ defects are frequently severe and life-threatening.

A newborn's numerous malformations indicate a possible chromosomal abnormality. Trisomy 13 is confirmed by examining the infant's chromosomal pattern through karyotyping or another procedure. Trisomy 13 is detectable during pregnancy through the use of ultrasonography, **amniocentesis**, and **chorionic villus sampling**.

See also Chromosomal mutations and abnormalities; Prenatal diagnostic techniques

PATERNITY AND PARENTAGE TESTING

Paternity or parentage identification is based on the ability to establish the genetic relationship between the parent(s)biological offspring. From the beginning of the twentieth century, several methods have been developed and utilized for this purpose. The utilization of these techniques has followed the developing knowledge about the characteristics of the human genotype as well as having, in the last two decades, incorporated the development of new techniques such as molecular biology. With the knowledge that each human is genetically unique as result of genetic **polymorphism**, blood group antigen typing and other **immunological techniques** including the human histocompatibility leukocyte antigen (HLA) typing, became widely used for paternity testing and forensic science application until the end of the 1970s.

In the mid-1980s, these techniques were replaced by direct analysis of the **DNA** polymorphisms. The first of such techniques, developed by Alec Jeffries utilized multilocus **DNA probes**. This technique is known as the **restriction fragment length polymorphism** (**RFLP**) testing. RFLP techniques are based on variable number of **tandem repeats** (VNTR), which are sequences of 10–60 bp (base pairs) of length, that lie adjacent to each other in the same chromosomal orientation (minisatellites). The number of repeats can be several hundred and the total size of the sequence is usually bigger than 10 kb (kilobases).

Soon after, several other laboratories developed similar methodology and in the late 1980s, the FBI adopted this technique. A second variation of this strategy was then introduced. It used a combination of single locus probes (SLP) to achieve a similar exclusion power, instead of the two MLP probes previously used.

Methodologically, both techniques are based on the digestion of the genomic DNA with restriction endonucleases, separation of the fragments by electrophoresis, followed by X transfer to a nylon membrane and finally, detection by hybridization with either a radioactive or chemiluminescent probe. This technique is known as Southern blot analysis.

A further significant improvement in the analysis of the RFLP took place through the use of the PCR (**polymerase chain reaction**) technique, where a certain region of DNA is amplified, producing millions of copies of the fragment of interest. This reaction is carried on in a thermocycler machine, and the products of amplification are separated by electrophoresis and may be visualized and documented on a UV light box. Briefly, the main advantages of such techniques include the great discriminatory power of each loci, the ability to process mixed samples and the rich experience that was developed in the last decades through the utilization of these techniques. By the other hand, 50 ng or more of DNA material is required to obtain clear results, and degraded DNA samples can pose a significant limitation to the process. Furthermore, the process is time consuming, taking up to a number of weeks to be completed.

The next big change in the analysis of the DNA for paternity (and forensic) analysis incorporated the PCR amplification of microsatellites instead of minisatellites. Microsatellites are also formed by tandem repeats but consist of 2–5 nucleotides per repeat units. Each repeat unit does have smaller sizes than minisatellites and can vary from 100–800 bp. This means that the amplification requires less DNA ($<$ than 1 ng) and the quality of the material may be less than ideal as even degraded DNA can be analyzed. The potential number of loci is very large and the process is rapid; it may be completed in a day or two. This system also has the benefit of

lending itself to **multiplexing** and automation. In addition, several kits are available, and for some multiplexes inexpensive silver stain materials may be employed without expensive equipment.

In brief, the limitation of the method includes a smaller number of alleles and less heterozygosity per locus; the possibility of contamination from stray DNA is increased because of the amplification process. The amplification process also may lead to the formation of "stutter bands," artifacts or preferential amplification, leading to imprecise interpretation of the result. For some automated uses, the equipment is relatively expensive if high-throughput analysis of fluorescent labeled multiplex systems are undertaken.

With the recent automation and miniaturization of DNA typing methods, the analysis of polymorphism began to revisit the single **nucleotide** polymorphism analysis (SNPs), another form of loci polymorphism. They were first described about in the 1980s, but only recently are being studied for paternity analysis. They are abundant in the **genome** and perceived as being more stable than STRs due to lower **mutation** rates.

In parallel to that, two other techniques are being utilized with specific objectives. They include **mitochondrial DNA** and **Y chromosome** DNA analysis. Mitochondrial DNA (mtDNA) is a double-stranded DNA found in the **mitochondria**, transmitted only by the egg. Therefore, mtDNA is particularly useful in the study of individuals related through the female line. Alternatively, the Y **chromosome** is transmitted from the father only, so DNA on the Y chromosome can be used to trace the male lineage. Y markers are particularly useful in resolving DNA from different males, such as with sexual assault mixtures.

See also Biochemical analysis techniques; Tandem repeat sequences

PAULING, LINUS (1901-1994)
American chemist

American chemist Linus Pauling has made important contributions to a number of fields, including the structure of proteins and other biologically important molecules, mineralogy, quantum mechanics, the nature of mental disorders, nuclear structure, and nutrition. He determined the molecular structure of more than 225 substances using an electron diffraction technique. Pauling was the first to construct the alpha helical structure of a protein. Pauling developed the theory of complementarity and used it to explain enzyme reactions and how genes might act as templates for the formation of **enzymes**. Using this theory, he studied the relationship between molecular abnormality and hereditary, and was able to determine that **sickle cell anemia** is caused by the change of a single **amino acid** in a hemoglobin molecule. Linus Pauling is the only person ever to win two unshared Nobel Prizes. His 1954 Nobel Prize for chemistry was given in recognition of his work on the nature of the chemical bond, while the 1963 Nobel

Peace Prize was awarded for his efforts to bring about an end to the atmospheric testing of nuclear weapons.

Linus Carl Pauling was born in Portland, Oregon. In the fall of 1917, Pauling entered Oregon Agricultural College (OAC), now Oregon State University, in Corvallis. After graduation from OAC in 1922 with a B.S. in chemical engineering, Pauling entered the California Institute of Technology (Cal Tech) in Pasadena. Pauling was awarded his Ph.D in chemistry *summa cum laude* in 1925 and decided to continue his studies in Europe where he learned about the new field of quantum mechanics. The science of quantum mechanics, less than a decade old at the time, is based on the revolutionary concept that particles can sometimes have wave-like properties, and waves can sometimes best be described as if they consisted of massless particles.

He left Zurich in the summer of 1927, and returned to Cal Tech, where Pauling took up his new post as Assistant Professor of Theoretical Chemistry. During Pauling's first few years as a professor at the college, he worked on the x-ray analysis of crystals, but also began to spend more time on the quantum mechanical study of atoms and molecules. In the summer of 1930, Pauling traveled through Europe learning about the use of electron diffraction techniques to analyze crystalline materials. Upon his return to Cal Tech, Pauling showed one of his students, L. O. Brockway, how the technique worked and had him build an electron diffraction instrument. Over the next 25 years, Pauling, Brockway, and their colleagues used the diffraction technique to determine the molecular structure of more than 225 substances.

By the mid–1930s, Pauling began studying the structure of biological molecules. These molecules are complex substances that are found in living organisms and can contain thousands of atoms in each molecule rather than the relatively simple molecules that Pauling had studied previously that contained only twenty or thirty atoms. The first substance that attracted his attention was the hemoglobin molecule. Hemoglobin is the substance that transports oxygen through the bloodstream. Pauling's initial work with hemoglobin, carried out with a graduate student Charles Coryell, produced some fascinating results. Their research showed that the hemoglobin molecule undergoes significant structural change when it picks up or loses an oxygen atom. In order to continue his studies, Pauling decided he needed to know much more about the structure of hemoglobin, in particular, and proteins, in general.

Fortunately, he was already familiar with the primary technique by which this research could be done: x-ray diffraction analysis. The problem was that x-ray diffraction analysis of protein is far more difficult than it is for the crystalline minerals Pauling had earlier worked with. In fact, the only suitable x-ray pictures of protein available in the 1930s were those of the British crystallographer, William Astbury. Pauling decided, therefore, to see if the principles of quantum mechanics could be applied to Astbury's photographs to obtain the molecular structures of proteins.

The earliest efforts along these lines by Pauling and Coryell in 1937 were unsuccessful. None of the molecular structures they drew based on quantum mechanical principles could account for patterns like those in Astbury's photographs. It was not until eleven years later that Pauling finally realized what the problem was. The mathematical analysis and the models it produced were correct. It was Astbury's patterns that were incorrect. In the pictures Astbury had taken, protein molecules tilted slightly from the expected position. By the time Pauling recognized this problem, he had already developed a satisfactory molecular model for hemoglobin. The model was that of a helix, or spiral-staircase-like structure in which a chain of atoms is wrapped around a central axis. Pauling had developed the model by using a research technique on which he frequently depended—model building. He constructed atoms and groups of atoms out of pieces of paper and cardboard and then tried to fit them together in ways that would conform to quantum mechanical principles. Not surprisingly, Pauling's technique was also adopted by two contemporaries, **Francis Crick** and James Watson, in their solution of the **DNA** molecule puzzle, a problem that Pauling himself nearly solved.

Pauling also turned his attention to other problems of biological molecules. In 1939, for example, he developed the theory of complementarity and applied it the subject of enzyme reactions. He later used the same theory to explain how genes might act as templates for the formation of enzymes. In 1945, Pauling attacked and solved an important medical problem by using chemical theory. He demonstrated that the genetic disorder known as sickle **cell** anemia is caused by the change of a single amino acid in the hemoglobin molecule.

At the age of 65, Linus Pauling began to examine the possible therapeutic effects of vitamin C. Pauling was introduced to the potential value of vitamin C in preventing colds by biochemist Irwin Stone in 1966. He soon became intensely interested in the topic and summarized his views in a 1970 book, *Vitamin C and The Common Cold*. Before long, he became convinced that the vitamin was also helpful in preventing **cancer**.

Pauling's long association with Cal Tech ended in 1964, at least partly because of his active work in the peace movement. He accepted an appointment at the Study of Democratic Institutions in Santa Barbara for four years and then moved on to the University of California at San Diego for two more. In 1969, he moved to Stanford University where he remained until his compulsory retirement in 1974. In that year, Pauling and some colleagues and friends founded the Institute of Orthomolecular Medicine, later to be renamed the Linus Pauling Institute of Science and Medicine, in Palo Alto. Pauling died of cancer at his ranch in California in 1994.

See also Proteins and enzymes; X-ray crystallography

PEDIGREE ANALYSIS

In **genetics**, a pedigree is a diagram of a family tree showing the relationships between individuals together with relevant facts about their medical histories. A pedigree analysis is the interpretation of these data that allows a better understanding of the transmission of genes within the family. Usually, at least

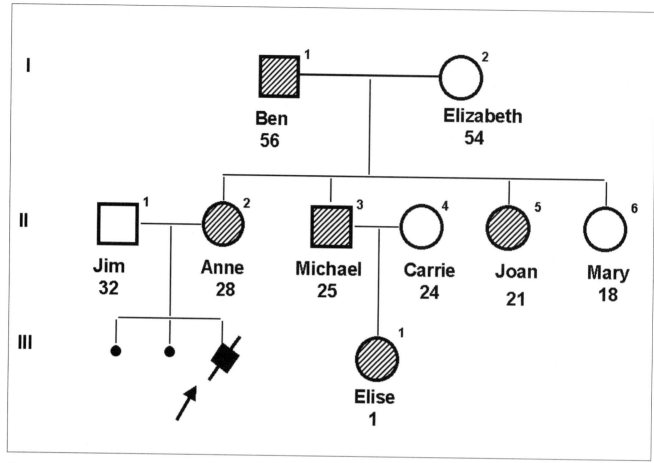

Three generation family pedigree analysis chart tracking the transmission of a chromosome rearrangement; all individuals with hatched symbols carry the translocation, squares are males, circles are females. The proband (deceased) is marked by an arrow. Small dark circles indicate spontaneous fetal losses. A Roman numeral at the left indicates the generations, and each individual has a specific number within the generation so that everyone can be uniquely identified, I.e., Carrie is individual II-4 (fourth person in generation II). *Courtesy of Dr. Constance Stein.*

one member of the family has a genetic disease, and by examining the pedigree, clues to the mode of **inheritance** of the disorder and the potential risk to other family members can be obtained.

The pedigree is initiated by using a symbol to represent the proband or individual seeking counseling. Immediate family members (parents, siblings, spouse, children) are added next, followed by aunts, uncles, cousins, grandparents, and others in the proper orientation. Males are indicated as squares and females as circles. The square or circle is filled in for any affected individuals to reflect their disease status. When two people marry or have children together, a single line is drawn between them. A vertical line descends from this marriage line and then connects to another horizontal line, the sibship line. Short vertical lines descend from the sibship line, one for each of the children of this union. All members of one generation are shown adjacent to one another in a row, with preceding generations above and later generations below. There are special symbols to denote consanguineous marriages (a double marriage line), identical **twins** (a single line from the sibship line that bifurcates for each twin), fraternal twins (an inverted

V drops from the sibship line), divorce and remarriage (cross hatches on the marriage line to show discontinuity between the divorced partners and a second marriage line to the new partner), and so on.

Each generation is labeled at the left with a Roman numeral beginning with the first generation. The members of each generation are consecutively numbered left to right with Arabic numbers, always starting each generation with one. In this way, each person can be specifically identified. For example, the second person in the first generation would be individual I-2, and the sixth person in the fourth generation would be IV-6.

Once the family members are properly arranged, important medical facts can be added. Proper interpretation of the pedigree is dependent upon obtaining accurate information about each individual in a pedigree. The first step in pedigree analysis is to observe the number and relationships of all individuals who express the same or similar clinical features. From this, it should be possible to determine if the disorder is dominant or recessive, autosomal or X-linked by looking for the typical patterns of inheritance. For example, an autosomal disease can usually be distinguished by seeing male-to-male transmission of the **mutation**, but since males pass only the **Y chromosome** to their sons, there should never be father to son transmission of an X-linked **gene**. Males will be most commonly affected in an X-linked disease, whereas males and females should be equally affected in autosomal disorders. In general, a dominant disease will be seen in approximately half of the individuals in each generation, but recessives occur very rarely. If the mutation is in the mitochondrial **genome**, affected mothers will pass the trait to all of their children, but none of the offspring of an affected male should have the disease.

Once the inheritance pattern of the disorder is determined, the status of family members in the pedigree can be evaluated. By carefully observing the position of affected individuals, mutation carriers may be identified. From this data, the risk of **carrier** status for other family members or the chance that a couple may have an affected child can be estimated.

Pedigrees are also maintained for many animals, though the purpose of pedigree analysis is somewhat different. The data contained in the pedigree are generally utilized to select individuals with specific characters for breeding purposes. Animals with unfavorable traits are eliminated from consideration so that the next generation will include individuals with more of the preferable traits. For each **species**, the characters of choice will vary. In the thoroughbred world, pedigree analysis tries to combine speed with stamina and a will to win that will yield winning racehorses. For cows, sheep, and pigs, such characteristics as high milk production, higher muscle content, or better wool are desirable. Even some plants have pedigrees as researchers strive to find drought and pest resistant species with high crop yields.

In medicine, pedigree analysis is an essential part of a complete medical work up for a genetic disease. The information obtained is an important aid in understanding the disorder and providing the best counseling to the family. For other plant and animal species, pedigree analysis is also a useful tool, though the goal is usually for gene **selection** rather than risk assessment.

See also Autosomes; Carrier; Dominance relations; Genetic Counseling; Population genetics; Recessive genes and traits

PENETRANCE • *see* GENE PENETRANCE

PETERMANN, MARY LOCKE (1908-1975)

American biochemist

Mary Locke Petermann isolated and worked out the structure of animal **ribosomes**, organelles that are now known as the sites of **protein synthesis** in cells. She began her original investigation of the particles (for a time they were known as "Petermann's particles") because they were interfering with her studies of **DNA** and **RNA**. Her work was fundamental and pioneering; her continued work established the importance of

ions in stabilizing ribosomes and elucidated ribosomal transformations.

Ribosomes are the organelle that are the site of protein synthesis in a **cell**. Petermann's research isolated several types of ribosomes and clarified their properties. She also pioneered the study of antibodies. This research later led to Rodney Porter winning a Nobel Prize in 1972 for his work on the structure of immunoglobulins.

Peterman was born in Laurium, Michigan, on February 25, 1908, one of three children and the only daughter of Albert Edward and Anna Mae Grierson Petermann. Her mother was a graduate of Ypsilanti State Teachers' College (now Eastern Michigan University). Her father, a graduate of Cornell University, became a lawyer for Calumet and Hecla Consolidated Copper Company in Calumet, Michigan, after World War I; he later was president and general manager. The Petermann family lived in a large company house and enjoyed high status in the community.

After graduating from Calumet High School in 1924, Petermann spent a year at a Massachusetts preparatory school before entering Smith College. In 1929, she graduated from Smith with high honors in chemistry and membership in Phi Beta Kappa. After a year at Yale University as a technician, she spent four years working at the Boston Psychopathic Hospital, investigating the acid-base balance of mental patients. In 1936 she entered the University of Wisconsin; she received a Ph.D. degree in physiological chemistry in 1939, with a thesis project on the role of the adrenal cortex in ion regulation.

In 1939 Petermann became the first woman chemist on the staff of the Department of Physical Chemistry at the University of Wisconsin. She remained at Wisconsin as a post-doctoral researcher until 1945. During these six years she and Alwin M. Pappenheimer began to investigate the physical chemisty of proteins. Petermann discovered what were at first called "Petermann's particles" but were named ribosomes at a meeting of the Biophysical Society in 1958.

After leaving the University of Wisconsin in 1945, Petermann accepted the position of research chemist at Memorial Hospital in New York City to explore the role of plasma proteins in **cancer**. In 1946, she was appointed Finney-Howell Foundation fellow at the newly founded Sloan-Kettering Institute, where she explored the role of nucleoproteins in cancer. She became an associate member of the institute in 1960, the first woman member in 1963, and member emeritus in 1973 when she retired. Concurrent with her work at Sloan-Kettering, she also taught biochemistry in the Sloan-Kettering Division, Graduate School of Medical Sciences, Cornell University. In 1966, she became the first woman appointed a full professor at Cornell. She authored or co-authored almost 100 scientific papers.

As the Sloan Award recipient in 1963, Petermann was honored for what the accompanying citation explained was her "many basic and distinguished contributions to the knowledge of the relevance of proteins and nucleoproteins in abnormal growth. An even greater contribution has been her fundamental work on the nature of the cell ribosome." Petermann used her award money to work for a year in the Swedish laboratory of Nobel laureate Arne Tiselius. She also lectured in several European countries, including England and France. In 1966 she received the Garvan Medal of the American Chemical Society, which honors contributions made by women scientists, an honorary doctorate from Smith College, and the Distinguished Service Award from the American Academy of Achievement.

Petermann never married. In 1974, the year before her death, she organized the Memorial Sloan-Kettering Cancer Center Association for Professional Women and served as its first president. She died in Philadelphia on December 13, 1975, of intestinal cancer. In 1976 the Educational Foundation of the Association for Women in Science named one of its graduate scholarships in her honor.

See also DNA (deoxyribonucleic acid); Messenger RNA (mRNA); Nucleic acid; Organelles and subcellular genetics; Transfer RNA (tRNA); Translation

PHAGE GENETICS

Bacteriophages, viruses that infect **bacteria**, are useful in the study of how genes function. The attributes of bacteriophages include their small size and simplicity of genetic organization.

The most intensively studied **bacteriophage** is the phage called lambda. It is an important model system for the latent infection of mammalian cells by retroviruses, and it has been widely used for **cloning** purposes. Lambda is the prototype of a group of phages that are able to infect a **cell** and redirect the cell to become a factory for the production of new **virus** particles. This process ultimately results in the destruction of the host cell (lysis). This process is called the lytic cycle. On the other hand, lambda can infect a cell, direct the integration of its **genome** into the **DNA** of the host, and then reside there. Each time the host genome replicates, the viral genome undergoes **replication**, until such time as it activates and produces new virus particles and lysis occurs. This process is called the lysogenic cycle.

Lambda and other phages, which can establish lytic or lysogenic cycles, are called temperate phages. Other examples of temperate phages are bacteriophage mu and P1. Mu inserts randomly into the host **chromosome** causing insertional mutations where intergrations take place. The P1 genome exists in the host cell as an autonomous, self-replicating plasmid.

Phage **gene expression** during the lytic and lysogenic cycles uses the host **RNA polymerase**, as do other viruses. However, lambda is unique in using a type of regulation called antitermination.

As host RNA polymerase transcribes the lambda genome, two proteins are produced. They are called cro (for "control of repressor and other things") and N. If the lytic pathway is followed, **transcription** of the remainder of the viral genes occurs, and assembly of the virus particles will occur. The N protein functions in this process, ensuring that transcription does not terminate.

The path to lysogeny occurs differently, involving a protein called cI. The protein is a repressor and its function is to bind to **operator** sequences and prevent transcription. Expression of cI will induce the phage genome to integrate into the host genome. When integrated, only the cI will be produced, so as to maintain the lysogenic state.

The virus adopts the lytic or lysogenic path early following infection of the host bacterium. The fate of the viral genetic material is governed by a **competition** between the cro and cI proteins. Both can bind to the same operator region. The region has three binding zones—cro and cI occupy these zones in reverse order. The protein, which is able to occupy the preferred regions of the operator first, stimulates its further synthesis and blocks synthesis of the other protein.

Analysis of the **genetics** of phage activity is routinely accomplished using a plaque assay. When a phage infects a lawn or layer of bacterial cells growing on a flat surface, a clear zone of lysis can occur. The clear area is called a plaque.

Aside from their utility in the study of **gene** expression, phage genetics has been put to practical use as well. Cloning of the human insulin gene in bacteria was accomplished using a bacteriophage as a vector. The phage delivered to the bacterium a recombinant plasmid containing the insulin gene. M13, a single-stranded filamentous DNA bacteriophage, has long been used as a cloning vehicle for molecular biology. It is also valuable for use in DNA **sequencing**, because the viral particle contains single-stranded DNA, which is an ideal template for sequencing. T7 phage, which infects *Escherichia coli*, and some strains of *Shigella* and *Pasteurella,* is a popular vehicle for cloning of complimentary DNA. Also, the T7 promoter and RNA polymerase are in widespread use as a system for regulatable or high-level gene expression.

See also Polymerase (DNA and RNA)

PHAGE THERAPY

Bacteriophage are well suited to deliver therapeutic payloads (i.e., deliver specific genes into a host organism). Characteristic of viruses, they require a host in which to make copies of their genetic material, and to assemble progeny **virus** particles. Bacteriophage are more specific in that they infect solely **bacteria**.

The use of phage to treat bacterial infections was popular early in the twentieth century, prior to the mainstream use of **antibiotics**. Doctors used phages as treatment for illnesses ranging from cholera to typhoid fevers. Sometimes, phage-containing liquid was poured into the wound. Oral, aerosol, and injection administrations were also used. With the advent of antibiotic therapy, the use of phage was abandoned. But now, the increasing resistance of bacteria to antibiotics has sparked a reassessment of phage therapy.

Lytic bacteriophage, which destroy the bacterial **cell** as part of their infectious process, are used in therapy. Much of the focus in the past fifteen years has been on nosocomial—

hospital acquired—infections, where multi-drug-resistant organisms have become a particularly lethal problem.

Bacteriophage offer several advantages as therapeutic agents. Their target specificity causes less disruption to the normal host bacterial flora, some **species** of which are vital in maintaining the ecological balance in various areas of the body, than does the administration of a relatively less specific antibiotic. Few side effects are evident with phage therapy, particularly allergic reactions, such as can occur to some antibiotics. Large numbers of phage can be prepared easily and inexpensively. Finally, for localized uses, phage have the special advantage that they can continue multiplying and penetrating deeper as long as the infection is present, rather than decreasing rapidly in concentration below the surface like antibiotics.

In addition to their specific lethal activity against target bacteria, the relatively new field of **gene therapy** has also utilized phage. Recombinant phage, in which carry a bit of non-viral genetic material has been incorporated into their **genome**, can deliver the recombinant **DNA** or **RNA** to the recipient genome. The prime use of this strategy to date has been the replacement of a defective or deleterious host **gene** with the copy carried by the phage. Presently, however, technical safety issues and ethical considerations have limited the potential of phage genetic therapy.

See also Bacteriophage; Gene insertion; Gene therapy

PHARMACOGENETICS

One of the newest subspecialties of **genetics** is pharmacogenetics, the science that deals with the relationship between inherited genes and the ability of the body to metabolize drugs. Medicine today relies on the use of therapeutic drugs to treat disease, but one of the longstanding problems has been the documented variation in patient response to drug therapy. The recommended dosage is usually established at a level shown to be effective in 50% of a test population, and based on the patient's initial response, the dosage may be increased, decreased, or discontinued. In rare situations, the patient may experience an adverse reaction to the drug and be shown to have a pharmacogenetic disorder. The unique feature of this group of diseases is that the problem does not occur until after the drug is given, so a person may have a pharmacogenetic defect and never know it if the specific drug required to trigger the reaction is never administered.

Consider the case of a 35-year-old male who is scheduled for surgical repair of a hernia. The patient is otherwise in excellent health and has no family history of any serious medical problems. After entering the operating theater, an inhalation anesthetic and/or muscle relaxant is administered to render the patient unconscious. Unexpectedly, there is a significant increase in body temperature, and the patient experiences sustained muscle contraction. If this condition is not reversed promptly, it can lead to death. Anesthesiologists are now familiar with this type of reaction. It occurs only rarely,

but it uniquely identifies the patient as having malignant hyperthermia, a rare autosomal dominant disorder that affects the body's ability to respond normally to anesthetics. Once diagnosed with malignant hyperthermia, it is quite easy to avoid future episodes by simply using a different type of anesthetic when surgery is necessary, but it often takes one negative, and potentially life threatening, experience to know the condition exists.

An incident that occurred in the 1950s further shows the diversity of pharmacogenetic disorders. During the Korean War, service personnel were deployed in a region of the world where they were at increased risk for malaria. To reduce the likelihood of acquiring that disease, the antimalarial drug primaquine was administered prophylactically. Shortly thereafter, approximately 10% of the African-American servicemen were diagnosed with acute anemia and a smaller percentage of soldiers of Mediterranean ancestry showed a more severe hemolytic anemia. Investigation revealed that the affected individuals had a **mutation** in the glucose 6-phosphate dehydrogenase (G6PD) **gene**. Functional G6PD is important in the maintenance of a balance between oxidized and reduced molecules in the cells, and, under normal circumstances, a mutation that eliminates the normal enzyme function can be compensated for by other cellular processes. However, mutation carriers are compromised when their cells are stressed, such as when the primaquine is administered. The system becomes overloaded, and the result is oxidative damage of the red blood cells and anemia. Clearly, both the medics who administered the primaquine and the men who took the drug were unaware of the potential consequences. Fortunately, once the drug treatment was discontinued, the individuals recovered.

Drugs are essential to modern medical practice, but, as in the cases of malignant hyperthermia and G6PD deficiency, it has become clear that not all individuals respond equally to each drug. Reactions can vary from positive improvement in the quality of life to life threatening episodes. Annually, in the United States, there are over two million reported cases of adverse drug reactions and a further 100,000 deaths per year because of drug treatments. The **Human Genome Project** and other research endeavors are now providing information that is allowing a better understanding of the underlying causes of pharmacogenetic anomalies with the hope that eventually the number of negative episodes can be reduced.

In particular, research on one enzyme family is beginning to revolutionize the concepts of drug therapy. The cytochrome P450 system is a group of related **enzymes** that are key components in the metabolic conversion of over 50% of all currently used drugs. Studies involving one member of this family, CYP2D6, have revealed the presence of several polymorphic genetic variations (poor, intermediate, extensive, and ultra) that result in different clinical phenotypes with respect to drug metabolism. For example, a poor metabolizer has difficulty in converting the therapeutic drug into a useable form, so the unmodified chemical will accumulate in the body and may cause a toxic overdose. To prevent this from happening, the prescribed dosage of the drug must be reduced. An ultra metabolizer, on the other hand, shows exceedingly rapid breakdown of the drug to the point that the substance may be destroyed so quickly that therapeutic levels may not be reached, and the patient may therefore never show any benefit from treatment. In these cases, switching to another type of drug that is not associated with CYP2D6 metabolism may prove more beneficial. The third phenotypic class, the extensive metabolizers, is less extreme than the ultra metabolism category, but nevertheless presents a relatively rapid turnover of drug that may require a higher than normal dosage to maintain a proper level within the cells. And, finally, the intermediate phenotype falls between the poor and extensive categories and gives reasonable metabolism and turnover of the drug. This is the group for whom most recommended drug dosages appear to be appropriate. However, the elucidation of the four different metabolic classes has clearly shown that the usual "one size fits all" recommended drug dose is not appropriate for all individuals. In the future, it will become increasingly necessary to know the patient's metabolic phenotype with respect to the drug being given to determine the most appropriate regimen of therapy for that individual.

At the present time, pharmacogenetics is still in its infancy with its full potential yet to be realized. Based on current studies, it is possible to envision many different applications in the future. In addition to providing patient specific drug therapies, pharmacogenetics will aid in the clinician's ability to predict adverse reactions before they occur and identify the potential for drug addiction or overdose. New tests will be developed to monitor the effects of drugs, and new medications will be found that will specifically target a particular genetic defect. Increased knowledge in this field should provide a better understanding of the metabolic effects of food additives, work related chemicals, and industrial byproducts. In time, these advances will improve the practice of medicine and become the standard of care.

PHENOTYPE • *see* GENOTYPE AND PHENOTYPE

PHENOTYPE AND PHENOTYPIC VARIATION

The word phenotype refers to the observable characters or attributes of individual organisms, including their morphology, physiology, behavior, and other traits. The phenotype of an organism is limited by the boundaries of its specific genetic complement (genotype), but is also influenced by environmental factors that impact the expression of genetic potential.

All organisms have unique genetic information, which is embodied in the particular **nucleotide** sequences of their **DNA** (**deoxyribonucleic acid**), the genetic biochemical of almost organisms, except for viruses and **bacteria** that utilize **RNA** as their genetic material). The genotype is fixed within an individual organism but is subject to change (mutations) from one generation to the next due to low rates of natural or spontaneous **mutation**). However, there is a certain degree of devel-

opmental flexibility in the phenotype, which is the actual or outward expression of the genetic information in terms of anatomy, behavior, and **biochemistry**. This flexibility can occur because the expression of genetic potential is affected by environmental conditions and other circumstances.

Consider, for example, genetically identical geranium seeds, with a fixed complement of genetic information, that produce geranium plants. If one geranium seed is grown under well-watered, fertile, non-crowded conditions, it will develop into a relatively tall, robust, and vigorously flowering specimen. However, if a genetically identical seed is grown under drier, less fertile, more competitive conditions, its productivity and growth is stunted. Such varying growth patterns of the same genotype are referred to as phenotypic plasticity. Some traits of organisms, however, are fixed genetically, and their expression is not affected by environmental conditions. For instance, the flower colour of individual geraniums (which can be white, red, or pink) is genetically fixed, and does not vary with the conditions of cultivation. Moreover, the ability of **species** to exhibit phenotypically plastic responses to environmental variations is itself, to a substantial degree, genetically determined. Therefore, phenotypic plasticity reflects both genetic capability and varying expression of that capability, depending on circumstances.

Phenotypic variation is essential for **evolution**. Without a discernable difference among individuals in a population there are no genetic **selection** pressures acting to alter the variety and types of alleles (forms of genes) present in a population. Accordingly, genetic mutations that do not result in phenotypic change are essentially masked from **evolutionary mechanisms**.

Phenetic similarity results when phenotypic differences among individuals are slight. In such cases, it may take a significant alteration in environmental conditions to produce significant selection pressure that results in more dramatic phenotypic differences. Phenotypic differences lead to differences in **fitness** and affect adaptation.

See also Adaptation and fitness; Biodiversity; Genetic variation; Genome; Genotype and phenotype; Phylogenetic; Variation of inherited characteristics; Wild-type

PHENYLKETONURIA

Phenylketonuria (PKU) is an inherited metabolic disorder in which an enzyme (phenylalanine hydroxylase) that is crucial to the appropriate processing of the **amino acid** phenylalanine, is absent or deficient. Normally, phenylalanine is converted to tyrosine in the body. When phenylalanine cannot be broken down, it accumulates in excess quantities throughout the body, causing mental retardation and other neurological complications. Treatment is usually started during babyhood; delaying such treatment results in a significantly lowered intelligence quotient (IQ) by age one. Because tyrosine is involved in the production of melanin (pigment), people with PKU usually have lighter skin and hair than other family members.

PKU is an autosomal recessive disorder, and is caused by mutations in both alleles of the **gene** responsible for phenylalanine hydroxylase, found on **chromosome** 12. Parents, both with the recessive gene **mutation** responsible for PKU, have a 25% chance with each pregnancy of producing a child affected with PKU. In the general population, the incidence of an adult **carrier** of PKU is approximately one in fifty people. PKU occurs in approximately one out of every 10,000 to 15,000 Caucasian or Asian births in the United States. The incidence of PKU in the African–American population is much less. Treatment involves a strict diet low in phenylalanine (warnings aimed at people with PKU can be found on cans of diet drinks containing the artificial sweetener aspartame, which is made from phenylalanine). **Gene therapy** offers the potential for permanently correcting the enzyme deficiencies of PKU and other **inborn errors of metabolism**. Current research focuses on the use of genetically altered adenoviruses (one of the causes of the common cold) to correct certain other inherited liver—based metabolic diseases. Understanding of the exact mechanisms of the neurological complications associated with PKU are, however, little understood, and knowledge of the precise genetic mutations responsible for PKU have yet to yield significant advances in treatment or prevention of PKU. Because it is vital to begin diet treatment immediately, most nations in the developed world require all that all infants be tested for the disease within the first week of life.

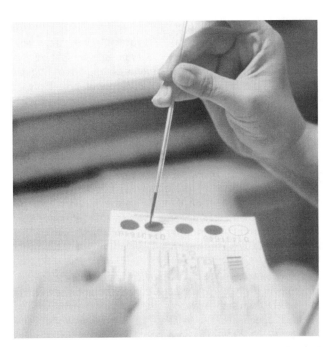

Technician performing test for phenylketonuria (PKU). *Custom Medical Stock Photo. Reproduced by permission.*

PHILOSOPHY OF GENETICS: NEO-DARWINISM AND THE MODERN SYNTHESIS

Neo-Darwinism is an attempt to reconcile gradual **evolution** by means of **natural selection** ("survival of the fittest") as proposed by **Charles Robert Darwin**, with developments in the field of **genetics**.

Darwinism asserted that evolution is a gradual process in which minute changes in organisms accumulate over time and eventually result in the appearance of new **species**. This transmutation was the result of breeding that favored the **replication** of specific traits that aided a species' overall adaptability, or "fitness."

However, Darwin only speculated about the means by which traits were passed on from generation to generation. Like many of his naturalist contemporaries, Darwin thought that paternal characteristics were somehow transmitted through the blood. This reliance on what was termed "blood theory" led Darwin to further conclude that unexpressed traits and variants could be diluted out of a population. This wrongly conceived explanation for the mechanisms of **heredity** became an obstacle to the continued acceptance of Darwinian evolution as it was originally proposed.

Seven years after the publication of Darwin's work on evolution, *On the Origin of Species*, German monk and geneticist, **Gregor Mendel**, proposed a new theory of biological **inheritance**, or heredity. Mendel's theory states that genetic material is transmitted as units. Each parent passes to their offspring a set of genes (units that represents each inheritable characteristic) that do not mix or blend, but remain separate and whole. Some genes are expressed, while others are dormant. Genes are thus unable to be diluted. Mendel's research remained in obscurity for three decades, but upon its rediscovery it challenged the prevailing interpretation and acceptance of Darwinism.

The scientific community in the late eighteenth century did not seek to actively discredit the notion of organic evolution that Darwin proposed. Rather, they criticized the mechanisms by which evolution was accomplished. Advocates of Mendelian genetics argued that the continual and infinitesimal transmutations central to Darwinism were antithetical to more broad changes observed by Mendel. Furthermore, geneticists were concerned with the seemingly infinite possibilities of variation that would result from constant, infinitesimal changes.

Also competing with Darwinism in he late 1800s was another theory of organic evolution known as neo-Lamarkism. The theory borrowed the concept of use and disuse, championed by French naturalist Jean-Baptiste Lamarck (1744–1829), to explain an organism's development of habits and adaptation to the environment. Neo-Lamarckism denied the role of natural **selection** as a means of electing traits of maximum adaptiveness, and favored the view that the environment itself directly acted on organisms, resulting in acquired traits that were also inheritable. Through his pioneering germ-plasm research German biologist August Weismann (1834–1914) proved that such traits of habit were not able to be passed on to offspring. Neo-Lamarckism largely fell out of favor, but the need for some sort of reconciliation between genetics and evolution theory was made clear. Again, scientists turned their attention to Darwinism. In light of his own research, Weismann became a champion of natural selection—but replaced Darwin's concepts of heredity with those of Mendel. Weismann and his followers became known as neo-Darwinists.

The actual synthesis of Darwinian evolution and genetic change was cemented by the work of Russian-born American naturalist and experimental geneticist **Theodosius Dobzhansky**. His book, *Genetics and the Origin of the Species*, published in 1937, presented a comprehensive explanation of evolution through genetic mechanisms. By 1960, more than a century after Darwin first introduced his theories, natural selection by genetic mechanisms was the most widely accepted theory of evolution in the scientific community.

Despite its general acceptance, many questions have arisen over the past 40 years regarding the comprehensiveness of neo-Darwinism. Advances in molecular biology have yielded conclusive observations of the microevoluntionary processes, changes in a **gene pool** over successive generations. However, macroevolution, or evolutionary trends on a large scale, is more problematic to research. **DNA** and enzyme analysis allows scientists to quantitatively study species divergence, the extent of **genetic variation** in populations, and the degree of genetic change in species formation. In 1968, Japanese geneticist **Motoo Kimura** (1924 –) advanced the neutrality theory, which states that at the foundation levels of life, such as DNA and protein sequences, some changes are "neutral" and do not affect the overall adaptiveness or function of an organism. He further theorized that the degree of difference between various amino acids sequences could estimate the time at which species diverged. This is known as the "molecular clock".

During the 1970s and '80s physical anthropologists made numerous discoveries of hominid fossils and had increasing success charting the progress of the evolution of man. Such discoveries in the natural sciences have made it gradually became that the molecular clock is not exact; nevertheless, it has continued to provide a reliable source of evidence for reconstructing a rough history of evolution.

Present-day the techniques of DNA **cloning** and **sequencing** have provided a new and more powerful means of investigating evolution at the molecular level and further assessing challenges levied against neo-Darwinism, but the exact relationship between microevolution and macroevolution remains unknown.

See also Darwinism; Selection; History of genetics: Ancient and classical views of heredity; History of genetics: Modern genetics

PHYLOGENETIC

The term phylogenetic (or phyletic) is used in reference to relationships among groups of organisms based on their shared evolutionary history (that is, on their **phylogeny**). One of the major goals of biological science (and of paleontology, the study of extinct organisms) is to understand the phylogenetic relations among living and extinct **species**, as well among higher taxa (this term refers to any taxonomic group, such as species, genus, family, class, or phylum). Phylogenetic knowledge is crucial to understanding the evolutionary history of life on Earth, including that of our own species, *Homo sapiens*.

Phylogenetic relationships among taxa are examined using various kinds of data, including information on the comparative anatomy of fossil species (and of higher taxa), and the anatomy, **biochemistry**, behavior, and ecology of living organisms. Complex sets of data on these sorts of variables are analyzed using certain mathematical procedures (such as cluster analysis and ordination) to determine groups of taxa based on their similarity and dissimilarity. In such analyses, a basic assumption is that taxa that are similar in terms of character variables (i.e., in their anatomy, biochemistry, behavior, and/or ecology) are likely to be more closely related than taxa that are dissimilar. However, it must be borne in mind that convergent **evolution** can result in non-related taxa having similar characters. In addition, homologous characters can evolve if closely related taxa are subject to different regimes of **natural selection** (i.e., a particular attribute may differ greatly between related taxa, as is the case of the wings of birds and the forelimbs of their reptilian ancestors).

The phylogeny of some groups of organisms has been worked out rather well by paleontologists. For example, an early progenitor of the lineage of modern horses was *Eohippus*, a relatively small, four-toed animal that lived in the Eocene epoch about 55 million years ago. The somewhat larger, three-toed *Mesohippus* lived during the Oligocene epoch (about 35 million years ago), apparently having evolved from the *Eohippus* lineage. Next to occur in this phylogenetic series was the larger, three-toed *Merychippus* of the Miocene epoch (25 million years ago), followed by the bigger, one-toed *Pliohippus* of the Pliocene (5 million years ago). Finally, this phylogenetic lineage led to the evolution of the even larger, one-toed, modern horses, of which there are several species, including the horse (*Equus caballus*), donkey (*E. asinus*), and zebras (including *E. zebra*).

See also Archaeogenetics: Genetic studies of ancient peoples; Evolutionary mechanisms; Phenotype

PHYLOGENY

Phylogeny is the inferred evolutionary history of a group of organisms. Paleontologists are interested in understanding life through time—not just at one time in the past or present, but over long periods of past time. Before they can attempt to reconstruct the forms, functions, and lives of once-living organisms, paleontologists have to place these organisms in context. The relationships of those organisms to each other are based on the ways they have branched out, or diverged, from a common ancestor. A phylogeny is usually represented as a

phylogenetic tree or cladogram, which are like genealogies of **species**.

Phylogenetics, the science of phylogeny, is one part of the larger field of systematics, which also includes taxonomy. Taxonomy is the science of naming and classifying the diversity of organisms. Not only is phylogeny important for understanding paleontology (study of fossils), but paleontology in turn contributes to phylogeny. Many groups of organisms are now extinct, and without their fossils we would not have as clear a picture of how modern life is interrelated.

There is an amazing diversity of life, both living and extinct. For scientists to communicate with each other about these many organisms, there must also be a classification of these organisms into groups. Ideally, the classification should be based on the evolutionary history of life, such that it predicts properties of newly discovered or poorly known organisms.

Phylogenetic systematics is an attempt to understand the evolutionary interrelationships of living things, trying to interpret the way in which life has diversified and changed over time. While classification is primarily the creation of names for groups, systematics goes beyond this to elucidate new theories of the mechanisms of **evolution**.

Cladistics is a particular method of hypothesizing relationships among organisms. Like other methods, it has its own set of assumptions, procedures, and limitations. Cladistics is now accepted as the best method available for phylogenetic analysis, for it provides an explicit and testable hypothesis of organismal relationships.

The basic idea behind cladistics is that members of a group share a common evolutionary history, and are "closely related," more so to members of the same group than to other organisms. These groups are recognized by sharing unique features which were not present in distant ancestors. These shared derived characteristics are called synapomorphies. Synapomorphies are the basis for cladistics.

In a cladistic analysis, one attempts to identify which organisms belong together in groups, or clades, by examining specific derived features or characters that those organisms share. For example, if a genus of plants has both red flowered and white flowered species, then flower color might be a useful character for determining the evolutionary relationships of those plants. If it were known that the white flowered form arose from the previously existing red flowered form (i.e., through a **mutation** that prevents formation of the red pigment), then it could be inferred that all of the white colored species arose from a single red-colored ancestor. Characters that define a clade (e.g., white flower color in the example above) are called synapomorphies. Characters that do not unite a clade because they are primitive (e.g., red flower color) are called plesiomorphies.

In a cladistic analysis, it is important to know which character states are primitive and which are derived (that is, evolved from the primitive state). A technique called outgroup comparison is commonly used to make this determination. In outgroup comparison, the individuals of interest (the ingroup) are compared with a close relative. If some of the individuals

of the ingroup possess the same character state as the outgroup, then that character state is assumed to be primitive. In the example discussed above, the outgroup has red flowers, so white is the derived state for flower color.

There are three basic assumptions in cladistics:

- 1. Any group of organisms are related by descent from a common ancestor.
- 2. There is a bifurcating pattern of cladogenesis.
- 3. Change in characteristics occurs in lineages over time.

The first assumption is a general assumption made for all evolutionary biology. It essentially means that life arose on earth only once, and therefore all organisms are related in one way or another. Because of this, scientists can take any collection of organisms and determine a meaningful pattern of relationships, provided they have the right kind of information.

The second assumption is that new kinds of organisms may arise when existing species or populations divide into exactly two groups. The final assumption, that characteristics of organisms change over time, is the most important assumption in cladistics. It is only when characteristics change that different lineages or groups are recognized. The convention is to call the "original" state of the characteristic plesiomorphic and the "changed" state apomorphic. The terms primitive and derived have also been used for these states, but they are often avoided by cladists, since those terms have been abused in the past.

Cladistics is useful for creating systems of classification. It is now the most commonly used method to classify organisms because it recognizes and employs evolutionary theory. Cladistics predicts the properties of organisms. It produces hypotheses about the relationships of organisms in a way that makes it possible to predict properties of the organisms. This can be especially important in cases when particular genes or biological compounds are being sought. Such genes and compounds are being sought all the time by companies interested in improving crop yield or disease resistance, and in the search for medicines. Only an hypothesis based on evolutionary theory, such as cladistic hypotheses, can be used for these endeavors.

As an example, consider the plant species *Taxus brevifolia*. This species produces a compound, taxol, which is useful for treating **cancer**. Unfortunately, large quantities of bark from this rare tree are required to produce enough taxol for a single patient. Through cladistic analysis, a phylogeny for the genus *Taxus* has been produced that shows *Taxus cuspidata*, a common ornamental shrub, to be a very close relative of *T. brevifolia*. *Taxus cuspidata*, then, may also produce large enough quantities of taxol to be useful. Having a classification based on evolutionary descent will allow scientists to select the species most likely to produce taxol.

Cladistics helps to elucidate mechanisms of evolution. Unlike previous systems of analyzing relationships, cladistics is explicitly evolutionary. Because of this, it is possible to examine the way characters change within groups over time—the direction in which characters change, and the relative frequency with which they change. It is also possible to compare the descendants of a single ancestor and observe patterns of origin and extinction in these groups, or to look at relative size and diversity of the groups. Perhaps the most important feature of cladistics is its use in testing long-standing hypotheses about adaptation.

See also Adaptation and fitness; Biochemical analysis techniques; Evolutionary mechanisms

PLANT BREEDING AND CROP IMPROVEMENT

Plant breeding began when early humans saved seeds and planted them. The cultural change from living as nomadic hunter-gatherers, to living in more settled communities, depended on the ability to cultivate plants for food. Present knowledge indicates that this transition occurred in several different parts of the world, about 10,000 years ago.

Today, there are literally thousands of different cultivated varieties (cultivars) of individual **species** of crop plants. As examples, there are more than 4,000 different peas (*Pisum sativum*), and more than 5,000 grape cultivars, adapted to a wide variety of soils and climates.

The methods by which this diversity of crops was achieved were little changed for many centuries, basically requiring observation, **selection**, and cultivation. However, for the past three centuries most new varieties have been generated by deliberate cross-pollination, followed by observation and further selection. The science of **genetics** has provided a great deal of information to guide breeding possibilities and directions. Most recently, the potential for plant breeding has advanced significantly, with the advent of methods for the incorporation of genes from other organisms into plants via recombinant DNA-techniques. This capacity is broadly termed "genetic engineering." These new techniques and their implications have given rise to commercial and ethical controversies about "ownership," which have not yet been resolved.

The genetic discoveries of Gregor Mendel with pea plants, first published in 1866, were revolutionary, although Mendel's work remained obscure until translated from German into English by **William Bateson** in 1903. Nevertheless, the relationship between pollen lodging on the stigma and subsequent fruit production was realized long before Mendel's work. The first **hybrid** produced by deliberate pollen transfer is credited to Thomas Fairchild, an eighteenth-century, English gardener. He crossed sweet William with the carnation in 1719, to produce a new horticultural plant.

Towards the end of that century, Thomas Andrew Knight, another Englishman, demonstrated the practical value of cross-pollination on an unprecedented scale. He produced hybrid fruit trees by cross-pollination, and then grafted shoots of their seedlings onto established, compatible root stalks. This had the effect of greatly shortening the time until fruit production, so that the horticultural success of the hybridization could be evaluated. After selecting the best fruit, the

hybrid seeds could be planted, and the process of grafting the seedlings and selection could be continued. The best hybrids, which were not necessarily stable through **sexual reproduction**, could be propagated by grafting. Thomas Knight was also responsible for the first breeding of wrinkled-seeded peas, the kind that provided Mendel with one of his seven key characters (round being dominant, with one allele sufficient for expression; wrinkled being recessive, requiring two copies of the allele for expression).

The concept of "diluting" hybrids by crossing them back to either parent also developed in the latter part of the nineteenth century. This strategy was introduced to ameliorate undesirable characters that were expressed too strongly. In genetic terms, there are two kinds of back-crossing. When one parent of a hybrid has many recessive characters, these are masked in the F_1 (first filial) hybrid generation by dominant alleles from the other parent. However, a cross of the F_1 hybrid with the recessive parent will allow the complete range of **genetic variation** to be expressed in the F_2 progeny. This is termed a test cross. A cross of the F_1 to the parent with more dominant characters is termed a back cross.

The broad aims of current plant breeding programs have changed little from those of the past. Improvements in yield, quality, plant hardiness, and pest resistance are actively being sought. In addition, the ability of plants to survive increasing intensities of ultraviolet radiation, due to damage in the ozone layer, and to respond favorably to elevated atmospheric concentrations of carbon dioxide are being assessed. To widen the available **gene** pools, collections of cultivars and wild relatives of major crop species have been organized at an international level. The United Nations' Food and Agriculture Organization (FAO) supported the formation of the International Board for Plant Genetic Resources in 1974. However, many cultivars popular in the nineteenth century have already fallen into disuse and been lost. The need to conserve remaining "heritage" varieties has been taken up by associations of enthusiasts in many countries, such as the Seed Savers' Exchange in the United States

Genetically identical plants, or clones, have been propagated from vegetative cuttings for thousands of years. Modern **cloning** techniques are used extensively to select for cultivars with particular characteristics, since there are limits to what can be achieved through direct hybridization. Some individual species or groups of cultivars cannot be genetically crossed. Sometimes this is because of natural polyploidy, when plant cells carry extra copies of some or all of the chromosomes, or because of inversions of **DNA** within chromosomes. In cases where cross-fertilization has occurred, "embryo rescue" may be used to remove hybrid embryos from the ovules and culture them on artificial media.

Pollen mother-cells in the anthers of some species have been treated with colchicine, to generate nuclei with double the **haploid chromosome** number, thus producing **diploid** plants that are genetically-identical to the haploid pollen. The use of colchicine to induce polyploidy in dividing vegetative cells first became popular in the 1940s, but tetraploids generated from diploids tend to mask recessive alleles. Generating diploids from haploids doubles all of the existing recessive alleles, and thereby guarantees the expression of the recessive characters of the pollen source.

In other difficult cases, the barriers to sexual crossing can sometimes be overcome by preparing protoplasts from vegetative (somatic) tissues from two sources. This involves treatment with cell-wall degrading **enzymes**, after which the protoplasts are encouraged to fuse by incubation in an optimal concentration of polyethylene glycol. A successful fusion of protoplasts from the two donors produces a new protoplast that is a somatic hybrid. Using tissue cultures, such cells can, in some cases, be induced to develop into new plants.

Somatic fusion is of particular interest for characters related to the chloroplast or mitochondrion. These plastids contain some genetic information in their specific, non-nuclear DNA, which is responsible for the synthesis of a number of essential proteins. In about two-thirds of the higher plants, plastids with their DNA are inherited in a "maternal" fashion—the cytoplasm of the male **gamete** is discarded after fusion of the egg and **sperm cells**. In contrast, in the minority of plants with biparental **inheritance** of plastid DNA, or when fusion of somatic protoplasts occurs, there is a mixing of the plastids from both parents. In this way, there is a potential for new plastid-nucleus combinations.

For chloroplasts, one application of plastid fusion is in the breeding of resistance to the effects of triazine herbicides. For **mitochondria**, an application relevant to plant breeding is in the imposition of male sterility. This is a convenient character when certain plants are to be employed as female parents for a hybrid cross. The transfer of male-sterile cytoplasm in a single step can avoid the need for several years of backcrosses to attain the same condition. Somatic hybridization has been used successfully to transfer male sterility in rice, carrot, rapeseed (canola), sugar beet, and citrus. However, this character can be a disadvantage in maize, where male sterility simultaneously confers sensitivity to the blight fungus, *Helminthosporium maydis*. This sensitivity can lead to serious losses of maize crops.

Replicate plant cells or protoplasts that are placed under identical conditions of tissue culture do not always grow and differentiate to produce identical progeny (clones). Frequently, the genetic material becomes destabilized and reorganized, so that previously concealed characters are expressed. In this way, the tissue-culture process has been used to develop varieties of sugar cane, maize, rapeseed, alfalfa, and tomato that are resistant to the toxins produced by a range of parasitic fungi. This process can be used repeatedly to generate plants with multiple disease resistance, combined with other desirable characters.

The identification of numerous mutations affecting plant morphology has allowed the construction of genetic linkage maps for all major cultivated species. These maps are constantly being refined. They serve as a guide to the physical location of individual genes on chromosomes.

DNA **sequencing** of plant genomes has shown that **gene expression** is controlled by distinct "promoter" regions of DNA. It is now possible to position genes under the control of a desired promoter, to ensure that the genes are expressed in the appropriate tissues. For example, the gene for a bacterial toxin (Bt) (from *Bacillus thuringiensis*) that kills insect larvae might be placed next to a leaf-development promoter sequence, so that the toxin will be synthesized in any developing leaf. Although the toxin might account for only a small proportion of the total protein produced in a leaf, it is capable of killing larvae that eat the genetically-modified leaves.

Agrobacterium tumefaciens and *A. rhizogenes* are soil **bacteria** that infect plant roots, causing crown gall or "hairy roots" diseases. Advantage has been taken of the natural ability of *Agrobacterium* to transfer plasmid DNA into the nuclei of susceptible plant cells. *Agrobacterium* cells with a genetically-modified plasmid, containing a gene for the desired trait and a marker gene, usually conferring antibiotic resistance, are incubated with protoplasts or small pieces of plant tissue. Plant cells that have been transformed by the plasmid can be selected on media containing the antibiotic, and then cultured to generate new, transgenic plants.

Many plant species have been transformed by this procedure, which is most useful for dicotyledonous plants. The gene encoding Bt, as well as genes conferring resistance to viral diseases, have been introduced into plants by this method.

Two methods have been developed for direct gene transfer into plant cells—electroporation and biolistics. Electroporation involves the use of high-voltage electric pulses to induce pore formation in the membranes of plant protoplasts. Pieces of DNA may enter through these temporary pores, and sometimes protoplasts will be transformed as the new DNA is stably incorporated (i.e., able to be transmitted in mitotic **cell** divisions). New plants are then derived from cultured protoplasts. This method has proven valuable for maize, rice, and sugar cane, species that are outside the host range for vector transfer by *Agrobacterium*.

Biolistics refers to the bombardment of plant tissues with microprojectiles of tungsten coated with the DNA intended for transfer. Surprisingly, this works. The size of the particles and the entry velocity must be optimized for each tissue, but avoiding the need to isolate protoplasts increases the potential for regenerating transformed plants. Species that cannot yet be regenerated from protoplasts are clear candidates for **transformation** by this method.

In 1992, a tomato with delayed ripening became the first genetically-modified (GM) commercial food crop. More than 40 different GM crops are now being grown commercially. GM corn and cotton contain bacterial genes that kill insects and confer herbicide-resistance on the crops. GM squash contains viral genes that confer resistance to viruses. Potatoes carry the Bt gene to kill the Colorado potato beetle and a viral gene that protects the potato from a **virus** spread by aphids. Mauve-colored carnations carry a petunia gene required for making blue pigment. In many cases, GM crops result in increased yields and reduced use of pesticides. New research is focused on producing GM foods containing increased vitamins and human or animal vaccines.

GM crops are controversial. There is concern that the widespread dissemination of the Bt gene will cause insects to become resistant. It has been reported that pollen from Bt corn is toxic to the caterpillars of monarch butterflies. It also is possible that GM crops will interbreed with wild plants, resulting in "superweeds" resistant to herbicides. There is also concern that the antibiotic-resistance genes, used as markers for gene transfer, may be passed from the plants to soil microorganisms or bacteria in humans who eat the food. Finally, the possibility of allergic reactions to the new compounds in food exists. Many countries have banned the production and importation of GM crops.

See also Agricultural genetics; Plant genetics

PLANT GENETICS

Plants have been central to **genetics**, dating back to their use by Gregor Mendel to establish the laws of **inheritance** of genetic traits. Since the discovery of **deoxyribonucleic acid** (**DNA**) and the abilities to transform **species**, including plants, with DNA derived elsewhere—recombinant DNA technology—the use of plants as vehicles to express traits encoded by recombinant DNA has exploded.

A plant breeder attempts to assemble a combination of genes in a crop plant which will make it as useful and productive as possible. Desirable traits include higher yield, improved quality, pest or disease resistance, or tolerance to environmental extremes. Obtaining the desired mix of traits can be lengthy and difficult. The ability to introduce foreign DNA into a plant via recombinant DNA technology has provided the means for identifying and isolating genes controlling specific characteristics in one kind of organism, and for moving those genes into another quite different organism, which will then express those characteristics. Modern plant genetics augments natural genetic **recombination**, but expands the natural limitations.

Plant genetics relies upon the universal presence of DNA in the cells of all living organisms. For most organisms, the sequence of events by which the information encoded in the DNA is expressed as proteins operates in a similar fashion. Thus, even species that are very different share similar genetic mechanisms, allowing DNA from one organism to be processed by a very different organism. Plant genetics relies upon the selective DNA sequence recognition and cutting abilities of a battery of **enzymes** called restriction endonucleases, and the re-splicing ability of other enzymes called ligases. The action of these enzymes makes the introduction of **exogenous DNA** possible. Identification of a single **gene** is usually not sufficient—regulation of the gene and its protein product is a complex process involving genetic and environmental factors. Current researchers expend much effort to rapidly sequence and determine the functions of genes of the most important crop species. Once genes having candidate potential are identified, their regulation must be defined.

Transformation of plants occurs mainly by two methods. The first is called the gene gun method. In this method, DNA attached to gold beads is ballistically introduced into the plant tissue. It has been especially useful in transforming monocot species like corn and rice. The second method relies on a bacterium called *Agrobacterium tumefaciens*. The bacterial method exploits the ability of *A. tumefaciens* to infect a plant with a piece of its own DNA, which commands the plant's cellular machinery to produce more **bacteria**. The bacteria accomplish this infection because it possesses a plasmid called the Ti (tumor-inducing) plasmid. Substances released through a wound in the plant activate genes on the plasmid, directing entry of the plasmid into plant tissues.

Engineered **plasmids** containing the wound-response genes and the genes of commercial or research interest have proved to be a successful means of transforming plants. Transformed plants can be subsequently detected by their growth in the presence of an inhibitory compound (since the plasmid also contains a gene(s) encoding resistance to the compound). Whole plants can then be obtained, seeds produced, and trials in the lab and the field performed to evaluate the performance of the transformed plant variety.

A popular target for transgenic technology has been the production of genetically engineered crops resistant to a target insect pest or herbicide. In 1999, almost 100 million acres of these crops were planted. In that year, the acreage devoted to transgenic varieties was approximately half of the U.S. soybean crop and about 25% of the U.S. corn crop. Herbicide resistant crops contained a gene encoding a protein that can chemically alter the herbicide so as to nullify its killing effect. Of the insect resistant crops, Bt insect-resistance has proved effective and commercially popular. Bt is short for *Bacillus thuringiensis,* a soil bacterium whose spores contain a crystalline protein. In the insect gut, the protein breaks down to release a toxin. The toxin binds to and creates pores in the intestinal lining, which is ultimately lethal for the insect. The use of Bt as a pest control is not new. However, its introduction to plants via recombinant DNA technology has been a novel application of pest control.

The use of Bt transgenic crops appears to be reducing the need for chemical pesticides, at least for cotton. The situation for other crops, like corn, is as yet unclear.

The production of drugs in genetically altered plants, biopharming, may represent the next wave in agricultural **biotechnology**. Until now, efforts have mainly been directed at protecting crops from pests and improving the taste and nutrition of the so-called genetically modified foods. About 20 companies worldwide are working on producing biopharmaceuticals in plants. A few are in human clinical trials, including vaccines for hepatitis B contained in potatoes and an antibody to prevent tooth decay, produced in genetically altered tobacco plants. The vaccines have been engineered from subunits of the disease-causing organism, rather than from the whole organism. Thus, the chances of developing an infection from the administration of the vaccine are thought to be much less frequent than traditional vaccination technologies. The production of compounds of industrial relevance in plants is also being investigated.

See also Agricultural genetics

PLANT REPRODUCTION

Plant reproduction may be sexual, in which two parents produce a genetically different individual; or asexual, involving the propagation of plants that are genetically identical to the parent.

Sexual reproduction in plants involves the fusion of two **haploid** gametes (or microspores; each has one set of chromosomes, signified as 1n, as a result of **meiosis**, or reduction division). The male gametes are found in pollen, which is produced by the anthers of plant flowers. The female gametes are contained within ovules, which are in the ovary of the pistil of plant flowers. **Pollination** is the process whereby pollen is transferred to the stigma of the pistil. This can occur by pollen dispersed into the atmosphere, or transferred by a pollinating insect, such as a bee or moth. The pollen germinates on the stigma and grows pollen tubes which penetrate through the style and into the ovary where **fertilization** of the ovule occurs, forming a **diploid** (or 2n) **zygote**, which develops into a seed. The ripe seed is then dispersed from the parent plant. If it encounters suitable environmental conditions the seed germinates to develop into a new individual that is closely related to its parents, but genetically different from both of them..

Asexual propagation can occur in various ways. Many plants produce genetically identical copies of themselves, through a mechanism referred to as **asexual reproduction**. However, botanists more properly refer to this mechanism as "sexual propagation" or "vegetative propagation" because many hold that only sexual reproduction should be referred to as true reproduction, since this is the only kind of propagation that results in the production of new, genetically unique individuals.

Another type of asexual propagation occurs when plants develop underground stems (or rhizomes) which grow outward, or new shoots which grow upward to form new shoots that are genetically identical to the parent. One such example is the trembling aspen (*Populus tremuloides*), which sometimes develops entire stands of trees growing out of the ground as seemingly individual stems, but are actually genetically identical and interconnected below-ground. Another example is the strawberry (*Fragaria virginiana*), although this **species** has its vegetative runners above-ground.

Other plants develop bulbils on their stems, which can detach, fall to the ground, and sprout to develop new plants that are genetically identical to the original one. One familiar species that does this is the tiger lily (*Lilium tigrinum.* Other plants can propagate from twigs or stem pieces that fall from the parent, then lodge into a suitable site and develop into a new plant. The crack willow (*Salix fragilis*) can spread itself along watercourses in this manner (as well as by disseminating seeds). Other non-sexual means of propagation include the production of underground bulbs, corms, and tubers that split into parts, each of which is capable of developing into a new plant. Plants that reproduce in this manner include irises and daffodils.

See also Agricultural genetics; Plant genetics

PLASMIDS

Plasmids are extra-chromosomal, covalently closed circular (CCC) molecules of double stranded (ds) **DNA** that are capable of autonomous **replication**. The prophages of certain bacterial phages and some dsRNA elements in **yeast** are also called plasmids, but most commonly plasmids refer to the extra-chromosomal CCC DNA in **bacteria**.

Plasmids are widely distributed in nature. They are dispensable to their host **cell**. They may contain genes for a variety of phenotypic traits, such as antibiotic resistance, virulence, or metabolic activities. The products plasmids encode may be useful in particular conditions of bacterial growth. Replication of plasmid DNA is carried out by subsets of **enzymes** used to duplicate the bacterial **chromosome** and is under the control of plasmid's own replicon. Some plasmids reach copy numbers as high as 700 per cell, whereas others are maintained at the minimal level of 1 plasmid per cell. One particular type of plasmid has the ability to transfer copies of itself to other bacterial stains or **species**. These plasmids have a tra operon. Tra **operon** encodes the protein that is the component of sex pili on the surface of the host bacteria. Once the sex pili contact with the recipient cells, one strand of the plasmid is transferred to the recipient cells. This plasmid can integrate into the host chromosomal DNA and transfer part of the host DNA to the recipient cells during the next DNA transfer process.

Plasmids are essential tools of genetic engineering. They are used as vectors in molecular biology studies. Ideally, plasmids as vectors should have three characteristics. First, they should have a multiple **cloning** site (MSC) which consists of multiple unique restriction enzyme sites and allows the insertion of foreign DNA. Second, they should have a relaxed replication control that allows sufficient plasmids to be produced. Last, plasmids should have selectable markers, such as **antibiotics** metabolite genes, which allow the identification of the transformed bacteria. Numerous plasmid vectors have been developed since the first plasmid vectors of the early 1970s. Some vectors have **bacteriophage** promoter sequences flanking the MSC that allows direct **sequencing** of the cloned DNA sequence. Some vectors have yeast or **virus** replication origin, which allows the plasmids to replicate in yeast and mammalian cells, hence enabling cloned cDNAs to express in these host cells. Many new features have and will be added into plasmids to make genetic engineering easier and faster.

See also Cloning vector; DNA replication

PLEIOTROPY

The products of many genes have several different effects or phenotypes in cells, a phenomenon known as pleiotropism or pleiotropy. Levels of pleiotropy vary from one **gene** to another, giving rise to expressions such as low-pleiotropy or great pleiotropy in the published scientific literature.

Genes encoding cytokines, polypeptide hormones used in cell-to-cell communication, may have many pleiotropic effects and, therefore, present great pleiotropy, as well as great redundancy because many functions of a given cytokine may also be performed by other different cytokines. For instance, several different cytokines may induce activated lymphocytes T (T cells) into proliferation. Examples of cytokines are as follows: Interferon alpha and beta (IFN-alfa and IFN-beta), **tumor** necrosis factor (TNF-alfa), interleukin–1 (IL–1alfa and IL–1beta).

Another example of great pleiotropy is given by GAPDH (enzyme glyceraldehyde-3–phosphate dehydrogenase). GAPDH was first known by its role in the metabolism of glucose, but several research groups gradually found that it played several different roles in different tissues, such as regulation of protein **translation**, export of RNAt from **cell nucleus** to cytoplasm, a role in **DNA repair**, a role in **DNA replication**, and another role in cell membrane during phagocytosis.

Some authors also use the term pleiotropism or pleiotropy when referring to synthetic molecules that present different effects in different disorders. For example, aspirin (acetyl salicylic acid) that has well-known analgesic (pain killing) and antipyretic (anti fever) effects, but also helps to reduce atheroma formation (plaques that sediment in veins and arteries, causing circulatory disorders), and more recently, has shown some anti tumor properties. Another example of a pleiotropic medicine is COX–2 (cyclooxygenase–2) inhibitors, first developed as an anti-inflammatory drug, and now under scrutiny by **cancer** researchers because of its anti tumor effects as an adjuvant treatment of cancer as well as a chemo preventive therapy. COX–2 is a member of a family of **enzymes**, the COX enzymes, and it is found overexpressed in colorectal tumors and other cancers. Cox genes are **housekeeping genes** that also present a great pleiotropism.

Thalidomide, first developed for treatment of nausea during pregnancy, caused thousands of cases of **birth defects** because it was unknown at that time that it is also an inhibitor of blood vessels' formation (angiogenesis). Presently, it is being prescribed as an antiangiogenic drug for cancer patients because it prevents tumor neovascularization (new vein formation), thus helping to control tumor growth. Several studies also indicate that some **antibiotics** of the molecular group of macrolides, such as chlarithromycin, do present a powerful antiangiogenic effect over some lung tumors.

The pleiotropic effect of several drugs, such as those mentioned above, may present an interesting field for biomedics, and a potential advantage to patients, because they are already marketed and tested as far as toxicity is concerned. However, they may end up as orphan drugs for other disorders, because they may never be adequately tested and standardized for new diseases, since it would require new investments in clinical trials.

See also Anticancer drugs; Cell proliferation; Congenital birth defects; DNA repair; Gene family; Housekeeping genes; Mitosis; Proteins and enzymes; Teratology and teratogens; Tumor

PLOIDY

Ploidy refers to the number of chromosomes found in a **cell** and is subdivided into two categories: euploidy and aneuploidy. Euploidy is one complete **chromosome** set or exact multiples of that set. Aneuploidy includes gain or loss of chromosomes that is less than a complete set. The optimum chromosome complement is fixed for each organism and can be described in terms of euploidy and aneuploidy, though the later is most often associated with variation or errors in the normal chromosome number.

Euploidy can refer to a broad range of chromosome findings. Haploidy, designated as the N number, is one complete chromosome set in which each chromosome is unique. Diploidy (2N) is two complete chromosome sets (pairs of chromosomes), triploidy (3N) is three sets, tetraploidy (4N) is four sets, and so on. Any cell with three or more chromosome sets is said to be polyploid. In humans, cells are **diploid**. They contain 46 chromosomes arranged in 23 pairs. One complete set (N) comes from the mother and the other N set from the father. Therefore, in diploid organisms, the **haploid** number equals the number of chromosomes in a **gamete**. In humans and most other animals, polyploidy is not compatible with life.

Some **species** of plants occur naturally as polyploids. The common dandelion, for example, is a triploid, as are several species of ferns. Phenotypic variation in some grains led to the finding that an increase in euploidy level within certain species may give rise to larger, more robust plants. In an attempt to improve crop yield, researchers designed crosses intended to generate plants with very high levels of polyploidy (5N, 6N, 7N and up). Larger plants with improved grain were obtained up to a level of polyploidy where excessive numbers of chromosomes began to affect the ability of the cell to divide accurately. From this point on, increasing the euploidy level actually results in smaller, less healthy plants.

Aneuploidy is most commonly the gain or loss of a single chromosome. For a diploid organism, gain of a single chromosome is known as trisomy (2N+1), i.e., all of the chromosomes are paired with the exception of one that has three copies. Monosomy (2N-1) is the loss of a single chromosome, so all chromosomes would be paired except one that would exist as a single chromosome. Occasionally, a cell may show a gain or loss of two (or more) separate chromosomes (double trisomy or double monosomy), but in general, the greater the gain or loss of chromosomes, the less viable the cell. If only a portion of a chromosome is lost, the cell is said to have a partial monosomy, and a **duplication** of only one chromosome region will result in a partial trisomy.

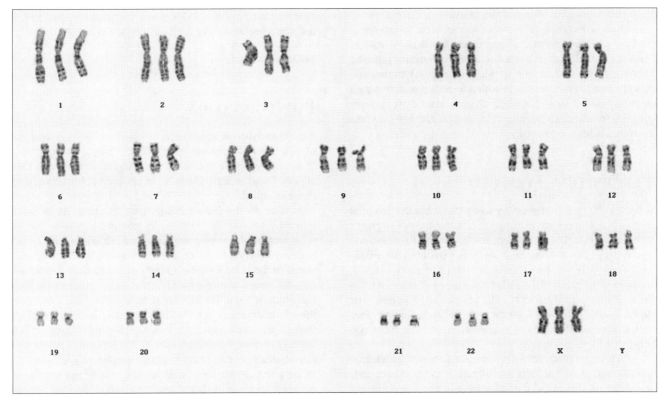

Human karyotype showing three copies of each chromosome as opposed to the normal pair of chromosomes. Triploidy is not compatible with life, but does occur as a result of cell division errors. *Courtesy of Dr. Constance Stein.*

In man, autosomal monosomy is not compatible with life, but there are three liveborn autosomal trisomies: trisomy 13 or **Patau syndrome**, trisomy 18 or Edward syndrome, and trisomy 21 or **Down syndrome**. **Sex chromosome** aneuploidies tend to better tolerated, probably because of the process of X inactivation. The only liveborn monosomy in humans is Turner syndrome: 45,X. Sex chromosome trisomies include: **Klinefelter syndrome** (47,XXY), triple X syndrome (47,XXX), and **XYY syndrome**. There are also some individuals who have multiple sex chromosomes (48,XXXX ; 49,XXXXX ; 48,XXYY, etc.), but all have varying abnormal phenotypes.

Trisomy or monosomy are usually the result of nondisjunction **cell division** errors in which one chromosome fails to move to the proper pole during division. This can occur in either **meiosis** or **mitosis**. Meiotic nondisjunction errors result in a gamete with one too many or one too few chromosomes that, after **fertilization**, gives rise to an individual with the aneuploid chromosome number. Mitotic nondisjunction errors in somatic cells of the body may give rise to an aneuploid cell line but usually have no phenotypic effect on the individual. However, should this type of mitotic division error occur in the **zygote** or early in fetal development, a second cell line could be established that would generate a **mosaic** chromosome complement, i.e., two cell lines would be present in the individual that would differ by a single chromosome change.

Because only chromosomes 13, 18 and 21 are seen in living trisomic individuals, the question is often asked if there is something about cell division that favors these three chromosomes when errors occur. Data from cytogenetic analysis of spontaneously aborted fetuses suggests that nondisjunction is random with respect to the chromosomes since nearly every other possible trisomy has been reported in abortus tissue. In fact, the most common trisomy seen in abortions is trisomy 16. The conclusion is that the total number and combination of genes present on the chromosomes must be the determining factor in **viability**. Chromosomes 13, 18, and 21 are some of the smallest of the chromosome set, so on the basis of size alone would be expected to contain a comparatively low number of genes that would lower the risk of lethality. This is supported by recent evidence from the **Human Genome Project** that suggests these three chromosomes have the lowest density of coding regions.

Although trisomies of chromosomes 13, 18 and 21 and monosomy X can produce liveborn children, most of the time these aneuploidies are not compatible with life. Of all 45,X conceptions, approximately 95% spontaneously terminate. Similarly, spontaneous termination occurs for 90% of trisomy 13 conceptions, 80% of trisomy 18 conceptions, and 65% of trisomy 21 conceptions. This implies that the allelic constitution of the trisomy is also important in determining viability, and that certain trisomic genotypes are not compatible with life. Because of this, only a relative few of all trisomic conceptions are liveborn.

Aneuploidy is also seen in **cancer** cell lines. Gain or loss of a few too many chromosomes can occur. Some aneuploidies are characteristic of particular disorders, as loss of chromosomes 5 and/or 7 is associated with myelodysplastic syndrome. Hyperdiploidy is seen in acute lymphocytic leukemia and is a condition in which there is a selective gain of chromosomes such that 50-55 chromosomes exist in a single cell. In advanced disease, random gain and loss of chromosomes is known to occur.

PLURIPOTENT STEM CELLS

A fertilized egg is totipotent, meaning that it has the potential to give rise to every type of **cell** needed for the development of the organism. The cell undergoes successive divisions, and at the beginning, the resulting identical cells are also totipotent. These cells form a hollow ball called a blastocyst, which contains a cluster of cells called the inner cell mass. At this point, some specialization of cells has already occurred. The outer layer of cells will form the extraembryonic membranes that will support the developing embryo, and the inner cell mass will form the embryo, itself.

The cells of the inner cell mass are known as pluripotent. Because they are incapable of giving rise to the supporting cells of the extraembryonic membranes, they are not sufficient for fetal development. Each of the inner-cell-mass cells does, however, have the potential to become any type of cell in the fetus, itself. As development proceeds, the pluripotent cells specialize further, so that each type, called a multipotent stem cell, has the ability to make only certain types of cells. Blood **stem cells**, for example, will make only red cells, white cells, and platelets. These multipotent cells are present throughout life.

Pluripotent stem cells have been isolated by one laboratory from the inner cell mass of human embryos in the blastocyst stage. These embryos were produced by *in vitro* **fertilization** and were in excess of those used in fertility treatment. In another laboratory, pluripotent stem cells were isolated from fetal tissue derived from terminated pregnancies. Another possible way to produce pluripotent stem cells is to fuse the **nucleus** of a somatic cell with an egg whose nucleus has been removed, forming a blastocyst from which innermast cells may be used.

There are many potential applications of pluripotent stem cells. Basic research on these cells will yield an understanding of how cell specialization occurs during normal human development. This will in turn help to explain what happens when the specialization process goes awry, for example in **cancer** and in **birth defects**. The knowledge may also be used in the development of therapies for conditions in which tissue degenerates or is destroyed, including Parkinson's and Alzheimer's diseases, spinal cord injury, burns, stroke, diabetes, and arthritis. Eventually, pluripotent cells may be manipulated to produce a particular type of tissue needed for therapy. For example, it may become possible to produce pancreatic islet cells to treat diabetes. Before that can occur, how-

ever, the changes that must occur for a pluripotent cell to become a pancreatic islet cell must be understood.

See also Embryology

POINT MUTATION

A point **mutation** is a change in a single base in the coding part of the **deoxyribonucleic acid** (**DNA**). Point mutations are changes in one base pair of the DNA and they can vary in their effects. They can range from being unobservable to being fatal to the organism.

Due to the inbuilt redundancy of the code of DNA (whereby several different triplet **codons** can code for the same **amino acid**) some point mutations are neutral, that is they have no observable effect, the same amino acid is still produced in the same location. Other point mutations can substitute the amino acid produced in the polypeptide chain. This can produce a non-functioning protein, which can then affect the whole organism even leading to death. In some cases the protein produced under the action of the point mutation can actually perform better than the original form. Where this beneficial effect occurs it may lead to a greater ability to survive in the given environment, thus leading to a greater number of offspring potentially being produced with the concomitant chance of a wider spread over the whole population for this new form of the **gene**. Some point mutations change the section of DNA affected from sense to nonsense. When nonsense DNA is found it is a signal for the **cell** to stop making the polypeptide chain. This is the production of a stop codon.

See also Chromosome structure and morphology; Mutations and mutagenesis

POLLINATION

Pollination is the transfer of pollen from the male reproductive organs to the female reproductive organs of a plant, and it precedes **fertilization**, the fusion of the male and the female sex cells. Pollination occurs in seed-producing plants, but not in the more primitive spore-producing plants, such as ferns and mosses. In plants such as pines, firs, and spruces (the gymnosperms), pollen is transferred from the male cone to the female cone. In flowering plants (the angiosperms), pollen is transferred from the flower's stamen (male organ) to the pistil (female organ). Many **species** of angiosperms have evolved elaborate structures or mechanisms to facilitate pollination of their flowers.

The Austrian monk and botanist **Johann Gregor Mendel** (1822-1884) conducted important pollination studies in Brno (now in the Czech Republic) in the mid-1800s. He studied **heredity** by performing controlled cross-pollinations of pea plants, thereby laying the foundation for the study of heredity and **genetics**. **Charles Darwin** continued the study of pollination in subsequent years. In 1876, Darwin wrote an important book on pollination biology, *The Effects of Cross and Self Fertilization in the Vegetable Kingdom.*

Botanists theorize that seed plants with morphologically distinct pollen (male) and ovules (female) evolved from ancestors with free-sporing heterospory, where the male and the female spores are also morphologically distinct.

The **evolution** of pollination coincided with the evolution of seed. Fossilized pollen grains of the seed ferns, an extinct group of seed-producing plants with fern-like leaves, have been dated to the late Carboniferous period (about 300 million years ago). These early seed plants relied upon wind to transport their pollen to the ovule. This was an advance over free-sporing plants, which were dependent upon water, as their sperm had to swim to reach the egg. The evolution of pollination therefore allowed seed plants to colonize terrestrial habitats.

It was once widely believed that insect pollination was the driving force in the evolutionary origin of angiosperms. However, paleobotanists have recently discovered pollen grains of early gymnosperms, which were too large to have been transported by wind. This and other evidence indicates that certain species of early gymnosperms were pollinated by insects millions of years before the angiosperms had originated.

Once the angiosperms had evolved, insect pollination became an important factor in their evolutionary diversification. By the late Cretaceous period (about 70 million years ago), the angiosperms had evolved flowers with complex and specific adaptations for pollination by insects and other animals. Furthermore, many flowers were clearly designed to ensure cross-pollination, exchange of pollen between different individuals. Cross-pollination is often beneficial because it produces offspring that have greater genetic heterogeneity, and are better able to endure environmental changes. This important point was also recognized by Darwin in his studies of pollination biology.

Most modern gymnosperms and many angiosperms are pollinated by wind. Wind-pollinated flowers, such as those of the grasses, usually have exposed stamens, so that the light pollen grains can be carried by the wind.

Wind pollination is a primitive condition, and large amounts of pollen are usually wasted, because it does not reach female reproductive organs. For this reason, most wind-pollinated plants are found in temperate regions, where individuals of the same species often grow close together. Conversely, there are very few wind pollinated plants in the tropics, where plants of the same species tend to be farther apart.

In general, pollination by insects and other animals is more efficient than pollination by wind. Typically, pollination benefits the animal pollinator by providing it with nectar, and benefits the plant by providing a direct transfer of pollen from one plant to the pistil of another plant. Angiosperm flowers are often highly adapted for pollination by insect and other animals.

Each taxonomic group of pollinating animals is typically associated with flowers which have particular characteristics. Thus, one can often determine which animal pollinates

a certain flower species by studying the morphology, color, and odor of the flower. For example, some flowers are pure red, or nearly pure red, and have very little odor. Birds, such as hummingbirds, serve as pollinators of most of these flowers, since birds have excellent vision in the red region of the spectrum, and a rather undeveloped sense of smell.

See also Plant genetics

POLYGENIC INHERITANCE AND DISORDERS

The **inheritance** of polygenic traits depends upon the interaction of two or more genes. This is sometimes confused with the idea of multiple alleles, which are just different forms of the same **gene**. However, the two concepts are related when discussing the probable phenotype resulting from a particular genotype. Such diseases as diabetes mellitus or heart disease are not the consequence of single gene inheritance, likewise physical traits such as height, weight or even behavior are all examples of quantitative traits whose expression depends upon several different factors. These include the number of genes involved, the number of alleles each gene has, and how much the phenotypic variability depends upon environmental interactions. Unlike qualitative traits such as blood type or other multiple allelic genes that show an unambiguous phenotype, quantitative polygenic traits show a range expression. The distribution of height, for instance, is expressed as short at one extreme and tall at the other extreme, with the bulk falling somewhere in the middle. Height can be used again when discussing the role environment plays in the expression of polygenic traits. Consider the change between the first generation of Japanese-Americans, to that of the second generation who ate an American diet. Another example would be the height of a tree grown in a temperate climate compared to the same one grown under drought conditions. The genotype would be the same, but the phenotype would be different due to lack of water.

Many different diseases show polygenic inheritance patterns. Hip dysplasia in dogs illustrates this expression well. Imagine two dogs with healthy hips are mated and produce offspring with the result that some of the puppies are completely crippled, some of them seem normal, but x-rays show they have hip dysplasia, and some of the puppies are not affected. This gradation from severely dysplastic to normal is the result of a cumulation of mutations that cause the disease to finally reach a threshold level and to then be expressed. Each of the parents has some of the mutations that cause the disease, but not enough to express the disease in themselves. However, when the two are bred, some of the their puppies get enough of the "bad" genes to have the severe form of the disease, some of the puppies get enough to show the moderate form of the disease and others were lucky and did not get enough of the deleterious genes to be affected, but are most likely carriers of some the disease genes as their parents.

Twin studies in humans have been useful in showing how the expression of polygenic traits is influenced by the environment. For example, two **twins** may have the potential to become diabetic, but because of their different diets, one becomes diabetic and the other does not. Because environmental factors can play such a large part in the expression of diseases that are polygenic, many people are not aware they carry these genes until their offspring accumulate enough of the defective genes to express the disease. This makes these types of disease hard to eradicate.

See also Phenotype

POLYMERASE CHAIN REACTION

Polymerase chain reaction is a technique in which cycles of denaturation, annealing with primer, and extension with **DNA** polymerase, are used to amplify the number of copies of a target DNA sequence by more than 106 times in a few hours. American molecular biologist **Kary Mullis** developed the idea of PCR in the 1970s. The idea was conceived while he was cruising in a Honda Civic on Highway 128 from San Francisco to Mendocino, California. For his ingenious invention, he was awarded a Nobel Prize in 1993.

PCR amplification of DNA is like any **DNA replication** by DNA polymerase *in vivo*. The difference is that PCR produces DNA in a test tube. For a PCR to happen, four components are necessary: template, primer, deoxyribonecleotides (adenine, thymine, cytosine, guanine) and DNA polymerase. In addition, part of the sequence of the targeted DNA has to be known in order to design the according **primers**. In the first step, the targeted double stranded DNA is heated to over 194°F (90°C) for denaturation. During this process, two stands of the targeted DNA are separated from each other and each stand is ready as template. The second step is carried out around 122°F (50°C). At this lowered temperature, the two primers annealed to their complementary sequence on each template. The DNA polymerase then extends the primer using the provided nucleotides. As a result, at the end of each cycle, the numbers of DNA molecules double. PCR is carried out manually in incubators of different temperatures for each step until the discovery of DNA polymerase from thermophilic **bacteria**. The famous bacterium *Thermus aquaticus* was found in Yellow Stone National Park. This bacterium lives in the hot springs at 203°F (95°C). The DNA polymerase from *T. aquaticus* keeps its activity at above 203°F over hours. Several heat-resistant DNA polymerases have been found since.

Genetically engineered heat resistant DNA polymerases, which have proofreading functions and make fewer mutations in the amplified DNA products, are available commercially. PCR reactions are carried out in different thermocyclers. Thermocyclers are designed to change temperatures automatically. Researchers set the temperatures and the time, and at the end of the procedure take the test tube out of the machine. The invention of PCR is a revolution to molecular biology. PCR is valuable to researchers because it allows them to multiply a unique DNA sequence to a large amount in a very short time. Researchers in the **Human Genome Project** are using PCR to look for markers in cloned DNA segments and to order DNA fragments in libraries. Molecular biologists use PCR in **cloning** DNA. PCR is used to produce biotin or other chemical-labeled probes. These probes are used in nucleic acid hybridization, *in situ* hybridization and other molecular biology procedures. PCR, coupled with fluorescence techniques and computer technology, allows the real time amplification of DNA. This enables quantitative detection of DNA molecules that exist in minute amounts. PCR is also used widely in clinical tests. Today, it has become a routine to use PCR in the diagnosis of infectious diseases such **AIDS**.

See also Hybridization of DNA; DNA synthesis

POLYMERASE (DNA AND RNA)

Polymerases are present in prokaryotes and **eukaryotes**, where they are essential to the manufacture of genetic material.

DNA polymerase functions to duplicate DNA (**deoxyribonucleic acid**). This process is called **replication**. Each time a **cell** divides, DNA polymerase duplicates its entire DNA, and the cell passes one copy to each daughter cell. This process ensures that genetic material is faithfully passed from generation to generation. DNA polymerase is extremely accurate in its task. Its error rate is less than one base in each billion duplicated.

The task of **DNA replication** requires a number of polymerases. Three distinct DNA polymerases (I, II, and III) have been isolated from *Escherichia coli*. **Arthur Kornberg** discovered the first DNA polymerase (I) in *E. coli* in 1956. Studies of mutant strains of *E. coli* led to the discovery of the other DNA polymerases II and III in the 1970s. Such studies also showed that the polymerase I is primarily involved in the repair of DNA damage, although it is also required for replication, and that polymerase III is the major replicative enzyme. The function of DNA polymerase II is still unknown.

Eukaryotes contain five DNA polymerases. Three are found in the **mitochondria** and are responsible for replication of **mitochondrial DNA**. Four (including two of the mitochondrial **species**) are active in the cell **nucleus**, where they function in replication.

In DNA replication, whether mitochondrial or nuclear, the involved polymerase functions to synthesize a new strand of DNA from each one of the old DNA strands, once unwinding of the old strands has occurred prior to replication. The polymerase begins synthesis at the site of **RNA primers**, which were previously synthesized on the two unwound DNA strands. Once synthesis of new DNA is complete, another DNA polymerase acts to remove the RNA primers and replace them with DNA, completing the synthetic process.

RNA polymerase functions to synthesize RNA using a DNA template and molecules called ribonucleotide triphosphates as the substrates. This process is called **transcription**. The DNA may code for a **messenger RNA**, which will be used for **protein synthesis**, for a **ribosomal RNA**, or for a **transfer RNA**. The latter two RNA species are used in the manufacture of protein. In prokaryotes, one RNA polymerase synthesizes all three of the RNAs. In eukaryotes, this function is performed by three separate RNA polymerases.

In a cell, DNA replication and DNA transcription will be occurring simultaneously, so both DNA and RNA polymerases will be in action at the same time. A mechanism must exist to coordinate the action of the polymerases in order to prevent a collision of the polymerase complexes. The best interpretation of available evidence suggests that the movement of the DNA polymerase through the RNA transcription complex accomplish this. The exact means by which this occurs remain unclear.

See also DNA structure

POLYMORPHISM

Polymorphism refers to the presence of two or more distinct forms which exist together within a single breeding population of a **species**. The forms are discontinuous, meaning that the population lacks individuals that are intermediate. Polymorphisms are found in many plants and animals and often exist at a frequency too high to be maintained solely by **mutation**. Polymorphisms are known to exist either as obvious physical variations easily detected by examination of the body of the organism, **enzymes**, or proteins of thoseorganisms, and as chromosomal or **deoxyribonucleic acid** (**DNA**) variants.

The term discontinuous is essential to an understanding of polymorphism. Most traits of organisms exist in a continuous gradation between extremes. Consider pigment and height in human populations. Skin pigment varies from a deep brown to various light hues. Similarly, height in humans varies from short to extremely tall. The existence of innumerable intermediate forms in both height and pigmentation qualifies those traits as being continuously variant. Other human traits have no intermediates. Consider the human ABO blood type polymorphism. Humans may be classified as blood type A, B, AB, or O. Intermediates of those blood types do not exist. Blood type is determined by **heredity** and this is true of other types of polymorphism. Environmental factors such as diet and ultraviolet radiation (uv) may affect height and pigment respectively, however environment does not change the genetic determinate for ABO blood types. Thus, they are not affected by what a human may or may not be exposed to.

Pigment pattern polymorphisms exist in populations of the northern leopard frog, *Rana pipiens*. The leopard frog ordinarily has spots on its back and limbs. There is a remarkable variation in the number and size and shape of the spots; it is appropriate to state that spotting is a continuously variable trait of the leopard frog. However, a spotless variant, known as the burnsi morph, exists in Minnesota and contiguous states. The spotless frog is not an albino. It has pigment cells, but the cells do not aggregate into the well recognized spots of this common frog. Genetic studies reveal that the burnsi morph differs from ordinary spotted frogs by the possession of a single dominant Mendelian allele. The burnsi morph of the leopard frog seems to persist at a relatively stable frequency in many Minnesota populations, creating a polymorphic population. It is difficult to understand the biological significance of the balanced stability where it occurs, however it seems that the burnsi **gene** might convey an advantage when relatively rare but not when it becomes more common.

The kandiyohi morph is another example of pigment pattern polymorphism in some populations of *R. pipiens*. Kandiyohi frogs are identified by small mottled patches of pigment between the usual dark spots that characterize leopard frogs. The kandiyohi morph differs from ordinary spotted frogs by possession of a single dominant Mendelian allele. Intermediates with pigment patterns between ordinary spotted frogs and the kandiyohi frog do not exist. They are sharply distinct, and thus, kandiyohi is another example of polymorphism in leopard frogs.

See also Phenotype

POLYPLOIDY • *see* PLOIDY

PONNAMPERUMA, CYRIL (1923-1994)
Sri Lankan-born American chemist

Cyril Ponnamperuma, an eminent researcher in the field of chemical evolution, rose through several National Aeronautics and Space Administration (NASA) divisions as a research chemist to head the Laboratory of Chemical Evolution at the University of Maryland, College Park. His career focused on explorations into the origin of life and the "primordial soup" that contained the precursors of life. In this search, Ponnamperuma took advantage of discoveries in such diverse fields as molecular biology and astrophysics.

Born in Galle, Ceylon (now Sri Lanka) on October 16, 1923, Cyril Andres Ponnamperuma was educated at the University of Madras (where he received a B.A. in Philosophy, 1948), the University of London (B.Sc., 1959), and the University of California at Berkeley (Ph.D., 1962). His interest in the origin of life began to take clear shape at the Birkbeck College of the University of London, where he studied with J. D. Bernal, a well-known crystallographer. In addition to his studies, Ponnamperuma also worked in London as a research chemist and radiochemist. He became a research associate at the Lawrence Radiation Laboratory at Berkeley, where he studied with Melvin Calvin, a Nobel laureate and experimenter in chemical evolution.

After receiving his Ph.D. in 1962, Ponnamperuma was awarded a fellowship from the National Academy of Sciences, and he spent one year in residence at NASA's Ames Research Center in Moffet Field, California. After the end of his associate year, he was hired as a research scientist at the center and became head of the chemical evolution branch in 1965.

Cyril Ponnamperuma. *AP/Wide World Photos, Inc. Reproduced by permission.*

During these years, Ponnamperuma began to develop his ideas about chemical evolution, which he explained in an article published in *Nature*. Chemical evolution, he explained, is a logical outgrowth of centuries of studies both in chemistry and biology, culminating in the groundbreaking 1953 discovery of the structure of **deoxyribonucleic acid (DNA)** by **James Watson** and **Francis Crick**. Evolutionist **Charles Darwin's** studies affirming the idea of the "unity of all life" for biology could be extended, logically, to a similar notion for chemistry: protein and nucleic acid, the essential elements of biological life, were, after all, chemical.

In the same year that Watson and Crick discovered DNA, two researchers from the University of Chicago, Stanley Lloyd Miller and **Harold Urey**, experimented with a primordial soup concocted of the elements thought to have made up earth's early atmosphere—methane, ammonia, hydrogen, and water. They sent electrical sparks through the mixture, simulating a lightening storm, and discovered trace amounts of amino acids.

During the early 1960s, Ponnamperuma began to delve into this primordial soup and set up variations of Miller and Urey's original experiment. Having changed the proportions of the elements from the original Miller-Urey specifications slightly, Ponnamperuma and his team sent first high-energy electrons, then ultraviolet light through the mixture, attempt-

ing to recreate the original conditions of the earth before life. They succeeded in creating large amounts of **adenosine triphosphate (ATP)**, an **amino acid** that fuels cells. In later experiments with the same concoction of primordial soup, the team was able to create the nucleotides that make up nucleic acid—the building blocks of DNA and **ribonucleic acid (RNA)**.

In addition to his work in prebiotic chemistry, Ponnamperuma became active in another growing field: exobiology, or the study of extraterrestrial life. Supported in this effort by NASA's interest in all matters related to outer space, he was able to conduct research on the possiblity of the evolution of life on other planets. Theorizing that life evolved from the interactions of chemicals present elsewhere in the universe, he saw the research possibilities of spaceflight. He experimented with lunar soil taken by the *Apollo 12* space mission in 1969. As a NASA investigator, he also studied information sent back from Mars by the unmanned Viking, Pioneer, and Voyager probes in the 1970s. These studies suggested to Ponnamperuma, as he stated in an 1985 interview with *Spaceworld,* that "Earth is the only place in the solar system where there is life."

In 1969, a meteorite fell to earth in Muchison, Australia. It was retrieved still warm, providing scientists with fresh, uncontaminated material from space for study. Ponnamperuma and other scientists examined pieces of the meteorite for its chemical make-up, discovering numerous amino acids. Most important, among those discovered were the five chemical **bases** that make up the nucleic acid found in living organisms. Further interesting findings provided tantalizing but puzzling clues about chemical evolution, including the observation that light reflects both to the left and to the right when beamed through a solution of the meteorite's amino acids, whereas light reflects only to the left when beamed through the amino acids of living matter on earth. "Who knows? God may be left-handed," Ponnamperuma speculated in a 1982 *New York Times* interview.

Ponnamperuma's association with NASA continued as he entered academia. In 1979, he became a professor of chemistry at the University of Maryland and director of the Laboratory of Chemical Evolution—established and supported in part by the National Science Foundation and by NASA. He continued active research and experimentation on meteorite material. Based on his findings from the Murchison meteorite experiments, Ponnamperuma reported the creation of all five chemical bases of living matter in a single experiment that consisted of bombarding a primordial soup mixture with electricity.

Ponnamperuma's contributions to scholarship include hundreds of articles. He wrote or edited numerous books, some in collaboration with other chemists or exobiologists, including annual collections of papers delivered at the College Park Colloquium on Chemical Evolution. He edited two journals, *Molecular Evolution* (from 1970 to 1972) and *Origins of Life* (from 1973 to 1983). In addition to traditional texts in the field of chemical evolution, he also co-authored a software program entitled "Origin of Life," a simulation model intended to introduce biology students to basic concepts of chemical evolution.

Although Ponnamperuma became an American citizen in 1967, he maintained close ties to his native Sri Lanka, even becoming an official governmental science advisor. His professional life has included several international appointments. He was a visiting professor of the Indian Atomic Energy Commission (1967); a member of the science faculty at the Sorbonne (1969); and director of the UNESCO Institute for Early Evolution in Ceylon (1970). His international work included the directorship of the Arthur C. Clarke center, founded by the science fiction writer, a Sri Lankan resident. The center has as one of its goals a Manhattan Project for food synthesis.

Ponnamperuma was a member of the Indian National Science Academy, the American Association for the Advancement of Science, the American Chemical Society, the Royal Society of Chemists, and the International Society for the Study of the Origin of Life, which awarded him the A. I. Oparin Gold Medal in 1980. In 1991, Ponnamperuma received a high French honor—he was made a *Chevalier des Arts et des Lettres.* Two years later, the Russian Academy of Creative Arts awarded him the first Harold Urey Prize. In October 1994, he was appointed to the Pontifical Academy of Sciences in Rome. Ponnamperuma died on December 20, 1994 at Washington, D.C.

See also Evolutionary mechanisms; Miller-Urey experiment; Origin of life; Replication

POPULATION GENETICS

Population **genetics** is the study of **gene** variations (alleles) in a population. Allele frequencies within populations can reveal the selective factors and breeding patterns of that population. To determine the true allele frequency of a population, all the individuals in that population must be tested, but this is not practical for most studies. Instead, allele frequencies are usually estimated by testing a sample of individuals from the population, which is representative of the population. How representative this sample is can be tested by mathematics and this sampling math is an important part of population genetics. A powerful tool within this area of research is that of mathematical modeling using computers. Mathematical models are used to predict what effects such things as **selection**, population size, **mutation**, and migration have upon the occurrence and frequency of both linked and unlinked genes.

For example, by measuring the allele frequencies in a population and comparing the observed proportion of homozygotes to the expected proportion, deviations from **Hardy-Weinberg equilibrium** can be detected. These deviations could be due to **inbreeding**, or selection, which both could reveal the behavior of the population and what evolutionary pressures it faces. If there is strong selection for an allele, for example, then within a few generations it will be the only version of that gene in the population—it will be fixed. The development of the mathematics of population genetics enabled it to answer more rigorous questions about **natural**

selection on populations. This was the work of three geneticists during the first half of the twentieth century: Sewell Wright, R.A. Fisher and J.B.S. Haldane. They also integrated mathematical population genetics (based on Mendel's concept of genetics) with Darwin's theory of **evolution** by natural selection, and this was called neo-Darwinism. Their theoretical models suggested that selection was more important than mutation and random genetic drift in changing allele frequencies in a population.

Population genetics is based on the Hardy-Weinberg equation, $p^2+2pq+q^2=1$, where p and q are the frequencies of two alleles of the same gene in a population, and vary between 0 and 1. The equation states that p^2 is the frequency of individuals having two copies of the a certain allele (p), q^2 is the frequency of individuals in the population having two q alleles, and 2pq is the frequency of heterozygotes. This only holds for **diploid** organisms (those with a full set of chromosomes) with no inbreeding. The Hardy-Weinberg equilibrium states that, in the absence of mutation, selection, and migration, the frequency of alleles in a population remains constant.

During the first half of the twentieth century, the selectionist ideal was prevalent in population genetics. This suggested that every allele had some selective effect on the individual, even if it was very slight. This was challenged by **Motoo Kimura** in the 1960s, who proposed what is called the neutral theory. He observed that there was more variation in populations than expected, if the selectionist model was correct. The selectionist model would have less variation because most alleles during the history of the population would have either died out or have become fixed thanks to selection. The neutral model said that most alleles were selectively neutral, and had no discernible effect on the individual. Therefore it was not selection but random events in breeding, population size and movement (termed genetic drift) that determined the frequency of alleles of most genes in a population. Later studies of **DNA** variation gave support to this idea by showing that most DNA polymorphisms were not shown in the proteins, and—because the structure of proteins greatly determines how the body works (**enzymes**, **cell** signaling etc.)—that there could be no selection for or against most of these DNA polymorphisms.

The reconciliation of selectionist and neutral theory is an important issue for modern population genetics. Using modern techniques such as DNA **sequencing** and high-throughput genotyping of DNA polymorphisms, investigators can now study more genes in larger samples for many different organisms. Generation of large amounts of data will undoubtedly resolve many issues. Studies of gene mutation rate and ecology cam also have an impact on population genetics. For example, genetic variablity in a single population will always be reduced due to fixation of alleles either due to mutation or to genetic drift. However, in the real world, new alleles will be generated by both mutation and migration from other populations. Therefore determining the rate of mutation for different genes, and the nature and frequency of population migration have important consequences for what alleles are in the population (**gene frequency** distribution). The use of cladistics and coalescent theory are also important aspects of modern population genetics. These use diagrammatic trees to deduce the relationship between populations and between **species** by using allele frequency and DNA sequence data.

See also Adaptation and fitness; Coalescence; Evolutionary mechanisms; Homozygous; Human population genetics; Outbreeding; Rare genotype advantage

POSITIONAL CLONING • *see* CLONES AND CLONING

POUCHET, FELIX-ARCHIMEDE (1800-1872)

French biologist

French-born scientist Felix-Archimede Pouchet was a biologist and popular science writer, and a leading proponent of **spontaneous generation**. While Pouchet contributed widely to advancing biological concepts, he is most widely remembered for his defeat in the widely known debates with **Louis Pasteur**.

Pouchet was born in Rouen, France, the son of a successful industrialist. Pouchet he became interested in biology at a young age, and graduated from the University of Rouen in 1827. At 27, he became director of the Museum of Natural History at Rouen, and stayed at that institution for the remainder of his life. Pouchet also taught in Rouen at the academy and the school of medicine, where he received many honors including the Legion of Honor in 1843.

Pouchet was a successful author who contributed widely to the literature of biology. He was recognized as a friend and supporter of **Charles Darwin** and his ideas of **natural selection** and **evolution**. Pouchet's many areas of interest included botany, zoology, physiology, and microbiology. His knowledge of history, as well as interest in a breadth of subjects, made him a respected scholar in his day.

While many science textbooks of the early nineteenth century appeared staid, Pouchet used a large number of illustrations in his 1865 general biology text, *The Universe*. Pouchet stated that his reason for writing was to help people enjoy and develop a taste for the natural sciences. Pouchet sought to popularize the study of science. His work received more acclaim for the English edition than that on the European continent.

It is in the controversy with Pasteur over spontaneous generation that Pouchet is most remembered. The fundamental question as framed by Pouchet was whether microorganisms could generate on their own. Underlying the proposal was the connection of spontaneous generation with the ideas of Darwin and materialism. Because France was a Catholic country and the Church condemned the materialism of evolution, a fierce philosophical battle developed during the years 1859–1864. Pouchet famed the raging flames with a logical, simply written work published in Paris in 1859, that detailed his stance on spontaneous generation. Pouchet held that there were three factors necessary for spontaneous generation:

putrescence, or decaying organic matter, air, and water. Electricity and sunlight aided in the growth. Red light helped the formation of animal proto-organisms and green light aided vegetable proto-organisms. Pouchet sought to disprove the idea of airborne particles, which he called atmospheric micrography. Using a device that passed air through distilled water, he obtained deposits and analyzed the presence of starch, fibers, carbon, and minerals. Occasionally, there would be a mold spore. According to Pouchet, airborne contamination was unlikely and obviously spontaneous generation would follow. Pouchet described how microorganisms would grow in nutrient media under laboratory conditions that made every effort to protect the solutions from the air.

Pasteur was a dedicated Catholic who condemned spontaneous generation as materialistic and a part of **Darwinism**. In 1861, Pasteur demonstrated that deposits from filtered air contained these microorganisms, and that no growth would occur in filtered media. In 1864, Pasteur was able to convince the Academy of Science with a variety of experiments. Pouchet tried to keep the controversy alive. With Pouchet's death in 1872, the proponents of spontaneous generation were slowly defeated, leading the way for microbiology to blossom in the latter two decades of the nineteenth Century.

See also Evolutionary mechanisms; Spontaneous generation theory

PRADER-WILLI SYNDROME

Prader-Willi syndrome (PWS) is a genetic condition caused by the absence of chromosomal material from **chromosome 15**. The genetic basis of PWS is complex. Characteristics of the syndrome include developmental delay, poor muscle tone, short stature, small hands and feet, incomplete sexual development, and unique facial features. Insatiable appetite is a classic feature of PWS. This uncontrollable appetite can lead to health problems and behavior disturbances.

Initially, scientists found that individuals with PWS have a portion of genetic material deleted (erased) from chromosome 15. In order to have PWS, the genetic material must be deleted from the chromosome 15 received from an individual's father. If the deletion is on the chromosome 15 inherited from the mother, a different syndrome develops. This was an important discovery because it demonstrated for the first time, that the genes inherited from an individual's mother can be expressed differently than the genes inherited from the father.

Over time, scientists realized that some individuals with PWS do not have a deletion of genetic material from chromosome 15. Further studies found that these patients, instead of inheriting one chromosome 15 from their mother and the other from their father, atypically inherited both copies of chromosome 15 from their mother. When a person receives both chromosomes from the same parent the condition is termed **uniparental disomy**. When a person receives both chromosomes from their mother it is called maternal uniparental disomy.

Scientists are still discovering other causes of PWS. A small number of patients with PWS have a change (**mutation**) in the genetic material on the chromosome 15 inherited from their father. This mutation prevents certain genes on chromosome 15 from working properly. PWS develops when these genes function normally.

Human carry 46 chromosomes in the cells of their body. Chromosomes contain genes that regulate the function and development of the body. The 46 chromosomes in the human body are divided into pairs. Each pair is assigned a number or a letter. Chromosomes are divided into pairs based on their physical characteristics. Most chromosomes have a constriction near the center called the centromere. The centromere separates the chromosome into long and short arms. The short arm of a chromosome is called the *p* arm. The long arm of a chromosome is called the *q* arm.

Chromosomes in the same pair contain the same genes. However, some genes may be expressed differently depending on whether they were inherited from the egg or the sperm. Sometimes, genes are silenced when inherited from the mother. Other times, genes are silenced when inherited from the father. When genes in a certain region on a chromosome are silenced, they are said to be imprinted.

Chromosome 15 contains many different genes. There are several genes found on the q arm of chromosome 15 that are imprinted. A **gene** called SNPRN is an example of one of these genes. It is normally imprinted, or silenced, if inherited from the mother. The imprinting of this group of maternal genes does not typically cause disease. The genes in this region are not normally imprinted if paternal in origin. Normal development depends on these paternal genes being present and active. If these genes are deleted, not inherited, or incorrectly imprinted, PWS develops.

Seventy percent of the cases of PWS are caused when a piece of material is deleted, or erased, from the paternal chromosome 15. This deletion happens in a specific region on the *q* arm of chromosome 15. The piece of chromosomal material that is deleted contains genes that must be present for normal development. These paternal genes must be working normally, because the same genes on the chromosome 15 inherited from the mother are imprinted. When these paternal genes are missing developmental abnormalities occur (e.g., the symptoms associated with PWS).

In 99% of the cases of PWS the deletion is sporadic. This means that it happens randomly and there is not an apparent cause. It does not run in the family. If a child has PWS due to a sporadic deletion in the paternal chromosome 15, the chance the parents could have another child with PWS is less than 1%. In less than 1% of the cases of PWS there is a chromosomal rearrangement in the family, which causes the deletion. This chromosomal rearrangement is termed a **translocation**. If a parent has a translocation the risk to have a child with PWS would be higher than 1%.

PWS can also develop if a child receives both chromosome 15s from their mother. This is observed in approximately 25% of the cases of PWS. Maternal uniparental disomy for chromosome 15 leads to PWS because the genes on the chromosome 15 that should have been inherited from the father are missing. These paternal genes must be present, because the same genes on both the chromosome 15s inherited from the mother are imprinted.

PWS caused by maternal uniparental is sporadic. This means that it happens randomly and there is not an apparent cause. If a child has PWS due to maternal uniparental disomy the chance the parents could have another child with PWS is less than 1%.

Approximately 3-4% of patients with PWS have a change (mutation) in a gene located on the *q* arm of chromosome 15. This mutation leads to incorrect imprinting. This mutation causes genes inherited from the father to be imprinted or silenced. These genes are not normally imprinted. If a child has PWS due to a mutation that changes imprinting, the chance the parents could have another child with PWS is approximately 5%.

It should be noted that if an individual has a deletion of the same material from the *q* arm of the maternal chromosome 15 a different syndrome develops. This syndrome is called **Angelman syndrome** and occurs if an individual receives both chromosome 15s from the father.

Newborns with PWS generally have poor muscle tone (hypotonia) and do not feed well. This can lead to poor weight gain and failure to thrive. Genitalia can be smaller than normal. Hands and feet are also typically smaller than normal. Some patients with PWS have unique facial characteristics.

As children with PWS age, development is typically slower than normal. Developmental milestones, such as crawling, walking, and talking occur later than usual. Developmental delay continues into adulthood for approximately 50% of individuals with PWS. At about one to two years of age, children with PWS develop an uncontrollable, insatiable appetite. Left to their own devices, individuals with PWS will eat until they suffer from life-threatening obesity. The desire to eat can lead to significant behavior problems.

The symptoms and features of PWS require life long support and care. If food intake is strictly monitored and various therapies provided, individuals with PWS have a normal life expectancy.

PWS affects approximately 1 in 10,000–25,000 live births. It is the most common genetic cause of life-threatening obesity. It affects both males and females. PWS can be seen in all races and ethnic groups.

Infants with PWS have weak muscle tone (hypotonia). This hypotonia causes problems with sucking and eating. Infants with PWS may have problems gaining weight. Some infants with PWS are diagnosed with "failure to thrive" due to slow growth and development. During infancy, babies with PWS may also sleep more than normal and have problems controlling their temperature.

Some of the unique physical features associated with PWS can be seen during infancy. Genitalia that is smaller than normal is common. This may be more evident in males with PWS. Hands and feet may also be smaller than average. The unique facial features seen in some patients with PWS may be difficult to detect in infancy. These facial features are very mild and do not cause physical problems.

As early as six months, but more commonly between one and two years of age, a compulsive desire to eat develops. This uncontrollable appetite is a classic feature of PWS. Individuals with PWS lack the ability to feel full or satiated. This uncontrollable desire to eat is thought to be related to a difference in the brain which controls hunger. Over-eating (hyperpahgia), a lack of a desire to exercise, and a slow metabolism places individuals with PWS at high risk for severe obesity. Some individuals with PWS may also have a reduced ability to vomit.

Behavior problems are a common feature of PWS. Some behavior problems develop from the desire to eat. Other reported problems include obsessive/compulsive behaviors, depression, and temper tantrums. Individuals with PWS may also pick their own skin (skin picking). This unusual behavior may be due to a reduced pain threshold.

Developmental delay, learning disabilities, and mental retardation are associated with PWS. Approximately 50% of individuals with PWS have developmental delay. The remaining 50% are described as having mild mental retardation. The mental retardation can occasionally be more severe. Infants and children with PWS are often delayed in development.

Puberty may occur early or late, but it is usually incomplete. In addition to the effects on sexual development and fertility, individuals do not undergo the normal adolescent growth spurt and may be short as adults. Muscles often remain underdeveloped and body fat increased.

During infancy the diagnosis of PWS may be suspected if poor muscle tone, feeding problems, small genitalia, or the unique facial features are present. If an infant has these features, testing for PWS should be performed. This testing should also be offered to children and adults who display features commonly seen in PWS (developmental delay, uncontrollable appetite, small genitalia, etc.). There are several different genetic tests that can detect PWS. All of these tests can be preformed from a blood sample.

Methylation testing detects 99% of the cases of PWS. Methylation testing can detect the absence of the paternal genes that should be normally active on chromosome 15. Although methylation testing can accurately diagnose PWS, it can not determine if the PWS is caused by a deletion, maternal uniparental disomy, or a mutation that disrupts imprinting. This information is important for genetic counseling. Therefore, additional testing should be performed.

Chromosome analysis can determine if the PWS is the result of a deletion in the q arm of chromosome 15. Chromosome analysis, also called karyotyping, involves staining the chromosomes and examining them under a microscope. In some cases the deletion of material from chromosome 15 can be easily seen. In other cases, further testing must be performed. **FISH** (fluorescence in-situ hybridization) is a special technique that detects small **deletions** that cause PWS.

More specialized **DNA** testing is required to detect maternal uniparental disomy or a mutation that disrupts imprinting. This DNA testing identifies unique DNA patterns in the mother and father. The unique DNA patterns are then compared with the DNA from the child with PWS.

PWS can be detected before birth if the mother undergoes **amniocentesis** testing or **chorionic villus sampling** (CVS). This testing would only be recommended if the mother or father is known to have a chromosome rearrangement, or if they already have a child with PWS syndrome.

There is currently not a cure for PWS. Treatment during infancy includes therapies to improve muscle tone. Some infants with PWS also require special nipples and feeding techniques to improve weight gain.

Treatment and management during childhood, adolescence, and adulthood is typically focused on weight control. Strict control of food intake is vital to prevent severe obesity. In many cases food must be made inaccessible. This may involve unconventional measures such as locking the refrigerator or kitchen cabinets. A lifelong restricted-calorie diet and regular exercise program are also suggested. Unfortunately, diet medications have not been shown to significantly prevent obesity in PWS. However, growth hormone therapy has been shown to improve the poor muscle tone and reduced height typically associated with PWS. Special education may be helpful in treating developmental delays and behavior problems. Individuals with PWS typically excel in highly structured environments. Life expectancy is normal and the prognosis good, if weight gain is well controlled.

See also Chromosomal mutations and abnormalities; Genetic disorders and diseases; Hereditary diseases; Inherited cancers; Mutations and mutagenesis

PREDICTIVE GENETIC TESTING

An immediate consequence of researchers' opening of the genetic book of life has been the opportunity to know about the human genetic makeup and, increasingly, the human genetic destiny. The use of genetic information to predict future onset of disease in an asymptomatic (presymptomatic) person is called predictive genetic testing.

Every aspect of our being is influenced by both genes and environment. In the future, a strategy for influencing development may be to alter genes. At present, the environment in which genes act can sometimes be changed, and thereby moderate their impact (taking medications or avoiding specific hazards, for example). Sometimes there is no known way to change the deterministic power of a **gene**, though with increased knowledge of its workings there is always hope for future interventions. Whether or not the course of a disease can be altered, predictive information is increasingly available, and some people choose knowledge over uncertainty.

For generations, people have used family information to anticipate outcomes for themselves. Insurers consider parental age and cause of death for actuarial tables. **Evolution** in knowledge has been from information with considerable associated uncertainty to that with greater predictive capacity. **Huntington** disease (HD) became the prototype for predictive testing and serves to illustrate.

HD is a neurological disease with onset of symptoms usually during adulthood. It is inherited as an autosomal dominant trait; someone with an affected parent has a 50/50 chance of eventually developing the disease. The HD gene was the first human gene to be linked to an otherwise anonymous **DNA** marker (a **restriction fragment length polymorphism**, called G8), and long before the gene itself was identified, this marker and others like it became powerful predictive tools. Families in which HD was segregating were studied to determine which variant of the marker was tracking with the mutant HD gene; once that relationship was established, the marker(s) could be used to test family members who wished to know their genetic status. This indirect approach to testing was associated with some probability of error, since the markers were only close to the gene, not within it. With discovery in 1993 of the gene responsible for HD, a direct assay was immediately possible, with or without access to samples from other family members, and results became highly predictive.

The laboratory advances made access to this information possible, but it was quickly recognized that great care would be needed in the application of such knowledge to individuals at risk. A large Canadian collaborative study of predictive testing for HD, initiated in the late 1980s, has been particularly informative for assessing the impact of such information on individuals and families and developing guidelines for the practice of predictive medicine, including the need for supportive counseling and follow-up. Lessons from experience with this relatively obscure disorder were soon applied to other late-onset diseases for which predisposing mutations were identified. Notable in this context are **inherited cancers** such as familial breast **cancer** or colon cancer, other neurological disorders such as spino-cerebellar ataxias (including Machado-Joseph Disease), and familial Alzheimer's disease. Common afflictions such as heart disease, diabetes, and arthritis will eventually be amenable to similar investigations.

The **Human Genome Project** recognized the need for ethical considerations to match scientific advances, and its mandate includes significant support for research into ethical, legal and social issues. This has set new standards for the application of knowledge, respecting public concerns about the implications of new technologies. The opportunity to know ones genetic destiny has potential risks that must be mitigated in order for the benefits to be realized. Once the predictive test for HD was available, it was soon apparent that not everyone at risk wished to be tested. The right not to know is a significant issue. The genetic nature of these diseases adds complication, since information revealed about one individual may secondarily imply information about other family members, and individual choices will impact others in the family network. Acting upon respect for individual autonomy, early guidelines have advised against the testing of children for late-onset disorders in the absence of preventive options. In countries without universal health care, insurance implications of predictive testing are huge. One question involves whether people be required to submit a clean genetic bill of health in order to secure health or life insurance.

Eventually, there will be effective therapeutic interventions for diseases such as HD and Alzheimer's disease, individually tailored to the needs of those at risk. Until then, there will be controversy over the practice of predictive testing, but many will continue to choose knowledge and maintain hope for the future.

See also Ethical issues of modern genetics; Genetic counseling; Genetic testing and screening; Molecular diagnostics; Risk factors

PRENATAL DIAGNOSTIC TECHNIQUES

The ability to detect genetic disorders before birth has improved dramatically during the last two decades. Although considered fairly safe, these procedures are only performed on those patients that have a higher risk for disease or have experienced previous birthing difficulties, as complications to both the mother and fetus can occur. With the advent of newer molecular biology techniques, the procedures for performing prenatal diagnosis keep improving as well as an increased understanding of the genetic mechanisms involved.

Amniocentesis is the removal of amniotic fluid via a needle inserted through the maternal abdomen into the uterus and amniotic sac, in order to gain information about the status of **fetal cells**. This technique is performed in the second trimester of pregnancy (between 15 to 18 weeks gestation). In 1966, researchers were first able to culture and analyze fetal cells form amniotic fluid. Soon thereafter it became possible to diagnosis the fetal karyotype anomalies. A karyotype is a pictorial representation of the complete set of cellular chromosomes at a particular stage of **mitosis**.

Cytogenetic diagnosis for chromosomal anomalies include biochemical investigations for metabolic diseases and screening methods as well as determining alpha-fetoproteins (AFP) levels that are elevated with anacephaly and **neural tube defects** such as **spina bifida**.

Indications for the use of chromosomal procedures include advanced maternal age (≥35 yrs at delivery), previous offspring affected by chromosomal anomalies, a parent affected with chromosomal anomalies, a positive screening test (biochemical and/or fetal nuchal translucency), or fetal ultrasound that indicates morphologic anomalies.

Indications for biochemical analysis also include a previous offspring affected by a congenital metabolic error, a parents who may be a carriers of a congenital metabolic error, or if the mother is a **carrier** of an X-linked congenital metabolic error. Other factors can include a history of recurrent miscarriage (spontaneous and natural abortion of the fetus), or *in vitro* **fertilization** because a higher rate of fetal aneuploidies is reported in such cases.

Clinicians utilize three common techniques to conduct these studies. Amniocentesis can be performed "freehand" after an ultrasound evaluation of the fetus in order to deter-

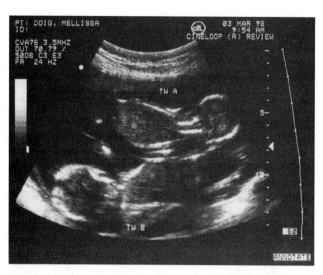

Multiple pregnancy, ultrasound, showing one baby on the bottom and the other at the top of the image. *Melissa Doig. Reproduced by permission.*

mine the site of the placenta, the position of the fetus, the amniotic fluid volume and placental pockets in the amniotic sac. The selected site is generally as high on the uterus as possible, away the placenta, and cephalic pole of the fetus (especially the face) and free of the umbilical cord. Making sure to monitor fetal movement, the physician marks the selected site with a soda-straw-like tube, which makes a little indentation on the skin. The area around the site is then cleaned with alcohol or iodophor (betadine) solution and a needle is then inserted, using the ultrasound to follow the path of the needle and guide it into the pocket of amniotic fluid.

Amniocentesis can also be performed "freehand" under direct ultrasound control with continuous needle tip visualisation. In this technique, the ultrasound probe (a linear probe is preferred) is used to withdraw the fluid. The needle is inserted in a parallel direction of the ultrasound scan plane, searching for an area of the uterus close to the end of the probe.

As a third alternative, amniocentesis may be performed under direct ultrasound control with probe adapter and continuous needle tip visualisation.

In all cases, after the needle is inserted, the first 0.0021 pint (1 ml) of fluid is discarded as it may contain the mother's skin cells that would contaminate the sample. Then 0.042 pint (20 ml) of amniotic fluid is withdraw—from which the fetal cells are then extracted for genetic analysis.

Although amniocentesis yields valuable genetic information, complications–including slight increased risk of miscarriage and infections (chorioamnionitis) demand a cautious approach to the use of the test.

Chorionic villus sampling (CVS) is placental villi biopsy performed between the tenth to twelfth week of gestation. Because the placental villi and the fetus are derived from the same tissue, CVS represents a source of genetic material identical to the fetus. Furthermore, these cells are actively growing and tissue cultures can be obtained more quickly

than those of the amniotic fluid cells. The indications for performing a CVS are similar to those for amniocentesis. Since it is a technique use din the first trimester (i.e. the first three months of a normal nine month pregnancy), it is possible to determine problems with the fetus earlier then with amniocentesis. One possible problem associated with its use is a possible discrepancy between fetal and trophoblast karyotype that can lead to false negative or false positive results. Another complication is that CVS has been associated with fetal limbs reduction.

CVS offers the diagnostic possibility for early cytogenetic based diagnosis of chromosomal anomalies and is part of the molecular analysis for Mendelian diseases.

CVS involves inserting a needle (abdominally) or a catheter (cervically) into the developing placenta but staying outside the amniotic sac. Then suction is applied with a syringe and about 10-15 milligrams of tissue are aspirated into the syringe. The first technique used was the transcervical (through the cervix), and then a transabdominal (through the abdomen) performance of CVS was suggested in order to avoid fetal contamination with **bacteria** coming from the cervix.

CVS obtained tissue is manually cleaned of maternal uterine tissue and then grown in culture, and a karyotype is determined as with amniocentesis. Three cytogenetic techniques are available for the diagnosis: direct preparation (using the spontaneous mitosis of the trophoblast, short term culture (incubation of 12-24 hours of the trophoblast cells) and long term culture, where mesenchimal cells are put in a medium. The diagnosis is available after 10-15 days.

CVS procedures carry a slight increased risk of miscarriage (approximately three times greater than with amniocentesis), infections (chorioamnionitis), leakage of amniotic fluid, fetal limb reduction. Contamination of tissues can also lead to false, or misleading diagnostic results.

Cordocentesis is a technique that allows the withdrawal of a fetal blood sample. It must be performed after the 18th week of gestation when the size of the umbilical vessels is large enough to insert a needle. Also, at this time there is sufficient fetal blood volume to obtain a sample without too much risk. In 1979, researchers were first able to obtain uncontaminated blood from the umbilical cord. The source of contamination usually includes both amniotic fluid and maternal blood.

Cordocentesis can also be utilized to provide rapid karyotyping in order to verify structural malformations of the fetus or as a confirmation of mosaicism detected by CVS and/or amniotic fluid examination. As with other prenatal diagnostic procedures there are risks of complications that include an increased risk of miscarriage (at a higher rate then that observed for amniocentesis and CVS), infections (chorioamnionitis), and fetal hemorrhage (bleeding).

See also Chromosomal mutations and abnormalities; Embryology; Hereditary diseases; Heterozygous advantage; Mutations and mutagenesis

PRÉVOST, JEAN-LOUIS (1790-1850)

Swiss physician

Swiss-born physiologist and physician Jean-Louis Prévost was a pioneer in the anatomy of the reproductive system, as well as hematology, or study of the blood. Prévost's discoveries in the early half of the nineteenth century paved the way for later revelations about reproduction and blood transfusion.

Born in Geneva, Switzerland, Prévost as a child had the privilege of knowing outstanding naturalists of the day, such as François and Jean-Pierre Huber and Jean Senebier. In the early years of the nineteenth century, it was not unusual for young men to begin studies in theology and later turn to medicine. One of the great educational centers for theology was the University of Paris, and Prévost attended there two years before transferring to the University at Edinburgh, Scotland. He attended Edinburgh from 1816–1818, where he received his doctor of medicine. The dire need for physicians in Ireland inspired Prévost to go to Dublin to begin his medical practice.

Returning to Geneva, Prévost set up a medical practice but soon became disenchanted with clinical medicine, and decided his true interest was in research. At the time, early microscopes were more refined, and chemistry was developing into a respectable science. Prévost organized a group of scientists to write important papers and memoirs. Included in the group were A. LeRoyer, H. Leber, Antoine Morin, and Jean-Baptist Dumas (1800–1884). These friends were also interested in care of the poor and indigent, and with the help of Louis Goses, a benefactor, the group founded a hospital to treat the poor without cost. This was one of the first outpatient hospitals in Europe.

Inspired by his friends, especially Dumas, Prévost studied sperm, showing the origin of sperm was in the testicles of the male. Dumas and Prévost wrote a book in 1821 *On the Sperm of Different Animals.* Their experiments were built on the work of Lorenzo Spallanzani (1729–1799) and others, but the two scientists were able to illuminate the subject because of the refined technology.

Next, Prévost turned to experimental **embryology**, continuing to publish his findings with Dumas in 1824. By studying frogs' eggs, they confirmed the classic laws of the development of fertilized eggs developed by Jan Swammerdam (1637–1680) and K.E. von Baer (1792–1876). For their work they received recognition and the Prix Montyon of the Paris Academy of Science.

Prévost then turned his attention to the process of digestion. Because he was interested in chemistry, Prévost was convinced chemical processes were the major components of digestion, and devised many experiments to prove his ideas. In this sense, Prévost became one of the first biochemists. Next, blood became Prévost's major interest. He approached hematology by analyzing circulation and composition of the blood, and also studying the embryological development of the heart in several animals. Prévost was one of the first to suggest the possibility of blood transfusion.

PRIMERS

A primer is an engineered piece of **deoxyribonucleic acid**. This short segment of **DNA**, typically 18–24 **bases**, is deliberately made to be complementary to a target stretch of DNA that lies immediately beside another stretch of DNA that is destined to be copied. The principle use of primers is used in the copying of DNA via the **polymerase chain reaction** (PCR).

The use of primers is vital to the performance of the **polymerase** chain reaction. The PCR technique itself has revolutionized molecular biology and areas such as forensic pathology, making the detection of even a single molecule of DNA achievable.

Primers are used following the selective cutting of the DNA. The two strands of nucleotides that comprise DNA can be selectively cut open using one of a variety of a class of **enzymes** called **restriction enzymes**. Each enzyme recognizes a **nucleotide** sequence at which to cut the DNA. Some restriction enzymes produce a staggered cut, in which a region of each single strand of DNA is exposed. The primer segments of DNA are complimentary to each of the two exposed single stranded regions. Because they are complimentary, the primer sequences are able to bind to the exposed single stranded regions of the restriction enzyme-treated DNA.

The restriction enzyme cutting of DNA can be performed at an elevated temperature, which discourages the DNA from annealing back together (rejoining). Once the temperature is reduced, annealing can occur. Then, the primer sequences, which are present in a large concentration, anneal to their target DNA strand. The primers provide an initiation site for the elongation of a complimentary DNA strand. Repeated cycles amplifies the quantity of the target region of DNA and forms the basis of PCR.

See also Complementary DNA (cDNA) and RNA (cRNA); Hybridization of DNA

PRIONS

A prion, short for "proteinaceous infectious articles," is a protein capable of causing both inheritable and communicable disease by inducing benign proteins to change their shape. This shape change renders the protein infectious. When the notion of prions was first suggested in the 1970s by **Stanley Prusiner**, a neurologist at the University of California at San Francisco, it was heretical. Dogma held that the conveyers of transmissible diseases required genetic material, composed of nucleic acid (**DNA or RNA**), in order to establish an infection in a host. Time and the accumulation of experimental evidence proved Prusiner correct. For his discovery of prions and the elucidation of their infectious mechanism, Prusiner received the 1997 Nobel Prize in medicine.

The known prion diseases, all fatal, are known as spongiform encephalopathies. They frequently cause the brain to become riddled with holes, and rod-shaped particles clumped together in large arrays are evident. The diseases are widespread in animals and develop slowly over the course of years, even decades in humans. The most common form is scrapie, found in sheep and goats. Afflicted animals lose coordination and become incapacitated, unable to stand. Other prion diseases of animals are called transmissible mink encephalopathy, chronic wasting disease of mule deer and elk, feline spongiform encephalopathy and bovine spongiform encephalopathy (also referred to as mad cow disease). The list of known human prion diseases is growing, and includes Kuru, Creutzfeldt-Jacob disease, Gerstmann-Straussler-Scheinker disease and fatal familial insomnia. Unproven, but suspected human diseases include **Alzheimer's disease**, Parkinson's disease and amyotrophic lateral sclerosis ("Lou Gehrig's disease").

Research by Prusiner and others in the 1970s and 1980s demonstrated that prions produced a single **gene** product, a protein called PrP (for prion protein). In the late 1980s, Prusiner and colleagues showed that PrP was present in the brains of people afflicted with Gerstmann-Straussler-Scheinker syndrome, a rare spongiform encephalopathy. The protein was also present in healthy individuals, but differed from the other Prp by a single amino acid— a **point mutation**. Analysis of a large number of patients with the syndrome established a genetic linkage between the **mutation** and the disease. Further research found scrapie to result from a similar mutation.

The mechanism of prion's is not fully understood. It is known that the disease is caused by a shape change of PrP. The mutated PrP adopts a sheet-like structure, whereas the normal form is more helical in shape. Yet, mutated and normal forms of PrP have been shown to have the same **amino acid** sequences. Somehow, the accumulation of the sheet-like forms in the brain causes damage. Preventing the conversion to the sheet format might be a useful therapy or preventative step in prion diseases.

The tranmissibility of prion from one **species** to another is controversial and still unresolved. PrPs from different species differ in their amino acid sequences. The differences can be slight to considerable; cows and human PrPs are very divergent, differing by over 30 amino acids. Prusiner and others are pursuing the notion that species barrier may be breached if the PrPs are very similar. Nonetheless, the development of Creutzfeld-Jacob disease in several farmers whose herds developed mad cow disease might be noteworthy.

See also Epidemiology and genetics

PROBABILITY AND STATISTICS

Probability is a quantitative statement of the likelihood or chance that some specified event has happened or will happen in the future. Probability is a number between zero and one, and is often abbreviated with the lowercase letter "p". Zero probability means that the event cannot happen (i.e., it is impossible). A probability of one means that an event is sure to happen (i.e., a certainty). Probability can be expressed as a fraction, where the probability p=0.01 is the same as saying p = 1/100.

The concepts of probability are perhaps best understood by way of examples. Common examples use dice, coins, or cards to illustrate probability in concrete ways. For example, what is the chance of rolling a four with a single roll of one die? If the die is conventional, with one number on each of its six faces, and the die is weighted evenly, so that no one surface has any advantage over the others to land face up, the chance is 1/6, because only one face of the six faces of the cube has the number four. To confirm the accuracy of this estimate of 1/6, one could roll the die many times and record the proportion of times that a four comes up. If it were found that a four does not come up 1/6 of the time, but rather some different number, one might question the fairness of the die.

If a certain outcome makes it impossible for another event to occur, the events are said to be mutually exclusive. For example, if a coin is tossed, and it comes up heads, it is not possible for it to also come up tails at the same time. Also, if a certain outcome in no way alters the chance of some different event, their probabilities are said to be independent. For example, suppose there was a large bucket with coins of various kinds (pennies, nickels, dimes, quarters), and one was to choose a coin and flip it to see the outcomes, the chance of obtaining heads on the flip is independent of the chance that one would one have selected a dime from the bucket. These are independent probabilities.

Probabilities of different events can be combined in certain ways. The chance that both of two events would occur is found by multiplying the chance for each event. This assumes the events are not mutually exclusive. The chance of either of two events occurring is found by adding their individual probabilities, and subtracting the chance that they both occurred. What is the chance of drawing the ace of spades from a standard deck of poker cards? This is the combined chance of drawing both a spade and an ace, so their probabilities are multiplied. There are 13 spades in a deck of 52, so that chance is 13/52, or 1/4. There are 4 aces in a deck of 52, so that chance is 4/52, or 1/13. The chance of getting the ace of spades is 1/13 X 1/4, or 1/52. What is the chance of drawing either an ace or a spade? Here, the second rule for adding probabilities is used. The chance would be 1/4 + 1/13 - 1/52, or 4/13.

Statistics can be described as the collection and analysis of data. In the example above of rolling a die many times to determine if it is fair, how is it determined whether the finding was significant or not? Consider a die rolled 600 times. It would be expected for the number four to be rolled 100 times. If it was rolled only 99 times, or 101 times, it is understood that this is close enough to the expected 1/6 to accept the die as fair. Likewise, the number four was rolled only 10 times out of 600, it would be easily concluded that the die was not a fair one. What if a four came up 95 times, or 90 times? At some point between the extremes, intuition fails, and an objective method of deciding where to draw the line is needed. Here, a statistical test can help maintain objectivity in deciding between different alterative interpretations of the data from our experiment. In the broader sense, statistics is the science of making decisions about the meaning and significance of the data we obtain, and for describing the way various data are distributed.

See also Genetic counseling: Risk calculations using Bayesian statistics; Mendelian genetics and Mendel's laws of heredity; Statistical analysis of genetic syndromes

PROBE · *see* DNA PROBES

PROGENY · *see* FILIAL GENERATION

PROGRAMMED CELL DEATH · *see* APOPTOSIS

PROKARYOTE

Prokaryotes are cells or organisms that lack a nuclear membrane and membrane bound organelles.

Prokaryotic cells differ from eukaryotic cells in important ways. Prokaryotic cells are usually much smaller than eukaryotic cells, and are able to grow and divide rapidly. **Eukaryotes**, with a true membrane-bound **nucleus** and discrete membrane-bound organelles, are able to compartmentalize functional activity into structural components and organelles such as **mitochondria**, chloroplasts, lysosomes, endoplasmic reticulum, and Golgi complex. Prokaryote cells carry out the essential activities that take place in eukaryotic organelles without the benefit of these specialized structures.

Prokaryotes and eukaryotes differ to some extent in their biochemical composition, particularly with respect to their lipid composition. These biochemical differences also allow prokaryotes to differ in many aspects of their metabolism.

Other than the morphological differences, perhaps the most significant of the differences between prokaryotes and eukaryotes involves the nature of genes and the manner in which genetic information is expressed. The genes of prokaryotes are simpler and do not contain the large amounts of material incidental to **protein synthesis** found in eukaryotes.

Prokaryotes have been present on Earth for much of its history, and in one form or another have occupied nearly all potential habitats. Although some prokaryotic **bacteria** cause disease in plants and animals, the overwhelming majority do not. Many prokaryotes are benign and necessary inhabitants of the larger ecosystem.

See also Bacterial genetics; Chromosome; Comparative genomics; Conservation genetics and biodiversity; Evolution, evidence of; Evolutionary mechanisms; Organelles and subcellular genetics; Origin of life

PROKARYOTIC GENOMES (CIRCULAR RNA AND DNA)

Life first appeared on Earth about 3.7 billion years ago in the form of unicellular organisms, i.e., single-cell organisms, termed prokaryotes. **Bacteria** and archaea are examples of ancient unicellular organisms that supposedly have evolved from a common primeval ancestral, through successive genetic variations, **DNA recombination** and **gene** divergence. Prokaryotes, (from the Greek, *pro* [before] and *karya* [nucleus]) are therefore unicellular organisms without an organized **nucleus** separating chromosomes from the cytoplasm.

Archaea and bacteria were first considered as one sole **species** because of their similar anatomies. Recent genetic analysis, however, had established archaea as a species apart from bacteria. This analysis revealed that only 44% of archaea's genes are similar to those found either in bacteria or in unicellular **eukaryotes** known as **yeast** cells; and the remaining 56% are genes exclusive of archaea, completely different from those found in the other two species.

Prokaryotes found in nature today store their encoded genetic information (i.e., genes) in circular chromosomes, mainly formed by double-stranded DNA. Scientists assume that the first prokaryotes had simpler cellular structures, storing their genetic information in circular **RNA** strands that enabled them to make only rudimentary proteins. This hypothesis states that only later, through evolutionary changes and due to the appearance of new catalysts (**enzymes**), the first DNA molecules were developed and used as a more efficient way of storing genes, as do the prokaryotes known at present. Once DNA was available, prokaryotic organisms began using RNA molecules for complementary functions during **protein synthesis** and **DNA replication**, while DNA became the permanent repository of genes. DNA opened the opportunity for more complex cellular functions to be developed because it confers a better genomic stability than does RNA. DNA and RNA are presently found in every **cell**, whether prokaryotic or eukariotic cells.

A great number of crucial proteins such as structural and functional proteins, enzymes, hormones, growth factors, etc., and many cellular functions such as chemical signaling among cells, energy production, enzymatic systems for DNA damage repair, etc., were first developed by prokaryotes. These proteins were further conserved in new different species throughout the evolutionary chain, from yeast cells to mammals. This phenomenon, known as evolutionary conservation, implies that certain genes were first developed by prokaryotes, then further conserved in higher species.

A well-studied **prokaryote** is the bacterium *Escherichia coli,* because of its simple structure and fast proliferation. *E. coli* has a rod-like shape, and its plasmatic membrane is similar to that found in eukaryotic cells, but the external face of the membrane is rigid and thicker than the internal face, thus sustaining the bacterial shape. The cytoplasm contains **ribosomes** bound to **messenger RNA** that form polyribosome complexes that take a crucial part in protein **translation**. DNA in *E. coli* is usually organized in two identical circular chromosomes, each located in different regions of cytoplasm, and attached to the internal face of the plasmatic membrane. The region where the **chromosome** is located is termed nucleoid, and although *E. coli* may have one, two, or sometimes even more chromosomes, they are always identical.

A peculiar trait of prokaryotes is the ability to transfer extra packages of genetic information, termed **plasmids**, among individuals of the same species and even among different groups. The plasmid is not an integral part of the prokaryotic **genome** and replicates independently of bacterial DNA. Plasmids are constituted by a short, ring-shaped double strand of extra genes. Individuals carrying one or more plasmids are better fitted to survive in stressful conditions, such as the presence of **antibiotics**, toxic compounds, high acidity medium, etc. Plasmids are also responsible for the well-known bacterial ability to acquire resistance against a number of antibiotics. This horizontal gene transfer among individuals, i.e., plasmid transfer, may occur through different approaches, such as **transformation** and conjugation.

During transformation, a bacterium finds free fragments of DNA delivered in the process of death of another bacterium. A protein complex, present on the outer membrane of bacteria, collects the free DNA pieces and digests one of the strands, releasing the nucleotides in the process, while the other strand is integrated to its chromosome. Transformation may also occur without digestion of one of the strands, when the particle taken is a plasmid, which is then partially integrated to the bacterial chromosome.

Conjugation may occur through two different ways. The donor cell shoots out extensions, known as pilli (pillus, in the singular), while the receptor cell presents one or more receptor docking sites. Sometimes, several donors conjugate with a single receptor, thus forming aggregates. Once the donators and the receptor aggregate, plasmids transfer through pilli to the receptor cell. The other way of plasmid transfer through conjugation is when the donor cells secretes chemical messengers that attract receptor cells and stimulate them to secrete adhesion proteins that bind both receptor and donator together. Once in contact, pores open in the membranes of both cells and the plasmids transfer.

Prokaryotic DNA also contains a class of genes termed transposons or "jumping genes" that can bind to other genes and transfer gene sequences from one site of the chromosome to another, or from the chromosome to a plasmid, also from a plasmid to the chromosome. There are three main types of transposons: insertion sequences of **bases** with length between 150-1500 bp (base pairs) that integrate to chromosome by homologous recombination; complex transposons, such as Tn5 and Tn3, that contain a central region of extra genes besides those necessary for **transposition**, and are mainly responsible for antibiotic resistance and other metabolic functions; and, phage-associated conjugative transposons. Phage-associated conjugative transposons are derived from the association of phage particles (proteins) and transposons elements. Phage proteins are resultant from **bacteriophage transduction** of viral proteins into the DNA of infected bacteria that are further integrated to bacterial genome through homologous recombination. Bacteriophages are viruses that infect bacteria and utilize their replicative system to synthesize its own viral genome and proteins.

Another kind of transposon gene, termed Type II transposons, do not have insertion sequences, but also carry extra genes, such as those that confer antibiotic resistance, and can be transferred horizontally among individuals. Instead of jumping, conjugative transposons perform replicative transposition; i.e., they replicate and transpose the copy to a plasmid that later is passed to another individual or will segregate in the daughter cells. This class of plasmid is known as resistance plasmids or R factors.

Although plasmid transfer is more commonly observed among bacterial individuals pertaining to the same group, such as gram-positive or gram-negative bacteria, genetic exchanges were also observed between individuals of both groups, as well as among different groups of archaea, or between bacteria and archaea. Even more surprising was the discovery that horizontal gene transfers also occur in nature among unrelated species, such as bacteria and unicellular eukaryotes (yeast cells and unicellular plants), and between archaea and yeast cells. This phenomenon is known as non-homologous or "illegitimate" DNA recombination because it may occur between chromosomes of species not related, and also between non-homologous DNA sequences. Evidence suggests that such horizontal gene transfers may also have contributed to foster the emergence of new species and biological diversity.

See also Bacterial genetics; Chromosome; Comparative genomics; Conservation genetics and biodiversity; Evolution, evidence of; Evolutionary mechanisms; Genome; Homologous chromosomes; Organelles and subcellular genetics; Origin of life; Recombinant DNA molecules; Transformation; Transposition

PROMOTER • *see* TRANSCRIPTION

PROPHAGE

A prophage is the **DNA** of a **bacteriophage** that is incorporated into the **genome** of a host **cell** during the lysogenic cycle. A bacteriophage is a **virus** that is only able to infect and reproduce in **bacteria**. The prophage enables the bacteriophage to replicate its own DNA without disruption of the host cell's DNA. After **replication**, the virus is able to exit the bacterial **chromosome** and further infect other cells.

Viral replication can occur by one of two cycles: the lysogenic cycle or the lytic cycle. The lysogenic cycle reproduces viral DNA usually without harming the host bacterial cell. Viruses that reproduce via the lysogenic cycle are called temperate viruses. In the lysogenic cycle, the bacteriophage attaches to the outer surface of a bacterium and injects its DNA into the bacterial chromosome. The bacteriophage then circularizes and combines its DNA with the bacterial chromosome and becomes a noninfectious prophage. The cell's new DNA, a combination of the phage and the host cell's genome, replicates producing identical bacterial cells. The prophage continues to spread as the bacterium reproduces regularly through **cell division** without harming the host cell. Sometimes the viral DNA is excised from the bacterial DNA. When this happens, the viral DNA begins to replicate without being incorporated into the host cell's DNA. Eventually, the viral DNA is assembled into new bacteriophages and causes the cell to lyse, killing the host and releasing the new viruses.

The lytic life cycle is similar to the lysogenic cycle however it does not express a prophage. Instead of combining its DNA into the bacterial chromosome of the host cell to form a prophage, the virus shuts down the DNA of the host and orchestrates the production of its own genome. Viruses that are solely disposed to the lytic cycle are called virulent viruses. Unlike the lysogenic cycle, a bacteriophage that reproduces via the lytic cycle ultimately destroys the host bacterial cell. Although viral reproduction through the lytic cycle does not display a prophage, it is an important process in the production and spreading of viruses.

See also Bacteria; Bacteriophage

PROPHASE · *see* CELL DIVISION

PROTEASE AND PROTEASE INHIBITORS

A protease is an enzyme that specifically recognizes and breaks apart a given protein or similar proteins. The destruction of the protein is accomplished by the ability of the protease to break the peptide bonds that link the constituent parts of the protein—the amino acids—together.

Proteases are present in every living **cell** of every living organism. As soon as the cells are disrupted, the proteases are released and can quickly degrade any given protein of interest. In experimental work this behavior is highly undesirable, as the protease destruction of other proteins greatly reduces the yield of the target protein in isolation experiments.

Fortunately, chemical substances called protease inhibitors can prevent protease activity. A protease inhibitor is any substance that partially or completely blocks functioning of the protease. This can be accomplished by the binding of the inhibitor to the **active site** within the protease at which the breakage of the peptide bonds occurs. Other protease inhibitors exert their influence by binding to some other place on the protease's structure that prevents, through a conformational (shape) change in the protein, the active site from functioning properly. Because of the binding specificity, a given inhibitor will be active against only one or a certain number of types of protease.

Protease inhibitors have become important clinically because of their promise as a treatment for AIDS—acquired immunodeficiency syndrome. They work by blocking the activity of a protease associated with the **AIDS virus**, the

human immunodeficiency virus (HIV). HIV protease clips a long viral protein into several differently sized smaller proteins. These smaller proteins are necessary for proper HIV assembly and for the ability of these newly-made virus particles to infect other cells once they are released from the infected cell. It is the processing of the long viral protein into the infection-critical smaller proteins that is blocked by protease inhibitors. The spread of the virus to new cells is subsequently halted.

To combat HIV, protease inhibitors are usually administered as a cocktail with other agents that block the **transcription** of the viral **RNA** into **DNA**. This treatment is costly and monitoring of the body's response is important, as resistance to protease inhibitors can develop.

See also Enzymes; Pharmacogenetics; Retroviruses

PROTEIN ELECTROPHORESIS

Protein electrophoresis is a sensitive analytical form of chromatography that allows the separation of charged molecules in a solution medium under the influence of an electric field. A wide range of molecules may be separated by electrophoresis, including, but not limited to **DNA**, **RNA** and protein molecules.

The degree of separation and rate of molecular migration of mixtures of molecules depends upon the size and shape of the molecules, the respective molecular charges, the strength of the electric field, the type of medium used (e.g., cellulose acetate, starch gels, paper, agarose, polyacrylamide gel, etc.) and the conditions of the medium (e.g., electrolyte concentration, pH, ionic strength, viscosity, temperature, etc.).

Some mediums (also known as support matrices) are porous gels that can also act as a physical sieve for macromolecules.

In general, the medium is mixed with buffers needed to carry the electric charge applied to the system. The medium/buffer matrix is placed in a tray. Samples of molecules to be separated are loaded into wells at one end of the matrix. As electrical current is applied to the tray, the matrix takes on this charge and develops positively and negatively charged ends. As a result, molecules such as DNA and RNA that are negatively charged, are pulled toward the positive end of the gel.

Because molecules have differing shapes, sizes, and charges they are pulled through the matrix at different rates and this, in turn, causes a separation of the molecules. Generally, the smaller and more charged a molecule, the faster the molecule moves through the matrix.

When DNA is subjected to electrophoresis, the DNA is first broken by what are termed **restriction enzymes** that act to cut the DNA is selected places. After being subjected to restriction **enzymes**, DNA molecules appear as bands (composed of similar length DNA molecules) in the electrophoresis matrix. Because nucleic acids always carry a negative charge, separation of nucleic acids occurs strictly by molecular size.

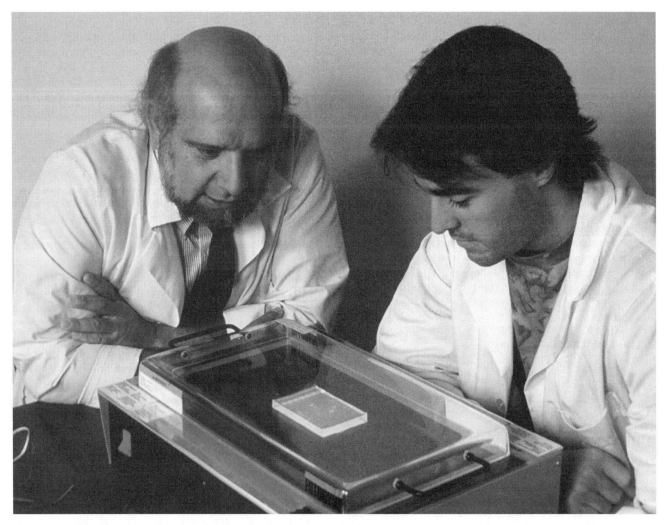

Molecular geneticists use protein gels to characterize DNA. This is a gel of Down syndrome DNA. *SIUBIOMED/Custom Medical Stock Photo. Reproduced by permission.*

Proteins have net charges determined by charged groups of amino acids from which they are constructed. Proteins can also be amphoteric compounds (a compound that can take on a negative or positive charge depending on the surrounding conditions.) A protein in one solution might carry a positive charge in a particular medium and thus migrate toward the negative end of the matrix. In another solution, the same protein might carry a negative charge and migrate toward the positive end of the matrix. For each protein there is an isoelectric point related to a pH characteristic for that protein where the protein molecule has no net charge. Thus, by varying pH in the matrix, additional refinements in separation are possible.

The advent of electrophoresis revolutionized the methods of protein analysis. Swedish biochemist Arne Tiselius was awarded the 1948 Nobel Prize in chemistry for his pioneering research in electrophoretic analysis. Tiselius studied the separation of serum proteins in a tube (subsequently named a Tiselius tube) that contained a solution subjected to an electric field.

Sodium dodecyl sulfate (SDS) polyacrylamide gel electrophoresis techniques pioneered in the 1960s provided a powerful means of protein fractionation (separation). Because the protein bands did not always clearly separate (i.e., there was often a great deal of overlap in the protein bands) only small numbers of molecules could be separated. The subsequent development in the 1970s of a two-dimensional electrophoresis technique allowed greater numbers of molecules to be separated.

Two-dimensional electrophoresis is actually the fusion of two separate separation procedures. The first separation (dimension) is achieved by isoelectric focusing (IEF) that separates protein polypeptide chains according to **amino acid** composition. IEF is based on the fact that proteins will, when subjected to a pH gradient, move to their isoelectric point. The second separation is achieved via SDS slab gel electrophoresis that separates the molecule by molecular size. Instead of broad, overlapping bands, the result of this two-step process is the formation of a two-dimensional pattern of spots, each comprised of a unique protein or protein fragment. These spots are subsequently subjected to staining and further analysis.

Some techniques involve the application of radioactive labels to the proteins. Protein fragments subsequently obtained from radioactively labels proteins may be studied my radiographic measures.

There are many variations on gel electrophoresis with wide-ranging applications. These specialized techniques include Southern, Northern, and Western Blotting. Blots are named according to the molecule under study. In Southern blots, DNA is cut with restriction enzymes then probed with radioactive DNA. In Northern blotting, RNA is probed with radioactive DNA or RNA. Western blots target proteins with radioactive or enzymatically-tagged antibodies.

Modern electrophoresis techniques now allow the identification of homologous DNA sequences and have become an integral part of research into **gene** structure, **gene expression**, and the diagnosis of heritable diseases. Electrophoretic analysis also allows the identification of bacterial and viral strains and is finding increasing acceptance as a powerful forensic tool.

See also Blotting analysis; DNA fingerprinting and forensics; DNA structure

PROTEIN SYNTHESIS

Protein synthesis is the process by which cells convert amino acids into long chain polymers called proteins. Proteins are molecules that have a variety of functions in cells such as providing structure, storing energy, providing movement, transporting other substances, catalyzing biological reactions, and protecting against disease. Proteins make up more than 50% of a cell's dry weight. Protein synthesis is programmed by **DNA**. During this process DNA is converted to **RNA** which is then translated into a protein by the **ribosomes**.

The theories that laid the foundation for modern understanding of protein synthesis began in 1909 with **Archibald Garrod**. He was the first to suggest that genes were chemically expressed through **enzymes** that catalyze specific chemical reactions in the **cell**. He even theorized that an inherited disease reflected a person's inability to make a particular enzyme. Unfortunately, his ideas about **inheritance** were ahead of their time, and it took several decades before they were supported by further research.

This research came in the 1930s when George Beadle and Edward Tatum established the relationship between genes and enzymes. They discovered that in certain bread mold **species**, there were mutants that required extra nutrients to grow on a plate. These different mutants were thought to be deficient of certain metabolic enzymes. Through their research, Beadle and Tatum established the **one gene-one enzyme** hypothesis, which states that the function of a **gene** is to dictate the production of a specific enzyme. This idea was later refined when it was found that genes also dictate the production of proteins.

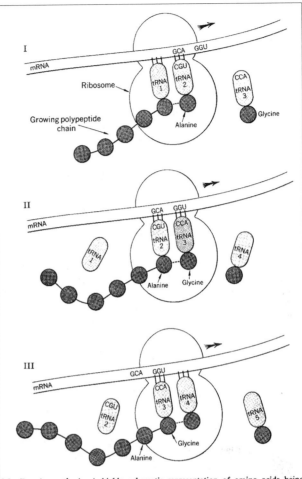

8-8 Protein synthesis. A highly schematic representation of amino acids being attached to a growing polypeptide chain. A single ribosome is shown at three different times.

Protein synthesis. *Oxford University Press, Inc. Used by permission of Oxford University Press, Inc.*

Genes are the genetic material that dictate the production of **proteins and enzymes**. The genes are located in the **nucleus** of the cell and are composed of DNA. The DNA is composed of nucleotides which are molecules made up of a sugar component, a phosphate group and a cyclic base. There are four different **nucleotide bases** in which all the information for making proteins is stored. A gene is typically hundreds or thousands of nucleotides long. Unlike DNA, proteins are made up of amino acids instead of nucleotides. To get from DNA to protein requires two steps, **transcription** and **translation**.

The first step in protein synthesis involves the transcription of DNA into **messenger RNA (mRNA)**. While DNA and RNA are similar, there are subtle differences. For example the sugar component of DNA has one less hydroxyl group than ribose, the sugar component of RNA. Also, RNA uses the nucleotide uracil instead of thymine.

During the process of transcription an enzyme known as RNA **polymerase** separates an area of the double strands of DNA. It first binds to a region of DNA known as a promoter region. It then begins to transcribe, or copy, the DNA into RNA at an initiation site. Only one strand is used as a template for making the mRNA. This occurs by the **base pairing** of the nucleotides. The polymerase than moves along the DNA and elongates the RNA molecule. As the polymerase travels along the DNA it both unwinds, catalyzes copying and rewinds the DNA. When the polymerase comes upon a certain sequence of DNA, called the termination sequence, it stops transcription and releases the RNA. The resulting RNA contains an exact copy of the gene. This RNA is further processed by **RNA splicing** to remove introns. The result is mRNA which is transported out of the nucleus into the cytoplasm where it can be translated into a protein.

Translation is the actual production of the protein or polypeptide. It is the process by which the mRNA nucleotide code is translated into amino acids. It turns out that different combinations of three nucleotides code for the 20 amino acids. For example, the base triplet GCA is translated into the **amino acid** alanine. Thus, the genetic code is in the form of triplet **codons**. Since it takes three nucleotides to code for 1 amino acid, each gene must have three times as many nucleotides as amino acids that make up the protein. For example, a strand of mRNA that has 300 nucleotides would code for a protein that is 100 amino acids long.

The process of translation involves cellular organelles called ribosomes. The ribosome are made up two subunits which are composed of proteins and **ribosomal RNA**. During translation the small subunit of the ribosome attaches to the mRNA at the initiation site. This is a sequence of three nucleotides which is the same in every gene. The large subunit then converges on the structure carrying with it another type of RNA called **transfer RNA** (RNA). The tRNA has a reading portion which recognizes three nucleotides. It interacts with an enzyme called aminoacyl-tRNA synthetase which attaches specific amino acids to the tRNA.

After initiation, the amino acids are combined to create a growing polypeptide chain. The ribosome travels along the mRNA in blocks of three nucleotides, or codons. The tRNA reads the codon and new amino acids are added one by one. When the ribosome gets to a certain codon, called the stop codon, it ceases translation. The polypeptide chain is then modified by other enzymes in the cytoplasm. These posttranslational modifications can include the attachment of sugars or lipids, the removal of certain amonio acids, or the joining of one polypeptide chain with another to make a multiple unit protein.

Cracking the genetic code was done during the 1960s. The first experiment was done by **Marshall Nirenberg**. He created an artificial RNA molecule which contained only uracill. He put this in a test tube with ribosomes and the other materials needed to synthesize proteins. The result was a protein that contained only phenylalanine. This suggested that a codon of UUU specified the amino acid phenylalanine. After this experiment, a series of similar ones were run until the entire genetic code was determined. Now when scientists get the nucleotide sequence of a gene, they know exactly the amino acid sequence of the protein it codes for. The genetic code is nearly universal. It is shared by organisms as diverse as **bacteria** to humans. This is important because it has allowed the incorporation of human genes in bacteria to produce human proteins.

See also Chromosome structure and morphology; DNA replication; Proteins and enzymes

PROTEINS AND ENZYMES

The building blocks of proteins and **enzymes** are molecules formed by carboxyl acids attached to amino groups ($-NH_2$), known as amino acids. Most protein structures consist of combinations of only about twenty of the most commonly found amino acids.

Amino acids bind to each other to form peptides and proteins. Conventionally, the term protein is used to designate chains of several peptides, known as polypeptides, with a molecular weight higher than thousands of Daltons. Peptides with a biological function go in length from dipeptides and tripeptides, up to polymers with thousands of Daltons.

Most proteins have well-defined structures and their specific biological functions depend upon the correct conformation of the molecular structure. For instance, the majority of soluble proteins of an organism, such as blood proteins, have globular structures, like small eggs. Some proteins are fiberlike and are associated in bundles, forming fibrils such as those of wool and hair. Myosin, the protein that makes muscles contract, has both globular and fibrous elements in its structure, whereas collagen, the protein of connective tissues, is constituted by three triple helices of fibrils that form super structures in the shape of a fibrous rope. Collagen represents one third of all proteins of the human body and together with elastin is responsible for both cohesion and elasticity of tissues.

Every enzyme is also a protein. Enzymes are proteins that function as catalysts of biochemical reactions. Most physiological activities in organisms are mediated by enzymes, from unicellular life forms to mammals. Enzymes speed up chemical reactions, allowing organic systems to reach equilibrium in a faster pace. For instance, every phase of the **cell** cycle is controlled by enzymes that alternately inhibit or stimulate specific cellular activities as well as **gene expression** or repression, hence affecting the time of specific physiological activities within each phase of the cell cycle. Enzymes are highly selective in their activities, with each enzyme acting over a specific substrate or group of substrates. Substrate is a term designating any molecule that suffers enzymatic action, whether being activated or inhibited.

The main property of catalyst molecules is that they are not altered by the chemical reactions they induce, although some rare exceptions are known where some enzymes are inactivated by the reactions they catalyze. Enzymatic catalysis involves the formation of protein complexes between substrate and enzyme, where the amount of enzymes is generally much greater than the amount of substrate.

Some families of enzymes play an important role during the process of **DNA replication**. For example, when **DNA synthesis** activates, helicases break hydrogen bridges and some topoisomerases separate the two **DNA** strands. DNA-polymerases synthesize the fragments of the new DNA strand, while topoisomerase III does the proofreading of the transcribed sequences, eliminating those containing errors. **Ribonuclease** H removes **RNA** sequences from polymers containing complexed RNA/DNA, and DNA-ligase unites the newly transcribed fragments, thus forming the new DNA strand.

In the last decade, researchers discovered that many proteins involved in intracellular communication are structured in a modular way. In other words, they are constituted by relatively short **amino acid** sequences of about 100 amino acids, and have the basic role of connecting one protein to another. Some proteins of such signaling pathways are entirely comprised of connecting modules and deprived of enzymatic activity. These non-enzymatic modules are termed protein dominium or protein modules, and they help enzymes in the transmission of signals to the cell **nucleus** in an orderly and controlled way. Proteins containing only connecting (or binding) modules, such as SH2 and SH3, act as important molecular adaptors to other proteins. While one of its modules binds to a signaling complex, such as a transmembrane tyrosine-kinase receptor, other binding modules permit the docking of other proteins that, once complexed, amplifies the signal to the nucleus. Such adaptor proteins also allow the cell to utilize certain enzymes that otherwise would not be activated in a given signaling pathway. The structure of adaptor proteins also displays binding sites that connect to DNA, where they recognize specific **nucleotide** sequences of a given **gene**, thus inducing **transcription**. In this case, the only enzyme in the cascade of signals to the nucleus is the receptor in the surface of the cell, and all the events that follow occur through the recognition among proteins and through the protein recognition of a locus in DNA.

Analysis of the communication circuitry in brain nerve cells revealed that some neuronal signaling pathways contain a great amount of protein binding modules, which are also known as scaffold proteins. Scaffold proteins sustain groups of signaling proteins that are gathered for prompt response near specific target proteins, thus amplifying the signal **transduction** in synapses. These findings suggest the existence of cellular signaling webs that depend upon permanent structures in cells that could be roughly compared to the hard disk of a computer that supports a variety of applications, while ensuring speed and fidelity in the transmission of information. An example of a scaffold protein is PSD-95 that plays a role in the learning process.

Proteins are encoded by genes. A gene usually encodes a nucleotide sequence that can be first transcribed in pre-messenger RNA, and then read and translated on the **ribosomes** into a group of similar proteins with different lengths and functions, known as protein isoforms. A single polypeptide may be translated and then cut by enzymes into different proteins of variable lengths and molecular weights.

During transcription, the non-coding DNA sequences (introns) are cut off, and the coding sequences (exons) are transcribed into pre-messenger RNA, which in turn is spliced to a continuous stretch of exons before protein **translation** begins. The spliced stretch subdivides in **codons**, where any of the four kinds of nucleotide may occupy one or more of the three positions, and each triplet codes for one specific amino acid. The sequence of codons is read on the ribosomes, three nucleotides at a time. The order of codons determine the sequence of amino acids in the protein molecule that is formed.

Introns may have a regulatory role of either the splicing or the translational process, and may even serve as exons to other genes. After translation, proteins may also undergo biochemical changes, a process known as post-translation processing. They may be either cut by enzymes or receive special bonds, such as disulfide bridges, in order to fold into a functional structure.

See also Biochemistry; DNA binding proteins; DNA repair; DNA synthesis; Gene family; Gene splicing; Non-coding DNA; Polymerase (DNA and RNA)

PROTEOMICS

Proteomics is the cataloging and analysis of proteins in the human body. With the decoding of **DNA** in the **genome**, the next logical step is finding the structure and function of proteins that are coded by DNA. Explaining proteins is much more complex.

In 1980, ten years before the inception of the **Human Genome Project**, Congress debated a Human Protein Index, reasoning that it would make sense to classify proteins which direct all biological functions and change during disease. The project was abandoned when considerations about the genome began to emerge.

In 1995, Australian post-doctoral researcher Marc Wilkins coined the term proteome. At that time few researchers envisioned a project of protein discovery similar to the effort of the human genome project. Unique tools have enabled scientists to probe the secrets of protein. Using **x-ray crystallography** and nuclear magnetic resonance (NMR), researchers have revealed structures of many proteins. For example, in 1998, Roderick MacKinnon at Rockefeller University obtained an image of a potassium channel, a concept important to neuroscience. To understand how proteins function, scientists must probe the entire area of the protein.

In late 2000, the National Institute of General Medical Sciences (NIGMS) invested 20 million dollars to detail 10,000 proteins in a ten-year span. Although these are few compared to the total estimated number of 50,000 to two million proteins, the number will probably cover most of the structures relevant to biology and medicine.

X-ray **crystallography** pictures of protein reveal folds and kinks; different colors represent different functions. In these grooves and folds, certain molecules fit like a key to a lock.

Specific proteins may have similar functions in worms, insects, or man, and the structural characteristics are shown in the genes that encode them. The science of naming the shapes of proteins is called structural proteomics. Scientists estimate there are about 1,000 shapes such as barrel shapes, doughnuts, spheres, or molecular zippers.

Using crystallized protein scatter x rays, scientists study the folds and kinks looking for their functions. The research is called **functional genomics**. Robots and powerful x-ray generators have increased the pace of discoveries. Like the genome project, private and government programs are working to decode the proteome.

See also Bioinformatics and computational biology; Functional genomics; Proteins and enzymes

PROTO-ONCOGENE

Proto-oncogenes are functional structural genes present in the human **genome** that have sequences very similar to oncogenes seen in viruses. Under normal conditions, the proto-oncogenes perform vital functions in the **cell**. They have been shown to be associated with control of cell growth, **differentiation**, and proliferation including regulation of secreted growth factors, cell surface receptors, signal **transduction** systems, **transcription** factors, and the cell cycle.

However, proto-oncogenes can be altered to an oncogenic form by **mutation**. This most commonly occurs by chromosomal rearrangement, usually a **translocation**, although point mutations and **gene amplification** have also been reported. Once transformed, or activated, the proto-oncogene loses its normal function causing a disruption in cell regulation patterns, and potentially leading to carcinogenesis. Each proto-oncogene appears to have a unique mutation that causes this change in state, i.e., every time a particular proto-oncogene is modified to the oncogenic form, the same mutation must occur. nly a single mutation in one of a pair of proto-oncogenes is necessary for disease to occur, so the resultant disease is considered to be dominant.

The first **chromosome** rearrangement involving a proto-oncogene was detected in chronic myelogenous leukemia (CML). The proto-oncogene c-*abl* is located on chromosome 9 and produces a tyrosine kinase necessary for cellular function. The **gene** is feedback inhibited such that only the amount of enzyme that can be utilized by the cell is normally produced. A translocation between c-*abl* and another protein kinase gene on chromosome 22 creates a structurally abnormal chromosome 22 known as the Philadelphia chromosome (named for the city in which it was identified). Biochemically, there is a change to constitutive synthesis of tyrosine kinase resulting in an overproduction of the enzyme, and this has been directly correlated with the disease CML. In cell lines, if the translocation is repaired and the excess of tyrosine kinase removed, the cells will revert to their normal state. eintroduction of the translocation once again causes the cells to express a **cancer** phenotype.

Over 50 proto-oncogenes have been identified in mammalian systems. Since the c-*abl* rearrangement was first described, additional similar proto-oncogene translocations have been identified that give rise to a variety of other leukemias and lymphomas. Mutations in proto-oncogenes have also been shown to be involved in solid **tumor** formation. Fortunately, many checks and balances exist within cells to protect against high levels of mutation, and since the mutation events that trans-

form the proto-oncogenes are highly specific, cases of cancer are relatively rare events. However, it is known that this type of mutation can be influenced by environmental factors, so, as with **tumor suppressor genes**, within a given population, some individuals will have a higher risk of disease than others.

See also Cancer genetics; Oncogene; Proto-oncogenes; Tumor suppressor genes

PROVIRUS HYPOTHESIS

American virologist **Howard Temin** proposed the provirus hypothesis. The hypothesis arose from Temin's observations that the Rous sarcoma **virus**, which has **ribonucleic acid (RNA)** as its core genetic material, could not infect a **cell** once the cell's synthesis of **deoxyribonucleic acid (DNA)** was stopped. Temin formed a possible explanation, the provirus hypothesis, claiming that the RNA of the invading virus is somehow copied or translated into the DNA of the host cell. As a result, the reproductive activity of the cell would be altered and the cell would become cancerous.

Temin's provirus hypothesis met with skepticism, as the prevailing dogma at the time was that genetic information passed only from DNA to RNA. However, working independently, Temin and **David Baltimore** proved, in 1970, that the hypothesis was correct. They identified the viral enzyme dubbed **reverse transcriptase**, which functions to pass genetic information to DNA from RNA.

This work has been useful in **cloning**, genetic engineering and research on **AIDS**, which is caused by a **retrovirus**. For the formulation of the provirus hypothesis and for sharing in its validation, Temin shared the 1975 Nobel Prize for physiology or medicine with Baltimore and **Renato Dulbecco**.

PRUSINER, STANLEY (1942-)
American physician

Stanley Prusiner performed seminal research in the field of neurogenetics, identifying the prion, a unique infectious protein agent containing no **DNA** or **RNA**.

Prusiner was born on in Des Moines, Iowa. His father, Lawrence, served in the United States Navy, moving the family briefly to Boston where Lawrence Prusiner enrolled in Naval officer training school before being sent to the South Pacific. During his father's absence, the young Stanley lived with his mother in Cincinnati, Ohio. Shortly after the end of World War II, the family returned to Des Moines where Stanley attended primary school and where his brother, Paul, was born. In 1952, the family returned to Ohio where Lawrence Prusiner worked as a successful architect.

In Ohio, Prusiner attended the Walnut Hills High School, before being accepted by the University of Pennsylvania where he majored in chemistry. At the University, besides numerous science courses, he also had the opportunity to broaden his studies in subjects such as philosophy, the history of architecture, economics, and Russian history. During the summer of 1963, between his junior and senior years, he began a research project on hypothermia with Sidnez Wolfson in the Department of Surgery. He worked on the project throughout his senior year and then decided to stay on at the University to train for medical school. During his second year of medicine, Prusiner decided to study the surface fluorescence of brown adipose tissue (fatty tissue) in Syrian golden hamsters as they arose from hibernation. This research allowed him to spend much of his fourth study year at the Wenner-Gren Institute in Stockholm working on the metabolism of isolated brown adipocytes. At this he began to seriously consider pursuing a career in biomedical research.

Early in 1968, Prusiner returned to the U.S. to complete his medical studies. The previous spring, he had been given a position at the NIH on completing an internship in medicine at the University of California San Francisco (UCSF). During that year, he met his wife, Sandy Turk, who was teaching mathematics to high school students. At the NIH, he worked on the glutaminase family of **enzymes** in *Escherichia coli* and as the end of his time at the NIH began to near, he examined the possibility of taking up a postdoctoral fellowships in neurobiology. Eventually, however, he decided that a residency in neurology was a better route to developing a rewarding career in research as it offered him direct contact with patients and therefore an opportunity to learn about both the normal and abnormal nervous system. In July 1972, Prusiner began a residency at UCSF in the Department of Neurology. Two months later, he admitted a female patient who was exhibiting progressive loss of memory and difficulty performing some routine tasks. This was his first encounter with a Creutzfeldt-Jakob disease (CJD) patient and was the beginning of the work to which he subsequently dedicated most of his professional life.

In 1974, Prusiner accepted the offer of an assistant professor position from Robert Fishman, the Chair of Neurology at UCSF, and began to set up a laboratory to study scrapie, a parallel disease of human CJD found in sheep. Early on in this endeavor, he collaborated with William Hadlow and Carl Eklund at the Rocky Mountain Laboratory in Hamilton, Montana, from whom he learnt much about the techniques of handling the scrapie agent. Although the agent was first believed to be a **virus**, data from the very beginning suggested that this was a novel infectious agent, which contained no nucleic acid. It confirmed the conclusions of Tikvah Alper and J.S. Griffith who had originally proposed the idea of an infectious protein in the 1960s. The idea had been given little credence at that time. At the beginning of his research into prion diseases, Prusiner's work was fraught with technical difficulties and he had to stand up to the skepticism of his colleagues. Eventually he was informed by the Howard Hughes Medical Institute (HHMI) that they would not renew their financial support and by UCSF that he would not be promoted to tenure. The tenure decision was eventually reversed, however, enabling Prusiner to continue his work with financial support from other sources. As the data for the protein nature of the scrapie agent accumulated, Prusiner grew more confident that

his findings were not artifacts and decided to summarize his work in a paper, published in 1982. There he introduced the term "prion," derived from "proteinaceous" and "infectious" particle and challenged the scientific community to attempt to find an associated nucleic acid. Despite the strong convictions of many, none was ever found.

In 1983, the protein of the prion was found in Prusiner's laboratory and the following year, a portion of the **amino acid** sequence was determined by Leroy Hood. With that knowledge, molecular biological studies of **prions** ensued and an explosion of new information followed. Prusiner collaborated with Charles Weissmann on the molecular **cloning** of the **gene** encoding the prion protein (PrP). Work was also done on linking the PrP gene to the control of scrapie incubation times in mice and on the discovery that mutations within the protein itself caused different incubation times. Antibodies that provided an extremely valuable tool for prion research were first raised in Prusiner's lab and used in the discovery of the normal form of PrP protein. By the early 1990s, the existence of prions as causative agents of diseases like CJD in humans and bovine spongiform encephalopathy (BSE) in cows, came to be accepted in many quarters of the scientific community. As prions gained wider acceptance among scientists, Prusiner received many scientific prizes. In 1997, Prusiner was awarded the Nobel Prize for medicine.

See also Viral genetics

PSEUDOGENE

A pseudogene is a sequence of **DNA** that is very similar to a normal **gene** but that has been altered slightly so that it is not expressed. A pseudogene was probably once functional. However, the accumulation of mutations over time rendered it incapable of encoding a protein product. The mutations in a pseudogene typically take the form of stop signals to the **transcription** process.

Typically, a functional gene from which the pseudogene drifted over time exits in addition to the pseudogene. There can be many copies, up to one thousand, of the pseudogene present in the **genome** in addition to the functional copy. The copies form part of the repeated DNA in the genome.

The functional significance of pseudogenes is not yet clear. Several speculations concerning the significance of pseudogenes have been proposed. These speculations include: they are remnants of evolutionary mistakes; conversely, they are genes in the process of evolving, and; they somehow serve a vital role in the expression of their normal counterparts.

Pseudogenes can be found in the genome proper and in the genome of **mitochondria**. Furthermore, pseudogenes are widespread in nature, being found in the DNA of fungi, plants, metazoans, arthropods, chordata, and mammals including man.

A beneficial of pseudogenes has been their use as markers to trace the evolutionary pathway of mitochondrial genomes of various organisms and in studies to ascertain the evolutionary relatedness of two organisms. One possible deleterious effect of pseudogenes may be their potential for complications in human identity tests that are based on **mitochondrial DNA** sequences. Pseudogenes that are identical to sequences in mitochondrial DNA have been detected outside of the mitochondria, in the DNA in the cell's **nucleus**, and in a variety of organisms. Accordingly, tests by forensic scientists designed to probe specifically for mitochondrial sequences would not be reliable without further modifications.

See also Gene mutation; Mutation; Repeated sequences

PSEUDOGENE · *see* GENE NAMES AND FUNCTIONS

PTASHNE, MARK STEVEN (1940-)
American molecular biologist

American molecular biologist Mark Steven Ptashne is head of the **gene regulation** research program at the Sloan-Kettering Institute in New York City.

After undergraduate study at Reed College, Ptashne received a Ph.D. in Molecular Biology from Harvard University in 1968. After receiving his doctorate, Ptashne stayed at Harvard to continue his research in molecular biology and **biochemistry**. Ptashne was awarded a Guggenheim Fellowship in 1973–1974 and later served as the head of Harvard's Department of Biochemistry and Molecular Biology from 1980 until leaving Harvard in 1983 to start the Genetics Institute. In 1986, Ptashne wrote a highly acclaimed book titled *A Genetic Switch* that explained mechanisms of genetic regulation. Subsequently, Ptashne accepted the Herchel Smith Professor of Molecular Biology at Harvard in 1993. In 1997, Ptashne moved his research to the Sloan-Kettering Institute.

Ptashne's research has centered on the fundamental mechanisms of **gene** regulation, especially the mechanisms of **transcription** regulation. Independent of the similar work done by his Harvard colleague, **Walter Gilbert**, Ptashne gained an international reputation for his identification and explanation of the operation of repressor genes.

Ptashne's work extends into concepts of regulated localization and the interactions of cells with extracellular molecular signals. Although much of the information underlying Ptashne's explanations of gene regulation comes from work done with **bacteria** and **bacteriophage**, Ptashne's research also extends to mechanisms of control in higher *Drosophila* and animal cells.

Recent work by Ptashne and his colleagues include investigations into the relation between **telomere** looping and gene activation; activation of **gene expression** by small molecule transcription factors; and various methods of transcriptional activation in mammalian cells. Ptashne continues work in measuring and describing the **evolution** of systems of genetic regulation.

Among his many honors and awards, Ptashne received the 1977 Lasker Award for Basic Research. In addition to his scientific research, Ptashne also studied music at the New England Conservatory of Music in Boston and plays the violin.

See also DNA (deoxyribonucleic acid); Gene induction; Gene regulation; Operon

PUNCTUATED EQUILIBRIUM

Punctuated equilibrium is an evolutionary theory put forward by **Niles Eldredge** and **Stephen Gould** in 1972 that runs counter to **Charles Darwin's** evolutionary theory of gradualism. Darwin's evolutionary theory maintains that adaptation in **species** are the result of a continuous process of gradual change. Within the tenants of gradualism, an adaptation would take many millions of years and all intermediate forms would exist for a long period, leaving a large likelihood for intermediate fossil types to exist. Surprisingly, intermediate forms are rare within the fossil record, and this lack of intermediate fossil evidence lead Eldredge and Gould to formulate the theory of punctuated equilibrium.

Contrary to Darwin's gradualism, punctuated equilibrium suggests that new species may have arisen rapidly over only a few thousand years and then remained unchanged for many millions of years before the next period of adaptation. Rather than occurring over the majority of the population, punctuated equilibrium postulates that change occurred in only a small part of the population. With change occurring rapidly within a small portion of the population, intermediate species would not likely be represented in the fossil record. Adaptation would be swift and profound to have such a major effect.

The sort of macro mutations that Gould and Eldredge envisioned include mutations within regulator genes which simultaneously affect a whole **operon** and, thus, drastically affect the development of the organism. The incipient species would have to then supplant the original species and **natural selection** would eliminate intermediate types.

Darwin himself had recognized that such events may occur and contrary to some peoples beliefs punctuated equilibrium can be quite easily explained by and accepted into the system of natural **selection**.

See also Darwinism; Evolutionary mechanisms

PUNNETT, R. C. (1875-1967)
English geneticist

R. C. Punnett was instrumental in introducing the field of **genetics** to lay audiences, especially to commercial breeders of livestock. His contributions significantly advanced knowledge of the genetics of fowl, ducks, rabbits, sweet pea plants, and humans; his research served as the foundation for poultry genetics for decades. Punnett was among the pioneering investigators who helped revolutionize scientific thought in the field of genetics after the rediscovery of **Gregor Mendel's** work with genetics and **heredity**.

Reginald Crundall Punnett, the eldest of three children, was born at Tonbridge in Kent, England, to George Punnett, the head of a Tonbridge building firm, and Emily Crundall. He suffered from appendicitis as a child. During the treatment, which consisted of applying leeches to the lower stomach, and the daily bedrest required afterwards, he spent his time reading Jardine's *Naturalist's Library* and discovered a strong liking for natural history. Punnett attended Cambridge University, where he developed an interest in human anatomy, human physiology, and zoology, and decided to pursue a career in zoology, not medicine as he originally had intended. As part of his zoological studies, Punnett observed sharks at the Zoological Station in Naples, Italy, for six months before graduating with first class honors from Cambridge's Caius College in 1898. He received his master's degree in 1902.

In 1899, Punnett was offered the position of demonstrator and part-time lecturer in the Natural History Department of St. Andrews University, where he stayed for three years. In 1901, he was elected a fellow of Caius College. During this time, his appendix had been troubling him sporadically and he decided to have it removed. As the scientific thought at the time was that worms caused appendicitis, he dissected the organ after the surgery, but found no worms. After he recovered from the operation, Punnett became unhappy with his teaching accommodations at St. Andrews and began a search for a new job. In 1902, he returned to Cambridge as a demonstrator in morphology in the Department of Zoology and remained in this position until 1904. While a demonstrator, Punnett had plenty of time to perform research and publish a number of papers on nemertines, a type of worm.

Punnett then turned from the study of nemertines to genetics, working with **William Bateson**, an investigator also researching Mendelian principles, during a six-year period in which the pair produced noteworthy and lasting advances in Mendelian genetics. Gregor Mendel, an Augustinian monk, had used pea plants to demonstrate how genetic traits are inherited. Using the sweet-pea plant or fowl for their studies, Punnett and Bateson determined several basic classical Mendelian genetic principles, including the Mendelian explanations of sex-determination, sex-linkage, complementary factors and factor interaction, and the first example of autosomal linkage. It was early in his relationship with Bateson that Punnett was awarded the Balfour Studentship in the Department of Zoology, and he resigned as demonstrator. He held the Balfour Studentship position until 1908. Also, during this year he was awarded the Thurston medal of Gonville and Caius College. Punnett wrote *Mendelism,* the first published textbook on the subject of genetics, in 1905. As a reflection on his research, he was appointed superintendent of the Museum of Zoology in 1909. A year later, Punnett became professor of biology at the University of Cambridge. Punnett and Bateson started the *Journal of Genetics* in 1911 and edited it jointly until Bateson died in 1926. Punnett continued to edit the journal, credited with drawing numerous new students to the field of genetics, for twenty more years.

In 1912, the University of Cambridge changed the name of the chair of biology to the chair of zoology, and offered the position to Punnett, making him the first Arthur Balfour Professor of Genetics, a position that was the first of its kind in Great Britain. He held this prestigious position until his retirement at the age of sixty-five. Also in 1912, Punnett was elected a fellow of the Royal Society, and in 1922, he was awarded its Darwin Medal. Punnett's interests in genetics led him to a founding membership in the British Genetical Society; he served as one of the group's secretaries from 1919 to 1930, when he then became president.

During World War I, Punnett bought his expertise in poultry breeding to a position with the Food Production Department of the Board of Agriculture. He suggested that hens' plumage color could be used to determine the sex of the birds much earlier than previously was possible. This enabled the breeders to destroy most of the unwanted males, which were not used for food, and save precious resources for raising females for consumption. After the war ended, Punnett produced the first breed of poultry in which a trait is demonstrated uniquely in one sex or the other. This first auto-sexing breed, the Cambar, was followed a decade later by a second breed, the Legbar. Punnett's work laid the foundation for poultry breeding research for several more decades.

Punnett was married to Eveline Maude Froude at age forty-one; they had no children. In his leisure time, he enjoyed playing bridge, participated in many sports including cricket and tennis, and collected Japanese color prints, Chinese porcelain and old and rare biological and medical texts. Upon his death, his collection of Japanese prints was acquired by the Bristol Corporation for the city art gallery. Punnett died during a game of bridge in Bilbrook, Somerset, England, at age 91.

See also Agricultural genetics; Mendelian genetics and Mendel's laws of heredity

PUNNETT SQUARE

A Punnett square is a checkerboard representation of predictable genotypes from a second generation offspring of an experimental breeding. Named for geneticist Reginald C. Punnett, who originally used the method to compute the results of a cross using only one **gene**. Punnett squares are now occasionally used for considering two genes and their alleles, however they quickly become unwieldy when used to calculate offspring of more than two alleles.

With a Punnett square, one gene is considered and the **genotype and phenotype** are predicted. Along the top of the checkerboard are the alleles found in the gametes of one parent (usually the male by convention) and the alleles found in the other parent are written down the left-hand side. The products of the possible matings are then placed in the four boxes in the middle of the checkerboard. The relative numbers of the different progenies can then be calculated. For example a man heterozygous for "A" can produce gametes of "A" and "a." If

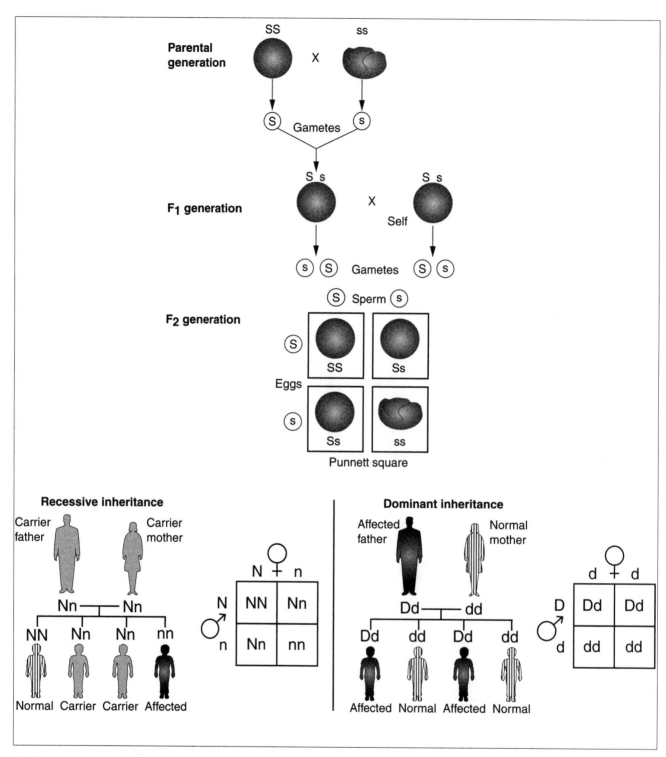

The Punnett square showing recessive and dominant inheritance.

the man mates with a homozygous recessive female, who can only produce gametes of type "a," their resulting offspring genotypes can be calculated using a Punnett square. The genotypes for the resultant offspring would either be Aa or aa. If

"A" was dominant to "a", then phenotypically half of the offspring would be "A" and half would be "a". If two heterozygotes mated, the phenotypical offspring would be produced in the ratio of three dominant to one recessive. In the latter case, the appropriate Punnett square would genotypically illustrate one quarter AA, one half Aa, and one quarter aa.

A Punnett square can also be used to calculate the relative frequencies of a mating when two different genes are observed. The same rules are used, but there are four different gametes produced along the top and side, giving 16 different possibilities for the offspring. If both parents are AaBb, then four different gametes are possible—AB, Ab, aB, and ab. Genotypically, the Punnett square shows us that in the 16 offspring one would be dominant homozygous, one would be recessive homozygous, and the remainder would be heterozygous for at least one of the gene pairs. If "A" coded for brown eyes, "a" for blue eyes, "B" coded for black hair, and "b" for blond hair, then the results would be in a 9:3:3:1 ratio (nine brown eyed and black haired, three brown eyed and blond haired, three blue eyed and black haired, and one blue eyed and blond haired).

A Punnett square considering two genes assumes the **law of independent assortment** is followed. First formulated by Gregor Johann Mendel, independent assortment is the random distribution of alleles to gametes during **meiosis**. If one particular allele goes to one **gamete**, this has no influence on the likelihood of any other allele going to the same gamete. All various allele combinations have an equal chance of representation in a gamete.

See also Cross breeding; Probability and statistics

PURINES • *see* NUCLEIC ACID

PYRIMIDINES • *see* NUCLEIC ACID

Q

QUANTITATIVE GENETICS • *see* STATISTICAL
ANALYSIS OF GENETIC SYNDROMES

QUANTUM SPECIATION

Quantum speciation, also known as saltational speciation, is the process by which a small population of **species** rapidly diverges into more than one species that is reproductively isolated from the original population. Scientists theorize that quantum speciation occurs because of genetic drift that results from the **founder effect**. Genetic drift refers to random genetic changes within a population rather than **natural selection**. The founder effect occurs because a few individuals from a population colonize a new area and become isolated from the rest of the species. The genes of these individuals have enough **genetic variation** that there is a change in allele frequency from the original population. Through **random mating**, genes may be mutated and alleles may become lost or fixed in the new population. Within a few generations, **chromosome** mutations cause changes in observable physical traits and the new species can no longer breed with individuals from the original population. This process can occur in large populations but is more pronounced in smaller ones. Ultimately, quantum speciation is dependent on isolation, both genetic and physical, and chromosome **mutation** in order to occur and allow the newly formed species to survive.

Evidence from the fossil record suggests that quantum speciation can occur in one or two generations because of polyploidy, the increase in number of chromosome sets. Usually organisms contain two sets of chromosomes, but sometimes polyploidy can occur during hybridization and produce three or more sets of chromosomes. Plants provide the best evidence of accelerated speciation by means of polyploidy. Almost 40% of angiosperms (flowering plants) are polyploids that evolved by this mechanism. For example, scientists theorize that the polyploid sequoia evolved from a **diploid** ancestor. Another example of quantum speciation, although not by polyploidy, is the origin of bats 50 million years ago. The sudden appearance of winged mammals in the fossil record began the new lineage that gave rise to the order Chiroptera.

Some scientists believe quantum speciation to be part of the **punctuated equilibrium** model that is characterized by slow and fast periods of change. This evolutionary theory states that long periods of time with relatively little genetic change are followed by sudden, rapid changes in **genetics** that result in the formation of new species. However, other scientists argue that rapid genetic change may lead to some divergence within a population but is not accompanied by speciation.

See also Evolutionary mechanisms; Ploidy

R

RADIATION MUTAGENESIS

Mutations are caused by **DNA** damage and genetic alterations that may occur spontaneously at a very low rate. The frequency of these mutations can be increased by using special agents called mutagens. Ionizing radiation was the first **mutagen** that efficiently and reproducibly induced mutations in a multicellular organism. Direct damage to the **cell nucleus** is believed to be responsible for both mutations and other radiation mediated genotoxic effects like chromosomal aberrations and lethality. Free radicals generated by irradiation of the cytoplasm are also believed to induce **gene** mutations even in the non-irradiated nucleus.

There are many kinds of radiations that can increase mutations. Radiation is often classified as ionizing or non-ionizing depending on whether ions are emitted in the penetrated tissues or not. X rays, gamma rays (γ), beta particle radiation (β), and alpha particle (α) radiation (also known as alpha rays) are ionizing form of radiation. On the other hand, UV radiation, such as that experience by exposure to Sunlight is non-ionizing. Biologically, the differences between types of radiation effects fundamentally involve the way energy is distributed in irradiated cell populations and tissues. With alpha radiation, ionizations occur every 0.2-0.5 nm leading to an intense localized deposition of energy. Accordingly, alpha radiation particles will travel only a short distance before expending of their energy. Primary ionization in x rays or gamma radiation occurs at intervals of 100 nm or more and traverses some centimeters deeper into tissues. This penetration leads to a more even distribution of energy as opposed to the more concentrated or localized alpha rays.

This principle has been used experimentally to deliver radiation to specific cellular components. A cumulative effect of radiation has been observed in animal models. This means that if a population is repeatedly exposed to radiation, a higher frequency of mutations is observed that is due to additive effect. Intensive efforts to determine the mutagenic risk of low dose exposure to ionizing radiation have been an ongoing concern because of the use of nuclear energy and especially because of the exposure to radon gas in some indoor environments. Radon is estimated by the United States Environmental Protection Agency to be the cause of more than 20,000 cases of lung cancer annually.

The relative efficiencies of the different types of radiations in producing mutations are usually deduced from experiments in a *Drosophila model*. The mutagenic effect of radiation is generally assumed to be due to direct damage to DNA, but the identity of the specific lesions remains uncertain.

Investigation of radiation's mutagenic effects on different tissues, cells, and subcellular compartments is becoming possible by the availability of techniques and tools that allow the precise delivery of small doses of radiation and that provide better monitoring of effects. Reactive oxygen **species** released in irradiated cells are believed to act directly on nuclear DNA and indirectly by modifying **bases** that will be incorporated in DNA, or inactivating **DNA repair enzymes**. Novel microbeam alpha irradiation techniques have allowed researchers to investigate radiation-induced mutations in non-irradiated DNA. There is evidence that radiation induces changes in the cytosol that are transmitted to the nucleus and even to neighboring cells. Direct measurement of DNA damage caused by ionizing radiation is performed by examining micronucleus formation or analysis of DNA fragments on agarose gels following treatment with specific endonucleases such as those that only cleave at certain sites. The **polymerase chain reaction** (PCR) is also used to detect the loss of some marker genes by large **deletions**. The effect of ionizing radiation on cells can also be measured by evaluating the expression level of the stress inducible p21 protein.

Critical lesions leading to mutations or killing of a cell include induction of DNA strand breaks, damaged bases, and production of abasic sites (where a single base is deleted), as well as large chromosomal deletions. Except for large deletions, most of these lesions can be repaired to a certain extent, and the lethal and mutagenic effect of radiation is assumed to result principally from incompletely or incorrectly repaired DNA. This view is supported by experimental studies which showed that mice given a single radiation dose, called acute dose, develop significantly higher level of mutations than mice given the same dose of radiation over a period of weeks or months. The rapid activation of the DNA-repair pathway through p53 protein and the stress-inducible p21 protein as well as the extreme sensitivity of cells with genetic defects in DNA repair machinery support the view that the ability of the cell to repair irradiation-induced DNA damage is a limiting factor in deciding the extent of the mutagenic effects.

See also Alleles and allotype; Carcinogenic; Chemical mutagenesis; Chromosomal mutations and abnormalities; Chromosome identification and banding; Evolutionary mechanisms; Gene mutation; Gene names and functions; Gene penetrance; Neutral selection; Point mutation; Population genetics

RANDOM GENETIC SYNDROME

A syndrome is a pattern of multiple **birth defects** or malformations, that are all the result of a single underlying cause. There are more than 1,000 different genetic syndromes known, although many are rare. Apert syndrome is chosen here as an example of a syndrome to illustrate the concept of syndromes and how the different features may be related to a single defective **gene**.

Apert syndrome is an autosomal dominant genetic disease affecting approximately one in every 100,000 live births which is characterized by a flat facial profile, a tall flat forehead, and fusion of the digits of the hands and/or feet. Examination of the growth plates of the bones in the skull shows irregular and premature fusion of the bones with constriction of the space normally available for brain growth. The constriction of space within the skull generally leads to mental retardation or developmental delay, and protrusion of the globes of the eyes unless it is effectively relieved through surgery. Although the condition is inherited as an autosomal dominant condition, the birth of an affected child to two healthy parents frequently occurs as a result of a new **mutation** in the **DNA**. Individuals affected with Apert syndrome have a 50% chance of passing the disease-causing mutation to each of their children.

The underlying genetic defect in Apert syndrome has been shown to be a defective fibroblast growth factor receptor gene called FGFR2. There are at least 19 different fibroblast growth factors which have been identified in humans influencing varied aspects of **cell** growth, tissue **differentiation**, and cell migration. The mutation of FGFR2 that causes Apert syndrome misinterprets signals from the fibroblast growth factors during fetal development and promotes premature fusion of the growth plates in bones, giving rise to the unusually shaped skull and the mitten-like appearance of the hands and feet.

Different mutations in the same FGFR2 gene can have a different influence on the development of the fetus. Other syndromes which are clinically similar to Apert syndrome but distinct enough to be identified as different syndromes includes Pfeiffer syndrome, Crouzon syndrome, Beare-Stevenson syndrome and Jackson-Weiss syndrome.

See also Developmental genetics; Mendelian genetics and Mendel's laws of heredity

RANDOM MATING

In **sexual reproduction**, random mating is the idea that mates are chosen at random from the pool of all possible mates, with no preference given to any genetic characteristic, ancestry or trait. When pine trees mate, for example, and the males release pollen into the wind, the chance that any given male is the one to produce the pollen that fertilizes a certain female is essentially random. There is no choosing on the part of the female, but rather, the pollen lands wherever it lands.

In contrast, nonrandom mating means that certain mates from a given pool of potential mates have a greater chance than others of mating with a specific individual of the opposite sex. The ways in which nonrandom mating actually occurs are varied. There are three primary situations which result in nonrandom mating: stratification, assortive mating, and consanguinity.

Stratification means that the population is subdivided into a number of subgroups. In a population of frogs in a certain region, this might mean that there are distinct ponds or pools that the subgroups tend to remain a part of during mate **selection** even though they are a part of the larger population. In humans, the subgroups might be bound by ethnicity, religious community, language, or other barriers.

Assortive mating occurs when a mate is chosen because he or she possesses some particular characteristic. For example, one might select a mate based on some musical or athletic talent, intellectual ability, or some other physical characteristic. People within the deaf community often choose their mates from the deaf population. Positive assortive mating means that people tend to select mates who possess a certain trait while negative assortive mating means that they tend not to select a mate who possesses a certain trait.

Consanguinity refers to matings between blood relatives. Consanguinity is generally considered to be matings between known relatives such as mother-son, father-daughter, uncle-niece, aunt-nephew, brother-sister, first cousins, second cousins, and so forth.

Random mating is particularly important with regard to calculating the chance for specific autosomal recessive diseases to occur in a family. The Hardy-Weinberg equation is often used to determine the chance that a person carries a certain **mutation**, but this method assumes that all matings are random. The concept of random mating also enters into discussions of why certain **gene** mutations seem to cluster in certain subgroups within a population. For example, **cystic fibrosis** tends to occur more often in Caucasians of European descent, and **Tay-Sachs disease** is more common among Jewish individuals of Eastern European descent.

See also Genetic implications of mating and marriage customs; Hardy-Weinberg equilibrium; Population genetics

RANDOMLY AMPLIFIED POLYMORPHIC DNA

Randomly amplified polymorphic **DNA** (RAPD) is a technique that is useful as a rapid means of comparing and discriminating the DNA from different individuals. RAPD is based on random priming—the use of short pieces of DNA that bind to many random places in the **genome**. The DNA is then fragmented using **restriction enzymes**. Because of the small size of the probe DNA and the many fragments that can be generated, a very large number of reaction products can be obtained. The products can be separated based on their size by electrophoresis and detected by staining of the gel or, if the **primers** are radioactive, by the technique of **autoradiography**.

RAPD has been combined with the **polymerase chain reaction** (PCR). The use of PCR allows the DNA to the probe bound to be amplified in number to detectable levels, and so eliminated the need for a large amount of DNA at the commencement of an experiment. Older techniques, like **restriction fragment length polymorphism**, require large amounts of genomic DNA, are difficult to automate and require much time to complete.

RAPD has found a place as a technique for the rapid and preliminary examination of DNA. While useful in this regard, PCR-based fingerprinting techniques relying on random primers are not robust and generally are unsuitable for use as population markers, particularly in critical or demanding situations such as human diagnostics or courtroom evidence. RAPD has proved useful in discriminating genotypes of various prokaryotic and eukaryotic **species**.

See also Amplification; Amplified Fragment Length Polymorphism (AFLP)

RARE GENOTYPE ADVANTAGE

Rare genotype advantage is the evolutionary theory that genotypes that have been rare in the recent past should have particular advantages over common genotypes under certain conditions. This can be best illustrated by a host-parasite interaction. Successful parasites are those carrying genotypes that allow them to infect the most common host genotype in a population. Thus, hosts with rare genotypes, those that do not allow for infection by the pathogen, have an advantage because they are less likely to become infected by the common-host pathogen genotypes. This advantage is transient, as the numbers of this genotype will increase along with the numbers of pathogens that infect this formerly rare host. The pattern then repeats. This idea is tightly linked to the so-called Red Queen Hypothesis first suggested in 1982 by evolutionary biologist Graham Bell (1949–) (so named after the Red Queen's famous remark to Alice in Lewis Carroll's *Through*

the Looking Glass: "Now here, you see, you have to run as fast as you can to stay in the same place."). In other words, **genetic variation** represents an opportunity for hosts to produce offspring to which pathogens are not adapted. Then, sex, **mutation**, and genetic **recombination** provide a moving target for the **evolution** of virulence by pathogens. Thus, hosts continually change to stay one step ahead of their pathogens.

This reasoning also works in favor of pathogens. An example can be derived from the use of **antibiotics** on bacterial populations. Bacterial genomes harbor genes conferring resistance to particular antibiotics. Bacterial populations tend to maintain a high level of variation of these genes, even when they seem to offer no particular advantage. The variation becomes critical, however, when the **bacteria** are first exposed to an antibiotic. Under those conditions, the high amount of variation increases the likelihood that there will be one rare genotype that will confer resistance to the new antibiotic. That genotype then offers a great advantage to those individuals. As a result, the bacteria with the rare genotype will survive and reproduce, and their genotype will become more common in future generations. Thus, the rare genotype had an advantage over the most common bacterial genotype, which was susceptible to the drug.

See also Adaptation and fitness; Hardy-Weinberg equilibrium; Selection; Sexual reproduction; Spontaneous mutations and reversions

RÉAUMUR, RENÉ ANTOINE FERCHAULT DE (1683-1757)

French scientist

René Antoine Ferchault de Réaumur was a French scientist who made outstanding contributions to several branches of science, including a well-regarded and influential study of insects that proved valuable to early twentieth century geneticists during the development of **chromosome** theory. Réaumur is perhaps best recognized for his work which established how the process of digestion works in animals. Réaumur as born in 1683 in La Rochelle, France. Although there is little record of his early years or his education, it is known that at the age of 20 he moved to Paris. In 1708, he became a member of the French Academy of Sciences.

Réaumur applied his diverse expertise to a variety of fields. He compiled his knowledge of insects in a six volume series of memoirs known collectively as *Mémoires pour servir à l'histoire des insectes* (which translates as Memoirs for Following Insect Study). This work, published between 1734 and 1742, is generally considered to be the first comprehensive publication describing insects. Réaumur's insect study was relied upon by **Carolus Linnaeus**, in his classic work *Systema Naturae*. Subsequently these publications provided an important database for the construction of evolutionary theory.

In the world of biology, Réaumur is also recognized for his studies of digestion in birds and animals. His research showed that digestion in higher animals was a chemical process and not strictly done by mechanical agitation. Réaumur showed that the stomach acted on food chemically by feeding meat-filled metal cylinders to a hawk and then examining the cylinders after the bird regurgitated them. He was eventually able to isolate these digestive chemicals by forcing the hawk to swallow a sponge and then squeezing the retrieved stomach liquid out of the regurgitated sponge. Réaumur continued his experiments until his death in 1757.

Early in his career Réaumur was put in charge of a government sponsored project known as *Description des arts et méitiers,* which required that he collect information on all of France's arts, industries and professions. This experience gave Réaumur a diverse background in science and technology which he put to good use. For example, his chemical knowledge helped him recognize the importance of carbon in the composition of steel. Using this fact, in 1720 he developed an improved method for manufacturing iron and steel which became known as the cupola furnace.

Réaumur also, he developed a new form of opaque porcelain, known as Réaumur's porcelain. He even devised the Réaumur temperature scale using a thermometer he had invented. This thermometer used a mixture of alcohol and water to register the freezing point of water as zero and boiling point as 80 degrees. He used 80 as the number of points on his scale because he believed it was "a number convenient to divide into parts." The Réaumur scale was once widely used in Europe, particularly in France, but it has now been replaced by the Celsius scale.

See also Chromosome; *Drosophila melanogaster;* Genotype and phenotype

READING FRAME • *see* FRAME SHIFTS

RECESSIVE GENES AND TRAITS

In **diploid** organisms, the attribute of a specified character (or phenotype) is said to be recessive if it is masked when in the presence of a dominant allele. A recessive phenotype is expressed only in homozygotes (or, for X-linked traits, also in hemizygotes). Strictly speaking, it is the phenotype that is recessive (or dominant), not the allele; however, the term "recessive gene" is widely, if loosely used.

Gregor Mendel recognized the dominant and recessive aspects of the traits he studied in garden peas in the mid nineteenth century. Those traits that become latent in the process of the hybridization he described as recessive, whereas those that constituted the character of the **hybrid** were seen as dominant. Mendel studied seven characters of peas, each with two recognizable traits; for example, round or wrinkled seeds, dwarf or tall stems, and several others. Starting with pure strains for each trait, he created hybrid crosses (called F1) and subsequently crosses of the hybrids to one another (called F2). For each character, one of the traits disappeared in the F1, but reappeared in the F2 generation in about one quarter of the plants.

He described the agent responsible for each trait as a "factor"; that factor (now called a **gene**) was responsible for the phenomenon of a recessive trait that could be hidden but not destroyed. The term atavism, though rarely used today, refers to this reappearance of a character after several generations.

In **experimental organisms**, the study of two characters simultaneously is called a dihybrid cross. An organism with the recessive phenotype for both is called a double recessive organism.

The concept of dominance and recessivity is an operational one, and the distinction is not absolute. Moving from Mendel's peas to examples from human **genetics**, a trait may be clinically undetectable in heterozygotes (also called carriers), and thus defined as recessive at that level of the clinical phenotype. At the cellular, biochemical or molecular level, the trait might be seen as recessive, incompletely recessive, or even dominant, depending on the aspect being studied. Consider the example of the **mutation** in the beta-globin gene that, when homozygous, causes the disease **sickle cell anemia**. The heterozygous individual is not anemic under normal circumstances, therefore the mutant gene would be considered recessive. Carriers, however, may demonstrate some sickle-shaped red blood cells under conditions of reduced oxygen tension, though not to the extent seen in the homozygotes, so at the cellular level, the trait is incompletely recessive. At the molecular level, both normal and sickle hemoglobin can be found within the cells, and the trait could be described as co-dominant at this level.

In family studies, autosomal recessive traits have certain distinctive **inheritance** characteristics. Traits that are lethal or deleterious are usually rare, and the inheritance patterns described apply to such rare alleles. First, a recessive trait often appears sporadically, with no evidence of it in either parent or collateral relatives. Those parents, however, are then obligate carriers of the recessive allele, and any of their subsequent offspring have a one in four chance of being affected by the disease or trait in question. Second, the more rare the recessive gene, the more likely the parents are to be related to one another. Consanguinity in the parents' relationship often alerts a clinician to consider a recessive basis for a diagnosis.

Recessive traits may also result from genes carried on the **X chromosome**. In this case, **hemizygous** males reveal the recessive phenotype due to the lack of any dominant allele to mask it.

At a molecular level, recessive traits are frequently the result of mutations in **enzymes** rather than in structural genes, and are more likely to result from the absence of a gene product that an altered one. The reason for this is that a trait will be recessive if the alternate allele provides enough product for full cellular function. Enzymes are typically produced in huge excess by cells, so that a 50% reduction is often not deleterious. This characteristic is a rule-of-thumb, but certainly not absolute.

Genetic **selection** acts on the phenotype of an individual. Carriers of recessive traits, by definition, are not detectable phenotypically, will not be selected against, and can, therefore, harbor and transmit their recessive alleles throughout generations.

See also Dominance relations; Filial generation; Genetic implications of mating and marriage customs; Genetic load; Genotype and phenotype; Inbreeding; Mendelian genetics and Mendel's laws of heredity

RECIPROCAL CROSS

A reciprocal cross is one of a pair of matings in which two opposite sexes are coupled with each of two different genotypes and mated in opposite combinations. For example, a female of a certain genotype A is first crossed with a male of genotype B. Then, in the reciprocal cross, a female of genotype B is crossed with a male of genotype A. Reciprocal crosses are used to determine whether maternal or paternal factors influence the **inheritance** of the characteristic. They are used to detect sex-linkage, **maternal inheritance**, or **cytoplasmic inheritance**.

If all crosses give the same results when made reciprocally, the observed phenotypes and proportions will be the same for sons and daughters. This occurs when genes for the trait are carried on **autosomes**, not on the sex chromosomes. However, in the case of sex-linkage, differences arise in reciprocal crosses.

Eye color in *Drosophila melanogaster,* commonly known as the fruit fly, provides a classic example of a trait identified as sex-linked on the basis of reciprocal crosses. **Wild-type** eye color in fruit flies is dull red and is denoted by w+. A recessive mutant allele of w+ that causes white-eye color is denoted by w. Using pure lines, crossing a female with red eyes by a white-eyed male produces all red-eyed sons and daughters. This is the expected result based on simple Mendelian **genetics**, because w+ is dominant over w. However, the reciprocal cross, that of a white-eyed female with a red-eyed male, produces only white-eyed sons and red-eyed daughters. From this result, it can be can concluded that w+ and its alleles are not carried on autosomes and are therefore sex-linked.

Genetically, the results described above can be explained by the fact that the alleles specifying eye color are carried on the **X chromosome**. Individual flies with two X chromosomes are females, while flies with one X and one **Y chromosome** are males. The Y **chromosome** lacks an allele of the eye color **gene**. Therefore, each son will express phenotypically whatever allele is on the single X chromosome he has received from his mother. The daughters, on the other hand, received two alleles specifying eye color, one on each X chromosome. They therefore present the phenotype of the dominant w+ allele.

See also Extranuclear Inheritance; Hemizygous; Monohybrid cross; Recessive genes and traits

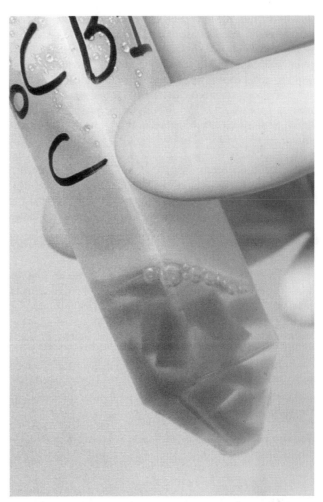

Researcher holds a tube of agarose "plugs" containing recombinant human and yeast DNA. *Photograph by © Klaus Guldbrandsen/Science Photo Library. Photo Researchers. Reproduced by permission.*

RECOMBINANT CLONE • *see* RECOMBINANT DNA
TECHNOLOGY AND GENETIC ENGINEERING

RECOMBINANT DNA MOLECULES

Recombinant **deoxyribonucleic acid** (**DNA**) is genetic material from different organisms that has been chemically bonded together to form a single macromolecule. The **recombination** can involve the DNA from two eukaryotic organisms, two prokaryotic organisms, or between an eukaryote and a **prokaryote**. An example of the latter is the production of human insulin by the bacterium *Escherichia coli,* which has been achieved by splicing the **gene** for insulin into the *E. coli* **genome** such that the insulin gene is expressed and the protein product formed.

The splicing of DNA from one genome to another is done using two classes of **enzymes**. Isolation of the target DNA sequence is done using **restriction enzymes**. There are well over a hundred restriction enzymes, each cutting in a very precise way a specific base of the DNA molecule. Used singly or in combination, the enzymes allow target segments of DNA to be isolated. Insertion of the isolated DNA into the recipient genome is done using an enzyme called DNA ligase.

Typically, the recombinant DNA forms part of the DNA making up a plasmid. The mobility of the plasmid facilitates the easy transfer of the recombinant DNA from the host organism to the recipient organism.

Paul Berg of Stanford University first achieved the manufacture of recombinant DNA in 1972. Berg isolated a gene from a human cancer-causing monkey **virus**, and then ligated the **oncogene** into the genome of the bacterial virus lambda. For this and subsequent recombinant DNA studies (which followed a voluntary one-year moratorium from his research while safety issues were addressed) he was awarded the 1980 Nobel Prize in chemistry.

In 1973, Stanley Cohen and Herbert Boyer created the first recombinant DNA organism, by adding recombinant **plasmids** to *E. coli*. Since that time, advances in molecular biology techniques, in particular the development of the **polymerase chain reaction**, have made the construction of recombinant DNA swifter and easier.

Recombinant DNA has been of fundamental importance in furthering the understanding of genetic regulatory processes and shows great potential in the genetic design of therapeutic strategies.

See also Bacterial genetics; Plasmids; Polymerase chain reaction (PCR)

RECOMBINANT DNA TECHNOLOGY AND GENETIC ENGINEERING

Recombinant **deoxyribonucleic acid (DNA)** is DNA that has been created artificially. DNA from two or more sources is incorporated into a single recombinant molecule. The technologies used to accomplish this are collectively referred to as recombinant DNA technology.

Recombinant DNA technology relies upon restriction endonucleases—enzymes that recognize certain sequences within the DNA, and cut the DNA at the particular site. Many restriction endonucleases exist that recognize many sequences of DNA **bases**. Some of the **enzymes** create a staggered cut, where one strand of DNA is longer than the complimentary strand. Because of the staggered cuts, treatment of different DNA with the **restriction enzymes** will generate regions, which compliment one another. These regions will anneal together. These behaviors allow foreign DNA to be inserted into target DNA. This is the basis of recombinant DNA technology.

To be useful, the recombinant molecule must be replicated many times to provide material for subsequent experimental manipulations. Producing many identical copies of the same recombinant molecule is called **cloning**. Cloning can be done in the laboratory, using the technique called the **polymerase chain reaction**.

Recombinant DNA technology also uses living organisms, such as **bacteria**, **yeast** and mammalian cells to achieve cloning. The DNA is taken up by the **cell** when incorporated in a vector. Viruses and **plasmids** typically are utilized as vectors. The bacterium *Agrobacterium tumefaciens* is a popular means of transforming plants. The DNA shuttled by the vector to the host typically contains a **gene** that encodes resistance to an antibiotic, or other feature, such as the ability to fluoresce when exposed to ultraviolet light. These act as indicators to permit a means of monitoring the success of DNA integration and protein expression. For example, exposure of the bacteria *Escherichia coli* to vectors containing the recombinant DNA will produce some cells that have successfully incorporated and expressed the DNA. If the recombinant indicator is an antibiotic resistance gene, then the successful cells can be detected when they form colonies on a growth substrate containing the antibiotic. Each colony represents a clone of transformed cells. Selecting recombinant plants can be done in a similar fashion, with the recombinant insert containing a herbicide-resistance gene and the growth medium containing the herbicide.

Recombinant DNA technology is used in human therapy. Many human genes have been cloned in *E. coli* or in yeast. This has made it possible to produce unlimited quantities of human protein in the laboratory. Cultured cells transformed with the human gene are being used to manufacture insulin for diabetes, factor VIII for males suffering from **hemophilia** A, factor IX for hemophilia B, human growth hormone, erythropoietin for the treatment of anemia, interferons, interleukins, factors for stimulating bone marrow after a bone marrow transplant, tissue plasminogen activator for dissolving blood clots, and adenosine deaminase for the treatment of some forms of severe combined immune deficiency.

Production of transgenic organisms has also created a great number of possibilities using plants. Transgenic plants have been obtained that express antibodies and recombinant pharmaceuticals. Molecular techniques combined with the traditional techniques of plant breeding are being utilized by the forestry industry to generate trees of commercial interest that are capable of faster and straighter growth. Transgenic crop plants, engineered to resist pests or targeted herbicides, have proved very popular. In 1999, the area planted to transgenic varieties was approximately half the U.S. soybean crop and about 25% of the country's corn crop.

Future applications of recombinant DNA technology include food crops engineered to produce edible vaccines. This strategy would make vaccination more readily available to children worldwide. Because of their use across many cultures and their ability to adapt to tropical and subtropical environments, bananas have be the object of considerable research effort. Transgenic bananas containing inactivated viruses that cause cholera, hepatitis B and diarrhea has been produced and is undergoing evaluation.

See also Plasmids; Recombination

RECOMBINATION

Recombination, is a process during which genetic material is shuffled during reproduction to form new combinations. This mixing is an important from an evolutionary standpoint because it allows the expression of different traits between generations. The process involves a physical exchange of nucleotides between duplicate strands of **deoxyribonucleic acid (DNA)**.

There are three types of recombination; homologous recombination, specific recombination and **transposition**. Each type occurs under different circumstances. Homologous recombination occurs in **eukaryotes**, typically during the first phase of the meiotic **cell division** cycle. In most eukaryotic cells, genetic material is organized as chromosomes in the **nucleus**. A nick is made on the chromosomal DNA of corresponding strands and the broken strands cross over, or exchange, with each other. The recombinant region is extended until a whole **gene** is transferred. At this point, further recombination can occur or it can be stopped. Both processes require the creation of another break in the DNA strand and subsequent sealing of the nicks by special **enzymes**.

Site specific recombination typically occurs in prokaryotes. It is the mechanism by which viral genetic material is incorporated into bacterial chromosomes. The event is site-specific, as the incorporation (integration) of viral genetic material occurs at a specific location on the bacterial **genome**, called the attachment site, which is homologous with the phage genome. Under appropriate conditions alignment and merging of the viral and bacterial genomes occurs.

Transposition is a third type of recombination. It involves transposable elements called transposons. These are short segments of DNA found in both prokaryotes and eukaryotes, which contain the information enabling their movement from one genome to another, as well as genes encoding other functions. The movement of a transposon, a process of transposition, is initiated when an enzyme cuts DNA at a target site. This leaves a section that has unpaired nucleotides. Another enzyme called transposase facilitates insertion of the transposon at this site. Transposition is important in genetic engineering, as other genes can be relocated along with the transposon DNA. As well, transposition is of natural significance. For example, the rapid reshuffling of genetic information possible with transposition enables immunocytes to manufacture the millions of different antibodies required to protect eukaryotes from infection.

See also Meiosis; Bacterial genetics

REGULATION GENE

Regulation genes are a type of **gene** that provides the instructions for creating proteins which help control the expression of other structural genes. They are a key part of the overall system of **gene expression**.

Gene expression is the process by which cells translate their genetic code into proteins. These proteins have a variety of functions such as providing structure, storing energy, providing movement, transporting other substances, catalyzing and controlling biological reactions, and protecting against disease. Depending on the type of **cell** and the environment, at any given time only a handful of genes are expressed in the cell. The other genes are suppressed by regulation genes.

Regulation genes were first discovered in 1961 by **François Jacob** and **Jacques Monod** in Paris. These scientists found that some of the genes in a bacterial **genome** were turned on or off depending on the environmental conditions. Their studies led to the development of the **operon** model. According to this model, regulation genes are present on the bacterial **chromosome** which control the expression of other structural genes. For example, certain **bacteria** have a set of genes which can produce the **amino acid** tryptophan through a variety of biosynthetic processes. When tryptophan is present in the cell's environment, it binds with the protein product of a regulation gene. This protein-tryptophan complex then binds to the **DNA** of the genes that synthesize tryptophan and turns them off. When tryptophan is depleted from the environment, the regulation protein then ceases to bind to the DNA, thereby turning the tryptophan-synthesizing genes back on. This type of control system allows the cell to respond rapidly to a changing environment.

The proteins that are produced from regulation genes control the **transcription** of other genes by binding to particular sites on the DNA. This binding action can affect the target gene in either a positive or negative way by turning it on or off. The tryptophan operon is an example of negative control. Regulation genes are typically located upstream of the structural genes that they regulate. While the binding site of most regulation proteins vary, depending on the protein type, most bind upstream of the target genes.

While operon systems are common in **prokaryote** genomes, they are not generally found in **eukaryotes**. Instead, the regulation genes in a eukaryote may control genes in much more subtle ways. Evidence shows that regulation proteins can affect gene expression even when they bind to DNA that is thousands of base pairs away from the gene. Regulation proteins may also control gene expression in other ways such as binding to **mRNA**, reducing the speed at which mRNA is degraded, and self regulating. While a large amount of information has been gathered on the subject of eukaryotic gene expression, much is still unknown. Scientists hope that one day this information could lead to practical cures for **cancer**.

See also Transcription

REGULATORY REGION OR SEQUENCE

A regulatory region refers to a region of a **gene** to which proteins bind. The region, typically located just before the stretch of **DNA** that encodes the protein product of the gene, stimulates or abrogates the expression of the protein.

In a gene, **transcription** of DNA to **messenger RNA**, which in turn provides the blueprint for the manufacture of a protein, begins upon the binding of transcription molecules to the promoter. The promoter is a regulatory region, determining in a positive sense whether the particular gene is transcribed or not. Another stretch of DNA called the **operator**, also functions as a regulatory region, but in a negative sense. The binding of repressor molecules to the operator can prevent transcription from occurring. The enhancer also functions as a regulatory region.

Regulation of **gene expression** via the binding of proteins is a ubiquitous mechanism, being found throughout the eukaryotic and prokaryotic world. The so-called master, or homeotic, genes, which coordinate the expression of genes at the proper times and in the proper places in organisms ranging from the fruit fly *Drosophilae* to **bacteria** such as *Escherichia coli,* depend upon the existence of regulatory regions. The homeotic gene products are very similar in **amino acid** sequence and three-dimensional shape to regulatory proteins that bind to promoter and enhancer regions.

In **eukaryotes**, a sequence of nucleotides called the Hogness box, or TATA box (denoting the base sequence), is an important regulatory region. The TATA box lies some 25 **bases** upstream from the promoter. In prokaryotic genes the Pribnow box, or TATAAT box, plays a similar role, but it is located closer to the gene sequence, about 10 nucleotides away.

The operons of prokaryotic organisms, which have been well studied and which are classic examples of genetic regulation, operate based on regulatory regions. In the lactose **operon**, a repressor gene encodes a repressor protein that can bind to a specific region of the DNA called the operator. The binding of the repressor inhibits transcription of the structural genes in the operon. On another level, however, the binding of the repressor can be prevented by the occupation of the regulatory region by another molecule, termed an inducer. This is referred to as negative control. Positive control, also involving the regulatory region, exists, in response to levels of specific metabolizable compounds in the environment.

See also Gene regulation; Homeobox; Master gene; Operon; Promoter; Upstream, downstream sites

REPEATED SEQUENCES

The mammalian **genome** can roughly be divided into two classes of **DNA**: unique and repeated. Unique sequence DNA occurs only once per **haploid** genome and codes for structural genes. Repeated sequence DNA refers to multiple copies of particular sequences that are found more than once, ranging from a few (ten or less) to many (greater than a thousand) times in the genome. Depending on the organism, repeated sequences can account for 20–80% of the total DNA. This diverse group can be further subdivided into moderate and highly repetitive elements. The moderately repetitive category is comprised primarily of the tRNA (**transfer RNA**) and rRNA (**ribosomal RNA**) genes. In humans, the rRNA genes are clus-

tered on the short arm of only 10 chromosomes, but the tRNA genes are more widely distributed in at least 50 different sites in the genome. These **RNA** genes are transcribed and the products are used in **translation**. The highly repetitive sequences are quite different in that they are usually non-coding, and due to this, were once thought to be nonfunctional and termed "junk" DNA. However, although highly repeated sequences do not produce a product, it has been shown that many have important roles in the **cell**.

The majority of the highly repeated sequences are arranged in **tandem repeats**. Of the remaining repeated sequences in the genome, one group is the pseudogenes, inactive regions that have a sequence similar to an active **structural gene**. Some **gene** families include several different pseudogenes that are thought to be evolutionary variants of the functional gene.

Interspersed repeats are another type of repeated sequence that are widely distributed throughout the genome, but tend to occur singly rather than in clusters. The two primary classes are : 1) SINES, short interspersed nuclear elements, and 2) LINES, long interspersed nuclear elements such as *LINE-1* or *L1*. SINES show some homology to RNA genes. The *Alu* repeats are the best studied of this group and are the most common sequence in the human genome. They are approximately 280 bp long and are generally found in regions adjacent to genes. The LINES are much larger, up to 7 kb in length. LINES may be transcribed but are not translated. Together, SINES and LINES are estimated to comprise approximately 10% of mammalian genomes. The function of these regions is unknown.

Transposable elements, LTRs (long terminal repeats), and retroviral-like elements (RTLV) make up another small group of repeated sequences. Analysis has shown a similarity between these elements and repeated sequences that occur in **bacteria** and viruses. In addition, some bacterial **plasmids** have repeated sequences known as iterons that are important in regulating the binding of a protein important in plasmid **replication**.

REPIFERMIN

The **Human Genome Project** (HGP) has bolstered the development of new technologies that utilize the projects comprehensive construction of genetic maps, physical maps, and **DNA** sequence data. Scientists apply this information, combined with bioinformatics and high **throughput screening** methods, to identify and develop novel drug candidates. One of the first drugs to come from the HGP and be used in clinical trials, Repifermin, was identified by a genomics company called Human **Genome** Sciences (HGSI) of Rockville, Maryland. HGSI uses **sequencing** data to identify genes like Repifermin. Repifermin, or keratinocyte Growth Factor–2 (KGF–2), is a human protein believed to be activated in response to epithelial wound damage. Its function is to attract cells involved in wound healing such as fibroblasts, collagen and connective tissue to site of the wound for tissue repair and regrowth. It was discovered while screening 10,000 different genes. HGSI found more than 300 full-length genes corresponding to potential therapeutic protein candidates. Scientists used recombinant expression systems to test each **gene** and protein in 70 cell-based screening assays that measured the role of each protein in wound repair. Repifermin is one of six new members of the fibroblast growth factor family discovered by HGSI scientists.

By making cells in skin and other part of the body migrate to the site of a wound, the therapeutic potential of Repifermin is enormous. For example, it can be used during routine surgery intravenously or by spraying it on tissues after surgical incisions to promote healing. It has been shown to promote healing of not only skin wounds but also internal wounds in the mucous membranes that line the gastrointestinal tract and, therefore, may be used to treat chronic skin ulcers, skin injuries, ulcerative colitis or inflammatory bowel disease. Chronic painful sores (that sometimes do not heal) and typically plague elderly patients might be treatable with Repifermin. Additionally, it may be helpful in treating **cancer** patients suffering from mucositis (mucositis is caused by inflammation of the mucous membranes lining oral cavities or the digestive tract and can be a side effect of chemotherapy treatment).

Repifermin has successfully completed both Phase I and Phase IIa clinical trials and is considered to be effective for topical indications in humans including; burns, acute surgical wounds, chronic wounds such as pressure ulcers and venous ulcers. As of 2001, Repifermin is being investigated in Phase IIb clinical trials. Repifermin is one of the first of many drugs to be developed using the sequencing data generated by HGP.

REPLICATION

Replication is the process by which nucleic acids such as **deoxyribonucleic acid** (**DNA**) or **ribonucleic acid** (**RNA**) are copied. The current method of replication was first proposed by **James Watson** and **Francis Crick** in 1953. Watson and Crick suggested that **DNA** was replicated by a semiconservative method in which each strand of DNA served as a template for new strands. In the late 1950s, this theory was confirmed through experiments performed by Matthew Meselson and Franklin Stahl. In their experiment, they grew *E. coli* **bacteria** in a medium, containing a heavy isotope of nitrogen. The bacteria naturally incorporated the heavy nitrogen into their DNA. The bacterial cultures were then placed in a lighter nitrogen, which enabled Meselson and Stahl to follow the production of new DNA and confirm the semiconservative replication model.

DNA replication begins with a double strand of DNA. This structure, or parent molecule, is made up of two complementary strands of DNA that have paired nucleotides. Each **nucleotide** is paired with a specific partner, adenine (A) with thymine (T), and guanine (G) with cytosine (C). Replication begins at specific sites known as the origin of replication. These sites have a nucleotide sequence that is recognized by certain proteins. When replication is initiated, a short segment of DNA is untwisted by various enzmes called helicases at the origin. Next, the two DNA strands are separated, creating a replication bubble. The replication bubble is stabilized by another set of proteins that bind to DNA. This leaves segments of DNA that have a certain length of unpaired nucleotides. These unpaired nucleotides provide the template needed for the replication of the DNA strands. At each end of the replication bubble, there is a section called the replication fork where the new strands of DNA emerge as they are elongated.

During the replication process, key **enzymes** known as DNA polymerases, bring new nucleotides onto each strand of DNA following the **base pairing** rules. DNA polymerases have the limitation of only being able to replicate DNA in one direction, namely 5'–3'. A double strand of DNA is antiparallel, however, meaning the nucleotides on one strand are oriented in the opposite way as nucleotides on the other strand. One strand, called the leading strand, can be replicated easily by the **polymerase**. The other strand, called the lagging strand, employs short strands of replication called Okazaki fragments. These fragments are connected into a single DNA strand by another enzyme called DNA ligase.

Another limitation of DNA polymerase is that it requires a primer to initiate DNA replication. These **primers** are short stretches of RNA that are joined to the DNA by a primase enzyme. In **eukaryotes**, the primer is about 10 nucleotides long. With the primer attached to the DNA, the polymerase begins replication from that point. After replication, one DNA molecule has been made into two. Each of these DNA strands is made up of an old strand and a new one. The rate at which replication occurs can be as fast as 500 nucleotides per second in bacteria. In eukaryotes, the rate is more like 50 nucleotides per second.

DNA replication occurs during the S phase of a cell's life cycle. At this time, duplicates of the chromosomes in the **cell** are visible. **Mitosis** separates the duplicate chromosomes and DNA is distributed between the two daughter cells. In this way, replication allows the transfer of genetic material from one cell to the next.

See also DNA replication

REPLICATION PROCESS MANIPULATION

Replication of genomes is a complex process that involves the interaction of many proteins and enzyme activities. It initiates at specific locations called origin of replication, which are sites of assembly and binding of protein and enzyme complexes termed origin recognition complex (ORC).

The progress and completion of replication is regulated both in space and in time in such a way that it only starts and terminates at the specific origin and termination sites and at predetermined time points of the **cell division** cycle. The protein and enzyme complexes responsible for the special and temporal control of the three major stages of initiation, elongation, and termination of **DNA replication** have been characterized in considerable detail. Biochemical, genetic, and *in vitro* complementation approaches have been used to reveal common sequence and structural features of the replication machinery such as binding sites for replication proteins, **DNA** unwinding elements, and sites for the attachment to the nuclear matrix and scaffold structures. The detailed mechanisms of how these different elements specify the origin of replication within the whole **chromosome** is still poorly understood, especially in the mammalian genomes where replication starts at many sites on each chromosome.

Different conditional mutants in **bacteria** and **yeast**, which are usually temperature sensitive mutants, have helped to identify the components of the ORC and identify consensus sequences determining the initiation site. DNA **methylation** of sites in the origin of replication has also been identified as critical in controlling the initiation of replication. Special **enzymes** called Dam methylases add methyl groups to these sites. During the semiconservative synthesis of DNA, only the old strand is methylated, the new (novel) strand remains non-methylated for a period of time after each replication cycle. The hemimethylated state (i.e., one strand methylated, the other not methylated) of these sites lead to the sequestration of the origin of replication to the bacterial membrane where a membrane bound inhibitor is thought to prevent the binding of the first component of the ORC. Bacterial mutants that lack the Dam methylase (Dam-) are used in recombinant DNA technology. If the copy number of a plasmid vector inside a bacteria needs to be controlled, one introduce this vector in a Dam negative strain and it will only replicate once.

The temporal regulation of the initiation of replication in eukaryotic cells is studied using complementation assays. These are controlled systems where a substrate DNA, is allowed to undergo replication under the control of factors from the cytosol of cells that are at different stages of the **cell** cycle. *Xenopus* eggs, which are known to contain replication factors that are sufficient to support several rounds of replication without new **protein synthesis**, are used to support the replication of DNA in nuclei from another cell type. When a **nucleus** is introduced into the egg and the nuclear membrane is prevented from disintegrating, the DNA in the injected nucleus will replicate only once. If the nuclear membrane is disrupted, the DNA will undergo several rounds of replication. This indicates the existence of a factor or factors responsible for initiating replication that were depleted from the intact nucleus.

The factors needed to initiate replication, and supplied in this case by the *Xenopus* cytosol, are called licensing factors. Their depletion from the nucleus after initiation of the first replication prevents further round of replication from occurring. This process, cytosolic licensing can not be imported into the nucleus and are only takes place when the nuclear membrane breaks down during **mitosis.**

The elongation of replication is maintained by the integrity of the protein and enzyme complex that is formed at the origin of replication. This complex moves along the DNA substrate until it reaches a termination site where it disassembles. The ORC complex in the yeast is an exception and does not disassemble. It is a constitutive **chromatin** bound complex. The initiation of replication or its inhibition depends, in this case, on the interaction of protein factors, called Cdc6 and Mcm, with the ORC and the origin of replication.

Control of replication termination is very poorly understood in eukaryotic and mammalian systems. The existence of replication termini in **plasmids** and bacteria was established in the 1980s. As of 2001, however, studies have identified only a few of the number of factors involved in the control of termination in these replicons.

See also Eukaryotic genetics; Proteins and enzymes; Regulator gene; Regulatory region or sequence; Somatic cell genetics

REPLISOMES

Replisome refers to a complex of proteins that are engaged in the elongation of the newly synthesized strand of **deoxyribonucleic acid**. The replisome assembles at the **replication fork**, the region where the new **DNA** stand is being made.

The replisome complex assumes three principle functions, DNA **polymerase**, DNA primase and DNA helicase. The helicase, a donut-shaped enzyme, initiates replication by unwinding the two parental DNA strands. The structure of the helicase was elucidated in 1999 using the technique of **x-ray crystallography**. It has been determined that the helicase moves along the DNA strand very quickly, 300 paired nucleotides per second. Once the DNA strands are unwound, single stranded **DNA binding proteins** attach to each unwound strand, preventing the strands from winding back together. Subsequently, DNA polymerase catalyzes the elongation of each strand. The polymerase is able to operate in a continuous fashion on the leading parental strand that is unwinding. On the other strand, the DNA primase molecule builds **primers**, which are eventually connected to the other replicated DNA strand.

See also DNA replication

REPORTER GENE · *see* GENE NAMES AND FUNCTIONS

REPRESSION · *see* GENETIC REGULATION

RESOLUTION

The amount or degree of detail that can be expressed on a physical genetic map of **DNA** is referred to as the resolution of the map or the resolution of study.

The greater the resolution on a physical map the more readily identifiable are the genetic elements designated on the map. Of course, such resolution reflects the precision of the studies used to characterize the section of DNA in question and of the ability of researchers to identify genetic landmarks (e.g., intergenic nitrogenous base sequences, genes, bands, markers, **cutting sites**, etc) within the region of interest.

The degree of resolution is often express in terms of base pairs (bp). The maximum resolution obtainable is the goal of the Human **Genome** Project—the complete and accurate **sequencing** of all the nucleotides comprising the human genome.

There are low and high resolution genetic maps (physical maps). Low resolution chromosomal maps carry an average resolution of about 10-12 Mb (1 M = 1,000,000 base pairs). The resolution may be increased depending on the tightness of DNA binding. Accordingly, DNA extracted from loose chromatin-like chromosomal DNA at interphase may be resolve down to 100,000 to 200,000 base pairs.

Higher resolution maps (e.g., connoting and macrorestriction maps) can be generated using restriction **enzymes**, **polymerase chain reaction** (PCR), and **protein electrophoresis** techniques. With macrorestriction maps (top- down mapping) **restriction enzymes** cut chromosomes into large pieces a single **chromosome** is cut (with rare- cutter restriction enzymes) into large pieces, which are ordered, further subdivided, and studied often yield resolutions of 100,000 bp to 1Mb. Contig maps (bottom-up maps) result from the cutting of chromosomes that are then cloned and ordered into contiguous blocks (hence a contig map). Such contig maps provide higher resolution usually from 1Mb down to 10,000 bp.

See also Restriction fragment length polymorphism (RFLP)

RESTRICTION ENZYMES

Restriction **enzymes** are proteins that are produced by **bacteria** as a defense mechanism against viruses that infect the bacteria (bacterial phages). Most bacteria have restriction modification systems that consist of methylases and restriction enzymes. In such systems a bacteria's own **DNA** is modified by **methylation** (the addition of a methyl group, CH_3) at a specific location determined by a specific pattern of **nucleotide** residue and protected from degradation by specialized enzymes termed endonucleases.

The names of restriction enzymes are created from the first letter of the bacterial genus followed by the fist two letters of the **species** plus a Roman numeral if more than one restriction enzyme has been identified in a particular species. Thus, the fifth restriction enzyme from *E. coli* is called EcoRV. Besides **cloning**, restriction enzymes are used in **genetic mapping** techniques, linking the **genome** directly to a conventional **genetic marker**.

Any DNA molecule, from viruses to humans, contains restriction-enzyme target sites purely by chance and, therefore, may be cut into defined fragments of size suitable for cloning. Restriction sites are not relevant to the function of the organism, nor would they be cut *in vivo,* because most organisms do not have restriction enzymes.

There are three types of restriction endonucleases in bacteria. Type I cuts unmodified DNA at a non-specific site 1000 base pairs beyond the recognition site. Type III recognizes a short asymmetric sequence and cuts at a site 24-26 base pairs from the recognition site. Type II recognizes short DNA of four to eight nucleotides. Type II restriction enzymes are widely used in molecular biology. Type II restriction enzymes have two properties useful in recombinant DNA technology. First, they cut DNA into fragments of a size suitable for cloning. Second, many restriction enzymes make staggered cuts generating single-stranded ends conducive to the formation of recombinant DNA. Hamilton Smith identified the first type II restriction enzyme, HindII, in 1970 at Johns Hopkins University.

Most type II restriction endonucleases cut DNA into staggered ends. For example, restriction enzyme EcoRI (from the bacterium *Escherichia coli*) recognizes the following six-nucleotide-pair sequence in the DNA of any organism: 5'-GAATTC-3', 3'-CTTAAG-5'. This type of segment is called a DNA **palindrome**, which means that both strands have the same nucleotide sequence but in antiparallel orientation. EcoRI cuts in the six-base-pair DNA between the G and the A nucleotides. This staggered cut leaves a pair of identical single stranded ends. Some enzymes cut DNA at the same position of both strands, leaving both ends blunt.

See also Cloning vector; Microbial genetics

RESTRICTION FRAGMENT LENGTH POLYMORPHISM (RFLP)

As **DNA** changes are not restricted to those that affect phenotype, restriction fragment length polymorphisms (RFLP) analysis is a powerful technique for the characterization of DNA at the molecular level. These markers are inherited in the same manner as genes that code for visible phenotypes. The **recombination** frequency between an RFLP and a detectable phenotype can be measured. Thus, genetic maps can be constructed to include both genotypic and phenotypic markers. RFLPs can thereby provide a link between genes that lie far apart. In 1980, an RFLP map was created for the human **genome**.

Restriction maps that result from different patterns of distribution of restriction sites in the DNA of individuals within a population of organisms are called restriction fragment length polymorphisms (RFLPs). Differences in individual base pairs between comparable sequences of any two individual chromosomes occur at a frequency of greater than one change per kilobase. Highly polymorphic regions are usually located between genes, where small variations in sequence do not affect **gene expression**. The polymorphisms can be identified by the digestion of genomic DNA with a restriction enzyme (endonuclease). Differences in sequences that result in the gain or loss of a restriction site cause variations in the lengths of the fragments produced. Polymorphisms can also result from the insertion or deletion of stretches of DNA between two restriction sites.

To visualize an RFLP, Southern blotting techniques are used to identify fragments of various lengths based on the location of the sites of a particular restriction enzyme. Probes of known sequence that highlight restriction fragments that often vary in length among different individuals are used to generate clearly discernible patterns. A probe specific for a portion of a particular **chromosome** will reveal differences within that region of the genome from one individual to another.

Because restriction polymorphisms should occur near any particular target **gene**, RFLPs can be identified that show tight linkage with a mutant (or disease) phenotype. Comparison of restriction maps of patients suffering from a particular disease with those of unaffected individuals can reveal specific bands that are always present or absent in affected individuals.

After the identification of an RFLP that is tightly linked with a disease, it may be used as a molecular procedure to detect the disease, either in a prenatal screen or after birth. For instance, an RFLP has been identified that is associated with the genetic disease **sickle cell anemia**. Sickle **cell** anemia is caused by a **mutation** in the alpha-globin gene and results in an abnormal form of hemoglobin. Digestion of DNA with the restriction enzyme Hpa1 and Southern analysis using a probe specific for the alpha-globin gene results in the production of a 7–kb fragment in normal individuals. In contrast, patients with sickle cell anemia display a fragment of 13–kb. Carriers of the disease, those who do not have sickle-cell anemia, but who have inherited one copy of the mutant gene from a parent, can also be identified by RFLP analysis. They will produce both the 7– and 13–kb fragments. It is important to note that the change in DNA sequence that causes this RFLP is not the same change that causes sickle cell anemia itself. Instead, it is tightly linked to the alpha-globin gene. Other human genetic diseases that can be detected via RFLP analysis include Huntington's Chorea, **phenylketonuria**, and **cystic fibrosis**.

RFLP analysis is particularly useful for diagnosis of disease because it assays directly for a genotype (DNA sequence) and does not depend on expression of a gene or even phenotypic expression of the disease itself. Thus, a disease can be identified in an individual before symptoms of the disease are apparent. Additionally, a fetus can be monitored for diseases before birth.

Additionally, RFLPs provide a beginning point for the isolation of the gene responsible for a disease. If the mutation that causes an RFLP in fact lies within the gene responsible, an RFLP at this gene must occur in all cases of the disease. Therefore, it is difficult to prove that a defect in a particular gene is in fact responsible for causing a disease. However, mapping and RFLP analysis can exclude certain genes as candidates. They may also provide a point from which researchers may proceed along the DNA to identify the causal gene itself.

The technique termed DNA fingerprinting utilizes RFLPs and other polymorphic markers to identify individuals based on their particular patterns. Through this technique, hair, blood or other bodily fluids found at a crime scene and blood obtained from a suspect can be used to compare DNA fingerprint patterns. A match between the two patterns pro-

vides strong evidence against the suspect. RFLP patterns are also used to establish parent-child relationships by comparison of the map of a suitable region of the chromosome between potential parents and the child.

See also Amplified fragment length polymorphism (AFLP); Blotting analysis; Molecular diagnostics; Paternity and paternity testing

RETINOBLASTOMA

Retinoblastoma is a rare malignant **tumor** which usually appears in infants or young children. It occurs at a frequency of about one in every 15,000 births. In some cases, there is a family (familial) history of the disease.

The genetic cause of retinoblastoma has been extensively studied. It is described as a "two-hit" process. Normally, individuals have two good copies of the retinoblastoma **gene** (RB-1) on **chromosome** 13. The disease develops in individuals in which **mutation** has occurred in both copies of RB-1. It appears that about 40% of patients are born with a defective copy (first "hit"), inherited from one parent. The second copy is rendered defective by a separate mutation (second "hit") that occurs in the eye. Individuals with an inherited RB-1 defect have high likelihood of developing retinoblastoma in both eyes. For these individuals, diagnosis usually occurs by age one. These patients also have increased risk of developing other types of **cancer**.

The other 60% of retinoblastoma patients inherit two normal copies of RB-1 and develop the disease only after each copy experiences an independent mutation. The likelihood of two independent "hits" is lower, and these individuals are less likely to develop retinoblastoma in both eyes. For these individuals, average age of diagnosis is 2.1 years. Chances of developing retinoblastoma decline sharply after age five. For those individuals who have had retinoblastoma in one eye, there is some possibility of the disease appearing in the other eye at any age into adulthood.

In cases with a family history of retinoblastoma, the child inherits a defective chromosome 13 from one parent. The defective alternative (allele) behaves dominantly, in that the victim needs only to inherit one defective gene. The tumor arises, however, only after a second, spontaneous mutation occurs in one of the cells of the retina; therefore, a situation then exists in which both copies of chromosome 13 carry defective alleles. In the majority of cases, spontaneous mutations appear to occur in both copies of chromosome 13.

The RB-1 gene carries the information for making a protein called pRB. This protein regulates **cell division**. When pRB is absent or defective due to defective copies of the gene, uncontrolled **cell** division occurs, and cancer results. pRB appears to be involved in many types of cancer besides retinoblastoma.

Diagnosis is usually made in early childhood. A white reflection in the pupil of the eye is often the first sign of the disease. The presence of a tumor can be confirmed by an oph-

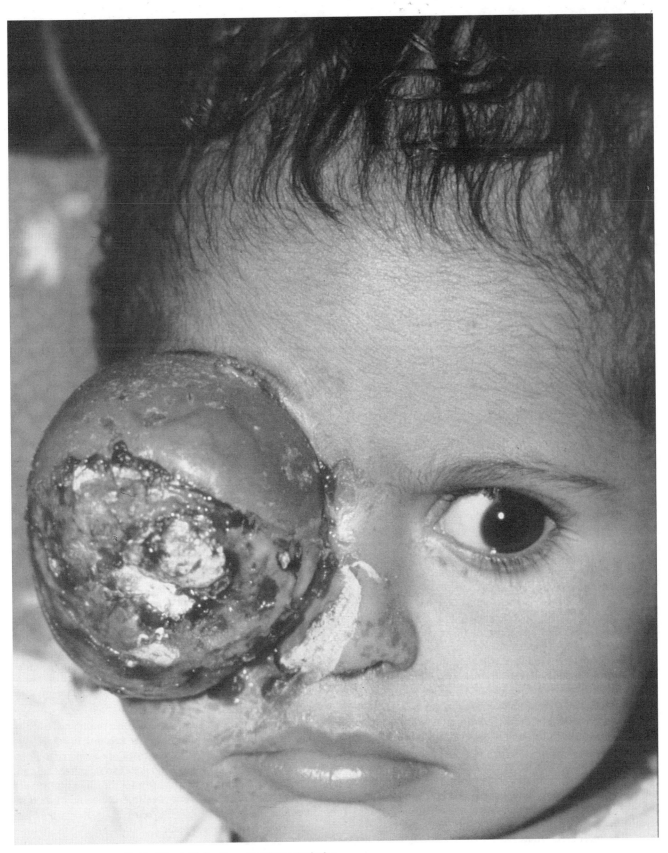

Retinablastoma. *Custom Medical Stock Photo. Reproduced by permission.*

thalmologist directly examining the retina through the pupil. Genetic testing may be recommended to see if a defective gene has been inherited.

Treatment for retinoblastoma depends on the size and number of tumor locations (foci) in the eye, as well as whether the disease is found in one or both eyes. When only one eye is involved, it is surgically removed (enucleated). If both eyes are involved, one eye can sometimes be saved by treating the tumor with radiation, photocoagulation (use of intense laser light to destroy cancer cells), or cryotherapy (use of intense cold to kill cancer cells). Chemotherapy is increasingly used as a follow-up to one of these treatments. However, some forms of radiation therapy have been shown to promote other cancers, especially of the bone.

Because many patients have a strong predisposition to this disease, frequent eye exams are recommended. Even after successful treatment, retinoblastoma patients receive close monitoring in order to get the earliest possible warning if the disease recurs. Individuals with familial tumors in only one eye have a high incidence (70%) of recurrence in the other eye. Some patients experience secondary tumors in other non-ocular tissues of the body.

No preventative measures are possible for genetic conditions such as retinoblastoma. When expectant parents are related to anyone who has had retinoblastoma, they often choose to receive genetic counseling.

See also Cancer genetics

RETROPOSONS AND TRANSPOSABLE ELEMENTS

Transposable elements are relatively long **DNA** sequences in prokaryotic and eukaryotic genomes that act as mobile genetic elements. These elements, which represent a large part of the genomes of many **species** transpose by a mechanism that involves **DNA synthesis** followed by random integration at a new target site in the **genome.**

All transposable elements encode for transposase, the special enzyme activity that helps in the insertion of transposons at a new site, and most of them contain inverted repeats at their ends. The major difference between bacterial transposable elements and their eukaryotic counterparts is the mechanism of **transposition.** Only eukaryotic genomes contain a special type of transposable elements, called retroposons, which use **reverse transcriptase** to transpose through an **RNA** intermediate.

Transposition may result in splicing of DNA fragments into or out of the genome. During replicative transposition, the transposon is first replicated giving a new copy that is transferred to a new site, with the old copy being left at the original site. Nonreplicative transposition however describes the movement of a transposon that is excised from a donor site, usually generating a double, and is integrated in a new site.

The most basic transposable elements in **bacteria** are insertion sequences, which encode only for one enzyme, the transposase. Longer bacterial transposons contain at least one more protein-coding **gene**, which most frequently is an antibiotic resistance gene. In **eukaryotes**, retroposons are more common than transposons. They are either retroviral or nonviral. Viral retroposons encode for the **enzymes** reverse transcriptase and integrase and are flanked by long terminal repeats (LTRs) in the same way as retroviruses. The typical and most abundant nonviral retroposons are the short interspersed elements (SINEs) and the long interspersed elements (LINEs), which are usually repeated, many times in the mammalian genome. Both SINEs and LINEs lack LTRs and are thought to transpose through a special retrotransposition mechanism that involves **transcription** of one strand of the retroposon into RNA. This RNA undergoes conformation change (looping) and provides a primer for the synthesis of single stranded **cDNA**. The cDNA later serve as template for the synthesis of a double stranded DNA that is inserted in the genome by yet unknown mechanisms.

Transposons and retroposons seem to play a role in **evolution** and biology by promoting rearrangement and restructuring of genomes. Transposition may directly cause both deletion and **inversion** mutagenesis. Furthermore, transposable elements mediate the movement of host DNA sequences to new locations, enrich the genome with identical sequences positioned at different locations, and promote homologous **recombination**. Such recombination may eventually result in **deletions**, inversions, and translocations.

Transposons usually influence the expression of the genes in proximity of their insertion sites. They have therefore been extensively used as tools to create random insertion mutants in bacteria, **yeast** and higher eukaryotes. They are also used in large-scale functional genomic studies. They are valuable both during the **cloning** of genes and the generation of transgenic animals.

See also Chromosomal mutations and abnormalities; Clones and cloning; Gene mutation; Mutation

RETROVIRUS

Retroviruses are viruses in which the genetic material consists of **ribonucleic acid** (**RNA**) instead of the usual **deoxyribonucleic acid** (**DNA**). Retroviruses produce an enzyme known as **reverse transcriptase** that can transform RNA into DNA, which can then be permanently integrated into the DNA of the infected host cells.

Many **gene therapy** treatments and experiments use disabled mouse retroviruses as a **carrier** (vector) to inject new genes into the host DNA. Retroviruses are rendered safe by adding, mutating, or deleting viral genes so that the **virus** cannot reproduce after acting as a vector for the intended delivery of new genes. Although viruses are not normally affected by **antibiotics**, genes can be added to retroviruses that make them susceptible to specific antibiotics.

As of 2001, researchers have discovered only a handful of retroviruses that infect humans. Human immunodeficiency virus (HIV), the virus that causes acquired immune deficiency syndrome (**AIDS**), is a retrovirus. Another human retrovirus, human T-cell leukemia virus (HTLV), was discovered three years prior to the discovery of HIV. Both HTLV and HIV attack human immune cells called T cells. T cells are the linchpin of the human immune response. When T cells are infected by these retroviruses, the immune system is disabled and several serious illnesses result. HTLV causes a fatal form of **cancer** called adult T **cell** leukemia. HTLV infection of T cells changes the way the T cells work in the body, causing cancer. HIV infection of T cells, however, eventually kills T cells, rendering the immune system powerless to stave off infections from microorganisms.

Retroviruses are sphere-shaped viruses that contain a single strand or a couple of strands of RNA. The sphere-shaped capsule of the virus consists of various proteins. The capsule is studded on the outside with proteins called receptor proteins. In HIV, these receptor proteins bind to special proteins on T cells called CD4 receptors. CD4 stands for cluster of **differentiation**, and CD type 4 is found on specific T cells called helper cells. The human retroviruses discovered so far bind only to CD4 receptors, which makes their affinity for T helper cells highly specific.

The retrovirus receptor docks with a CD4 receptor on a T cell, and enters the T cell through the T cell membrane. Once inside, the retrovirus begins to replicate. But because the retrovirus's genetic material consists of RNA, not DNA, **replication** is more complicated in a retrovirus than it is for a virus that contains DNA.

In all living things, DNA is the template by which RNA is transcribed. DNA is a double-stranded molecule that is located within the **nucleus** of cells. Within the nucleus, DNA transcribes RNA, a single-stranded nucleic acid. The RNA leaves the nucleus through tiny pores and enters the cytoplasm, where it directs the synthesis of proteins. This process has been called the "central dogma" of molecular biology. No life form has been found that violates this central dogma—except retroviruses. In retroviruses, the RNA is used to transcribe DNA, which is exactly opposite to the way genetic material is transcribed in all other living things. This reversal is why they are named retrograde, or backwards, viruses.

In addition to RNA, retroviruses contain an enzyme called reverse transcriptase. This is the enzyme that allows the retrovirus to make a DNA copy from RNA. Once this DNA copy is made, the DNA inserts itself into the T cell's DNA. The inserted DNA then begins to produce large numbers of viral RNA that are identical to the infecting virus's RNA. This new RNA is then transcribed into the proteins that make up the infecting retrovirus. In effect, the T cell is transformed into a factory that produces more retroviruses. Because reverse transcriptase enzyme is unique to retroviruses, drugs that inhibit the action of this enzyme are used to treat retroviral infection, such as HIV. Reverse transcriptase is vital for retrovirus replication, but not for human cell replication. Therefore, modern reverse transcriptase inhibitor drugs are specific for retroviruses. Often, reverse transcriptase inhibitors are used in combination with other drugs to treat HIV infection.

Retroviruses are especially lethal to humans because they cause a permanent change in the T cell's DNA. Other viruses merely commandeer their host cell's cytoplasm and chemical resources to make more viruses; unlike retroviruses, they do not insert their DNA into the host cell's DNA. Nor do most viruses attack the body's T cells. Most people's cells, therefore, can recover from an attack from a virus. Eventually, the body's immune system discovers the infection and neutralizes the viruses that have been produced. Any cells that contain viruses are not permanently changed by the viral infection. Because retroviruses affect a permanent change within important cells of the immune system, cellular recovery from a retrovirus infection does not occur.

In 1980, researchers headed by Robert Gallo at the National Cancer Institute discovered the first human retrovirus. They found the virus within leukemic T cells of patients with an aggressive form of T cell cancer. These patients were from the southern United States, Japan, and the Caribbean. Almost all patients with this form of cancer were found to have antibodies (immune system proteins made in response to an infection) to HTLV.

HIV is perhaps the most famous retrovirus. Discovered independently by several researchers in 1983, HIV is now known to be the causative agent of AIDS. People with AIDS test positive for HIV antibodies, and the virus itself has been isolated from people with the disease.

HIV attacks T cells by docking with the CD4 receptor on its surface. Once inside the cell, HIV begins to transcribe its RNA into DNA, and the DNA is inserted into the T cell's DNA. However, new HIV is not released from the T cell right away. Instead, the virus stays latent within the cell, sometimes for ten years or more. For reasons that are not yet clear, at some point the virus again becomes active within the T cell, and HIV particles are made within the cell. The new HIV particles bud out from the cell membrane and attack other T cells. Soon, all of the T cells of the body are infected and die. This infection cycle explains why very few virus particles are found in people with the HIV infection (those who do not yet have AIDS); many particles are found in people who have fulminate AIDS.

No cure has yet been found for AIDS. Researchers are still unsure about many aspects of HIV infection, and research into the immune system is still a relatively new science. Several anti-retroviral drugs, such as AZT, ddI, and ddC, have been administered to people with AIDS. These drugs do not cure HIV infection; they merely postpone the development of AIDS. AIDS is almost invariably fatal.

Simian immunodeficiency virus (SIV) is the primate version of HIV. In fact, monkeys infected with SIV are used to test AIDS drugs for humans. Rous sarcoma virus (RSV) causes cancer in chickens and was the first retrovirus identified. Feline leukemia virus (FELV) causes feline leukemia in cats and is characterized by symptoms similar to AIDS. Feline leukemia is a serious disease that, like AIDS, is fatal. Unlike AIDS, a vaccine has been developed to prevent this disease.

See also Bacterial genetics; Cancer genetics; Epidemiology and genetics; Immunogenetics; Viral genetics

RETT SYNDROME

Rett syndrome is a progressive neurological disorder seen almost exclusively in females. The most common symptoms include decreased speech, mental retardation, severe lack of coordination, small head size, and unusual hand movements.

Dr. Andreas Rett first reported females with the symptoms of Rett syndrome in 1966. Females with this X-linked dominant genetic condition are healthy and of average size at birth. During infancy, head growth is abnormally slow and microcephaly (small head size) develops. Babies with Rett syndrome initially have normal development. At approximately one year of age, development slows and eventually stops. Patients with Rett syndrome develop autistic features. Involuntary hand movements are a classic feature of Rett syndrome.

Females with Rett syndrome may also develop seizures, curvature of the spine (scoliosis), irregular breathing patterns, swallowing problems, constipation, and difficulties walking. Some females with Rett syndrome are unable to walk. There is currently no cure for Rett syndrome. Most girls with Rett syndrome live until adulthood. The **gene** responsible for Rett syndrome has been identified and genetic testing is available.

Rett syndrome is an X-linked condition. This means that the **mutation** (genetic change) responsible for Rett syndrome affects a gene located on the **X chromosome**. The affected gene is the methyl CpG-binding protein 2 (MECP2) gene. This gene makes a protein that regulates other genes. When there is a mutation in MECP2, the protein it makes does not work properly. This is thought to prevent normal neuron (nerve **cell**) development.

Rett syndrome is considered to be X-linked dominant in nature. Males have one X **chromosome** and one **Y chromosome**. Females have two X chromosomes. Males with a mutation in their MECP2 gene typically die as infants or are miscarried before birth. Rett syndrome is usually considered fatal in males because the Y chromosome cannot compensate for the MECP2 mutation on the X chromosome. Females with a mutation in the MECP2 gene develop Rett syndrome, but the presence of the second X chromosome in females carrying a normal MECP2 gene enables them to survive.

The severity of the syndrome in females is related to the type of mutation in the MECP2 gene and the activity of the X chromosomes. Normally, both X chromosomes have the same activity. However, the activity can be unequal. If the X chromosome with the mutation in the MECP2 gene is more active than the X chromosome without the mutation, the female is more severely affected. The reverse is also true. If the X chromosome without the mutation is more active than the X chromosome with the mutation, the female is less severely affected.

If a woman has a mutation in her MECP2 gene, she has a 50% risk with any pregnancy to pass on her X chromosome with the mutation. However, it is uncommon for women with Rett syndrome to have children due to the severity of the disorder.

The incidence of Rett syndrome is thought to be between 1 in 10,000 and 1 in 20,000 live births. It is seen almost exclusively in females. The vast majority of cases of Rett syndrome are sporadic in nature. Therefore, the risk of a family having more than one affected daughter is typically very low.

Infants with Rett syndrome are typically a normal size at birth. They develop normally until approximately 6-18 months of age. Development then slows, eventually stops, and soon regresses. Affected individuals are unable to do things they were once able to do. Girls with Rett syndrome lose the ability to speak, become uninterested in interacting with others, and stop voluntarily using their hands. The loss of language and eye contact causes girls with Rett syndrome to appear to be autistic. Between one and three years of age girls with Rett syndrome develop the unusual hand movements that are associated with the disease. Patients wring their hands, clap their hands, and put their hands in their mouth involuntarily. Some patients with Rett syndrome also lose the ability to walk. If the ability to walk is maintained, the gait is very ataxic (uncoordinated, clumsy).

By preschool age the developmental deterioration of girls with Rett syndrome stops, but they continue to have lack of speech, inability to understand language, poor eye contact, mental retardation, ataxia, and apraxia (inability to make purposeful movements). Other common symptoms associated with Rett syndrome include seizures, constipation, irregular breathing, scoliosis, swallowing problems, teeth grinding, sleep disturbances, and poor circulation. As patients with Rett syndrome get older, their ability to move decreases and spasticity (rigidity of muscles) increases.

The diagnosis of Rett syndrome is made when the majority of the symptoms associated with the disease are present. If a physician suspects an individual has Rett syndrome, **DNA** testing is recommended. Approximately 75% of patients with Rett syndrome have a mutation in the MECP2 gene. DNA testing can be performed on a blood sample, or other types of tissue from the body. If a mutation is found in the MECP2 gene, the diagnosis of Rett syndrome is confirmed.

As of 2001, there is not a cure for Rett syndrome. Treatment of patients with Rett syndrome focuses on the symptoms present. Treatment may include medications that inhibit seizures, medications that reduce spasticity, and medications that prevent sleep disturbances. Nutrition is monitored in females with Rett syndrome due to their small stature and the constipation associated with the disorder. In the absence of severe medical problems, most patients with Rett syndrome live into adulthood.

See also Chromosomal mutations and abnormalities; Genetic disorders and diseases; Hereditary diseases; Inherited cancers; Mutations and mutagenesis

REVERSE TRANSCRIPTASE

Reverse transcriptase, also known as RNA-directed **DNA polymerase**, was independently discovered by **Howard Temin** and **David Baltimore** in 1970. Reverse transcriptase acts like a regular DNA polymerase in that it synthesizes DNA in the 5' to 3' direction from primed templates. The difference is that reverse transcriptase uses **RNA** as its template.

Reverse transcriptase was first discovered in retroviruses. Retroviruses are RNA containing viruses such as certain **tumor** viruses and human immunodeficiency **virus** (HIV, the causative agent of **AIDS**). In the infection cycle of **retrovirus**, the virus attaches to the host **cell** surface and releases its RNA **genome** and the prepackaged reverse transcriptase into the host cell. Reverse transcriptase then starts to retro-transcribe RNA into double stranded DNA. This event happens at the cytoplasma. The double stranded viral DNA genome is then translocated into the **nucleus**. The linear copy of the retroviral genome is inserted into the host chromosomal DNA with the help of other proteins to form a provirus. The provirus achieves the status of a cellular **gene** and is expressed through the agency of cellular RNA polymerase, and replicated by cellular **enzymes** in concert with chromosomal DNA. At the end of the virus life cycle, retroviral RNA and reverse transcriptase, together with other components, are packed into a viral particle and released out of the cell.

The discovery of reverse transcriptase was a surprise to the scientific world. The central dogma of molecular biology was that genetic information always flows from DNA to RNA to protein. Since 1970, reverse transcriptases have been found in insect and plant viruses. It is now believed that **reverse transcription** is not a singular odd exception, but rather a paradigm for a process that is shared by viral and nonviral genetic elements occurring widely in nature. The elements of **yeast** and the copia and Ulysses elements of *Drosophila* (fruit fly) all resemble retrovirus in that they all encode reverse transcriptases.

Reverse transcriptase has been a useful tool in genetic engineering because of its ability to transcribe mRNAs to complementary stands of DNA (**cDNA**). cDNAs can then be cloned, sequenced, or expressed for further studies. Murine and avian virus reverse transcriptases are the mostly used reverse transcriptases in genetic engineering. Research of reverse transcriptase also has clinical significance. Studies on inhibitors of reverse transcriptases are promising in the search for a cure to AIDS and other retrovirus related diseases.

See also DNA replication; Gene expression

REVERSE TRANSCRIPTION

Reverse **transcription** is an atypical method of synthesizing **DNA** from a template of **RNA**. Typically, transcription is a unidirectional process in which the **nucleotide** sequence of one strand of DNA is copied, thereby creating a single strand of RNA with a nearly identical sequence. This RNA strand is then utilized by the **cell** for a variety of reasons related to the synthesis of **proteins and enzymes**.

In reverse transcription, the process is reversed. RNA is used as a template to make DNA. This process is performed by certain retroviruses whose genetic code is made up of single-stranded RNA molecules. It also requires a special enzyme

known as a **reverse transcriptase** enzyme. When these viruses infect a cell, they inject it with their RNA. Instead of being utilized in **protein synthesis**, this RNA goes through the process of reverse transcription, and is converted into a single-stranded DNA molecule. This single-stranded DNA is further converted into a double-stranded DNA that then becomes integrated into the cell's **genome**. When these foreign genes are expressed, the cell's normal functions are altered and it becomes a manufacturing site for more viruses.

The existence of reverse transcription establishes the general rule that information stored in nucleic acid sequences as either RNA or DNA can be converted between either type. However, reverse transcription does not generally occur in the normal operations of a cell.

See also Viral genetics

RH FACTOR • *see* BLOOD GROUP GENETICS

RIBONUCLEASE

Ribonuclease is a type of enzyme which catalyzes the breakdown of bonds in **ribonucleic acid** (**RNA**). It plays an important role in the regulation of protein production in cells. There are a wide variety of ribonucleases found throughout nature. The most thoroughly studied of these is bovine pancreatic ribonuclease. It was the first enzyme whose two dimensional and three dimensional structures were determined.

Bovine pancreatic ribonuclease has been extensively studied because it is a relatively small enzyme, abundant, and heat stable. It was first discovered in 1920 by W. Jones. He demonstrated that an extract from a bovine pancreas could hydrolyze **yeast** nucleic acid. In 1940, this ribonuclease was crystallized and isolated from the extract. It became the first enzyme for which structural information was obtained. Its primary structure was determined in 1962 and its three dimensional crystal structure was resolved by **x-ray crystallography** in 1967.

Ribonucleases help degrade ribonucleic acid (RNA). RNA is the material in a **cell** that codes for different proteins. When ribonuclease is present in the cytoplasm, RNA is degraded. This is important because it allows the cell to control the amount of protein produced by any RNA and to reuse the nucleotides to create new RNA. While this process is designed to degrade unnecessary RNA, ribonucleases will degrade any type of RNA so a large amount is not present in the cell at any given time.

Bovine ribonuclease contains four disulfide bonds. It is the classic example of a protein whose activity is dependant on it higher order structure. When it is denatured, it loses its catalytic activity. This means that it no longer degrades RNA. When the denaturing process is reversed, it regains its activity. It is known as a ribonucleoprotein because it consists of both an RNA molecule and a protein. **Sidney Altman** discovered that the RNA component alone possesses enzymatic activity. It can interact to degrade **transfer RNA**. The protein portion helps maintain structure and is important for the proper function of the ribonuclease.

There are numerous types of ribonucleases distributed throughout nature, found in both plant and animal **species**. There are three different types of ribonucleases which are generally called ribozymes. A ribozyme is any RNA that has catalytic activity.

The different ribonucleases work in similar ways. Pancreatic ribonuclease catalyzes the hydrolysis of phosphodiester bonds between RNA chains. This is thought to occur in two steps. First, the enzyme breaks a bond between a phosphorus and oxygen on an RNA molecule. Next, water reacts with the free nucleoside permanently removing it from the RNA. This causes a breakdown of the RNA at specific sites, namely at pyrimidine **bases**. Other ribonucleases hydrolyze RNA at other sites. For example bacterial ribonucleases interact with purines. Typically, ribonucleases react with bases on hairpin loop structures in the RNA. Since ribonucleases breakdown RNA at different places, they have become important tools in the analysis of the sequence and structure of RNA.

See also Enzymes, genetic manipulation of; RNA (ribonucleic acid)

RIBONUCLEIC ACID (RNA) • *see* RNA

RIBONUCLEOTIDE • *see* NUCLEOTIDE

RIBOSOMAL RNA

Although it is **DNA (deoxyribonucleic acid)** that contains the instructions for directing the synthesis of specific structural and enzymatic proteins, several types of **RNA** actually carry out the processes required to produce these proteins. These include **messenger RNA (mRNA)**, ribosomal RNA (rRNA) and **transfer RNA** (tRNA). Further processing of the various RNA's is carried out by another type of RNA called small nuclear RNA (snRNA). The structure of RNA (ribonucleic acid) is very similar to that of DNA, however, instead of the base thymine, RNA contains the base uracil. In addition, the pentose sugar ribose is missing an oxygen atom at position two in DNA, hence the name deoxy-.

The first step in **protein synthesis** is the **transcription** of DNA into mRNA. The mRNA exits the nuclear membrane through special pores and enters the cytoplasm. It then delivers its coded message to tiny protein factories called **ribosomes** that consist of two unequal sized subunits. Some of these ribosomes are found floating free in the cytosol, but most of them are located on a structure called rough endoplasmic reticulum (rER). It is thought that the free-floating ribosomes manufacture proteins for use within the **cell (cell proliferation)**, while those found on the rER produce proteins for export out of the cell or those that are associated with the cell membrane.

Ribosomes are composed of ribosomal RNA (as much as 50%) and special proteins called ribonucleoproteins. In **eukaryotes** (an organism whose cells have chromosomes with nucleosomal structure and are separated from the cytoplasm by a two membrane nuclear envelope and whose functions are compartmentalized into distinct cytoplasmic organelles), there are actually four different types of rRNA. One of these molecules is called 18SrRNA and along with some 30–plus different proteins, it makes up the small subunit of the ribosome. The other three types of rRNA are called 28S, 5.8S, and 5S rRNA. One of each of these molecules, along with some 45 different proteins is used to make the large subunit of the ribosome. There are also two rRNAs exclusive to the mitochondrial (a circular molecule of some 16,569 base pairs in the human) **genome**. These are called 12S and 16S. A **mutation** in the 12SrRNA has been implicated in non-syndromic hearing loss. Ribosomal RNA's have these names because of their molecular weight. When rRNA is spun down by ultracentrifuge, these molecules sediment out at different rates because they have different weights. The larger the number, the larger the molecule.

The larger subunit appears to be mainly involved in such biochemical processes as catalyzing the reactions of polypeptide chain elongation and has two major binding sites. Binding sites are those parts of large molecule that actively participate in its specific combination with another molecule. One is called the aminoacyl site and the other is called the peptidyl site. Ribosomes attach their peptidyl sites to the membrane surface of the rER. The aminoacyl site has been associated with binding transfer RNA. The smaller subunit appears to be concerned with ribosomal recognition processes such as mRNA. It is involved with the binding of tRNA also. The smaller subunit combines with mRNA and the first "charged" tRNA to form the initiation complex for **translation** of the RNA sequence into the final polypeptide.

The precursor of the 28S, 18S and the 5.83S molecules are transcribed by RNA **polymerase** I (Pol I) and the 5S rRNA is transcribed by RNA polymerase III (PoIII). Pol I is the most active of all the RNA polynmerases, and is one indication of how important these structures are to cellular function.

Ribosomal RNAs fold in very complex ways. Their structure is an important clue to the evolutionary relationships found between different kinds of organisms. Sequence comparisons of the various rRNAs across various **species** show that even though their base sequences vary widely, **evolution** has conserved their secondary structures, therefore, organization must be important for their function.

See also Nucleic acid

RIBOSOMES

Ribosomes are organelles that play a key role in the manufacture of proteins. Found throughout the **cell**, ribosomes are composed of ribosomal **ribonucleic acid** (rRNA) and proteins. They are the sites of **protein synthesis**.

Although Robert Hooke first used a light microscope to look at cells in 1665, it was only in the last few decades that the cell's organelles were found. This is primarily because light microscopes do not have the magnifying power required to see these tiny structures. Using an **electron microscope**, scientists have been able to see most of the cells substructures, including the ribosomes.

Ribosomes are composed of a variety of proteins and rRNA. They are organized in two functional subunits that are constructed in the cell's nucleolus. One is a small subunit that has a squashed shape, while another is a large subunit that is spherical in shape. The large subunit is about twice as big as the small unit. The subunits usually exist separately, but join when they are attached to a **messenger RNA** (**mRNA**). This initiates protein synthesis.

Production of a protein begins with initiation. In this step, the ribosomal small subunit binds to the mRNA along with the first **transfer RNA** (tRNA). The next step is elongation, where the ribosome moves along the mRNA and strings together the amino acids one by one. Finally, the ribosome encounters a stop sequence and the two subunits release the mRNA, the polypeptide chain, and the tRNA.

Protein synthesis occurs at specific sites within the ribosome. The P site of a ribosome contains the growing protein chain. The A site holds the tRNA that has the next **amino acid**. The two sites are held close together and a chemical reaction occurs. When the stop signal is present on the mRNA, protein synthesis halts. The polypeptide chain is released and the ribosome subunits are returned to the pool of ribosome units in the cytoplasm.

Ribosomes are found in two locations in the cell. Free ribosomes are dispersed throughout the cytoplasm. Bound ribosomes are attached to a membranous structure called the endoplasmic reticulum. Most cell proteins are made by the free ribosomes. Bound ribosomes are instrumental in producing proteins that function within or across the cell membrane. Depending on the cell type, there can be as many as a few million ribosomes in a single cell.

Since most cells contain a large number of ribosomes, rRNA is the most abundant type of **RNA** and rRNA plays an active role in ribosome function. It interacts with both the mRNA and tRNA and helps maintain the necessary structure. Transfer RNA is the molecule that interacts with the mRNA during protein synthesis and is able to read a three amino acid sequence. On the opposite end of the tRNAs, amino acids are bonded on a growing polypeptide chain. Generally, it takes about a minute for a single ribosome to make an average sized protein. However, several ribosomes can work on a single mRNA at the same time. This allows the cell to make many copies of a single protein rapidly. Sometimes these multiple ribosomes, or polysomes, can become so large that they can be seen with a light microscope.

The ribosomes in **eukaryotes** and prokaryotes are slightly different. Eukaryotic ribosomes are generally larger and are made up of more proteins. Since many diseases are caused by prokaryotes, these slight differences have important medical implications. Drugs have been developed that can inhibit the function of a prokaryotic ribosome butleave the eukaryotic ribosome unaffected. One example is the antibiotic tetracycline.

See also Cell; DNA synthesis

RIDDLE, OSCAR (1877-1968)
American zoologist

Oscar Riddle is known for his research on **evolution**, reproduction, **heredity**, and endocrinology. Riddle is also responsible for isolating the hormone prolactin.

Riddle conducted his research in both the United States and Europe, but spent most of his career at the Station for Experimental Evolution at Cold Spring Harbor, Long Island. His extensive study of birds included investigations into the physiology of reproduction, the basis of sex, as well as breeding, heredity, and evolution. In 1932, while working with pigeons at Cold Spring Harbor's Biological Laboratory, he isolated the hormone, prolactin, which is responsible for stimulating the mammary glands to produce milk in females.

Riddle wrote several papers on the physiology of sex, heredity, and endocrinology, as well as development and reproduction issues. One of his most famous treatises, "Any Hereditary Character and the Kinds of Things We Need to Know About It," discusses the diversity of traits contained in one hereditary characteristic. The paper also questions the limitations faced by the scientists in the field of **genetics** during his time, especially when dealing with matters of heredity and evolution. Riddle asserted that in order to gain understanding in those areas, it was necessary to thoroughly understand the nature of one individual hereditary character.

Riddle was a member of several biological societies in South America, India, France, England, and the United States. In 1958, he was given the Humanist of the Year award by the American Humanist Association. Riddle was president of the American Rationalist Federation from 1959 to 1960.

See also Darwinism

RIMOIN, DAVID (1936-)
Canadian-American geneticist

David Rimoin is one of the pioneers in the field of **medical genetics**. His work in describing the genetic and molecular triggers of skeletal abnormalities led to the discovery of several new syndromes, and helped develop new methods for preventing and treating dwarfism.

Rimoin completed his undergraduate work and received his M.D. from McGill University in Montreal, Canada. He went on to receive his Ph.D. in Human Genetics from Johns Hopkins University. In 1970, Rimoin was appointed chief at the University of California at Los Angeles (UCLA) Medical Center's division of medical genetics. Three years later, Rimoin became a professor of pediatrics and medicine at the

UCLA School of Medicine. In 1986, Rimoin was made chairman of the department of pediatrics and director of the Medical Genetics-Birth Defects Center at Cedars-Sinai Medical Center in Los Angeles.

Rimoin's research is focused on the genetic basis of skeletal dysplasias, or skeletal abnormalities. Rimoin led some of the earliest studies on disorders of growth hormone metabolism, and investigated new treatments and methods for preventing short stature. Rimoin is also credited with increasing public awareness and understanding of dwarfism.

In 1979, Rimoin became the founding president of the American Board of Medical Genetics. He was also the founding president of the American College of Medical Genetics and president of the American Society of Human Genetics.

Rimoin has published over 350 papers and edited 11 books, including *Principles and Practice of Medical Genetics* (with Alan E.H. Emery), considered one of the most comprehensive texts on the science of genetics. Dr. Rimoin also played an integral role in developing the Skeletal Dysplasia Center at UCLA.

See also Genetic defect; Genetic disorders and diseases; Molecular biology and molecular genetics

RISK FACTORS

Several biological, environmental, and behavioral patterns may contribute to increase the incidence of disease or death in a given population, and such deleterious patterns are generically termed risk factors. For instance, several recent studies points to tobacco addiction as the leading cause of premature death in the United States, accounting for approximately 500,000 deaths (i.e., about 20% of all deaths) every year. Smoking, therefore, is a serious risk behavioral factor contributing to the development of coronary disease, **cancer**, stroke, pulmonary emphysema, and other respiratory and circulatory disorders.

Biological risk factors may be either an inherited **gene mutation** that increases the predisposition to one or more diseases, such as cancer, diabetes mellitus, or familial hypercholesterolemia; or they may be an infection by viruses or **bacteria** that induce **DNA** mutations in the cells of the host, such as hepatitis **virus** (liver necrosis or liver cancer), *Helicobacter pylori* (gastric cancer).

Life styles and nutritional habits may also pose risks to both mental and physical health. A stressful life style and the inability to manage daily time, generating anxiety and conflicts, work overload, inadequate sleeping and malnutrition, and lack of daily leisure and relax time, increases the risk of gastric disturbance, mental illness, high blood pressure, depression and even cancer. Obesity and lack of physical exercise, for example, are risky behaviors that account for about 15% of all deaths in the United States, and it is also an important risk factor associated with chronic diseases affecting one and a half million of Americans.

The risk of cancer, heart attack, stroke, and infections is increased by a diet rich in fat and poor in crucial micronutrients, such as vitamins, essential amino acids, and minerals. Alcoholism, i.e., the ingestion of alcoholic beverages on a daily basis, also increases risk for liver conditions, brain deterioration, vitamin B depletion, and cancer, as well as the risk of accidents. Excess of salt in foods are linked by researchers not only to high blood pressure later in life, but it is also considered a risk factor of gastric cancers, with a high incidence among Japanese population and other Asian cultures whose cuisine includes salty sauces, such as shoyu and oyster sauce.

Some genetic conditions, such as diabetes mellitus, also increase the risk of occurrence of other illnesses, such as stroke and myocardial infarction. Age and gender are also risk factors for some diseases. For instance, men is in a greater risk of infarction than women, and in men heart attacks occur earlier than in women. After menopause, women has an increased risk of heart attacks, high blood pressure and stroke; but even then, women's death incidence from these conditions is not as great as men's.

Environmental and occupational exposure to toxic pollutants, such as asbestos, polycyclic aromatic hydrocarbons (PAHs), ultra violet radiation, benzene, insecticides, paint solvents, dyes containing heavy metals (mercury, lead, etc.), constitutes risk factors of a wide range of illnesses, from chronic to lethal ones.

The accumulative effects of several risk factors may increase exponentially the incidence of certain diseases, such as respiratory disorders, leukemia and other cancers, as well as chronic diseases.

Most of the public health risk factors known today are preventable through the adoption of health educational campaigns and public policies aiming at environmental pollution control and occupational protection against exposure. Groups at risk should also be oriented to undergo regular medical check ups in order to prevent disease onset and allow early diagnosis, thus benefiting from better chances of treatment of some otherwise incurable diseases.

See also Amino acid; Carcinogenic; Chemical mutagenesis; Epidemiology and genetics; Genetic disorders and diseases; Genetic dyslipidemias; Hereditary diseases; Radiation mutagenesis; Transduction

RNA FUNCTION

Ribonucleic acid (RNA) is a nucleic acid that is similar, but not identical, to a single strand of **deoxyribonucleic acid (DNA)**. In RNA, the base uracil is substituted for the base thymine in the genetic code. While DNA and RNA are very similar in their composition, RNA has a much more versatile role. It both conveys genetic information and catalyzes reactions.

There are five major types of RNA that are found in eukaryotic cells. The types are known as heterogeneous nuclear RNA (hnRNA), **messenger RNA (mRNA)**, **transfer RNA** (tRNA), **ribosomal RNA** (rRNA), and small nuclear

RNA. Depending on the type of RNA, it can function as a **carrier** for genetic information, a catalyst for biochemical reactions, an adapter molecule in **protein synthesis**, and a structural molecule in **cell** organelles.

RNA functions principally to convey the genetic information contained in DNA into the **gene** product, a protein. In the **nucleus**, a series of enzymatic reactions produces hnRNA from DNA. **Enzymes** are molecules that help (facilitate) a reaction but that are not altered by the reaction. Accordingly, they are not a part of the final product of a reaction. This RNA **species** is a direct copy of DNA, and so contains the coding and noncoding regions of the parent molecule. The extraneous noncoding regions are spliced out as the hnRNA is converted to mRNA. The mRNA is directed to the cytoplasm, in this way, it functions as the carrier for information from the DNA to the **ribosomes**, the cell's protein synthesizing organelles.

The ribosomes interact with the mRNA and construct a protein based on the instructions in the **nucleotide** sequence. A component of the ribosome is rRNA. It provides the structure and shape for the catalytic areas of the ribosome, and acts as a catalyst (or a ribozyme) for the action of another ribosomal RNA, the tRNA. Protein synthesis is contingent upon tRNA. It is the functional bridge between the coding nucleotides on the mRNA and the amino acids used to construct the protein. During protein synthesis, one end of the tRNA interacts with three nucleotides on the mRNA. The other end of the tRNA carries an **amino acid**. The amino acid is transferred and bonded to the growing chain of amino acids that have resulted from the action of other tRNA species.

Another form of RNA that acts as a ribozyme is snRNA. During splicing of the hnRNA, snRNA catalyzes reactions in the spliceosome, a group of biomolecules that accomplish the splicing.

RNA is the sole genetic material present in viruses known as single strand RNA viruses. One class, the Retroviridae, contains an enzyme called **reverse transcriptase** that uses the RNA as a template for the synthesis of viral DNA. The viral DNA integrates into the host cell **genome** to provide the template for viral RNA synthesis by host-derived mechanisms. Retroviruses are also noteworthy because the group includes viruses associated with leukemia and human immunodeficiencies.

RNA (RIBONUCLEIC ACID)

Ribonucleic acid (RNA) conveys genetic information and catalyzes important biochemical reactions. Similar, but not identical, to a single strand of **deoxyribonucleic acid** (**DNA**), in some lower organisms, RNA replaces DNA as the genetic material. As with DNA, RNA follows specific **base pairing** rules, except that in RNA the base uracil replaces the base thymine (i.e., instead of an adenine-thymine or A-T pairing, there is an adenine-uracil or A-U pairing). Accordingly, when RNA acts as a **carrier** of genetic information, uracil replaces thymine in the genetic code.

In humans, **messenger RNA** (**mRNA**) is the product of **transcription** and acts to convey genetic information from the **nucleus** to the protein assembly complex at the ribosome. The ribosome is composed of **ribosomal RNA** (rRNA) and other proteins. Transfer RNAs (tRNA) act to catalyze the **translation** process by acting as carriers of specific amino acids. Because tRNAs bind to specific sites on the strand of mRNA, the sequence of amino acids subsequently inserted into the synthesized protein is both specific and genetically determined by the **nucleotide** sequence in DNA from which the mRNA strand was originally transcribed.

Other forms of RNA perform important roles in other biochemical reactions. Regardless of function, RNA is a biopolymer made up of ribonucleotide units and is present in all living cells and some viruses. The chemical units of RNA are ribonucleotide monomers consisting of a ribose sugar $(C_5H_{10}O_5)$, phosphorylated at the third carbon (C3) and linked to one of four **bases** through a type of chemical linkage formed between a sugar and a base by a condensation reaction (glycosidic bond). The four bases found in RNA are adenine (A), guanine (G), cytosine (C), and uracil (U). Other bases may also be found, although they are generally modified versions of these four (e.g., methylated bases are found in parts of transfer tRNA).

The single nucleotides (monomers) of RNA form a linear chain by linking their phosphate groups and sugars in phosphodiester bonds. RNA does not form a double stranded alpha-helix as does DNA. In some parts of the RNA molecule, there is a folding into alpha-helical-like regions. Corresponding to their unique functions, messenger RNA (mRNA), ribosomal RNA (rRNA), and **transfer RNA** (tRNA) all have different three-dimensional structures. In higher eukaryotic organisms, different RNAs are found distributed throughout the cell—in the nucleus, cytoplasm, and also in cytoplasmic organelles such as **mitochondria** and, in plants, chloroplasts.

The nucleus is the chief site of RNA synthesis and the source of all cytoplasmic RNA, while mitochondria and chloroplasts synthesize their RNA from their own DNA. rRNA is synthesized by the nucleoli within the nucleus, while the high molecular weight precursor to cytoplasmic mRNA, sometimes termed heterogeneous nuclear or hnRNA, is transcribed on the DNA **chromatin**. Low molecular weight RNA also occurs in the nucleus and consists partly of tRNA and partly of RNA, which has a regulatory function in **gene** activation. The cytoplasm contains tRNA and rRNA in the **ribosomes** and mRNA in polysomes, or polyribosomes. The latter are the structural units of protein biosynthesis, consisting of several ribosomes attached to a strand of mRNA.

The function of mRNA is to transcribe the information held in DNA. In the cells of eukaryotic organisms, the first transcriptional product is the long, heterogenous nuclear RNA, or hnRNA. This contains both the nucleotide sequences eventually transcribed into polypeptides and large tracts of sequences not translated. Non-translated sequences are termed introns (or intervening sequences). Removal of introns, and other untranslated portions of the molecule, edits hnRNA into mRNA molecules. After editing removes as much as 90 % of

Molecular structure of RNA. S stands for sugar; P for phosphate.

hnRNA, the resulting mRNA molecules are transported into the cytoplasm.

rRNA is located within ribosomes, the sites of protein biosynthesis. Ribosomes are large ellipsoid cytoplasmic organelles consisting of RNA and protein.

tRNA, the smallest known functional RNA, is essential for protein biosynthesis. Its purpose is to transfer a specific **amino acid** from the cytoplasm and incorporate it into the growing polypeptide chain on the polysome. Different tRNAs contain between 70 and 85 nucleotides. The most characteristic feature of tRNA is that it contains the anticodon, a sequence of three nucleotides specific for the mRNA codon sequence. There is at least one tRNA per **cell** bearing the anticodon for each of the 20 amino acids. The aminoacyl-tRNA (the tRNA carrying the amino acid) becomes bound to the large subunit of a ribosome, where antiparallel basepairing occurs between the anticodon of the tRNA and the complementary codon of the associated mRNA. The specificity of this base pairing ensures that the amino acid inserts into the correct position in the growing protein polypeptide chain. During translation, the deacylated tRNA (i.e., with its amino acid removed) is released from the ribosome and becomes available once again for recharging with its amino acid.

DNA-dependent RNA synthesis is the process of RNA sythesis on a template of DNA. According to the rules of base pairing, the base sequence of DNA determines the synthesis of a complementary base sequence in RNA. Assisted (catalyzed) by the enzyme RNA **polymerase**, the growing RNA chain releases from the template so that the process can start again, even before the previous molecule is complete. Termination **codons** and a termination factor known as rho-factor end the synthesis process. In certain viruses, RNA-dependent RNA synthesis occurs, with the viral RNA acting as a template for the synthesis of new RNA.

See also RNA function; Viral genetics

RNA SPLICING

RNA splicing is a biological reaction in which introns are removed from a transcribed RNA to create **mRNA**. This process occurs in conjunction with the **transcription of DNA** to mRNA.

Before 1977, scientists were not aware that eukaryote genes were dramatically different than prokaryotes. It was known, however, that **eukaryotes** had significantly more DNA than prokaryotes. This difference, called the C-value paradox, led to the discovery that eukaryotes had interrupted genes. These are genes containing **exons and introns**; **nucleotide** sequences that are both coding and non-coding. Evidence for RNA splicing was obtained when nuclear RNA was compared to mRNA. It was found that nuclear RNA was much longer than mRNA suggesting mRNA was further processed before being transported to the cytoplasm.

RNA splicing is one step in the overall process in which the genetic code is transcribed into mRNA and then translated into proteins. It is known to occur in the **nucleus** of the **cell** where DNA transcription takes place. During the process of transcription, each nucleotide from the specific **gene** is translated into RNA. This results in an RNA molecule that contains sequence of both the exons and the introns. Since the introns do not code for proteins, they are removed by RNA splicing.

There are several types of known splicing systems. One system involves a spliceosome which is an array of proteins that function together. The human spliceosome has been found to contain 44 different components. Another type of system involves excision of introns by the RNA itself. Still another type involves the removal of introns by tRNA.

The spliceosome system has been one of the most thoroughly studied splicing system. It is responsible for removing an intron that is located between two exons. The process begins with a set of **enzymes** that recognize and bind to the splice sites. These sites are located at the exon-intron boundaries and are made up of four nucleotides, two at each end of the intron. The bond between the left exon and intron is cut by enzymes in the spliceosome. This is a transesterification chemical reaction. Next, the intron is folded over on itself to the right, forming a molecular lariat structure. Another set of enzymes cuts the right end of the intron and finally joins the two exons.

The existence of a processing system for changing nuclear RNA to mRNA was surprising to scientists at first. Scientists considered it strange that a cell would waste the energy required to maintain an amount of genetic material that does not directly aid in the production of proteins. To explain this phenomena, researchers have attempted to find a function for introns. Current theories suggested that introns play a regulatory role in **gene expression**. Also, they may help cells produce multiple proteins from a single gene.

See also Chromosome structure and morphology; Gene regulation

ROBERTS, RICHARD J. (1943-)
English molecular biologist

For decades scientists assumed that genes are continuous segments within **deoxyribonucleic acid (DNA)**, the chemical template of **heredity**. In 1977, however, Richard J. Roberts, a thirty-four year old British scientist working with adenovirus, the same **virus** that is one cause of the common cold, discovered that genes (the functional units of heredity) can be com-

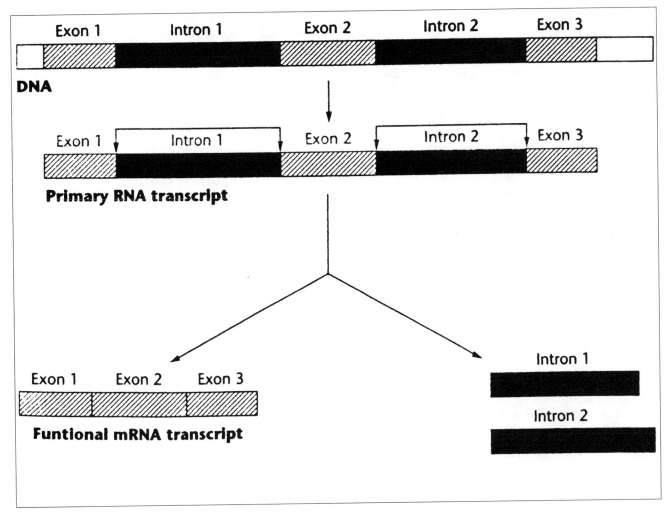

The splicing of a eukaryotic primary RNA transcript. *Fitzgerald Science Press. Reproduced by permission.*

posed of several separate segments rather than of a single chain along the DNA strand. For his discovery of split genes, Roberts, along with colleague **Phillip A. Sharp**, was awarded the Nobel Prize for physiology or medicine in 1993.

Richard John Roberts was born in Derby, England, a mid-sized industrial city about forty miles northeast of Birmingham. His father, John Roberts, was a motor mechanic, while his mother, Edna (Allsop) Roberts, took care of the family and served as Richard's first tutor. In 1947, the Roberts family moved to Bath, where Roberts spent his formative years. At St. Stephen's junior school, Roberts encountered his first real mentor, the school's headmaster known only to the students as Mr. Broakes. Here he was exposed to a variety of mentally stimulating games, ranging from crossword to logical puzzles. "Most importantly, I learned that logic and mathematics are fun!" Roberts wrote in a brief autobiography for the Nobel Foundation.

At the City of Bath Boys School (now Beechen Cliff School), Roberts became enamored with the life and literature of detectives, as they represented the ultimate puzzle solvers. His young career path changed abruptly, however, when he

received a chemistry set from his parents. His supportive father had a large chemistry cabinet constructed and, with the aid of a local chemist who supplied the myriad chemicals he needed, Roberts soon discovered how to assemble fireworks and other concoctions not found in a beginner's chemistry manual. "Luckily I survived those years with no serious injuries or burns. I knew I had to be a chemist," he wrote in the Nobel Foundation autobiography.

At the age of seventeen, Roberts entered Sheffield University, where he concentrated in chemistry. His initial introduction to **biochemistry** was negative, he recalled in his autobiography: "I loathed it. The lectures merely required rote learning and the laboratory consisted of the most dull experiments imaginable." After graduating with honors in 1965, Roberts remained at Sheffield to study for his doctoral degree under David Ollis, his undergraduate professor of organic chemistry. But the direction of Roberts' scientific interests were profoundly altered after reading a book by John Kendrew on **crystallography** and molecular biology. Roberts became hooked on molecular biology and was later invited to conduct his postdoctoral work as part of a research team assembled by his colleague, Jack Strominger, a professor of biochemistry and molecular biology at Harvard University.

In 1969, Roberts left the English countryside and moved to Cambridge, Massachusetts, where he spent the next four years deciphering the sequence of nucleotides in a form of **ribonucleic acid** known as tRNA. Using a new method devised by English biochemist **Frederick Sanger** at Cambridge, he was able to sequence the **RNA** molecule, while teaching other scientists Sanger's technique. His creative work with tRNA led to the publication of two papers in *Nature* and an invitation by genetic pioneer and Nobel laureate, **James Watson**, to join his laboratory in Cold Spring Harbor, Long Island, New York.

In 1972, Roberts moved to Long Island to research ways to sequence DNA. American microbiologists **Daniel Nathans** and **Hamilton Smith** had shown that a restriction enzyme, Endonuclease R, could split DNA into specific segments. Roberts thought that such small segments could be used for DNA **sequencing** and began looking for other new **restriction enzymes** to expand the repertoire. (**Enzymes** are complex proteins that catalyze specific biochemical reactions.) He noted in his autobiography that his laboratory was responsible for discovering or characterizing three-quarters of the world's first restriction enzymes. In 1977, he developed a series of biological experiments to "map" the location of various genes in adenovirus and found that one end of a messenger ribonucleic acid (**mRNA**) did not react as expected. With the use of an **electron microscope**, Roberts and his colleagues observed that genes could be present in several, well-separated DNA segments. As he told the *New York Times,* "Everybody thought that genes were laid out in exactly the same way, and so it came as a tremendous surprise that they were different in higher organisms, such as humans."

In 1986, Roberts married his second wife, Jean. He is the father of four children, Alison, Andrew, Christopher and Amanda. He returned to Massachusetts in 1992 to join New England Biolabs, a small, private company in Beverly, Massachusetts, involved in making research reagents, particularly restriction enzymes. He serves as joint research director. In 1993, Roberts was awarded the Nobel Prize for his discovery of split genes. The Nobel Committee stated that, "The discovery of split genes has been of fundamental importance for today's basic research in biology, as well as for more medically oriented research concerning the development of **cancer** and other diseases."

See also DNA structure; Gene splicing; Sequencing

ROUS, PEYTON (1879-1970)

American physician

Francis Peyton Rous was a physician-scientist at the Rockefeller Institute for Medical Research (later the Rockefeller University) for over sixty years. In 1966, Rous won the Nobel Prize for his 1910 discovery that a virus can cause **cancer** tumors. His other contributions to scientific medicine include creating the first blood bank, determining

major Functions of the liver and gall bladder, and identifying factors that initiate and promote malignancy in normal cells.

Rous was born on October 5, 1879, in Baltimore, Maryland, to Charles Rous, a grain exporter, and Frances Wood, the daughter of a Texas judge. His father died when Rous was eleven, and his mother chose to stay in Baltimore to ensure that her three children would have the best possible education. His sisters were professionally successful, one a musician, the other a painter.

Rous, whose interest in natural science was apparent at an early age, wrote a "flower of the month" column for the *Baltimore Sun*. He pursued his biological interests at Johns Hopkins University, receiving a B.A. in 1900 and an M.D. in 1905. After a medical internship at Johns Hopkins, however, he decided (as recorded in *Les Prix Nobel en 1966*) that he was "unfit to be a real doctor" and chose instead to concentrate on research and the natural history of disease. This led to a full year of studying lymphocytes with Aldred Warthin at the University of Michigan and a summer in Germany learning morbid anatomy at a Dresden hospital.

After Rous returned to the United States, he developed pulmonary tuberculosis and spent a year recovering in an Adirondacks sanatorium. In 1909, Simon Flexner, director of the newly-founded Rockefeller Institute in New York City, asked Rous to take over cancer research in his laboratory. A few months later, a poultry breeder brought a Plymouth Rock chicken with a large breast **tumor** to the Institute and Rous, after conducting numerous experiments, determined that the tumor was a spindle-cell sarcoma. When Rous transferred a cell-free filtrate from the tumor into healthy chickens of the same flock, they developed identical tumors. Moreover, after injecting a filtrate from the new tumors into other chickens, a malignancy exactly like the original formed. Further studies revealed that this filterable agent was a virus, although Rous carefully avoided this word. Now called the Rous sarcoma virus RSV) and classed as an **RNA retrovirus**, it remains a prototype of animal tumor viruses and a favorite laboratory model for studying the role of genes in cancer.

Rous's discovery was received with considerable disbelief, both in the United States and in the rest of the world. His viral theory of cancer challenged all assumptions, going back to Hippocrates, that cancer was not infectious but rather a spontaneous, uncontrolled growth of cells and many scientists dismissed his finding as a disease peculiar to chickens. Discouraged by his failed attempts to cultivate viruses from mammal cancers, Rous abandoned work on the sarcoma in 1915. Nearly two decades passed before he returned to cancer research.

After the onset of World War I, Rous, J. R. Turner, and O. H. Robertson began a search for emergency blood transfusion fluids. Nothing could be found that worked without red blood corpuscles so they developed a citrate-sugar solution that preserved blood for weeks as well as a method to transfuse the suspended cells. Later, behind the front lines in Belgium and France, they created the world's first blood bank from donations by army personnel. This solution was used again in World War II, when half a million Rous-Turner blood units were shipped by air to London during the Blitz.

During the 1920s, Rous made several contributions to physiology. With P. D. McMaster, Rous demonstrated the concentrating activity of bile in the gall bladder, the acid-alkaline balance in living tissues, the increasing permeability along capillaries in muscle and skin, and the nature of gallstone formation. In conducting these studies, Rous devised culture techniques that have become standard for studying living tissues in the laboratory. He originated the method for growing viruses on chicken embryos, now used on a mass scale for producing viral vaccines, and found a way to isolate single cells from solid tissues by using the enzyme trypsin. Moreover, Rous developed an ingenious method for obtaining pure cultures of Kupffer cells by taking advantage of their phagocytic ability; he injected iron particles in animals and then used a magnet to separate these iron-laden liver cells from suspensions.

In 1933, a Rockefeller colleague's report stimulated Rous to renew his work on cancer. Richard Shope discovered a **virus** that caused warts on the skin of wild rabbits. Within a year, Rous established that this papilloma had characteristics of a true tumor. His work on mammalian cancer kept his viral theory of cancer alive. However, another twenty years passed before scientists identified viruses that cause human cancers and learned that viruses act by invading genes of normal cells. These findings finally advanced Rous's 1910 discovery to a dominant place in cancer research.

Meanwhile, Rous and his colleagues spent three decades studying the Shope papilloma in an effort to understand the role of viruses in causing cancer in mammals. Careful observations, over long periods of time, of the changing shapes, colors, and sizes of cells revealed that normal cells become malignant in progressive steps. **Cell** changes in tumors were observed as always evolving in a single direction toward malignancy.

The researchers demonstrated how viruses collaborate with carcinogens such as tar, radiation, or chemicals to elicit and enhance tumors. In a report co-authored by W. F. Friedewald, Rous proposed a two-stage mechanism of carcinogenesis. He further explained that a virus can be induced by carcinogens or it can hasten the growth and transform benign tumors into cancerous ones. For tumors having no apparent trace of virus, Rous cautiously postulated that these spontaneous growths might contain a virus that persists in a masked or latent state, causing no harm until its cellular environment is disturbed.

Rous eventually ceased his research on this project due to the technical complexities involved with pursuing the interaction of viral and environmental factors. He then analyzed different types of cells and their nature in an attempt to understand why tumors go from bad to worse.

Rous maintained a rigorous workday schedule at Rockefeller. His meticulous editing and writing, both scientific and literary, took place during several hours of solitude at the beginning and end of each day. At midday, he spent two intense hours discussing science with colleagues in the Institute's dining room. Rous then returned to work in his laboratory on experiments that often lasted into the early evening.

Rous was appointed a full member of the Rockefeller Institute in 1920 and member emeritus in 1945. Though officially retired, he remained active at his lab bench until the age of ninety, adding sixty papers to the nearly three hundred he published. He was elected to the National Academy of Sciences in 1927, the American Philosophical Society in 1939, and the Royal Society in 1940. In addition to the 1966 Nobel Prize for Medicine, Rous received many honorary degrees and awards for his work in viral oncology, including the 1956 Kovalenko Medal of the National Academy of Sciences, the 1958 Lasker Award of the American Public Health Association, and the 1966 National Medal of Science.

As editor of the *Journal of Experimental Medicine,* a periodical renowned for its precise language and scientific excellence, Rous dominated the recording of forty-eight years of American medical research. He died of abdominal cancer on February 16, 1970, in New York City, just six weeks after he retired as editor.

See also Aging and life expectancy; Chromosomal mutations and abnormalities; Gene mutation; Viral genetics

ROUX, WILHELM (1850-1924)

German embryologist

Wilhelm Roux, the father of experimental **embryology**, was born in Jena, Germany. His teachers were some of the finest scientists of the time, namely Ernst Haeckel at Jana, Rudolf Virchow at Berlin, and Freidrich von Recklinghausen (1833–1910) at Strasbourg. After receiving his university education, Roux taught at the University of Breslau 1879 until 1889, he became a professor at Innsbruck, Austria in 1889, and a professor and director of the anatomical institute at Halle, a position he held from 1889 until he retired in 1921.

Roux was a tireless worker who studied embryos seeking a causal connection between function and form. Rather than simply looking at embryos as his predecessors did, he often manipulated embryos to seek understanding of developmental patterns. Ultimately, he sought to describe development in physical and chemical terms. He focused on frog eggs and embryos and investigated many phenomena. An example: he studied the orientation of the plane of the first cleavage, i.e., he sought to ascertain where the cleavage furrow appears as the **zygote** begins its first **cell division**. Roux noted that when a sperm enters an egg, it leaves behind a track known as the copulation path. Roux noted that the track was often the site of the first cleavage furrow. This appears to be true much of the time, but there are exceptions. Roux was the first to describe the gray crescent in the eggs of *Rana*. This area of intermediate pigmentation (i.e., not as darkly pigmented as the animal hemisphere nor as light as the vegetal hemisphere, hence "gray") occurs after **fertilization** and is opposite the site of sperm entry. The gray crescent is the site of blastopore formation in gastrulation. Prior to fertilization, it may be said that the frog egg is radially symmetric; after fertilization and especially obvious with gray crescent formation, the egg is bilaterally symmetrical. The establishment of a left and right side of an embryo is crucial for future development and Roux demonstrated his penetrating insight by seeking understanding of this change in an egg.

It was, however, the half embryos for which Roux is best remembered. Roux had marked eggs by pricking with a microneedle prior to his most famous experiment. Superficial egg pricking permits a tiny exovate (a small bleb of cytoplasm at the site of the prick) to form and this was a mode of marking a **cell** and following it in its development. In his most famous experiment, Roux plunged a hot needle into one of the blastomeres (a blastomere is an early embryonic cell) of a zygote at the two cell stage. The hot needle killed the operated cell. The question to be resolved was what developmental pattern would the remaining living cell follow. Roux reported that a half embryo was formed. He reasoned that genetic determinants were parcelled out with cell division and each of the two cells had received only the determinants to form a half embryo. Roux had studied the precision of nuclear division in **mitosis**. He reasoned that this precise mitotic mechanism functioned to divide the **nucleus** into qualitatively unequal halves. Mitosis thus segregated genetic determinants to daughter cells which were then unequal. His half embryo experiment supported this notion. Roux was incorrect. Hans Driesch separated sea urchin eggs at early cleavage stages. He got two whole but smaller larvae developing from each blastomere. Roux had not separated his frog blastomeres. It was later believed that the living frog blastomere developed only a half embryo because of the inhibiting effect of the dead blastomere next to it. Driesch attempted to separate frog eggs at the two cell stage but was unable to do so. However, several workers were successful in isolating frog blastomeres with results similar to those of Driesch with the sea urchin—these results are also obtained with mouse embryos at the two cell stage.

While Roux's most famous experiment led to an incorrect conclusion, it is nevertheless extraordinarily notable. Earlier embryologists looked but did not experiment. A notable example of this was William Harvey who described the development of the chick embryo in detail. However, Roux went to the embryo and sought answers to developmental questions by experimental means. He began the modern era of analytical experimental embryology. Roux is remembered also because he founded the first scientific journal devoted to experimental embryology, *Archiv füur Entwicklungsmechanik*, which is published at this time as *Development, Genes and Evolution*.

See also Embryology

ROWLEY, JANET DAVISON (1925-)
American cytogeneticist

Janet D. Rowley's research on **chromosome** abnormalities in a type of leukemia have introduced new diagnostic tools for oncologists—doctors specializing in cancer—and have also opened new avenues of inquiry into possible **gene** therapies for **cancer**. A specialist in cytogenetics (the investigation of the role of cells in **evolution** and **heredity**), Rowley has helped to pinpoint cancer gene locations and correlate them to chromosome aberrations.

Janet Rowley was born in New York City, the daughter of Hurford Henry and Ethel Mary (Ballantyne) Davison. She attended the University of Chicago where she earned a B.S. degree in 1946 and her M.D. in 1948. During the latter year she married Donald A. Rowley. The couple eventually had four sons. Rowley's professional career took her from a research assistant job at the University of Chicago in 1949 and 1950 to an internship at Marine Hospital in Chicago, a residency at Cook County Hospital, and a clinical instructor position in neurology at the University of Illinois Medical School before returning to the University of Chicago Medical School in 1962. From 1962 to 1969, she was a research associate both in the department of medicine and at Argonne Cancer Research Hospital. Then from 1969 to 1977, she became an associate professor, and in 1977, was made a full professor at the medical school and at Franklin McLean Memorial Research Institute. In 1984, Rowley became the Blum-Riese Distinguished Service professor in the department of medicine and in the department of molecular **genetics** and **cell** biology at Franklin McLean Memorial Research Institute.

During her work at the University of Chicago, Rowley has committed her research to understanding the cytogenetic causes of cancer. She developed the use of quinacrine and Giemsa staining to identify chromosomes in cloned cells. Once the chromosomes were easily identifiable, she could then study abnormalities that occur in some chromosomes in certain cancers. With the discovery of oncogenes, or cancer-inducing genes, Rowley focused on chromosome rearrangements that occur in a form of blood cancer known as chronic myeloid leukemia (CML). Studying the so-called Philadelphia chromosome, Rowley was able in 1972 to show a consistent chromosome **translocation** or shifting of genetic material in CML cells. This was the first recurring translocation to be discovered in any **species**. Since that time, more than seventy such translocations have been detected in human malignant cells. In general, Rowley's research indicates that both translocations and **deletions** of genetic material occur in malignancy, and that cancer is caused by a complex series of events within a single cell, making some genes overactive (**tumor** producing) and eliminating other genes that would normally suppress growth. Any cell, according to Rowley's research, is therefore potentially cancerous. What is needed to activate malignant growth is this complex series of events, including translocation and deletion.

Co-founder and co-editor of the journal *Genes, Chromosomes and Cancer,* Rowley has received numerous honors and awards, including the Esther Langer Award in 1983; the Kuwait Cancer Prize in 1984; the A. Cressy Morrison Award from the New York Academy of Sciences in 1985; the Judd Memorial Award from the Sloan-Kettering Cancer Center, the Charles S. Mott Prize from the General Motors Research Foundation, and the G. H. A. Clowes Memorial Award from the American Association for Cancer Research, all in 1989; and the Robert de Villiers award from the Leukemia Society of America in 1993.

See also Cancer genetics; Cell proliferation; Oncogenetic research

S

SAME-SENSE MUTATION

A same-sense **mutation** is form of **silent mutation** where a change in the base sequence of a **gene** does not result in a change in the **amino acid** sequence of the protein for which the gene carries the genetic instructions. Because the genetic code contains multiple codes for the same instructions, the genetic code is considered a **degenerate code**. Because the genetic code is degenerate, a same-sense mutation produces **codons** that carry the same instructions for the synthesis of proteins as do the normal, non-mutated, codons.

Changes in nucleic acid (**DNA** or **RNA**) **sequencing** are natural and those that occur during **replication** are usually repaired by specialized repair mechanisms. When these mechanisms fail to correct a change (revert the sequence of **bases** back to there normal state), a mutation results. As a result of a same-sense mutation, however, there is no change in the insertion of the proper amino acid into the chain of amino acids bonded together to form a protein. Because there is no change in the amino acid sequence, there is no observable change in the structure and function of the protein.

A same-sense mutation occurs when, for example, there is a substitution of adenine for cytosine at final position in a codon determining triplet sequence of a gene. Such a **point mutation** results in a cytosine-cytosine-adenine (CCA) sequence instead of a cytosine-cytosine-cytosine (CCC) sequence. Because both CCA and CCA carry the instruction to insert the amino acid proline during **protein synthesis**, the protein translated from the codons produced by the mutation is unaltered.

Using another sequence as an example, at the molecular level, if the base sequence of the coding stand of DNA is guanine-adenine-thymine (GAT), because of restricted base paring (e.g., A-T and C-G) the anticoding DNA strand used as a **transcription** template contains a cytosine-thymine-adenine (CTA) sequence. Because uracil (U) replaces thymine in **mRNA**, the DNA coding strand sequence (the strand that has the same sequence as does the mRNA transcript) results in a mRNA codon sequence of GAU. This codon sequence is ultimately interpreted to insert aspartic acid into the translated protein's sequence of amino acids. A same-sense mutation results from a substitution of cytosine (C) for thymine (T) in the coding strand because the mutation results in a DNA coding strand sequence of GAC that corresponds to a CTG anticoding sequence that is ultimately transcribed into a GAC sequence mRNA codon. Both the normal GAU and the mutant produced GAC codons carry the instructions to insert the amino acid aspartic acid into the lengthening amino acid chain during the **translation** process.

Same-sense mutations are also called synonymous base pair changes and are a form of silent mutation (mutations that do not result in an identifiable change in visible expression of the genes). Although same-sense mutations reflect molecular level changes in a gene or **chromosome**, the effects of same-sense mutations are not reflected in the translated protein. Because such silent mutations result in no observable phenotypic changes, they would not be expected to provide a basis for the mechanisms of **natural selection** to operate. Although there are no easily identifiable affects on elements of **fitness** (e.g., protein function, developmental **viability**, fertility, etc.), there are some models of **selection** based upon an organism's preferred use of selected codons (codon usage bias) that may allow selection to act based on differing rates of translation among codons. Although evidence for such codon selection bias has been found in studies of **bacteria** and insects, as of 2001, there is no evidence that such a mechanism operates in humans.

See also Base pairing (bp); Deletions; DNA repair; Frame shifts

SANGER, FREDERICK (1918-)
English biochemist

Frederick Sanger's important work in **biochemistry** has been recognized by two Nobel Prizes for chemistry. In 1958, Sanger

Frederick Sanger.

received the award for determining the arrangement of the amino acids that make up insulin, becoming the first person to identify a protein molecule. In 1980, Sanger shared the award with two other scientists, and was cited for his work in determining the sequences of nucleic acids in **deoxyribonucleic acid (DNA)** molecules. This research has had important implications for genetic research, and taken in conjunction with Sanger's earlier work on the structure of insulin, represent considerable contributions to combating a number of diseases.

Frederick Sanger was born in Rendcombe, Gloucestershire, England, on August 3, 1918. His father, also named Frederick, was a medical doctor, and his mother, Cicely Crewsdon Sanger, was the daughter of a prosperous cotton manufacturer. Young Frederick attended the Bryanston School in Blandford, Dorset, from 1932 to 1936 and was then accepted at St. John's College, Cambridge. By his own admission, Sanger was not a particularly apt student. Later in life Sanger wrote in *Annual Review of Biochemistry* that "I was not academically brilliant. I never won scholarships and would probably not have been able to attend Cambridge University if my parents had not been fairly rich."

Upon arriving at Cambridge and laying out his schedule of courses, Sanger found that he needed one more half—

course in science. In looking through the choices available, Sanger came across a subject of which he had never heard— biochemistry—but that sounded appealing to him. "The idea that biology could be explained in terms of chemistry," he later wrote in *Annual Review of Biochemistry,* "seemed an exciting one." Sanger followed the introductory course with an advanced one and eventually earned a first–class degree in the subject.

Sanger rapidly discovered his strengths and weaknesses in science. Although he was not particularly interested in or skilled at theoretical analysis, Sanger was a superb experimentalist. After receiving his bachelor's degree from St. John's in 1939, Sanger decided to continue his work in biochemistry. Though World War II had just begun, Sanger avoided service in the English army because his strong Quaker pacifist beliefs qualified him as a conscientious objector. Instead, Sanger began looking for a biochemistry laboratory where he could serve as an apprentice and begin work on his Ph.D. The first position he found was in the laboratory of protein specialist, N. W. Pirie. Pirie assigned Sanger a project involving the extraction of edible protein from grass. That project did not last long as Pirie left Cambridge, and Sanger was reassigned to Albert Neuberger. Neuberger changed Sanger's assignment to the study of lysine, an **amino acid**. By 1943, Sanger had completed his research and was awarded his Ph.D. for his study on the metabolism of lysine.

After receiving his degree, Sanger decided to stay on at Cambridge, where he was offered an opportunity to work in the laboratory of A. C. Chibnall, the new Professor of Biochemistry. Chibnall's special field of interest was the analysis of amino acids in protein, a subject in which Sanger also became involved. The structure of proteins had been a topic of considerable dispute among chemists for many years. On the one hand, some chemists were convinced that proteins consisted of some complex, amorphous material that could never be determined chemically. Conversely, other chemists believed that, while protein molecules might be complex, they did have a structure that could eventually be unraveled and understood.

Probably the most influential theory of protein structure at the time of Sanger's research was that of the German chemist Emil Fischer. In 1902, Fischer had suggested that proteins consist of long chains of amino acids, joined to each other head to tail. Since it was known that each amino acid has two reactive groups, an amino group and a carboxyl group, it made sense that amino acids might join to each other in a continuous chain. The task facing researchers like Sanger was to first determine what amino acids were present in any particular protein, and then to learn in what sequence those amino acids were arranged. The first of these steps was fairly simple and straight-forward, achievable by conventional chemical means. The second was not.

The protein on which Sanger did his research was insulin. The reason for this choice was that insulin—used in the treatment of diabetes—was one of the most readily available of all proteins, and one that could be obtained in very high purity. Sanger's choice of insulin for study was a fortuitous one. As proteins go, insulin has a relatively simple structure. Had he, by chance, started with a more complex protein, his research would almost certainly have stretched far beyond the ten years it required.

In 1945, Sanger made an important technological break-through that made possible his later **sequencing** work on amino acids. He discovered that the compound dinitrophenol (DNP) will bond tightly to one end of an amino acid and that this bond is stronger than the one formed by two amino acids bonding with one another. This fact made it possible for Sanger to use DNP to take apart the insulin molecule one amino acid at a time. Each amino acid could then be identified by the newly discovered process of paper chromatography. This was a slow process, requiring Sanger to examine the stains left by the amino acids after they were strained through paper filters, but the technique resulted in the eventual identi-fication of all amino acid groups in the insulin molecule.

Sanger's next objective was to determine the sequence of the amino acids present in insulin, but this work was made more difficult by the fact that the insulin molecule actually consists of two separate chains of amino acids joined to each other at two points by sulfur-sulfur bonds. In addition, a third sulfur-sulfur bond occurs within the shorter of the two strands. Despite these difficulties, Sanger, in 1955, announced the results of his work: he had determined the total structure of insulin molecule, the first protein to be analyzed in this way. Sanger's work in this area was considered important because it involved proteins, "the most important substances in the human body," as Sanger described them in a *New York Times* report on his work. Proteins are integral elements in both the viruses and toxins that cause diseases and in the antibodies that prevent them. Sanger's research, in laying the ground-work for future work on proteins, greatly increased scientists' ability to combat diseases. For his important work on proteins, Sanger was awarded the Nobel Prize in chemistry in 1958.

In 1962, Sanger joined the newly established Medical Research Council (MRC) Laboratory of Molecular Biology at Cambridge, a center for research that included such scientists as Max Perutz, **Francis Crick**, and **Sydney Brenner**. This move marked an important turning point in Sanger's career. The presence of his new colleagues—and Crick, in particular—sparked Sanger's interest in the subject of nucleic acids. Prior to joining the MRC lab, Sanger had had little interest in this subject, but he now became convinced of their importance. His work soon concentrated on the ways in which his protein-sequencing experiences might be used to determine the sequencing of nucleic acids.

The latter task was to be far more difficult than the for-mer, however. While proteins may consist of as few as 50 amino acids, nucleic acids contain hundreds or thousands of basic units, called nucleotides. The first successful sequencing of a nucleic acid, a **transfer RNA** molecule known as alanine, was announced by **Robert William Holley** in 1965. Sanger had followed Holley's work and decided to try a somewhat differ-ent approach. In his method, Sanger broke apart a nucleic acid molecule in smaller parts, sequenced each part, and then deter-mined the way in which the parts were attached to each other.

In 1967, Sanger and his colleagues reported on the structure of an **RNA** molecule known as 5S using this technique.

When Sanger went on to the even more challenging structures of DNA molecules, he invented yet another new sequencing technique. In this method, a single-stranded DNA molecule is allowed to replicate itself but stopped at various stages of **replication**. Depending on the chemical used to stop replication, the researcher can then determine the **nucleotide** present at the end of the molecule. Repeated applications of this process allowed Sanger to reconstruct the sequence of nucleotides present in a DNA molecule.

Successful application of the technique made it possible for Sanger and his colleagues to report on a 12 nucleotide sequence of DNA from **bacteriophage** λ in 1968. Ten years later, a similar approach was used to sequence a 5,386 nucleotide sequence of another form of bacteriophage. In recognition of his sequencing work on nucleic acids, Sanger was awarded his second Nobel Prize in chemistry in 1980, shares of which also went to **Walter Gilbert** and **Paul Berg**. Their work has been lauded for its application to the research of congenital defects and **hereditary diseases** and has proved vitally important in producing the artificial genes that go into the manufacture of insulin and interferon, two substances used to treat diseases.

In 1983, at the age of 65, Sanger retired from research. He began to be concerned, he said in the *Annual Review of Biochemistry,* about "occupying space that could have been available to a younger person." He soon found that he very much enjoyed retirement, which allowed him to do many things for which he had never had time before. Among these were gardening and sailing. He also had more time to spend with his wife, the former Margaret Joan Howe, whom he had married in 1940, and his three children, Robin, Peter Frederick, and Sally Joan.

During his career, Sanger received many honors in addi-tion to his two Nobel Prizes. In 1954, he was elected to the Royal Society and in 1963, he was made a Commander of the Order of the British Empire, among other honors.

SANGER, MARGARET LOUISA HIGGINS (1879-1966)
American nurse and social reformer

Margaret Sanger opened the first birth control clinic in America, spearheaded the birth control movement, and founded both organizations that later merged to form the Planned Parenthood Federation of America.

"Maggie" Higgins was born in Corning, New York, the sixth of eleven children of a freethinking Irish immigrant stonecutter, Michael Hennessy Higgins, and his wife, Anne Purcell Higgins, a sickly, passive, woman. Margaret never accepted the patriarchal family structure, and mourned her mother's early death, hastened by repeated pregnancies. Although she considered men in the style of her father to be sexual tyrants, Margaret admired her father's his leftist icono-

Margaret Sanger. *Planned Parenthood Federation of America.*
Reproduced by permission.

clastic ideas, and shared many political and ideological heroes with him.

The family fortunes declined after 1894 when Michael Hennessy alienated the local Roman Catholic constituency by engaging the atheist socialist Robert Green Ingersoll (1833–1899) to speak at a public meeting in Corning. Margaret always longed to escape from what she considered Corning's provincialism. She endured the taunts of her teachers and classmates at the parish school of St. Mary's Church until 1896, when, with financial help from her older sisters, she was able transfer to Claverack College and Hudson River Institute, a Protestant boarding school far from her detested Corning. For the first time in her life, she had regular access to secular books.

After graduating in 1900, she enrolled in the nursing program at the White Plains, New York, Hospital, unable to afford to study medicine leading to an M.D. degree. In 1902, she received her nursing credentials and married Jewish architect William Sanger (1874–1961). The Sangers were already involved with both the Socialist Party and the International Workers of the World when they moved to Manhattan in 1910, and became involved in a prominent socialist circle including Emma Goldman (1869–1940), John Reed (1887–1920), and Upton Sinclair (1878–1968).

Margaret worked on the Lower East Side as a visiting nurse midwife for women suffering from too many children, inadequate reproductive health care, frequent miscarriages,

sexually transmitted diseases, and abortion. She was so affected by the misery of these mothers that she quit nursing in 1912 to devote full time to the cause of freeing women from the medical and economic afflictions of unwanted pregnancy. She became a prolific propagandist and publicist. In 1912, Sanger began writing a sex education column for *The New York Call*. Her first monthly issue of *The Woman Rebel*, a journal subtitled "No Gods, No Masters," appeared in March 1914, but it was soon suppressed. In 1917 Sanger founded *Birth Control Review*.

Sanger opened her clinic in Brooklyn, New York, on October 16, 1916, to distribute family planning literature and contraceptive devices. Because sex education was considered legally obscene, the police almost immediately raided and closed the clinic. She was arrested and convicted.

She divorced Sanger in 1920, and two years later married a self-made millionaire, J. Noah H. Slee (1860–1943). He liberally funded the birth control movement. In 1922, Sanger founded the American Birth Control League and in 1923, the Birth Control Clinical Research Bureau, which in 1939 merged into the Birth Control Federation of America. She retired in 1942 when, against her wishes, it changed its name to the Planned Parenthood Federation of America. Some historians suggest Sanger was forced out of the movement by moderates who believed that she offended mainstream citizens.

Sanger died of atherosclerosis at home in Tucson, Arizona.

See also Eugenics; Family; Genetic implications of mating and marriage customs; Sociobiology

SARGENT, THOMAS DEAN (1953-   )
American geneticist

Thomas Dean Sargent made his first major contribution to molecular **genetics** while he was a postdoctoral fellow in Igor David's laboratory at the National Institute of Health (NIH). He was involved in the production of the complex complementary **DNA** (**cDNA**) libraries and later helped to develop the first subtracted cDNA libraries. This is a powerful methodology that allows the selective detection and characterization of differentially expressed **RNA** molecules.

Using subtracted cDNA libraries in combination with other **cell** biological techniques, Sargent and colleagues isolated classes of RNAs that are expressed at different stages of the embryonic development of the frog *Xenopus laevis*. They identified and characterized several genes involved in cell adhesion and development, and established the requirement for cell-cell interaction and communication in the induction of mesoderm formation in frog embryos. In addition, they also discovered the role of the **transcription** factor AP-2 in the regulation of epidermal **gene expression**. Subtracted cDNA libraries has been used by other groups to discover differentially expressed genes. Sargent's laboratory uses the mouse model to study the **differentiation** of the epidermis using transgenic and **gene** knockout approaches. The focus is especially on the Distal-less class of the **homeobox** genes, which seem to be involved in skin development. The *Xenopus* embryo, which was the first animal model used, continues to be investigated with a special interest in the role the Eph-class of receptor tyrosine kinases play in adhesion and cell-cell communication. The laboratory is also involved in the investigation of embryonic neuronal development using the zebrafish as a model organism.

Sargent is now the head of the Section on Vertebrate Development, Laboratory of Molecular Genetics, the National Institute of Child Health and Human Development (NICHD), NIH. Apart from his position at the NIH, Dr. Sargent holds the position of adjunct Professor of Genetics at the Graduate Genetics program at George Washington University in Washington D.C. He received his Bachelor of Arts degree in 1975 from Indiana University and his Ph.D. in Biochemistry from Caltech in 1981. He was a postdoctoral fellow at the NIH and has been a tenured scientist in the (NICHD) since 1989.

See also Bases; Transcription; Translation

SCHLEIDEN, MATTHIAS JACOB (1804-1881)

German botanist

Matthias Schleiden is credited, along with Theodor Schwann, with articulating the **cell** theory. Born in Hamburg, he began his career as a lawyer. He met with no great success in law, and, becoming increasingly depressed, attempted suicide. After recovering from the failed attempt, he returned to school to study medicine, specializing in botany.

Schleiden served as a professor first at Jena and later at Dorpat, then resigned and moved frequently from town to town until he died in 1881. Possibly as a result of his previous career, Schleiden was impulsive, sharp and scornful of his opposition. He rejected the botanist as a glorified scientific librarian, opting for a focus on the anatomy and physiology of plants. "Most people of the world, even the most enlightened," he said, "are still in the habit of regarding the botanist as a dealer in barbarous Latin names, as a man who gathers flowers, names them, dries them, and wraps them in paper, and all of whose wisdom consists in determining and classifying this hay which he has collected with such great pains."

Schleiden's chief contribution to the cell theory was elaborated in an 1838 essay on the origins of the cell. First, he concluded that plants structure was based on cells and that these cells were created in a common fashion. Schleiden argued that the cell developed from the growth of the **nucleus**, which he called the "cytoblast". He stated—and Schwann accepted his position—that the nucleus was spontaneously generated out of the cytoplasm or other unformed organic substances. Once the cell was fully formed, Schleiden attested, the nucleus dissolved.

That theory of cell formation was refuted by Robert Remak in 1852, who insisted—as now understand—that cells are created by the division of other cells.

Despite its flaws, Schleiden's paper was extremely important to the world of biology. His conclusion that plants consist entirely of cells or cell products focused attention on the cell as the basic unit of living organisms. Also, what Schleiden lacked in rigorous scientific foundation he made up for in forceful argument and ardent conclusions which, while many were later found to be wrong, laid the foundation for Schwann's broader, more comprehensive work on the cell theory. Together their work produced one of the most critical biological developments of their time. Schleiden and Schwann wedded empirical microscopic observations with the more speculative conclusions of natural philosophers to create a unifying theory on the structural similarity of plants and animals.

Matthias Schleiden also published a textbook on botany in 1842, outlining some of his own theories on natural science and criticizing other botanists of the age. Much of the book repeated general theories of the time, including his own work on cells, but it did attempt in its methodology to initiate comparitive investigations into plant **evolution**.

After his initial foray into cell theory, however, Schleiden chose not pursue the subject to any greater degree.

See also Cell proliferation; Embryology, the history of developmental and generational theory; History of genetics: ancient and classical views of heredity

SEEDS AND SEED SAVING

Seed saving, the storage of seeds for future use in agriculture or research, has been practiced since the dawn of human horticulture. Each year's crop was planted with seeds saved from the previous year, or found in the local environment. Eventually seeds crossed regions, by nature or through human migration and trade, thus introducing new **species** into various local environments. such dispersal of plant genetic materials resulted in origin species producing numerous different varieties as successive generations were hybridized with local flora. With the development of the field of **genetics** in the late 1800s, scientists began in earnest research seed saving, seed sharing, and plant hybridization. They amassed large specimen collections of plant genetic material in the form of seeds. Seeds were not only the most convenient means by which to store various species, but they permitted scientists greater control over specimens and research variables, and allowed for the observation of plants in various developmental stages.

Most of the early research in **plant genetics** involved the production of sustainable (e.g. able to reproduce) hybrids. Large seed collections were necessary reservoirs of genetic material for hybridization and modification of species. In turn, hybrids that were able to reproduce added their viable seed to collection. In the early twentieth century, scientists were eagerly sought practical applications for plant genetics, seed saving, and seed sharing. Agriculture was in the midst of a revolution of mechanization, but new farming methods soon demonstrated the need for better adapted crops. Individual farmers had long been controlling the genetic character of their crops by selecting strong, healthy plants that exhibit characteristics they desired from which to harvest seeds. Over time, however, this practice potentially limited the number of varieties of a given crop in certain areas. While these crops may have been well adapted to their growing environment, limited variation left whole harvests vulnerable to disease. For instance, a lack of species variation may have contributed to the massive crop failures during the Irish Potato Famine in 1845–1847. Some varieties of potato may have been less susceptible to the suspected guilty pathogen, *Phytophthora*. However, much of the original tuber and seed stock imported from Ireland was from a relatively small region in South America.

Seed saving is also an important part of efforts to understand plant **evolution** and to preserve rare and endangered plants whose natural habitats are threatened. As habitat destruction increases globally, conservation—geneticists rely on genebanks, facilities that store seeds and other genetic materials, for preservation and research specimen. **Gene** banks are also a valuable when conservation—geneticists wish to introduce certain plants to a location in order to better balance a damaged ecosystem. Bio-conservationists are currently developing new ways to treat and store seeds in order keep them viable them for longer periods of time. Cold, low humidity storage can preserve viable seeds for a century; chemically freezing specimen may permit even longer storage. Long-term preservation methods allow time for ecosystems to fully recover from disasters like fire or flood, before endangered plants, which may not reappear on their own, are reintroduced back into their habitat.

Agricultural restoration after a disaster is also possible. A farmer in a drought-affected region could utilize a gene bank to plant a wheat crop more suited to arid conditions than his usual variety. Since seeds allotted for agricultural use are usually only stored for short periods of time, many agricultural seed banks are run relatively locally, such as in farm co-operatives. However, disaster relief and agricultural reform are also implemented internationally. hough the origin points (the places where various types of plants exhibit their greatest genetic diversity) of many plants are in developing countries, industrial nations hold most of the world's seeds and plant genetic materials. Thus, agricultural development projects, disaster relief, and even research in plant genetics requires not only co-operation in the scientific community, but also a system of rules governing the purchase and responsible usage of seeds. Like any other natural resource, seed are regulated by local, national, and even international laws.

In some cases, modern genetics and traditional agricultural practices regarding seed saving clash because of legal patent rights regarding genetically modified crops. Courts around the world are hearing increasing numbers of cases regarding the rights and responsibilities of farmers who use genetically modified crops.

See also Agricultural genetics; Biotechnology; Shotgun method; Plant genetics

SELECTION

Selection refers to an evolutionary pressure that is the result of a combination of environmental and genetic pressures that affect the ability of an organism to live and, equally importantly, to raise their own reproductively successful offspring.

As implied, **natural selection** involves the natural (but often complex) pressures present in an organism's environment. Artificial selection is the conscious manipulation of mating, manipulation, and fusion of genetic material to produce a desired result.

Evolution requires **genetic variation**, and these variations or changes (mutations) are usually deleterious because environmental factors already support the extent genetic distribution within a population.

Natural selection is based upon expressed differences in the ability of organisms to thrive and produce biologically successful offspring. Importantly, selection can only act to exert influence (drive) on those differences in genotype that appear as phenotypic differences. In a very real sense, evolutionary pressures act blindly.

There are three basic types of natural selection: **directional selection** favoring an extreme phenotype; **stabilizing selection** favoring a phenotype with characteristics intermediate to an extreme phenotype (i.e., normalizing selection); and **disruptive selection** that favors extreme phenotypes over intermediate genotypes.

The evolution of pesticide resistance provides a vivid example of directional selection, wherein the selective agent (in this case DDT) creates an apparent force in one direction, producing a corresponding change (improved resistance) in the affected organisms. Directional selection is also evident in the efforts of human beings to produce desired traits in many kinds of domestic animals and plants. The many breeds of dogs, from dachshunds to shepherds, are all descendants of a single, wolf-like ancestor, and are the products of careful selection and breeding for the unique characteristics favored by human breeders.

Not all selective effects are directional, however. Selection can also produce results that are stabilizing or disruptive. Stabilizing selection occurs when significant changes in the traits of organisms are selected against. An example of this is birth weight in humans. Babies that are much heavier or lighter than average do not survive as well as those that are nearer the mean (average) weight.

On the other hand, selection is said to be disruptive if the extremes of some trait become favored over the intermediate values. Perhaps one of the more obvious examples of disruptive selection is sexual dimorphism, wherein males and females of the same **species** look noticeably different from each other. One sex may be larger, have bright, showy plumage, bear horns, or display some kind of ornament that the other lacks. The male peacock, for instance, has deep green and sapphire blue plumage and an enormous, fanning tail, while the female is a drab brown, with no elaborate tail.

Sexual dimorphism is considered the result of sexual selection, the process in which members of a species compete for access to mates. Sexual selection and natural selection may often operate in opposing directions, producing the two distinct sex phenotypes. Males, who are typically the primary contestants in the **competition** for mating partners, usually bear the ornaments such as showy plumage in spite of the potential costs of these ornaments, such as increased visibility to predators, and attacks from rival males. Females are less often involved in direct competition for mates, and they are not generally subject to the forces of sexual selection (although there are role reversals in a few species). Females are believed to play a critical role in the evolution of many elaborate male traits, however, because if the female preference has a genetic basis, female choice of particular males as mating partners will cause those male traits to spread in subsequent generations.

Sometimes the **fitness** of a phenotype in some environment depends on how common (or rare) it is; this is known as frequency-dependent selection. Perhaps an animal enjoys an increased advantage if it conforms to the majority phenotype in the population; this occurs when, for example, predators learn to avoid distasteful butterfly prey, because the butterflies have evolved to advertise their noxious taste by conforming to a particular wing color and pattern. Butterflies that deviate too much from the "warning" pattern are not as easily recognized by their predators, and are eaten in greater numbers. Interestingly, frequency-dependent selection has enabled butterflies who are not distasteful to mimic the appearance of their noxious brethren and thus avoid the same predators. Conversely, a phenotype could be favored if it is rare, and its alternatives are in the majority. Many predators tend to form a "search image" of their prey, favoring the most common phenotypes, and ignoring the rarer phenotypes. Frequency-dependent selection provides an interesting case in which the **gene frequency** itself alters the selective environment in which the genotype exists.

Many people attribute the phrase "survival of the fittest" to Darwin, but in fact, it originated from another naturalist/philosopher, Herbert Spencer (1820–1903). Recently, many recent evolutionary biologists have asked: **Survival of the fittest** what? At what organismal level is selection most powerful? What is the biological unit of natural selection-the species, the individual, or even the **gene**?

Although it seems rational that organisms might exhibit parental behavior or other traits "for the good of the species". In his 1962 book *Animal Dispersion in Relation to Social Behaviour*, behavioral biologist V. C. Wynne-Edwards proposed that animals would restrain their reproduction in times of resource shortages, so as to avoid extinguishing the local supply, and thus maintain the "balance of nature." However, Wynne-Edwards was criticized because all such instances of apparent group-level selection can be explained by selection acting at the level of individual organisms. A mother cat who suckles her kittens is not doing so for the benefit of the species; her behavior has evolved because it enhances her kittens' fitness, and ultimately her own as well, since they carry her genes.

Under most conditions, group selection will not be very powerful, because the rate of change in gene frequencies when one individual replaces another in the population is greater than that occurring when one group replaces another group. The number of individuals present is generally greater than the number of groups present in the environment, and individual turnover is greater. In addition, it is difficult to imagine that individuals could evolve to sacrifice their reproduction for the good of the group; a more selfish alternative could easily invade and spread in such a group.

However, there are some possible exceptions; one of these is reduced virulence in parasites, who depend on the survival of their hosts for their own survival. The *myxoma* **virus**, introduced in Australia to control imported European rabbits (*Oryctolagus cuniculus*), at first caused the deaths of many individuals. However, within a few years, the mortality rate was much lower, partly because the rabbits became resistant to the pathogen, but also partly because the virus had evolved a lower virulence. The reduction in the virulence is thought to have been aided because the virus is transmitted by a mosquito, from one living rabbit to another. The less deadly viral strain is maintained in the rabbit host population because rabbits afflicted with the more virulent strain would die before passing on the virus. Thus, the viral genes for reduced virulence could spread by group selection. Of course, reduced virulence is also in the interest of every individual virus, if it is to persist in its host. Scientists argue that one would not expect to observe evolution by group selection when individual selection is acting strongly in an opposing direction.

Some biologists, most notably Richard Dawkins (1941–), have argued that the gene itself is the true unit of selection. If one genetic alternative, or allele, provides its bearer with an adaptive advantage over some other individual who carries a different allele then the more beneficial allele will be replicated more times, as its bearer enjoys greater fitness. In his book *The Selfish Gene*, Dawkins argues that genes help to build the bodies that aid in their transmission; individual organisms are merely the "survival machines" that genes require to make more copies of themselves.

This argument has been criticized because natural selection cannot "see" the individual genes that reside in an organism's **genome**, but rather selects among phenotypes, the outward manifestation of all the genes that organisms possess. Some genetic combinations may confer very high fitness, but they may reside with genes having negative effects in the same individual. When an individual reproduces, its "bad" genes are replicated along with its "good" genes; if it fails to do so, even its most advantageous genes will not be transmitted into the next generation. Although the focus among most evolutionary biologists has been on selection at the level of the individual, this example raises the possibility that individual genes in genomes are under a kind of group selection. The success of single genes in being transmitted to subsequent generations will depend on their functioning well together, collectively building the best possible organism in a given environment.

When selective change is brought about by human effort, it is known as artificial selection. By allowing only a selected minority of individuals to reproduce, breeders can produce new generations of organisms featuring particular traits, including greater milk production in dairy cows, greater oil content in corn, or a rainbow of colors in commercial flowers. The repeated artificial selection and breeding of individuals with the most extreme values of the desired traits may continue until all the available genetic variation has been exhausted, and no further selection is possible. It is likely that dairy breeders have encountered the limit for milk production in cattle—eventually, a cow's milk production will increase more slowly for a given increase in feed-but the limit has not yet been reached for corn oil content, which continues to increase under artificial selection.

Seemingly regardless of the trait or characteristic involved (e.g. zygotic selection), there is almost always a way to construct a selectionist explanation of the manifest phenotype.

See also Adaptation and fitness; Chromosomal mutations and abnormalities; Darwinism; Evolution, evidence of; Evolutionary mechanisms

SELF-INCOMPATIBILITY

The failure of gametes from the same plant to form a viable embryo is known as self-incompatibility (SI). The process was initially studied in ornamental tobacco, where it was noticed that pollen grains that fell on the stigma of the same plant failed to grow down the style. However, the pollen could grow successfully down the style of other plants of the same **species**. In addition to tobacco, SI mechanisms have evolved in many diverse angiosperms families, with different families utilizing different proteins to accomplish the same goal. SI increases outcrossing and heterozygosity and prevents **inbreeding**.

SI systems operate prior to **fertilization** during interactions between the male gametophyte (pollen or pollen tube) and the female sporophytic tissue (pistil). Incompatibility can be achieved by prevention of pollen germination, retardation of the growth or disorientation of the pollen tube, or failure of nuclear fusion. These processes are usually controlled by a single genetic locus (the S locus) with multiple alleles. The S locus contains one or more genes expressed in the male or female reproductive tissues. Differences in proteins encoded by these genes are the basis for the recognition of self or non-self pollen; when the pollen genotype matches either of the two S alleles of the pistil, the pollen cannot grow.

The two major types of SI can be described as gametophytic and sporophytic SI systems. Gametophytic SI (GSI), in which the behavior of the pollen is determined by the **haploid** pollen genotype at the S locus, is the more common system, found in more than 60 families of plants. In many cases of GSI, the pollen tube is able to initiate growth through the style before growth is arrested. GSI has been well studied in many

species of the family Solanaceae, including petunia, tobacco, and tomato. The proteins expressed in the female tissues necessary for GSI have been identified and are ribonucleases. They were named the S-RNases because they are encoded by genes found on the S locus. The pollen-expressed male component that interacts with these S-RNases is still unknown, however, as is the mechanism by which S-RNases mediate the incompatibility response.

In contrast to GSI, the less common sporophytic SI (SSI) is determined by the genotype of the female tissue, not the pollen. SSI responses occur early in pollen-pistil interactions and often block hydration of the pollen grain or emergence of the pollen tube at the stigma surface. SSI has been studied in the family Brassicaceae, where the stigma proteins involved have been identified. They include a protein secreted into the **cell** wall of certain stigma cells and a second protein that appears to be a receptor. Both of these proteins are likely required in order to recognize the pollen SSI component, which has been shown to be a small peptide found on the surface of the pollen grain.

See also Xenogamy; Plant genetics; Plant reproduction; Sexual reproduction

SELFISH DNA

Also known as junk **deoxyribonucleic acid** (junk **DNA**), selfish DNA are areas of DNA that have no apparent function, but are passed on from generation to generation. In some cases, the sequences of selfish DNA are repeated (repetitive DNA) and use the "host" organism as a means for survival (survival machine). This phenomena is documented in the Selfish DNA theory, which states that the eukaryotic organisms carrying the replicating DNA are nothing more than survival machines that allow the DNA to survive and reproduce. This type of DNA is generally repetitive in its composition and it is typical of such regions as spacer DNA and satellite DNA.

Recently, the possibility that selfish DNA includes functional genes that are coded with viable characteristics has been suggested, and the selfish DNA theory has been expanded to the selfish DNA **gene** theory. The expanded theory proposes that the "host" organism is not merely a survival machine, but an organism that is affected by the selfish DNA/genes. The selfish DNA is functional, and therefore it can influence its own and its host's survival.

Selfish DNA is one example of a class of highly repetitive DNAs with unknown functions. The ability of the host to regulate selfish DNA appears to be limited. There is a high correlation between the amount of selfish DNA and the underlying cellular complexity of the host. Higher animal cells generally contain greater amounts of selfish DNA. There is also a strong correlation between the amount of cellular **differentiation** (e.g., of the type observed in higher plant and animal **species**) and the amount of selfish DNA carried by those species.

Selfish genes are transmitted at a higher rate than other genes. They can often be distributed in abnormal patterns during **meiosis** by a process termed meiotic drive. In such cases the selfish DNA may begin to work to the detriment of the host species, becoming such a large percentage of the **genome** that it may be possible that the species faces a substantial risk of extinction.

See also Gene frequency

SELKOE, DENNIS JAMES (1943-   )

American neurologist

Dennis Selkoe was, according to the Institute of Scientific Information (ISI), one of the scientists who produced the hottest and most highly cited papers in neuroscience in the past decade.

Selkoe's laboratory is investigating the **biochemistry** and molecular **cell** biology of neuronal degeneration during aging of the mammalian brain. He was among the first to discover that abnormal amyloid protein deposits in the human brain causes certain types of Alzheimer's diseases. The invention of assays and cell-culture systems to analyze the synthesis and metabolism of amyloid protein has been well received by researchers in the neuroscience field. These assays and methods were very useful as diagnostic tools for testing predisposition to **Alzheimer's disease** and stimulated research to find drug targets for neurodegenerative diseases.

Together with other research groups, Selkoe and colleagues demonstrated the role that amyloid peptide beta accumulation played in the development of Alzheimer's disease. Selkoe and colleagues also investigated the role two membrane proteins called presenilin one and two play in the production and accumulation of the amyloid peptide beta polymers. Presenilin was first thought to be involved, in association with a number of proteolytic **enzymes** called secretases, in the degradation of amyloid parent protein (APP) and the production of amyloid peptide beta peptide. The latter is secreted in the extracellular matrix of the brain where it accumulates and then polymerizes to produce the plaques characteristic of Alzheimer's diseases.

Presenilin has been shown to play an important physiological role in the normal development of the fly embryo. Its most important and still controversial feature, discovered by Selkoe and colleagues, is that it is identical to the secretases and is therefore the first protease described in biology that is active deep inside a cell membrane. The researchers in Selkoe's group believe that it is possible to design drugs that will act through blocking the enzymatic activity of presenilin and thus control the progression of Alzheimer's disease. Because of the importance of presenilin in embryonic development and its role in the pathogenesis of Alzheimer's disease, Selkoe proposes that presenilin has evolved as a helpful molecule that incidentally cuts APP and causes Alzheimer's disease.

Born in New York City, Selkoe received a BA from Columbia University and an MD from the University of Virginia school of Medicine in 1969. He became Professor of neurology at Harvard Medical School in 1990. He is also the director of the Center for Neurological Diseases at Brigham and Women's Hospital in Boston.

See also Alzheimers disease; Cytogenetics; Medical genetics

SEMI-LETHAL GENES · *see* LETHAL AND SEMI-LETHAL GENES

SEQUENCE TAGGED SITES

Sequence-tagged sites (STSs) are short stretches of **DNA** sequence, usually a few hundred **bases** long, which can be amplified by the **polymerase chain reaction**, and are regarded as landmarks in a **genome**. Presence of a specific STS in a genomic clone or DNA from a specific **chromosome** can be tested by PCR. STSs are most useful when they have been mapped to a specific chromosome using somatic **cell** hybrids and to a specific region by screening cell hybrids containing fragments of human chromosomes. An STS map is a collection of STSs ordered along a chromosome with their physical position known. They are used to orientate and order genomic clones such as BACs and YACs in order to generate physical maps. STS maps have been generated for the human genome as part of the **human genome project**. They have also been generated for other organisms such as the mouse and the zebrafish, usually as a precursor for full genomic **sequencing**.

STSs can be generated in different ways. Many STSs were generated for the human genome by sequencing a few hundred bases of both ends of BACs. Other STSs were generated by randomly **cloning** and sequencing small fragments of the human genome. STSs were also generated for the human genome when simple tandem repeat sequence markers (microsatellites) were isolated for **genetic mapping**. These STSs are therefore used as physical markers and as genetic markers, providing a link between physical and genetic maps.

Clones containing a certain STS can be isolated by screening genomic libraries by PCR. This is an alternative to screening genomic libraries by hybridisation with the STS used as a radioactively labeled probe. The PCR-based screening method is labor-intensive and false negatives are more common in PCR compared to hybridization, so the hybridization method is more commonly used.

See also Bacterial artificial chromosome (BAC); Clones and cloning; DNA probes; Genetic mapping; Genomic library; Polymerase Chain Reaction (PCR); Tandem repeat sequences; Yeast Artificial Chromosome (YAC)

SEQUENCING

Sequencing refers to the techniques that determine the order of the constituents of **deoxyribonucleic acid (DNA)** or protein. DNA sequencing determines the order of the constituent bases—adenine, thymine, guanine and cytosine. Protein sequencing determines the order of the constituent amino acids. DNA is typically sequenced for several reasons: to determine the sequence of the protein encoded by the DNA, the location of sites at which **restriction enzymes** can cut the DNA, the location of DNA sequence elements that regulate the production of **messenger RNA**, or to detect alterations in the DNA that might be important in genetic diseases, such as **cystic fibrosis** or **cancer**.

In recent years, the best known example of sequencing has been the effort to sequence the human **genome**, the complete set of genetic instructions carried within each **cell** of an organism. Decoding the DNA sequence of the human genome should pave the way towards a better understanding of the genetic basis of certain diseases.

The sequencing of DNA is accomplished by stopping the lengthening of a DNA chain at a known base and at a known location in the DNA. Practically, this can be done in two ways. In the first method, called the Sanger-Coulson procedure, a small amount of a specific so-called dideoxynucleoside base is incorporated in along with a mixture of the other four normal **bases**. This base is slightly different from the normal base and is also radioactively labeled. The radioactive base becomes incorporated into the growing DNA chain instead of the normal base, growth of the DNA stops. This stoppage is done four times, each time using one of the four different dideoxynucleosides. This generates four collections of DNA molecule. Also, since **replication** of the DNA always begins at the same point, and because the amount of altered base added is low, for each reaction many DNA pieces of different length will be generated. When the sample is used for gel electrophoresis, the different sized pieces can be resolved as radioactive bands in the gel. Then, knowing the location of the bases, the sequence of the DNA can be deduced. The second DNA sequencing technique is known as the Maxam-Gilbert technique, after its co-discoverers. In this technique, both strands of double-stranded DNA are radioactively labeled using radioactive phosphorus. Upon heating, the DNA strands separate and can be physically distinguished from each other, as one strand is heavier than the other is. Both strands are then cut up using specific **enzymes** and the different sized fragments of DNA are separated by gel electrophoresis. Based on the pattern of fragments the DNA sequence is determined.

The Sanger-Coulsom is the more popular method. Various modifications have been developed and it has been automated for very large-scale sequencing, such as the **Human Genome Project**. In the sequencing of the human genome, a sequencing method called shotgun sequencing was very successfully employed. Shotgun sequencing refers to a method that uses enzymes to cut DNA into hundreds or thousands of random bits. So many bits are necessary since automated sequencing machines can only decipher relatively short fragments of DNA about 500 bases long. The many sequences are then pieced back together using computers to generate the entire DNA genome sequence.

Protein sequencing involves determining the arrangement of the **amino acid** building blocks of the protein. It is common to sequence a protein by the DNA sequence encoding the protein. This, however, is only possible if a cloned **gene** is available. It still is often the case that chemical protein sequencing, as described subsequently, must be performed in order to manufacture an oligonucleotide probe that can then be used to locate the target gene. The most popular direct protein chemical sequencing technique in use today is the Edman degradation procedure. This is a series of chemical reactions, which remove one amino acid at a time from a certain end of the protein (the amino terminus). Each amino acid that is released has been chemically modified in the release reaction, allowing the released product to be detected using a technique called reverse phase chromatography. The identity of the released amino acids is sequentially determined, producing the amino acid sequence of the protein.

Another protein sequencing technique is called fast atom bombardment mass spectrometry, or FAB-MS. This is a powerful technique in which the sample is bombarded with a stream of fast atoms, such as argon. The protein becomes charged and fragmented in a sequence-specific manner. The fragments can be detected and their identify determined. The expense and relative scarcity of the necessary equipment can be a limitation to the technique.

Still another protein sequencing strategy is the digestion of the protein with specialized protein-degrading enzymes called proteases. The shorter fragments that are generated, called peptides, can then be sequenced. The problem then is to order the peptides. This is done by the use of two proteases that cut the protein at different points, generating overlapping peptides. The peptides are separated and sequenced, and the patterns of overlap and the resulting protein sequence can be deduced.

See also Genome; Genomic library

SERIAL ANALYSIS OF GENE EXPRESSION (SAGE)

Serial analysis of **gene expression** (SAGE) is a technique that allows the quantitave evaluation of the level of **gene** expression in a **cell**. Developed in the laboratory of Bert Vogelstein in 1995, this technique is based on the production and **sequencing**, and count of small **cDNA** pieces called tags which are subsequently assigned to corresponding genes. Extensive sequencing, deep bioinformatic knowledge, and powerful computer software are required to assemble and analyze results from SAGE experiments. This is most likely one of the reasons of the limited use of this sensitive technique in academic research laboratories.

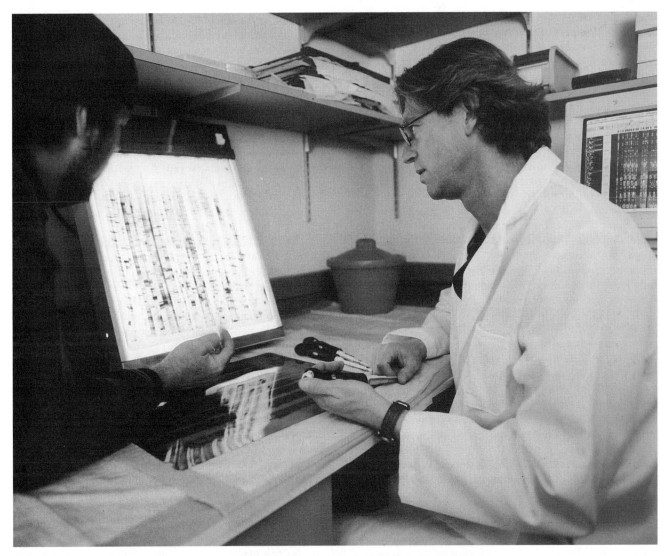

Robot designed for DNA mapping and sequencing. *Custom Medical Stock Photo. Reproduced by permission.*

In a typical SAGE experiment, **messenger RNA (mRNA)** that are naturally tagged by poly (A), are isolated from control and experimental cells and converted into double-stranded cDNAs using biotin-labeled oligo (dT) **primers**. The cDNAs are digested with specific **restriction enzymes** to produce small biotin labeled **DNA** fragments that can be bound to straptavidin beads as a result of the strong interaction between straptavidin and biotin. An oligonucleotide linker containing a type II restriction endonuclease site is then ligated to the free (non-biotinilated) ends of all the cDNA fragments. Type IIS endonucleases are restriction **enzymes** that cut at a defined number of **bases** downstream of their recognition sites. Subsequent digestion of the biotin tagged cDNA fragments by a type IIS restriction enzyme will result in the release of small, biotin free pieces of cDNA of an average length of 20 base pairs (bp). These short DNA pieces called tags should represent all cDNAs and, therefore, mRNA originally expressed in the studied cell. All these DNA tags are ligated together to form large fragments of DNA that usually contain multiple tagged inserts. The large fragments are PCR amplified and cloned in a sequencing vector. Sequencing will identify each tag that will be representative of an mRNA molecule. Computer analysis is performed to match sequence tags to known cDNAs or **expressed sequence tags** (**ESTs**) and to evaluate the frequency of each tag. The frequency of a specific tag is related to the abundance of the corresponding mRNA in the cell. SAGE, therefore, provides a measure of the relative expression levels of mRNA **species** under specific cellular condition. Because representative fragments of all mRNA species present in a cell are sequenced, this method allows the identification of both known and novel genes being expressed under the conditions studied.

One major disadvantage of the SAGE technique is that it does not measure the actual expression level of a gene, but counts the frequency of a tag that is assigned to an mRNA species, the **transcription** product of a single gene. Currently, the average size of a tag produced during SAGE analysis is ten bases and this makes it difficult to assign a tag to a specific transcript with accuracy. In fact, two different genes could have the same tag and the same gene that is alternatively spliced could have different tags at the 3' ends. Assigning each tag to an mRNA transcript could be made even more difficult and ambiguous if sequencing errors are also introduced in the process.

It is estimated that only about 10% of all genes are simultaneously expressed in a cell. The level of expression of particular genes determines which proteins will be made and consequently decides how the cell is functioning and how it is responding to the environment. The same cell will usually express different sets of genes during exposure to different drugs, toxic agents or pathogens. Using the SAGE technique, researchers were able to identify many genes that are specifically expressed in **cancer** and which promise to be useful as diagnostic markers and target for therapy.

See also Biochemical analysis techniques

SEVERE COMBINED IMMUNODEFICIENCY

Severe combined immunodeficiency (SCID) is a rare genetic disease that is actually a group of inherited disorders characterized by a lack of immune response, usually occurring in infants less than six months old. SCID is the result of a combination of defects of both T-lymphocytes and B-lymphocytes. Lymphocytes are white blood cells that are made in bone marrow, and many move to the thymus gland where they become specialized immune T and B cells. In healthy individuals, T cells attack antigens while B cells make plasma cells that produce antibodies (immunoglobulins). However, this immune response in SCID patients is absent making them very susceptible to invading diseases, and thus children with untreated SCID rarely live to the age of two years.

SCID is characterized by three main features. The helper T-lymphocytes are functioning poorly or are absent, the thymus gland may be small and functioning poorly or is absent, and the **stem cells** in bone marrow, from which mature T- and B-lymphocytes arise, are absent or defective in their function. In all of these situations, little or no antibodies are produced. If, for example, T-lymphocytes are never fully developed, then the immune system can never function normally. Moreover, the results of these defects include the following: impairment of normal functioning T- and B-lymphocytes, negative effects on the maturation process for T-helper and T-suppressor cells, and elimination and damage of the original source of the lymphocytes.

The immune disorders characterized in SCID arise because of the **inheritance** of abnormal genes from one or both parents. The most common form of SCID is linked to the **X chromosome** inherited from the mother; this makes SCID more common among males. The second most common defect

is caused by the inheritance of both parents' abnormally inactive genes governing the production of a particular enzyme that is needed for the development of immunity, called adenosine deaminase (**ADA**). Although many defective genes for other forms of SCID have been identified in the last few years, scientists do not fully understand all of the forms of the disease.

There are many specific clinical signs that are associated with SCID. After birth, an infant with SCID is initially protected by the temporarily active maternal immune cells; however, as the child ages, his or her immune system fails to take over as the maternal cells become inactive. Pulmonary problems such as pneumonia, non-productive coughs, inflammation around the bronchial tubes, and low alveolar oxygen levels can affect the diseased infant repetitively. Chronic diarrhea is not uncommon, and can lead to severe weight loss, malnutrition, and other gastrointestinal problem. Infants with the disease have an unusual number of bacterial, fungal, viral, or protozoal infections that are much more resistant to treatment than in healthy children. Mouth thrush and **yeast** infections, both fungal, appear in SCID patients and are very resistant to treatment. Additionally, chronic bacterial and fungal skin infections and several abnormalities of the blood cells can persist.

Severe combined immunodeficiency is a disease that can be successfully treated if it is identified early. The most effective treatment has been hematopoietic stem **cell** transplants that are best done with the bone marrow of a sister or brother; however, the parent's marrow is acceptable if the infant is less than three months old. Early treatment can also help to avoid pre-transplant chemotherapy often necessary to prevent rejection of the marrow in older children. This is especially advantageous because chemotherapy can leave the patient even more susceptible to invading bodies. When successful, treatment for SCID corrects the patient's immune system defect, and recent success rates have been shown to be nearly 80% for the bone marrow transplant.

Gene therapy is the subject of ongoing research, and shows promise as a treatment for SCID. Researchers remove T cells of SCID patients and expose those cells to the ADA **gene** for ten days, and then return the cells intravenously. Although it was successful in one case, this treatment of SCID is still very much in the experimental stage. Nevertheless, these and other treatments hold potential for the development of a cure for SCID.

See also Adenosine deaminase deficiency (ADA); Immunogenetics

SEX CELLS · *see* GAMETOGENESIS

SEX CHROMOSOME

Sex chromosomes are the chromosomes within a **cell** that carry the genes important in the determination of the sex of an organism. In the cells of all **diploid** organisms, with the exception of the sperm and eggs and the gametophyte generation in plants, chromosomes are arranged in pairs called homologous pairs. Each **chromosome** in the homologous pair comes from a different parent. A cell that has both chromosomes of each homologous pair is a diploid cell. In the human body, diploid cells have 46 chromosomes arranged in 23 homologous pairs.

In humans, each diploid cell has two sex chromosomes that make up one of the 23 homologous pairs of chromosomes. There are two types of sex chromosomes, the **X chromosome** and the **Y chromosome**. The X chromosome is larger than the Y chromosome, and, therefore, has spaces for genes that are not present on the Y chromosome.

In humans and many other **species**, the sex chromosomes of females are identical. Females have two X chromosomes (symbolized as XX). The chromosomes that make up the pair of sex chromosomes in males are different. Males have one X chromosome and one Y chromosome (symbolized as XY). In birds, moths, and butterflies the sex chromosomes are reversed; males are XX and females are XY. In some insects, the Y chromosome is absent. Thus, a female would be XX, but a male would be simply X.

The sex of an individual (based on what sex chromosomes it has) is determined at the time of **fertilization**, when the male's sperm joins with the female's egg. Because all human female cells have two X chromosomes, when the egg forms by **meiosis**, it must contain an X chromosome. However, since all human male cells have both an X and a Y chromosome, when the sperm form by meiosis, they can contain either an X chromosome or a Y chromosome. Thus, during fertilization, the egg always contributes an X chromosome and the sperm can contribute either an X or a Y chromosome. If the sperm has an X chromosome, the resulting offspring will have two X chromosomes and will be a female (XX). If, on the other hand, the sperm contributes a Y chromosome, the **zygote** that forms as a result of fertilization will have one X and one Y chromosome and will be a male (XY). In humans and many other species, it is the gametic contribution of the male that leads to the determination of sex of the offspring, depending on which sex chromosome he contributes in his sperm.

See also Sex determination; Sex-linked genes; Sexual reproduction

SEX DETERMINATION

The sex of an individual is usually determined by the sex chromosomes that the organism has. Each **diploid cell** in the organism has one pair of sex chromosomes that differs from the other homologous pairs of chromosomes. For example, in humans, one of the 23 pairs of **homologous chromosomes** makes up the sex chromosomes. However, unlike all of the other homologous pairs, the sex chromosomes may not be the same as each other in size. There are two types of sex chromosomes, the **X chromosome** and the **Y chromosome**. The X

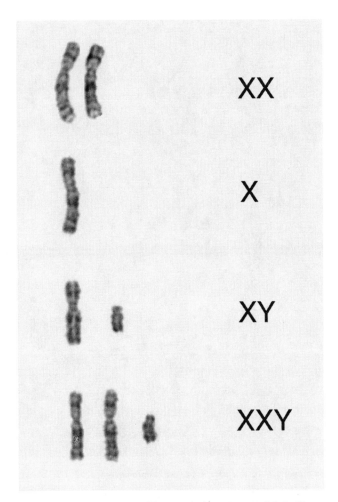

Sex chromosomes showing different complements associated with different phenotypes. A) Two X chromosomes—normal female; B) An X and a Y chromosome—normal male; C) One X chromosome—Turner syndrome female; D) Two X and one Y chromosomes—Klinefelter syndrome. *Courtesy of Dr. Constance Stein.*

chromosome is larger than the Y chromosome, and therefore has spaces for genes that are not present on the Y chromosome.

In humans and many other **species**, females have two X chromosomes (symbolized as XX). However, the chromosomes that make up the pair of sex chromosomes in males are different. Males have one X chromosome and one Y chromosome (symbolized as XY). In some species, such as birds, moths and butterflies the sex chromosomes are reversed; males are XX and females are XY. In some insects the Y chromosome is absent altogether. Thus, a female would be XX, but a male would be XO (the "O" indicates that the Y chromosome is absent).

The sex of an individual is based on its sex chromosomes. This is determined at the time of **fertilization**, when the male's sperm joins with the female's egg. For example, because all human female cells have two X chromosomes, when the egg forms by **meiosis**, it must contain an X chromosome. However, since all human male cells have both an X and a Y chromosome, when the sperm form by meiosis, they can

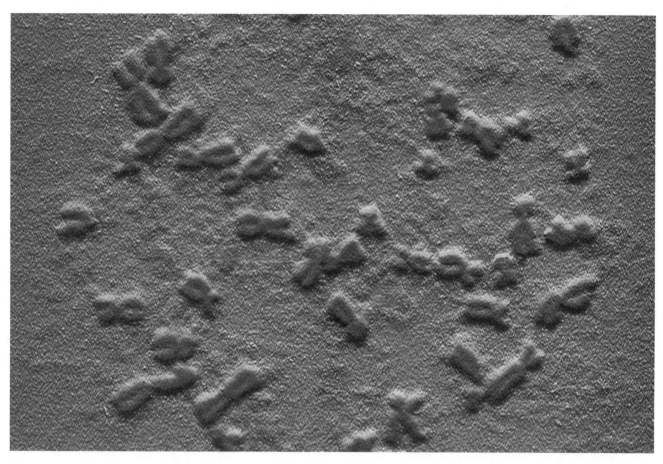

In humans, normal females carry two X chromosomes. *J.L. Carson/Custom Medical Stock Photo. Reproduced by permission.*

contain either an X chromosome or a Y chromosome. Thus, during fertilization, the egg always contributes an X chromosome and the sperm can contribute either an X or a Y chromosome. If the sperm has an X chromosome, the resulting offspring will have two X chromosomes and will be a female (XX). If, on the other hand, the sperm contributes a Y chromosome, the **zygote** that forms as a result of fertilization will have one X and one Y chromosome and will be a male (XY). In humans and many other species, it is the **sex chromosome** contributed by the male that is responsible for determining the sex of the offspring. Since the sperm have a 50% chance of receiving an X chromosome and a 50% chance of receiving a Y chromosome, there are equal numbers of X and Y sperm. As a result, there is an almost equal chance of having a male or female offspring.

In some organisms, sex chromosomes are not wholly responsible for sex determination. Environmental conditions can also play a role. For example, in marine turtles and some other reptiles, temperature is an important factor. In some marine turtles, when eggs are incubated at higher temperatures, females will be produced. Under lower incubation temperatures, males will be produced. In some coral reef fish, such as wrasses, individuals can actually change sex. These fish live in small groups of females, with one dominant male.

When the male dies, one of the females becomes a male. Her color and behavior change, and she begins to produce sperm. In addition, many types of organisms are hermaphroditic, possessing both male and female sex organs. Thus, while sex is usually determined by the sex chromosomes an organism possesses, there are exceptions.

See also Embryology; Sex chromosomes

SEX-LINKED TRAITS

Sex-linked genes are genes that are physically located on either the X or the **Y chromosome**. Because there are very few genes located on the Y **chromosome**, and none that are considered to be Mendelian disease genes, the overwhelming majority of sex-linked genes are located on the **X chromosome**. Therefore, sex-linked traits and X-linked traits are essentially synonymous.

The distinctive nature of sex-linked traits compared to traits located on any of the 22 **autosomes** is related to the differences in **gene** copy number for sex-linked genes between males and females. Females possess two X-chromosomes and no copy of the Y chromosome. For X-linked genes, females have two copies of each (dizygosity) just as is the case for the autosomal genes and they lack genes encoded on the Y-chromosome altogether. Males, however, possess one X chromosome and one Y chromosome and they therefore have but a single copy (hemizygosity) for all genes on both of the sex chromosomes.

Females inherit one copy of the X chromosome at conception from both the mother and the father. Because a single active copy appears to be optimal for sex-linked genes, genetic balance is maintained in the female for genes encoded on the X chromosome through inactivation of the second X chromosome. Although some genes appear to escape inactivation, most genes on the second X chromosome of the female are silenced in the process. X inactivation occurs early in fetal development, and the X that is chosen for inactivation is chosen at random. Half of the cells will select the copy of the X chromosome inherited from the mother, and half will choose the one from the father. Should one of the X chromosomes carry a gene **mutation**, it will be expressed in only half of the cells with the normal gene on the other X being expressed in the other half. For recessive disease genes, the cells that express the normal gene compensate for the cells that express the mutated gene, and health is normal. For dominant genes, the mutation causes the disease even though it is expressed in only half of the cells.

For males, their X chromosome is inherited from the mother, and the Y chromosome is inherited from the father. The same X must be active in every **cell** and therefore any mutations inherited on the X chromosome will be expressed as genetic disease. X-linked recessive diseases are far more frequently observed in males than in females. In contrast, X linked dominant diseases are about half as common in males than in females.

The most common X-linked recessive genetic diseases include Duchenne muscular dystrophy, Becker muscular dystrophy (a milder form than Duchenne), **hemophilia** A, hemophilia B, and some forms of color blindness. The classic pattern of **inheritance** is one in which affected males are related to one another through healthy **carrier** females.

There are only a few examples of X-linked dominant diseases in humans, the best known of which is fragile X mental retardation syndrome. The inheritance pattern is similar to an autosomal dominant pattern in that females pass the trait to half of their offspring regardless of their sex. It is distinctive from autosomal dominant inheritance in that there is no father to son transmission.

Sex-influenced traits are traits whose expression is influenced by the sex of the individual. These traits are generally not coded by genes located on either of the sex chromosomes, and are therefore different in principle than sex-linked traits. Often, these traits involve hormonal or physiological elements that are different in males than in females. Examples include the autosomal dominant male pattern baldness trait that causes early baldness in males with a milder pattern of hair thinning in females, and the autosomal recessive iron storage related disease hemochromatosis. Sex-limited traits also

are usually coded by genes located on the autosomes and are different from sex-influenced traits as they are expressed exclusively in either males or females.

See also Mendelian genetics and Mendel's laws of heredity; Sex determination

SEXUAL REPRODUCTION

Sexual reproduction is the process whereby two parents produce offspring genetically different from themselves that have new combinations of the parents' characteristics. This contrasts with **asexual reproduction**, where one parent produces offspring genetically identical to itself. During sexual reproduction, each parent contributes one **haploid gamete** (a sex **cell** with half the normal number of chromosomes). The two sex cells fuse during **fertilization** and form a **diploid zygote** (which has the normal number of chromosomes). **Recombination**, which is the production of variations in **gene** combinations, occurs at fertilization, bringing together new combinations of alleles. **Crossing over**, the exchange of pieces of chromosomes by two **homologous chromosomes**, also brings about **genetic variation** during sexual reproduction. Sexual reproduction is advantageous because it generates variations in characters that can adapt a **species** over time and improve its chances of survival.

Sexual reproduction occurs in practically all forms of life. Even **bacteria**, which are always haploid, exchange genetic material. **Eukaryotes**, organisms possessing a nuclear membrane, generally produce haploid gametes (or sex cells). A gamete, such as an egg or a sperm, possesses half the normal number of chromosomes, and is produced by **meiosis**, which is reduction **cell division**, which reduces the number of chromosomes from diploid in the parent cell to haploid in the gametes. When the gametes fuse at fertilization, they restore the normal number of chromosomes. Conjugation, alternation of generations, and animal reproduction illustrate various modes of sexual reproduction.

Conjugation is a process of genetic recombination that occurs between two organisms (such as bacteria) in addition to asexual reproduction. Conjugation only occurs between cells of different mating types. In bacteria, cells designated F+ and F- lie close together, and a narrow bridge of cytoplasm forms between them. F+ cells contain a plasmid or reproductive factor that is made of **DNA**, which replicates within the bacterial cell. A copy is transferred from a donor F+ cell to a recipient F-. *Spirogyra*, a freshwater filamentous alga, also exhibits conjugation, where two nearby filaments develop extensions that contact each other. The walls between the connecting channels disintegrate, and one cell moves through the conjugation tube into the other. The cells fuse to form a diploid zygote, the only diploid stage in the life of *Spirogyra*. The black bread mold, *Rhizopus*, reproduces asexually by spores and sexually by conjugation. During conjugation, the tips of short hyphae act as gametes, and fuse. The resulting zygote develops a protective wall and becomes dormant. Finally, meiosis occurs, and a haploid bread mold germinates and grows spore-producing sporangia.

During sexual reproduction in animals, a haploid sperm and unites with a haploid egg cell to form a diploid zygote. The zygote divides mitotically and differentiates into an embryo. The embryo grows and matures. After birth or hatching, the animal develops into a mature adult capable of reproduction. Some invertebrates reproduce by self-fertilization, in which an animal's sperm fertilizes its own eggs. Self-fertilization is common in tapeworms and other internal parasites, which lack the opportunity to find a mate. Most animals, however, use cross-fertilization, in which different individuals donate the egg and the sperm. Even hermaphrodites animals (such as the earthworms) that produce both types of gametes use cross-fertilization.

Animals exhibit two patterns for bringing sperm and eggs together. One is external fertilization, whereby animals shed eggs and sperm into the surrounding water. The flagellated sperm need an aquatic environment to swim to the eggs, the eggs require water to prevent drying out. Most aquatic invertebrates, most fish, and some amphibians use external fertilization. These animals release large numbers of sperm and eggs, thereby overcoming large losses of gametes in the water. In addition, courting behavior in some species brings about the simultaneous release of the gametes, which helps insure that sperm and egg meet.

The other pattern of sexual reproduction is internal fertilization, whereby the male introduces sperm inside the female's reproductive tract where the eggs are fertilized. Internal fertilization is an adaptation for life on land, for it reduces the loss of gametes that occurs during external fertilization. Sperms are provided with a fluid (semen) that provides an aquatic medium for the sperm to swim when inside the male's body. Mating behavior and reproductive readiness are coordinated and controlled by hormones so that sperm and egg are brought together at the appropriate time.

After internal fertilization, most reptiles and all birds lay eggs that are surrounded by a tough membrane or a shell. Their eggs have four membranes, the amnion, the allantois, the yolk sac and the chorion. The amnion contains the fluid surrounding the embryo; the allantois stores the embryo's urinary wastes and contains blood vessels that bring the embryo oxygen and take away carbon dioxide. The yolk sac holds stored food, and the chorion surrounds the embryo and the other membranes. After the mother lays her eggs, the young hatch.

Mammals employ internal fertilization, but except for the Australian montremes such as the duckbill platypus and the echidna, mammals do not lay eggs. The fertilized eggs of mammals implant in the uterus, which develops into the placenta, where the growth and **differentiation** of the embryo occur. Embryonic nutrition and respiration occur by diffusion from the maternal bloodstream through the placenta. When development is complete, the birth process takes place.

In plants, sexual and asexual reproduction unite in a single cycle called alternation of generations. During alternation of generations, a gametophyte, (a haploid gamete-producing plant), alternates with a sporophyte (a diploid spore-producing plant). In *Ectocarpus*, a brown aquatic alga, the two generations are equally prominent, whereas in mosses, the gametophyte generation dominates. In ferns and seed plants, the sporophyte dominate, because the sporophyte generation is better adapted to survive on land.

Mosses are small plants that lack vascular tissue and do not produce seeds, and depend on a moist environment to survive. The green leafy ground cover of mosses that we are familiar with is the haploid gametophyte. The gametophyte develops sex organs, a male antheridium and a female archegonium on the same or different plants. The antheridium produces flagellated **sperm cells** that swim to the egg cells in the archegonium. After fertilization, the zygote grows into a diploid sporophyte. The sporophyte consists of a foot, stalk, and capsule. It remains attached to the gametophyte. Cells in the capsule undergo meiosis and develop into haploid spores. When released, spores grow into gametophytes with root like, leaf like and stem like parts.

Ferns, in the form of the familiar green leafy plants, represent the diploid sporophyte generation. Ferns have a vascular system and true roots, stems, and leaves, but they do not produce seeds. Sporangia, or spore cases, develop on the leaves of ferns, and produce haploid spores by means of meiosis. The spores germinate into haploid green gametophytes. The fern gametophyte is a tiny heart-shaped structure that bears antheridia and archegonia. Flagellated sperm swim to the eggs in a layer of ground water. Although the sporophyte is adapted to land life, this need for water limits the gametophyte. After fertilization, the diploid zygote develops into the sporophyte.

In flowering plants, the diploid sporophytes are plants with roots, leaves, stems, flowers and seeds. Anthers within the flower contain four sporangia. Cells in the sporangia undergo meiosis and produce haploid microspores. The wall of each microspore thickens, and the haploid **nucleus** of the microspore divides by **mitosis** into a generative nucleus and a tube nucleus. These microspores are now called pollen, and each pollen grain is an immature male gametophyte. **Pollination** occurs when pollen escapes from the anthers and lands on the stigma of a flower, either of the same plant or a different plant. There, a pollen tube begins to grow down the style toward the ovary of the pistil, and the two nuclei move into the pollen tube. The generative nucleus divides to form two haploid sperm cells. The germinated pollen grain is now a mature male gametophyte. Finally, the pollen tube penetrates the ovary and the sperm enter. The ovary contains sporangia called ovules. Meiosis occurs within each ovule forming four haploid megaspores. Three disintegrate, and the remaining megaspore undergoes repeated mitosis to form the female gametophyte. The female gametophyte is a haploid seven-celled structure. One of the seven cells is an egg cell. Another of the seven cells contains two nuclei called polar nuclei. When the two sperm cells enter, double fertilization occurs. One sperm fertilizes the egg, forming a zygote that develops into a diploid embryo sporophyte. The two polar nuclei fuse and their product unites with the second sperm forming a triploid endosperm. The endosperm serves as stored food for the embryo sporophyte. After fertilization, the ovule matures into a seed, consisting of embryo, stored food, and seed coat. In angiosperms, the ovary usually enlarges to become a fruit. Upon germination, the seed develops into a mature diploid sporophyte plant. Internal fertilization and seeds help adapt flowering plants to life on land.

See also Crossing over; Embryology; Fertilization; Gametogenesis; Genetic implications of mating and marriage customs; Ovum; Sperm cells

SEXUAL SELECTION · *see* SELECTION

SHARP, PHILLIP A. (1944-)
American molecular biologist

Phillip A. Sharp has conducted research into the structure of **deoxyribonucleic acid** (DNA—the chemical blueprint that synthesizes proteins) which has altered previous views on the mechanism of genetic change. For his work in this area, Sharp was presented with the 1977 Nobel Prize in medicine along with **Richard J. Roberts**. In addition to the Nobel Prize, Sharp has received honors from the American Cancer Society and the National Academy of Science, and is the recipient of the Howard Ricketts Award, the Alfred P. Sloan Jr. Prize, the Albert Lasker Basic Medical Research Award, and the Dickson Prize.

Born in Falmouth, Kentucky, Sharp grew up on a small agricultural farm owned by his parents, Katherin Colvin and Joseph Walter Sharp. Earnings from tobacco land given to him by his parents allowed him to attend Union College in Barbourville, Kentucky, where he received a B.A. degree in chemistry and mathematics in 1966. Sharp earned his Ph.D. degree from the University of Illinois in 1969.

Sharp and Roberts discovered in 1977 that, in some higher organisms, genes may be comprised of more than one segment, separated by material which apparently plays no part in the creation of the proteins. Previously, most scientists believed that genes were continuous sections of **DNA** and that the string of coding information that makes up each **gene** was a single, linear unit. Sharp and Roberts, however, distinguished between the exons, the sequences that contain the vital information needed to create the protein, and the introns, incoherent biochemical information that interrupts the protein-manufacturing instructions. Each gene is apparently composed of fifteen to twenty exons, in between which introns may be located. During **protein synthesis**, exons are copied and spliced together, creating complete sequences, while the introns are ignored.

This discovery had not been made earlier largely because scientists had conducted most of their genetic research on prokaryotic organisms, such as **bacteria**, which do not have their genetic material located in clearly defined nuclei. Studies of bacteria had indicated that gene activity resulted in the **transcription** of double-stranded DNA into single-stranded messenger **ribonucleic acid** (**mRNA**); this is translated to the corresponding protein by **ribosomes**. Prokaryotic organisms have no introns, however, and therefore could not supply evidence for the existence, or the significance, of noncoding regions of DNA. Roberts and Sharp carried their research out on adenoviruses, the **virus** responsible for the common cold in humans. Although these are also prokaryotic organisms, Roberts and Sharp were able to take advantage of the fact that viruses reproduce themselves using the mechanisms of eukaryotic cells. Since their **genome** has some similarities to the genetic material in human cells, their protein synthesis was therefore relevant to the study of the cells of higher organisms.

In their experiments, Sharp's team created **hybrid** molecules in which they could observe mRNA strands binding to their complementary DNA strands. Electron micrographs allowed the scientists to identify which parts of the viral genomes had produced the mature mRNA molecules. What they discovered was that substantial sections of DNA were ignored in producing the final mRNA. This unexpected result gave evidence of a greater complexity of mRNA synthesis in eukaryotic organisms than in prokaryotic ones. Further research indicated that the mRNAs of eukaryotic organisms are synthesized as large mRNA precursor molecules; the introns are spliced out by means of enzyme activity to produce the mature mRNA that manufactures proteins. They found that a single gene could produce a variety of proteins—some defective—as a result of different splicing patterns.

Further research has indicated that the introns, rather than simply being "junk" DNA, contain some information that is necessary for the production of proteins, but the nature of this information is not yet understood.

It is now believed that many **hereditary diseases** are caused by imperfect splicing of the genetic material, leading to the creation of faulty proteins. This may occur if the copying and splicing of the exons is not carried out accurately. One such disease is beta-thalassemia, a form of anemia prevalent in some Mediterranean areas that is caused by a faulty protein responsible for the formation of hemoglobin. Because of the insight Sharp's and Roberts's research has produced into the mechanisms of **cell** reproduction, it has important ramifications for research on malignant tumors and the viruses responsible for their development. It has also led to an investigation of methods for stopping the **replication** of the human immunodeficiency virus type 1 (HIV–1), with potential benefits in the search for a treatment for **AIDS**.

Sharp and Roberts's work has also led to new theories on the nature of evolutionary change; rather than being the cumulative effect of genetic **mutation** over time, it is now believed that it may be the result of the shuffling of large segments of DNA into new combinations to produce new proteins.

In 1990, before his earlier work had led to his Nobel Prize, Sharp was offered, and accepted, the presidency of the

Massachusetts Institute of Technology. A short time later, he decided not to accept the position in order to devote his time exclusively to research. He has remained active in the field of academic administration, however, and has lobbied for research funding. He has also been active in industry; he was one of the founders of Biogen, a corporation started in Switzerland and now operating in Cambridge, Massachusetts, that has employed techniques developed in genetic engineering to produce the drug interferon.

See also DNA synthesis; Selfish DNA; Viral genetics

SHOCKLEY, WILLIAM (1910-1989)
American physicist

William Shockley was an American physicist who became involved in the study of the genetic basis of intelligence and who advanced now discredited ideas regarding race and intelligence. During the 1960s, he argued, in a series of articles and speeches, that people of African descent have genetically inferior mental capacity when compared to those of Caucasian ancestry. Using data taken primarily from U.S. Army pre-induction IQ tests, Shockley came to the conclusion that the genetic component of a person's intelligence was based on racial heritage. He also surmised that the more white genes a person of African descent carried, the more closely his or her intelligence corresponded to that of the general white population. Shockley ignited further controversy with his suggestion that inferior individuals (those whose IQ numbered below 100) receive payment to undergo voluntary sterilization.

The social implications of Shockley's theories were, and still are profound. Many scholars regarded Shockley's whole analysis as flawed, and they rejected his conclusions. Others were outraged that such views were even expressed publicly.

William Bradford Shockley was born in London, England, on February 13, 1910, and moved to California in 1913. Shockley spent a year at the University of California at Los Angeles (UCLA) before attending the California Institute of Technology (Cal Tech), where he earned a bachelor's degree in physics in 1932. He continued his education at the Massachusetts Institute of Technology (MIT) where he was awarded his Ph.D. in 1936.

Upon graduation from MIT, Shockley accepted an offer to work at the Bell Telephone Laboratories in Murray Hill, New Jersey. In 1945, Shockley became director of its research program on solid-state physics. Together with John Bardeen, a theoretical physicist, and Walter Brattain, an experimental physicist, Shockley researched semiconductors as a means of amplification. After more than a year of failed trials, Bardeen suggested that the movement of electric current was being hampered by electrons trapped within a semiconductor's surface layer. That suggestion caused Shockley's team to suspend temporarily its efforts to build an amplification device and to concentrate instead on improving their understanding of the nature of semiconductors.

William Shockley.

By 1947, Bardeen and Brattain had learned enough about semiconductors to make another attempt at building Shockley's device. This time they were successful. Their device consisted of a piece of germanium with two gold contacts on one side and a tungsten contact on the opposite side. When an electrical current was fed into one of the gold contacts, it appeared in a greatly amplified form on the other side. The device was given the name point contact transistor.

Shockley's work in the development of the transistor led to a Nobel Prize. By the late 1950s, his company, the Shockley Transistor Corporation, was part of a rapidly growing industry created as a direct result of his contributions to the field. Shockley shared the 1956 Nobel Prize in physics with John Bardeen and Walter Brattain, both of whom collaborated with him on developing the point contact transistor.

In 1963, Shockley embarked on a new career, accepting an appointment at Stanford University as its first Alexander M. Poniatoff Professor of Engineering and Applied Science. Here he became interested in **genetics** and the origins of human intelligence, in particular, the relationship between race and the Intelligence Quotient (IQ).

Shockley remained at Stanford until retirement in 1975, when he was appointed Emeritus Professor of Electrical Engineering. He died in San Francisco on August 12, 1989, of cancer.

See also Eugenics

Shotgun Method

The shotgun method (also known as shotgun **cloning**) is a method in cloning genomic **DNA**. It involves taking the DNA to be cloned and cutting it either using a restriction enzyme or randomly using a physical method to smash the DNA into small pieces. These fragments are then taken together and cloned into a vector. The original DNA can be either genomic DNA (whole **genome** shotgun cloning) or a clone such as a **YAC** (**yeast** artificial **chromosome**) that contains a large piece of genomic DNA needing to be split into fragments.

If the DNA needs to be in a certain cloning vector, but the vector can only carry small amounts of DNA, then the shotgun method can be used. More commonly, the method is used to generate small fragments of DNA for **sequencing**. DNA sequence can be generated at about 600 **bases** at a time, so if a DNA fragment of about 1100kb is cloned, then it can be sequenced in two steps, with 600 bases from each end, and a hundred base overlap. The sequencing can always be primed with known sequence from the vector and so any prior knowledge of the sequence that has been cloned is not necessary. This approach of shotgun cloning followed by DNA sequencing from both ends of the vector is called shotgun sequencing.

Shotgun sequencing was initially used to sequence small genomes such as that of the cauliflower **mosaic virus** (CMV), which is 8kb long. More recently, it has been applied to more complex genomes. Usually this involves creating a physical map and a contig (line of overlapping clones) of clones containing a large amount of DNA in a vector such as a YAC, which are then shotgun clone into smaller vectors and sequenced. However, a whole genome shotgun approach has been used to sequence the mouse, fly and human genomes by the private company Celera. This involves shotgun cloning the whole genome and sequencing the clones without creating a physical map. It is faster and cheaper than creating a physical **gene** map and sequencing clones one by one, but the reliability of reassembling all the sequences of the small fragments into one genomic sequence has been doubted. For example, a part of the fly genome was sequenced by the one-by-one approach and the whole genome shotgun method. The two sequences were compared, and showed differences. 60% of the genes were identical, 31% showed minor differences and 9% showed major differences. The whole genome shotgun method generated the sequence much more quickly, but the one-by-one approach is probably more accurate because the genes were studied in more detail.

See also Clones and cloning; Cloning vector; Genetic mapping; Sequencing

Shull, George Harrison (1874-1954)
American botanist and geneticist

Without George Shull's concept of **hybrid** vigor, the development of high-yield hybrid corn would not have been possible. Shull used maize to demonstrate that repeated self-fertilization of pure lines resulted in inferior ears of corn, but hybrid crosses between different lines produced vigorous plants. Shull originally introduced the concept as "heterosis" at a lecture in Göttingen in 1914. Although both Shull and American geneticist Edward Murray East (1879–1938) realized the potential agricultural uses of hybrid corn, Shull was responsible for urging American seed companies to develop commercial hybrid lines. The use of hybrid agricultural plant lines has dramatically increased agricultural yields, first in the United States and later around the world.

Born on a farm near North Hampton, Ohio, Shull received a largely informal and unorthodox education. In spite of this, Shull published his first article, in *American Garden,* at the age of 17. After teaching school in rural Ohio for several years, he entered Antioch College at age 23. Shull then held a series of federal positions, including appointments with the United States Bureau of Forestry, the National Herbarium in Washington, and the Bureau of Plant Industry within the Department of Agriculture. While working as a federal naturalist, Shull simultaneously began graduate studies in plant physiology at the University of Chicago where he met American geneticist and later eugenicist Charles Davenport (1866–1944). Davenport, an avid Mendelian, first introduced Shull to the potential of maize as an experimental organism. Most maize geneticists, such as East or American geneticist Rollins Emerson (1873–1947), started with maize and adopted Mendel's laws as an analytical tool to understand the **inheritance** of this agriculturally useful plant. Shull, on the other hand, began with the concepts of Mendelian inheritance and thought of maize as an experimental tool rather than as a focus of his research. Ironically, Shull's theoretical contributions to plant breeding transformed the field of American agriculture.

Although Shull is most remembered for his contributions to hybrid breeding, he also founded the journal *Genetics* in 1916. As editor of the journal for its first ten years, Shull oversaw the publication of many important early articles in **genetics** research. Shull's professional networks varied widely, including membership in the National Academy of Science, the Ecological Society of America, the American Philosophical Society, the American Association for the Advancement of Science, and the **Eugenics** Society of America. In 1917, he served as the president of the American Society of Naturalists. Most of his honors, however, such as the 1940 DeKalb Agricultural Association Medal, stemmed from his work on hybrid corn.

See also Agricultural genetics; Hybrid; Hybridization of plants; Mendelian genetics and Mendel's laws of heredity; Plant breeding and crop improvement

Sibley, Charles G. (1917-1998)
American biologist

Charles G. Sibley advanced a very controversial methodology to compare the **DNA** of differing **species**. Sibley initially attempted to determine the extent of genetic differences between bird species—but his work was ultimately expanded to characterize the differences between chimps, gorillas, and humans. In contrast to previous studies that indicated that chimps, gorillas, and humans were all equally distant from each other (each was a genetically distinct line from a common ancestor) within phylogenic classifications. Sibley's methodology indicated that humans were closer to chimps in terms of DNA or **gene** content.

Sibley initially developed a genetically based system of bird taxonomy. In conjunction with Jon Ahlquist, he pioneered a new technique to create DNA-DNA hybridizations in order to determine the evolutionary relationships among birds. Sibley's objective was ultimately to devise a methodology to determine how similar the DNA of one species is to that of another species. The more similar the DNA, the more evolutionary related are the two species (i.e, more closely related they are to a common ancestor).

Sibley obtained DNA from a species and separated the **double helix** into its complementary strands by applying heat. The strands were then labeled with a radioactive isotope and combined with strands from another species. The mixture was incubated at 140° Fahrenheit for 120 hours causing complementary strands to bond forming DNA **hybrid** molecules. Sibley's hypothesis depended on measuring how much heat was necessary to separate the DNA hybrids and hypothesized that that measure of heat determined how closely related the two species of birds were. DNA molecules from the same species had the highest melting temperature. In addition, dissimilar strands melt at a lower temperature because the less bonding occurs due to lack of complementary nitrogenous **bases**. The lower the melting temperature required to separate the hybrid, the less similarity existed between the DNA of the two species, and the greater the measure of genetic distance.

Sibley and Ahlquist compared 26,000 DNA hybrids from 1700 species of birds and constructed a genetically based **phylogeny** from their findings.

Sibley's theories were ultimately disproved because he assumed equal rates of change among lineages when he constructed the phylogeny; however, time between generations of birds varies from species to species and thus, affects the rate of DNA **evolution**. In addition, the distance measures from the data were pooled together and the calculated statistics did not account for variation in the data. In effect, his classification system was argued to be mathematically unreliable.

Sibley was born in Fresno, California. He earned his Ph.D. in zoology from the University of California, Berkeley.

See also Anneal; Bases; Transcription; Translation

SICKLE CELL ANEMIA

Sickle **cell** anemia is an inherited blood disorder that arises from a single **amino acid** substitution in one of the component proteins of hemoglobin. The component protein, or globin, that contains the substitution is defective. Hemoglobin molecules constructed with such proteins have a tendency to stick to one another, forming strands of hemoglobin within the red blood cells. The cells that contain these strands become stiff and elongated—that is, sickle shaped.

Normal hemoglobin is composed of a heme molecule and two pairs of proteins called globins. Humans have the genes to create six different types of globins—alpha, beta, gamma, delta, epsilon, and zeta—but do not use all of them at once. Which genes are expressed depends on the stage of development: embryonic, fetal, or adult. Virtually all of the hemoglobin produced in humans from ages 2-3 months onward contains a pair of alpha-globin and beta-globin molecules.

A change, or **mutation**, in a **gene** can alter the formation or function of its product. In the case of sickle cell hemoglobin, the gene that carries the blueprint for beta-globin has a minute alteration that makes it different from the normal gene. This mutation affects a single nucleic acid along the entire **DNA** strand that makes up the beta-globin gene. (Nucleic acids are the chemicals that make up **deoxyribonucleic acid**, known more familiarly as DNA.) Specifically, the nucleic acid, adenine, is replaced by a different nucleic acid called thymine.

Because of this seemingly slight mutation, called a **point mutation**, the finished beta-globin molecule has an amino acid substitution: valine occupies the spot normally taken by glutamic acid. (Amino acids are the building blocks of all proteins.) This substitution creates a beta-globin molecule—and eventually a hemoglobin molecule—that does not function normally.

Sickle-shaped cells—also called sickle cells—die much more rapidly than normal red blood cells, and the body cannot create replacements fast enough. Anemia develops due to the chronic shortage of red blood cells. Further complications arise because sickle cells do not fit well through small blood vessels, and can become trapped. The trapped sickle cells form blockages that prevent oxygenated blood from reaching associated tissues and organs. Considerable pain results in addition to damage to the tissues and organs. This damage can lead to serious complications, including stroke and an impaired immune system. Sickle cell anemia primarily affects people with African, Mediterranean, Middle Eastern, and Indian ancestry. In the United States, African Americans are particularly affected.

Genes are inherited in pairs, one copy from each parent. Therefore, each person has two copies of the gene that makes beta-globin. As long as a person inherits one normal beta-globin gene, the body can produce sufficient quantities of normal beta-globin. A person who inherits a copy each of the normal and abnormal beta-globin genes is referred to as a **carrier** of the sickle cell trait. Generally, carriers do not have symptoms, but their red blood cells contain some hemoglobin S.

A child who inherits the sickle cell trait from both parents—a 25% possibility if both parents are carriers—will develop sickle cell anemia. Sickle cell anemia is characterized by the formation of stiff and elongated red blood cells, called sickle cells. These cells have a decreased life span in comparison to normal red blood cells. Normal red blood cells survive for approximately 120 days in the bloodstream; sickle cells last only 10-12 days. As a result, the bloodstream is chronically short of red blood cells and the affected individual develops anemia.

However, the severity of the symptoms cannot be predicted based solely on the genetic **inheritance**. Some individuals with sickle cell anemia develop health- or life-threatening problems in infancy, but others may have only mild symptoms throughout their lives. For example, genetic factors, such as the continued production of fetal hemoglobin after birth, can modify the course of the disease. Fetal hemoglobin contains gamma-globin in place of beta-globin; if enough of it is produced, the potential interactions between hemoglobin S molecules are reduced.

Worldwide, millions of people carry the sickle cell trait. Individuals whose ancestors lived in sub-Saharan Africa, the Middle East, India, or the Mediterranean region are the most likely to have the trait. The areas of the world associated with the sickle cell trait are also strongly affected by malaria, a disease caused by blood-borne parasites transmitted through mosquito bites. According to a widely accepted theory, the genetic mutation associated with the sickle cell trait occurred thousands of years ago. Coincidentally, this mutation increased the likelihood that carriers would survive malaria outbreaks. Survivors then passed the mutation on to their offspring, and the trait became established throughout areas where malaria was common.

Although modern medicine offers drug therapies for malaria, the sickle cell trait endures. Approximately 2 million Americans are carriers of the sickle cell trait. Individuals who have African ancestry are particularly affected; one in 12 African Americans are carriers. An additional 72,000 Americans have sickle cell anemia, meaning they have inherited the trait from both parents. Among African Americans, approximately one in every 500 babies is diagnosed with sickle cell anemia. Hispanic Americans are also heavily affected; sickle cell anemia occurs in one of every 1,000-1,400 births. Worldwide, it has been estimated that 250,000 children are born each year with sickle cell anemia.

Symptoms typically appear during the first year or two of life, if the diagnosis has not been made at or before birth. However, some individuals do not develop symptoms until adulthood and may not be aware that they have the genetic inheritance for sickle cell anemia.

To confirm a diagnosis of the sickle cell trait or sickle cell anemia, another laboratory test called gel electrophoresis is performed. This test uses an electric field applied across a slab of gel-like material to separate protein molecules based on their size, shape, or electrical charge. Although hemoglobin S (sickle) and hemoglobin A (normal) differ by only one amino acid, they can be clearly separated using gel electrophoresis. If both types of hemoglobin are identified, the individual is a carrier of the sickle cell trait; if only hemoglobin S is present, the person most likely has sickle cell anemia.

The sickle cell trait is a genetically linked, inherited condition. Inheritance cannot be prevented, but it may be predicted. Screening is recommended for individuals in high-risk populations; in the United States, African Americans and Hispanic Americans have the highest risk of being carriers.

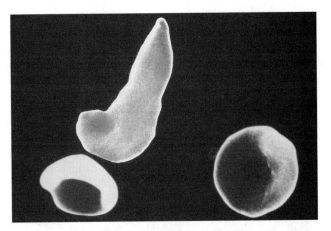

Scanning electron micrograph (SEM) showing healthy red blood cells (round) and a diseased sickle-shaped cell. *Photograph by Dr. Gopal Murti. National Audubon Society Collection/Photo Researchers, Inc. Reproduced by permission.*

The gel electrophoresis test is also used as a screening method for identifying the sickle cell trait in newborns. More than 40 states screen newborns in order to identify carriers and individuals who have inherited the trait from both parents.

Early identification of sickle cell anemia can prevent many problems. The highest death rates occur during the first year of life due to infection, aplastic anemia, and acute chest syndrome. If anticipated, steps can be taken to avert these crises. With regard to long-term treatment, prevention of complications remains a main goal. Sickle cell anemia cannot be cured—other than through a bone marrow transplant—but treatments are available for symptoms.

Screening at birth offers the opportunity for early intervention; more than 40 states include sickle cell screening as part of the usual battery of blood tests done for newborns. Pregnant women and couples planning to have children may also wish to be screened to determine their carrier status. Carriers have a 50% chance of passing the trait to their offspring. Children born to two carriers have a 25% chance of inheriting the trait from both parents and having sickle cell anemia. Carriers may consider genetic counseling to assess any risks to their offspring. The sickle cell trait can also be identified through prenatal testing; specifically through use of amniotic fluid testing or **chorionic villus sampling**.

See also Genetic testing and screening; Point mutation

SILENT MUTATION

A silent **mutation** is any change in **DNA** or **RNA** (mutation) that does not result in a change in the function of the protein. With silent mutations, the function of the protein coded for by the **gene** undergoing the mutational event remains unaltered.

In structural genes, a silent mutation can result from a change in the base sequence of DNA (or, in prokaryotes, RNA) that does not result in a change in the **amino acid** sequence of the protein for which the gene carries instructions. The genetic code is termed a **degenerate code** because more than one base sequence (and the **codons** produced from them) may code for a particular amino acid. Accordingly, a change from one sequence to another does not result in a change in the amino acid sequence of the protein. There are, for example, six different **nucleotide** sequences and codons that direct the insertion of the amino acid arginine (Arg) during **translation**. In RNA comprised codons, the base sequences are composed of the nitrogenous **bases** adenine (A), uracil (U), cytosine (C), and guanine (G). Codon sequences CGU (cytosine-guanine-uracil), CGC, CGA, CGG, AGA, and AGG all instruct (code for) the insertion of arginine into the growing amino acid chain during **protein synthesis**. Accordingly, a mutational event that results in the substitution of cytosine (C) for uracil (U) in the third position of a CGU codon does not result in a structural or functional change in the protein for which it codes.

Silent mutations can also occur even if there is a change in the amino acid sequence of a protein, as long as the substitution has no effect on the functioning of the protein. Some regions of a protein are more critical to its function than other regions. Mutations that result in the substitution of amino acids in non-critical areas of the protein molecule may have no discernable effect on the shape or function of that protein with regard to its biochemical reactions. Silent mutations can also result from a change in nucleotide or codon sequence that is reversed by another mutation. Such reversing mutations are termed suppressor mutations.

Regulatory genes may also exhibit silent mutations. In regulatory genes, a silent mutation results from a change in the DNA or RNA base sequence that does not change the regulatory function or activity of a regulatory gene.

At the level of the organism, silent mutations are those mutations that result in no observable changes in organism, even when the organism can exist in a **haploid** (half the genetic complement) state or, as with humans, when the organism may be homozygous for the mutation (e.g., carries the mutation on both alleles situated on homologous (corresponding) **chromosome** pairs).

See also Base pairing (bp); Bases; Deletions; DNA repair; Frame shifts; Point mutation; Transcription

SINGER, MAXINE (1931-)

American biochemist and geneticist

Maxine Singer, a leading scientist in the field of human **genetics**, is also a staunch advocate of responsible use of biochemical genetics research. During the height of the controversy in the 1970s over the use of recombinant **deoxyribonucleic acid** (**DNA**) techniques to alter genetic characteristics, Singer advocated a cautious approach. She helped develop guidelines to balance calls for unfettered genetics research as a means of making medically valuable discoveries with demands for restrictions on research to protect the public from possible harm. After the DNA controversy waned, Singer continued to

contribute to the field of genetics, researching cures for **cancer, hemophilia,** and other diseases related to genetics.

Singer was born in New York City, to Hyman Frank, an attorney, and Henrietta (Perlowitz) Frank, a hospital admissions officer, children's camp director, and model. Singer received her B.A. from Swarthmore College in Pennsylvania in 1952, and earned her Ph.D. in biochemistry from Yale in 1957. From 1956 to 1958, she worked as a U.S. Public Health Service postdoctoral fellow at National Institute for Arthritis, Metabolism and Digestive Diseases (NIAMD), National Institutes of Health (NIH), in Bethesda, Maryland. Singer then became a research chemist on the staff of the section on **enzymes** and cellular biochemistry from 1958 to 1974. There she conducted DNA research on tumor-causing viruses as well as on **ribonucleic acid** (**RNA**). In the early 1970s, Singer also served as a visiting scientist with the department of genetics of the Weizman Institute of Science in Rehovot, Israel.

While Singer was working at NIH, scientists learned how to take DNA fragments from one organism and insert them into the living cells of another. This "recombinant DNA "could direct the production of proteins in the foreign organism as if the DNA was still in its original home. This technique had the potential of creating completely new types of organisms, a cause of concern for some in the scientific and general community. Recombinant DNA technologies also bring unprecedented opportunities to discover cures for serious diseases, to develop new crops, and otherwise to benefit humanity.

In 1972, one of Singer's colleagues and personal friends, **Paul Berg** of Stanford University, was the first to create **recombinant DNA molecules**. He later voluntarily stopped conducting related experiments involving DNA manipulation in the genes of tumor-causing viruses because of some scientists' fears that a **virus** of unknown properties might escape from the laboratory and spread into the general population.

Although Berg's self-restraint was significant, the catalyst for the debate over gene-splicing was the 1973 Gordon Conference, an annual high-level research meeting. Singer, who was co-chair of the event, was approached by several nucleic acid scientists with the suggestion that the conference include consideration of safety issues. Singer agreed. She opened the discussion with an acknowledgment that DNA manipulation could assist in battling health problems, yet such experimentation brought to bear a number of moral and ethical concerns.

The scientists present decided, by ballot, to send a public letter about the safety risks of recombinant DNA research to the president of the National Academy of Sciences, and asked *Science* magazine to publish it. Singer and her co-chair, Dieter Söll of Yale University, wrote the letter warning that organisms of an unpredictable nature could result from the new technique, and suggested that the National Academy of Sciences study the problem and recommend guidelines. Concern generated by this letter led to another meeting at the Asilomer Conference Center in Pacific Grove, California, where a debate ensued. Such proceedings—to consider the ethical issues arising from the new DNA research—were unprecedented in the scientific community. Immediately after the Asilomer Conference concluded, a NIH committee began formulating guidelines for recombinant DNA research.

In helping develop the guidelines, Singer advocated a careful analytic approach. In 1976, she presented four principles to the committee to be used in drafting the guidelines. Singer advised that certain experiments posed such serious hazards that they should be banned altogether; that experiments with lesser or no potential hazards should be permitted if their benefits are unobtainable through conventional methods and if they are properly safeguarded; that the more risk in an experiment, the stricter the safeguards should be; and that the guidelines should be reviewed annually.

During her career, Singer has also served on the editorial board of *Science* magazine and has contributed numerous articles. In her writing for that publication about recombinant DNA research, she stressed the benefits to humanity that recombinant DNA techniques could bring, especially in increasing the understanding of serious and incurable disease. After the NIH guidelines were implemented, she told *Science* readers that "under the Guidelines work has proceeded safely and research accomplishments have been spectacular." By 1980, when public near-hysteria had waned, Singer called for a "celebration" of the progress in molecular genetics. In *Science* she wrote: "The manufacture of important biological agents like insulin and interferon by recombinant DNA procedures," as well as the failure of any "novel hazards" to emerge, was evidence of the value of the cautious continuation of DNA research.

In 1974, Singer accepted a new position at NIH as chief of the Section of Nucleic Acid Enzymology, Division of Cancer Biology and Diagnosis (DCBD) at the National Cancer Institute in Bethesda, Maryland. In 1980, Singer became chief of the DCBD's Laboratory of Biochemistry. She held this post until 1988, when she became president of the Carnegie Institution in Washington, D.C. Singer remains affiliated with the National Cancer Institute, however, as scientist emeritus, where she continues her research in human genetics.

In addition to her laboratory research, Singer has devoted considerable time and energy to other scientific and professional pursuits. In 1981, she taught in the biochemistry department at the University of California at Berkeley. A prolific writer, she has issued more than one hundred books, articles, and papers. Most are highly technical, including numerous articles published in scientific journals. Singer also compiled a graduate-level textbook with Paul Berg on molecular genetics called *Genes and Genomes: A Changing Perspective.* Singer's peers gave the work high praise for its clear presentation of difficult concepts.

Singer has also written extensively on less technical aspects of science. Singer and Berg authored a book for laypeople on genetic engineering, and she continued to promote the benefits of recombinant DNA techniques and battle public suspicion and fear long after the controversy peaked in the 1970s. In the early 1990s, for example, Singer issued an article encouraging the public to try the first genetically engineered food to reach American supermarket shelves. In describing the harmlessness of the "Flavr Savr" tomato, she

Maxine Singer. *The Bettmann Archive/Newsphotos, Inc. Reproduced by permission.*

decried public objections that eating it was dangerous, unnatural, or immoral to readers of the *Asbury Park Press.* Pointing out that "almost all the foods we eat are the product of previous genetic engineering by cross-breeding," Singer said that the small amount of extra DNA in the tomato would be destroyed in the digestive tract, and that people already consume the DNA present in the other foods in their diets. Moreover, she said the decision to eat a genetically altered tomato did not reduce her admiration for nature's creations.

In addition to her writing and lecturing, Singer has served on numerous advisory boards in the United States and abroad, including science institutes in Naples, Italy, Bangkok, Thailand, and Rehovot, Israel. She also has served on an advisory board to the Pope and as a consultant to the Committee on Human Values of the National Conference of Catholic Bishops. She worked on a Yale committee that investigated the university's South African investments, and serves on Johnson and Johnson's Board of Directors. Concerned about the quality of science education in the United States, she started First Light, a science program for inner-city children.

Singer travels extensively and maintains long workweeks to accommodate all her activities. She married Daniel Singer in 1952; the couple has four children. Singer is the recipient of more than forty honors and awards, including some ten honorary doctor of science degrees and numerous commendations from NIH.

See also Cancer genetics; DNA synthesis; Ethical issues of modern genetics

SINGLE NUCLEOTIDE POLYMORPHISM

A single **nucleotide polymorphism** (SNP) is a single **DNA** nucleotide that is polymorphic within a **species**. Each SNP has two alleles, represented by the nucleotide at that site (A,C,G or T). Some SNPs are in genes, and they can change the **amino acid** in the protein that the **gene** codes for, but most are in non-coding DNA.

SNPs can be identified in several different ways. Any **mutation** detection technique, such as **denaturing gradient gel electrophoresis** (DGGE) or denaturing high-performance liquid chromotography (DHPLC) can be used to identify sequence changes, which can be confirmed as polymorphic after observing that sequence change in DNA from several unrelated individuals. DHPLC was used to isolate SNPs on the human **Y chromosome**. Many SNPs in the human **genome** have been identified by comparison of clone sequence overlaps. If a certain sequence is covered by two or more probes and those clones are sequenced, then any sequence differences are likely to be SNPs, although they could also be **cloning** or **sequencing** artefact. Confirming that the sequence change is indeed a SNP involves developing a **polymerase chain reaction** (PCR)-based assay for each specific sequence change and identifying the two alleles in a sample of individuals. Ideally, Mendelian **inheritance** should also be shown by analysis the SNP in a family. These tests would also rule out any sequence variation cause by **duplication** of very similar sequences so individuals would appear to have two alleles at a single position.

Determining the allele at a SNP is commonly called typing a SNP. This can be done by several methods, but most involve amplifying the DNA containing the SNP of interest by PCR, then testing that amplified DNA. The technique used to identify the SNP such as DGGE and DHPLC can also be used to type the SNP. More commonly, a specific assay is developed involving a restriction enzyme. If the SNP alters a restriction enzyme cut site, the two alleles can be distinguished by whether the enzyme cuts or not, which can be seen by gel electrophoresis. If a restriction site is not altered, an artificial restriction enzyme site can often be created by amplifying with a PCR primer with a base different from the original DNA sequence. This causes the amplified DNA to have a restriction site, which can be tested in the manner described above. Alleles at a SNP tested by restriction enzyme digestion are sometimes called + and -, defined by whether the enzyme cuts or not, instead of the type of nucleotide at the site.

Instead of amplifying the DNA and testing for an allele at a SNP, it is possible to amplify the DNA only if the DNA contains a specific allele. This is called allele specific **polymerase** chain reaction (AS-PCR). Normally the DNA is detected by analysis by gel electrophoresis after amplification, but recently technology has allowed this detection to be automated by detecting the DNA amplification product being produced during the amplification. This technology allows a fluorescent molecule to be produced during amplification, and measurement of fluorescence allows the measurement of how much DNA has been amplified.

SNPs are considered to be useful genetic markers in the human genome, because they are so common—by comparing two random sequences a SMP is found about every thousand **bases**, with 1.4 million found as of 2001. Because of their high density, they may be useful in mapping genes which have an influence in many common human diseases.

See also Allele frequency; Bases; Denaturing gradient gel electrophoresis; DNA sequence; Heteroduplex analysis; Polymerase chain reaction (PCR); Restriction enzymes; Sequencing

SINGLE-GENE DISORDERS

Single-gene disorders, also called monogenic disorders, are defects caused by a mutant allele of a single **gene** that results in the functional loss of a protein. Single-gene disorders are generally easy to trace because they follow the rules of classical Medelian **genetics**. Single-gene disorders can be autosomal or X-linked, dominant or recessive and usually become evident by studying a particular family's pedigree. Common single-gene disorders include **sickle cell anemia**, familial hypercholesterolemia, Duchenne muscular dystrophy, and **cystic fibrosis**. Some speech and language disorders are also caused by the interference of a single gene.

Mutations in **DNA (deoxyribonucleic acid)** arise because of a spontaneous chemical change that results in a substitution, deletion, or insertion of a **nucleotide** base pair. As a result, many mutations cause physical or mental disorders. Sickle **cell** anemia is caused by a nucleotide base substitution that results in a defective hemoglobin molecule. Because sickle cell anemia is an autosomal recessive disorder, individuals who have the disease must have two abnormal copies of the gene, one from each parent. Familial hypercholesterolemia (FHC) is due to a defective gene on **chromosome** 19 that codes for an abnormal receptor protein unable to mediate the uptake of cholesterol into cells. As a result, there is a cholesterol buildup in the blood that may eventually lead to arteriosclerosis or coronary heart disease. FHC is an autosomal dominant disorder where only one abnormal allele is necessary to have the disease. Duchenne muscular dystrophy is due to an X-linked recessive, abnormal gene that is much more frequent among males than females. Males only need one mutant allele to have the disease because they only have one **X chromosome**; females do not express the disease unless both X chromosomes are affected. The disease is caused by a lack of dystrophin, a protein found in the membranes of muscle fibers. Hypophosphatemia is an X-linked dominant disorder characterized by low levels of phosphorus in the blood. Because of a defective protein, the kidneys are unable to reabsorb phosphorus or calcium resulting in abnormal, fragile bones. Hypophosphatemia is a relatively rare disorder that affects both males and females.

In most single-gene disorders, adding a functional protein to affected cells would cure the disorder. Currently, doctors attempt to do this through **gene therapy**. Theoretically, if doctors can locate the mutant gene, they can replace it with a

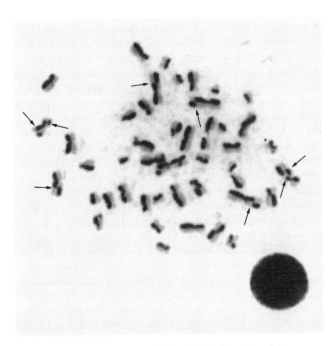

Metaphase stage cell stained to differentiate between the sister chromatids of each chromosome (one dark and one light staining). Arrows indicate sister chromatid exchanges—an exchange between the chromatid arms results in a harlequin appearance of each chromatid which is partly black and partly light. *Courtesy of Dr. Constance Stein.*

conducted a series of investigations into the physical and genetic characteristics of a **bacteriophage** called Phi X 174. These breakthrough studies illuminated the viral genetic processes. Sinsheimer and his colleagues also succeeded for the first time in isolating, purifying and synthetically replicating viral **DNA**.

In 1977, Sinsheimer left California Institute of Technology to become a chancellor of the University of California, Santa Cruz. One reason the position of chancellor appealed to him was that it provided a forum to address his concerns that had developed concerning the social implications and potential hazards of recombinant DNA technology and **cloning** methods. Sinsheimer was one of the first scientists to question the potential of molecular biology and the ethical implications of the developing technologies. He had become committed to promoting scientific literacy among non-scientists.

His early years at Santa Cruz were challenging. During his tenure the university re-established itself as a seat of research and academic excellence. Some of Sinsheimer's accomplishments included the establishment of the Kleck telescope, the establishment of programs in agroecology, applied economics, seismological studies, and a major in computer engineering.

Sinsheimer also participated fundamentally in the genesis of the **Human Genome Project**. In May of 1985, Sinsheimer organized a conference at Santa Cruz to consider the benefits of **sequencing** the human **genome**. From these and other such deliberations arose the Human Genome Project.

See also Ethical issues of modern genetics

SISTER CHROMATID EXCHANGE

During the S-phase of the **cell** cycle, **DNA** is replicated and each **chromosome** is present in a duplicated state with the two genetically identical chromatids joined together at the centromere. These two sister chromatids are readily apparent in late prophase or early metaphase of **mitosis**.

Sister chromatid exchange (SCE) is the process wherein the two sister chromatids break and rejoin with one another, physically switching positions on the chromosome. Because the exchanges occur with tremendous precision with respect to the DNA sequence, and the sister chromatids are genetically identical, no information is altered during the exchange. Such exchanges are natural events during cell **replication** with each cell typically undergoing three to four SCEs during each replication cycle.

SCEs can be visualized by growing cells in a medium, which contains the DNA base analog, bromodeoxyuridine (BrdUrd). BrdUrd closely resembles thymidine, and gets incorporated into the growing DNA strand during replication. Because replication of DNA is semiconservative, all of the BrdUrd is taken up into the newly synthesized strand, with the template strands remaining free of BrdUrd. During a second round of growth in BrdUrd medium, the two sister chromatids differ in the amount of BrdUrd present. The sister chromatid which has the original template strand of DNA has one strand

normal gene that will produce functional proteins, curing single-gene disorders such as cystic fibrosis (CF). Normal CF genes are manufactured in the laboratory and delivered into defective cells by microscopic fat capsules called liposomes. Scientists also use viruses as vehicles for transporting healthy genes into affected cells. However, most gene therapies are still in the experimental stage as scientists continue to work towards a cure for single-gene disorders.

The underlying genetics of speech and language are believed to be due to several genes. However, one such speech and language disorder has been linked to the disruption of a single gene. This gene abnormality exists on chromosome seven, where a gene linked to autism also exists. Autism is a polygenic disorder that affects the brain's ability to communicate with others and react well with the environment. Polygenic disorders involve the combination of alleles in more than one gene. Although these disorders are also inherited, they are far more complex than single-gene disorders.

See also Duchenne muscular dystrophy; Gene therapy

SINSHEIMER, ROBERT LOUIS (1920-)
American molecular biologist and biophysicist

Robert Sinsheimer was born in Washington, the first scientist in is family. He was professor of biophysics at California Institute of Technology from 1957 to 1977, and was Chairman of the Division of Biology from 1968 to 1977. During this period he of normal DNA and one which contains BrdUrd. The other sister chromatid has BrdUrd incorporated into both strands. With careful chemical and ultraviolet light treatment, the two sister chromatids can be distinguished from each other. SCEs can then be identified by differences in staining patterns.

SCE is useful for **genetic toxicology** studies as chemicals, that cause mutations, will often increase the frequency of SCEs. SCE may be used as a diagnostic test for the rare genetic disease called Bloom syndrome.

See also Mitosis; Replication

SITE • *see* LOCUS (LOCI)

SLOW VIRUSES

Historically, the term "slow **virus** infections" was coined for a poorly defined group of seemingly viral diseases which were later found to be caused by several quite different conventional viruses, also unconventional infectious agents. They nevertheless shared the properties of causing diseases with long incubation periods and a protracted course of illness, affecting largely the central nervous and/or the lymph system and usually culminating in death. The slow virus concept was first introduced by the Icelandic physician Bjorn Sigurdsson (1913–1959) in 1954. He and his co-workers had made pioneering studies on slow diseases in sheep including maedi-visna and scrapie. Maedi is a slowly progressive interstitial pneumonia of adult sheep while visna is a slow, progressive encephalomyelitis and the same virus, belonging, to the lentivirus subgroup of retroviruses, was found to be responsible for both conditions.

Since the original isolation of the maedi-visna virus, concern with and knowledge of slow viral infections, both in animals and in humans, has grown. Research on sheep lentiviruses and their pathogenesis has continued to this day and received an important impetus in the 1980s with the recognition of the devastating condition in humans known as acquired immunodeficiency syndrome (**AIDS**). AIDS shared many of the attributes of slow virus infections in animals and led virologists to suspect, then to identify, the lentivirus causing AIDS: the human immunodeficiency virus or HIV. Questions posed by Bjorn Sigurdsson's work on maedi-visna also became the central pathogenic questions of HIV disease. For example: how and where does HIV persist despite an initially robust and long-sustained immune response? How does HIV actually destroy the tissues it infects? Why do these events unfold so slowly? Final answers to all these questions have still not been found and there is much research still to be done on the lentiviruses but Sigurdsson's contribution to HIV research through the study of maedi-visna should be acknowledged.

Other slow virus infections of humans due by conventional viruses include progressive multifocal leukoencephalopathy (PML) caused by the JC papovavirus. This is an opportunistic infection in hosts that have defective **cell** mediated immunity and the majority of human cases now occur in HIV 1 infected individuals. Patients present with progressive multifocal signs including visual loss, aphasia (difficulty speaking), seizures, dementia, personality changes, gait problems and less commonly, cerebellar, brain stem, and spinal cord features. Death occurs within weeks to months of clinical onset. Subacute sclerosing panencephalitis (SSPE), another slow infection, has been identified as a rare consequence of chronic persistant infection by the measles (rubella) virus, causeing an insidious syndrome of behavioral changes in young children. Patients develop motor abnormalities, in particular myoclonic jerks. and ultimately become mute, quadriplegic, and in rigid stupor. SSPE is found worldwide with a frequency of one case per million per year. Progressive rubella panencephalitis is another very rare slow virus infection of children and young people caused by the same virus. Most patients have a history of congenital or acquired rubella and the clinical course is more protracted than in SSPE with progressive neurologic deficit occurring over several years. A third slow virus of humans that has had some publicity in recent years is the human T-cell leukaemia virus (HTLV) types 1 and 2 which are associated with adult T-cell leukaemia. It was initially thought that the causative agent of AIDS was related to HTLV though it later became clear that whereas HTLV 1 and 2 are both oncogenic ("cancer producing") retroviruses, HIV belongs to the lentivirus sub-group.

An unconventional agent causing slow infections has now been identified as a non-viral "proteinaceous, infectious" agent, or prion. **Prions** give rise to the group of diseases now called transmissible spongiform encephalopathies. In animals these include scrapie in sheep and bovine spongiform encephalopathy (BSE) in cows. Human prion infections include rare dementing diseases like kuru, Creutzfeldt-Jakob disease (CJD), Gerstmann-Straussler-Scheinker (GSS) syndrome and fatal familial insomnia (FFI). The prion agent is replicated without provoking any antibody response, appears not to have any recognizable nucleic acid component and is resistant to conventional inactivation techniques for infectious agents. Current evidence suggests that the prion protein is an abnormal isoform of a normal host encoded protein known as PrP, which is coded on the short arm of **chromosome** 20. Prions appear to "replicate" by a novel form of protein-protein information transfer, the abnormal PrP seemingly inducing the normal protein to undergo a structural change into the abnormal form. Most pathological changes observed with transmissible spongiform encephalopathies are confined to the brain; however, scrapie-induced disorders of the pancreas have also been described. The neuropathology sometimes shows a dramatic spongiform disruption of brain tissue but may also be subtle and non-characteristic, even at the terminal stages of the disease. In the latter cases, diagnosis has to rely on features like clinical signs, transmissibility, detection of abnormal PrP or identification of mutations in the PrP **gene**.

In humans, prion diseases may be sporadic, acquired or inherited. Iatrogenic transmissions of the prion have occurred following medical procedures such as pituitary growth hormone injections, where the hormone source was contaminated with prion tissue, or corneal transplants, where a patient accidentally received an infected cornea. The first recognized human prion disease was kuru, which emerged among the South Fore people of New Guinea and is now generally thought to have been transmitted by the practice of ritual cannibalism. CJD is today the most common human prion disease occuring worldwide with a frequency of about one per million per year. The peak incidence occurs in older people between the ages 55 and 65 although recently a "new variant" has emerged in the U.K., which affects individuals at a much earlier age. It is widely believed that the new variant CJD is closely related to the variety causing BSE in cattle and may be contracted by the ingestion of infected beef.

Inherited prion disease can arise from specific point mutations in the PrP gene. Perhaps 10–15% of CJD cases are familial with an autosomal dominant pattern of **inheritance**. Gerstmann-Straussler-Scheinker syndrome is another rare familial condition that is vertically transmitted in an apparently autosomal dominant way. As with other prion diseases it can be horizontally transmitted to non-human primates and rodents through intracerebral inoculation of brain homogenates from patients with the disease. The exact incidence of the syndrome is unknown but is estimated to be between one and ten per hundred million per year and the condition appears to be an allelic variant of familial Creutzfeldt-Jakob disease. Fatal familial insomnia is the third most common inherited human prion disease. The region of the brain most affected in this condition is the thalamus which monitors sleep patterns. The symptoms of the disease are characterized by progressive insomnia and, as with other prion diseases, eventual motor signs.

See also Viral genetics

SMITH, HAMILTON O. (1931-)
American molecular biologist

Hamilton O. Smith shared the 1978 Nobel Prize in physiology or medicine with fellow biologists **Werner Arber** and **Daniel Nathans** for the set of linked discoveries that started off the boom in **biotechnology**. Because of these discoveries, researchers can more easily elucidate the structure and coding of **deoxyribonucleic acid** (**DNA**) molecules (the basic genetic map of an organism), and they hope to correct many genetic illnesses in the future. It is also possible to design new organisms, a controversial but potentially beneficial technology. Smith purified and explained the activity of the first restriction enzyme, which became the principal tool used by genetic engineers to selectively cut up DNA. (Arber had linked restriction and modification to DNA, and predicted the existence of **restriction enzymes**. Nathans, under Smith's encouragement at Johns Hopkins, developed techniques that enabled their practical use.)

Hamilton Othanel Smith was born in New York City, to Bunnie (Othanel) Smith and Tommie Harkey Smith. His father, an assistant professor of education at the University of Florida, finished his Ph.D. at Columbia University in 1937 and took a new teaching job at the University of Illinois. The family then moved to Urbana, where Smith and his brother attended public school. It was here that Smith's interest in science began. He and his brother even equipped a laboratory in their basement with money from their paper routes. Smith graduated from University High School in three years, enrolling at a local university in1948.

Smith came to the study of **genetics** by way of medicine. Initially a mathematics major at the University of Illinois, he transferred to the University of California at Berkeley in 1950 to study biology and graduated with a bachelor's degree in 1952. He obtained a medical degree from the Johns Hopkins School of Medicine in 1956. The following year, Smith married Elizabeth Anne Bolton, a nurse. They eventually had four sons and a daughter. During the years 1956 to 1962, he held various posts, including an internship at Washington University in St. Louis, Missouri, a two-year Navy stint in San Diego, California, and a residency at Henry Ford Hospital in Detroit, Michigan. He gradually taught himself genetics and molecular biology in his spare time. In 1962 he began a research career at the University of Michigan on a postdoctoral fellowship from the National Institutes of Health, before finally returning to Johns Hopkins in 1965 as a research associate in the microbiology department. He was named a full professor of microbiology in 1973, and professor of molecular biology and genetics in 1981. In 1975, Smith was awarded a Guggenheim Fellowship for a year of study at the University of Zurich in Switzerland.

DNA, the genetic material in all cells, is a long, chain-like molecule encoded along its structure for individual genes and thereby individual proteins. Each link in the chain is one of four possible nucleotides (adenine, guanine, cytosine, and thymine) arranged in varying sequences. An individual gene's function is coded by the order of the links, as a word's meaning is coded by the order of the letters. Much of the fundamental research in biotechnology was accomplished by studying DNA from both **bacteria** and the viruses—also called bacteriophages—that can infect bacteria. If a bacterium can break up invading viral DNA without harming its own DNA, it destroys the **virus** and thus resists infection. Bacteria that resist infection chemically modify parts of their DNA, usually with a methylating enzyme called a methylase, so it cannot be cut. Then they damage the viral DNA with a specific restriction enzyme (also called a restriction endonuclease). The bacterial DNA remains undamaged because of its chemical alteration, while the viral DNA is cut apart by the endonuclease.

Restriction **enzymes** are classified as Class I or Class II. Class I enzymes recognize specific DNA sequences, but they do not cut DNA only at those locations. Each Class II endonuclease, however, cuts only between two specific sequences of nucleotides, and no others. The mechanism of this type of bacterial resistance is called restriction-modification and involves the matched set of Class II endonuclease—methylase enzymes. Restriction refers to cutting DNA at a specific location, and modification refers to the enzyme-driven chemical change in the DNA which prevents such cutting. (Of course, viruses eventually can evolve modifications of their own which prevent their destruction by restriction enzymes; an appropriately modified virus can thus infect a formerly resistant strain of bacteria.)

Smith had been interested in bacterial genetics for many years, and spent 1966 working in Geneva, Switzerland, with Werner Arber. After his return to the U.S., Smith purified the first Type II restriction endonuclease, which he obtained from the bacterium *Hemophilus influenzae,* and identified the **nucleotide** sequence which the enzyme would cut. He gave a supply of the enzyme to Daniel Nathans, who used it in his own work. The three men eventually won the 1978 Nobel Prize. The presenter of the prize noted that Smith proved Arbor's hypothesis about restriction enzymes, pointing the way for future research.

The exacting specificity of Class II restriction enzymes makes them useful because biotechnologists can now cut DNA apart selectively. Then, they can add and subtract specific nucleotides, and reproducibly weld (recombine) the links back together in a new order. This new piece of DNA now codes for a different protein. The current and potential uses of these procedures are enormous. Biotechnologists can genetically engineer bacteria that produce a particular chemical; human insulin, for the treatment of diabetes is now made by such recombinant bacteria. Other bacteria have been designed to chew up oil slicks. One of the tasks that biotechnologists would like to accomplish is the eradication of genetic illness by correcting the mistaken DNA codes that cause it. Sickle-cell anemia, for example, is a life-threatening and incurable disease resulting from a mistake in two of the genes that code for the blood protein hemoglobin. Currently, although some **gene** therapies are experimental and controversial, several promising and successful human trials and treatment protocols are underway.

See also Gene therapy

SMITH, MICHAEL (1932-)

English-Canadian biochemist

Michael Smith began his professional research career in salmon physiology and endocrinology, but returned to the chemical synthesis that had been his first interest, including the chemical synthesis of **deoxyribonucleic acid (DNA)**. Smith experimented with isolating genes and invented site-directed mutagenesis, a technique for deliberately altering **gene** sequences. Smith's work was hailed as having tremendous implications for genetic studies and the understanding of how individual genes function, and already has been applied in the study of disease-producing viruses. In 1993, Smith shared the Nobel Prize in Chemistry independently with Kary Mullis. The Royal Swedish Academy of Sciences credited Smith and Mullis with having revolutionized basic research and saluted the possibilities offered by their research toward the cure of **hereditary diseases.**

Smith was born in Blackpool, England, to Rowland Smith, a market gardener, and Mary Agnes Armstead Smith, a bookkeeper who also helped with the market gardening. Smith was admitted to Arnold School, the local private secondary school, with a scholarship he earned based on his examination results (this examination was taken, at the time, by all English children when they finished their primary education). While at Arnold School, Smith became involved in scouting, which eventually led to a life-long interest in camping and other outdoor activities.

After graduating from Arnold School in 1950, Smith enrolled at the University of Manchester in order to study chemistry, realizing a natural inclination toward the sciences. He moved rapidly through school, receiving a B.Sc. in 1953, and a Ph.D. in chemistry in 1956, both sponsored by scholarship. Afterward, Smith was accepted into biochemist Har Gobind Khorana's laboratory in Vancouver, Canada. Smith's original plan in migrating to Canada was to work for a year, then return to England and work for a chemical company. However, his experience working with Khorana, who would win the Nobel Prize in 1968 for his contributions to **genetics**, changed his plans. Smith decided university research was the path he wanted to take and that British Columbia, with its natural beauty, would be his home. Smith is now a Canadian citizen.

Smith stayed with the Khorana group and moved with it in 1960 to the Institute for Enzyme Research at the University of Wisconsin. Until then, Smith's work in Canada had been in several different areas of chemical synthesis. In 1961, Smith decided it was time for a change and decided to re-locate to the West Coast. Smith accepted a position as head of the chemistry section of the Vancouver Laboratory of the Fisheries Research Board of Canada. His work there was mainly in salmon physiology and endocrinology, but he also continued to work in chemical synthesis.

In 1966, Smith entered the academic field, taking an appointment as associate professor of biochemistry and molecular biology at the University of British Columbia (UBC), and bringing with him an interest in chemically synthesized DNA (the molecule of **heredity**). Also beginning in 1966 Smith held a concurrent position as medical research associate of the Medical Research Council of Canada. He was made full professor in 1970, and has continued his teaching duties ever since. In 1986 he was asked to establish a **biotechnology** laboratory on the campus of UBC, which he has headed since that time.

Smith has taken three sabbaticals from his duties at the University of British Columbia, spending three months in 1971 at Rockefeller University in New York, one year during 1975 and 1976 at the Medical Research Council laboratory in Cambridge, and eight months in 1982 at Yale University. The middle excursion was spent in English biochemist Frederick Sanger's laboratory learning about DNA sequence determination, essential to Smith's later research.

Smith was first able to isolate genes using chemical synthesis in 1974. Slowly he developed what became known as site-directed mutagenesis, a technique that allows gene sequences to be altered deliberately. More specifically, it involves separating one strand of a piece of DNA and producing a mirror image of it. This mirror image can then be used as a probe into a gene. It can also be used with chemical enzymes—proteins that act as catalysts in biochemical reactions—that are able to cut and splice DNA in living cells. Jeffrey Fox, editor of *Bioscience,* called this process the "intellectual bombshell that triggered protein engineering," as quoted in the Toronto *Globe and Mail.* Smith's findings were published in 1978 in *Journal of Biological Chemistry.* This paper lays the foundation of the research Smith has done since. The paper concludes, "This new method of mutagenesis has considerable potential in genetic studies. Thus, it will be possible to change and define the role of regions of DNA sequence whose function is as yet incompletely understood."

Smith, in demonstrating that biological systems are chemical, has allowed scientists to tinker systematically with genes, altering properties one at a time to see what effect each alteration may have on the gene's functioning. Genes are the building blocks for countless proteins that make up skin, muscles, bone, and hormones. Changes in the expression of these proteins reveal to the scientist how his or her tinkering has altered the gene function. This process has been used specifically to study disease-producing viruses, such as those that cause **cancer**. The eventual goal is to uncover the functioning of the genes, so drugs to combat the viruses can be developed.

After being several times a nominee, Smith was awarded the Nobel Prize in Chemistry in 1993 jointly with Kary Mullis from California. Their work was not collaborative, though both dealt with biotechnology. Announcing the award, the Royal Swedish Academy of Sciences credited Smith for having "revolutionized basic research and entirely changed researchers' way of performing their experiments," as quoted in the Toronto *Globe and Mail.*The academy further said Smith's work holds great promise for the future with the "possibilities of **gene therapy**, curing hereditary diseases by specifically correcting mutated code words in the genetic material."

With the Nobel Prize proceeds, Smith established an endowment fund, half of which will be earmarked to aid research on molecular genetics of the central nervous system, specifically in relation to schizophrenia research. The other half is to be divided between general science awareness projects and the Society for Canadian Women in Science and Technology in an effort to induce more women to pursue careers in science. He also convinced both the provincial and federal governments to contribute to his funds.

In addition to his receipt of the Nobel Prize, Smith has garnered numerous other honors in the course of his career, including the Gairdner Foundation International Award in 1986, and the Genetics Society of Canada's Award of Excellence in 1988. He has assumed several administrative responsibilities, including becoming acting director of the Biomedical Research Center, a privately funded research institute, in 1991, and is a member of the Canadian Biochemical Society, the Genetics Society of America, and the American Association for the Advancement of Science. He is a fellow of the Chemical Society of London, the Royal Society of Canada, and the Royal Society of London, and has served on several medical committees, such as the advisory committee on

research for the National Cancer Institute of Canada. He is a popular speaker, and has delivered over 150 addresses throughout the world during the course of his career. His scientific research articles number more than two hundred.

See also Gene splicing; Gene therapy

SMITH-FINEMAN-MYERS SYNDROME

Smith-Fineman-Myers syndrome (SFMS) is a rare and severe type of X-linked inherited mental retardation.

Smith-Fineman-Myers syndrome is also known as Smith-Fineman-Myers type mental retardation and Smith-Fineman-Myers type X-linked mental retardation. SFMS results in severe mental retardation along with characteristic facial features and skeletal differences.

Smith-Fineman-Myers syndrome is an X-linked disease. X-linked diseases map to the human **X chromosome**, a **sex chromosome**. Females have two X chromosomes, whereas males have one X **chromosome** and one **Y chromosome**. Because males have only one X chromosome, they require only one copy of an abnormal X-linked **gene** to display disease. Because females have two X chromosomes, the effect of one X-linked recessive disease gene is masked by the disease gene's normal counterpart on her other X chromosome.

In classic X-linked **inheritance** males are affected, presenting full clinical symptoms of the disease. Females are not affected. Affected fathers can never pass X-linked diseases to their sons. However, affected fathers always pass X-linked disease genes to their daughters. Females who inherit the faulty gene but do not show the disease are known as carriers. Female carriers of SFMS have a 50% chance to pass the disease-causing gene to each of their children. Each of a female carrier's sons has a 50% chance to display the symptoms of SFMS. None of a female carrier's daughters would display symptoms of SFMS.

Some patients with SFMS have been found to have a **mutation** in the ATRX gene, on the X chromosome at a location designated as Xq13. ATRX is also the disease gene for several other forms of X-linked mental retardation. Mutations in ATRX are associated with X-linked Alpha-thalassemia/mental retardation syndrome, Carpenter syndrome, Juberg-Marsidi syndrome, and X-linked mental retardation with spastic paraplegia. It is possible that some patients with SFMS have X-linked Alpha-thalassemia/mental retardation syndrome without the hemoglobin H effects that lead to Alpha-thalassemia in the traditionally recognized disease.

SFMS affects only males and is very rare. As of early 2001, only 12 cases have been reported in the medical literature. SFMS has been reported in brothers of affected boys.

SFMS visibly affects the skeletal and nervous systems and results in an unusual facial appearance. The genitals may also show effects ranging from mild (e.g. undescended testes) to severe (leading to female gender assignment).

Boys with SFMS have short stature and a thin body build. Their heads are small and may also be unusually shaped.

Scoliosis and chest deformities have been reported to occur with SFMS. X rays may show that their bones have characteristics of the bones of people younger than they are. Hands are often short with unusual palm creases and short, unusually shaped fingers. Fingernails may be abnormal. Foot deformities and shortened or fused toes have also been reported.

Boys with SFMS exhibit severe mental retardation. Restlessness, behavior problems, seizures, and severe delay in language development are common. Boys with SFMS may be self-absorbed with reduced ability to socialize with others. Affected boys show reduced muscle tone as infants and young children. Later, muscle tone and reflexes are abnormally increased causing spasticity.

Boys with SFMS may display cortical atrophy, or degeneration of the brain's outer layer, on brain imaging studies. Cortical atrophy is commonly found in older normal people. When cortical atrophy is found in younger people it is typically due to a serious brain injury. Brain biopsies of two patients with SFMS have been normal.

SFMS is associated with unusual facial features including a large mouth with a drooping lower lip, prominent upper jaw and front teeth, and an underdeveloped chin. Cleft palate has been reported in one set of affected **twins**. Eyes are widely spaced with drooping eyelids. Skin may be lightly pigmented with multiple freckles.

Assessments for any type of mental retardation usually include a detailed family history and thorough physical exam. Brain and skeletal imaging through CT scans or x rays may be helpful. A chromosome study and certain other genetic and biochemical tests help to rule out other possible causes of mental retardation.

Diagnosis of SFMS has traditionally been based on the visible and measurable symptoms of the disease. Until 2000, SFMS was not known to be associated with any particular gene. As of 2001, scientists do not yet know if other genes may be involved in some cases of this rare disease. Genetic analysis of the ATRX gene may, however, prove to be helpful in diagnosis of SFMS.

Treatment for SFMS is based on the symptoms each individual displays. Seizures are controlled with anticonvulsants. Medications and behavioral modification routines may help to control behavioral problems. SFMS usually manifests severe retardation with a normal lifespan.

See also Chromosomal mutations and abnormalities; Genetic disorders and diseases; Hereditary diseases; Inherited cancers; Mutations and mutagenesis

SNELL, GEORGE DAVID (1903-1996)

American immunogeneticist

Geneticist George David Snell's pioneering research on the immune system in the 1930s and 1940s enabled medical science to develop the process of organ transplantation. Through skin grafts performed on mice at the Jackson Hole Laboratory, he discovered the factor (known as histocompatibility) that enables doctors to determine whether organs and tissues can be successfully transplanted from one body to another. Snell's research earned him the 1980 Nobel Prize for medicine or physiology.

One of three children, Snell was born in Bradford, Massachusetts, to Cullen Snell and the former Kathleen Davis. Snell's father developed and manufactured many inventions, including a mechanism for starting motorboat engines. In Snell's fifth year, the family moved to Brookline. Snell's interests while growing up were varied, and included science, math, sports, and music.

After enrolling at New Hampshire's Dartmouth College in 1922, Snell was influenced to major in biology after taking a **genetics** course taught by Professor John Gerould. He obtained a B.S. degree in that subject in 1926 and enrolled at Harvard that same year to study genetics under the renowned biologist William Castle, who was among the first American scientists to delve into the biological laws of **inheritance** regarding mammals. Snell received a Ph.D. in 1930 after completing his dissertation on linkage, the means by which two or more genes on a **chromosome** are interrelated. That same year he became an instructor of zoology at Rhode Island's Brown University, only to leave in 1931 to work at the University of Texas at Austin following receipt of a National Research Council Fellowship.

Snell's decision to accept the fellowship turned out to be a momentous one, as he began work for the famed geneticist **Hermann Joseph Muller**, whose research with fruit flies led to the discovery that x rays could produce mutations in genes. At the university, Snell experimented with mice, showing that x rays could produce mutations in rodents as well. Although Snell left the University of Texas in 1933 to serve as assistant professor at the University of Washington, he ventured to the Jackson Laboratory in Bar Harbor, Maine, in 1935 to return to research work. The laboratory, specialized in mammalian genetics, and was well-known for its work in spite of its small size.

After continuing his work with x rays and mice, Snell decided to embark on a new study. Snell's project was concerned with the notion of transplants. Earlier scientific research had indicated that certain genes are responsible for whether a body would accept or reject a transplant. The precise genes responsible had not, however, then been identified.

Snell began his experiments by performing transplants between mice with certain physical characteristics. He quickly discovered those mice with certain identical characteristics—in particular a twisted tail—tended to accept each other's skin grafts. In 1948, Peter Gorer came to Jackson Laboratory from London, England. Gorer, who had also conducted experiments on mice, developed an antiserum. He had discovered the existence of a certain antigen (foreign protein) in the blood of mice which induced an immune reaction when injected into other mice. Gorer had called this type of substance "Antigen II."

In collaboration, Snell and Gorer proved that Antigen II was present in mice with twisted tails, indicating that the genetics code for Gorer's antigen and the code found by Snell to be vital for tissue acceptance were identical. They called their discovery of this factor "H–2," for "Histocompatibility Two" (a term invented by Snell to describe whether a transplant would be accepted or rejected).

Later research revealed that instead of only a single **gene** being responsible for this factor, a number of closely related genes controlled histocompatibility. As a result, this was subsequently designated as the **Major Histocompatibility Complex** (**MHC**). The discovery of the MHC, and subsequent research by other scientists in the 1950s which proved it also existed in humans, made widespread organ transplantation possible. Donors and recipients could be matched (as had been done with blood types) to see if they were compatible.

Eventually Snell was able to produce what he called "congenic mice"—animals that are genetically identical except for one particular genetic characteristic. Unfortunately, the first strains of these mice were destroyed in a 1947 forest fire, which burned down the laboratory. However, Snell's tenacity and dedication enabled him to rebound from this setback. Within three years, Snell created three strains of mice which differed genetically only in their ability to accept tissue grafts. The development of congenic strains of mice opened up a new field for experimental research, with Jackson Laboratory eventually being able to supply annually tens of thousands of these mice to other laboratories.

In 1952, Snell became staff scientific director and, in 1957, staff scientist at Jackson Laboratories. In those capacities, Snell continued his research, particularly on the role that MHC plays in relation to **cancer**. Experiments he conducted with congenic mice found that on some occasions the mice rejected tumors, which had been transplanted from their genetic **twins**. This "hybrid resistance" indicated that some tumors provoke an immune response, causing the body to produce antibodies to fight the **tumor**. This discovery could eventually be of great importance in developing weapons to fight cancer.

Although he retired in 1968, Snell continued to visit the lab, discuss scientific and medical matters with colleagues, and write articles and books. Elected to the American Academy of Arts and Sciences in 1952 and to the National Academy of Science in 1970, he was also a member of international scientific societies, including the French Academy of Science and the British Transplantation Society. Snell won numerous awards during the 1960s and 1970s, such as the Hectoen Silver Medal from the American Medical Association, the Gregor Mendel Award for genetic research, and a career award from the National Cancer Institute. This culminated in his winning the 1980 Nobel Prize in medicine or physiology for his work on histocompatibility. He shared this with two other immunogeneticists, **Jean Dausset** and **Baruj Benacerraf**. After being told of the Nobel committee's decision, Snell said there should have been a fourth recipient—his colleague Peter Gorer who died in 1962 and was thus ineligible to receive the prize.

Married in 1937 to the former Rhoda Carson, Snell and his wife had three sons. He died at his home in Bel Harbor, Maine, at the age of 92.

See also Cancer genetics; Immunogenetics; Transplantation genetics

SOCIAL CUSTOMS · *see* GENETIC IMPLICATIONS OF MATING AND MARRIAGE CUSTOMS

SOCIAL DARWINISM

Social **Darwinism** is the theory that persons, groups, and races are subject to the same laws of **natural selection** as Charles Darwin had perceived in plants and animals in nature. According to the theory, which was popular in the late nineteenth and early twentieth centuries, the weak were diminished and their cultures delimited, while the strong grew in power and in cultural influence over the weak. Social Darwinists held that the life of humans in society was a struggle for existence ruled by "survival of the fittest," a phrase proposed by the British philosopher and scientist Herbert Spencer.

The theory of **evolution** by natural **selection** was proposed by Charles Darwin and **Alfred Russel Wallace** in 1858. They argued that **species** with useful adaptations to the environment are more likely to survive and produce progeny than are those with less useful adaptations, thereby increasing the frequency with which useful adaptations occur over the generations. The limited resources available in an environment promotes **competition** in which organisms of the same or different species struggle to survive. In the competition for food, space, and mates that occurs, the less well-adapted individuals must die or fail to reproduce, and those who are better adapted do survive and reproduce. In the absence of competition between organisms, natural selection may be due to purely environmental factors, such as inclement weather or seasonal variations.

The social Darwinists, notably Spencer and Walter Bagehot in England and William Graham Sumner in the United States, believed that the process of natural selection acting on variations in the population would result in the survival of the best competitors and in continuing improvement in the population. Societies, like individuals, were viewed as organisms that evolve in this manner.

The theory was used to support laissez-faire capitalism and political conservatism. Class stratification was justified on the basis of "natural" inequalities among individuals, for the control of property was said to be a correlate of superior and inherent moral attributes such as industriousness, temperance, and frugality. Attempts to reform society through state intervention or other means would, therefore, interfere with natural processes; unrestricted competition and defense of the status quo were in accord with biological selection. The poor were the "unfit" and should not be aided; in the struggle for existence, wealth was a sign of success. At the societal level, social Darwinism was used as a philosophical rationalization for imperialist, colonialist, and racist policies, sustaining belief in Anglo-Saxon or Aryan cultural and biological superiority.

Social Darwinism declined during the twentieth century as an expanded knowledge of biological, social, and cultural phenomena undermined, rather than supported, its basic tenets.

See also Adaptation and fitness; Darwinism; Evolutionary mechanisms; Natural selection

SOCIOBIOLOGY

Sociobiology, also called behavioral ecology, is the study of the **evolution** of social behavior in all organisms, including human beings. The highly complex behaviors of individual animals become even more intricate when interactions among groups of animals are considered. Animal behavior within groups is known as *social* behavior. Sociobiology asks about the evolutionary advantages contributed by social behavior and describes a *biological* basis for such behavior. It is theory that uses biology and **genetics** to explain why people (and animals) behave the way they do.

Sociobiology is a relatively new science. In the 1970s, Edward O. Wilson, now a distinguished professor of biology at Harvard University, pioneered the subject. In his groundbreaking and controversial book, *Sociobiology: The New Synthesis*, Dr. Wilson introduced for the first time the idea that behavior is likely the product of an interaction between an individual's genetic makeup and the environment (or culture in the case of human beings). Wilson's new ideas rekindled the debate of "Nature vs. Nurture," wherein nature refers to genes and nurture refers to environment.

Sociobiology is often subdivided into three categories: narrow, broad, and pop sociobiology. Narrow sociobiology studies the function of specific behaviors, primarily in nonhuman animals. Broad sociobiology examines the biological basis and evolution of general social behavior. Pop sociobiology is concerned specifically with the evolution of human social behavior.

Sociobiologists focus on reproductive behaviors because reproduction is the mechanism by which genes are passed on to future generations. It is believed that behavior, physically grounded in an individual's **genome** (or genes), can be acted upon by **natural selection**. Natural **selection** exerts its influence based upon the **fitness** of an organism. Individuals that are *fit* are better suited (genetically) to their environment and therefore reproduce more successfully. An organism that is fit has more offspring than an individual that is unfit. Also, fitness requires that the resulting offspring must survive long enough to themselves reproduce. Because sociobiologists believe that social behavior is genetically based, they also believe that behavior is heritable and can therefore contribute to (or detract from) an individual's fitness. Examples of the kinds of reproductive interactions in which sociobiologists are interested include courtship, mating systems like monogamy (staying with one mate), polygamy (maintaining more than one female mate), and polyandry (maintaining more than one

male mate), and the ability to attract a mate (called *sexual selection*.)

Sociobiology also examines behavior that indirectly contributes to reproduction. An example is the theory of optimal foraging which explains how animals use the least amount of energy to get the maximum amount of food. Another example is altruistic behavior (**altruism** means selfless). Dominance hierarchies, territoriality, ritualistic (or symbolic) behavior, communication (transmitting information to others through displays), and instinct versus learning are also topics interpreted by sociobiology.

Sociobiology applied to human behavior involves the idea that the human brain evolved to encourage social behaviors that increase reproductive fitness. For example, the capacity for learning in human beings is a powerful characteristic. It allows people to teach their relatives (or others) important life skills that are passed-down from generation to generation. However, the ability to learn is also a variable trait. That is, not every person learns as quickly or as well as every other person. A sociobiologist would explain that individuals who learn faster and more easily have increased fitness. Another example is smiling. The act of smiling in response to pleasurable experiences is a universal social behavior among people. Smiling is observed in every culture. Furthermore, smiling is an example of an instinct that is modified by experience. Therefore, because the behavior is instinctual, it has a genetic and inheritable basis. Because it is altered by experience, the behavior is socially relevant. Sociobiologists might speculate, then, that since smiling is a visual cue to other individuals that you are pleased, people who tend to smile more easily are more likely to attract a suitable mate, and are therefore more fit.

The discipline of sociobiology is an important set of ideas because nearly every animal **species** spends at least part of its life cycle in close association with other animals. However, it is also riddled with debate, principally because it attempts to not only explain the behavior of animals but also of human beings. More dangerously, it tries to describe "human nature." The idea that human behavior is subject to genetic control has been used in the past to justify racism, sexism, and class injustices. In this respect, sociobiology is similar to **Social Darwinism**. For this reason, sociobiology remains a controversial discipline. Further criticisms include the observation that sociobiology contains an inappropriate amount of anthropomorphism (giving human characteristics to animals) and it excessively generalizes from individuals to whole groups of organisms. Despite criticism, however, sociobiology is an enlightening new aspect of biology which, taken in context, can bridge the gap between life science and the humanities.

See also Altruism; Darwinism; Evolutionary mechanisms; Survival of the fittest

SOMATIC CELL GENETICS

Eukaryotic cells may be classified broadly as either germ cells (sex cells, or gametes) or all the other body cells, known as somatic cells. Somatic **cell genetics**, strictly speaking, means genetic studies of cells other than germ cells, but it is usually used more restrictively, referring to mammalian cells in tissue culture.

Key characteristics of somatic cells are that they are **diploid**, and differentiated. Though all somatic cells from one individual are (in principle) genetically identical, they evolve different tissue-specific characteristics during the process of **differentiation**, through regulatory and epigenetic changes.

The opportunity to study cells *in vitro* (isolated from the whole organism; literally, in glass) creates several advantages: one can assay cells in test tubes in ways that would be inappropriate to undertake on an individual, the cellular environment can be defined and manipulated, and rare genotypes can be maintained for study beyond the lifespan of the individual.

Most differentiated cells do not grow readily in culture; those that do tend to lose differentiated properties, and only have a finite lifespan (though they may be frozen for longer-term storage). Those that have arisen from tumors, on the other hand, or that are experimentally transformed *in vitro* may become permanent lines with indefinite lifespan. The most common source of tissue for non-transformed lines is skin fibroblasts. Lymphocytes can be cultured (short-term) from peripheral blood following stimulation by phytohemagglutinin, still the method used in diagnostic cytogenetics. Permanent lymphoblast lines are created with Epstein-Barr **virus transformation**, providing infinite supplies of otherwise rare genetic materials.

Perhaps the greatest contribution of somatic cell genetics was in the first stages of mapping the human **genome**. It was found in the 1960s that hybrids of human and rodent cells could be cultured, but selectively retained rodent and discarded human chromosomes. This phenomenon was taken to advantage, through selective culture conditions and identification of human chromosomes retained by the hybrids, allowing assignment of specific **gene** functions to specific chromosomes, starting with thymidine kinase to human **chromosome** 17. These early markers were the anchors for all later gene mapping studies.

In its broader sense, somatic cell genetics also encompasses the study of **cancer** cells, since the process of malignant transformation is one in which the genetic makeup of the **tumor** is altered relative to the otherwise stable constitution of the cells of origin.

See also Cancer genetics; Cell differentiation; Cell line; Chromosome mapping and sequencing; Clones and cloning; Gene regulation; Oncogenetic research

SONNEBORN, TRACY MORTON
(1905-1981)
American zoologist

Tracy Sonneborn's breakthrough research on mating types in Paramecium allowed geneticists to conduct cross-breeding analyses of unicellular organisms. Before Sonneborn identi-fied the mating types, researchers knew that the protozoa reproduced both asexually, through budding, and sexually, through conjugation, but could not control the reproductive path taken by the organism. Sonneborn's findings added protozoa to the growing number of model organisms available to **genetics** researchers, including maize, *Drosophila, Chlamydomonas,* and *Neurospora.* Paramecium's complex **sexual reproduction** cycle, involving both a macronucleus, two macronuclei, and cytoplasmic factors, led Sonneborn to argue for the importance of cytoplasmic and non-nuclear **inheritance**. This placed him in opposition to many of the genetic leaders of the day, especially those trained by American T. H. Morgan (1866–1945) at Columbia University, who believed that all hereditary information was stored in the chromosomes of the **nucleus**. Sonneborn's first major piece of evidence for **cytoplasmic inheritance** was the Kappa factor. The Kappa factor resulted in the production of a substance toxic to other varieties of Paramecium, was inherited through the cytoplasm, and responded to environmental conditions; however, it was later determined that the Kappa factor was a symbiotic microorganism rather than an inherited trait. Sonneborn's later studies on the immunological properties of Paramecium were more successful in demonstrating the presence of environmentally sensitive traits inherited through the cytoplasm. Many mainstream geneticists resisted such findings, partially out of fear that this sort of research might offer support to the theories of Soviet agriculturist T. D. Lysenko (1898–1976). Geneticists now recognize a number of non-nuclear modes of inheritance, including many of the structural attributes of the **cell**, the autonomous reproduction of some organelles (i.e., **mitochondria**) and the presence of some symbiotic microorganisms.

Born to a traditional Jewish family in Baltimore, Sonneborn briefly considered becoming a rabbi. He quickly changed direction, however, and began studying biology at the Johns Hopkins University in 1922. There, he followed the advice of his mentor, American biologist Herbert Spencer Jennings (1868–1947), and concentrated his research on a single organism. Although the microscopic flatworm *Stenostomum* had been adequate for Sonneborn's doctoral research, he turned to other organisms after receiving his Ph.D. in 1928. Sonneborn eventually settled on Paramecium, and he continued to work on that protozoa for the rest of his career. He spend most of the 1930s as a research associate in Jennings' lab, occasionally funded by outside agencies such as the National Research Council. His identification of mating types in Paramecium propelled him to national fame, and he accepted a position as associate professor of zoology at Indiana University in 1939. Although Sonneborn received offers for positions at other prestigious institutions throughout his career, he remained at Indiana University for the rest of his professional life. Indiana briefly became a major center for genetics in the 1940s, attracting such leading geneticists as Americans H. J. Muller (1890–1967), Marcus Rhoades (1903–1991), and the Italian-born naturalized American Salvador Luria (1912–1991). Indiana's distance from other major centers of genetics research, such as the California

Institute of Technology, the Massachusetts Institute of Technology, and Columbia University, surely helped Sonneborn maintain his unorthodox interest in cytoplasmic inheritance.

Sonneborn was recognized as an outstanding teacher both at Indiana University and within the genetic community. For years, he and his wife, Ruth, opened their home weekly for an informal seminar where graduate students and visiting researchers gathered to exchange ideas. As a leader of the professional biological community, he served terms as the President of the American Society of Naturalists and played a major role in establishing the American Institute of Biological Sciences. His presidency of the Genetics Society in the late 1940s located him squarely in the middle of the debates on **Lysenkoism**. He was a member of the National Academy of Sciences, the American Academy of Arts and Sciences, and a foreign members of the Royal Society of London. Sonneborn received numerous prizes and awards throughout his career, including the Newcombe Cleveland Research Prize of the American Association for the Advancement of Science (1946) and the Kimber Genetics Award of the National Academy of Sciences (1939).

See also Lysenkoism; Microbial genetics; Nucleus

SOUTHERN BLOTTING ANALYSIS · *see*

BLOTTING ANALYSIS

SPECIES

A species is a population of individual organisms that can interbreed in nature, mating and producing fertile offspring in a natural setting. Species are organisms that share the same **gene pool**, and therefore genetic and morphological similarities.

All organisms are given two names (a binomial name); the first is the genus name and the second is the species name, for example *Homo sapiens*, the name for humans. The Linnaean classification system places all organisms into a hierarchy of ranked groups. The genus includes one or more related species, while a group of similar genera are placed in the same family. Similar families are grouped into the same order, similar orders in the same class, and similar classes in the same phylum.

Organisms are assigned to the higher ranks of the Linnaean classification scheme largely on the basis of shared similarities (syna pomorphisus). Species are identified on the basis of an organism's ability to interbreed, in addition to its morphological, behavioral, and biochemical characters. Although species are defined as interbreeding populations, taxonomists rarely have information on an organism's breeding behavior and therefore often infer interbreeding groups on the basis of reproductive system morphology, and other shared characters.

In the last 20 years, modern molecular techniques such as **DNA hybridization** have allowed biologists to gain extensive information on the genetic distance between organisms, which they use to construct hypotheses about the relatedness of organisms. From this information researchers hypothesize as to whether or not the populations are genetically close enough to interbreed.

While the biological species concept has historically been the most widely used definition of a species, more recently the **phylogenetic** and ecological species concepts have taken the forefront as a more inclusive and useful definition. Whereas the biological species concept defines a species as a group of organisms that are reproductively isolated (able to successfully breed only within the group), the phylogenetic species concept considers tangible (and measurable) differences in characteristics. This idea, also called the cladistic species concept, examines the degree of genetic similarity between groups of related individuals (called clades) as well as their similarities in physical characteristics. For instance, the biological species concept might group coyotes and wolves together as one species because they can successfully breed with one another. In contrast, the phylogenetic concept would definitively split coyotes and wolves into two species based upon the degree of divergence in genetic characters and larger observable traits (like coat color, for instance). In contrast to these, the ecological species concept might classify wolves and coyotes as different species by comparing the differing environmental resources that they exploit, called adaptive zones. Currently, the precise definition of a species is a topic under constant scientific debate and likely will never fully be resolved. Rather, the definition may change with the perspectives and needs of each sub-discipline within biology (ecology versus zoology, for example). A pluralist approach combines some or all of these species concepts to arrive at a more inclusive definition.

Speciation is the process whereby a single species develops over time into two distinct reproductively isolated species. Speciation events are of two types—either allopatric or sympatric. Allopatric speciation results from the division of a population of organisms by a geographical barrier. The isolation of each of the two populations slowly results in differences in the **gene** pools until the two populations are unable to interbreed either because of changes in mating behavior or because of incompatibility of the **DNA** from the two populations. The early stages of allopatric speciation are often evident when one examines the same species of fish from different ponds. Fish from the two ponds may not appear to be morphologically different, but there may be slight differences in the gene pools of each population. If the two fish populations remain separated for enough generations, they may eventually become two separate reproductively isolated species.

Sympatric speciation is less frequent than allopatric speciation and occurs when a group of individuals becomes reproductively isolated from the larger population occupying the same range. This type of speciation may result from genetic changes (or mutations) occurring within individuals that inhibits them from interbreeding with others, except those in

The Giant Panda, a species indigenous to China, is one of the many endangered species. *JLM Visuals. Reproduced by permission.*

which the same **mutation** has occurred. Polyploid plant species, that is, species with more than two copies of each **chromosome**, are thought to have arisen by sympatric speciation.

More than 1.5 million species have been described, and it is estimated that there are between 10-50 million species currently inhabiting Earth.

See also Ecological and environmental genetics; Evolution; Evolutionary mechanisms; Survival of the fittest

SPEMANN, HANS (1869-1941)

German embryologist

Hans Spemann was recognized for his research into the development of embryos, and in particular for his studies into the causes behind the specialization and **differentiation** of embryonic cells. In the mid–1930s, Spemann discovered "organizers"—regions within developing embryos that cause undifferentiated tissue to evolve in a specific way. For this finding he was awarded the 1935 Nobel Prize for physiology or medicine. In addition to these achievements, Spemann is credited with founding the early techniques of microsurgery, the minute manipulations of tissue or living structure.

The son of a well known book publisher, Spemann was born in Stuttgart, Germany. He was the eldest of four children of Johann Wilhelm Spemann and the former Lisinka Hoffman. The family, which was socially and culturally active, lived in a large home that was well stocked with books, which helped shape the young Spemann's intellect. Upon entering the Eberhard Ludwig Gymnasium, Spemann first wished to study the classics. Although he later turned to **embryology**—the branch of biology that focuses on embryos and their development—he never relinquished his love of artistic endeavors; throughout his lifetime he organized evening gatherings of friends to discuss art, literature, and philosophy.

Before entering the University of Heidelberg in 1891 to study medicine, Spemann worked at his father's business and served a tour of duty in the Kassel hussars. His strict interest in medicine lasted only until he met German biologist and psychologist Gustav Wolff at the University of Heidelberg. Only a few years older than Spemann, Wolff had begun experiments on the embryological developments of newts and had shown how, if the lens of an embryological newt's eye is removed, it regenerates. Spemann remained interested and intrigued by both Wolff's finding and also in the newt, on which he based much of his future work. But more than the regeneration phenomenon, Spemann was interested in how the eye develops from the start. He devoted his scientific career to the study of how embryological cells become specialized and differentiated in the process of forming a complete organism.

Spemann left Heidelberg in the mid–1890s to continue his studies at the University of Munich; he then transferred to the University of Würzberg's Zoological Institute to study under the well-known embryologist Theodor Boveri. Spemann quickly became Boveri's prize student, and completed his doctorate in botany, zoology, and physics in 1895.

Shortly thereafter he married Clara Binder; the couple eventually had two sons. Spemann stayed at Würzburg until 1908, when he accepted a post as professor at the University of Rostock. During World War I, he served as director of the Kaiser Wilhelm Institute of Biology (now the Max Planck Institute) in Berlin-Dahlem, and following the war, in 1919, Spemann took a professorship at the University of Freiburg.

By the time Spemann began research at the Zoological Institute in Würzburg, he had already developed a keen facility and reputation for conducting well-designed experiments that centered on highly focused questions. His early research followed Wolff's closely. The eye of a newt is formed when an outgrowth of the brain, called the optic cup, reaches the surface layer of embryonic tissue (the ectoderm). The cells of the ectoderm then form into an eye. In removing the tissue over where the eye would form and replacing it with tissue from an entirely different region, Spemann found that the embryo still formed a normal eye, leading him to believe that the optic cup exerted an influence on the cells of the ectoderm, inducing them to form into an eye. To complete this experiment, as well as others, Spemann had to develop a precise experimental technique for operating on objects often less than two millimeters in diameter. In doing so, he is credited with founding the techniques of modern microsurgery, which is considered one of his greatest contributions in biology. Some of his methods and instruments are still used by embryologists and neurobiologists today.

In another series of experiments—conducted in the 1920s, Spemann used a method less technically demanding to make an even more critical discovery. By tying a thin hair around the jelly-like egg of a newt early in embryogenesis (embryo development), he could split the egg entirely, or squeeze it into a dumbbell shape. When the egg halves matured, Spemann found that the split egg would produce either a whole larva and an undifferentiated mass of cells, or two whole larva (although smaller than normal size). The split egg never produced half an embryo. In the case of the egg squeezed into a dumbbell shape, the egg formed into an embryo with a single tail and two heads. Spemann's primary finding in these experiments was that if an egg is split early in embryogenesis, the two halves do not form into two halves of an embryo; they either become two whole embryos, or an embryo and a mass of cells.

This led Spemann to the conclusion that at a certain stage of development, the future roles of the different parts of the embryo have not been fixed, which supported his experiments with the newt's eye. In an experiment conducted on older eggs, however, Spemann found that the future role of some parts of the embryo had been decided, meaning that somewhere in between, a process he called "determination" must have taken place to fix the "developmental fate" of the cells.

One of Spemann's greatest contributions to embryology—and the one for which he won the 1935 Nobel Prize in physiology or medicine—was his discovery of what he called the "organizer" effect. In experimenting with transplanting tissue, Spemann found that when an area containing an organizer is transplanted into an undifferentiated host embryo, this transplanted area can induce the host embryo to develop in a certain way, or into an entirely new embryo. Spemann called these transplanted cells organizers, and they include the precursors to the central nervous system. In vertebrates, they are the first cells in a long series of differentiations of which the end product is a fully formed fetus.

Spemann remained at the University of Freiburg until his retirement in the mid–1930s. When not busy with his scientific endeavors, he cultivated his love of the liberal arts. He died at his home near Freiburg at age 72.

See also Embryology, the history of developmental and generational theory

SPERM CELLS

Sperm cells are **haploid** sex cells of the male. They are also known as spermatozoa. Unlike eggs which are large, nonmotile, and generally few in number, sperm are tiny, motile, and produced in huge numbers. While the human sperm length (0.002 in [600 cm]) is relatively great due to its long tail, the volume of an entire sperm, tail and all, is only 1/85,000 of the egg.

Reproduction in humans occurs when the sperm are deposited in the vagina of a female near the cervical opening of the uterus. The haploid sperm move into the uterus and up the Fallopian (uterine) tube where an egg may be encountered. With **fertilization**, the egg finishes its second meiotic division to become haploid. The haploid sperm and mature egg together form a **diploid zygote** which is the beginning of a new individual.

The male gonad (testis singular, testes plural; testicle is derived from the diminutive of testis and perhaps is best used to describe the gonads of a sexually immature boy) produces the hormone testosterone and sex cells. Early in embryonic development, primordial germ cells, which are diploid, migrate to the embryonic gonad. The primordial germ cells give rise to the diploid **stem cells** of the testis, known as spermatogonia. Each of the many spermatogonia, after the first meiotic division, form two primary spermatocytes which in turn, after the second meiotic division, form four haploid spermatids. In the process of forming mature sperm the spermatids lose much of their cytoplasm and develop a long, propulsive tail.

Motility of the sperm is due to the long tail which is a modified flagellum. Cilia and flagella, from protozoa through humans, all have a basically similar structure which has been intensively investigated since first described in early **electron microscope** studies. Microtubules that run the length of the sperm tail are arranged in a ring of nine pairs surrounding a pair in the center. Ciliary dynein is associated with each of the nine microtubule pairs. It is the interaction of the dynein with the microtubules which causes flagellar bending and thus propulsion.

As stated, sperm occur in large numbers. It is estimated that a quarter of a billion sperm are released in a single ejaculate in a healthy male human.

See also Ovum; Sexual reproduction

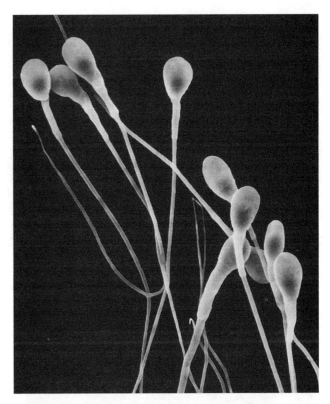

Scanning electron micrograph (SEM) of human spermatazoa, magnified 760 times. *Scanning electron micrograph (SEM) by Dr. Dennis Kunkel. © Dr. Dennis Kunkel/Phototake. Reproduced by permission.*

SPERMATOGENESIS • *see* GAMETOGENESIS

SPINA BIFIDA

Spina bifida is the common name for a range of **birth defects** caused by problems with the early development of the spine. The term comes from the Latin "spina," meaning thorn, or "spine," and "bifida," meaning split into two parts. The main defect of spina bifida involves an abnormal opening in the bony column through which the spinal cord passes, called the vertebral column. In spina bifida, there is an abnormal opening somewhere along the vertebral column, which leaves the spinal cord unprotected, and vulnerable to either mechanical injury or invasion by infection.

Spina bifida occurs in one out of every 700 births to whites in North America, but in only one in every 3,000 births to African-Americans. In some areas of Great Britain, the occurrence of spina bifida is as high as one in every 200 births. Spina bifida tends to run in families, leading researchers to suspect a genetic as well as environmental cause of the disease. Current research indicates that more than one **gene** may be responsible for causing spina bifida, and that the occurrence of the disease can be dramatically reduced by women of childbearing age consuming 0.4 milligrams daily of folic acid, a B-vitamin, prior to conceiving a child.

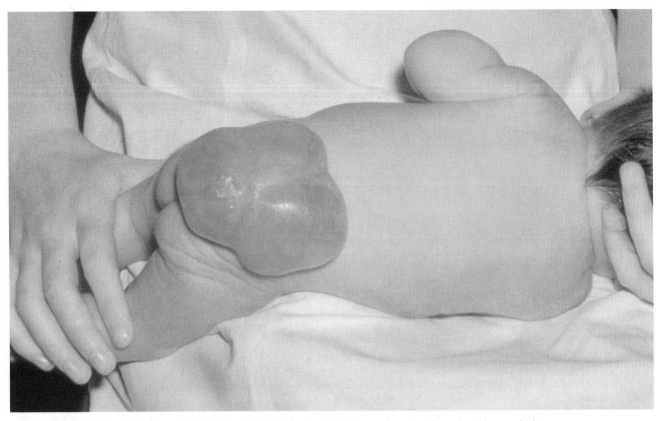

Spina bifida. *Biophoto Associates, National Audubon Society Collection/Photo Researchers, Inc. Reproduced by permission.*

Genes currently under study for association with spina bifida are: genes involved with the metabolism of folic acid, genes known to cause **neural tube defects** in rats and mice, genes involved with glucose metabolism, HOX and PAX genes which help determine body symmetry, and the proto-oncogenes which may contribute to physical development as well as **cancer**.

The classic defect of spina bifida cystica is an opening in the spine, obvious at birth, out of which protrudes a fluid-filled sac. This sac may include either just the meninges, those membranes which cover the spinal cord (a meningocele)or may include both the meninges and some part of the actual spinal cord (myelomeningocele). Often, the spinal cord itself has not developed properly. In spina bifida occulta, there may be some opening in the vertebrae, but no protruding sac. The entire defect may be covered with skin. At the other end of the spectrum of severity is rachischisis, in which the entire length of the spine may be open.

The problems caused by spina bifida depend on a number of factors, including the location of the spinal defect and what degree of disorganization of the spinal cord exists. Because different levels of the spinal cord are responsible for different functions, the location and the size of the defect in spina bifida will affect what kind of disabilities an individual will experience. Most patients with clinically identifiable spina bifida have some degree of weakness in the legs. This can be so severe as to be complete paralysis, depending on the

spinal cord condition. The higher up in the spine the defect occurs, the more severe the disabilities.

People with spina bifida frequently face severe problems with both bladder and bowel function, because complete emptying of both bladder and bowels requires an intact spinal cord. Difficulty in completely emptying the bladder can result in severe, repeated infections, ultimately causing kidney damage, which can be life-threatening.

Other types of problems may accompany spina bifida, including changes in the architecture and arrangement of brain structures, resulting in the accumulation of fluid on the brain. Many people with spina bifida have other orthopedic complications, including clubfeet and hip dislocations, and abnormal curves and bends in the spine, resulting in a hunchbacked or twisted appearance. Intelligence in children with spina bifida varies widely, and certainly depends on the severity of the spinal defect and the presence of other associated defects. Children with brain abnormalities, as well as those who have had meningitis, are most likely to have intellectual deficits. With aggressive surgery, **antibiotics**, and other therapies, many victims of spina bifida live meaningful and productive lives; 60% of those with spina bifida survive past their sixteenth birthday.

See also Birth defects

SPONTANEOUS GENERATION

Spontaneous generation, which is also called biopoesis or abiogenesis, is the process of living organisms arising *de novo* from non-living material. Until the nineteenth century, it was believed that spontaneous generation was the process by which many living creatures arose. Examples of spontaneous generation included mites arising from piles of non-living dust and fungi arising from dead wood, instead of merely taking advantage of the fact the dead wood is there and moving in to colonize it. Ancient Egyptians believed snakes arose from mud and the Greeks believed rats came from garbage. These misguided theories were logically based on physical observation of the relationship between the organism and its habitat.

By the time of the Renaissance, scientists did not believe that larger animals arose from spontaneous generation, but they still clung to the belief that **bacteria** arose spontaneously. It was not until 1862, when **Louis Pasteur** offered proof to the contrary, that spontaneous generation was discarded. Pasteur showed that microorganisms were not spontaneously generated from non-living matter, but arose from preexisting microbes. By killing these preexisting microbes and not allowing new ones to grow, Pasteur illustrated that no microorganisms arose, even though the conditions prevailing had previously given rise to them. This is part of the germ theory of disease.

See also Evolution; Evolutionary mechanisms

SPONTANEOUS MUTATIONS AND REVERSIONS

Mutations and reversions are events that change the expression of traits of an organism. They change the organism's phenotype. The change can result from a change in a base—a constituent of the deoxyribonucleic acid—or a change in one of the **amino acid** constituents of the **ribonucleic acid** that is used as a blueprint to make a protein.

Each time **DNA** replicates, mistakes can occur. If the **mutation** is deleterious, the organism does not survive. But, if the mutation changes the organism's phenotype in a nonlethal fashion, the mutation can persist. The accuracy of **DNA replication**, the occurrence of damage to the DNA, and the effectiveness of mechanisms for **DNA repair** determine the rate at which mutations occur. **Bacteria** and other organisms exhibit natural mutation and reversion rates, where mutations and reversions can occur spontaneously.

Salvador Luria and **Max Delbrück** provided the demonstration of spontaneous mutations. They showed that the rate of acquisition of resistance to a **virus** by bacteria was not dependent on the number of cells used in the experiment. Rather, they showed that some mutations increased depending on the time that had elapsed between the appearance of the mutation and contact with the virus. For these studies, Luria and Delbrück received a Nobel Prize. Another milestone was reached when **Joshua Lederberg** isolated virus-resistant cells from populations that have never encountered the virus. These experiments confirmed the existence of spontaneous mutations—mutations that arise without contact with a selecting agent.

Cells have evolved several repair mechanisms to deal with spontaneous mutations. Specialized **enzymes** called DNA repair nucleases can recognize an altered base, break the bonds holding the base in position so that it is removed from the remainder of the DNA chain, and then fill in the missing base pairs. A second enzyme, DNA ligase, seals the nick in the DNA strand, cementing the inserted base in position. Another method of mutation repair is called the SOS response. More than fifteen genes are activated by the detection of DNA damage due to mutations, including spontaneous mutations.

Reversions can also occur spontaneously. Reversions, or reverse mutations, change the new phenotype back to the original, or so-called wild-type. Bacterial strains that contain reversions are known as revertants. True revertants are identical to the original organism in both DNA sequence and phenotypic expression of the encoded information. The second, reversion mutation has restored the molecular change that precipitated the initial change in phenotype.

Reversion is different from suppression, where the second mutation does not restore the exact **nucleotide** sequence.

See also Mutations and mutagenesis

SRY GENE

In humans, early studies comparing normal males and females to individuals with a structurally aberrant **Y chromosome** determined that a **gene** key to the development of the testes was located on the short arm of the Y **chromosome**. The name given at that time was TDF, testis-determining factor. Ongoing research narrowed the area of interest to a 35 kb segment just below the pseudoautosomal region and eventually the relevant gene was identified and named SRY, for sex determining region of the Y chromosome. Cross **species** comparisons showed that this gene is very similar to the mouse Sry gene and that it is also represented in other mammals and marsupials.

As its name implies, SRY plays a major role in **sex determination**, a complex, multistep pathway that gives rise to either a male or female individual. In humans, female development appears to be the default, so in order for a male to develop, a switch to the male pathway must occur. The activator for this is SRY. **Transcription** and **translation** of the gene results in the development of the testes that in turn produce testosterone and launch **differentiation** of the male sex genitalia. At the same time, another gene, MIF—Müllerian inhibitory factor, is stimulated to shut down the female developmental process. The entire series of events requires a large number of genes on many different chromosomes.

As with all genes, it is possible for mutations to occur which change its function. **Deletions** and a variety of point mutations have been described in SRY that result in blockage of the male developmental pathway and give rise to individuals with gonadal dysgenesis, which is clinically, a phenotypic female with a male XY **sex chromosome** complement. Although outwardly female, these individuals usually have undifferentiated or abnormal internal sexual organs and are therefore infertile.

Translocations between the X and Y chromosomes in which the SRY gene is moved from the Y to the tip of one X have also been reported. The change in position does not affect the function of SRY, so in the absence of any other factors, male development will proceed. However, the individual will have an apparent female chromosome complement with two X chromosomes.

The key aspect of the SRY gene is, therefore, its importance as the initiator of the male developmental pathway.

See also Sex chromosomes; Sex determination; X chromosome; Y chromosome

STABILIZING SELECTION

Stabilizing **selection** (also called optimizing selection or normalizing selection) is one of three main forms of **natural selection**. In stabilizing selection, selection acts to maintain a well-adapted balance with regard to **fitness** among the forms of a **gene** (alleles) that exist in a given population. According, stabilizing selection conserves existing genetic balances and continues to do so until there is a sufficient change in environment to cause directional or **disruptive selection**. As with directional and disruptive selection, stabilizing selection is based upon differences in greater reproductive success of individuals within the population. Stabilizing selection specifically acts on visible characters or traits that result in difference in reproductive success (differential reproduction).

The existing fossil record provides abundant evidence for the dominance of stabilizing selection over the majority of the history of life on Earth. The other main mechanisms of natural selection include **directional selection** (when an extreme phenotype is fittest) and disruptive selection (two extreme phenotypes are fitter than an intermediate phenotype).

Stabilizing selection acts to conserve or maintain the existing frequencies of alleles, including their percentage distribution, within a population. This form of genetic conservation is characteristic of well-adapted populations existing in areas of environmental stability. Stabilizing selection acts to reduce numbers of individuals with extreme phenotypes within the population. At the genetic level simple stabilizing selection occurs when heterozygotes (**diploid** organisms with two different alleles) are fitter than homozygous organisms.

Because stabilizing selection tends to reduce variation visible characteristics or traits (phenotype) within a population, organisms reflecting the exiting phenotype distribution tend to be numerically favored and have higher fitness than individuals that exhibit different characters or traits.

Stabilizing selection, reduces phenotypic deviation to maintain a particular set of characteristics within a population.

At a more complex level, stabilizing selection acts on such phenotypic traits as human birth weight. Low weight newborns have a significantly reduced chance of survival. Correspondingly, very high birth weight babies suffer high mortality rates due to a number of birth process (e.g., the limitations placed on size by pelvic clearance). Although the causes differ, light infants and heavy infants die during childbirth at increased rates when compared to average weight infants.

Although stabilizing selection is a form of natural selection; the fundamental principles can be utilized during artificial selection. By culling out extremes of phenotypes, breeders essentially remove genetic or allele stock from the population and thereby restrict the numbers and types of alleles that can be passed to the next generation.

See also Animal husbandry; Evolution; Evolutionary mechanisms; Genome; Genotype and phenotype; Neutral selection; Plant breeding and crop improvement; Polymorphism; Selection

STAHL, FRANKLIN W. (1929-)

American geneticist

Franklin W. Stahl, in collaboration with Matthew Meselson, discovered direct evidence for the semiconservative nature of **deoxyribonucleic acid (DNA) replication** in **bacteria**. In experiments, Stahl and Meselson showed that when a double stranded DNA molecule is duplicated, the double strands are separated and a new strand is copied from each parent strand forming two new double stranded DNA molecules. The new double stranded DNA molecules contain one conserved parent strand and one new daughter strand. Therefore, the replication of a DNA molecule is semiconservative: it retains some of the original material while creating some new material. The understanding of the semiconservative nature of DNA in replication was a major advancement in the field of molecular biology.

Franklin William Stahl, the youngest of three children, was born in Boston, Massachusetts, to Oscar Stahl, an equipment specialist with New England Telephone and Telegraph, and Eleanor Condon Stahl, a homemaker. Stahl received a baccalaureate degree from Harvard University in 1951 in the area of biological sciences, and continued with graduate studies in the field of biology, earning a Ph.D. degree at the University of Rochester in New York, in 1956. From 1955 to 1958, Stahl was a research fellow at the California Institute of Technology, where he collaborated with Matthew Meselson on the semiconservative replication experiment.

In 1952, having just graduated from Harvard, Stahl attended a course at Cold Spring Harbor Laboratories in New York given by A. H. Doermann. Doermann was well known for research on bacteriophages, microscopic agents that destroy disease-producing bacteria in a living organism. This course gave Stahl his first exposure to the **genetics** of bacteriophages. The subject so fascinated him that he spent his summers in the laboratory of Dr. Doermann while working on his doctorate at the University of Rochester during the school year. **Bacteriophage** genetics would later become the major focus of his own laboratory's scientific research. Stahl would also come to teach the same course at Cold Spring Harbor.

After receiving his doctorate in biology, Stahl moved to California to work in the laboratory of **Max Delbrück** at the California Institute of Technology as a postdoctoral fellow. While at Cal Tech, he began a collaboration with graduate student Meselson to design an experiment to describe the nature of **DNA replication** from parent to offspring using bacteriophages. The idea was to add the substance 5-bromouracil, which would become incorporated into the DNA of a T4 bacteriophage upon its replication during a few rounds of reproduction. Phage samples could then be isolated by a **density gradient centrifugation** procedure which was originally designed by Stahl, Meselson, and Jerome Vinograd. It was thought that the phage samples containing the incorporated 5–bromouracil would separate in the density gradient centrifugation to a measurable degree based on the length of the new strands of DNA acquired during replication. Several attempts to obtain measurable results were unsuccessful. Despite these first setbacks, Stahl had confidence in the theory of the experiment. After further contemplation, Stahl and Meselson decided to abandon the use of the T4 bacteriophage and the labeling substance 5–bromouracil and turned to the use of a bacteria, *Escherichia coli,* with the heavy nitrogen isotope 15N as the labeling substance. This time, when the same experimental steps were performed using the new substitutions, the analysis of the density gradient centrifugation samples showed three distinct types of bacterial DNA, two from the original parent strands of DNA and one from the new offspring. Analysis of the new offspring showed each strand of DNA came from a different parent. Thus the theory of semiconservative replication of DNA had been proven.

After spending 1958 at the University of Missouri as an associate professor of zoology, Stahl took a position as associate professor of biology and research associate at the Institute of Molecular Biology, located at the University of Oregon in Eugene, Oregon. In 1963, he was awarded status of professor; he was appointed acting director of the Institute from 1973 to 1975. Stahl has held a concurrent position as resident research professor of molecular genetics at the American Cancer Society.

Stahl set up his own laboratory at the Institute, contributing further to the scientific research and understanding in the area of bacteriophage genetics, as well as the genetic **recombination** of bacteriophages and fungi. In the early years at Eugene, he continued to focus his research in the area of genetic recombination and replication in bacteriophages using the techniques of density gradient and equilibrium centrifugations. Through the years, he was able to map the **DNA structure** of the T4 bacteriophage. The experiments involved T4 bacteriophages inactivated by decay of DNA incorporated radionucleotides or by x-irradiation of the DNA that would cause breaks in the DNA sequence. By performing reactiva-

tion-crosses of these bacteriophage, Stahl studied the patterns in which markers on the DNA were "knocked out" or lost. Although the inactivated phages are unable to produce offspring themselves, they can contribute particular markers to their offspring when they are grown in the presence of rescuing phages that supply the functions necessary for phage development. By the pattern of markers seen in the offspring of these reactivation crosses, a map can be constructed. From the map constructed, the correlated knockout markers reflected a linkage relationship in the form of a circle. With this map in hand, particular DNA sequences could be shown to be important for various functions of the bacteriophage.

Much of Stahl's later work focused on the bacteriophage Lambda, which has a more complex structure than bacteriophage T4, and its replication inside of a bacterial **cell**. He determined particular "hot spots" in the DNA sequence that were susceptible to various mutations or recombinations during the process of replication. These "hot spots" were particular sites in the DNA sequence of the phage that tended to show **crossing over** between two DNA strands of the **chromosome**. The resulting mutations (Chi mutations), which occurred at four or five particular sites in the Lambda phage, conferred a particular large plaque forming character by accelerating the rate of crossing over at these sites. These mutations affected the overall function of the bacteriophage, sometimes causing complete inactivation. Through further studies of genetic recombination in bacteriophages, Stahl became known to the scientific world as an expert on their structure and life cycle.

From 1964–85, Stahl held several year-long positions as visiting professor or volunteer scientist in various universities throughout the world. He was the volunteer scientist in the Division of Molecular Genetics for the Medical Research Council in Cambridge, England. He took a sabbatical leave from Oregon and conducted research in the Medical Research Council Unit of Molecular Genetics at the University of Edinburgh, Scotland, as well as at the Laboratory of International Genetics and Biophysics in Naples, Italy. He held the position of Lady Davis Visiting Professor in the Genetics Department at Hebrew University in Jerusalem, Israel. Stahl also taught courses on bacterial viruses at Cold Spring Harbor Laboratories and in Naples. He is a member of the National Academy of Sciences, the American Academy of Arts and Sciences, and the Viral Study Section of the National Institutes of Health.

See also Crossing over; DNA replication

STANLEY, WENDELL MEREDITH (1904-1971)

American biochemist

Wendell Meredith Stanley was a biochemist who was the first to isolate, purify, and characterize the crystalline form of a **virus**. During World War II, he led a team of scientists in developing a vaccine for viral influenza. His efforts have paved the way for understanding the molecular basis of **hered-**

ity and formed the foundation for the new scientific field of molecular biology. For his work in crystallizing the **tobacco mosaic virus**, Stanley shared the 1946 Nobel Prize in chemistry with John Howard Northrop and James B. Sumner.

Stanley was born in the small community of Ridgeville, Indiana. His parents, James and Claire Plessinger Stanley, were publishers of a local newspaper. As a boy, Stanley helped the business by collecting news, setting type, and delivering papers. After graduating from high school he enrolled in Earlham College, a liberal arts school in Richmond, Indiana, where he majored in chemistry and mathematics. He played football as an undergraduate, and in his senior year he became team captain and was chosen to play end on the Indiana All-State team. In June of 1926 Stanley graduated with a bachelor of science degree. His ambition was to become a football coach, but the course of his life was changed forever when an Earlham chemistry professor invited him on a trip to Illinois State University. Here, he was introduced to Roger Adams, an organic chemist, who inspired him to seek a career in chemical research. Stanley applied and was accepted as a graduate assistant in the fall of 1926.

In graduate school, Stanley worked under Adams, and his first project involved finding the stereochemical characteristics of biphenyl, a molecule containing carbon and hydrogen atoms. His second assignment was more practical; Adams was interested in finding chemicals to treat leprosy, and Stanley set out to prepare and purify compounds that would destroy the disease-causing pathogen. Stanley received his master's degree in 1927 and two years later was awarded his Ph.D. In the summer of 1930, he was awarded a National Research Council Fellowship to do postdoctoral studies with Heinrich Wieland at the University of Munich in Germany. Under Wieland's tutelage, Stanley extended his knowledge of experimental **biochemistry** by characterizing the properties of some **yeast** compounds.

Stanley returned to the United States in 1931 to accept the post of research assistant at the Rockefeller Institute in New York City. Stanley was assigned to work with W. J. V. Osterhout, who was studying how living cells absorb potassium ions from seawater. Stanley was asked to find a suitable chemical model that would simulate how a marine plant called *Valonia* functions. Stanley discovered a way of using a water-insoluble solution sandwiched between two layers of water to model the way the plant exchanged ions with its environment. The work on *Valonia* served to extend Stanley's knowledge of biophysical systems, and it introduced him to current problems in biological chemistry.

In 1932, Stanley moved to the Rockefeller Institute's Division of Plant Pathology in Princeton, New Jersey. He was primarily interested in studying viruses. Viruses were known to cause diseases in plants and animals, but little was known about how they functioned. Stanley's assignment was to characterize viruses and determine their composition and structure.

Stanley began work on a virus that had long been associated with the field of virology. In 1892, D. Ivanovsky, a Russian scientist, had studied tobacco **mosaic** disease, in which infected tobacco plants develop a characteristic mosaic pattern of dark and light spots. He found that the tobacco plant juice retained its ability to cause infection even after it was passed through a filter. Six years later M. Beijerinck, a Dutch scientist, realized the significance of Ivanovsky's discovery: the filtration technique used by Ivanovsky would have filtered out all known **bacteria**, and the fact that the filtered juice remained infectious must have meant that something smaller than a bacterium and invisible to the ordinary light microscope was responsible for the disease. Beijerinck concluded that tobacco mosaic disease was caused by a previously undiscovered type of infective agent, a virus.

Stanley was aware of recent techniques used to precipitate the tobacco mosaic virus (TMV) with common chemicals. These results led him to believe that the virus might be a protein susceptible to the reagents used in protein chemistry. He set out to isolate, purify, and concentrate the tobacco mosaic virus. He planted Turkish tobacco plants, and when the plants were about six inches tall, he rubbed the leaves with a swab of linen dipped in TMV solution. After a few days the heavily infected plants were chopped and frozen. Later, he ground and mashed the frozen plants to obtain a thick, dark liquid. He then subjected the TMV liquid to various **enzymes** and found that some would inactivate the virus and concluded that TMV must be a protein or something similar. After exposing the liquid to more than 100 different chemicals, Stanley determined that the virus was inactivated by the same chemicals that typically inactivated proteins, and this suggested to him, as well as others, that TMV was protein-like in nature.

Stanley then turned his attention to obtaining a pure sample of the virus. He decanted, filtered, precipitated, and evaporated the tobacco juice many times. With each chemical operation, the juice became more clear and the solution more infectious. The end result of two-and-one-half years of work was a clear concentrated solution of TMV which began to form into crystals when stirred. Stanley filtered and collected the tiny, white crystals and discovered that they retained their ability to produce the characteristic lesions of tobacco mosaic disease.

After successfully crystallizing TMV, Stanley's work turned toward characterizing its properties. In 1936, two English scientists at Cambridge University confirmed Stanley's work by isolating TMV crystals. They discovered that the virus consisted of ninety-four percent protein and six percent nucleic acid, and they concluded that TMV was a nucleoprotein. Stanley was skeptical at first. Later studies, however, showed that the virus became inactivated upon removal of the nucleic acid, and this work convinced him that TMV was indeed a nucleoprotein. In addition to chemical evidence, the first **electron microscope** pictures of TMV were produced by researchers in Germany. The pictures showed the crystals to have a distinct rod-like shape. For his work in crystallizing the tobacco mosaic virus, Stanley shared the 1946 Nobel prize in chemistry with John Howard Northrop and James Sumner.

During World War II, Stanley was asked to participate in efforts to prevent viral diseases, and he joined the Office of

Scientific Research and Development in Washington D.C. Here, he worked on the problem of finding a vaccine effective against viral influenza. Such a substance would change the virus so that the body's immune system could build up defenses without causing the disease. Using fertilized hen eggs as a source, he proceeded to grow, isolate, and purify the virus. After many attempts, he discovered that formaldehyde, the chemical used as a biological preservative, would inactivate the virus but still induce the body to produce antibodies. The first flu vaccine was tested and found to be remarkably effective against viral influenza. For his work in developing large-scale methods of preparing vaccines, he was awarded the Presidential Certificate of Merit in 1948.

In 1948, Stanley moved to the University of California in Berkeley, where he became director of a new virology laboratory and chair of the department of biochemistry. In five years Stanley assembled an impressive team of scientists and technicians who reopened the study of plant viruses and began an intensive effort to characterize large, biologically important molecules. In 1955 **Heinz Fraenkel-Conrat**, a protein chemist, and R. C. Williams, an electron microscopist, took TMV apart and reassembled the viral **RNA**, thus proving that RNA was the infectious component. In addition, their work indicated that the protein component of TMV served only as a protective cover. Other workers in the virus laboratory succeeded in isolating and crystallizing the virus responsible for polio, and in 1960 Stanley led a group that determined the complete **amino acid** sequence of TMV protein. In the early 1960s, Stanley became interested in a possible link between viruses and **cancer**.

Stanley was an advocate of academic freedom. In the 1950s, when his university was embroiled in the politics of McCarthyism, members of the faculty were asked to sign oaths of loyalty to the United States. Although Stanley signed the oath of loyalty, he publicly defended those who chose not to, and his actions led to court decisions which eventually invalidated the requirement.

Stanley received many awards, including the Alder Prize from Harvard University in 1938, the Nichols Medal of the American Chemical Society in 1946, and the Scientific Achievement Award of the American Medical Association in 1966. He held honorary doctorates from many colleges and universities. He was a prolific author of more than 150 publications and he co-edited a three volume compendium entitled *The Viruses.* By lecturing, writing, and appearing on television he helped bring important scientific issues before the public. He served on many boards and commissions, including the National Institute of Health, the World Health Organization, and the National Cancer Institute.

Stanley married Marian Staples Jay on June 25, 1929. The two met at the University of Illinois, when they both were graduate students in chemistry. They co-authored a scientific paper together with Adams, which was published the same year they were married. The Stanleys had three daughters and one son. While attending a conference on biochemistry in Spain, Stanley died from a heart attack at the age of 66.

STASIGENESIS

Stasigenesis is the apparent absence of observable change (**evolution**) in a population over time. In terms of **genetics**, a population that remains observably stable (stasigenic) over time is a phenotypically stable population. Although a population may be phenotypically stable there are still underlying changes that constantly occur at the genetic molecular level. When a population is stasigenic, the **evolutionary mechanisms** that cause genetic changes (or result in changes in the **gene pool** of a population) do not result observable (phenotypic) changes upon which **natural selection** acts.

The existing fossil record reflects extended periods of stasigenesis for the majority of **species**. In fact, stasigenesis is the most common macroevolutionary pattern found in the fossil record. This evolutionary stagnation (stasis) is often explained by evolutionary biologists as a result of a lack of variation in traits upon which the mechanisms of natural **selection** can act to drive genetic and phenotypic change.

Stasigenesis is also linked to environmental change. Genetic changes, acted upon by selection, result in species well adapted to their environments. If a population is well adapted to its environment, and that environment remains stable, the random mechanisms of genetic change may cancel out, produce changes (mutations) that remain at very low levels within a population, or not be sufficient to produce observable change. As a result, once a species is well-adapted, changes must occur as a result of environmental change or as a result of random genetic changes (e.g., mutations).

Within a stasigenic population, where there is stability in the phenotype from one generation to the next, a genetic **mutation** resulting in a phenotypic difference may be subjected to tremendous selective pressure (i.e., the change may be so great that it is lethal to the individual or otherwise has a significant and adverse impact on reproductive success). In this case, the evolutionary mechanisms of selection are stabilizing in that they act to conserve the existing genes (**genome**).

Stasigenesis is a type of macroevolution that contrasts with gradual changes in phyla caused by small but measurable differences in observable (phenotypic) traits and with more rapid changes (**punctuated equilibrium**). Macroevolutionary changes ultimately reflect genetic changes significant to produce changes in structure and function (e.g., changes to wing structure and function). In accord with the theory of punctuated equilibrium, populations may remain stasigenic (stable) for extended periods between periods of rapid change that can result in genetically incompatible species and or extinctions.

See also Founder effect

STATISTICAL ANALYSIS OF GENETIC SYNDROMES

Because of computer technologies, **genetics** has benefited during the past 20 years from specific and complex explanations of the mechanisms leading to genetic syndromes and diseases.

Genetic syndromes are complex occurrences with genotype aberrations, with or without observable (phenotypic) characteristics, that can seriously affect the normal life of individuals. Embryological and fetal anomalies (abnormalities) that may or may not include genetic defects, are classified as individual anomalies (e.g., malformation, disruption, deformation, dysplasia), and pattern anomalies (e.g., associations, sequences, syndromes, and developmental field defects). With computers, statistical analyses of such diseases can be accomplished easily, including evaluation of the frequency of a single malformation in the presence of a genetic disease, or in complex processes such as genetic transmission analysis (linkage studies), and multivariate analyses of patterns and **risk factors**.

Linkage is the occurrence of two or more genes (genetic loci) with a higher probability to segregate together rather than independently during **meiosis**. Linkage occurs because **crossing over** does not usually take place between loci that are close to each other. The unit used to express how close two linked genes are is the centimorgan (cM), or percent **recombination**. The statistical method of measuring linkage is the logarithm of the odds (Lod score) that expresses the odds in favor of finding the observed combination of alleles at the loci being studied if they are linked at the given distance, rather than being unlinked. A specific cut-off point of Lod score of +3 is considered strong evidence of linkage. Linkage allows the determination of the likely genotype of an individual, and the patterns of **inheritance** of a specific form of a heterogenetic disease. Linkage is also useful in determining the role of genetic factors in heterogeneous conditions, as well as **cleft lip and palate** and insulin-dependent diabetes. Clinical uses of linkage include prenatal diagnosis, **carrier** detection and pre-symptomatic diagnosis.

Cluster analysis is a wide set of multivariate techniques that attempts to identify relatively homogeneous groups of cases (or variables) based on selected characteristics, as well as phenotype patterns, using an algorithm that starts with each case (or variable) in a separate cluster and combines clusters until only one is left. In the analysis of genetic syndromes, cluster analysis can be useful in finding new syndromes based upon a recurrent pattern of phenotypic characteristics. Cluster analysis also yields the accuracy of the statistical model to properly classify the subjects, if the dependent variable (the ascertained genetic disease, for example) is known. A new application of cluster analysis is represented by the microarray. This technology promises to monitor, by means of sophisticated software, the whole **genome** on a single chip so that researchers have a better picture of the simultaneous interactions among thousands of genes. A stepwise **selection** of **gene expression** that has a different pattern in pathologic cases when compared to controls (experimental standards) allows a proper prediction model able to swiftly classify the subjects.

Latent class analysis is another sophisticated model that can be used for assessing the validity of discrete measurement, such as categorical results of some non-definitive diagnostic criteria. Initially, latent class analysis was mainly used for studies in psychology. In this statistical approach, the observed data (e.g., phenotype anomalies) are considered indicators of a non-directly observable variable for a genetic disease. Therefore, different realizations correspond to different patterns of observations. Latent class analysis is able to identify these latent types of patterns that can also be used to estimate the conditional probability for a subject belonging to the assigned class. Latent class analysis is also used to evaluate how a pattern of malformations can correctly detect a specific phenotype associated with a karyotype abnormality.

Non-invasive prenatal genetic research is a growing field that has as its main aim the selection of a population of pregnant women with a higher risk of carrying a fetus affected by genetic disease. One statistical method used is the Bart test, which calculates the probability of the diseases using quantitative biochemical markers (or a sonographic marker called nuchal translucency) able to adjust the age-specific probability of a syndrome for an appropriate correction factor for the individual called the likelihood ratio. The likelihood ratio is approximately homogenous with maternal age probability, and their combination yields an adjusted and more correct probability of the disease (also called odds of being affected given a positive result, or OAPR). An analogue of the Bart test for qualitative markers (as well as sonographic findings) is logistic regression, that evaluates — by means of a slightly different factor of correction named the odds ratio — the association of a positive marker (also called risk factor) to the pathology (also called outcome). The odds ratio is a very useful method to understand the magnitude of an association between a qualitative variable with an event. It is expressed as the ratio between the risk of observing a disease with and without a risk factor. The logistic regression output is usually a bit less reliable than the Bart test, and it must, therefore, be used and evaluated with care.

Several new models are available for the evaluation of genetic diseases. The use of such techniques is strictly relegated to their clinical applicability, and only after after careful patient counselling. In fact, because of the relative rarity of genetic syndromes and the enormous span of phenotype patterns, it is difficult to collect a population to produce "robust" results unaffected by bias and inaccurate predictive values.

See also Genetic counseling: Risk calculations using Bayesian statistics; Prenatal diagnostic techniques

STEBBINS, JR., GEORGE LEDYARD (1906-2000)
American botanist

Considered the founder of evolutionary botany, George Ledyard Stebbins, Jr., was the first scientist to apply modern synthetic evolutionary theory to the plant kingdom. Stebbins was one of the twentieth-century architects who developed the evolutionary synthesis by considering analysis of organic fossils and genetic information with respect to Darwinian theories, specifically **natural selection**. Stebbins' ideas established and advanced evolutionary biology techniques that examined the processes and mechanisms of genetic **mutation**, **recombination**, and **chromosome** structure and quantity in plants. By applying evolutionary concepts to plants, Stebbins became the first person to synthesize artificially a plant **species** that survived in a natural environment. His contributions to plant **evolution** guided other researchers who elaborated on his research, and his findings, especially concerning plant speciation, became accepted as a foundation for botanical investigations. Stebbins' familiarity with plants enhanced his theoretical insights which have been cited as some of the most significant twentieth-century scientific achievements.

Stebbins was born at Lawrence, New York, to affluent parents who encouraged him to explore and understand nature. Because his mother had tuberculosis, the family moved to California where Stebbins enjoyed examining regional plants. At the family's summer home near Seal Harbor, Maine, Stebbins learned about the flora and fauna of tide pools. Enrolling at Harvard University in 1924, Stebbins considered a career as an attorney or musician until Professor Merritt Lyndon Fernald, a botany expert, convinced him to earn botanical degrees. Stebbins completed a doctorate in 1931. While attending graduate school, Stebbins became interested in applying **genetics** to botanical research, but the use of innovative chromosomal analysis techniques divided botanists, and conflicts between professors delayed Stebbins's academic progress. Stebbins attended the International Botanical Congress at Cambridge, England, 1930, meeting eminent botanists who inspired him.

Conducting cytogenetic investigations, Stebbins taught at Colgate University in Hamilton, New York, from 1931 to 1935 when he accepted a position at the University of California at Berkeley where he intensified his evolution research. He became acquainted with **Theodosius Dobzhansky**, a California Institute of Technology geneticist who researched fruit flies. Both men were interested in studying the role of chromosomes in evolution, collected specimens together during horseback rides, and debated evolution research. Stebbins helped establish the Biosystematists, a group of scientists in the San Francisco Bay area committed to developing evolutionary methodology.

Working with Ernest B. Babcock, Stebbins experimented with flowering plant species. As early as the 1940s, Stebbins produced fertile hybrids by using artificially induced polyploidy to double the number of plants' chromosomes and form new species. He created wild grass polyploids, including *Ehrharta erecta,* a new species grown in a natural setting. Stebbins actively participated in the Society for the Study of Evolution, serving as that group's third president in 1948. Because most evolutionary researchers were zoologists, Stebbins voiced his concerns that the society's journal, *Evolution,* focused more on animals than plants, alienating botanists.

In 1947, Stebbins lectured at Columbia University, and his presentations were the basis of his 1950 book, *Variation and Evolution in Plants,* considered a classic text. Stebbins discussed his pioneering evolutionary botanical research and development of plant species, stressing that plants, like animals, undergo evolution but in unique ways because their cells differ. After his book was published, Stebbins moved to the University of California's campus at Davis to create a genetics department.

Stebbins wrote *Processes of Organic Evolution* (1966), *The Basis of Progressive Evolution* (1969), and *Chromosomal Evolution in Higher Plants* (1971) prior to retiring in 1973. He was a visiting professor at the University of Chile and several American universities. Stebbins later published *Flowering Plants: Evolution Above the Species Level* (1974) and *Darwin to DNA, Molecules to Humanity* (1982). He was a co-author of the textbook *Evolution* (1977) with Theodosius Dobzhansky, Francisco Ayala, and James Valentine. Stebbins promoted conservation and sought preservation of a Monterey Peninsula beach while he was president of the California Native Plant Society in 1967. He encouraged the protection of indigenous plants, especially rare species, and their habitats.

Elected a member of the National Academy of Sciences in 1952, Stebbins was also awarded a 1979 National Medal of Science in addition to receiving other significant medals, prizes, and fellowships presented by international scientific societies and honorary degrees from educational institutions. He was president of the Botanical Society of America and secretary general of the Union of Biological Sciences. In 1980, the University of California at Davis designated the Stebbins Cold Canyon Reserve to recognize his commitment to botany. National symposiums featured papers about Stebbins's theories and work. Stebbins died at Davis, California.

See also Agricultural genetics; Biotechnology

STEIN, WILLIAM HOWARD (1911-1980)

American biochemist

William Howard Stein, in partnership with **Stanford Moore**, was a pioneer in the field of protein chemistry. Although other scientists had previously established that proteins could play such roles as that of **enzymes**, antibodies, hormones, and oxygen carriers, almost nothing was known of their chemical makeup. Stein and Moore, during some forty years of collaboration, were not only able to provide information about the inner workings of protein molecules, but also invented the mechanical means by which that information could be extracted. Their discovery of how protein amino acids function was accomplished through a study of **ribonuclease** (RNase), a pancreatic enzyme that assists in the digestion of food by catalyzing the breakdown of nucleic acids. But their work could not have been accomplished without the development of a technology to assist them in collecting and separating the amino acids contained in ribonuclease. Their invention of the fraction collector and an automated system for analyzing amino acids was of great importance in furthering protein research, and these devices have become standard laboratory equipment.

Stein and Moore began their collective work in the late 1930s under Max Bergmann at the Rockefeller Institute (now

Rockefeller University). After Bergmann's death in 1944, the pair developed the protein chemistry program at the Institute and began their research into enzyme analysis. Except for a brief period during World War II when Moore served with the Office of Scientific Research and Development in Washington D.C., and the two years when Stein taught at the University of Chicago and Harvard University, the partnership continued uninterrupted until Stein's death in 1980. Their joint inventions and co-authorship of most of their scientific papers were said to make it impossible to separate their individual accomplishments. Their combined efforts were acknowledged in 1972 with the Nobel Prize in chemistry. According to Moore, writing about Stein in the *Journal of Biological Chemistry* in 1980, they received the award "for contributions to the knowledge of the chemical structure and catalytic function of bovine pancreatic ribonuclease." Christian Anfinsen shared the Nobel Prize with Stein and Moore for related research.

The son of community-minded parents, Stein was born in New York City on June 25, 1911. He was the second of three children. His father, Fred M. Stein, was involved in business and retired at an early age to lend his services to various health care associations in the community. The scientist's mother, Beatrice Borg Stein, worked to improve recreational and educational conditions for underprivileged children. >From an early age, Stein was encouraged by his parents to develop an interest in science. He received a progressive education from grade school on, attending the Lincoln School of the Teacher's College of Columbia University, transferring at sixteen years of age to Phillips Exeter Academy for his college preparatory studies. He graduated from Harvard University in 1933, then took a year of graduate study in organic chemistry there. Finding that his real interest was **biochemistry**, he completed his graduate studies at the College of Physicians and Surgeons of Columbia University, receiving his Ph.D. in 1938. His dissertation concerned the **amino acid** composition of elastin, a protein found in the walls of veins and arteries. This work marks the beginning of his long search to understand the chemical function of proteins.

The successful research being done at the Rockefeller Institute under the direction of Max Bergmann caught Stein's attention. He pursued post-graduate studies there in 1938, spending his time improving analytical techniques for purifying amino acids. Moore joined Bergmann's group in 1939. There, he and Stein began work in developing the methodology for analyzing the amino acids glycine and leucine. Their work was interrupted when the United States entered World War II. Then, Bergmann's laboratory was given over to the study of the physiological effects of mustard gases, in the hope of finding a counteractant.

The group's efforts to find accurate tools and methods for the study of amino acid structure increased in importance when they assumed the responsibility of establishing the Institute's first program in protein chemistry. Looking for ways to improve the separation process of amino acids, they turned to partition chromatography, a filtering technique developed during the war by the English biochemists A. J. P. Martin and Richard Synge. Building on this technology, as

well as that of English biochemist Frederick Sanger's column chromatography and the ion-exchange technique of Werner Hirs, Stein and Moore went on to invent the automatic fraction collector and develop the automated system by which amino acids could be quickly analyzed. This automated system replaced the tedious two-week sequence that was previously required to differentiate and separate each amino acid.

From then on, the isolation and study of amino acid structure was advanced through these new analytical tools. Ribonuclease was the first enzyme for which the biochemical function was determined. The discovery that the amino acid sequence was a three-dimensional, chain-like structure that folds and bends to cause a catalytic reaction was a beginning for understanding the complex nature of enzyme catalysis. Stein and Moore were certain that this understanding would result in crucial medical advances. By 1972, the year Stein and Moore shared the Nobel Prize, other enzymes had been analyzed using their methods.

Because he was extremely eager to see that research done in laboratories all over the country be disseminated as widely and as quickly as possible, Stein devoted many years in various editorial positions to the *Journal of Biological Chemistry*. Under his leadership, the journal became a leading biochemistry publication. He had joined the editorial board in 1962 and became editor in 1968. He only held the latter post for one year, however. While attending an international meeting in Denmark, he contracted Guillain-Barré Syndrome, a rare disease often causing temporary paralysis. In grave danger of dying, he managed to recover somewhat. The illness left him a quadriplegic, confined to a wheelchair for the rest of his life. Although he remained involved with the work of his colleagues both in the laboratory and at the *Journal,* he was unable to participate actively.

In addition to the Nobel Prize, Stein shared with Moore the 1964 Award in Chromatography and Electrophoresis and the 1972 Theodore Richard Williams Medal of the American Chemical Society. He served as chairperson of the U.S. National Committee for Biochemistry from 1968 to 1969, as trustee of Montefiore Hospital, and as board member of the Hebrew University medical school. He married Phoebe L. Hockstader on June 22, 1936. They had three sons: William Howard, Jr., David, and Robert. Stein died in Manhattan on February 2, 1980.

See also Proteins and enzymes; Translation

STEM CELL RESEARCH

Stem cells are cells that can divide for an infinite period of time when being grown outside of the body, and which can differentiate into various types of specialized cells. When **fertilization** of an egg with sperm occurs, the resulting fertilized **cell** has the capability to form an entire organism. The cell is described as being totipotent (having total potential). After some time, as rounds of **cell division** occur, specialization of cells occurs. But, early in fetal development, before the developing mass of cells attaches itself to the wall of the uterus, some cells still retain the ability to form virtually every type of cell in the body. These cells are pluripotent (capable of differentiating into many types of cells but not all types required for fetal development). With continued fetal development, further specialization of **pluripotent stem cells** results in multipotent stem cells—cells that give rise to cells having a particular function, such as blood cells and various types of skin cells. Indeed, life would be impossible without blood stem cells, which function to replenish blood cell supply throughout life.

Stem cell research is concerned primarily with the pluripotent cells. The field is relatively new. James Thomson reported in *Science* in late 1998 his success in maintaining undifferentiated embryonic stem cells in their undifferentiated state in lab culture.

Stem cells can be obtained from human embryos at the so-called blastocyst stage (a stage very early in fetal development, only a few division cycles after fertilization). As well, cells can be obtained from fetal tissue from terminated pregnancies. The latter procedure has precipitated much discourse. In August 2001, United States president George W. Bush announced that he would support very limited federal funding of research using stem cells from human embryos. It was a compromise that did not completely satisfy parties on either side of the controversial issue.

Another potential means of obtaining pluripotent stem cells may be a technique called somatic cell nuclear transfer. In the technique involves the physical removal of its **nucleus** from an egg cell. The nucleus is the specialized area of the cell that contains the organized pieces of genetic material called the chromosomes. The material left behind in the egg cell contains nutrients and other energy-producing materials necessary for development of the embryo. Then, a somatic cell—any cell other than an egg or a sperm cell—is placed next to the denucleated egg cell, and the two cells are chemically fused together. After a requisite number of cell divisions, pluripotent stem cells can, at least in theory thus far, be recovered and used.

Pluripotent stem cells are important to science and to advances in health care. At the most fundamental level, study of these cells could advance the understanding of the processes of cellular development, such as the orchestrated mechanisms by which genes are turned on and off during development and growth. Some of the most serious medical conditions, such as **cancer** and **birth defects**, are due to abnormal cell specialization and cell division. Pluripotent stem cells could also be used to screen new drugs, eliminating the need to use living subjects for the early phases of drug discovery. The most far-reaching potential application of the stem cells is the generation of cells and tissue that could be used for so-called cell therapies. Potentially stem cells may function as a kind of universal human donor cell, which could serve as raw material for whatever diseased cell requires replacing. Such donor cells would have to be genetically engineered so as not to form the cell surface molecules that would alert the recipient's immune system. The cells could be used for replacement of defective or diseased cells without the danger of transplantation rejection that occurs presently.

Potential applications include the replacement of defective heart tissue and replacement of malfunctioning insulin producing cells in Type I diabetes. In the last several years, several lines of research have produced concrete results showing the potential of stem cells in **cell therapy**. Genetic engineering of stem cells may be promising as a cancer eradication strategy. In rats, neural stem cells genetically engineered to convert a compound into a cancer-killing agent have been found to selectively target and destroy cancerous cells in the brain. Elsewhere, neural stem cells have also been shown capable of integration into the diseased retina of rats and of taking on some of the characteristics of retinal cells. This holds the promise that stem cell therapy may aid in repairing retinal damage. Other researchers have demonstrated, again in rats, that stem cells in the brain were able to repair damaged areas and restore function when stimulated by a growth-inducing protein. If replicated in humans, then stem cell treatments for stroke, nervous system and spinal cord injury and diseases such as Parkinson's and Alzheimer's that are marked by degeneration of nerve cells.

Another application of stem cells has been to form a chimera—an animal that grows from an embryo in which stem cells from another animal have been inserted. Some of the chimera's cells have one set of parents, and some cells have another set of parents. "Knockout" mice, research animals lacking specific genes, are chimeras. While theoretically conceivable, human chimeras are not contemplated.

Researchers have claimed success at reprogramming multipotent cells for a function other than that they were programmed for. Specifically, adult skin cells from cattle were reverted to stem cells and then transformed into heart cells. Other studies involved neural stem cells from mice and bone marrow cells from rats have also indicated that functional reprogramming of adult cells may be feasible. These breakthrough studies hold forth the potential of using cells from adults to treat diseases, rather than extracting embryonic cells. For the present, however, there are barriers to the use of adult stem cells. Such cells may not be present in all tissues of the body. More knowledge of the locations of adult stem cells is still required. Secondly, adult stem cells are present in minute quantities, are difficult to isolate and their number decreases with age. The time necessary to locate, harvest, and grow the cells to usable numbers may be too long for practical purposes. Finally, adult stem cells may contain **DNA** abnormalities, which have accumulated as a result of a lifetime of exposure to DNA-altering agents such as sunlight and toxic chemicals. Further research may overcome these limitations, allowing stem cells obtained from adults to be used in cell therapy.

See also Cell differentiation

STEM CELLS

Stem cells are undifferentiated, multipotential somatic cells from which other, specialized cells arise by **differentiation**. Stem cells are cells which have the ability to divide infinitely, and to develop into the huge variety of specialized cells required for the creation of an organism. Most cells have a specific function (e.g., liver cells, skin cells, brain cells, etc.) and once they have taken on this function, in a process called differentiation, they cannot be adapted for any other function. Stem cells, however, have not gone through the differentiation process.

By isolating stem cells in a laboratory, scientists theoretically could grow new heart cells to repair damage from heart attacks, new liver cells to treat hepatitis, or new red blood cells for **cancer** patients. As of 2001 the methods by which customize stem cells with respect to function is as yet unknown.

A number of different types of stem cells have been identified. Totipotent (total potency) stem cells represent the **cell** created during the fusion of sperm and egg in conception, and the cells created in the next few **cell division** cycles. Every cell contains all the genetic information necessary to create an entire being. With time these totipotent cells begin to specialize into pluripotent cells, that have the potential to create many but not all of the types of tissue necessary for the development of the entire being.

Pluripotent stem cells develop into more specialized stem cells committed to the generation of a specific cell line (examples include skin stem cell, liver stem cell, blood stem cell, nerve stem cell). Specialization is a one-way process and a specific stem cell line cannot become less specialized.

Human pluripotent stem cells are important to science and to advances in health care. At the most fundamental level, stem cells could further the understanding of the myriad complex events that occur during human development. Of primary importance would be the identification of factors involved in the cellular decision-making process that results in cell specialization. Conditions such as cancer and **birth defects** are due to abnormal cell specialization and cell division. A better understanding of normal cell processes will allow scientists to better understand the fundamental errors that cause these diseases.

Stem cell research could also change the way drugs are developed and tested. Testing in animals and humans could be reduced, as stem cell lines could be used in the initial screening of therapeutic agents. Only those showing potential would be carried through to animal or human testing.

The most far-reaching potential of human pluripotent stem cells is the generation of cells and tissue that could be used for so-called cell therapies. Many diseases and disorders result from disruption of cellular function or destruction of tissues of the body. The organs and tissues available for transplantation is far outnumbered by the number of people in need. Stem cells, stimulated to develop into specialized cells, offer the possibility of a renewable source of replacement cells and tissue to treat diseases and conditions such as Parkinson's

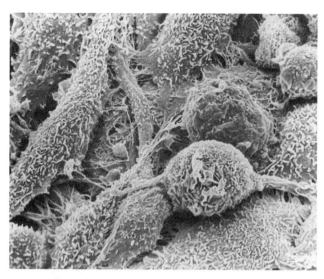

Scanning electron micrograph of cultured mammalian embryonic stem cells. *Illustration by Yorgos Nikas. Photo Researchers. Reproduced by permission.*

disease, **Alzheimer's disease**, spinal cord injury, stoke, burns, heart disease, diabetes, osteoarthritis and rheumatoid arthritis.

Blood stem cells are capable of differentiating into red blood cells, platelets, and the many types of white blood cells necessary for the immune system. Stem cells have been utilized to treat cancer patients whose immune systems have been destroyed by chemotherapy.

Another stem cell therapeutic procedure involves the administration of bone marrow cells. Anticancer strategies that target the rapidly growing cancerous cells also compromise other fast growing cells, such as bone marrow cells. Transplantation of bone marrow cells, themselves a type of stem cells, or a peripheral blood stem cell transplant, can restore the vital functions lost by the therapeutic destruction of the original marrow population.

In early 1999, there were several reports of the successful laboratory growth and differentiation of stem cells into multiple cell lines. Such a strategy could provide a ready supply of transplantation and material. However, ethical concerns over the embryonic source of the stem cells need to be addressed before the potential of this stem cell technology can be fully realized. Adult stem cells do exist. However, their small number would require cell culture techniques to increase their numbers to therapeutically useful levels. The time required to do this might not allow for fully effective therapy.

More generally, the potential of stem cell transplantation therapy will require a solution to the problem of immune rejection. Because pluripotent stem cells obtained from embryos or fetal tissue would be genetically different from the recipient, future research would need to focus on modifying human stem cells to minimize tissue incompatibility or to create tissue banks with the most common tissue-type profiles.

See also Cell differentiation; Developmental genetics; Gene therapy; Stem cell research

STEPTOE, PATRICK (1913-1988)

English gynecologist

Patrick Steptoe, an English gynecologist and medical researcher, helped develop the technique of *in vitro* **fertilization**. In this process, a mature egg is removed from the female ovary and is fertilized in a test tube. After a short incubation period, the fertilized egg is implanted in the uterus, where it develops as in a typical pregnancy. This procedure gave women whose fallopian tubes were damaged or missing, and were thus unable to become pregnant, the hope that they too could conceive children. Steptoe and his colleague, English physiologist Robert G. Edwards, received international recognition—both positive and negative—when the first so-called test tube baby was born in 1978.

Patrick Christopher Steptoe was born in Oxfordshire, England. His father was a church organist, while his mother served as a social worker. Steptoe studied medicine at the University of London's St. George Hospital Medical School and, after being licensed in 1939, became a member of the Royal College of Surgeons. His medical career was interrupted for a time by World War II, when Steptoe volunteered as a naval surgeon. Following the war, Steptoe completed additional studies in obstetrics and gynecology. In 1948, he became a member of the Royal College of Obstetricians and Gynecologists and moved to Manchester to set up a private practice. In 1951, Steptoe began working at Oldham General and District Hospital in northeast England.

While at Oldham General and District Hospital, Steptoe pursued his interest in fertility problems. He developed a method of procuring human eggs from the ovaries by using a laparoscope, a long thin telescope replete with fiber optics light. After inserting the device—through a small incision in the navel—into the inflated abdominal cavity, Steptoe was able to observe the reproductive tract. Eventually the laparoscope would become widely used in various types of surgery, including those associated with sterility. But, at first, Steptoe had trouble convincing others in the medical profession of the merits of laparoscopy; observers from the Royal College of Obstetricians and Gynecologists considered the technique fraught with difficulties. Five years passed before Steptoe published his first paper on laparoscopic surgery.

In 1966 Steptoe teamed with Cambridge University physiologist Robert G. Edwards to propel his work with fertility problems. Utilizing ovaries removed for medical reasons, Edwards had pioneered the fertilization of eggs outside of the body. With his laparoscope, Steptoe added the dimension of being able to secure mature eggs at the appropriate moment in the monthly cycle when fertilization would normally occur. A breakthrough for the duo came in 1968 when Edwards successfully fertilized an egg that Steptoe had extracted. Not until 1970, however, was an egg able to reach the stage of **cell** division—into about 100 cells—when it generally moves to the uterus. In 1972 the pair attempted the first implantation, but the embryo failed to lodge in the uterus. Indeed, none of the women with implanted embryos carried them for a full trimester.

As their work progressed and word of it leaked out, the researchers faced criticism from scientific and religious circles concerning the ethical and moral issues relating to tampering with the creation of human life. Some opponents considered the duo's work akin to the scenario in Aldous Huxley's 1932 work, *Brave New World,* in which babies were conceived in the laboratory, cloned, and manipulated for society's use. Members of Parliament demanded an investigation and sources of funds were withdrawn. A *Time* reporter quoted Steptoe as saying, "All I am interested in is how to help women who are denied a baby because their tubes are incapable of doing their small part." Undaunted, Steptoe and Edwards continued their work at Kershaw's Cottage Hospital in Oldham, with Steptoe financing the research by performing legal abortions. Disturbed with the criticism, Steptoe and Edwards became more secretive, which made the speculation and criticism more intense.

In 1976 Steptoe met thirty-year-old Leslie Brown, who experienced problems with her fallopian tubes. Steptoe removed a mature egg from her ovary, and Edwards fertilized the egg using her husband Gilbert's sperm. The fertilized egg—implanted after two days—thrived, and on July 25, 1978, Joy Louise Brown, a healthy five pound twelve ounce girl was born in Oldham District and General Hospital. Even before the birth, reporters and cameramen congregated outside of the four story brick hospital, hoping for a glimpse of the expectant mother.

Steptoe and Edwards were reluctant to discuss the procedures in press conferences and did not immediately publish their findings in a medical journal. The procedures were fully presented at the January 26, 1979 meeting of the Royal College of Obstetricians and Gynecologists and at the conference of the American Fertility Society in San Francisco. Steptoe reported that with modified techniques, ten percent of the *in vitro* fertilization attempts could succeed. He further predicted that there could one day be a 50% success rate for the procedure.

In the aftermath of the first successful test tube baby, Steptoe received thousands of letters from couples seeking help in conception. He retired from the British National Health Service and constructed a new clinic near Cambridge. For their efforts, Steptoe and Edwards were both named Commanders of the British Empire, and in 1987 Steptoe was honored with fellowship in the Royal Society. Steptoe and his wife, a former actress, had one son and one daughter. His interests outside of medicine included piano and organ, cricket, plays, and opera. Steptoe died of cancer in Canterbury at age 74. Since the birth of baby Brown and the pioneering techniques of Steptoe, thousands of couples throughout the world with certain infertility problems have been able to have children through *in vitro* fertilization.

See also Embryology; In vitro fertilization

STERN, CURT (1902-1981)

German geneticist

Curt Stern furthered the understanding of the mechanism of **heredity** and fundamentally contributed to establishing the relevance of genetic studies to medicine. Stern was born in Hamburg, Germany, and early in life he displayed a strong interest in natural history. His interest led him to enroll at the University of Berlin in 1920. Stern's doctoral studies in protozoology at the Kaiser Wilhelm Institute led to his receiving a Ph.D. in 1923, at the age of 21. At the time, he was the youngest person to receive a degree from the university.

After graduation, Stern changed research direction, having the opportunity to study *Drosophila* (fruit fly) genetics at Columbia University. Stern arrived at Columbia in 1924. In the next 12 years, he published several seminal papers detailing chromosomal-linked **inheritance** of physical characteristics and crossing-over of genetic material between chromosomes.

Due to return to Germany, Stern remained in the United States because of Hitler's rise to power and the implications for German Jews. After a short stint at Western Reserve University in Cleveland, Ohio, he moved to the University of Rochester, where he remained until 1947. From 1941 to 1947, Stern served as chairman of both the Department of Zoology and the Division of Biological Sciences. While at Rochester, Stern made fundamental contributions to *Drosophila* genetics. He discovered and coined the term isoallele, to describe the range of **genetic variation** possible from certain regions of the **chromosome**. In other research, Stern demonstrated the ability of low levels of radiation to induce genetic mutations via changes in the **DNA**.

In 1947, Stern joined the faculty of the University of California, Berkeley, where he became professor of zoology and, 11 years later, professor of genetics as well. Until his retirement in 1970, Stern investigated the genetic control of **differentiation** in *Drosophila*. Stern also contributed to human genetics by alerting physicians, through his three editions of the book *Principles of Human Genetics,* that indistinguishable phenotypes can have different genetic **bases.**

Stern received many honors and awards in recognition of his scientific achievements. In 1970, Stern was diagnosed with Parkinson's disease, which contributed to his death from cardiac failure eleven years later.

See also Gene regulation; Transcription; Translation

STEVENS, NETTIE MARIA (1861-1912)

American biologist

Nettie Maria Stevens was a biologist and cytogeneticist and one of the first American women to be recognized for her contributions to scientific research. "She...produced new data and new theories," wrote Marilyn Bailey Ogilvie in *Women in Science,* "yet beyond these accomplishments passed along her expertise to a new generation.... illustrat[ing] the importance of the women's colleges in the education of women scientists." Although Stevens started her research career when she was in her thirties, she successfully expanded the fields of **embryology** and cytogenetics (the branch of biology which focuses on the study of **heredity**), particularly in the study of histology (a branch of anatomy dealing with plant and animal tissues) and of regenerative processes in invertebrates such as hydras and flatworms. She is best known for her role in genetics—her research contributed greatly to the understanding of chromosomes and heredity. She theorized that the sex of an organism was determined by the **inheritance** of a specific chromosome—X or Y—and performed experiments to confirm this hypothesis.

Stevens, the third of four children and the first daughter, was born in Cavendish, Vermont to Ephraim Stevens, a carpenter of English descent, and Julia Adams Stevens. Historians know little about her family or her early life, except that she was educated in the public schools in Westford, Massachusetts, and displayed exceptional scholastic abilities. Upon graduation, Stevens taught Latin, English, mathematics, physiology and zoology at the high school in Lebanon, New Hampshire. As a teacher, she had a great zeal for learning that she tried to impart both to her students and colleagues. Between 1881 and 1883, Stevens attended the Normal School at Westfield, Massachusetts, consistently achieving the highest scores in her class from the time she started until she graduated. She worked as a school teacher, and then as a librarian for a number of years after she graduated. Stevens then she enrolled at Stanford University in 1896, at the age of 35.

At Stanford, Stevens studied physiology under professor Oliver Peebles Jenkins. She spent summers studying at the Hopkins Seaside Laboratory, Pacific Grove, California, and pursuing her love of learning and of biology. During this time, Stevens decided to switch careers to focus on research, instead of teaching. While at Hopkins she performed research on the life cycle of *Boveria,* a protozoan parasite of sea cucumbers. Her findings were published in 1901 in the *Proceedings of the California Academy of Sciences.* After obtaining her masters degree—a highly unusual accomplishment for a woman in that era—Stevens returned to the East to study at Bryn Mawr College, Pennsylvania, as a graduate biology student in 1900. She was such an exceptional student that she was awarded a fellowship enabling her to study at the Zoological Station in Naples, Italy, and then at the Zoological Institute of the University of Würzburg, Germany. Back at Bryn Mawr, Stevens obtained her doctorate in 1903. At this time, she was made a research fellow in biology at Bryn Mawr and then was promoted to a reader in experimental morphology in 1904. From 1903 until 1905, her research was funded by a grant from the Carnegie Institution. In 1905, she was promoted again to associate in experimental morphology, a position she held until her death in 1912.

While Stevens' early research focused on morphology and taxonomy, then later expanded to cytology, her most important research was with chromosomes and their relation to heredity. Because of the pioneering studies performed by the renowned monk Gregor Mendel (showing how pea plant genetic traits are inherited), scientists of the time had some early information about how chromosomes acted during **cell division** and maturation of germ cells. However, no inherited trait had been traced from the parents' chromosomes to those of the offspring. In addition, no scientific studies had yet linked one **chromosome** with a specific characteristic. Stevens, and the well-known biologist **Edmund Beecher Wilson**, who worked independently on this type of research, were the first to demonstrate that the sex of an organism was determined by a particular chromosome; moreover, they proved that gender is inherited in accordance with Mendel's laws of **genetics**. Together, their research confirmed, and therefore established, a chromosomal basis for heredity. Working with the meal worm, *Tenebrio molitor,* Stevens determined that the male produced two kinds of sperm—one with a large X chromosome, and the other with a small **Y chromosome**. Unfertilized eggs, however, were alike, and had only X chromosomes. Stevens theorized that sex, in some organisms, may result from chromosomal inheritance. She suggested that eggs fertilized by sperm carrying X chromosomes produced females, and those by sperm carrying the Y chromosome resulted in males. She performed further research to prove this phenomenon, expanding her studies to other **species**. Although this theory was not accepted by all scientists at the time, it was profoundly important in the **evolution** of the field of genetics and to an understanding of determination of gender.

Stevens was a prolific author, publishing some thirty-eight papers in eleven years. For her paper, "A Study of the Germ Cells of *Aphis rosae* and *Aphis oenotherae,*" Stevens was awarded the Ellen Richards Research Prize in 1905, given to promote scientific research by women. Stevens died of breast **cancer** at the age of 50, before she could occupy the research professorship created for her by the Bryn Mawr trustees. Much later, **Thomas Hunt Morgan**, a 1933 Nobel Prize recipient for his work in genetics, recognized the importance of Stevens' ground-breaking experiments, as quoted by Ogilvie in the *Proceedings of the American Philosophical Society,* "Stevens had a share in a discovery of importance and her name will be remembered for this, when the minutiae of detailed investigations that she carried out have become incorporated in the general body of the subject."

See also Sex chromosomes; Sex determination; Sex-linked genes

STOPES, MARIE CHARLOTTE CARMICHAEL (1880-1958)

Scottish paleobotanist and social reformer

Inspired by meeting **Margaret Sanger** (1879–1966) in 1915, Marie Stopes began crusading for sexual freedom and birth control. With her second husband, Humphrey Verdon Roe (1878–1949), she opened the first birth control clinic in Great Britain, "The Mothers' Clinic" in Holloway, North London, on 17 March 1921.

Marie Stopes. *Photo by Underwood & Underwood. The Library of Congress.*

Stopes was born in Edinburgh, Scotland to the English architect, archeologist, and geologist Henry Stopes (1852–1902) and his feminist wife, Charlotte Carmichael (1841–1929), one of the first women to attend a Scottish university. Stopes and her younger sister, Winnie, were raised in London in a curious mixture of socially progressive scientific thought and stern Scottish Protestantism. Her authoritarian mother trusted the Bible, but supported woman suffrage, clothing reform, and free thought. Stope's father cared mainly for science. As a young girl Stopes met many of her father's friends in the British Association for the Advancement of Science, including **Francis Galton** (1822–1911), **Thomas Henry Huxley** (1825–1895), Norman McColl (1843–1904), and Charles Edward Sayle (1864–1924). Through them came Stope's interest in **Charles Darwin**, **evolution**, and eventually, **eugenics**.

Stopes enrolled at University College, London, in 1900 on a science scholarship, graduating B.Sc. in 1902 with honors in botany and geology. She did graduate work there until 1903, then at the University of Munich, where she received her Ph.D. in paleobotany in June 1904. In October of the same year, Stopes became the first woman scientist on the faculty of the University of Manchester. In 1905, University College made her the youngest Briton of either gender to earn the D.Sc. She studied at the Imperial University of Tokyo from 1907 to 1908, then returned to Manchester in 1909. Stopes married botanist and geneticist Reginald Ruggles Gates (1882–1962) in 1911, but obtained an annulment five years later.

Devoted to eugenics, Stopes founded the Society for Constructive Birth Control and Racial Progress in 1921, and after 1937 was a Life Fellow of the British Eugenics Society. Stopes advocated the involuntary sterilization of anyone she deemed unfit for parenthood, including the mentally impaired, addicts, subversives, criminals, and those of mixed racial origin. At one time, Stopes persecuted her son, Harry Stopes Roe (b. 1924), for marrying a woman with bad eyesight. While Sanger's expressed main motivation in promoting birth control was to relieve the misery of the poor, Stopes campaigned vigorously and often flamboyantly for birth control to prevent "inferior" women from reproducing, and to allow all women to lead fulfilling lives without fear of pregnancy. Stopes made enemies on all sides of the issue. Sanger, Havelock Ellis (1859–1939), and other left-leaning rivals within the birth control movement accused her of anti-Semitism, political conservatism, and egomania. Stope's strongest opposition came from the Roman Catholic Church, especially because, unlike most other early advocates of birth control, she did not oppose abortion.

By her own account Stopes had three distinct careers, a scientist until about 1914, a social reformer until the late 1930s, and a poet thereafter. Among her books are *Married Love: A New Contribution to the Solution of Sex Difficulties* (1918), *Wise Parenthood: A Sequel to "Married Love": A Book for Married People* (1919), *Radiant Motherhood: A Book for Those who are Creating the Future* (1920), *Contraception (Birth Control): Its Theory, History and Practice* (1923), *The Human Body* (1926), *Sex and the Young* (1926), *Enduring Passion: Further New Contributions to the Solution of Sex Difficulties* (1928), *Mother England: A Contemporary History* (1929), *Roman Catholic Methods of Birth Control* (1933), *Birth Control To-Day* (1934), *Marriage in my Time* (1935), *Change of Life in Men and Women* (1936), and *Your Baby's First Year* (1939). Stopes died quietly at her home near Dorking, Surrey, England.

See also Birth defects; Eugenics; Family; Genetic concepts of race and ethnicity; Genetic implications of mating and marriage customs; Inbreeding; Social Darwinism; Sociobiology

STRASBURGER, EDUARD ADOLF (1844-1912)

German botanist

Eduard Adolf Strasburger (not Eduard Gottlieb as is sometimes reported) was born in Warsaw, Poland, then governed by Russia. After attending the Sorbonne in Paris from 1862 to 1864, he studied botany under Hermann Schacht (1814–1864) at the University of Bonn, Germany, where he also heard botanical lectures by Julius von Sachs (1832–1897), who was visiting from Poppelsdorf, and Nathanael Pringsheim (1823–1894), who was visiting from Jena. When Schacht died

suddenly, Strasburger transferred to the University of Jena and became Pringsheim's assistant and protege.

Strasburger was also the protege of **Ernst Haeckel** (1834–1919) at Jena. Through Haeckel, Strasburger became a devotee of **Charles Darwin's** (1809–1882) evolutionary theory. Haeckel's influence gained Strasburger a Jena professorship in 1869 when Pringsheim retired. Strasburger also served as director of the botanical gardens in Jena. His vacations to the Riviera with Haeckel combined science and pleasure.

In 1881, Strasburger accepted a teaching and research position at the University of Bonn, where he spent the rest of his life. Under his leadership, the Botanical Institute and Botanical Garden of the University of Bonn became the world center for plant cytology studies. His laboratory at the Poppelsdorfer Palace, formerly the residence of the electors of Cologne, became a Mecca for foreign graduate students of plant cytology, especially Americans.

The plant **cell nucleus** was the general focus of Strasburger's research. Naturally this interest led him to questions of sex and **inheritance**. He disagreed with the idioplasm theory of **Carl Wilhelm von Nägeli** (1817–1891), believing instead that nuclear **chromatin** performs the inheritance transmission function that Nägeli assigned to hereditary cell material not specific to the nucleus. Strasburger further concluded that free cells never arise, and that cell formation always depends upon either nuclear division, as in the production of daughter cells, or nuclear fusion, as in the production of zygotes. Among the terms he coined are cytoplasm, **diploid**, and **haploid**.

The observations of German anatomist **Walther Flemming** (1843–1905) and Strasburger in the late 1870s and early 1880s were the beginning of the discovery of chromosomes. From the orderly division of these rod like structures, Strasburger inferred that the cell nucleus is the locus of **heredity**. His progressive elucidation of the mechanics and principles of **mitosis** made possible the further **chromosome** studies of Belgian cytologist Edouard Joseph Louis-Marie van Beneden (1846–1910), German biologist August Weismann (1834–1914), and German anatomist **Heinrich Wilhelm Gottfried von Waldeyer-Hartz** (1836–1921), who coined the term chromosome in 1888.

Having spent nearly his entire life with his work, Strasburger died in Poppelsdorf while still fully engaged in research. His *Textbook of Botany* (Lehrbuch der Botanik), first published in 1894, appeared in its thirty-fourth edition in 1998.

See also Chromatin; Chromosome; Diploid; Evolution; Haploid; Phylogeny; Plant genetics; Sex determination

STRUCTURAL GENE

A structural **gene** is a gene whose product is an enzyme, or a protein that is involved in structural functions (i.e., tRNA, rRNA, etc.). It is a concept relative to regulatory genes. A regulatory genes' function is to regulate the expression of other genes through its own gene product such as RNAs or proteins.

The overwhelming majority of bacterial genes are structural genes. In **bacteria**, a series of structural genes that are often, but not necessarily related in function, tend to cluster. These genes may encode **enzymes** in certain metabolic pathways. They use a common promoter and transcribe as a single **transcription** unit and hence, are coordinately regulated. The regulator gene generally has its own promoter, and may be expressed constitutively or may be subject to autogenous regulation. The expression of structural genes is under two modes of control. The structural genes are said to be under negative control if they are expressed until turned off by the regulator protein, and under positive control if they are not expressed until turned on by the regulator protein. The entire system, including the structural genes and other elements that control their expression forms an **operon**.

In a positively controlled operon, the regulatory gene encodes a protein called an activator that binds to a region of the operon (the initiator), thereby turning on the transcription initiation. In a negatively controlled operon, the regulatory gene encodes a protein called repressor protein that binds to a specific region of the **DNA** (the **operator**) upstream of the structural genes, thereby preventing the initiation of transcription. The classic mode of control in bacteria is negative. In **eukaryotes**, one structural gene transcribes as a single unit. The regulation of the structural gene transcription is much more complicated in eukaryotes. It involves a large number of factors that bind to variety of cis-acting elements. Termination of the transcript happens far beyond the end of the coding sequence. The common mode of eukaryotic transcription is positive. Regulation by specific repression of a promoter is rare.

See also Gene names and functions

STURTEVANT, ALFRED HENRY (1891-1970)

American geneticist

A. H. Sturtevant, an influential geneticist and winner of the National Medal of Science in 1968, is best known for his demonstrations of the principles of **gene** mapping. This discovery had a profound effect on the field of **genetics** and led to projects to map both animal and human chromosomes. He is the unacknowledged father of the **Human Genome Project**, which mapped all of man's 30,000–plus chromosomes by the year 2000. Sturtevant's later work in the field of genetics led to discovery of the first reparable gene defect as well as the position effect, which showed that the effect of a gene is dependent on its position relative to other genes. He was a member of Columbia University's "*Drosophila* Group," whose studies of the genetics of fruit flies advanced new theories of genetics and **evolution**.

Alfred Henry Sturtevant, the youngest of six children, was born in Jacksonville, Illinois, to Alfred and Harriet (Morse) Sturtevant. Sturtevant's father taught at Illinois College briefly, but later chose farming as a profession. When Alfred Sturtevant was seven, his family moved to a farm in southern Alabama. He attended high school in Mobile, which was 14 miles from his home and accessible only by train.

Sturtevant enrolled in Columbia University in New York City in 1908, boarding with his older brother, Edgar, who taught linguistics at Columbia's Barnard College. Edgar and his wife played a significant role in young Sturtevant's life. They sent him Columbia's entrance examination, found him a scholarship, and welcomed him into their home in Edgewater, New Jersey, for four years. Edgar was also responsible for steering his brother toward a career in the sciences. The young Sturtevant had discovered genetic theory at an early age and often drew pedigrees of his family and of his father's horses. Edgar encouraged him to write a paper on the subject of color **heredity** in horses and to submit the draft to Columbia University's **Thomas Hunt Morgan**, the future Nobel Laureate geneticist. The paper used the recently rediscovered theories of Gregor Mendel, the 19th-century Austrian monk and founder of genetics, to explain certain coat-color **inheritance** patterns in horses. Sturtevant somehow mastered this subject in spite of his color-blindness.

As a result of his paper on horses, which was published in 1910, Sturtevant was given a desk in Morgan's famous "fly room," a small laboratory dedicated to genetic research using *Drosophila* (fruit flies) as subjects. Fruit flies are ideal subjects for genetic research. They mature in ten days, are less than one-eighth inch long, can live by the hundreds in small vials, require nothing more substantial than **yeast** for food, and have only four pairs of chromosomes.

Morgan's early work focused on the phenomenon of crossing-over in the fruit fly. By 1910, he had already described the sex-limited inheritance of white-eye. From this observation, Morgan postulated the idea that genes were linked because they were carried by the same **chromosome** and that genes in close proximity to one another would be linked more frequently than those that were farther apart. Sometimes, dominant linked traits, such as eye color and wing size, became unlinked in offspring. Sturtevant studied the process of **crossing over** of **sex-linked traits**, which are carried on the **X chromosome**. Female fruit flies have two X chromosomes. In addition to one X chromosome, males have a **Y chromosome**, which carries very few genes. Sturtevant correctly hypothesized that the exchange between X chromosomes probably occurred early on in the process of egg formation, when the paired chromosomes lie parallel to each other.

Morgan assumed that the relative distance between genes could be measured if the crossing over frequencies could be determined. From this lead, Sturtevant developed a practical method for determining this frequency rate. He began by studying six sex-linked traits and measured the occurrence of this related trait. The more frequently the traits occurred, Sturtevant reasoned, the closer the genes must be. He then calculated the percentages of crossing over between the various traits. From these percentages, he determined the relative distance between the genes on the chromosome, the first instance of gene mapping. This major discovery, which Sturtevant published in 1913 at the age of 22, eventually enabled scientists to map human and animal genes. It is often considered to be the starting point of modern genetics.

In 1914, Sturtevant received his Ph.D. from Columbia and stayed on in Morgan's lab as an investigator for the Carnegie Institution of Washington, D.C. Along with C. B. Bridges, **Hermann Joseph Muller**, and Morgan, he formed part of an influential research team that made significant contributions to the fields of genetics and entomology. He later described the lab as highly democratic and occasionally argumentative, with ideas being heatedly debated. The 16 x 24-foot lab had no desks, no separate offices, one general telephone, and very few graduate assistants. Sturtevant thrived in this environment. He worked seven days a week, reserving his mornings for *Drosophila* research and his afternoons for reading the scientific literature and consulting with colleagues. Sturtevant possessed a near photographic memory and wide-ranging interests. His only shortcoming as a researcher was his incessant pipe smoking, which often left flakes of tobacco ash mixed in with the samples of fruit flies. In spite of this minor flaw, the fly-room group raised research standards and elevated research writing to an art form. They also perfected the practice of chromosome mapping, using Sturtevant's methods to develop a chromosome map of *Drosophila,* detailing the relative positions of fifty genes.

Sturtevant published a paper in 1914 that documented cases of double crossing over, in which chromosomes that had already crossed over broke with one another and recrossed again. His next major paper, published in 1915, concerned the sexual behavior of fruit flies and concentrated on six specific mutant genes that altered eye or body color, two factors that played important roles in sexual **selection**. He then showed that specific genes were responsible for selective intersexuality. In later years, Sturtevant discovered a gene that caused an almost complete sex change in fruit flies, miraculously transforming females into near males. In subsequent years, researchers identified other sex genes in many animals, as well as in humans. These discoveries led to the development of the uniquely twentieth-century view of sex as a gene-controlled trait which is subject to variability.

During the 1920s, Sturtevant and Morgan examined the unstable bar-eye trait in *Drosophila*. Most geneticists at that time assumed that bar eye did not follow the rules of Mendelian heredity. In 1925, Sturtevant showed that bar eye involved a **recombination** of genes rather than a **mutation**, and that the position of the gene on the chromosome had an effect on its action. This discovery, known as the position effect, contributed greatly to the understanding of the action of the gene.

In 1928, Morgan received an offer from the California Institute of Technology to develop a new Division of Biological Sciences. Sturtevant followed his mentor to California, where he became Caltech's first professor of genetics. The new genetics group set up shop in Caltech's Kerckhoff Laboratory. Sturtevant continued working with fruit flies and conducted genetic investigations of other animals and plants, including snails, rabbits, moths, rats, and the evening primrose, *Oenothera.*

In 1929, Sturtevant discovered a sex ratio gene that caused male flies to produce X sperm almost exclusively, instead of X and Y sperm. As a result, these flies' offspring were almost always females. In the early 1930s, giant chromosomes were discovered in the salivary glands of fruit flies. Under magnification, these chromosomes revealed cross patterns that were correlated to specific genes. The so-called physical map derived from these giant chromosomes did not exactly match Sturtevant's relative location maps. In the physical map, some of the genes tended to cluster toward one end of the chromosome and the distances between genes was not uniform. But the linear order of the genes on the chromosome matched Sturtevant's relative maps gene for gene. This discovery confirmed that Sturtevant had been correct in his assumptions about chromosomal linearity.

In 1932, Sturtevant took a sabbatical leave and spent the year in England and Germany as a visiting professor of the Carnegie Endowment for International Peace. When he returned to America, he collaborated with his Caltech colleague **Theodosius Dobzhansky**, a Russian-born geneticist, on a study of inversions in the third chromosome of *Drosophila pseudoobscura*. In the 1940s, Sturtevant studied all of the known gene mutations in *Drosophila* and their various effects on the development of the **species**. From 1947 to 1962, he served as the Thomas Hunt Morgan Professor of Biology at Caltech. His most significant scientific contribution during that time occurred in 1951, when he unveiled his chromosome map of the indescribably small fourth chromosome of the fruit fly, a genetic problem that had puzzled scientists for decades.

During the 1950s and 1960s, Sturtevant turned his attention to the iris, and authored numerous papers on the subject of evolution. He became concerned with the potential dangers of genetics research and wrote several papers on the social significance of human genetics. In a 1954 speech to the Pacific Division of the American Association for the Advancement of Science, Sturtevant described the possible genetic consequences of nuclear war and argued that the public should be made aware of these possible cataclysmic hazards before any further bomb testing was performed. One of his last published journal articles, written in 1956, described a mutation in fruit flies that, by itself, was harmless but which proved lethal in combination with another specific mutant gene.

Sturtevant was named professor emeritus at Caltech in 1962. He spent the better part of the early 1960s writing his major work, *A History of Genetics,* which was published in 1965. In 1968, he received the prestigious National Medal of Science for his achievements in genetics. Sturtevant married Phoebe Curtis Reed in 1923, and the couple had three children. Sturtevant died at the age of 78.

See also Crossing over; Genetic mapping

SUBSTITUTION MUTATION · *see* MUTATION

SUICIDE GENE · *see* GENE NAMES AND FUNCTIONS

SUPERFEMALES AND SUPERMALES • *see* SEX
DETERMINATION

SUPERNUMERARY CHROMOSOMES

Supernumerary or B-chromosomes are small, extra chromosomes found in a large number of plant and animal **species**. They usually have a centromere and possibly coding **DNA**, but they usually produce no functional protein. They are not essential to normal cellular function, and do not follow normal patterns of **replication** and segregation during **cell division**. Supernumerary chromosomes may be unstable and lost from some cells, or accumulate in others. In some organisms, supernumerary chromosomes are normal variants in the population, but even in these situations, not all member of a species or all cells within a given individual will have the supernumerary chromosomes. An example of a positive effect of the presence of B-chromosomes has been reported in foxes. Male foxes with supernumerary chromosomes tend to eat better and be more fit overall than foxes lacking the extra chromosomes. In general, the origin and function of these structures are not well understood, but they appear to be more common in plants than in animals.

In higher organisms, small chromosomes derived from one **chromosome** of the normal set are often classed as supernumeraries. In humans, it has been estimated that supernumerary chromosomes exist in the population at a frequency of 1 in 2,000 to 1 in 3,500. he majority of these are bisatellited chromosomes probably derived from an error in the division of an acrocentric chromosome. As long as such structures do not carry any genes, they tend to be non-deleterious and do not affect the individual's physical appearance. However, if a bisatellited marker includes active genes, it can have negative consequences (partial trisomy or tetrasomy) if the extra copies of the genes are expressed.

Occasionally, a supernumerary chromosome is identified during a prenatal cytogenetic karyotype analysis. These may be bisatellited or have a single or no satellite. If one of the parents can be shown to have the same chromosome, it is unlikely that the child will be adversely affected. However, if the supernumerary is *de novo* (present for the first time in this child), the risk for abnormalities is higher since the extra chromosome could be due to an error resulting in a partial trisomy in the child. Genetic counseling is important to provide the family with the relative risks of the finding.

See also Chromosome structure and morphology; Ploidy

SURVIVAL OF THE FITTEST

The term "survival of the fittest" was first used by the Victorian naturalist Herbert Spencer as a metaphor to help explain **natural selection**, the central element of Charles Darwin's revolutionary theory of evolutionary change, first published in 1859 in his famous book, *The Origin of Species by Means of Natural Selection.*

In this extremely influential and important book, Darwin reasoned that all **species** are capable of producing an enormously larger number of offspring than actually survive. Darwin held that the survival of progeny was not a random process. (In fact, he described it as a "struggle for existence.") Rather, Darwin suggested that those progeny that were better adapted to coping with the opportunities and risks presented by environmental circumstances would have a better chance of surviving, and of passing on their favorable traits to subsequent generations. These better-adapted individuals, which contribute disproportionately to the genetic complement of subsequent generations of their population, are said to have greater reproductive **fitness**. Hence the use, and popularization, of the phrase: "survival of the fittest." Darwin also used another, more awkward expression to explain the same thing: the "preservation of favored races in the struggle for life." In fact, this is the subtitle that he used for *The Origin of Species by Means of Natural Selection.*

Darwin's theory of **evolution** by natural **selection** is one of the most important concepts and organizing principles of modern biology. The differential survival of individuals that are more-fit, for reasons that are genetically heritable, is believed to be one of the most important mechanisms of evolution. And because of its clarity, the phrase "survival of the fittest" is still widely used to explain natural selection to people interested in understanding the evolution of life on Earth.

See also Adaptation and fitness; Darwinism; Evolutionary mechanisms

SUTTON, WALTER STANBOROUGH (1877-1916)
American physician

Walter Stanborough Sutton was a surgeon and a biologist who advanced the findings and confirmed the genetic theories of Gregor Mendel. A physician in private practice for the last half of his life, Sutton discovered the role of chromosomes in **meiosis** (sex **cell division**) and their relationship to Mendel's laws of **heredity**. From his research with grasshoppers collected at his parent's farm in the summer of 1899, he went on to make a major contribution to the understanding of the workings of chromosomes in **sexual reproduction**.

The fifth of six sons, Sutton was born in Utica, New York, to William Bell Sutton and Agnes Black Sutton. At age ten, Sutton moved with his family to Russell County, Kansas, where he attended public schools. He studied engineering at the University of Kansas, Lawrence, beginning in 1896. Following his younger brother's death from typhoid in 1897, however, he made a pivotal change in the course of his education that would eventually lead him to the study of medicine and to his discoveries in **genetics**.

Sutton earned a bachelor's degree from the University of Kansas School of Arts in 1900. While an undergraduate, he met **Clarence Erwin McClung**, a zoology instructor. Their four-year association would greatly influence Sutton's later work. McClung persuaded Sutton to study histology, which led the young student to other areas of inquiry, including cytological examinations of the lubber (grass)hopper (*Brachystola magna*).

Sutton's careful camera lucida drawings of the stages of spermatogenesis in the lubber hopper were described in his first paper, "The Spermatogonial Divisions in *Brachystola Magna*," which appeared in the *Kansas University Quarterly* in 1900 and served as his master's thesis the following year. Following this work, McClung and other faculty members encouraged Sutton to pursue his doctoral studies under biologist **Edmund Beecher Wilson** at Columbia University. In 1901, Sutton began work there. He continued his cytological studies on chromosomal division in germ cells, and in 1902, detailed his research in "On the Morphology of the **Chromosome** Group in *Brachystola Magna*." Earlier that year, based on his readings of work done by British biologist **William Bateson** on the relationship between meiosis in germ cells and body characteristics, Sutton made the connection between cytology and heredity, and thus opened the field of cytogenetics. Sutton's hypotheses and generalizations were later published in what has become a landmark work, "The Chromosomes in Heredity" in the *Biological Bulletin*, 1903.

In these papers, Sutton explained that through his observation of meiosis he found that all chromosomes exist in pairs very similar to each other; that each **gamete**, or sex **cell**, contributes one chromosome of each pair, or reduces to one-half its genetic material, in the creation of a new offspring cell during meiosis; that each fertilized egg contains the sum of chromosomes of both parent cells; that these pairs control heredity; and that each particular chromosome's pair is based on independent assortment, that is, the maternal and paternal chromosomes separate independently of each other. The result, Sutton found, was that an individual in a **species** may posses any number of random combinations of different pairs of maternal and paternal chromosomes. Sutton also hypothesized that each chromosome carries in it groups of genes, each of which represents a biological characteristic—a thought that contradicted the then prevalent theory that ascribed one inherited trait to each chromosome.

At the time Sutton's paper was published, Austrian scientist Theodor Boveri claimed he had reached the same conclusions. As a result, the biological generalizations of the association of paternal and maternal chromosomes in pairs and their subsequent separation, which makes up the physical basis of the Mendelian law of heredity, is called the Sutton-Boveri Hypothesis. This 1903 discovery, however, was to be Sutton's last in cytology; due to unknown reasons, he never completed his course of study at Columbia.

Returning to Kansas after his laboratory research, Sutton worked as a foreman in the Chautauqua County oil fields until 1905. While there he used his abilities to solve technical problems. He developed the first technique for starting large gas engines with high-pressure gas. In 1907, he patented a device to raise an oil pump mechanism from a well when worn valve components required replacement. Sutton also began a design to use electric motors to run drilling devices, but did not complete it.

Still fascinated with the intricacies of the life sciences, and at the request of his father, Sutton returned to the Columbia University College of Physicians and Surgeons. He earned his medical degree in 1907 and for the next two years served as surgical house officer at Roosevelt Hospital in New York City. During that time, Sutton designed and built a device to deliver rectal anesthesia to patients unable to inhale ether. In 1909, he moved back to Kansas to practice privately in Kansas City, and to teach in the department of surgery at the University of Kansas. Six years later, he took a leave of absence from the University to work at the American Ambulance Hospital in Juilly, France, during World War I. His experiences of working on injured soldiers led to a book chapter on wound surgery. In addition, Sutton developed a method of finding and removing foreign bodies in soft tissue involving the use of fluoroscopy and a simple device made from a hooked piece of wire. After the war, Sutton returned to his private practice and his teaching duties in Kansas, continuing his work there until his death from a ruptured appendix at the age of 39.

See also Chromosome structure and morphology; Mendelian genetics and Mendel's laws of heredity; Sexual reproduction

SYNAPSIS

Synapsis is the formation of a structure termed a bivalent (a pairing of **homologous chromosomes**) during meiotic division. The chromotids of each of the homologous chromosomes align, essentially **gene** for gene to form a synaptonemal complex that allows crossover (exchange of genetic material).

In contrast to **mitosis**, **meiosis** results in the reduction of the cellular **genome** to a **haploid** state. This reduction of **chromosome** number is achieved by a single round of **DNA replication** followed by two sequential rounds of nuclear and cellular division (meiosis I and meiosis II). Like mitosis, meiosis is initiated after the synthetic phase is completed and the parental chromosomes have replicated to produce identical sister chromatids. Once the cellular **DNA** has replicated, the chromosomes condense into daughter chromatids of the duplicated chromosome. However, the patterning of chromosomal segregation during meiosis I differs from that of mitosis in that during meiosis I homologous chromosomes first pair up with each other and then segregate to different daughter cells.

Synapsis is the specific pairing up of these chromatids with their homologous chromosomes along the midline of the **cell**. Not only is the pairing of homologous after DNA **replication** a key event in the meiotic segregation of chromosomes, but also this lining up allows a process called **crossing over** to occur. Crossing over involves the symmetrical breaking of one maternal and one paternal chromosome and allows the reciprocal exchange of genetic information by breaking followed by the crosswise rejoining of the chromosomes.

As a result of synapsis and crossing over the daughter cells contain a mixture of parental types called the **recombination** form.

Synapsis takes place during an extended prophase of meiosis I that is further divided into five stages. These five stages are named leptotene, zygotene, pachytene, diplotene, and diakinesis (this naming is based on the physical appearance of the chromosomes during these various stages). It is thought that the initial association of homologous chromosomes is most likely the result of **base pairing** between complementary DNA strands but true synapsis occurs during the zygotene stage. A base-for-base zipper-like protein structure called the synaptomenal complex is formed along the length of the chromosome. This complex keeps the chromosomes aligned with one another and tightly associated through the pachytene stage where crossing over occurs.

Some congenital defects are associated with synaptic misalignment that can cause chromosomal anomalies to occur. One of these anomalies is called chromosomal nondisjunction and it results in aneuploidy. Aneuploidy refers to any condition where the normal number of chromosomes is not present, either too few to too many. In humans, **Down syndrome** is an example of aneuploidy. Other types of chromosomal anomalies are translocations and inversions.

See also Cell division; Cell proliferation; Centromere; Chromatin; Chromosome; Eukaryotic genetics; Gamete; Gametogenesis; Sister-chromatid exchange

SYNGANMEON • *see* EVOLUTIONARY MECHANISMS

SZOSTAK, JACK WILLIAM (1952-)
Canadian biochemist

Jack W. Szostak, a professor of **genetics** and a biochemist at Massachusetts General Hospital developed the **Yeast** Artificial **Chromosome** (**YAC**), in collaboration with Harvard biochemist Andrew Murray. The YAC, developed in 1980s, was the first artificial (synthetic) chromosome. YAC chromosomes are used by scientists to study and map chromosomes. In addition, YACs are now used as **cloning vectors**.

Szostak is also a principal investigator affiliated with the Howard Hughes Medical Institute.

Szostak's work also sheds new light upon the biological processes responsible for the origin of life on Earth. In a continuation of the type of work started by American scientists Stanley Miller and **Harold Urey**, who, in the 1950's, were able to form amino acids from inorganic materials present in Earth's early atmosphere. Szostak's work further advanced such studies by creating a synthetic entity capable of **replication**. More recent work by Szostak has focused upon the role of **RNA** as an enzyme in biochemical reactions. Such studies may shed light on the role primitive RNAs may have played in primitive **cell** organization and **biochemistry**.

By creating random RNA, Szostak is able to test those RNAs and the genetic code sequences they contain against different environmental conditions that, in effect, reflect differing **selection** pressures. Szostak and his coworkers are then able to evaluate the binding functions and reactions of the synthetic RNAs. By making a large number of copies of the RNA, Szostak is then able to mimic natural **mutation** driven alterations to the RNA in an effort to evaluate how changes in the RNA sequences affect their role in cellular reactions and functions. Szostak can selectively study the role of RNAs in reactions by developing specific RNAs that function as **enzymes** in those reactive pathways. Once again, the role of the RNAs can be evaluated in terms of the effect of mutations on previously observed reactions.

The degree to which primitive RNAs were able to catalyze reactions is critical. If, Szostak argues, RNAs acting as enzymes (ribozymes) were able to catalyze a large number of reactions, then this ability supports hypotheses that utilize RNA as a primitive organizer in early cell forms. Such organizer RNA could have originated from random assortments of nucleotides in the primitive Earth seas or pools formed by runoff. Any random assembly of nucleotides that was capable of replication could have easily become widespread.

See also Genomic library; Origin of life; RNA function

T

TANDEM REPEATS

Tandem repeats are repeated **DNA** sequences arranged together one after another. The total number of iterations can vary from site to site and person to person. In general, this group of repeats can be subdivided by size into four categories; megasatellite, satellite, minisatellite, and microsatellite. Some types are localized to specific sites, but others are dispersed throughout the **genome**.

Megasatellite DNA repeats range in size from 3–4.7 kb (kilobases) and have been identified on chromosomes 4, 8, 19 and the X. Their function is unknown.

The name satellite DNA was originally derived from the early buoyant **density gradient centrifugation** technique used to separate different size DNAs. This group of sequences formed a minor peak or satellite in the gradients. It is now known that the satellite DNA is primarily derived from the centromeric region of chromosomes and includes several different types of repeats ranging in size from 5-171 base pairs. This is the portion of the **chromosome** to which the spindle fibers attach during **cell division**, and it is the last DNA to replicate, occurring just prior to separation of the chromosomes. There is significant conservation of sequences among different **species** probably due to the similar function the centromere performs in different organisms.

Minisatellite DNA is principally found at the ends of linear chromosomes (telomeres and subtelomeres). These are smaller units than the classic satellite DNA and range from 6-64 bp (base pairs) in size. The telomeric sequences function to stabilize the chromosome and prevent interaction between the ends of the DNA strands. This sequence is also highly conserved across species. In humans, it is a 6 base pair repeat, $(TTAGGG)_n$, spanning an area of 0.1-20 kb. Over time, the repeats are lost during **replication**, and when they have been totally depleted, the chromosome destablizes with an increased likelihood of breakage, rearrangement, or loss of the chromosome. This has been shown to be an important mechanism in aging. In **cancer**, activation of telomerase, an enzyme that replaces the telomeric repeat sequences as they are lost, causes an abnormal immortalization of the chromosomes in the disease **cell** line.

The larger forms of the minisatellite group are usually classified as hypervariable minisatellite DNA and though commonly found in the subtelomeric region, may also occur in other locations of the genome. The actual function of these sequences is unknown, but they have become quite useful in DNA fingerprinting applications.

As one would expect, microsatellite repeats are the smallest by size, 1-4 bp. They are found throughout the genome and the reason for their existence is mostly unknown. In humans, trinucleotide (3 bp) repeats have been implicated in **mutation** leading to disease. Tandem repeats of three base pair units exist normally throughout the genome both within genes and in non-coding regions, but, in some individuals under certain conditions, amplification may occur that increases the total number of copies of the repeated sequence. This abnormal amplification has been shown to lead to a number of diseases of the central or peripheral nervous systems that result in neurodegenerative (**Huntington disease**, myotonic dystrophy, Friedrich ataxia, etc.) or mental deficiency (**fragile X syndrome**) disorders.

Tandem repeats have also been recognized in some genes. This is demonstrated by the alphafetoprotein (AFP) and serum albumin genes in mice. The two genes are linked in tandem and though they are currently dissimilar, their structure indicates they were probably both derived from a single original **gene**. It is hypothesized that this gene tandemly duplicated and over time, separate genes with related but different functions arose. A similar mechanism may explain the similarity and close proximity of the genes necessary for collagen production.

As times goes on, more information is being acquired about the different types of repeat sequences. The **Human Genome Project** may provide further insights into the function of these diverse sequences.

See also Huntington disease; Fragile X syndrome; Centromere; Telomere

TANZI, RUDOLPH EMILE (1958-)
American geneticist

Rudolph Emile Tanzi, Unit director of the Genetics and Aging Unit at Massachusetts General Hospital (MGH) and professor of neurology (neuroscience) at Harvard Medical School, specializes in the genetic causes of **Alzheimer's disease** (AD), a debilitating and fatal neurological disorder that causes its victims to suffer from dementia. Working with different research teams, Tanzi has identified several genes implicated in AD and in other neurodegenerative disorders. He is the recipient of numerous awards, including the Potamkin Prize and the Metropolitan Life Foundation Award.

Upon completing his undergraduate work in microbiology and history at the University of Rochester in 1980, Tanzi accepted a position in a genetics research lab at MGH and was part of the team led by **James Gusella** that in 1983 located the **gene** that causes Huntington's disease, a fatal neurological disorder. In Gusella's lab, Tanzi also studied **chromosome** 21 in an attempt to locate the gene(s) involved in **Down syndrome**; Tanzi published his map of chromosome 21 in 1988.

Because of a striking similarity between Down syndrome and AD, the study of chromosome 21 led Tanzi to research the genetic causes of AD. Both Down syndrome and AD result in amyloid deposits or plaques in the brains of its victims; amyloid is a gooey substance made of a protein fragment called amyloid beta, or A-beta, and other proteins. In 1986, Tanzi was among the first to isolate on chromosome 21 the amyloid beta protein precursor (APP) gene that causes early-onset AD (early-onset AD accounts for about 5% of AD victims and affects people under the age of 60; most AD victims have late-onset Alzheimer's).

After Tanzi received his Ph.D. in neurobiology from Harvard University in 1990, he continued his genetic research of AD at MGH and focused on the role of amyloid in AD. Tanzi is a proponent of the amyloid hypothesis that states that amyloid deposits are directly related to the destruction of neurons in AD, and that drugs can and should be designed based on this hypothesis. Opponents of the amyloid hypothesis look instead to the neurofibrillary tangles of a protein called tau; these tangles, like amyloid, are present in the brains of all AD victims. Tangles twist in and around neurons, destroying the cells.

Tanzi worked with the research groups that identified two more early-onset AD genes in 1995. He was on the team with Peter Hyslop that discovered presenilin (PS1), on chromosome 14. Subsequent research by Tanzi found several of the more than 80 mutations of this gene. Working with Wilma Wasco at MGH and Jerry Schellenberg, Tanzi also discovered the second presenilin gene (PS2) on chromosome 1. Scientists argue that mutations in the genes APP, PS1, and PS2 explain about half of all early-onset AD.

The discovery of the first late-onset gene, apoliprotein E (APOE) was announced in 1992. Four years later, Tanzi's lab at MGH, renamed the Genetic and Aging Unit, was chosen by the National Institute of Mental Health (NIMH) to help set up a **genome** screen of more than 300 families with familial AD. Important results of the screening include the discovery of the gene Alpha-2-macroglobulin (A2M) and of its location on chromosome 12. Tanzi also learned more about amyloid: while A2M binds amyloid and removes it from the body, APOE appears to promote the formation of amyloid.

In the late 1990s, Tanzi and another researcher at MGH studied the link of certain metals, specifically copper, zinc, and iron, to the amyloid plaques in AD. In early 1999, a metal-binding compound developed by Tanzi and his colleague was tested on transgenic mice; results showed a reduction of brain amyloid by 40% and an elimination of the amyloid in 25% of the mice. The compound is one of two drug approaches that may demonstrate that AD is reversible by dissolving existing amyloid plaques. By 2000, the metal-reducing drug was being cleared for human trials that would last at least two years.

Tanzi's ongoing research at MGH targets the perplexing questions of how early-onset genetic mutations cause AD, how A-beta and tau are related, and whether A-beta is the main cause of neuron destruction. His work also focuses on the development of drugs that will either inhibit the production of A-beta or accelerate the breakdown of A-beta in the brains of AD victims.

See also Neurological and opthalmological genetics; Pharmacogenetics

TATUM, EDWARD LAWRIE (1909-1975)
American biochemist

Edward Lawrie Tatum's experiments with simple organisms demonstrated that **cell** processes can be studied as chemical reactions and that such reactions are governed by genes. With George Beadle, he offered conclusive proof in 1941 that each biochemical reaction in the cell is controlled via a catalyzing enzyme by a specific **gene**. The "one gene-one enzyme" theory changed the face of biology and gave it a new chemical expression. Tatum, collaborating with **Joshua Lederberg**, demonstrated in 1947 that **bacteria** reproduce sexually, thus introducing a new experimental organism into the study of molecular **genetics**. Spurred by Tatum's discoveries, other scientists worked to understand the precise chemical nature of the unit of **heredity** called the gene. This study culminated in 1953, with the description by **James Watson** and **Francis Crick** of the structure of **DNA**. Tatum's use of microorganisms and laboratory mutations for the study of biochemical genetics led directly to the **biotechnology** revolution of the 1980s. Tatum and Beadle shared the 1958 Nobel Prize in physiology or medicine with Joshua Lederberg for ushering in the new era of modern biology.

Tatum was born in Boulder, Colorado, to Arthur Lawrie Tatum and Mabel Webb Tatum. He was the first of three children. Tatum's father held two degrees, an M.D. and a Ph.D. in pharmacology. Edward's mother was one of the first women to graduate from the University of Colorado. As a boy, Edward played the French horn and trumpet; his interest in music lasted his whole life.

Tatum earned his A.B. degree in chemistry from the University of Wisconsin in 1931, where his father had moved the family in order to accept as position as professor in 1931. In 1932, Tatum earned his master's degree in microbiology. Two years later, in 1934, he received a Ph.D. in biochemistry for a dissertation on the cellular biochemistry and nutritional needs of a bacterium. Understanding the biochemistry of microorganisms such as bacteria, **yeast**, and molds would persist at the heart of Tatum's career.

In 1937, Tatum was appointed a research associate at Stanford University in the department of biological sciences. There he embarked on the *Drosophila* (fruit fly) project with geneticist **George Beadle**, successfully determining that kynurenine was the enzyme responsible for the fly's eye color, and that it was controlled by one of the eye-pigment genes. This and other observations led them to postulate several theories about the relationship between genes and biochemical reactions. Yet, the scienyists realized that *Drosophila* was not an ideal experimental organism on which to continue their work.

Tatum and Beadle began searching for a suitable organism. After some discussion and a review of the literature, they settled on a pink mold that commonly grows on bread known as *Neurospora crassa*. The advantages of working with *Neurospora* were many: it reproduced very quickly, its nutritional needs and biochemical pathways were already well known, and it had the useful capability of being able to reproduce both sexually and asexually. This last characteristic made it possible to grow cultures that were genetically identical, and also to grow cultures that were the result of a cross between two different parent strains. With *Neurospora*, Tatum and Beadle were ready to demonstrate the effect of genes on cellular biochemistry.

The two scientists began their *Neurospora* experiments in March 1941. At that time, scientists spoke of "genes" as the units of heredity without fully understanding what a gene might look like or how it might act. Although they realized that genes were located on the chromosomes, they didn't know what the chemical nature of such a substance might be. An understanding of DNA (**deoxyribonucleic acid**, the molecule of heredity) was still twelve years in the future. Nevertheless, geneticists in the 1940s had accepted **Gregor Mendel's** work with **inheritance** patterns in pea plants. Mendel's theory, rediscovered by three independent investigators in 1900, states that an inherited characteristic is determined by the combination of two hereditary units (genes), one each contributed by the parental cells. A dominant gene is expressed even when it is carried by only one of a pair of chromosomes, while a recessive gene must be carried by both chromosomes to be expressed. With *Drosophila*, Tatum and Beadle had taken genetic mutants—flies that inherited a variant form of eye color—and tried to work out the biochemical steps that led to the abnormal eye color. Their goal was to identify the variant enzyme, presumably governed by a single gene that controlled

the variant eye color. This proved technically difficult, and as luck would have it, another lab announced the discovery of kynurenine's role before theirs did. With the *Neurospora* experiments, they set out to prove their **one gene-one enzyme** theory another way.

The two investigators began with biochemical processes they understood well: the nutritional needs of *Neurospora*. By exposing cultures of *Neurospora* to x rays, they would cause genetic damage to some bread mold genes. If their theory was right, and genes did indeed control biochemical reactions, the genetically damaged strains of mold would show changes in their ability to produce nutrients. If supplied with some basic salts and sugars, normal *Neurospora* can make all the amino acids and vitamins it needs to live except for one (biotin).

This is exactly what happened. In the course of their research, the men created, with x-ray bombardment, a number of mutated strains that each lacked the ability to produce a particular **amino acid** or vitamin. The first strain they identified, after 299 attempts to determine its mutation, lacked the ability to make vitamin B_6. By crossing this strain with a normal strain, the offspring inherited the defect as a recessive gene according to the inheritance patterns described by Mendel. This proved that the **mutation** was a genetic defect, capable of being passed to successive generations and causing the same nutritional mutation in those offspring. The x-ray bombardment had altered the gene governing the enzyme needed to promote the production of vitamin B_6.

This simple experiment heralded the dawn of a new age in biology, one in which molecular genetics would soon dominate. Nearly forty years later, on Tatum's death, Joshua Lederberg told the *New York Times* that this experiment "gave impetus and morale" to scientists who strived to understand how genes directed the processes of life. For the first time, biologists believed that it might be possible to understand and quantify the living cell's processes.

Tatum and Beadle were not the first, as it turned out, to postulate the one gene-one enzyme theory. By 1942, the work of English physician **Archibald Garrod**, long ignored, had been rediscovered. In his study of people suffering from a particular inherited enzyme deficiency, Garrod had noticed the disease seemed to be inherited as a Mendelian recessive. This suggested a link between one gene and one enzyme. Yet Tatum and Beadle were the first to offer extensive experimental evidence for the theory. Their use of laboratory methods, like x rays, to create genetic mutations also introduced a powerful tool for future experiments in biochemical genetics.

During World War II, the methods Tatum and Beadle had developed in their work with pink bread mold were used to produce large amounts of penicillin, another mold. In 1945, at the end of the war, Tatum accepted an appointment at Yale University as an associate professor of botany with the promise of establishing a program of biochemical microbiology within that department. In 1946. Tatum did indeed create a new program at Yale and became a professor of microbiology. In work begun at Stanford and continued at Yale, he demonstrated that the one gene-one enzyme theory applied to yeast and bacteria as well as molds.

In a second fruitful collaboration, Tatum began working with Joshua Lederberg in March 1946. Lederberg, a Columbia University medical student fifteen years younger than Tatum, was at Yale during a break in the medical school curriculum. Tatum and Lederberg began studying the bacterium *Escherichia coli*. At that time, it was believed that *E. coli* reproduced asexually. The two scientists proved otherwise. When cultures of two different mutant bacteria were mixed, a third strain, one showing characteristics taken from each parent, resulted. This discovery of biparental inheritance in bacteria, which Tatum called genetic **recombination**, provided geneticists with a new experimental organism. Again, Tatum's methods had altered the practices of experimental biology. Lederberg never returned to medical school, earning instead a Ph.D. from Yale.

In 1948 Tatum returned to Stanford as professor of biology. A new administration at Stanford and its department of biology had invited him to return in a position suited to his expertise and ability. While in this second residence at Stanford, Tatum helped establish the department of biochemistry. In 1956, he became a professor of biochemistry and head of the department. Increasingly, Tatum's talents were devoted to promoting science at an administrative level. He was instrumental in relocating the Stanford Medical School from San Francisco to the university campus in Palo Alto. In that year Tatum also was divorced from his wife June. On December 16, 1956, he married Viola Kantor in New York City. Tatum left the West coast and took a position at the Rockefeller Institute for Medical Research (now Rockefeller University) in January 1957. There he continued to work through institutional channels to support young scientists, and served on various national committees. Unlike some other administrators, he emphasized nurturing individual investigators rather than specific kinds of projects. His own research continued in efforts to understand the genetics of *Neurospora* and the nucleic acid metabolism of mammalian cells in culture.

In 1958, together with Beadle and Lederberg, Tatum received the Nobel Prize in physiology or medicine. The Nobel Committee awarded the prize to the three investigators for their work demonstrating that genes regulate the chemical processes of the cell. Tatum and Beadle shared one-half the prize and Lederberg received the other half for work done separately from Tatum. Lederberg later paid tribute to Tatum for his role in Lederberg's decision to study the effects of x-ray-induced mutation. In his Nobel lecture, Tatum predicted that "with real understanding of the roles of heredity and environment, together with the consequent improvement in man's physical capacities and greater freedom from physical disease, will come an improvement in his approach to, and understanding of, sociological and economic problems."

Tatum's second wife, Viola, died on April 21, 1974. Tatum married Elsie Bergland later in 1974 and she survived his death the following year, in 1975. Tatum died at his home on East Sixty-third Street in New York City after an extended illness, at age 65.

See also Bacterial genetics; One gene-one enzyme; Enzymes, genetic manipulation of

TAY-SACHS DISEASE

Tay-Sachs disease is a genetic disorder that causes progressive neurological degeneration. Although the disease primarily affects infants, juvenile and adult forms of Tay-Sachs disease also exist. Tay-Sachs disease is incurable, and infants with the disease usually die by age four. The defective genes responsible for Tay-Sachs disease are especially prevalent in Jewish people of Eastern European descent, but non-Jewish cases of Tay-Sachs have also been identified. Genetic screening can be performed on persons whose ethnic descent or family history place them at high risk for carrying the defective genes.

Tay-Sachs disease is named for two researchers who first noticed the characteristic symptoms in the 1880s. Tay, an opthamologist, noted the typical eye disorders that occur late in the disease in several infants of Jewish descent. Sachs, a neurologist, identified the neurological disturbances in infants with the disease. By the 1940s, researchers had uncovered the physiological basis of the disease, a deadly accumulation of fatty acids in the brain that destroys nerve cells. In the 1960s, researchers discovered that decreased levels of an important enzyme caused the ganglioside accumulation in the brain. By the 1970s, screenings were conducted for carriers of the defective **gene**, and in 1989, the actual genes responsible for Tay-Sachs were discovered.

Tay-Sachs disease has been traced to a defect in the HEXA gene, the gene that encodes an enzyme called hexosaminidase A. **Enzymes** are proteins that break down substances in the body. Hexosaminidase A breaks down a group of fatty acids called the gangliosides. In Tay-Sachs disease, however, the defective HEXA genes do not function correctly, leading to low levels of hexosaminidase A. Without proper amounts of this enzyme, gangliosides accumulate in the brain and destroy nerve cells.

Tay-Sachs is a homozygous recessive genetic disorder. In homozygous recessive genetic disorders, two defective alleles, or copies, of the gene, one from each parent, must combine to produce the defective gene. If two people who each carry a defective allele have a child, chances are one in four that the child will have Tay-Sachs disease.

About 1 in 27 people of Eastern European Jewish descent are thought to be carriers, and about 1 in 3600 Jewish infants are born with this disease, accounting for approximately 90% of all Tay-Sachs cases world-wide. Among the general population, about one in 300 people are carriers of the defective allele. Carriers do not have symptoms of Tay-Sachs, although their levels of the hexosaminidase A enzyme may be reduced as much as 50%. This reduction, however, is not sufficient to produce symptoms.

Because no cure exists for Tay-Sachs disease, efforts to identify genetic carriers are essential in controlling the disease. Carriers are at risk for having children with Tay-Sachs if they have children with another **carrier**. In 1969, a nationwide screening program was initiated in the United States. With the help of religious and community leaders, screenings for low levels of the hexosaminidase A enzyme (a sign of carrier status) were conducted in synagogues, community centers, and on college campuses. Carriers were told their status and informed of their options. The results of this screening and educational program were extraordinarily successful. In 1970, the number of children born in the United States with Tay-Sachs disease dropped from 100 to 50; in 1980, the number dropped even further, to 13. Many carriers identified in this screening program opted not to have children. However, carriers could also reduce their risk of having children with the disease if they made sure their potential partners were also tested for carrier status.

An improved screening test is now available. Instead of testing for hexosaminidase A enzyme deficiency, the accuracy of which is sometimes affected by pregnancy and other factors, researchers now use a genetic test that searches for evidence of the defective gene in a blood sample. In 1989, researchers identified three genetic changes, or mutations, responsible for Tay-Sachs disease. Two of these mutations cause the infantile form of the disease. By testing for the existence of these mutations in a person's blood, carriers are more accurately identified. The test is more specific than the enzyme test and is also easier to perform.

In the future, researchers hope that **gene therapy** may cure Tay-Sachs disease. In gene therapy, cells that have been infected with viruses carrying normal genes are injected into the body. The healthy genes would then encode enough hexosaminidase A to break down the accumulating gangliosides. Because the defective genes responsible for Tay-Sachs are known, this therapy is a possibility. Researchers, however, must solve many technical difficulties before gene therapy for Tay-Sachs is a reality.

See also Gene therapy

TECHNOLOGY TRANSFER • *see* ECONOMICS AND

GENETIC TECHNOLOGY TRANSFER

TELOMERE

Telomeres are specialized **DNA** regions that consist of short, tandemly **repeated sequences** at the ends of linear chromosomes. Telomeres are found in all **eukaryotes** and are essential for normal **cell** growth. During **DNA replication**, telomeres complete the synthesis of chromosomes that otherwise would have been lost or recombined with other chromosomes during **replication**. However, there is still some shortening of the telomere with each DNA replication that is theorized to be equivalent to cell aging.

In 1973, Aleksei M. Olovnikov, a scientist at the Institute of Biochemical Physics in the Russian Academy of Sciences in Moscow, proposed the telomere theory of aging. Olovnikov claimed that during **DNA synthesis**, DNA **polymerase** fails to replicate all of the nucleic acids resulting in

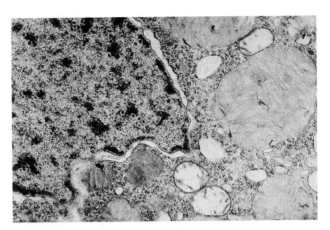

Cerebral sphingolipidosis, indicative of Tay-Sachs disease. © 1992 IMS Creative. Custom Medical Stock Photo. Reproduced by permission.

shortened chromosomes with each successive generation. Eventually, the cell will no longer divide and after enough critical regions have been deleted, the cell cycle will arrest.

Telomeres are synthesized by the enzyme telomerase, a **reverse transcriptase** that contains its own **RNA** template complementary to the telomere. After DNA replication, the RNA primer is removed leaving a gap at the end of the newly synthesized strand of DNA. In order to fill in the gap, the enzyme telomerase catalyzes the extension of the template strand at the end. This extension of DNA forms a hairpin loop that attaches to the end **nucleotide** on the newly synthesized strand, thus completing the telomere.

The existence and persistence of telomerase in normal cells is under active study. Telomerase is one of the factors believed to control the length of telomeres that act as a biological clock for the cell. Telomerase is not active all the time, nor is it found in every cell of the body. Researchers have discovered that if the action of telomerase is interrupted, the telomere will be abnormally short and the aging process of the cell accelerates, resulting in cell death much earlier than normal. Likewise, scientists theorize that the uninhibited action of telomerase can lead to cell immortality. **Cancer** cells have been found to have active telomerase that increases **cell proliferation** by extending telomeres. Researchers claim that when newly formed **tumor** cells arise, they turn on a **gene** for telomerase. However, some cancer cells have been found to extend their telomeres without turning on this gene. By understanding the role of telomeres and telomerase, scientists hope to one day predict and treat the onset of disease.

See also DNA structure

TELOPHASE • *see* CELL DIVISION

TEMIN, HOWARD (1934-1994)

American virologist

Howard Temin is an American virologist who revolutionized molecular biology in 1965, when he found that genetic information in the form of **ribonucleic acid** (**RNA**) can be copied into **deoxyribonucleic acid** (**DNA**). This process, called **reverse transcriptase**, contradicted accepted beliefs of molecular biology at that time, which stipulated that DNA always passed on genetic information through RNA. Temin's research also contributed to a better understanding of the role viruses play in the onset of **cancer**. For this, he was featured on the cover of *Newsweek* in 1971, which hailed his discovery as the most important advancement in cancer research in sixty years. In addition, Temin shared the 1975 Nobel Prize in physiology or medicine for his work on the Rous sarcoma **virus**. His discovery of the reverse transcriptase process contributed greatly to the eventual identification of the human immunodeficiency virus (HIV). Temin's later research focused on genetic engineering techniques. A vehement anti-smoker, Temin took every opportunity to warn against the dangers of tobacco, even in his acceptance speech for the Nobel Prize.

Howard Martin Temin was born in Philadelphia to Henry Temin, a lawyer, and Annette (Lehman) Temin. The second of three sons, Temin showed an early aptitude for science and first set foot in a laboratory when he was only fourteen years old. As a student at Central High School in Philadelphia, he was drawn to biological research and attended special student summer sessions at the Jackson Laboratory in Bar Harbor, Maine. After graduation from high school, Temin enrolled at Swarthmore College in Pennsylvania where he majored and minored in biology in the school's honors program. He published his first scientific paper at the age of eighteen and was described in his college yearbook as "one of the future giants in experimental biology."

After graduating from Swarthmore in 1955, Temin spent the summer at the Jackson Laboratory and enrolled for the fall term at the California Institute of Technology in Pasadena. For the first year and a half, he majored in experimental **embryology** but then changed his major to animal virology. Temin studied under **Renato Dulbecco**, a renowned biologist who worked on perfecting techniques for studying virus growth in tissue and developed the first plaque assay (a chemical test to determine the composition of a substance) for an animal virus. Temin received his Ph.D. in biology in 1959 and worked for another year in Dulbecco's laboratory. In 1960 he joined the McArdle Laboratory for Cancer Research at the University of Wisconsin—Madison, where he spent the remainder of his career as the Harold P. Rusch Professor of Cancer Research and the Steenbock Professor of Biological Sciences.

Temin began studying the Rous sarcoma virus (RSV) while still a graduate student in California. First identified in the early twentieth century by **Peyton Rous**, RSV is found in some **species** of hens and was one of the first viruses known to cause tumors. In 1958 Temin and Harry Rubin, a postdoctoral fellow, developed the first reproducible assay *in vitro* for the quantitative measuring of virus growth. Accepting an appointment as assistant professor of oncology at Wisconsin in 1960, Temin continued his research with RSV. Using the assay method he and Rubin developed, Temin focused on delineating the differences between normal and **tumor** cells. In 1965, Temin announced his theory that some viruses cause cancer through a startling method of information transfer.

Scientists at the time thought that genetic information could only be passed from DNA to RNA. DNA is a long molecule comprised of two chains of nucleic units containing the sugar deoxyribose. RNA is a molecule composed of a chain of nucleic units containing the sugar ribose. For years, many of Temin's colleagues rejected his theory that some viruses actually reverse this mode of transmitting genetic information, and they cited a lack of direct evidence to support it. Temin, however, was convinced that RNA sometimes played the role of DNA and passed on the genetic codes that made a normal **cell** a tumor cell.

It took Temin several years, however, to prove his theory. Despite making further inroads in gathering evidence implicating **DNA synthesis** in RSV infection, many of his colleagues remained skeptical. Finally, in 1970, Temin, working with Satoshi Mitzutani, discovered a viral enzyme able to copy RNA into DNA. Dubbed a reverse transcriptase virus, this enzyme passed on hereditary information by seizing control of the cell and making a reverse transcript of the host DNA. In other words, the enzyme synthesized a DNA virus that contained all the genetic information of the RNA virus. This discovery was made simultaneously by biologist **David Baltimore** at his laboratory at the Salk Institute in La Jolla, California.

The work of Temin and Baltimore led to a number of impressive developments in molecular biology and recombinant DNA experimentation over the next twenty years, including characterizing retroviruses, a family of viruses that cause tumors in vertebrates by adding a specific **gene** for cancer cells. In 1975, Temin shared the Nobel Prize in physiology or medicine with his former mentor, Renato Dulbecco, and David Baltimore. These three scientists' research illustrated how separate avenues of scientific research could converge to produce significant advances in biology and medicine. Eventually, interdisciplinary research was to become a mainstay of modern science.

In 1987, Temin reflected on his discovery of viral roles in causing cancer. "I measure [my discovery's importance] by comparing what I taught in the experimental oncology course 25 years ago," said Temin in a University of Wisconsin press release, pointing out that the topic of viral carcinogenesis (the viral link to cancer) was rarely the focus of any lectures at that time. "Now, in the course we're teaching, between a third and half of the lectures are related directly or indirectly to viral carcinogenesis."

Temin's continuing work into the role viruses play in carcinogenesis had an important impact on acquired immunodeficiency syndrome (**AIDS**) research. Temin's discovery of reverse transcriptase provided scientists with the means to find and identify the AIDS virus. His interest in genetic engineering and the causes of cancer eventually led him to another exciting discovery. He found a way to measure the **mutation** rate in retroviruses (viruses that engage in reverse transcriptase), which led to insights on the variation of cancer genes and viruses, such as AIDS. Determining the speed at which genes and viruses change provided vital information for devising attempts to vaccinate or treat viral diseases. Temin's discovery of reverse transcriptase also led to the development of standard tools used by biologists to prepare radioactive **DNA probes** to study the genetic makeup of viral and malignant cells. Another genetic engineering technique that arose from this research was the ability to make DNA copies of **messenger RNA**, which could be isolated and purified for later study.

Temin was also interested in such areas as **gene therapy**, which uses **gene splicing** techniques to genetically improve the host organism. As he began to apply genetic engineering techniques to his research, Temin recognized legitimate concerns about producing pathogens (microorganisms that cause disease) that could escape into the environment. He also served on a committee that drew up federal guidelines in human gene therapy trials.

Temin's research convinced him that science was making progress in the fight against cancer. Temin said in a 1984 United Press International release, "We know the names of some of the genes which are apparently involved in cancer. If past history is a guide, this understanding will lead to improvement in diagnosis, therapy, and perhaps prevention."

Throughout his career, Temin continued to teach general virology courses for graduates and undergraduates. He also worked with students in his laboratory. "I get satisfaction from a number of things—from discovering new phenomena, from understanding old phenomena, from designing clever experiments—and from seeing students and postdoctoral fellows develop into independent and outstanding scientists," he stated in a University of Wisconsin press release.

A scientist and family man who shunned the spotlight after winning the Nobel Prize (which he kept in the bottom drawer of a file cabinet), Temin was committed to quietly searching for clues into the mysteries of cancer-causing viruses. Temin married Rayla Greenberg, also a geneticist, in 1962, and the couple had two daughters, Miriam and Sarah. A familiar site on the Wisconsin-Madison campus, Temin bicycled to work every day on his mountain bike.

Temin spoke out against cigarette smoking. During the award ceremonies for the Nobel Prize, Temin told the audience that he was "outraged" that people continued to smoke even though cigarettes were proven to contain carcinogens. He instructed that 80% of all cancers were preventable because they resulted from environmental factors, such as smoking. "It was the most important general statement I could make about human cancer," he said later in a *People* magazine interview. "And I realized the Nobel Prize would give me an opportunity to speak out that a person does not ordinarily have." Temin went on to testify before the Wisconsin legislature and congress in support of antismoking bills. His research efforts in AIDS led him to urge the federal government to increase funding for further research into the AIDS epidemic. Despite living a lifestyle designed to minimize the risk of cancer, Temin, who never smoked, developed lung cancer in 1992. He died of this disease on February 9, 1994. In addition to the Nobel Prize, Temin received many other awards for his research, including the prestigious Albert Lasker Award in Basic Medical Research in 1974 and the National Medal of Science in 1992.

See also Viral genetics

TERATOLOGY

Teratology is the study of serious deviations from normal growth and development in organisms. The word teratogenesis is derived from the Greek *gennan*, meaning to produce, and *terata*, meaning monster. In humans, teratology is the study of chemicals, drugs, medications, alcohol, disease, or other environmental agents in relation to fetal abnormalities and **birth defects**. To be classified as a teratogen, an agent must cause either low birth weight, intrauterine growth retardation, or small size for gestational age; stillbirth or miscarriage; structural abnormalities; or functional defects (mental or developmental retardation). To cause birth defects, a teratogen must reach the developing fetus. Teratogens can also cause premature birth and fetal/newborn addiction to pharmacological agents. Chemicals such as thalidomide, tetracycline, valproic acid, cocaine, alcohol; metals such as lead and mercury; viruses such as syphilis, herpes, and rubella can all be regarded as teratogens.

Birth defects and abnormalities have occurred throughout history. Rock carvings and drawings depict individuals with deformities and malformations; writings from ancient Babylon describe them, as do clay tablets believed to date back to 2000 B.C. found in the library of an Assyrian king around 700 B.C..

Before the advent of the modern study of birth defects in the eighteenth century, congenital abnormalities were viewed with superstition or as supernatural—perhaps punishment from the gods or the work of the devil. The first person to articulate the theory of "developmental arrest" was William Harvey in 1651, when he observed that a cleft lip in a newborn infant closely resembled the normal condition of a fetus at a certain stage of development. Albrecht von Haller (1768) and Kaspar Friedrich Wolff (1759) followed a similar line of thinking. Modern studies show that teratogens act by disrupting the normal developmental pathways of a growing fetus. So a small disruption of **cell** signalling by a teratogenic chemical can alter the development of a morphological feature. A teratogen may also be a **mutagen**, a substance which increase the chance of a **DNA mutation**, which may lead to birth defects.

Approximately 3-5% of all infants worldwide are born with some form of birth defect. Approximately 5% of those appear to be due to teratogens, a small percentage of which are preventable through pre-pregnancy counselling and the assessment of **risk factors** in certain populations. In many countries, educational and informational organizations are working toward reducing the number teratogen-related birth defects. For example, recent evidence indicates a woman can reduce her chances of bearing a child with **spina bifida**, one of the most common types of birth defects, by 50% simply by consuming at least 400 micrograms of folic acid daily for least one month prior to conception.

See also Biological warfare; Cleft lip/palate

TEST-CROSS • *see* CROSSING

TETRAD ANALYSIS

Tetrads, the four **haploid cell** products of a single meiotic division, can be used to study the behavior of chromosomes and genes during **meiosis**. This kind of analysis is only possible in haploid organisms, such as some fungi or single-celled algae, in which the products of each meiosis are held together in a kind of bag (or ascus). Thus, the four products of a single meiosis are recoverable and testable. Tetrads can take different forms in different organisms. In some **species**, four sexual spores containing the four nuclei are formed as products of meiosis e.g. in baker's **yeast**, *Saccharomyces cerevisiae*. In other species, e.g. the red bread mould, *Neurospora crassa*, eight sexual spores are formed: in these cases each of the four meiotic product nuclei undergoes a further mitotic division, producing eight nuclei that are then enclosed in eight spores. These groups of eight may be called octads, but most geneticists also call them tetrads because they simply represent double tetrads.

Tetrad analysis has proved very powerful in testing some of the assumptions of the **chromosome** theory of **heredity** directly. In many experiments in Mendelian **genetics**, individuals are studied on the basis of random meiotic product analysis. For example, a testcross of *Aa* to *aa* produces a 1:1 ratio, from which the equal segregation of *A* and *a* in a single meiosis is inferred. The use of tetrads provides a more direct test of this notion because the products of meiosis can be examined directly.

In some species, tetrads can be jumbled and unordered within the ascus while in others they can be arranged linearly. The latter are particularly interesting because they can be used to estimate the distance of the locus under study from the centromere (the region of the chromosome where spindle fibers attach). The linear array of sexual spores is a result of the lack of spindle overlap in the meiotic divisions (or in any subsequent mitotic divisions). Because no spindle overlap occurs, the nuclei do not pass each other in the long ascus and a linear array of haploid spores is produced. Using marker genes *A* and *a*, it is possible to cross *A* and *a* cultures

and then isolate the tetrads. Typical results for a cross *A* X *a* are shown below:

A	*a*	*A*	*a*	*A*	*a*
A	*a*	*A*	*a*	*A*	*a*
A	*a*	*a*	*A*	*a*	*A*
A	*a*	*a*	*A*	*a*	*A*
a	*A*	*A*	*a*	*a*	*A*
a	*A*	*A*	*a*	*a*	*A*
a	*A*	*a*	*A*	*A*	*a*
a	*A*	*a*	*A*	*A*	*a*

Columns representing spores within asci.

The first two types represented by the two left-most columns are called first division segregation, or M_I, patterns. These asci result from meioses in which there has been no detectable crossover between marker locus and the corresponding centromere locus of that chromosome. Because of this, the two alleles *A* and *a* segregate into separate nuclei at the first meiotic division. The four ascus types on the right, however, show second division segregation, or M_{II}, patterns. These asci arise from meioses in which there has been a crossover between the marker locus and centromere locus. As a result of the crossover, *A* and *a* alleles appear in the same **nucleus** at the end of the first meiotic division, and they do not segregate until the second division. The frequency of second meiotic division segregation spore patterns for a particular **gene** locus is proportional to distance of that locus from the centromere. It is possible to calculate this distance in map units by dividing the percentage of asci showing a second-division segregation pattern for that locus by two. The analysis can be further extended to map more than one locus.

See also Fungal genetics; Mendelian genetics and Mendel's laws of heredity; Non-Mendelian inheritance

THALASSEMIA

Thalassemia is an inherited disorder that affects the production of hemoglobin and causes anemia. Hemoglobin is the substance in red blood cells that enables them to transport oxygen throughout the body. It is composed of a heme molecule and protein molecules called globins. Owing to an inherited genetic trait, lower-than-normal amounts of globins are manufactured in the bone marrow. If the trait is inherited from both parents, a globin may be entirely absent. Thalassemia causes varying degrees of anemia, which can range from insignificant to life threatening. People of Mediterranean, Middle Eastern, African, and Asian descent are at higher risk of carrying the genes for thalassemia.

Hemoglobin molecules are vital for transporting oxygen throughout the body. Hemoglobin and red blood **cell** production is carried out in the bone marrow, a spongy tissue found within certain bones, such as the hips, skull, and breast bone. The hemoglobin molecule is actually made up of five smaller component molecules: a heme molecule and four protein molecules called globins. Normal hemoglobin molecules contain two pairs of different globin molecules.

Humans have the genes to construct six types of globins, but do not use all six at once. Different globins are produced depending on the stage of development: embryonic, fetal, or adult. During embryonic development, hemoglobin contains two zeta-globin molecules and two epsilon-globin molecules. At the fetal stage, the body switches to alpha-globin and gamma-globin production. Within weeks after birth, an infant continues to produce alpha-globin, but gamma-globin is replaced by beta-globin and a very minor amount of delta-globin. After the first two to three months of life, most hemoglobin in the body is composed of two alpha-globins and two beta-globins; approximately 0.5% is composed of two alpha-globins and two delta-globins.

There is a different **gene** for each type of globin, with the exception of alpha-globin, which has two genes. (Genes are inherited in pairs, one copy from each parent.) A gene **mutation** may lead to inadequate levels of the related globin, reduced hemoglobin formation, and anemia. Such mutations are the underlying cause of thalassemia.

Thalassemia is classified according to the globin that is affected. The most common types of thalassemia are beta-thalassemia and alpha-thalassemia. Beta-thalassemia is caused by a mutation in the gene responsible for beta-globin. If a mutated beta-globin gene is inherited from both parents, the result is beta-thalassemia major, a severe, potentially life-threatening anemia. Beta-thalassemia major may also be referred to as Cooley's anemia or erythroblastic anemia. If only one mutated copy of the beta-globin gene is inherited, mild-to-nonexistent symptoms may appear; this condition is called beta-thalassemia minor. A person with one mutated copy of the beta-globin gene is referred to as a **carrier** of the beta-thalassemia trait.

The alpha-thalassemias are more complex because a person inherits two alpha-globin genes from each parent, yielding a total of two pairs of alpha-globin genes. Mutations in these genes can give rise to a range of symptoms. As long as adequate levels of alpha-globins are produced, the person—called the carrier of the alpha-thalassemia trait—will have few, if any symptoms. In cases in which alpha-globin is severely reduced, or not produced at all, the consequences can be fatal during fetal development or shortly after birth.

People with Mediterranean (including North African), Middle Eastern, or Southeast Asian ancestry are at higher risk of being carriers of or developing beta-thalassemia than are other populations. Alpha-thalassemia also is more likely to affect people of Mediterranean, African, Middle Eastern, and Southeast Asian descent. In some areas, 1 in 150-200 children are born with thalassemia major. It has been estimated that 2 million people in America carry the thalassemia trait. When two carriers of the same type thalassemia produce a child, there is a 25% possibility that the child will inherit moderate or severe thalassemia.

Thalassemia arises from mutations in one or more globin genes, leading to a reduction or absence of the associated globin. The severity of symptoms is directly related to whether one or both genes in a pair is mutated. However, symptoms may be modified by other genetic or environmental factors.

The **inheritance** of one mutated beta-globin gene is accompanied by few, if any symptoms. A person may have mild anemia, but may not even be aware of being a carrier unless tested. Anemia may only appear during pregnancy or following severe infections.

An inheritance of two mutated beta-globin genes causes a potential cascade of symptoms. Some individuals develop mild symptoms that do not require treatment, but others experience life-threatening anemia and complications. At the earliest, symptoms appear several weeks after birth; that is, when fetal gamma-globin production gives way to beta-globin production. Alpha-globin is produced normally and, relative to beta-globin, it is over-produced. This excess alpha-globin precipitates in the immature red blood cells, destroying them.

Similar to beta-thalassemia, if enough alpha-globin is being produced, symptoms may be nonexistent. If the alpha-globin levels are low enough to produce symptoms, onset may begin early in life as alpha-globin production starts during the fetal stage.

Thalassemia may be diagnosed from the symptoms; however, with proper medical treatment, a diagnosis may be made before symptoms become life- or health-threatening. Basic information that is used in diagnosis includes race and ethnic background, family history, and age. Unexpectedly slow development, along with pallor, jaundice, enlarged spleen or liver, or deformed bones can be common signs of thalassemia. Laboratory blood tests are used to confirm a diagnosis and determine the type of thalassemia. These tests can also be used to identify carriers.

Thalassemia cannot be cured; therapy focuses on managing symptoms. Treatment is not necessary for individuals who are unaffected or only develop mild symptoms. The mainstays of thalassemia management are blood transfusions and iron chelation therapy. Blood transfusions are typically given every 6-8 weeks, but may be more frequent in some cases. These transfusions have two purposes: to keep hemoglobin at or near normal levels and to prevent the bone marrow from producing ineffective red blood cells. Thalassemia has been treated with bone marrow transplantation. However, bone marrow transplants are strictly limited by several factors, including the general health of the marrow recipient and whether a donor with compatible marrow can be found.

An individual's outlook depends on the exact type of thalassemia and the associated genes. In the carrier state, a person may never develop symptoms. Other carriers develop mild anemia at times of extreme stress, such as pregnancy or illness. Some forms of alpha-thalassemia, in which alpha-globin is severely reduced or absent, are nearly always fatal. Affected

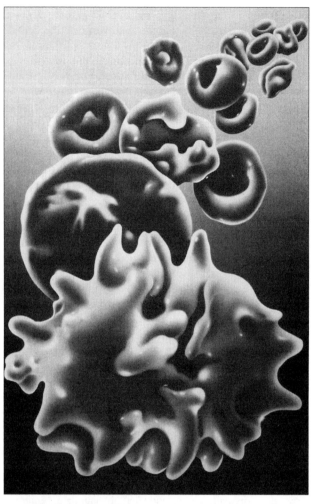

Illustration of distorted red blood cells in the inherited blood disorder of thalassemia. *Photograph by John Bavosi. Photo Researchers, Inc. Reproduced by permission.*

nucleotide changes, or small insertions or **deletions**, in **DNA**. The duplex DNA, usually between 100 and 300 nucleotides long and generated by the **polymerase chain reaction** (PCR), is forced through a gel that contains an increasing temperature gradient. TGGE is essentially the same technique as **Denaturing Gradient Gel Electrophoresis** (DGGE) where, instead of a temperature gradient, a chemical gradient is used. Although TGGE requires more specialist equipment, it is often preferred by scientists because the chemicals used in DGGE are toxic. As the DNA moves through the gel, it encounters a temperature, which makes the duplex melt, or denature. This denatured DNA is essentially arrested at that point in the gel, and forms a sharp band. The point at which the DNA duplex denatures can be affected by a single nucleotide change in the DNA, and so the band is formed in a different position.

For TGGE to work correctly to detect mutations, it is important that the DNA denatures from one end of the duplex rather than denaturing in the middle first or denaturing at both ends at the same time. The physical properties of a piece of DNA when it denatures can be determined from its nucleotide sequence, and is usually calculated by a computer program, such as Melt87. Like DGGE, the main advantage of TGGE is its sensitivity—it can detect virtually all mutations in a given piece of DNA. Because of this, it has been used in human **genetics** to screen exons of a disease **gene** for mutations in patients with this disease. An example is screening the human prion protein for disease-causing mutations. It is also commonly used to analyse DNA sequence differences in bacterial populations.

Although TGGE is mainly used for **mutation** detection, it can be used for other purposes. Studies on the physical nature of DNA have used TGGE. For example, the thermal stability of DNA duplexes with a single base pair mismatch (one base does not have its complementary partner) has been studied. It was found that the base that is mismatched and the **bases** that immediately surround it influence when the DNA duplex melts in a thermal gradient. TGGE can also analyse how a protein binds to a section of DNA, and how the shape (conformation) of a protein can vary with temperature. The thermal behaviour of alpha-amylase (an enzyme which digests the amylose in starch) from the fungus *Aspergillus oryzae* was studied using this technique. It can also be used to select for bacterial mutants that produce a certain enzyme which works at high temperatures (increased thermal stability), which may be important for the suitability of the enzyme in industrial processes operating at higher temperatures.

See also DNA binding proteins; Gene Mutation; Mutation; Prions

fetuses often die during weeks 34-40 of a pregnancy; some infants are born alive, but die within hours of birth.

Once the genes that determine thalassemia are inherited, the disease cannot be prevented. Screening offers the opportunity of identifying thalassemia carriers. Carriers may decide to undergo genetic counseling to assess potential risks to their children. Finally, prenatal testing, usually **chorionic villus** sampling or amniotic fluid testing, allows identification of thalassemia in unborn children.

See also Prenatal diagnostic techniques

THERMAL GRADIENT GEL ELECTROPHORESIS

Thermal or Temperature Gradient Gel Electrophoresis (TGGE) is a variant of gel electrophoresis that detects single

THREE-DIMENSIONAL ARRAY · *see*

DIMENSIONAL ARRAYS

THROUGHPUT SCREENING

Throughput screening is the process whereby many compounds can be tested for their target activity in a single experiment. The experiments, also called assays, are automated both with respect to the performance of the assay and detection of activity. Such automated assays are also referred to as high throughput screening.

The technique of throughput screening has proved a particularly powerful tool in pharmaceutical research. Many compounds can be screened for activity in a short time. Promising candidates can then be analyzed with more vigor to assess their therapeutic and commercial potential.

Throughput screening was first done using animals. However, the expense of this procedure resulted in the development of a test tube-type of assay system. Now, in a typical assay each compound to be tested is put into a well in a plastic plate containing 96 wells. The assay reagent, a solution containing the molecules needed to stimulate a reaction for the test compounds, is added to each well. After a set time, each well is monitored for the target reaction. Inhibition or activation are two typical reactions. High throughput screening thus functions to rapidly weed out potentially useful compounds or cells—tens of thousands can be analyzed each day—from those that are not useful for the particular purpose.

The advent of laboratory automation and robotic technology in the mid–1990s made throughput screening possible. In a further **evolution**, the assay mixture now can incorporate cells that contain genetically modified genes. The well plates used in this technique contain over 1500 wells. The characteristic positive reaction, such as fluorescence, is easier to detect and requires a smaller volume of test substance. This advance, designated ultra-high throughput screening, permits up to 200,000 substances a day to be screened.

Typically, out of a million compounds tested by throughput screening, three to five candidates emerge. One in twenty candidates end up being developed into a new drug. Without the speed and economy of afforded by throughput screening, drug development would be extremely slow and extraordinarily expensive. In the human **genome**, estimates are that thousands of genes may be potential targets of drug therapy. There are around 10^{63} chemical molecules that can potentially be used against these target genes. Without throughput screening, identification of therapeutic molecules would be virtually impossible.

See also Fluorescent dyes

THYMINE (T) · *see* BASES

TIJO, JOE-HIN (1919-   )
Chinese Cytogeneticist

Joe-Hin Tijo was born on February 11, 1919 in Java, once a part of the Dutch East Indies. In college Tijo studied a branch of agriculture called agronomy that deals with field-crop production and soil management. He became interested in breeding potatoes and attempted to produce hybrids resistant to disease. However, during World War II, Tijo was taken prisoner and held in a concentration camp by the Japanese Imperial Army until the war ended. After release from the concentration camp, Tijo was taken to Holland and granted a fellowship by the government to study plant breeding in Copenhagen. Shortly after, Tijo began research on cytogenetics with Albert Levan, head of the Institute of Genetics at the University of Lund in Sweden. From 1948 to 1959, Tijo worked in Zaragoza, Spain as director of cytogenetics on a plant improvement program but continued research at the University of Lund during holidays and summers.

In 1956, Joe-Hin Tijo and Albert Levan were working with human embryonic tissue when they discovered that there was 46 chromosomes, not 48, which scientists had believed to be the case for over 30 years. Previously, scientists were unable to distinguish the correct number of chromosomes because techniques for preparing microscope slides had not yet been perfected. However, Tijo and Levan's methods for separating chromosomes on microscope slides proved successful when they was able to count 46 chromosomes per **cell** during observations of 261 embryonic cells.

Tijo traveled to the United States in 1957 to pursue his PhD at the University of Colorado. After earning his doctorate, Tijo was invited to join the National Institutes of Health (NIH) and began working for the National Institute of Arthritis and Metabolic Diseases' Laboratory of Experimental Pathology. He continued studying chromosomes with special emphasis on leukemia and mental retardation. In 1962, Tijo and Whang discovered **chromosome** abnormalities in leukemias after producing a method for aspirating bone marrow cells and obtaining a view of the cell's chromosomes during the metaphase stage of **mitosis** called a metaphase spread. Tijo retired from NIH in 1992 at the age of 73.

See also Chromosome structure and morphology; History of genetics: Modern genetics; Karyotype and karyotype analysis

TOBACCO MOSAIC VIRUS

Tobacco **mosaic virus** (TMV), also known as tobamovirus, is a rod-shaped virus with **ribonucleic acid** (**RNA**) surrounded by a coat of protein that causes mosaic-like symptoms in plants. Mosaic-like symptoms are characterized by mottled patches of green or yellow color on the leaves of infected plants. The virus causes abnormal cellular function that usually does not kill the plant but stunts growth. Infected plants may have brittle stems, abnormally small, curled leaves, and unripened fruit.

Tobacco mosaic virus is capable of infecting many kinds of plants, not just tobacco plants. TMV is spread through small wounds caused by handling, insects, or broken leaf hairs that result from leaves rubbing together. The virus attaches to the **cell** wall, injects its RNA into the host cell, and forces the host cell to produce new viral RNA and proteins. Finally, the viral

RNA and proteins assemble into new viruses and infect other cells by passing through small openings called plasmodesmata that connect adjacent plant cells. This process allows the virus to take over metabolic processes without killing cells.

Tobacco mosaic virus is highly infectious and can survive for many years in dried plant parts. Currently, there is no vaccine to protect plants from TMV, nor is there any treatment to eliminate the virus from infected plants. However, seeds that carry TMV externally can be treated by acid extraction or trisodium phosphate and seeds that carry the virus internally can receive dry heat treatments.

The discovery of viruses came about in the late 1800's when scientists were looking for the **bacteria** responsible for damaging tobacco plants. During one experiment in 1892, Russian biologist Dimitri Ivanovsky concluded that the disease in tobacco plants could not be caused by bacteria because it passed through a fine-pored filter that is too small for bacteria to pass through. In 1933, American biologist Wendell Stanley of the Rockefeller Institute discovered that the infectious agent formed crystals when purified. The purified extract continued to cause infection when applied to healthy tobacco plants and therefore, could not be a living organism. Soon after, scientists were able to break down the virus into its constituent parts. Today, it is known that the infectious agent that causes the disease in tobacco plants is a virus, not bacteria.

See also Agricultural genetics; Viral genetics

TODD, ALEXANDER ROBERTUS (1907-1997)

English chemist

Alexander Todd was awarded the 1957 Nobel Prize in chemistry for his work on the chemistry of nucleotides. He was also influential in synthesizing vitamins for commercial application. In addition, Todd invesitgated active ingredients in cannabis and hashish, and helped develop efficient means of producing chemical weapons.

Alexander Todd was born in Glasgow, Scotland, to Alexander and Jane Lowrie Todd. In his autobiography, *A Time to Remember,* Todd recalls how, through hard work, his parents rose to the lower middle class despite having no more than an elementary education, and how determined they were that their children should have an education at any cost.

In 1918, Todd gained admission to the Allan Glen's School in Glasgow, a science high school. His interest in chemistry, which first arose when he was given a chemistry set at the age of eight or nine, developed rapidly. On graduation, six years later, he entered the University of Glasgow instead of taking a recommended additional year at Allan Glen's. In his final year at university, Todd did a thesis on the reaction of phosphorus pentachloride with ethyl tartrate and its diacetyl derivative under the direction of T. E. Patterson, resulting in his first publication.

After receiving his B.Sc. degree in chemistry with first-class honors in 1928, Todd was awarded a Carnegie research scholarship and stayed on for another year working for Patterson on optical rotatory dispersion. Deciding that this line of research was neither to his taste nor likely to be fruitful, he traveled to Germany to do graduate work at the University of Frankfurt, studying natural products. Todd preferred Jöns Berzelius' definition of organic chemistry as the chemistry of substances found in living organisms to Gmelin's definition of it (in the *Gmelin Handbook of Organometallic Chemistry*) as the chemistry of carbon compounds.

At Frankfurt he studied the chemistry of apocholic acid, one of the bile acids (compounds produced in the liver and having a structure related to that of cholesterol and the steroids). In 1931, he returned to England with his doctorate. He applied for and received an 1851 Exhibition Senior Studentship, which allowed him to enter Oxford University to work under Robert Robinson, who would receive the Nobel Prize in chemistry in 1947. Todd received his Ph.D. from Oxford in 1934. His research at Oxford dealt first with the synthesis of several anthocyanins, the coloring matter of flowers, and then with a study of the red pigments from some molds.

After leaving Oxford, Todd went to the University of Edinburgh on a Medical Research Council grant to study the structure of vitamin B_1 (thiamine, or the anti-beriberi vitamin). The appointment came about when George Barger, professor of medical chemistry at Edinburgh, sought Robinson's advice about working with B_1. At that time, only a few milligrams of the substance were available, and Robinson suggested Todd because of his interest in natural products and his knowledge of microchemical techniques acquired in Germany. Although Todd and his team were beaten in the race to synthesize B_1 by competing German and American groups, their synthesis was more elegant and better suited for industrial application. It was at Edinburgh that Todd met and became engaged to Alison Dale—daughter of Nobel Prize laureate Henry Hallett Dale—who was doing postgraduate research in the pharmacology department. They were married in January of 1937, shortly after Todd had moved to the Lister Institute where he was reader (or lecturer) in biochemistry.

Toward the end of his stay at Edinburgh, Todd began to investigate the chemistry of vitamin E (a group of related compounds called tocopherols), which is an antioxidant—that is, it inhibits loss of electrons. He continued this line of research at the Lister Institute and also started an investigation of the active ingredients of the *Cannabis sativa* plant (marijuana) that showed that cannabinol, the major product isolated from the plant resin, was pharmacologically inactive.

In 1938, Todd was offered a professorship at Manchester which he accepted, becoming Sir Samuel Hall Professor of Chemistry and director of the chemical laboratories of the University of Manchester. At Manchester, Todd was able to continue his research with little interruption. During his first year there, he finished the work on vitamin E with the total synthesis of alpha-tocopherol and its analogs. Attempts to isolate and identify the active ingredients in cannabis resin failed because the separation procedures available at the time were inadequate; however, Todd's synthesis of cannabinol involved an intermediate, tetrahydrocannabinol (THC), that had an effect much like that of hashish on rabbits and suggested to him that the effects of hashish were due to one of the isomeric tetrahydrocannabinols. This view was later proven correct, but by others, because the outbreak of World War II forced Todd to abandon this line of research for work more directly related to the war.

As a member, and then chair, of the Chemical Committee, which was responsible for developing and producing chemical warfare agents, Todd developed an efficient method of producing diphenylamine chloroarsine (a sneeze gas), and designed a pilot plant for producing nitrogen mustards (blistering agents). He also had a group working on penicillin research and another trying to isolate and identify the "hatching factor" of a parasite that attacks potatoes.

Late in 1943 Todd was offered the chair in biochemistry at Cambridge University, which he refused. Shortly thereafter he was offered the chair in organic chemistry, which he accepted, choosing to affiliate with Christ's College. From 1963 to 1978, he served as master of the college. As professor of organic chemistry at Cambridge, Todd reorganized and revitalized the department and oversaw the modernization of the laboratories (they were still lighted by gas in 1944) and, eventually, the construction of a new laboratory building.

Before the war, his interest in vitamins and their mode of action had led Todd to start work on nucleosides and nucleotides. Nucleosides are compounds made up of a sugar (ribose or deoxyribose) linked to one of four heterocyclic (that is, containing rings with more than one kind of atom) nitrogen compounds derived either from purine (adenine and guanine) or pyrimidine (uracil and cytosine). When a phosphate group is attached to the sugar portion of the molecule, a nucleoside becomes a **nucleotide**. The nucleic acids (**DNA** and **RNA**), found in **cell** nuclei as constituents of the chromosomes, are chains of nucleotides. While still at Manchester, Todd had worked out techniques for synthesizing nucleosides and then attaching the phosphate group to them (a process called phosphorylating) to form nucleotides. Later, at Cambridge, he worked out the structures of the nucleotides obtained by the degradation of nucleic acid and synthesized them. This information was a necessary prerequisite to **James Watson** and **Francis Crick's** formulation of the double-helix structure of DNA two years later.

Todd had found the nucleoside adenosine in some coenzymes, relatively small molecules that combine with a protein to form an enzyme, which can act as a catalyst for a particular biochemical process. He knew from his work with the B vitamins that B_1 (thiamine), B_2 (riboflavin) and B_3 (niacin) were essential components of coenzymes involved in respiration and oxygen utilization. By 1949 he had succeeded in synthesizing adenosine—a triumph in itself—and had gone on to synthesize adenosine di- and triphosphate (ADP and **ATP**). These compounds are nucleotides responsible for energy production and energy storage in muscles and in plants. In 1952, he established the structure of flavin adenine dinucleotide (FAD), a coenzyme involved in breaking down carbohydrates so that they can be oxidized, releasing energy for an organism to use. For his pioneering work on nucleotides and nucleotide **enzymes**, Todd was awarded the 1957 Nobel Prize in chemistry.

In 1952, Todd became chairman of the advisory council on scientific policy to the British government, a post he held until 1964. He was knighted in 1954 by Queen Elizabeth for distinguished service to the government. Named Baron Todd of Trumpington in 1962, he was made a member of the Order of Merit in 1977. In 1955, he became a foreign associate of the United States' National Academy of Sciences. He traveled extensively and been a visiting professor at the University of Sydney (Australia), the California Institute of Technology, the Massachusetts Institute of Technology, the University of Chicago, and Notre Dame University.

A Fellow of the Royal Society since 1942, Todd served as its president from 1975 to 1980. He increased the role of the society in advising the government on the scientific aspects of policy and strengthened its international relations. Extracts from his five anniversary addresses to the society dealing with these concerns are given as appendices to his autobiography. In the forward to his autobiography, Todd reports that in preparing biographical sketches of a number of members of the Royal Society he was struck by the lack of information available about their lives and careers and that this, in part, led him to write *A Time to Remember.* Todd died in his home city of Cambridge, England, at age 89.

See also DNA structure

TONEGAWA, SUSUMU (1939-   )
Japanese molecular biologist

Japanese molecular biologist Susumu Tonegawa is a professor at the Massachusetts Institute of Technology (MIT). Tonegawa's work made important advances into understanding of the genetic mechanisms of immunological systems.

Tonegawa received his doctorate from the University of California at San Diego in 1969. In 1971, Tonegawa became a of member of the Basel Institute for Immunology in Switzerland where he conducted research until accepting the professorship at MIT in 1981.

Tonegawa work in immunogenetics showed distinct relationships between antibodies and the genes responsible for their production and regulation. Tonegawa reported that there were alterations in patterns of chromosomal **recombination** that allowed the genes responsible for antibody production to move closer to one another on chromosomes. Tonegawa's worked gained a Nobel Prize in physiology or medicine in 1987.

The selective recombination mechanism allows organisms to enhance the production of antibodies. For example, although the human body has a limited number of chromosomes and a finite amount of **DNA** (only a portion of which is related to immune system function), cells are able to produce highly specific antibodies to a vast number of antigens. Tonegawa's work established that **gene** rearrangements allowed for increased variety in the production of antibodies.

Born in Nagoya, Japan, Tonegawa took his undergraduate studies in chemistry at Kyoto University in Japan in 1963 moving to the United States to undertake his graduate studies

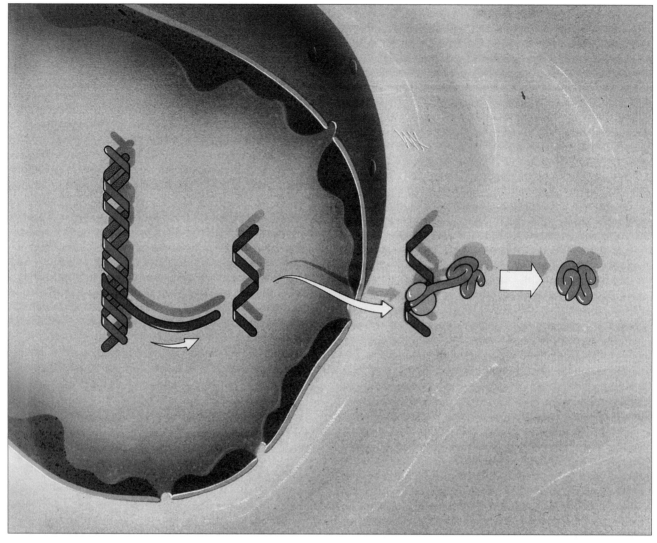

Diagram depicting the transcription of the DNA code into a code to mRNA. The transportation of mRNA into the cytoplasm and translation at the ribosomal complex. *M. Hartstock, Medical Art Co./Custom Medical Stock Photo. Repr*

at the University of California. Tonegawa's Nobel Prize winning research was conducted at Basel Institute for Immunology in Basel, Switzerland.

See also Antibody, monoclonal; Autoimmune diseases; Chromosome structure and morphology; Immunogenetics; Transplantation genetics

TRAIT • *see* CHARACTERS AND TRAITS

TRANSCRIPTION

The genetic information that is passed on from parent to offspring is carried by the DNA of a **cell**. The genes on the **DNA** code for specific proteins that determines appearance, differ-

ent facets of personality, health etc. In order for the genes to produce the proteins, it must first be transcribed from DNA to **RNA** in a process known as transcription. Thus, transcription is defined as the transfer of genetic information from the DNA to the RNA.

The process of transcription occurs in the **nucleus** of the cell. There are three different phases involved: initiation, elongation, and termination.

To initiate the process of information transfer, one of the strands of the double stranded DNA serves as a template for the synthesis of a single strand of RNA that is complementary to the DNA strand. The enzyme RNA **polymerase** binds to a particular region of the DNA that is termed as the "promoter." The promoter is a particular unidirectional sequence that appears at the beginning of the genes, and tells the enzyme where to start the synthesis and which strand to synthesize. Once the enzyme is bound to the promoter, it unwinds the

DNA and starts to make a strand of RNA with a base sequence complementary to the DNA template that is downstream of the RNA polymerase binding site. The strand from which it copies is known as the template or the antisense strand, while the other strand to which it is identical is called the sense or the coding strand.

After initiation, is the process of elongation. The substrates for RNA polymerase are nucleoside triphosphates. The RNA polymerase matches a base on the DNA to an RNA **nucleotide** (by complementary base pair binding) and then adds that nucleotide to the elongating RNA strand. As a new ribonucleotide triphosphate forms a bond with the 3'-hydroxyl end of the growing strand, a pyrophosphate is given off. The energy that is needed for synthesizing RNA is derived from splitting up of the triphosphate into a monophosphate and releasing the other two inorganic phosphates.

The next phase is called termination. Termination occurs when the RNA polymerase reaches a signal on the DNA template strand that tells it to stop. Once this termination signal is recognized by the RNA polymerase, it releases the DNA and transcription ceases. The newly synthesized RNA strand now undergoes "post-transcriptional processing."

Eukaryotic genes are not continuous. A typical **gene** consists of both coding sequences (exons) and non-coding sequences (introns). The primary transcript that is formed at the end of the transcription is actually known as hnRNA and is an exact copy of the gene with both introns and exons. A process called **RNA splicing** occurs and the introns are removed. The remaining exons are joined together to form the final **mRNA** product which codes for a single protein. This post-transcriptional RNA processing take place in the nucleus. Besides splicing, the hnRNA strand also has to be capped and poly-adenylated before being transported to the cytoplasm for **translation** into proteins. The 5' capping of the hnRNA occurs soon after the beginning of transcription. A methylated G nucleotide that is believed to play an important role in the initiation of **protein synthesis** is added to the 5' end. For the addition of a poly A tail, one hundred to two hundred residues of adenylic acid are added by an enzyme known as poly-A polymerase. This tail helps to guard the RNA transcript against degradation and enables the transcript to exit from the nucleus to the cytoplasm where it can be translated into proteins.

See also Chromosome structure and morphology; DNA (deoxyribonucleic acid); DNA replication

TRANSDUCTION

Transduction is defined as the transfer of genetic information between cells using a type of **virus** particle called a **bacteriophage**. The virus contains genetic material from one **cell**, which is introduced into the other cell upon virus infection of the second cell. Transduction does not, therefore, require cell to cell contact and is resistant to **enzymes** that can degrade **DNA**.

Bacteriophage can infect the recipient cell and commandeer the host's **replication** machinery to produce more copies of itself. This is referred to as the lytic cycle. Alternatively, the phage genetic material can integrate into the host DNA where it can replicate undetected along with the host until such time as an activation signal stimulates the production of new virus particles. This is referred to as the lysogenic cycle. Transduction relies on the establishment of the lysogenic cycle, with the bacterial DNA becoming incorporated into the recipient cell **chromosome** along with the phage DNA. This means of transferring **bacteria** DNA has been exploited for genetic research with bacteria like *Escherichia coli, Salmonella typhimurium,* and *Bacillus subtilis,* which are specifically targeted by certain types of bacteriophage.

There are two types of transduction: generalized transduction and specialized transduction. In generalized transduction, the packaging of bacterial DNA inside the phage particle that subsequently infects another bacterial cell occurs due to error. The error rate is about one phage particle in 1,000. Experimental mutants of phage have been engineered where the error rate is higher. Once the bacterial DNA has been injected inside the second bacterium, there is approximately a 10 percent change that the DNA will be stably incorporated into the chromosome of the recipient. A successful integration changes the **genotype and phenotype** of the recipient, which is called a transductant. A transductant will arise for about every 10^6 phage particles that contain bacterial DNA.

Specialized transduction utilizes specialized phage, in which some of the phage genetic material has been replaced by other genetic material, typically the bacterial chromosome. All of the phage particles carry the same portion of the bacterial chromosome. The phage can introduce their DNA into the recipient bacterium as above or via **recombination** between the chromosomal DNA carried by the phage and the chromosome itself.

Transduction has proved to be a useful means of transferring genetic traits from one bacterial cell to another.

See also Bacterial genetics; DNA replication

TRANSFECTION

Transfection is the process of delivering foreign molecules such as **DNA** into eucaryotic cells. After **cloning** a **gene**, scientists analyze its characteristics by reintroducing it into various **cell** types. In order to study the regulation of **gene expression**, the relevant DNA sequence can be mutated and transfected into cells, and its activities can be studied under different physiological conditions. Cell lines that express proteins can be established. The protein can be purified for further biomedical research. Large-scale production of a protein can be used as a drug.

There are two types of transfections that are routinely done in mammalian systems—transient and stable, or permanent transfections. In transient transfection, the **plasmids** get into the cell **nucleus**, but are not integrated into the chromosomes. There can be several copies of plasmids in the cells. As a result, the expression level is high. The transfected gene can be analyzed between one to four days after induction of the

DNA, depending on the vectors. Supercoiled plasmid DNA is used in transient transfection to achieve high efficiency. In stable transfection, the transfected DNA integrates into the chromosomes or exists as an episome. The cells that have integrated, or episomal foreign DNA, can be distinguished by selectable markers located on the plasmids. The commonly used markers include genes encoding aminoglycoside phosphotransferase and hygromycin B-phosphotransferase, among others. Linear DNA is normally used in stable transfections to facilitate the optimal integration of the foreign DNA into the host **genome**.

Over the years, many methods have been developed to introduce DNA into mammalian cells. The four basic methods include: calcium phosphate transfection, DEAE-dextran transfection, electroporation, and liposome-mediated transfection. In the first two methods, the treatment of cells resulted in DNA attaching to the cell surface. The DNA is then endocytosed by uncharacterized pathways. Electroporation uses electric field to produce holes on cell membrane. The DNA enters the cells through these holes. This method is versatile in transfecting various cell types. The mechanisms of the liposome-mediated transfection is not well understood. Presumably, negatively charged phosphate groups on DNA bind to the positively charged surface of the liposome, and the residual positive charge then mediates binding to negatively charged sialic acid residues on the cell surface.

See also Transformation

Transfer RNA

Transfer **RNA** (tRNA) is often referred to as the "Rosetta Stone" of **genetics**, as it translates the instructions encoded by **DNA**, by way of **messenger RNA (mRNA)**, into specific sequences of amino acids that form proteins and polypeptides. This class of small globular RNA is only 75 to 90 nucleotides long, and there is at least one tRNA for every **amino acid**. The job of tRNA is to transport free amino acids within the **cell** and attach them to the growing polypeptide chain. First, an amino acid molecule is attached to its particular tRNA. This process is catalyzed by an enzyme called aminoacyl—tRNA synthetase that binds to the inside of the tRNA molecule. The molecule is now charged. The next step, joining the amino acid to the polypeptide chain, is carried out inside the ribosome. Each amino acid is specified by a particular sequence of three **nucleotide bases** called **codons**. There are four different kinds of nucleotides in mRNA. This makes possible 64 different codons (4^3). Two of these codons are called STOP codons; one of these is the START codon (AUG). With only 20 different amino acids, it is clear that some amino acids have more then one codon. This is referred to as the degeneracy of the genetic code. On the other end of the tRNA molecule are three special nucleotide bases called the anticodon. These interact with three complimentary codon bases in the mRNA by way of hydrogen bonds. These weak directional bonds are also the force that holds together the double strands of DNA.

In order to understand how this happens it was necessary to first understand the three dimensional structure (conformation) of the tRNA molecule. This was first attempted in 1965, where the two-dimensional folding pattern was deduced from the sequence of nucleotides found in **yeast** alanine tRNA. Later work (1974), using x-ray diffraction analysis, was able to reveal the conformation of yeast phenylalanine tRNA. The molecule is shaped like an upside-down L. The vertical portion is made up of the D stem and the anti-codon stem, and the horizontal arm of the L is made up of the acceptor stem and the T stem. Thus, the **translation** depends entirely upon the physical structure. At one end of each tRNA is a structure that recognizes the genetic code, and at the other end is the particular amino acid for that code. Amazingly, this unusual shape is conserved between **bacteria** plants and animals.

Another unusual thing about tRNA is that it contains some unusual bases. The other classes of nucleic acids can undergo the simple modification of adding a methyl (CH3-) group. However, tRNA is unique in that it undergoes a range of modifications from **methylation** to total restructuring of the purine ring. (There are two types of bases, purines and pyrimidines.) These modifications occur in all parts of the tRNA molecule, and increase its structural integrity and versatility.

Transformation

Transformation is a process in which **exogenous DNA** is taken up by a (recipient) **cell**, sphaeroplast, or protoplast. In order to take up **DNA**, the cells must be competent. Competence is a state of bacterial cells during which the usually rigid cell wall can transport a relatively large DNA macromolecule. This is a highly unusual process, for **bacteria** normally lack the ability to transport macromolecules across the rigid cell wall and through the cyotplasmic membrane. Several bacteria, such as *Bacillus, Haemophilis, Neisseria,* and *Streptococcus,* possess natural competence because their cells do not require special treatment to take up DNA. This process is transient and occurs only in special growth phases, typically toward the end of log phase.

The demonstration of DNA transformation was a landmark in the history of **genetics**. In 1944, Oswald Avery, Colin MacLeod, and **Maclyn McCarty** conducted famous *Streptococcus pneumoniae* transformation experiments. Bacterial pneumonia is caused by the S strain of *S. pneumoniae.* The S strain synthesizes a slimy capsule around each cell. The capsule is composed of a polysaccharide that protects the bacterium from the immune response of the infected animal and enables the bacterium to cause the disease. The colonies of the S strain appear smooth because of the capsule formation. The strain that does not synthesize the polysaccharide, hence does not have the capsule, is called R strain because the surface of the colonies looks rough. The R strain does not cause the disease. When heat-killed S strain was mixed with live R strain, cultured, and spread on to a solid medium, a few S strain colonies appeared. When S cell extract was treated with RNase or proteinase and mixed with the live

R strain, R colonies and a few S colonies appeared. When the S strain cell extract was treated with DNase and mixed with live R strain, there were only R strain colonies growing on the agar plates. These experiments proved fundamentally that DNA is the genetic material that carries genes.

Transformation is widely used in DNA manipulation in molecular biology. For most bacteria that do not possess natural competency, special treatment, such as calcium chloride treatment, can render the cells competent. This is one of the most important techniques for introducing **recombinant DNA molecules** into bacteria and **yeast** cells in genetic engineering.

See also Conjugation; Transfection

TRANSGENICS

The term transgenics refers to the process of transferring genetic information from one organism to another. By introducing new genetic material into a **cell** or individual, a transgenic organism is created that has new characteristics it did not have before. The genes transferred from one organism or cell to another are called transgenes. The development of biotechnological techniques has led to the creation of transgenic **bacteria**, pants, and animals that have great advantages over their natural counterparts and sometimes act as living machines to create pharmaceutical therapies for the treatment of disease. Despite the advantages of transgenics, some people have great concern regarding the use of transgenic plants as food, and with the possibility of transgenic organisms escaping into the environment where they may upset ecosystem balance.

DNA is a complex and long molecule composed of a sequence of smaller molecules, called nucleotides, linked together. Nucleotides are nitrogen-containing molecules, called **bases** that are combined with sugar and phosphate. There are four different kinds of nucleotides in DNA. Each **nucleotide** has a unique base component. The sequence of nucleotides, and therefore of bases, within an organism's DNA is unique. In other words, no two organisms have exactly the same sequence of nucleotides in their DNA, even if they belong to the same **species** or are related. DNA holds within its nucleotide sequence information that directs the activities of the cell. Groups, or sets of nucleotide sequences that instruct a single function are called genes.

Much of the genetic material, or DNA, of organisms is coiled into compact forms called chromosomes. Chromosomes are highly organized compilations of DNA and protein that make the long molecules of DNA more manageable during **cell division**. In many organisms, including human beings, chromosomes are found within the **nucleus** of a cell. The nucleus is the central compartment of the cell that houses genetic information and acts as a control center for the cell. In other organisms, such as bacteria, DNA is not found within a nucleus. Instead, chromosomes are free within the cell. Additionally, many cells have extrachromosomal DNA—DNA that is not found within chromosomes. The **mitochondria** of cells, and the chloroplasts of plant cells have extrachromosomal DNA that help direct the activities of these organelles independent from the activities of the nucleus where the chromosomes are found. **Plasmids** are circular pieces of extrachromosomal DNA found in bacteria that are extensively used in transgenics.

DNA, whether in chromosomes or in extrachromosomal molecules, uses the same code to direct cell activities. The genetic code is the sequence of nucleotides in genes that is defined by sets of three nucleotides. The genetic code itself is universal, meaning it is interpreted the same way in all living things. Therefore, all cells use the same code to store information in DNA, but have different amounts and kinds of information. The entire set of DNA found within a cell (and all of the identical cells of a multicellular organism) is called the **genome** of that cell or organism.

The DNA of chromosomes within the cellular genome is responsible for the production of proteins. The universal genetic code simply tells cells which proteins to make. Proteins, in turn have many varied and important functions, and in fact help determine the major characteristics of cells and whole organisms. Proteins have structural functions, which give shape and strength to cells and tissues and they also act as hormones. Insulin, for example, is a protein hormone. Proteins can act as neurotransmitters, enabling nerve cells to relay signals. As **enzymes**, proteins carry out thousands of kinds of chemical reactions that make life possible. Proteins also act as cell receptors and signal molecules, which enable cells to communicate with one another, to coordinate growth and other activities important for wound healing and development. Proteins also have contractile functions involved in motion. Thus, many of the vital activities and characteristics that define a cell are really the result of the proteins that are present. The proteins, in turn, are determined by the genome of the organism.

Because the genetic code with genes is the same for all known organisms, and because genes determine characteristics of organisms, the characteristics of one kind of organism can be transferred to another. If genes from an insect, for example, are placed into a plant in such a way that they are functional, the plant will gain characteristics of the insect. The insect's DNA provides information on how to make insect proteins within the plant because the genetic code is interpreted in the same way. That is, the insect genes give new characteristics to the plant. This very process has already been performed with firefly genes and tobacco plants. Firefly genes were spliced into tobacco plants, which created new tobacco plants that could glow in the dark. This amazing artificial genetic mixing, called recombinant **biotechnology**, is the crux of transgenics. The organisms that are created from mixing genes from different sources are transgenic. The glow-in-the-dark tobacco plants in the previous example, then, are transgenic tobacco plants.

One of the major obstacles in the creation of transgenic organisms is the problem of physically transferring DNA from one organism or cell into another. It was observed early on that bacteria resistant to **antibiotics** transferred the resistance characteristic to other nearby bacterial cells that were not previously resistant. It was eventually discovered that the resistant bacterial cells were actually exchanging plasmid DNA carrying resistance genes. The plasmids traveled between resistant and susceptible cells. In this way, susceptible bacterial cells were transformed into resistant cells.

The permanent modification of a genome by the external application of DNA from a cell of different genotype is called **transformation**. Transformed cells can pass on the new characteristics to new cells when they reproduce because copies of the foreign transgenes are replicated during cell division. Transformation can be either naturally occurring or the result of transgenics. Scientists mimic the natural uptake of plasmids by bacterial cells for use in creating transgenic cells. Certain chemicals make transgenic cells more willing to take-up genetically engineered plasmids. Electroporation is a process where cells are induced by an electric current to take up pieces of foreign DNA. Transgenes are also introduced via engineered viruses. In a procedure called **transfection**, viruses that infect bacterial cells are used to inject the foreign pieces of DNA. DNA can also be transferred using microinjection, which uses microscopic needles to insert DNA to the inside of cells. A new technique to introduce transgenes into cells uses liposomes. Liposomes are microscopic spheres filled with DNA that fuse to cells. When liposomes merge with host cells, they deliver the transgenes to the new cell. Liposomes are composed of lipids very similar to the lipids that make up cell membranes, which gives them the ability to fuse with cells.

Humankind has been beneficially manipulating the genomes of organisms for a long time. Historically, the manipulations involved interbreeding of plants and animals to give new variations of useful characteristics. Such a practice is called artificial selection—the purposeful manipulation of reproduction to select for a particular genetic characteristic. It is the intentional counterpart to the more random process of **natural selection**. Examples of artificial **selection** used to create organisms with new and desirable characteristics include unusual breeds of dogs for pets, larger breeds of horses for work, and new corn varieties developed to withstand dry conditions or improve productivity. With the aid of new scientific knowledge, scientists can now use transgenics to accomplish the same results as selective breeding.

By recombining genes, bacteria that metabolize petroleum products are created to clean-up the environment, antibiotics are made by transgenic bacteria on mass industrial scales, and new protein drugs are produced. By creating transgenic plants, food crops have enhanced productivity. Transgenic corn, wheat, and soy with herbicide resistance, for example, are able to grow in areas treated with herbicide that kills weeds. Transgenic tomato plants produce larger, more colorful tomatoes in greater abundance. Transgenics is also used to create influenza immunizations and other vaccines.

Despite their incredible utility, there are concerns regarding trangenics. The **Human Genome Project** is a large collaborative effort among scientists worldwide that announced the determination of the sequence of the entire human genome in 2000. In doing this, the creation of transgenic humans could become more of a reality, which could lead to serious ramifications. Also, transgenic plants used as genetically modified food is a topic of debate.

See also Agricultural genetics; Animal husbandry; Bacterial genetics; Genetic engineering technology; Immunogenetics; Pharmacogenetics

TRANSLATION

Translation is the process in which genetic information, carried by **messenger RNA (mRNA)**, directs the synthesis of proteins from amino acids, whereby the primary structure of the protein is determined by the **nucleotide** sequence in the mRNA.

A molecule known as the ribosome is the site of the **protein synthesis**. The ribosome is protein bound to a second **species** of **RNA** known as **ribosomal RNA** (rRNA). Several **ribosomes** may attach to a single mRNA molecule, so that many polypeptide chains are synthesized from the same mRNA. The ribosome binds to a very specific region of the mRNA called the promoter region. The promoter is upstream of the sequence that will be translated into protein.

The nucleotide sequence on the mRNA is translated into the **amino acid** sequence of a protein by adaptor molecules composed of a third type of RNA known as transfer RNAs (tRNAs). There are many different species of tRNAs, with each species binding a particular type of amino acid. In protein synthesis, the nucleotide sequence on the mRNA does not specify an amino acid directly, rather, it specifies a particular species of tRNA. Complementary tRNAs match up on the strand of mRNA every three **bases** and add an amino acid onto the lengthening protein chain. The three base sequence on the mRNA are known as "codons," while the complementary sequence on the tRNA are the "anti-codons."

The ribosomal RNA has two subunits, a large subunit and a small subunit. When the small subunit encounters the mRNA, the process of translation to protein begins. There are two sites in the large subunit, an "A" site, and a "P" site. The start signal for translation is the codon ATG that codes for methionine. A tRNA charged with methionine binds to the translation start signal. After the first tRNA bearing the amino acid appears in the "A" site, the ribosome shifts so that the tRNA is now in the "P" site. A new tRNA molecule corresponding to the codon of the mRNA enters the "A" site. A peptide bond is formed between the amino acid brought in by the second tRNA and the amino acid carried by the first tRNA. The first tRNA is now released and the ribosome again shifts. The second tRNA bearing two amino acids is now in the "P" site, and a third tRNA can then bind to the "A" site. The process of the tRNA binding to the mRNA aligns the amino acids in a specific order. This long chain of amino acids constitutes a protein. Therefore, the sequence of nucleotides on the mRNA molecule directs the order of the amino acids in a given protein. The process of adding amino acids to the growing chain occurs along the length of the mRNA until the ribosome comes to a sequence of bases that is known as a "stopcodon." When that happens, no tRNA binds to the empty "A" site. This is the signal for the ribosome to release the polypeptide chain and the mRNA.

After being released from the tRNA, some proteins may undergo post-translational modifications. They may be cleaved by a proteolytic (protein cutting) enzyme at a specific site. Alternatively, they may have some of their amino acids biochemically modified. After such modifications, the polypeptide forms into its native shape and starts acting as a functional protein in the **cell**.

In RNA there are four different nucleotides, A, U, G and C. If they are taken three at a time (to specify a codon, and thus, indirectly specify an amino acid), 64 **codons** could be specified. However, there are only 20 different amino acids. Therefore, several triplets code for the same amino acid; for example UAU and UAC both code for the amino acid tyrosine. In addition, some codons do not code for amino acids, but code for polypeptide chain initiation and termination. The genetic code is non-overlapping, i.e., the nucleotide in one codon is never part of the adjacent codon. The code also seems to be universal in all living organisms.

See also DNA synthesis

TRANSLOCATION

In **genetics**, chromosomal translocation refers to type of interchange of **chromosome** pieces of **DNA** following breakage, in which segments are transferred between nonhomologous chromosomes. When this exchange occurs without a net loss or gain of genetic material, it is called a balanced, or reciprocal, translocation, and there is no phenotypic change in the individual. When the exchange results in a deletion or **duplication** of chromosomal material, in gametes or somatic cells, severe phenotypic changes may result.

With chromosomal translocations, there is a physical movement of genes located on one segment of DNA to another location (locus) on another chromosomal segment of DNA. Because translocations involve changes to genes the are also termed translocation mutations.

In **cell** biology, translocation is a term also used to describe the movement of proteins or other substances through membranes. Translocation is also said to occur at the level of the ribosome during **translation (protein synthesis)** and involves the step-by-step movement along **mRNA codons** as each codon is read to determine which **amino acid** to insert into the lengthening protein chain. At the site of translation a translocation is said to occur as the mRNA moves one codon at a time through the ribosome complex (i.e., through the A site).

The process of ribosomal translocation involves specialized **enzymes** termed translocase enzymes.

Unbalanced chromosomal translocations, where the deleted segment, fragment, of portion of a chromosome attaches to another chromosome are rare in humans.

See also Chromosomal mutations and abnormalities; Gene mutation; Molecular biology and molecular genetics

TRANSPLANTATION GENETICS

There are several different types of transplantation. An autograft is a graft from one part of the body to another site on the same individual. An isograft is one between individuals that are genetically alike, as in identical **twins**. An allograft is a graft between members of the same **species** but who are not genetically alike. A xenograft is one between members of different species. The allograft we are most familiar with is that of a blood transfusion. Nonetheless, the replacement of diseased organs by transplantation of healthy tissues has frustrated medical science because the immune system of the recipient recognizes that the donor organ is not "self" and rejects the new organ.

The ability to discriminate between self and nonself is vital to the functioning of the immune system so it can protect the body from disease and invading microorganisms. However, the same immune response that serves well against foreign proteins prevents the use of organs needed for life saving operations. Virtually every **cell** in the body carries distinctive proteins found on the outside of the cell that identify it as self. Central to this ability is a group of genes that are called the **Major Histocompatibility Complex (MHC)**. The genes that code for those proteins in humans are called the HLA or Human Leukocyte Antigen. These are broken down to class I (HLA-A, B and Cw), class II (HLA-DR, DQ and DP) and class III (no HLA genes).

The MHC was discovered during **tumor** transplantation studies in mice by Peter Gorer in 1937 at the Lister Institute, and was so named because "histo" stands for tissue in medical terminology. The genes that compose the MHC are unique in that they rarely undergo **recombination** and so are inherited as a haplotype, one from each parent. They are also highly polymorphic. This means that the genes and the molecules they code for vary widely in their structure from one individual to another and so transplants are likely to be identified as foreign and rejected by the body. Scientists have also noted that this area of the **genome** undergoes more mutational events then other regions, which probably accounts for some of its high degree of **polymorphism**. As previously mentioned, there are several classes of the MHC. The role of the MHC Class I is to make those proteins that identify the cells of the body as "self," and they are found on nearly every cell in the body that has **nucleus**. Nonself proteins are called antigens and the body first learns to identify self from nonself just before birth, in a **selection** process that weeds out all the immature T-cells that lack self-tolerance. Normally, this process continues throughout the lifespan of the organism. A breakdown in this process leads to allergies and at the extreme, results in such autoimmune diseases as multiple sclerosis, rheumatoid arthritis, and systemic lupus erythematosus. The job of the Class I proteins is to alert killer T cells that certain cells in the body have somehow been transformed, either by a viral infection or **cancer**, and they need to be eliminated. Killer T-cells will only attack cells that have the same Class I glycoproteins that they carry themselves. The Class II MHC molecules are found on another immunocompetent cell called the B-cells. These cells mature into the cells that make antibodies against foreign proteins. The class II molecules are also found on macrophages and other cells that present foreign antigens to T-helper cells. The Class II antigens combine with the foreign antigen and form a complex with the antibody, which is subsequently recognized and then eliminated by the body.

The ability of killer T-cells to respond only to those transformed cells that carry Class I antigen, and the ability of helper T-cells to respond to foreign antigens that carry Class II antigen, is called MHC restriction. This is what is tested for when tissues are typed for transplantation. Most transplantation occurs with allogeneic organs, which by definition are those that do not share the same MHC locus. The most sensitive type of transplantation with respect to this are those involving the bone marrow (Haematopoietic Stem Cell Transplantation) HLA matching is an absolute requirement so its use is limited to HLA-matched donors, usually a brother or sister. The major complications include graft-versus-host disease (GvHD is an attack of immunocompetant donor cells to immunosuppressed recipient cells) and rejection, which is the reverse of GvHD. The least sensitive are corneal lens transplantation, probably because of lack of vascularisation in the cornea and its relative immunological privilege. Drugs like cyclosporin A have made transplant surgery much easier, although the long term consequences of suppressing immune function are not yet clear. This antirejection drug is widely used in transplant surgery and to prevent and treat rejection and graft-versus-host disease in bone marrow transplant patients by suppressing their normal immune system. Newer strategies, including **gene therapy**, are being developed to prevent the acute and chronic rejection of transplanted tissues by introducing new genes that are important in preventing rejection. One promising aspect is the delivery of genes that encode foreign donor antigens (alloantigens). This might be an effective means of inducing immunological tolerance in the recipient and eliminate the need for whole-body immunosuppression.

See also Antibody and antigen; Immunogenetics; Immunological analysis techniques

TRANSPOSITION

A transposition is a physical movement of genetic material (i.e., **DNA**) within a **genome** or the movement of DNA across genomes (i.e., from one genome to another). Because these segments of genetic material contain genes, transpositions resulting in changes of the loci (location) or arrangements of genes are mutations. Transposition mutations occur in a wide range of organisms. Transposons occur in bacteria— and transposable elements have been demonstrated to operate in higher eukaryotic organisms, including mammalian systems.

Transposition mutations may only occur if the DNA being moved—termed the transposon—contains intact inverted repeats at its ends (terminus). In addition, functional tranposase **enzymes** must be present.

There are two types or mechanisms of transposition. Replicative transpositions involve the copying of the segment of section DNA to be moved (transposable element) before the segment is actually moved. Accordingly, with replicative transposition, the original section of DNA remains at its original location and only the copy is moved and inserted into its new position. In contrast, with conservative transpositions, the segment of DNA to be moved is physically cut from its original location and then inserted into a new location. The DNA from which the tranposon is removed is termed the donor DNA and the DNA to which the transposon is added is termed the receptor DNA.

Transposons are not passive participants in transposition. Transposons carry the genes that code for the enzymes needed for transpositon. In essence, they carry the mechanisms of transposition with them as they move or jump (hence Barbara McClintok's original designation of "jumping genes") throughout or across genomes. Transposons carry special insertion sequences (IS elements) that carry the genetic information to code for the enzyme transposase that is required to accomplish transposition mutations. One of the most important mechanisms of transposase is that they are the enzymes responsible for cutting the receptor DNA to allow the insertion of the transposon.

Transitions are a radical mutational mechanism. The physical removal of both DNA and genes can severely damage or impair the function of genes located in the transposons (especially those near either terminus). Correspondingly, the donor molecules suffer a deletion of material that may also render the remaining genes inoperative or highly impaired with regard to function.

McClintok's discovery of transposons, also termed "jumping genes" in the late 1940s (before the formation of the Waston-Crick model of DNA resulted in her subsequent award of a Nobel Prize for Medicine or Physiology.

Transposition segments termed retrotransposons may also utilize an **RNA** intermediate complimentary copy to accomplish their transposition.

Transposition can radically and seriously affect phenotypic characteristics including transfer of antibiotic resistance in bacterium. Following insertion, transposed genetic elements usually generate multiple copies of the genes transferred—further increasing their disruption to both the genotype and phenotypic expression.

See also Antibiotics; Bacterial genetics; Chromosomal mutations and abnormalities; Clones and cloning; Gene mutation; Mutation; Rare genotype advantage

TRANSVERSIONS

Tranversions are mutations that result in a change from a pyrimidine to a purine, or from a purine to a pyrimidine in **deoxyribonucleic acid (DNA)** or **ribonucleic acid (RNA)**.

DNA nucleotides, each contain one of four nitrogenous (nitrogen containing) **bases**: adenine (A), thymine (T), guanine (G), and cytosine (C). RNA molecules contain uracil (U) in the place of thymine. In DNA, the pyrimidines, single-ring nitrogenous basic compounds, are cytosine and thymine. Because uracil substitutes for thymine in RNA, the pyrimidines in RNA are cytosine and uracil. For both DNA and RNA the purines, nitrogenous double-ringed basic compounds are adenine and guanine.

The bonding between DNA strands, or between an RNA molecule and a DNA strand, or between two RNA molecules is very specific. Adenine always bonds with thymine in DNA strands (A-T **base pairing**) or, in the alternative, adenine bonds with the substituting uracil in RNA molecules to form a stable A-U bond. In both DNA and RNA, cytosine always bonds with guanine (C-G). Such pairs are pairing between a purine and a pyrimidine. This is critical to the structural integrity of the DNA helix because only a bond between a purine and a pyrimidine is chemically stable enough and structurally the right size to create the **double helix**.

Mutations are changes **nucleotide** sequence of the **genome**. If the normal sequence of a DNA triplet (a set of three bases upon which **codons** are constructed that ultimately direct the order of insertion of amino acids in the polypeptide chain or direct other processes during **translation**) was ATC the normal codon for that sequence would have a UAG sequence. The UAG sequence normally instructs the termination of **protein synthesis**. For example, if a transversion **mutation** were to take place in the second position of the ATC sequence, either the purine, adenine, or guanine would be substituted for the pyrimidine thymine. Such a transversion mutation would result in either a sequence of AAC or AGC. Instead of creating a stop codon, these mutated sequences would, respectively, produce UUG and UCG sequences. Instead of ordering the end of translation, the UUG sequence would order the insertion of the **amino acid** leucine and the UCG codon would order the insertion of amino acid serine. In either case, the protein structure would be altered. The alteration of one amino acid for another, or the production of a protein chain that is too long or too short can have a dramatic effect on the functional ability of the protein product.

Transitions (mutations involving the substitution of one purine for another purine, e.g., thymine substituting for cytosine in a DNA molecule or adenine substituting for guanine) are far more common than are transversions.

See also Chemical mutagenesis; Chromosomal mutations and abnormalities; DNA (deoxyribonucleic acid); DNA repair; Genetic code; Gene mutation; Imprinting; Polymerase chain reaction (PCR); Restriction enzymes

TRINUCLEOTIDE EXPANSION

Trinucleotide repeats are tandem repetitions of polymorphic (many forms) triplets of nucleotides. The spreading out of these repeats from a low to a high copy number is referred to as trinucleotide expansion, a process that leads to genetic

instability and is most frequently associated with the emergence of serious diseases.

Simple **nucleotide** repeats (dinucleotide, trinucleotide, and tetranucleotide repeats) make up a substantial fraction of the eukaryotic **genome**. The high level of polymorphisms in these repeats has been very useful in genome analysis. The structures known as microsatellites and minisatellites established by these repeats helped facilitate forensic studies, mapping and positional **cloning** of disease related genes, and analysis of **phylogenetic** data. The full significance of these repeats, besides their involvement in certain diseases, has not been understood. Trinucleotide repeats are most frequently involved in disease association in human. The distribution of trinucleotide repeats in the eukaryotic genome seems to depend on the nature of the triplet nucleotide. The CAG triplet is most frequently found in the coding regions of genes and is translated as polyglutamine homopeptides.

The molecular mechanisms of triplet repeats expansion are not well understood but are thought to involve events that occur during both normal genome **replication** and **DNA repair**. Expansion could result from slippage of repeats in the nascent (growing) lagging strand, coupled with the formation of hairpin or triplex. It is also likely that a newly synthesized fragment leads to the formation of a single stranded **DNA** fragment that is partially free from the growing DNA double strand. The free, single stranded DNA could later convert to a hairpin or triplex. These hairpins and triplexes could also result from nicks or gaps in a repeat tract. During repair of these gaps, slippage and subsequent hairpin formation may lead to repeat expansion. Although the normal trinucleotide repeats are stable in number when transmitted through generation, abnormally expanded ones become unstable with predilection to expand further resulting in copy numbers of up to thousands through future generation. Unstable trinucleotide repeats are usually inherited in a dominant pattern. Instability of the trinucleotide repeats is influenced by the rate of genetic **recombination** and is enhanced during meiotic events. It is thought that the instability of expanded repeats is due to the fact that they become adjacent, through expansion, to recombination hotspots.

Trinucleotide expansion and the resulting genetic instability have been linked to many human diseases through different mechanisms. These disorders typically involve genetic anticipation, which refers to the increasing severity and penetrance and decreasing age of onset with each new generation as the expansion of repeats increases. In **fragile X syndrome**, which is caused by expansion of CAG triplets (also known as polyglutamine expansion), the disease is caused by loss of function of the FMR1 **gene**, which seems to get silenced because of hypermethylation of adjacent **chromatin**. In the case of some neurological disorders, however, the expansion of CAG trinucleotide repeats result in a larger protein, which gets extended by excessively long polyglutamine sequences. These proteins acquire new function that interferes with the life cycle of the **cell** and induce **apoptosis**. In other cases, like in **Huntington disease**, proteolytic degradation products of polyglutamine containing proteins are believed to play a major role in the development and manifestation of diseases.

Expansion of another trinucleotide repeat, polyCTG, causes the autosomal dominant disease myotonic dystrophy, which is the most common myotonic dystrophy in adults. The expansion of polyCTG seems to affect the regulation of expression of myotonic dystrophy associated homeodomain protein gene (DMHAP). The downregulation of the expression of this gene is mediated through chromatin condensation at the enhancer region leading to the inhibition of any interaction with other factors (e.g. corresponding trans-acting factors).

See also Chromosomal mutations and abnormalities; Genetic disorders and diseases; Hereditary diseases; Inherited cancers; Mutations and mutagenesis

TRIPLET CODE · *see* GENETIC CODE

TRISOMY · *see* PLOIDY

TRUE-BREEDING · *see* BREEDING

Lap-Chee Tsui. *Reproduced by permission of Lap-Chee Tsui.*

TSUI, LAP-CHEE (1950-)

Chinese-Canadian molecular geneticist

Lap-Chee Tsui is most recognized for leading the team that, in 1989, found the **gene** responsible for **cystic fibrosis** (called CFTR, for cystic fibrosis transmembrane regulator gene). The large team involved the laboratories at Toronto's Hospital for Sick Children of Tsui, Jack Riordan, and Manuel Buchwald, and spanned the border to the University of Michigan laboratory of Francis Collins (now Director of the National Human **Genome** Research Institute of NIH). By today's perspective, reports of gene discovery may seem so commonplace as to be taken for granted; however, this discovery was noteworthy for two reasons. It was the first success of the approach known as positional **cloning**, meaning that it was found, not because of any clues about its function, but merely by knowing its location and a little about its genetic neighbors. Second, cystic fibrosis is the most common single-gene disease among Caucasians, with symptoms usually starting in childhood, which are eventually fatal. Much anticipation awaited this particular discovery, and it had its share of international competition in the process.

Tsui evolved gradually into his **genetics** career. With a significant artistic bent and inclination to precision and detail, he aspired to be an architect, but was assigned to study biology during his early years at the Chinese University of Hong Kong. He moved to the U.S. to finish a Ph.D. at the University of Pittsburgh, to Tennessee to the Oak Ridge National Laboratory, and then in 1981 to the Hospital for Sick Children for postdoctoral work, where he was soon offered a staff position. Tsui's career continued to flourish; he holds the Sellers

Chair of Cystic Fibrosis Research, and in 1996 he became the hospital's Geneticist-in-Chief. Tsui is a Professor of Molecular and **Medical Genetics** at the University of Toronto.

In his research endeavors, Tsui has not rested on his laurels. Immediately after the discovery of CFTR, he moved on to establish an international consortium of CF researchers and an organized mechanism to report and disseminate information about the hundreds of different mutations that would be found in the CFTR gene. His laboratory went on to make significant contributions to the mapping and cloning of other genes on **chromosome** seven, to knowledge of the structure-function relationships between mutant and normal versions of CFTR, and to the understanding of other genes that interact with CFTR and influence the phenotype of cystic fibrosis. His generosity in the support of students, postdoctoral fellows and charitable causes is widely recognized by all who know him.

In the international sphere, Tsui has also been generous of his time and expertise. He travels around the world many times each year in support of universities, research institutes and programs, and causes, including in Europe and Asia. For 2000-2001, he is president of the international Human Genome Organization (HUGO). On the national scene, was a founder of the Canadian Genetic Disease Network, and he lead the formation of Canada's funding organization for genomics research, now known as Genome Canada. Tsui oversees the development of The Center for Applied Genomics in Toronto.

Tsui's contribution to science, to his adopted country, and to international relations have all been appropriately recognized by awards far too numerous to list. Tsui is an Officer of the

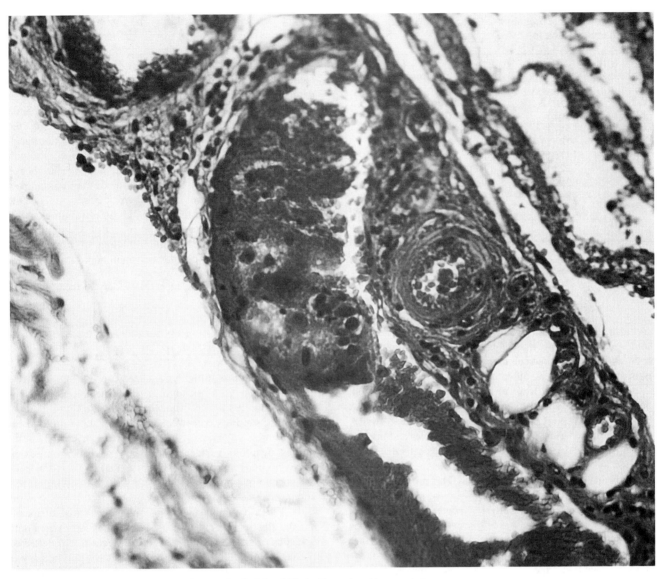

A malignant giant cell soft tissue tumor. *Custom Medical Stock Photo. Reproduced by permission.*

Order of Canada, a Fellow of the Royal Society of Canada and a Fellow of the Royal Society of London. His other major awards include the Gairdner International Award, the Franklin Institute Cresson Medal, the Canadian Medical Association Medal of Honor, the Medical Research Council of Canada's Distinguished Scientist Award and the Paul di Sant'Agnese Distinguished Scientific Achievement Award of the Cystic Fibrosis Foundation. He has several honorary degrees.

Tsui's architectural talent has not been wasted. In 1989, as the team of scientists struggled to interpret the information revealed by sequence of the CFTR gene, he created the conceptual drawings of the protein structure that would soon be seen around the world as the discovery was revealed.

See also Clones and cloning; Cloning: Applications to biological problems; Genetic mapping

TUMOR

A tumor (also known as a neoplasm) is an abnormal tissue growth. Neoplasm means new formation. Tumors can be either malignant (cancerous) or nonmalignant (benign), but either type may require therapy to remove or reduce its size. In either case, the tumor's growth is unregulated by normal genetic and somatic body control mechanisms. Usually the growth is not beneficial to the organ in which it is developing. **Mutation** in normal functioning genes may lead to the expression of oncogenes and affect expression of **tumor suppressor genes** to produce cancerous tumors.

Normally, cells are generated at a rate needed to replace those that die or are needed for an individual's growth and development. Moreover, cells become differentiated into specialized **cell** forms (muscle cells, bone cells). Genetic controls

Tumor Suppressor Gene	Chromosome Location	Disease or Tumor
Rb1	13q14	Retinoblastoma
p53	17p13.1	Li Fraumeni syndrome
WT-1	11p13	Wilms' tumor
APC	5q21	Colon Ca (FAP)
MCC	5q21	Colon Ca (FAP)
DCC	18q21-qter	
NF1	17q11.2	Neurofibromatosis 1
Merlin	22q12	Neurofibromatosis 2
VHL	3p25-26	von Hippel Lindau
MTS1	9p21	Melanoma
BRCA1	17q21	Breast and ovarian cancer
BRCA2	13q12-13	Breast cancer

Tumor suppressor genes, their chromosomal location, and the type of disease or tumor most commonly associated with them. This list is only a fraction of the known tumor supressors. *Courtesy of Dr. Constance Stein.*

modulate the formation of any given cells. The process of some cells becoming muscle cells, some becoming nerve cells, and so on is called cell **differentiation**. Tumor formation is an abnormality in cell differentiation.

A benign tumor is a well-defined growth with smooth boundaries. This type of tumor simply grows in diameter. A benign growth compresses adjacent tissues as it grows. A malignant tumor usually has irregular boundaries and invades the surrounding tissue. This **cancer** also sheds cells that travel through the bloodstream implanting themselves elsewhere in the body and starting new tumor growth. This process is called metastasis.

It is important that the physician determine which kind of tumor is present when one is discovered. In some cases this is not a simple matter. It is difficult to determine whether the growth is benign without taking a sample of it and studying the tissue under the microscope. This sampling is called a biopsy. Biopsy tissue can be frozen quickly, sliced thinly, and observed without staining (this is called a frozen section); or it can be sliced, stained with dyes, and observed under the microscope. Cancer tissue is distinctly different from benign tissue.

A benign tumor can be lethal if it compresses the surrounding tissue against an immovable obstacle. A benign brain tumor compresses brain tissue against the skull or the bony floor of the cranium and results in paralysis, loss of hearing or sight, dizziness, or loss of control of the extremities. A tumor growing in the abdomen can compress the intestine and interfere with digestion. It also can prevent proper liver or pancreatic function. The benign tumor usually grows at a relatively slow pace and may stop growing for a time when it reaches a certain size.

A cancer may grow quite rapidly or slowly, but usually is irregular in shape. It invades the neighboring tissue instead of pressing it aside. Most importantly, a cancer sheds cells, that is, metastasizes, so that new cancer growths can spring up in areas

distant from the original cancer. The cancerous cells also can establish a cancer in tissue that is different from the original cancer. A breast cancer could spread to bone tissue or to liver.

A benign tumor can be removed surgically if it is in a location that a surgeon can reach. A tumor growing in an unreachable area of the brain can be treated using radiation. It can also be treated by inserting thin probes through the brain tissue into the tumor and circulating liquid nitrogen through the probe to freeze the tumor. This operation is called cryosurgery.

A malignancy requires steps to remove it but consideration must be given to the possibility that the tumor has begun to metastasize. The main or primary tumor may be removed surgically but if the tumor has been growing for some time the patient also may require treatment with powerful drugs to kill any stray cells. This treatment is called chemotherapy. Chemotherapy allows the antitumor drug to be circulated throughout the body to counter any small tumor growths.

See also Breast/Ovarian Cancer, Hereditary; Cancer genetics; Carcinogenic

TUMOR SUPPRESSOR GENES

Cell proliferation is a highly regulated process in which two classes of genes play crucial roles, proto-oncogenes and **tumor** suppressor genes. Proto-oncogenes induce **cell** proliferation, and tumor suppressor genes control the levels of proto-oncogenic expression, or the period of activity of proteins transcribed by proto-oncogenes, thus keeping cell proliferation within normal rates. Since uncontrolled cell proliferation leads to tumor formation, genes that prevent abnormal cell proliferation are termed tumor suppressor genes.

When a tumor suppressor **gene** is mutated or inactivated, cells become either displastic or malignant. When only one copy or allele of a tumor suppressor gene is mutated, dysplasia, or benign overgrowth, usually occurs, due to a lower level of expression of anti proliferative proteins. When a second **mutation** occurs in the other allele of the same gene, the onset of **cancer** triggers. Some examples of tumor suppressor genes that are found mutated or inactivated in both hereditary and sporadic cancers are as follows: RB1 (resulting in hereditary **retinoblastoma** or osteosarcomas), p53 (resulting in Li-Fraumeni syndrome, breast cancer, brain tumors, and leukemias), p16 (resulting in melanomas and pancreatic cancer), and BRCA1 and BRCA2 (resulting in hereditary breast, ovarian and pancreatic cancers).

Additional functions of tumor suppressor genes are more complex and wider than the short definition given above suggests. There are about 12 gene suppressor genes identified today, whose functions are well studied, revealing the fundamental role these genes play in protecting body tissues against abnormal transformations that can lead to cancer. Some products of such genes (i.e., proteins or **enzymes**) antagonize directly or indirectly the activity of proto-oncogenes. Such proteins are found in every mature tissue of the body in roughly equivalent levels.

The most-studied tumor suppressor gene is p53, that controls phase G1 of the cell cycle and prevents the transition to phase S until **DNA** damage is repaired. In the case of irreparable DNA damage, p53 activates programmed cell death, also known as **apoptosis**.

Some oncogenes (i.e., the mutated counterpart of proto-oncogenes) transcribe proteins that inhibit p53 expression or inactivate p53 proteins. Consequently, the DNA-damaged cell escapes apoptosis and respond to mitotic signals, following to **cell division**. The resultant daughter cells will inherit such mutations and start a lineage of mutated cell clones that will eventually form a tumor. Inactivation of either a tumor suppressor gene or its proteins is also an important step in the development of tumor resistance against **anticancer drugs**. Cells that have lost the capacity for apoptosis are known as immortalized cells.

See also Anticancer drugs; Cancer genetics; Cell proliferation; DNA repair; Gene names and functions; Inherited cancers; Somatic cell genetics; Transformation

TURNER'S SYNDROME

Turner's syndrome (also referred to as gonadal dysgenesis) is a relatively common genetic disorder of females, which affects many body systems. Approximately one in every 2,000-5,000 female babies has Turner's syndrome. About 98-99% of pregnancies with Turner's syndrome abort spontaneously, usually during the first trimester of pregnancy. Approximately 10% of fetuses from pregnancies that have spontaneously aborted have Turner's Syndrome.

Described by Dr. Henry Turner in 1938, this disorder in due to a deficiency in the amount of genetic material on the **X chromosome**, one of the two sex chromosomes. Diagnosis of Turner's syndrome is made with a **chromosome** analysis. Turner's syndrome is not related to advanced maternal age and is a sporadic event, with the risk of recurrence not increased for subsequent pregnancies.

Turner's syndrome is associated with short stature and failure to mature sexually. Other problems may include learning difficulties, skeletal abnormalities (e.g., webbed neck, low posterior hair line), lymphedema (swelling of a part of the body due to an obstruction or deficiency of the lymphatic drainage system), heart and kidney abnormalities, infertility, obesity, formation of keloids (thick scars), and thyroid gland dysfunction (hypothyroidism). The type and amount of missing genetic material influence which specific organ abnormalities will be present, as well as the person's potential for growth. Because distinctive features do not always accompany the syndrome, Turner's syndrome is often not diagnosed during infancy, but may be suspected during childhood because of the short stature of a child. During teenage years, Turner's syndrome may be discovered due to delayed puberty and menarche, while in adult women, anovulation and infertility may indicate Turner's syndrome.

Short stature is usually present in females with Turner's syndrome. The causes are intrauterine growth retardation, a gradual decline in growth rate during childhood, and the absence of a pubertal growth spurt. Females with Turner's syndrome have abnormal body proportions characterized by markedly shortened lower extremities. The ultimate height range is 55–58 inches, although familial height may play a role in determining the ultimate adult height. Recent studies have indicated that much of the growth deficit in females with Turner's syndrome can be restored by injections of human growth hormone.

Normal pubertal development and spontaneous menstrual periods do not occur in the majority of females with Turner's syndrome. Though 10% of females with Turner's syndrome will go through puberty spontaneously, most will require the use of female hormone therapy for development of secondary sexual characteristics and menstruation. The time of initiation of therapy varies with each female, but usually begins when she expresses concern about the onset of puberty, usually around 15 years of age. Delaying the use of estrogen therapy is recommended to maximize the height the female will achieve. Various estrogenic and progestational agents and schedules have been used as hormone therapy. Although infertility cannot be altered, pregnancy may be possible through *in vitro* **fertilization**.

Renal abnormalities occur in 1/3 to 1/4 of females with Turner's syndrome. The most common abnormality is a horse shoe kidney. Cardiac abnormalities are also common, with the coarctation of the aorta being the most common. There is an increased incidence of thyroid dysfunction, diabetes mellitus, and carbohydrate intolerance. Prevalence of mental retardation appears to be no greater than in the general population, but Turner's syndrome females may exhibit learning disabilities, especially with regard to spatial perception, visual-motor coordination, and mathematics. As a result, the nonverbal IQ in Turner's syndrome tends to be lower than the verbal IQ. Females with Turner's syndrome may also be socially immature for their age and may need support in developing independence and social relationships.

See also Developmental genetics; Obesity, genetic factors

TWIN STUDIES

The scientific study of human **twins** began in the 1870s when Sir **Francis Galton** (1822–1911) published a series of articles arguing that **heredity** (nature) was a stronger factor than environment (nurture) in determining the respective characteristics of twins. He later suggested that identical twins might come from a single egg while non-identical twins might come from two separate eggs, simultaneously fertilized and implanted. This guess, published in *Inquiries into Human Faculty and its Development* (1883), was later proved correct, but neither Galton nor his contemporaries had any rigorous evidence to support it.

Human twin studies are important for genetic and psychological research because twins provide a natural control for experiments. Because respect for each twin's feelings, privacy, and personhood is easy for even the best-intentioned scientist to compromise, and because twin research readily evokes **eugenics**, some twin studies will probably always remain controversial, both ethically and scientifically.

Twin studies ramify into genetic, embryological, biochemical, immunological, behavioral, anthropological, psychological, and sociological aspects. Among the questions asked are: What genetic, biological, or environmental factors cause twinning? How do the inherited and acquired traits of twins differ from those of singletons? How do the traits of identical (monozygotic) twins differ from those of non-identical (dizygotic) twins? Why and how does one twin typically dominate the other? What are the effects when monozygotic twins grow up apart? What are the effects of adoption?

American scientists such as Horatio Hackett Newman (b. 1875), Frank Nugent Freeman (b. 1880), and Karl John Holzinger (b. 1892) at the University of Chicago extended Galton's tradition of careful twin research. Their 1937 work, *Twins: A Study of Heredity and Environment*, is a landmark in the literature of nature versus nurture. In 1940, capitalizing on the widespread frenzy over the Dionne quintuplets (b. 1934), Newman wrote a semi-popular work, *Multiple Human Births: Twins, Triplets, Quadruplets, and Quintuplets*, speculating on the biological or genetic causes of their seemingly miraculous birth and survival.

Motivated primarily by eugenics, twin studies took a more sinister turn in Germany in the early twentieth century. Johannes Lange (1891–1938) argued that criminal propensities in one twin increased the likelihood of similar sociopathy in the other. His 1929 book, *Verbrechen als Schicksal: Studien an kriminellen Zwillingen* (Crime as Fate: Studies of Criminal Twins), was admired by the British eugenicist John Burdon Sanderson Haldane (1892–1964) and translated into English in 1930 by Haldane's wife, Charlotte (1894–1969), as *Crime and Destiny*.

German studies of twins, heredity, and **genetics** in the 1930s became inseparable from the politics of race. In 1935, Joseph Mengele, a wealthy Bavarian from a prominent family, a member of the Sturmabteilung (SA) received a Ph.D. in anthropology at the University of Munich with a dissertation entitled *Rassenmorphologische Untersuchungen des vorderen Unterkieferabschnitts bei vier rassischen Gruppen* (Race-Morphological Investigations of Sections of the Anterior Lower Jaw in Four Racial Groups). He then enrolled at the University of Frankfurt Institute for Hereditary Biology and Racial Hygiene to study under Otmar Freiherr von Verschuer (1896–1969). He joined the Nazi party in 1937 and its Schutzstaffel (SS) in 1938. His second dissertation, *Sippenuntersuchungen bei Lippen-Kiefer-Gaumenspalte* (Genus Investigations on Cleft Lip, Jaw, and Palate) gained him an M.D. in 1938. As an officer in the medical corps of the Waffen SS on the Russian front, he was twice decorated for bravery, seriously wounded, sent home in 1942 to recover, then assigned to Auschwitz as camp physician, where he arrived on May 30, 1943.

Mengele had a morbid and perverse fascination with twins, and saw in the concentration camp an endless supply of subjects for his experiments. Verschuer arranged for full funding of Mengele's research at Auschwitz, provided that Mengele would send the most significant data and specimens to Frankfurt. Mengele arranged to be the camp's principal "selector," deciding who would die and who would live. He ordered all twins lives spared for his purpose of study.

Mengele's twin studies were marked by arbitrariness, cruelty, and lack of scientific rigor. His work was mostly random trial and error, without hypotheses. Among his obsessions was trying to change eye color to blue. These attempts would often result in pain, infection, or blindness among the subjects. Convinced that the inmates at Auschwitz were less than human, Mengele kept his subjects naked so that he could measure and observe them more easily. He would inject, bleed, dismember, irradiate, or transfuse his twins, expose them to diseases, or perform unconventional surgical procedures on them without anesthesia. Mengele often killed one or both twin, dissected the bodies, and sent the results to Verschuer. Mengele's experiments involved over 1,500 pairs of twins. Only about 200 of these twins survived.

After World War II, Mengele escaped to South America. In 1959, prompted by Jewish "Nazi hunters," West Germany issued a warrant for Mengele's arrest. The Universities of Munich and Frankfurt revoked his degrees in 1964. A surviving pair of "Mengele Twins," Eva and Miriam Mozes (b. 1935), founded the international support group C.A.N.D.L.E.S. (Children of Auschwitz Nazi Deadly Laboratory Experiments Survivors) in 1985.

Not all twin research is as inhumane, illogical, or useless as Verschuer's and Mengele's. Yet, because of Auschwitz, reasonable and legitimate twin researchers after World War II experienced some difficulty in restoring its domain of inquiry to the level of approval enjoyed in Galton's time. In 1951, Italian geneticist Luigi Gedda (1902–2000) published the first significant postwar work on twins, *Studio dei gemelli* (A Study of Twins), a large book that was partially translated into English in 1961 as *Twins in History and Science*. Twin research gained momentum in the 1960s as geneticists revived Newman's interest in discovering the causes of higher-order multiple births.

In 2001, the leading scientific investigator of twins was Thomas Joseph Bouchard, Jr. (b. 1937), professor of psychology at the University of Minnesota and director of the Minnesota Center for Twin and Adoption Research. Beginning in 1990, his team published copiously on the results of the project they conducted from 1979 to 1999, the Minnesota Study of Twins Reared Apart, wherein they periodically administered batteries of psychological, educational, medical, and dental tests to a large population of monozygotic and dizygotic twins and their spouses. Data was collected and maintained by the Minnesota Twins Registry. From 1987 to 2001, the team did a longitudinal study on twin aging, using subjects between 24 and 66 years old at first appointment.

See also Acquired character; Ethical issues of modern genetics; Genetic implications of mating and marriage customs; Genetic transmission, patterns of; Social Darwinism

Three generations of female Russian twins. *Gerald Davis/Phototake NYC. Reproduced with permission.*

TWINS

There are two distinct types of twins in humans: identical and fraternal. Identical twins, also called monozygotic twins, result from the splitting of a single **zygote** into two distinct embryos. Because they arise from a single fertilized egg, monozygotic twins are always of the same sex, and being genetically identical, they tend to bear remarkable similarities in physical appearance. Identical twinning occurs in roughly 1/250 pregnancies worldwide, with relatively little variation between different populations. Evidence indicates that there is no tendency toward identical twin births in families (i.e., it does not "run in families").

Fraternal twins result from the **fertilization** of two different eggs by two different sperm during the same ovulatory cycle. Fraternal twins, also called dizygotic twins, can be of the same sex, or one of each sex, and are no more alike genetically than other siblings. Fraternal twinning varies in frequency around the world with the highest reported rates in Africa, where twins account for up to one in every eight pregnancies, and the lowest rates reported in Asian populations where twins account for less than one percent of all pregnancies. An increased chance for fraternal twinning can run in families. In these cases, the predisposition toward twinning appears to be due to a higher chance for the woman to produce multiple mature eggs in the same menstrual cycle, and thus the history of twinning tends to fall on the side of the mother.

At birth, both fraternal and identical twins can be very similar in appearance, and identifying which type of twinning has occurred is often difficult. Examination of the placenta and membranes is sometimes helpful. Twins that share a common amnion, chorion, or placenta are always identical twins. The reverse is not always true, however. Identical twins that split into two embryos early in development may develop their own placenta and membranes. At the other end of the spectrum, twins which separate from each other late in development may be conjoined.

Twinning in other mammals sometimes carries unexpected consequences. For example, when male/female fraternal twin pairs occur in cattle, the male hormones produced by the male twin causes the female to remain infertile after birth. These females are called "freemartins". Fortunately, human male/female twin pairs do not appear to interfere with each other's development.

Twins are of special significance in understanding the **genetics** of complex traits. Comparisons between monozygotic twins and dizygotic twins are the basis for determination of the degree to which these traits are influenced by genes.

See also Fertilization; Multifactorial or multigenic disorder; Multifactorial transmission

TWO-DIMENSIONAL ARRAY • *see* DIMENSIONAL ARRAYS

TWORT, FREDERICK WILLIAM (1877-1950)

English bacteriologist

As a pioneering bacteriologist, Frederick Twort was responsible for several important advances in his field. He discovered what would be known as bacteriophages, bacteria-attacking viruses. This discovery led to the advent of molecular biology. Twort was the first scientist to grow the organism that caused Jöhne's disease, a deadly cattle infection, and his efforts contributed to its elimination. Twort also discovered a nutritional element later identified as vitamin K.

Twort was born in Camberley, Surrey, England, the son of William Henry Twort, a physician. Frederick was the oldest of ten siblings. He studied medicine in London at St. Thomas's Hospital Medical School. Twort became qualified and licensed in 1900, although he never practiced clinical medicine. Soon after graduation, Twort began his work as an assistant to Louis Jenner in London's St. Thomas's Hospital, working in the clinical laboratory. In 1902, Twort found work with William Bullock as an anatomy instructor in London Hospital. It was here that Twort's first work in bacteriology began. He spent several years familiarizing himself with the bacteriology of hospitals and soon began his own experimentation.

Twort married Dorothy Nony Banister, who helped him with his work, and with her had a son and three daughters. His son, Antony, also became a doctor as well as his father's biographer.

Twort's own research became of primary importance to him in 1907. In that year, he published one of his earliest significant papers on **bacteria**. In it, he outlined how bacteria adapted and mutated. Two years later, in 1909, he published on bacterial growth and related growth agents. In what became a common occurrence, Twort's results were overlooked by the scientific community of the time, and found to be important only decades later.

Also in 1909, Twort was named superintendent of the Brown Institution at the University of London, an animal hospital. While working here, Twort was able to devote all his time to research. His work was limited, however, by funding and support problems, which plagued him throughout his career. Twort advanced the notion at Brown that all pathogenic, or disease-causing, bacteria developed from organisms that lived freely. Most of his contemporary bacteriologists believed pathogens originated in the body.

Twort's first important achievement was his in-depth study, with G.L.Y. Ingram, of Jöhne's Disease, the results of which were published in the early 1900s. Twort did the earliest cultures of the organism that caused the disease. He believed that there was a connection between tuberculosis and Jöhne's Disease, so he derived what he called his "essential substance" from dead tubercle bacilli. These bacilli, when incorporated in a culture medium, proved ideal for growing Jöhne's bacillus. Twort's study of Jöhne's disease directly led to the development of the Jöhnin test. His discovery also eventually proved important to **biochemistry**, specifically in the study of bacteria and their nutritional needs.

In 1915, Twort discovered what came to be known as bacteriophages. Twort's discovery was something of an accident. He spent several years using artificial media to grow viruses. Twort noticed that the bacteria infecting his plates kept becoming transparent. This was the earliest recorded proof of bacteriophages, though Twort called his discovery "transmissible lytic agent."

Twort published his results, but he was not certain about what he discovered. He made several guesses in his articles, but did not commit to any specific one, a hallmark of his career that lessened his findings in the eyes of his peers. Twort's experiments in this area were also overshadowed by World War I; he served in the Royal Army Medical Corps from 1915-18.

In 1917, Canadian bacteriologist Felix d'Hérelle made the same discovery, independent of Twort. D'Hérelle gave his findings their now common name, bacteriophages, which translates as bacteria eater. After d'Hérelle announced his findings, there was some controversy over who made the discovery first and when, in part because of Twort's published uncertainties. The results eventually carried both their names, and became known as the Twort-d'Hérelle phenomenon. Both scientists shared a life-long obsession with their discovery, hoping to fight diseases plaguing humans.

Before **antibiotics** were developed, scientists searched for ways to fight disease. Twort and d'Hérelle thought bacteriophages might be an answer, but the viruses did not work when used on human patients. The importance of the discovery of bacteriophages did not emerge until after Twort and d'Hérelle had died. In recent years, the idea has again come to light, as bacteria continue to develop resistance to antibiotics. In 1984, it was learned that certain illnesses in livestock and humans can be treated with modified bacteriophages.

Based on his many accomplishments, Twort was accorded honors. Among other distinctions, he was appointed professor of bacteriology at the University of London in 1919 and in 1929, he was elected a fellow of the Royal Society. Twort's peers found him a difficult and remote man, which perhaps limited the acceptance of him and his ideas. Still, his unique ability to work independently and at a high level of technical aptitude contributed to his capacities as a scientific explorer.

As Twort's research progressed, he became obsessed with proving, in more specific terms, that bacteria evolved from viruses, and that these viruses had evolved from more primary cellular forms. Though he spent years on this idea, he did not publish his findings. Twort's research was permanently interrupted in 1944 when the Brown Institution was destroyed by enemy fire during World War II. His laboratories and specimens were completely decimated. Twort spent his last years suffering greatly from this loss.

Twort died in the city of his birth, Camberley, on March 20, 1950. Posthumously, he was remembered for his scientific accomplishments, as well as his uncompromising belief that scientific funding should not be controlled by the government.

See also Microbial genetics; Viral genetics

U

UNIPARENTAL DISOMY

Normally, individuals inherit their chromosomes in a manner that each pair consists of one homologue from each parent. The genetic state in which both copies of a particular **chromosome** (or part of a chromosome) have been inherited from only one parent is named uniparental disomy (UPD). UPD is a term to describe a class of anomalies belonging to nontraditional **inheritance**. In most cases of UPD, the karyotype appears completely normal. **DNA** markers are the only way to determine the one parent origin for each chromosome. Molecular analysis shows that the disomy can be the result of two copies of the same chromosome (isodisomy) coming from one parent carrying identical genes (i.e., from one grandparent), or of a pair of chromosomes (heterodisomy) coming from one parent but carrying different genes (i.e., from both grandparents). The incidence of UPD is estimated to be as high as 2.8 to 16.5 per 10,000 conceptions.

A trisomic conception with post zygotic loss of chromosome (trisomic rescue) is the most frequent cause of UPD. Trisomy is a common cause of miscarriage unless the loss of the **supernumerary chromosomes** happens early in development. This mechanism leads to "rescue" of the conceptus due to mosaicism with a **diploid cell** line. Another cause of UPD is the **fertilization** of a nullisomic **gamete** by a disomic gamete (gametic complementation). This pathogenesis is due to the fertilization of an egg with an extra chromosome (disomy) by a sperm that is missing the same chromosome (nullisomy) or vice-versa.

At the present time, the postulated mechanisms for uniparental disomy include post-zygotic compensatory **duplication** of the single chromosome in a monosomic cell, and exchange of segments between chromosomes. The replacement of an absent rather than an abnormal chromosome through duplication of the remaining chromosome is probably the most frequent situation. The exchange of segments between chromosomes may manifest segmental UPD if the exchanged region is imprinted. This UPD arises from a normal cell line, and is associated with some cancers and perhaps complex age dependent diseases (atherosclerosis, Alzheimer disease).

UPD can result in complex clinical conditions by producing either homozygosity for recessive mutations or anomalous patterns of imprinting (defined as the differential expression of maternal and paternal genetic material). Imprinted genes show non-equivalence of the paternally and maternally inherited alleles, with one copy of each being required for normal function. Again, UPD is frequently found in conjunction with mosaicism, which can also contribute to phenotypic abnormalities.

As a consequence of isodisomy, where two copies of a recessive **mutation** are transmitted from a heterozygous parent, conditions such as **cystic fibrosis** and abnormally short stature have resulted (**maternal inheritance** of chromosome seven). Father to son transmission of **hemophilia** A has also been reported, due to the inheritance of both X and **Y chromosome** from the father. Abnormal imprinting has been associated with many clinical diseases. For example, maternal disomy of chromosome 15 has been reported in approximately 20–25% of the patients affected by Prader-Willis syndrome, while paternal disomy of the same chromosome is known in two to four percent of **Angelman syndrome** patients. Other clinical conditions include transient neonatal diabetes and Russell-Silver syndrome.

UPD is also likely to have a major role in the etiology of pregnancy loss and unexplained fetal growth restriction. This feature, however, is correlated to imprinting since the majority of imprinted genes identified so far seem to be involved in aspects of fetal growth. The risk of UPD is higher in the presence of trisomy mosaicism, mosaicism with an abnormal chromosome, isochromosome, and small additional chromosomal segments. Segmental UPD is likely to occur at an unknown rate, and may contribute to **cancer** and degenerative diseases, which have age-dependent risks.

See also Chromosomal mutations and abnormalities

URACIL (U) · *see* BASES

UREY, HAROLD (1893-1981)
American chemist and physicist

Already a scientist of great honor and achievement, Harold Urey's last great period of research brought together his interests and experiences in a number of fields of research to which he had devoted his life. The subject of that research was the origin of life on Earth.

Urey hypothesized that the Earth's primordial atmosphere consisted of reducing gases such as hydrogen, ammonia, and methane. The energy provided by electrical discharges in the atmosphere, he suggested, was sufficient to initiate chemical reactions among these gases, converting them to the simplest compounds of which living organisms are made, amino acids. In 1953, Urey's graduate student Stanley Lloyd Miller carried out a series of experiments to test this hypothesis. In these experiments, an electrical discharge passed through a glass tube containing only reducing gases resulted in the formation of amino acids.

The Miller-Urey expeiment is a classic experiment in molecular biology and **genetics**. The experiment established that the conditions that existed in Earth's primitive atmosphere were sufficient to produce amino acids, the subunits of proteins comprising and required by living organisms. In essence, the **Miller-Urey experiment** fundamentally established that Earth's primitive atmosphere was capable of producing the building blocks of life from inorganic materials.

The Miller-Urey experiment also remains the subject of scientific debate. Scientists continue to explore the nature and composition of Earth's primitive atmosphere and thus, continue to debate the relative closeness of the conditions of the experimental conditions to Earth's primitive atmosphere.

The Miller-Urey experiment was but one part of a distinguished scientific career for Urey. In 1934, Harold Urey was awarded the Nobel Prize in chemistry for his discovery of deuterium, an isotope, or **species**, of hydrogen in which the atoms weigh twice as much as those in ordinary hydrogen. Also known as heavy hydrogen, deuterium became profoundly important to future studies in many scientific fields, including chemistry, physics, and medicine. Urey continued his research on isotopes over the next three decades, and during World War II his experience with deuterium proved invaluable in efforts to separate isotopes of uranium from each other in the development of the first atomic bombs. Later, Urey's research on isotopes also led to a method for determining the Earth's atmospheric temperature at various periods in past history. This experimentation has become especially relevant because of concerns about the possibility of global climate change.

Urey was born in Walkerton, Indiana. His father, Samuel Clayton Urey, was a schoolteacher and lay minister in the Church of the Brethren. His mother was Cora Reinoehl Urey. After graduating from high school, Urey hoped to attend college but lacked the financial resources to do so. Instead, he accepted teaching jobs in country schools, first in Indiana (1911–1912) and then in Montana (1912–1914) before finally entering Montana State University in September of 1914 at the age of 21. Urey was initially interested in a career in biology, and the first original research he ever conducted involved a study of microorganisms in the Missoula River. In 1917 he was awarded his bachelor of science degree in zoology by Montana State.

The year Urey graduated also marked the entry of the United States into World War I. Although he had strong pacifist beliefs as a result of his early religious training, Urey acknowledged his obligation to participate in the nation's war effort. As a result, he accepted a job at the Barrett Chemical Company in Philadelphia and worked to develop high explosives. In his Nobel Prize acceptance speech, Urey said that this experience was instrumental in his move from industrial chemistry to academic life.

At the end of the war, Urey returned to Montana State University where he began teaching chemistry. In 1921 he decided to resume his college education and enrolled in the doctoral program in physical chemistry at the University of California at Berkeley. His faculty advisor at Berkeley was the great physical chemist Gilbert Newton Lewis. Urey received his doctorate in 1923 for research on the calculation of heat capacities and entropies (the degree of randomness in a system) of gases, based on information obtained through the use of a spectroscope. He then left for a year of postdoctoral study at the Institute for Theoretical Physics at the University of Copenhagen where Niels Bohr, a Danish physicist, was researching the structure of the atom. Urey's interest in Bohr's research had been cultivated while studying with Lewis, who had proposed many early theories on the nature of chemical bonding.

Upon his return to the United States in 1925, Urey accepted an appointment as an associate in chemistry at the Johns Hopkins University in Baltimore, a post he held until 1929. He interrupted his work at Johns Hopkins briefly to marry Frieda Daum in Lawrence, Kansas, on June 12, 1926. Daum was a bacteriologist and daughter of a prominent Lawrence educator. The Ureys later had four children., Gertrude Elizabeth, Frieda Rebecca, Mary Alice, and John Clayton.

In 1929, Urey left Johns Hopkins to become associate professor of chemistry at Columbia University, and in 1930 he published his first book, *Atoms, Molecules, and Quanta*, written with A. E. Ruark. Writing in the *Dictionary of Scientific Biography*, Joseph N. Tatarewicz called this work "the first comprehensive English language textbook on atomic structure and a major bridge between the new quantum physics and the field of chemistry." At this time he also began his search for an isotope of hydrogen. Since Frederick Soddy, an English chemist, discovered isotopes in 1913, scientists had been looking for isotopes of a number of elements. Urey believed that if an isotope of heavy hydrogen existed, one way to separate it from the ordinary hydrogen isotope would be through the vaporization of liquid hydrogen. Urey's subsequent isolation of deuterium made Urey famous in the scientific world, and only three years later he was awarded the Nobel Prize in chemistry for his discovery.

During the latter part of the 1930s, Urey extended his work on isotopes to other elements besides hydrogen. Urey found that the mass differences in isotopes can result in modest differences in their reaction rates

The practical consequences of this discovery became apparent during World War II. In 1939, word reached the United States about the discovery of nuclear fission by the German scientists Otto Hahn and Fritz Strassmann. The military consequences of the Hahn-Strassmann discovery were apparent to many scientists, including Urey. He was one of the first, therefore, to become involved in the U.S. effort to build a nuclear weapon, recognizing the threat posed by such a weapon in the hands of Nazi Germany. However, Urey was deeply concerned about the potential destructiveness of a fission weapon. Actively involved in political topics during the 1930s, Urey was a member of the Committee to Defend America by Aiding the Allies and worked vigorously against the fascist regimes in Germany, Italy, and Spain. He explained the importance of his political activism by saying that "no dictator knows enough to tell scientists what to do. Only in democratic nations can science flourish."

Urey worked on the Manhattan Project to build the nation's first atomic bomb. As a leading expert on the separation of isotopes, Urey made critical contributions to the solution of the Manhattan Project's single most difficult problem, the isolation of uranium–235.

At the conclusion of World War II, Urey left Columbia to join the Enrico Fermi Institute of Nuclear Studies at the University of Chicago where Urey continued to work on new applications of his isotope research. During the late 1940s and early 1950s, he explored the relationship between the isotopes of oxygen and past planetary climates. Since isotopes differ in the rate of chemical reactions, Urey said that the amount of each oxygen isotope in an organism is a result of atmospheric temperatures. During periods when the earth was warmer than normal, organisms would take in more of a lighter isotope of oxygen and less of a heavier isotope. During cool periods, the differences among isotopic concentrations would not be as great. Over a period of time, Urey was able to develop a scale, or an "oxygen thermometer," that related the relative concentrations of oxygen isotopes in the shells of sea animals with atmospheric temperatures. Some of those studies continue to be highly relevant in current research on the possibilities of global climate change.

In the early 1950s, Urey became interested in yet another subject: the chemistry of the universe and of the formation of the planets, including the earth. One of his first papers on this topic attempted to provide an estimate of the

Harold Urey.

relative abundance of the elements in the universe. Although these estimates have now been improved, they were remarkably close to the values modern chemists now accept.

In 1958, Urey left the University of Chicago to become Professor at Large at the University of California in San Diego at La Jolla. At La Jolla, his interests shifted from original scientific research to national scientific policy. He became extremely involved in the U.S. space program, serving as the first chairman of the Committee on Chemistry of Space and Exploration of the Moon and Planets of the National Academy of Science's Space Sciences Board. Even late in life, Urey continued to receive honors and awards from a grateful nation and admiring colleagues.

See also Amino acid; Ancient DNA; Archae; Bases; Biochemistry; Cell; Chargaff's rules; Chromosome; Codons; Comparative genomics; DNA structure; DNA synthesis; Double helix; Mitochondrial inheritance; Origin of life; RNA function

V

Variation of Inherited Characteristics

Although there are characteristics whose **inheritance** adheres strictly to Mendel's predictions, more often than not, that is not the case. Many different mechanisms can alter the phenotypic expression of an inherited genotype, complicating attempts to predict offspring phenotypes.

One such case is called incomplete penetrance. If a phenotype is determined by a dominant allele, it is expected that all individuals bearing the dominant allele will show the phenotype. Sometimes, the phenotype is not expressed in some individuals, or the phenotype seems to skip generations, only to reappear later. If all individuals in a population express the phenotype, its allele is said to have complete penetrance. The degree of penetrance is described on the level of population—in what percentage of individuals having the dominant allele the phenotype will be expressed. An example of a **gene** with incomplete penetrance is that causing polydactyly (extra fingers or toes).

Polydactyly serves also as an example of variable expressivity of a gene, or the degree to which a gene is expressed within an individual. Low expressivity of the polydactyly gene might result in an individual's having one extra finger, whereas high expressivity might result in extra digits on each hand and foot.

Another variation in the expression of inherited characteristics involves the individual's sex. Some characteristics, like pattern baldness, are sex-influenced. Depending on the sex of the individual, a given genotype may be expressed or not. In the baldness example, an individual of either sex who is homozygous dominant for the baldness gene will demonstrate hair loss, although it is more extreme and earlier in onset in males. In individual who is homozygous recessive for this gene will escape baldness. Males, but not females, who are heterozygous for the gene will lose their hair. This difference appears to be caused by the background of the individual's sex hormones. Females having adrenal gland tumors that cause overproduction of male hormones lose their hair if they are heterozygous for the baldness gene; but it grows back if the condition is corrected by surgical removal of the **tumor**.

Sex-limited traits can be passed through generations from one sex to another (from mother to son, for example), but are expressed in only one sex. These include breast development in females and beard growth in males.

In an interaction called epistasis, a gene at one locus can alter the phenotypic expression of a gene at another locus. For example, black fur is dominant over brown fur in mice, and inheritance of coat color in mice follows Mendelian predictions. However, there is another gene that determines whether any pigment at all will be deposited in the fur. If a mouse is recessive for this deposition gene, it will be albino, regardless of its genotype for the black/brown gene. Thus, the deposition gene is said to be epistatic to the color gene.

Certain genes, called suppressor genes, nullify the effect of other genes. In *Drosophila*, a **mutation** called hairy-winged shows in the phenotype of flies recessive for this gene. If they have the suppressor hairy-winged allele as well, however, their normal-winged phenotype is restored.

The position of a gene on a **chromosome** may also affect its expression. If a gene is moved from its original location due to a **translocation** or an **inversion**, it may not be expressed. This is true particularly if the gene is relocated to an area near **heterochromatin**, which appears to inhibit the expression of adjacent genes.

Temperature can effect a gene's expression. A well-known example is in the Siamese cat, whose nose, ears, and paws are more darkly colored than the rest of its body. A temperature-sensitive allele for pigment production works better at the slightly lower temperatures in the extremities than in the rest of the body, yielding the characteristic color pattern.

A phenomenon called genetic anticipation explains the observation that some conditions become more severe as they pass from one generation to the next. In anticipation, segments of **DNA**, such as trinucleotide repeats, are unstable and increase in number during **replication**; therefore, the number of repeats increases with each generation. The higher the number of repeats, the more severe the disease will be, and the earlier its onset. Anticipation has been observed in **Huntington disease**, fragile-X syndrome, and myotonic dystrophy.

In cases of **genomic imprinting**, phenotypic expression varies depending on the parental origin of the chromosome carrying a particular allele. Some unidentified chemical imprint, perhaps a methyl group, is added to chromsomes before or during **gamete** formation, leading to differentially imprinted genes or chromosome regions, the imprint indentifying the chromosome as being from the mother or the father. This imprint can be reversed from one generation to the next, as when a chromosome is passed from father to daughter. Probably the best-known example of imprinting in humans occurs in the q1 region of the 15th chromosome. A deletion in the paternal chromosome in this region causes Prader-Willis syndrome, whereas a corresponding deletion in the maternal chromosome causes **Angelman syndrome**. These two disorders are phenotypically distinct.

See also Epistasis; Genetic anticipation; Sex-linked genes

Varmus, Harold E. (1939-)
American virologist

When Harold E. Varmus was appointed director of the National Institutes of Health (NIH) in November, 1993, he was already famous throughout the world for his investigations into cancer-causing genes and other fundamental areas of biology, including the complex mechanisms of viruses. Varmus, who helped prove that there is a genetic component to **cancer**, was the co-recipient of a 1989 Nobel Prize for his research into oncogenes (genes with the capacity to turn normal cells into cancerous ones). Varmus' title of director of the NIH carries with it immense responsibilities, including the managing of a ten billion-dollar-plus NIH budget and the determination of grant awards for many types of medical research.

Harold Eliot Varmus was born in Oceanside, New York, to Frank and Beatrice (Barasch) Varmus. He attended Amherst College, graduating with a B.A. degree in 1961 (twenty-three years later, Amherst would award him with an honorary doctorate). Varmus went on to perform graduate work at Harvard University, receiving an M.A. degree in 1962, then he studied medicine at Columbia University, receiving an M.D. in 1966.

Varmus practiced medicine as an intern and resident at the Presbyterian Hospital of New York City between 1966 and 1968. He then worked as a clinical associate at the National Institutes of Health in Bethesda, Maryland, from 1968 to 1970. Moving to California, Varmus served as a lecturer in the department of microbiology at the University of California in San Francisco, becoming an associate professor in 1974—the same year that he was named associate editor of *Cell and Virology*—then, in 1979, he was promoted to full professor of microbiology, biochemistry and biophysics. During the 1980s,

Varmus began to accumulate a number of prestigious honors for his research, including the 1982 California Academic Scientist of the Year award and the 1983 Passano Foundation award; he was also the co-recipient of the Lasker Foundation award. In 1984, Varmus received both the Armand Hammer Cancer prize and the General Motors Alfred Sloan award, and the American Cancer Society made him an honorary professor of molecular virology. These honors were followed by the Shubitz Cancer prize and, in 1989, the Nobel Prize in physiology or medicine.

Varmus and **J. Michael Bishop**, his colleague from the University of California at San Francisco, were awarded the Nobel Prize in honor of their 1976 discovery that showed normal cells contain genes that can cause cancer. Varmus and Bishop, working with Dominique Stehelin and Peter Vogt, helped to prove the theory that cancer has a genetic component, demonstrating that oncogenes are actually normal genes that are altered in some way, perhaps due to carcinogen-induced mutations. Their research focused on Rous sarcoma, a **virus** which can produce tumors in chickens by attaching to a normal chicken **gene** as it duplicates within a **cell**. Since then, research has identified a number of additional "proto-oncogenes" which, when circumstances dictate, abandon their normal role of overseeing **cell division** and growth and turn potentially cancerous. Varmus's and Bishop's **oncogene** studies had a tremendous impact on the efforts to understand the genetic basis of cancer. The results of their work quickly found practical applications, especially in cancer diagnosis and prognosis.

Varmus was nominated by United States President Bill Clinton to the directorship of the National Institutes of Health and was confirmed in November, 1993. The director of the NIH plays a vital part in setting the course for biomedical research in the United States. Varmus's nomination was strongly supported by biomedical scientists, but there was some opposition from **AIDS** activists. They—as well as others who were concerned with the health of women and members of minority groups—were concerned that Varmus would be more interested in basic biomedical research than in applied studies and feared that the medical research related to their specific concerns might be neglected. Varmus has argued that basic research in science, especially investigations of the fundamental properties of cells, genes, and tissues, could eventually lead to cures for many diseases, such as AIDS and cancer. As director, Varmus is also interested in revitalizing the intramural research program at NIH. He believes that science education in the United States needs to be improved and that students should be exposed to a science curriculum sooner, in smaller classes, by better-informed teachers.

A licensed physician in the state of California, Varmus is a member of numerous professional and academic associations, including the National Academy of Sciences, the American Society of Microbiologists, the American Society of Virologists, and the American Academy of Arts and Sciences. He married Constance Louise Casey on October 25, 1969, and they have two sons, Jacob Carey and Christopher Isaac.

See also Cancer genetics; Viral genetics

VAVILOV, NIKOLAI IVANOVITCH (1887-1943)

Russian plant geneticist

Nicolai Ivanovich Vavilov made significant contributions to the field of modern **plant genetics**. His identification of plant centers of origin, and the relationship between cultivated plants and their wild cousins, revolutionized the means by which scientists evaluate plant populations. Devoting a great deal of his career to the cause of agricultural improvement, he was a consummate traveler, fervent researcher, and a passionate advocate of the practical application of genetic research.

Vavilov was born in Moscow. He studied **genetics** at Cambridge and the John Innes Horticultural Institution in London under the direction of **William Bateson** (1861–1926). He returned to Russia and took a position as a professor of botany at the University of Saratov. In 1921, he left the university to work for the government and became director of the Bureau of Applied Botany in St. Petersburg. His next position was as director of the All—Union V.I. Lenin Academy of Agricultural Sciences. During his tenure at the Academy of Agricultural Sciences, the government began programs to advance the scientific research. Vavilov aided in the establishment of nearly 400 research institutes throughout the Soviet Union.

Vavilov was an avid traveler. By 1933, he had completed numerous research expeditions, visiting over 40 countries and collecting some 80, 000 plant specimen, a third of which were various types of wheat. An especially difficult expedition to Afghanistan in 1924 earned him a gold medal from the Russian Geographic Society. On these voyages, Vavilov began to formulate his theories on plant populations based upon his observations in the field. He published the results of his study in *The Origin, Variation, Immunity and Breeding of Cultivated Plants*. Vavilov concluded that the place of origin for any cultivated plant could be found in the region where varieties of the plant's non—cultivated relatives were most prolific and best adapted. He later expanded his theory. Postulating that a plant's center of origin is where the genetic diversity of a plant **species** is greatest, he identified over a dozen such global points of origin. When published, Vavilov's conclusions became known respectively as the law of homologous series in variation (1920) and the theory of the centers of origin of cultivated plants (1926). The theories provided a structure for charting movement, adaptation, and change within plant populations.

Vavilov's theories of plant populations, and their origins, remain important today. As natural habitats of wild plant species are increasingly threatened on a global scale, conservation—geneticists are concerned with its impact on the genetic pools of both wild and cultivated plants.

Despite his frequent research expeditions, Vavilov remained committed to using plant—genetics to improve Soviet agriculture. Vavilov used his prominence to further promote the establishment of agricultural research institutes. To facilitate better communication among scientists who were studying the breeding and raising of plant–crops, Vavilov organized conferences, societies, and institutes. He was a foreign member of six national academies of science, and served as the director of the Soviet Genetics Institute.

Although Vavilov received great acclaim from the international scientific community for his contributions to the study and understanding of botanical populations, his academic standing in the Soviet Union was later shattered by followers of Trofim Desnovich Lysenko (1898–1976), the government's Director of Genetics. Lysenko did not recognize the validity of the laws of **heredity**, and instead advocated that a plant's genetics could be altered (without hybridization or breeding) simply by changing its environmental context. He sought to integrate his scientifically unsound ideas into Soviet agriculture, but Vavilov and handful other geneticists publicly opposed Lysenko's plan and recommended instead a farming model similar to the United States. Lysenko denounced Vavilov on several occasions, and in 1940 petitioned for his arrest. Vavilov was arrested and sent to a concentration camp for political prisoners near Saratov. He died, still imprisoned, in 1943. Vavilov's work regained prestige in the early 1960s after **Lysenkoism** was discredited.

See also Lysenkoism; Plant Genetics

VECTOR • *see* CLONING VECTOR

VENTER, JOHN CRAIG (1946-)

American molecular biologist

John Craig Venter, currently the President and Chief Executive Officer of Celera Genomics, is one of the central figures in the **Human Genome Project**. Celera, headquartered in Rockville MD, has succeeded in completed a draft of the human **genome**. Using a fast **sequencing** technique, Venter and his colleagues were able to sequence the human genome, and the genomes of other organisms, including the bacterium *Haemophilus infuenzae,* in only about a decade.

Venter was born in Salt Lake City. After high school he seemed destined for a career as a surfer rather than as a molecular biologist. But a tour of duty in Vietnam as a hospital corpsman precipitated a change in the direction of his life. He returned from Vietnam and entered university, earning a doctorate in physiology and pharmacology from the University of California at San Diego in six years. After graduation he commenced research at the National Institutes of Health. While at NIH, Venter became frustrated at the then slow pace of identifying and sequencing genes. He began to utilize a technology whereby the normal copies of genes made by living cells were obtained and to decode only a portion of the **DNA**. These partial transcripts, called **expressed sequence tags**, could then be used to identify the gene-coding regions on the DNA from which they came. The result was to speed up the identification of genes. Hundreds of genes could be discovered in only weeks using the method.

Supported by venture capital, Venter started the non-profit The Institute for Genomic Research (TIGR) In the mid–1990s. Here, thousands of the expressed sequence tag probes to the human genome were made.

Venter's success and technical insight attracted the interest of PE Biosystems, makers of automated DNA sequencers. With financial and equipment backing from PE Biosystems, Venter left TIGR and formed a private for-profit company, Celera (meaning 'swift' in Latin). The aim was to decode the human genome faster than the government effort that was underway. Celera commenced operations in May 1998.

Another of Venter's accomplishments was to use a non-traditional approach to quickly sequence DNA. At that time, DNA was typically sequenced by dividing it into several large pieces and then decoding each piece. Venter devised the so-called shotgun method, in which a genome was blown apart into many small bits and then to sequence them without regard to their position. Following sequencing, supercomputer power would reassemble the bits of sequence into the intact genome sequence. The technique, which was extremely controversial, was tried first on the genome of the fruit fly *Drosophila*. In only a year the fruit fly genome sequence was obtained. The sequencing of the genome of the bacterium *H. influenzae* followed.

Venter's accomplishments are considerable, both technically and as a catalyst to spur genome sequencing.

VERNALIZATION • *see* LYSENKOISM

VERY LARGE SCALE INTEGRATION (VLSI)

Very large scale integration, or VSLI, has two meanings, both of which are relevant to **genetics**. VSLI refers to the strategies involving a combination of computational and mathematical tools being deployed to develop quantitative ways of dealing with the many databases housing genomic and biological information.

The explosion of information resulting from **genome sequencing** programs, such as the **Human Genome Project**, has created a need for large databases with search and query capabilities. Incompatibilities among heterogeneous databases and computer programs has hindered the ability to query different sources of genomic information. Researchers are actively pursuing the large-scale integration of such databases, which can be structured differently and which encompass different aspects of biology, such as **gene** sequences and biochemical pathways. Strategies include a VSLI system that does not require a central controlling computer, as there is no centralized control. Such flexibility allows data to be exchanged without being tied to a single mode of representation. Thus, the information residing in varied databases can be retrieved and compared. This kind of exploratory power is well-suited to the needs of genomics researchers.

The potential biological applications of VSLI go beyond the meshing of data from various genomics databases. VSLI comparison and integration of data could permit modeling the role of genes in determining the **genetic variation** of disease causing organisms. There are likely to be many genes involved that are linked and interact with one another. Modeling of the joint dynamics of such complex systems could be possible with VSLI. Another biological application of VSLI is the study of genetically controlled patterns in plants. Such studies could lead to **selection** of tailored breeding and growth conditions that lead to a more desirable plant.

Integration of databases also has use in non-biological realms. VSLI is useful in the study of the **evolution** of communication, learning and group structure. This encompasses such diverse topics as the origins of culture, evolution of conformity, and how genetic and cultural evolution interact. Finally, VSLI is finding a role in **population genetics**. Simulation studies of how random change alters populations of increasing size are possible because of the database merging power of VSLI.

VSLI is also a computer-related term. In this context, VSLI refers to the development of integrated circuits that are capable of housing millions of transistors. The Intel Pentium chip is an example of an integrated circuit. The increasing computing power afforded by VSLI chips is finding use in bioinformatics, **medical genetics**, and population genetics. Semiconductor technology is advancing to the point where devices will have the complexity that is require to solve lower level perception tasks. In other words, synthetic neurobiology will be possible, analogous to modeling experiments that are now done in molecular biology and genetics. Silicon neurons—neuronal function on a computer chip—could be used to emulate intelligent circuits in the brain. Simulations of learning, memory, cognition and perhaps consciousness are conceivable.

See also Comparative genomics; Genomic library

VIABILITY

The probability of zygotes (i.e., cells formed by the fusion of the egg **cell** (in human, the **ovum**) and the spermatozoid) or seeds to survive and develop into an adult organism is known as viability.

Mutations in gametes (i.e., germ cells, egg or spermatozoid) may give origin to a **zygote** carrying a deleterious allele that can result in reduced **fitness**, implying a lower probability for development or survival up to a reproductive age. Since most mutated genes are recessive, a heterozygous **carrier** will usually not manifest the related disease phenotype. However, if such recessive **mutation** is also present in the other inherited allele, embryo inviability, or the onset of a hereditary lethal syndrome may occur in early or late stages of life, depending on the **gene** involved.

Recessive mutated alleles are therefore termed semilethal genes, because death will occur only in **homozygote** individuals. Nevertheless, if the allele mutation occurs in the dominant copy of the gene, it will suffice for disease onset or embryo inviability of the **heterozygote** carrier. When the allelic mutation causes death, whether at the time of **fertilization** or in any stage of life, it is termed lethal gene or lethal allele. An example of the action of a semi lethal gene is **sickle cell anemia**, a hereditary autosomal recessive blood disorder that usually causes death of homozygote carriers during late childhood or early adolescence, but is not lethal to heterozygote individuals. An instance of a lethal gene disorder is **Huntington disease** (HD), an inherited autosomal dominant neuronal disorder, that may occur in early adulthood (juvenile form) or around middle age (50 years or later), affecting both heterozygote and homozygote individuals. The juvenile form of the disease is inherited from the father in the vast majority of the studied cases, whereas in the late-onset form of HD more than twice as many cases are inherited from an affected mother.

The risk rate for certain hereditary syndromes, such as **Down syndrome**, increases with the age of the mother at the time of conception, such rates are related to the mean maternal age (in the U.S. the mean maternal age is 29.2 years). This is known as the maternal age effect and may be related to follicular atresia (i.e., degeneration of the follicles surrounding the ovum) or genetic factors.

Achondroplasia, the most common form of human dwarfism, is an autosomal dominant trait derived in seven-eighths of the cases from a new mutation or *de novo* mutation in the father's **gamete**. Paternal age effect is associated with Achondroplasia present in the offspring of normal parents and about 80% of achondroplastic individuals die during childhood.

Viability of the offspring may be affected either by inherited mutations present in the **genome** of one of the parents, such as Li-Fraumeni syndrome, or breast and ovarian hereditary cancers, or by new age-related mutations occurring in the gametes of one or both parents, as exemplified above.

With regard to **evolutionary mechanisms**, the minimum level of viability for an organism is the ability to live long enough to successfully produce offspring that are, in turn, capable of successful reproduction.

See also Autosomes; Birth defects; Carrier; Chromosomal mutations and abnormalities; Evolutionary mechanisms; Gametogenesis; Gene mutation; Genetic disorders and diseases; Genotype and phenotype; Germ cells and the germ cell line

VIRAL GENETICS

Viral **genetics**, the study of the genetic mechanisms that operate during the life cycle of viruses, utilizes biophysical, biological, and genetic analyses to study the viral **genome** and its variation. The **virus** genome consists of only one type of nucleic acid, which could be a single or double stranded **DNA** or **RNA**. Single stranded RNA viruses could contain positive-sense (+RNA), which serves directly as **mRNA** or negative-sense RNA (-RNA) that must use an RNA **polymerase** to

synthesize a complementary positive strand to serve as mRNA. Viruses are obligate parasites that are completely dependant on the host **cell** for the **replication** and **transcription** of their genomes as well as the **translation** of the mRNA transcripts into proteins. Viral proteins usually have a structural function, making up a shell around the genome, but may contain some **enzymes** that are necessary for the virus replication and life cycle in the host cell. Both bacterial virus (bacteriophages) and animal viruses play an important role as tools in molecular and cellular biology research.

Viruses are classified in two families depending on whether they have RNA or DNA genomes and whether these genomes are double or single stranded. Further subdivision into types takes into account whether the genome consists of a single RNA molecule or many molecules as in the case of segmented viruses. Four types of bacteriophages are widely used in biochemical and genetic research. These are the T phages, the temperate phages typified by **bacteriophage** lambda, the small DNA phages like M13, and the RNA phages. Animal viruses are subdivided in many classes and types. Class I viruses contain a single molecule of double stranded DNA and are exemplified by adenovirus, simian virus 40 (SV40), herpes viruses and human papillomaviruses. Class II viruses are also called parvoviruses and are made of single stranded DNA that is copied in to double stranded DNA before transcription in the host cell. Class III viruses are double stranded RNA viruses that have segmented genomes which means that they contain 10-12 separate double stranded RNA molecules. The negative strands serve as template for mRNA synthesis. Class IV viruses, typified by poliovirus, have single plus strand genomic RNA that serves as the mRNA. Class V viruses contain a single negative strand RNA which serves as the template for the production of mRNA by specific virus enzymes. Class VI viruses are also known as retroviruses and contain double stranded RNA genome. These viruses have an enzyme called **reverse transcriptase** that can both copy minus strand DNA from genomic RNA catalyze the synthesis of a complementary plus DNA strand. The resulting double stranded DNA is integrated in the host **chromosome** and is transcribed by the host own machinery. The resulting transcripts are either used to synthesize proteins or produce new viral particles. These new viruses are released by budding, usually without killing the host cell. Both HIV and HTLV viruses belong to this class of viruses.

Virus genetics is studied by either investigating genome mutations or exchange of genetic material during the life cycle of the virus. The frequency and types of genetic variations in the virus are influenced by the nature of the viral genome and its structure. Especially important are the type of the nucleic acid that influence the potential for the viral genome to integrate in the host, and the segmentation that influence exchange of genetic information through assortment and **recombination**.

Mutations in the virus genome could either occur spontaneously or be induced by physical and chemical means. Spontaneous mutations that arise naturally as a result of viral replication are either due to a defect in the genome replication machinery or to the incorporation of an analogous base instead of the normal one. Induced virus mutants are obtained by either using chemical mutants like nitrous oxide that acts directly on **bases** and modify them or by incorporating already modified bases in the virus genome by adding these bases as substrates during virus replication. Physical agents such as ultra-violet light and x-rays can also be used in inducing mutations. Genotypically, the induced mutations are usually point mutations, **deletions**, and rarely insertions. The phenotype of the induced mutants is usually varied. Some mutants are conditional lethal mutants. These could differ from the **wild type** virus by being sensitive to high or low temperature. A low temperature mutant would for example grow at 87.8°F (31°C) but not at 100.4°F (38°), while the wild type will grow at both temperatures. A mutant could also be obtained that grows better at elevated temperatures than the wild type virus. These mutants are called hot mutants and may be more dangerous for the host because fever, which usually slows the growth of wild type virus, is ineffective in controlling them. Other mutants that are usually generated are those that show drug resistance, enzyme deficiency or an altered pathogenicity or host range. Some of these mutants cause milder symptoms compared to the parental virulent virus and usually have potential in vaccine development as exemplified by some types of influenza vaccines.

Besides **mutation**, new genetic variants of viruses also arise through exchange of genetic material by recombination and reassortment. Classical recombination involves breaking of covalent bonds within the virus nucleic acid and exchange of some DNA segments followed by rejoining of the DNA break. This type of recombination is almost exclusively reserved to DNA viruses and retroviruses. RNA viruses that do not have a DNA phase rarely use this mechanism. Recombination usually enables a virus to pick up genetic material from similar viruses and even from unrelated viruses and the eukaryotic host cells. Exchange of genetic material with the host is especially common with retroviruses. Reassortment is a non-classical kind of recombination that occurs if two variants of a segmented virus infect the same cell. The resulting progeny virions may get some segments from one parent and some from the other. All known segmented virus that infect humans are RNA viruses. The process of reassortment is very efficient in the exchange of genetic material and is used in the generation of viral vaccines especially in the case of influenza live vaccines. The ability of viruses to exchange genetic information through recombination is the basis for virus-based vectors in recombinant DNA technology and hold great promises in the development of **gene therapy**. Viruses are attractive as vectors in **gene** therapy because they can be targeted to specific tissues in the organs that the virus usually infect and because viruses do not need special chemical reagents called transfectants that are used to target a plasmid vector to the genome of the host.

Genetic variants generated through mutations, recombination or reassortment could interact with each other if they infected the same host cell and prevent the appearance of any phenotype. This phenomenon, where each mutant provide the missing function of the other while both are still genotypically

mutant, is known as complementation. It is used as an efficient tool to determine if mutations are in a unique or in different genes and to reveal the minimum number of genes affecting a function. Temperature sensitive mutants that have the same mutation in the same gene will for example not be able to complement each other. It is important to distinguish complementation from multiplicity reactivation where a higher dose of inactivated mutants will be reactivated and infect a cell because these inactivated viruses cooperate in a poorly understood process. This reactivation probably involves both a complementation step that allows defective viruses to replicate and a recombination step resulting in new genotypes and sometimes regeneration of the wild type. The viruses that need complementation to achieve an infectious cycle are usually referred to as defective mutants and the complementing virus is the helper virus. In some cases, the defective virus may interfere with and reduce the infectivity of the helper virus by competing with it for some factors that are involved in the viral life cycle. These defective viruses called "defective interfering" are sometimes involved in modulating natural infections. Different wild type viruses that infect the same cell may exchange coat components without any exchange of genetic material. This phenomenon, known as phenotypic mixing is usually restricted to related viruses and may change both the morphology of the packaged virus and the tropism or tissue specificity of these infectious agents.

See also Bacterial genetics; Cancer genetics; Epidemiology and genetics; Immunogenetics

VIRUS

A virus is a small, infectious agent that consists of a core of genetic material (either **deoxyribonucleic acid** [DNA] or **ribonucleic acid** [RNA]) surrounded by a shell of protein. Viruses cause disease by infecting a host **cell** and commandeering the host cell's synthetic capabilities to produce more viruses. The newly made viruses then leave the host cell, sometimes killing it in the process, and proceed to infect other cells within the host. Because viruses invade cells, drug therapies have not yet been designed to kill viruses, although some have been developed to inhibit their growth. The human immune system is the main defense against a viral disease.

Viruses can infect both plants, **bacteria**, and animals. The **tobacco mosaic virus**, one of the most studied of all viruses, infects tobacco plants. Bacterial viruses, called bacteriophages, infect a variety of bacteria, such as *Escherichia coli,* a bacteria commonly found in the human digestive tract. Animal viruses cause a variety of fatal diseases. Acquired Immune Deficiency Syndrome (**AIDS**) is caused by the Human Immunodeficiency Virus (HIV); hepatitis and rabies are viral diseases; and the so-called hemorrhagic fevers, which are characterized by severe internal bleeding, are caused by filoviruses. Other animal viruses cause some of the most common human diseases. Often, these diseases strike in childhood.

Measles, mumps, and chickenpox are viral diseases. The common cold and influenza are also caused by viruses. Finally, some viruses can cause **cancer** and tumors. One such virus, Human T-cell Leukemia Virus (HTLV), was only recently discovered and its role in the development of a special kind of leukemia is still being elucidated.

Although viral structure varies considerably between the different types of viruses, all viruses share some common characteristics. All viruses contain either **RNA** or **DNA** surrounded by a protective protein shell called a capsid. Some viruses have a double strand of DNA, others a single strand of DNA. Other viruses have a double strand of RNA or a single strand of RNA. The size of the genetic material of viruses is often quite small. Compared to the 100,000 genes that exist within human DNA, viral genes number from 10 to about 200 genes.

Viruses contain such small amounts of genetic material because the only activity that they perform independently of a host cell is the synthesis of the protein capsid. In order to reproduce, a virus must infect a host cell and take over the host cell's synthetic machinery. This aspect of viruses—that the virus does not appear to be "alive" until it infects a host cell—has led to controversy in describing the nature of viruses. Are they living or non-living? When viruses are not inside a host cell, they do not appear to carry out many of the functions ascribed to living things, such as reproduction, metabolism, and movement. When they infect a host cell, they acquire these capabilities. Thus, viruses are both living and non-living. It was once acceptable to describe viruses as agents that exist on the boundary between living and non-living; however, a more accurate description of viruses is that they are either active or inactive, a description that leaves the question of life behind altogether.

The origin of viruses is also controversial. Some viruses, such as the pox viruses, are so complex that they appear to have been derived from some kind of living eukaryote or **prokaryote**. The origin of the poxvirus could therefore resemble that of **mitochondria** and chloroplasts, organelles within eukaryotic cells, which are thought to have once been independent organisms. On the other hand, some viruses are extremely simple in structure, leading to the conclusion that these viruses are derived from cellular genetic material that somehow acquired the capacity to exist independently. This possibility is much more likely for most viruses; however, scientists still believe that the poxvirus is the exception to this scenario.

All viruses consist of genetic material surrounded by a capsid, but within the broad range of virus types, variations exist within this basic structure. Studding the envelope of these viruses are protein "spikes." These spikes are clearly visible on some viruses, such as the influenza viruses; on other enveloped viruses, the spikes are extremely difficult to see. The spikes help the virus invade host cells. The influenza virus, for instance, has two types of spikes. One type, composed of hemagglutinin protein (HA), fuses with the host cell membrane, allowing the virus particle to enter the cell. The other type of spike, composed of the protein neuraminidase

(NA), helps the newly formed virus particles to bud out from the host cell membrane.

The capsid of viruses is relatively simple in structure, owing to the few genes that the virus contains to encode the capsid. Most viral capsids consist of a few repeating protein subunits. The capsid serves two functions: it protects the viral genetic material and it helps the virus introduce itself into the host cell. Many viruses are extremely specific, targeting only certain cells within the plant or animal body. HIV, for instance, targets a specific immune cell, the T helper cell. The cold virus targets respiratory cells, leaving the other cells in the body alone. How does a virus "know" which cells to target? The viral capsid has special receptors that match receptors on their targeted host cells. When the virus encounters the correct receptors on a host cell, it "docks" with this host cell and begins the process of infection and **replication**.

Most viruses are rod- or roughly sphere-shaped. Rod-shaped viruses include tobacco **mosaic** virus and the filoviruses. Although they look like rods under a microscope, these viral capsids are actually composed of protein molecules arranged in a helix. Other viruses are shaped somewhat like spheres, although many viruses are not actual spheres. The capsid of the adenovirus, which infects the respiratory tract of animals, consists of 20 triangular faces. This shape is called an icosahedron. HIV is a true sphere, as is the influenza virus.

Some viruses are neither rod- or sphere-shaped. The poxviruses are rectangular, looking somewhat like bricks. Parapoxviruses are ovoid. Bacteriophages are the most unusually shaped of all viruses. A **bacteriophage** consists of a head region attached to a sheath. Protruding from the sheath are tail fibers that dock with the host bacterium. The bacteriophage's structure is eminently suited to the way it infects cells. Instead of the entire virus entering the bacterium, the bacteriophage injects its genetic material into the cell, leaving an empty capsid on the surface of the bacterium.

Viruses are obligate intracellular parasites, meaning that in order to replicate, they need to be inside a host cell. Viruses lack the machinery and **enzymes** necessary to reproduce; the only synthetic activity they perform on their own is to synthesize their capsids.

The infection cycle of most viruses follows a basic pattern. Bacteriophages are unusual in that they can infect a bacterium in two ways (although other viruses may replicate in these two ways as well). In the lytic cycle of replication, the bacteriophage destroys the bacterium it infects. In the lysogenic cycle, however, the bacteriophage coexists with its bacterial host, and remains inside the bacterium throughout its life, reproducing only when the bacterium itself reproduces.

An example of a bacteriophage that undergoes lytic replication inside a bacterial host is the T4 bacteriophage which infects *E. coli*. T4 begins the infection cycle by docking with an *E. coli* bacterium. The tail fibers of the bacteriophage make contact with the cell wall of the bacterium, and the bacteriophage then injects its genetic material into the bacterium. Inside the bacterium, the viral genes are transcribed. One of the first products produced from the viral genes is an enzyme that destroys the bacterium's own genetic material. Now the virus can proceed in its replication unhampered by the bacterial genes. Parts of new bacteriophages are produced and assembled. The bacterium then bursts, and the new bacteriophages are freed to infect other bacteria. This entire process takes only 20-30 minutes.

In the lysogenic cycle, the bacteriophage reproduces its genetic material but does not destroy the host's genetic material. The bacteriophage called lambda, another *E. coli*-infecting virus, is an example of a bacteriophage that undergoes lysogenic replication within a bacterial host. After the viral DNA has been injected into the bacterial host, it assumes a circular shape. At this point, the replication cycle can either become lytic or lysogenic. In a lysogenic cycle, the circular DNA attaches to the host cell **genome** at a specific place. This combination host-viral genome is called a **prophage**. Most of the viral genes within the prophage are repressed by a special repressor protein, so they do not encode the production of new bacteriophages. However, each time the bacterium divides, the viral genes are replicated along with the host genes. The bacterial progeny are thus lysogenically infected with viral genes.

Interestingly, bacteria that contain prophages can be destroyed when the viral DNA is suddenly triggered to undergo lytic replication. Radiation and chemicals are often the triggers that initiate lytic replication. Another interesting aspect of prophages is the role they play in human diseases. The bacteria that cause diphtheria and botulism both harbor viruses. The viral genes encode powerful toxins that have devastating effects on the human body. Without the infecting viruses, these bacteria may well be innocuous. It is the presence of viruses that makes these bacterial diseases so lethal.

Scientists have classified viruses according to the type of genetic material they contain. Broad categories of viruses include double-stranded DNA viruses, single-stranded DNA viruses, double-stranded RNA viruses, and single stranded RNA viruses. For the description of virus types that follows, however, these categories are not used. Rather, viruses are described by the type of disease they cause.

Poxviruses are the most complex kind of viruses known. They have large amounts of genetic material and fibrils anchored to the outside of the viral capsid that assist in attachment to the host cell. Poxviruses contain a double strand of DNA.

Viruses cause a variety of human diseases, including smallpox and cowpox. Because of worldwide vaccination efforts, smallpox has virtually disappeared from the world, with the last known case appearing in Somalia in 1977. The only places on Earth where smallpox virus currently exists are two labs: the Centers for Disease Control in Atlanta and the Research Institute for Viral Preparation in Moscow. Prior to the eradication efforts begun by the World Health Organization in 1966, smallpox was one of the most devastating of human diseases. In 1707, for instance, an outbreak of smallpox killed 18,000 of Iceland's 50,000 residents. In Boston in 1721, smallpox struck 5,889 of the city's 12,000 inhabitants, killing 15% of those infected.

Edward Jenner (1749–1823) is credited with developing the first successful vaccine against a viral disease, and that disease was smallpox. A vaccine works by eliciting an immune response. During this immune response, specific immune cells, called memory cells, are produced that remain in the body long after the foreign microbe present in a vaccine has been destroyed. When the body again encounters the same kind of microbe, the memory cells quickly destroy the microbe. Vaccines contain either a live, altered version of a virus or bacteria, or they contain only parts of a virus or bacteria, enough to elicit an immune response.

In 1797, Jenner developed his smallpox vaccine by taking pus from a cowpox lesion on the hand of a milkmaid. Cowpox was a common disease of the era, transmitted through contact with an infected cow. Unlike smallpox, however, cowpox is a much milder disease. Using the cowpox pus, he inoculated an 8-year-old boy. Jenner continued his vaccination efforts through his lifetime. Until 1976, children were vaccinated with the smallpox vaccine, called vaccinia. Reactions to the introduction of the vaccine ranged from a mild fever to severe complications, including (although very rarely) death. In 1976, with the eradication of smallpox complete, vaccinia vaccinations for children were discontinued, although vaccinia continues to be used as a **carrier** for recombinant DNA techniques. In these techniques, foreign DNA is inserted in cells. Efforts to produce a vaccine for HIV, for instance, have used vaccinia as the vehicle that carries specific parts of HIV.

Herpesviruses are enveloped, double-stranded DNA viruses. Of the more than 50 herpes viruses that exist, only eight cause disease in humans. These include the human herpes virus types 1 and 2 that cause cold sores and genital herpes; human herpes virus 3, or varicella-zoster virus (VZV), that causes chicken pox and shingles; cytomegalovirus (CMV), a virus that in some individuals attacks the cells of the eye and leads to blindness; human herpes virus 4, or Epstein-Barr virus (EBV), which has been implicated in a cancer called Burkitt's lymphoma; and human herpes virus types 6 and 7, newly discovered viruses that infect white blood cells. In addition, herpes B virus is a virus that infects monkeys and can be transmitted to humans by handling infected monkeys.

Adenoviruses are viruses that attack respiratory, intestinal, and eye cells in animals. More than 40 kinds of human adenoviruses have been identified. Adenoviruses contain double-stranded DNA within a 20-faceted capsid. Adenoviruses that target respiratory cells cause bronchitis, pneumonia, and tonsillitis. Gastrointestinal illnesses caused by adenoviruses are usually characterized by diarrhea and are often accompanied by respiratory symptoms. Some forms of appendicitis are also caused by adenoviruses. Eye illnesses caused by adenoviruses include conjunctivitis, an infection of the eye tissues, as well as a disease called pharyngoconjunctival fever, a disease in which the virus is transmitted in poorly chlorinated swimming pools.

Human papoviruses include two groups: the papilloma viruses and the polyomaviruses. Human papilloma viruses (HPV) are the smallest double-stranded DNA viruses. They replicate within cells through both the lytic and the lysogenic replication cycles. Because of their lysogenic capabilities, HPV-containing cells can be produced through the replication

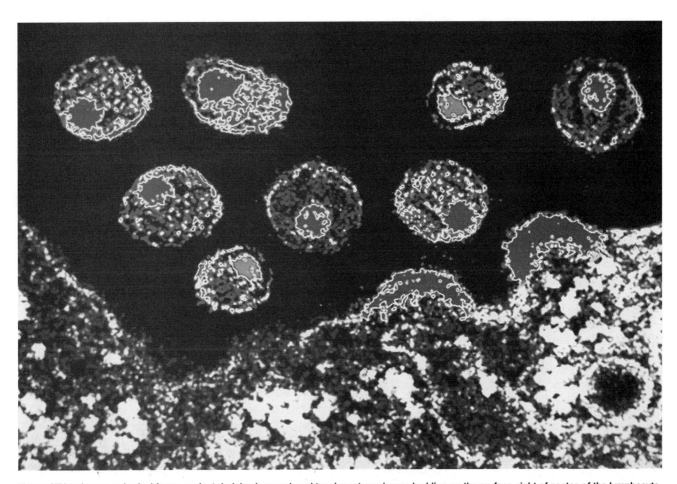

Mature HIV-1 viruses, spherical forms against dark background, and two immature viruses budding on the surface, right of center, of the lymphocyte from which the mature viruses emerged. *Photograph by Scott Camazinr. National Audubon Society Collection/Photo Researchers, Inc. Reproduced by permission.*

of those cells that HPV initially infects. In this way, HPV infects epithelial cells, such as the cells of the skin. HPVs cause several kinds of benign (non-cancerous) warts, including plantar warts (those that form on the soles of the feet) and genital warts. However, HPVs have also been implicated in a form of cervical cancer that accounts for 7% of all female cancers.

HPV is believed to contain oncogenes, or genes that encode for growth factors that initiate the uncontrolled growth of cells. This uncontrolled proliferation of cells is called cancer. When the HPV oncogenes within an epithelial cell are activated, they cause the epithelial cell to proliferate. In the cervix (the opening of the uterus), the **cell proliferation** manifests first as a condition called cervical neoplasia. In this condition, the cervical cells proliferate and begin to crowd together. Eventually, cervical neoplasia can lead to full-blown cancer.

Polyomaviruses are somewhat mysterious viruses. Studies of blood have revealed that 80% of children aged 5-9 years have antibodies to these viruses, indicating that they have at some point been exposed to polyomaviruses. However, it is not clear what disease this virus causes. Some evidence exists that a mild respiratory illness is present when the first antibodies to the virus are evident. The only disease that is certainly caused by polyomavirses is called progressive multifocal leukoencephalopathy (PML), a disease in which the virus infects specific brain cells called the oligodendrocytes. PML is a debilitating disease that is usually fatal, and is marked by progressive neurological degeneration. It usually occurs in people with suppressed immune systems, such as cancer patients and people with AIDS.

The hepadnaviruses cause several diseases, including hepatitis B. Hepatitis B is a chronic, debilitating disease of the liver and immune system. The disease is much more serious than hepatitis A for several reasons: it is chronic and long-lasting; it can cause cirrhosis and cancer of the liver; and many people who contract the disease become carriers of the virus, able to transmit the virus through body fluids such as blood, semen, and vaginal secretions.

The hepatitis B virus (HBV) infects liver cells and has one of the smallest viral genomes. A double-stranded DNA virus, HBV is able to integrate its genome into the host cell's genome. When this integration occurs, the viral genome is replicated each time the cell divides. Individuals who have integrated HBV into their cells become carriers of the disease. Recently, a vaccine against HBV was developed. The vaccine is especially recommended for health care workers who through exposure to patient's body fluids are at high risk for infection.

Parvoviruses are icosahedral, single-stranded DNA viruses that infect a wide variety of mammals. Each type of parvovirus has its own host. For instance, one type of parvovirus causes disease in humans; another type causes disease in cats; while still another type causes disease in dogs. The disease caused by parvovirus in humans is called erythremia infectiosum, a disease of the red blood cells that is relatively rare except for individuals who have the inherited disorder **sickle cell anemia**. Canine and feline parvovirus infections are fatal, but a vaccine against parvovirus is available for dogs and cats.

Orthomyxoviruses cause influenza ("flu"). This highly contagious viral infection can quickly assume epidemic proportions, given the right environmental conditions. An influenza outbreak is considered an epidemic when more than 10% of the population is infected. Itibodies that are made against one type of rhinovirus are often ineffective against other types of viruses. For this reason, most people are susceptible to colds from season to season.

These helical, enveloped, single-stranded RNA viruses cause pneumonia, croup, measles, and mumps in children. A vaccine against measles and mumps has greatly reduced the incidence of these diseases in the United States. In addition, a paramyxovirus called respiratory syncytial virus (RSV) causes bronchiolitis (an infection of the bronchioles) and pneumonia.

Flaviviruses (from the Latin word meaning "yellow") cause insect-carried diseases including yellow fever, an often fatal disease characterized by high fever and internal bleeding. Flaviviruses are single-stranded RNA viruses.

The two filoviruses, **Ebola virus** and Marburg virus, are among the most lethal of all human viruses. Both cause severe fevers accompanied by internal bleeding, which eventually kills the victim. The fatality rate of Marburg is about 60%, while the fatality rate of Ebolavirus approaches 90%. Both are transmitted through contact with body fluids. Marburg and Ebola also infect primates.

Rhabdoviruses are bullet-shaped, single-stranded RNA viruses. They are responsible for rabies, a fatal disease that affects dogs, rodents, and humans.

Retroviruses are unique viruses. They are double-stranded RNA viruses that contain an enzyme called **reverse transcriptase**. Within the host cell, the virus uses reverse transcriptase to make a DNA copy from its RNA genome. In all other organisms, RNA is synthesized from DNA. Cells infected with retroviruses are the only living things that reverse this process.

The first retroviruses discovered were viruses that infect chickens. The Rous sarcoma virus, discovered in the 1950s by **Peyton Rous** (1879–1970), was also the first virus that was linked to cancer. But it was not until 1980 that the first human **retrovirus** was discovered. Called Human T-cell Leukemia Virus (HTLV), this virus causes a form of leukemia called adult T-cell leukemia. In 1983, another human retrovirus, Human Immunodefiency Virus, the virus responsible for AIDS, was discovered independently by two researchers. Both HIV and HTLV are transmitted in body fluids.

See also Bacterial genetics; Cancer genetics; Epidemiology and genetics; Immunogenetics; Viral genetics

VON MOHL, HUGO (1805-1872)
German botanist

Hugo von Mohl made noteworthy advances in **cell** microscopy, plant microanatomy, and cell physiology. Mohl was born into a prosperous, intellectual family in Stuttgart, Germany. He showed an early predisposition toward both optics and botany, and became an expert microscopist while still a teenager. He earned a medical degree at the University of Tübingen in 1828 with a dissertation on plant pores. After four years of postdoctoral botanical research at Tübingen, at the University of Munich, and in the field, he was professor of physiology at the University of Bern, Switzerland, from 1832 to 1835 and professor of botany at Tübingen thereafter. At Tübingen he also directed the botanical gardens, founded in 1837 the Herbarium Tubingense, which in 2001 contained over a quarter million cataloged specimens, and co-founded in 1843 a major journal, *Botanische Zeitung*. He died peacefully in Tübingen.

With no clique of either students or colleagues, no wife or children, and no serious personal upheavals, Mohl led a singularly uneventful life. He was absolutely dedicated to his work, to which he brought technical skill, imaginative thinking, and rigorous empiricism. In writing, he just reported his observations clearly and directly, and refrained from drawing conclusions. He reacted strongly against German romanticism in general and the nature philosophy of Friedrich Wilhelm Joseph von Schelling (1775–1854) and **Lorenz Oken** (1779–1831) in particular. Both his anti-speculative philosophical stance and his lifelong scientific achievement foreshadowed the positivism that, in the wake of Auguste Comte (1798–1857), would dominate Western European thought throughout the latter half of the nineteenth century.

Mohl may have been the first to distinguish the cell wall from its contents. **Matthias Jacob Schleiden** (1804–1881) and **Carl Wilhelm von Nägeli** (1817–1891) pursued this line of research further. From his studies of algae, Mohl originated the hypothesis that cells reproduce by dividing. His most important discoveries are contained in his long article, *Die vegetabilische Zelle* (The vegetable cell) in Rudolph Wagner's standard 1853 reference work, *Handwörterbuch der Physiologie* (Dictionary of Physiology).

The Czech physiologist Jan Evangelista Purkinje (1787–1869) coined the term "protoplasm" in 1839 to apply to the embryonic substance of egg yolks. In 1846, Mohl used this name for the colloidal and granular complex of living substances in cells. Max Johann Sigismund Schultze (1825–1874) more narrowly defined protoplasm in 1861, but the concept of protoplasm in cytology is forever associated primarily with Mohl.

See also Cell; Cell division; Mitosis; Nucleus

VON NÄGELI, CARL WILHELM (1817-1891)
Swiss botanist

Carl Wilhelm von Nägeli, best known in the history of **genetics** as the scientist whose correspondence with Gregor Mendel (1822–1884) rescued Mendel's important work on **inheritance** from oblivion, also added to the knowledge of cells, sperm, and hybridization. Born in Kilchberg, Switzerland, Nägeli attended the Zürich Gymnasium. Because his father, a physician, intended him for a career in medicine, he entered the University of Zürich as a medical student in 1836, but upon hearing the lectures of **Lorenz Oken** (1779–1851), his interests shifted toward natural science. Influential too in changing Nägeli's plans was **Albert von Kölliker** (1817–1905), his schoolmate at both the gymnasium and the university. Nägeli transferred to the University of Geneva to study botany under Alphonse de Candolle (1806–1893), but received his Ph.D. at Zürich in 1840 with a dissertation on the Swiss **species** of the thistle Cirsium.

Nägeli went to the University of Berlin to study the philosophy of Georg Wilhelm Friedrich Hegel (1770–1831) at a very interesting time, the summer of 1841, when the suppression of Hegelian ideas by Prussian King Friedrich Wilhelm IV (1795–1861) and Minister of Education Johann Albrecht Friedrich Eichhorn (1779–1856), as well as the concomitant underground political activities of the Young Hegelians who eventually contributed to the 1848 Revolutions, were just beginning. Although Nägeli later claimed not to have been swayed by Hegelianism, his writings show a concern for wholeness, context, and ontology that is quite consistent with Hegel's overarching rational system.

From 1842 to 1844 Nägeli continued botanical study at the University of Jena, then toured England in 1845 to collect specimens. From 1844 to 1847 he co-published a journal, *Zeitschrift für wissenschaftliche Botanik* with the vitalist botanist, Matthias Schleiden (1804–1881). From 1845 to 1852 he taught at Zürich, where Carl Eduard Cramer (1831–1901) was among his students. He worked at the University of Freiburg in Breisgau with Cramer as collaborator until 1855, returned briefly to Zürich, then in 1857 accepted a distinguished professorship at the University of Munich, where he spent the rest of his life. With a colleague from Zürich, Simon Schwendener (1829–1919), he vastly improved the botanical gardens and laboratories in Munich. One of his Munich students was **Carl Erich Correns** (1864–1933).

Nägeli was the discoverer of **sperm cells** in ferns. Mendel corresponded with him because he was recognized as the world's leading expert on hybrids. In many ways his **cell** investigations anticipated the work of August Weismann (1834–1914). He famously disagreed with Eduard Strasburger (1844–1912) about **heredity**. Nägeli's "idioplasm" theory held that inheritance occurs in cell material not specific to the **nucleus**, while Strasburger argued that only nuclear **chromatin** performs that function.

See also Cell; Cell division; Heredity; Hybridization of plants; Inheritance; Mendelian genetics and Mendel's laws of heredity; Mitosis; Sperm cells

VON TSCHERMAK-SEYSENEGG, ERICH (1871-1962)
Austrian botanist

Austrian botanist Erich von Tschermak-Seysenegg was an important plant geneticist who applied Gregor Mendel's laws of **heredity** to develop several new disease-resistant crops, including wheat-rye and oat hybrids. Tschermak has also been credited, along with two other botanists, with the discovery in 1900 of missing work done by Mendel. Mendel's work in plant breeding was neither appreciated nor even understood by the scientific community until the rediscovery of his findings and the independent validation of those findings by Tschermak and the two other scientists.

Erich von Tschermak-Seysenegg was born in Vienna, Austria, in 1871, and received his doctorate from the University of Halle, Germany, in 1896. Tschermak accepted a teaching position in Vienna at the Academy of Agriculture in 1901, and became professor there five years later, in 1906.

Prior to joining the staff at the Academy of Agriculture in Vienna, Tschermak conducted breeding experiments with garden peas at the Botanical Garden of Ghent, Belgium, and then in a private garden at the Imperial Family's Foundation at Esslingen, near Vienna. Using 3,580 yellow peas and 1,190 green peas, Tschermak found that the F2 (second filial generation) ratio of dominant to recessive was 3.01:1. In the process of writing up his results, Tschermak discovered a reference to the garden pea experiments of Mendel. More than one hundred copies of Mendel's forty-six page report, "Experiments in Plant Hybridization", had been sent to various scientific libraries around the world, and Tschermak was able to obtain one of the original documents from the library at the University of Vienna. Reading Mendel's work, Tschermak realized that not only did his experiments, like those of Mendel, involve repeat crosses between yellow and green peas, but also that his results were identical to the 1866 results of Mendel.

Tschermak's discovery in 1900 of Mendel's work and duplication of Mendel's results coincided with the independent findings of two other scientists. Dutch botanist **Hugo De Vries** published his results in March 1900, and German botanist Carl Correns presented his report on April 24, 1900; like Tschermak, both botanists had found copies of Mendel's work and referred to it when writing the results of their experiments with plant breeding. Tschermak, De Vries, and Correns fully credited Mendel for first reaching the same conclusions and thus secured the historical impact of Mendel's research on the fields of **genetics** and heredity.

At his teaching post in Vienna, Tschermak's interest lay in the practical application of Mendelian **plant genetics**. Tschermak's innovative work with plant breeding led to pioneering breakthroughs in applied genetics. His contributions include the development of new cultivars in several important crop **species**, and the establishment of plant breeding stations and research facilities for plant genetics. He is likewise credited with establishing plant cultivation as an area of applied genetics. Erich Tschermak died in Vienna in 1962, at the age of 90.

See also Agricultural genetics; Mendelian genetics and Mendel's laws of heredity; Plant breeding and crop improvement

VON WALDEYER-HARTZ, HEINRICH WILHELM GOTTFRIED (1836-1921)

German anatomist

Heinrich Wilhelm Gottfried von Waldeyer-Hartz was a professor of anatomy and histology who coined the word "chromosome" in 1886. Waldeyer, noting the ability of thread-like structures in the **nucleus** to be stained by the dye fuchsin, named them "chromo" meaning color, and "soma", meaning body. Waldeyer also coined the term "neuron." In 1884, he described an area in the pharynx near the tonsils that has come to be known as Waldeyer's tonsillar ring. His contributions were critical to the science of neurology and to the understanding and treatment of **cancer**.

Waldeyer was born in Germany in Hehlen, a small village near Braunschweig. In 1856, at the age of twenty, he attended Göttingen University where he studied mathematics and then medicine, specializing in anatomy. Waldeyer then taught at the university in Königsberg, Germany, for two years and then at the University of Breslau, Germany, for eight years. In 1872, he became professor of anatomy at the University of Strasbourg (at that time in Germany), where he remained for eleven years. On October 1, 1883, Waldeyer joined the faculty at the University of Berlin, Germany, as professor of anatomy and later served as the director of the department of anatomy for more than thirty years. Additionally, Waldeyer served as rector at the University of Berlin from 1889-1899; he retired in October 1916.

Waldeyer's first main contribution to modern science was in 1867, when he postulated that abnormal **cell division** led to cancer. Waldeyer's theory countered the contemporary authority Rudolf Virchow, the German pathologist and founder of cellular pathology. Virchow had written in his three-volume work, *Die Krankhaften Geschwulste* (1863–67), that cancer originated in changes in the connective tissue. Waldeyer's explanation that cancer begins on a cellular level wasn't accepted until 1872, when Virchow's theory on the origin of cancer was finally recognized as erroneous and the legitimacy of Waldeyer's work was confirmed. Waldeyer also explained that the relocation of a single cancer **cell** could lead to the formation of secondary tumors elsewhere in the body. He wrote that early detection and treatment of cancer offered the best cure. His recommendations were especially significant when radiation and chemotherapy became standard cancer treatments.

In 1884 Waldeyer identified a ring of lymphoid tissue formed by the pharyngeal, palatine and lingual tonsils. This tissue, now referred to as Waldeyer's throat ring, forms a protective ring at the opening of the pharynx. The function of Waldeyer's ring is to provide immunity from certain antigens and to protect against oropharyngeal (throat and mouth) cancer. Waldeyer's ring is frequently involved in a form of cancer, termed non-Hodgkin lymphoma, which originates in the lymphatic system.

In 1888, Waldeyer suggested a name for the threads within the nucleus of a cell. German embryologist Oskar Hertwig had discovered the sphere of microscopic threads in the 1860's, and approximately twenty years later Waldeyer named the threadlike structures chromosomes.

Waldeyer's contributions to neurology followed upon the work of Camillo Golgi, an Italian pathologist. Golgi's work on the nervous system, which earned him the Nobel Prize in 1906, had laid the groundwork for further study of the nervous system. After Golgi's discovery of a particular type of nerve cell, now referred to as Golgi cells, Waldeyer was the first to hypothesize that the nervous system was comprised of individual cells. In 1891, he named these cells neurons and thus paved the way for the neuron theory, which is the cell theory of the nervous system.

W

WALLACE, ALFRED RUSSEL (1823-1913)

English naturalist

Alfred Russel Wallace and **Charles Darwin** (1809–1882) were the first two significant evolutionists, each discovering **natural selection** independently of the other, Darwin receiving most of the credit.

Wallace was born into a large, middle-class family in Usk, Monmouthshire, Wales. While still a teenager, he became devoted to the utopian socialism of Robert Owen (1771–1858). He developed an interest in natural history while apprenticed as a surveyor to his older brother William in the late 1830s. William hired him in 1839 but laid him off in 1843. He then taught mathematics, English, drafting, and surveying at the Collegiate School in Leicester until 1845, when he returned to the surveying business.

In 1844, Wallace met **Henry Walter Bates**, with whom he set sail on 25 April 1848 to gather specimens in the Amazon basin. Bates wanted merely to collect and explore, while, in addition to that, Wallace's goal was to find organic evidence for the Owenist theory of social **evolution**. Bates stayed until 1859, but Wallace left in 1852. On the return voyage his ship sank. He managed to survive in a leaky lifeboat for ten days, but lost all his collections. After seventeen months in England and Switzerland, frustrated by having nothing left to study, he set out for the field again.

From 1854 to 1862, Wallace was in Indonesia and Malaysia. In Sarawak in 1855 he wrote a key paper in the history of evolutionary theory, "On the Law Which has Regulated the Introduction of New Species. In 1856, he identified the imaginary "Wallace's line" of Asian/Australian faunal discontinuity between Bali and Borneo to the northwest and Lombok and Sulawesi to the southeast.

In February 1858, Wallace sent Darwin a copy of his recently written and still unpublished paper, "On the Tendency of Varieties to Depart Indefinitely from the Original Type," which outlined his ideas about natural **selection**. Darwin was then writing *The Origin of Species* and was afraid that Wallace might preempt him. He allegedly conspired with the geologist Charles Lyell (1797–1875) and the botanist Joseph Dalton Hooker (1817–1911) to minimize the impact of Wallace's contribution to the theory of evolution. Thus the Linnean Society of London heard both Wallace's and Darwin's views on natural selection at its July 1858 meeting without Wallace's permission.

Some historians of science have speculated that if Wallace had been in London in 1858–1859, then his name, not Darwin's, would be primarily and inextricably linked to the concept of evolution. Even though contemporary scientists generally knew that Wallace and Darwin had separately and almost simultaneously conceived of the idea of natural selection, Wallace did not seem to mind Darwin getting the glory. He and Darwin remained friends.

Wallace's and Darwin's theories of evolution grew apart. In 1864, Wallace presented "The Origin of Human Races Deduced from the Theory of 'Natural Selection'" to the Anthropological Society of London, which indicated that he was beginning to think **Darwinism** too materialist, unable to account for the higher psychological and spiritual facts about humans. In later years Wallace's forays into spiritualism, extraterrestrialism, the anti–vaccination movement, and other unusual enterprises and causes lessened his reputation as a scientist. He died at home in Old Orchard, Broadstone, Dorset.

See also Biodiversity; Evolution; Mimicry; Natural selection; Social Darwinism; Sociobiology

WALLACE, DOUGLAS C. (1946-)

American molecular geneticist

At a time when his colleagues were searching for genes in the nuclear **genome**, Douglas Wallace set himself apart by initiating a research program on mitochondrial genes. The mitochondrion is an organelle in the cytoplasm of all human cells that is known as the "powerhouse" of the **cell** since it is the primary source of the cell's energy. It contains a circular piece of **DNA** that is actually a second human genome. Although it is quite small and contains only genes that are specifically associated with the functions of the mitochondrion itself, this DNA has proven to be very important since mutations in the genes located there have been implicated in a variety of different human diseases. Wallace realized quite early that elucidation of the mitochondrial genome would be as critical to the complete understanding of the human body as were the studies on the nuclear genes.

One of his first tasks was to map the entire mitochondrial genome and identify all of the genes and their functions. The results of this research are contained in a searchable database, Mitomap, that has been established and is maintained by Wallace and colleagues at Emory University. This is an invaluable tool for clinicians and researchers alike who need information on mitochondrial genes. The data are updated on a weekly basis and can be freely accessed via the internet.

A variety of different disorders resulting from mutations in the mitochondrial genes have also been studied and disease-causing mutations have been identified. These include conditions involving vision loss, deafness, muscle weakness including some forms of cardiac disease, diabetes mellitus, and dementias. Identification of the particular causative defects have led to specific molecular tests that can be performed in clinical laboratories to aid in the diagnosis of the diseases. Counseling of patients has also benefited by the better understanding of the mode of **inheritance** and recurrence risks for each disorder.

A review of common features of **mitochondria** related diseases led to the suggestion that the mitochondria might play an important role in aging. Results of studies on the accumulation of mutations in mitochondria and the degeneration and death of mitochondria are being compared to information obtained on human aging to gain insights into the potential involvement of mitochondria in the aging process.

In addition to the disease related studies performed in his laboratory, Wallace has been interested in other applications for **mitochondrial DNA** technology. Interestingly enough, a connection was made with forensic medicine and pathology. Because the mitochondria are found in the cytoplasm, they are passed almost exclusively from mother to child, i.e., the father does not contribute cytoplasm to the **zygote** so cannot contribute his mitochondria. It is therefore possible to trace maternal lineages through the study of mitochondrial DNA. This is similar to the use of DNA from the **Y chromosome** in examining male lineages, except that with mitochondria, there is the added bonus of being able to show a connection between a mother and her sons as well as to her daughters. Because of

this, mitochondrial DNA analysis is being used to reunite mothers with their children in war ravaged areas of the world such as Bosnia. In another sphere, identification of the skeleton of a United States soldier was possible by comparing the mitochondrial DNA still present in his bones with putative female family members. On a larger scale, studies of changes in the mitochondria over time have facilitated research and understanding of how human ancestors may have migrated from Africa to Europe and Asia and eventually to the New World.

Wallace received his Ph.D from Yale University in 1975. He is currently the Robert W. Woodruff Professor of Molecular Genetics and Director of the Center for Molecular Medicine in the Department of Genetics and Molecular Medicine at Emory University, Atlanta, Georgia. As the head of the Center for Molecular Medicine, his responsibilities include overseeing a variety of applications of molecular genetics in addition to his own personal interests in the mitochondria. In the fall of 1999, the Center undertook a project to evaluate samples collected from several Egyptian mummies with the hopes of being able to establish a family relationship among them. Because all of the mummies being studied are male, the DNA of choice is from the Y **chromosome** looking at patrilineal inheritance. This is an interesting twist of fate, since Dr. Wallace's life work has focused primarily on the study of female lineages.

Douglas Wallace is currently the world's expert on mitochondrial DNA, and the laboratory he heads at Emory University is the most advanced site for **molecular diagnostics** of mitochondrial disorders.

See also Mitochondrial DNA; Molecular diagnostics; Y chromosome

WANG, JAMES C. (1936-)

Chinese-American biochemist

James C. Wang is a biochemist who trained as a chemical engineer before turning to biophysical chemistry and molecular biology. Wang discovered **deoxyribonucleic acid** (**DNA**) topoisomerases (or local **enzymes**) and proposed a mechanism for their operation in the 1970s. He also studied the configuration (or topology) of DNA, an approach that proved fruitful in helping to explain how the structure of the **double helix** coils and relaxes.

Wang once noted that his interest in DNA topology (shape and form) came about by chance. His training in engineering and chemistry had led him to study the physical basis of chemical processes. He began to think about the questions raised by the double helix structure of DNA soon after its discovery by molecular biologists **James Watson** and **Francis Crick**. His own study of DNA confirmed the unique structure. But it was not clear how the two tightly intertwined strands could unravel at the speed at which the biochemical processes were thought to occur. Topologically, it did not seem possible for the strands to unravel at all. Wang studied supercoiling in *E. coli* **bacteria** and found that the rotation speed of an unraveling DNA strand is 10,000 revolutions per minute. He also found that the same enzyme, a DNA topoisomerase, is responsible for both breaking and rejoining the DNA strands.

Wang believed that the topological characteristics of the double helix affected all its chemical transformations, including **transcription**, **replication**, and **recombination**. In a 1991 interview in *Cell Science,* Wang remarked that he relied on intuition to further his scientific understanding, particularly when trying to make sense of "bits and pieces of information" that didn't obviously fit together. "There are times," he told the interviewer, "when results that seem to make no sense are key to new advances."

Wang was born in mainland China on November 18, 1936. Less than a year later the Sino-Japanese War began. Wang lost his mother during the conflict, and shortly after it ended, his older sister also died. His father remarried, moving the family to Taiwan in 1949. Because of the war, Wang received only about two years of elementary education before starting junior high school in Taiwan. As a child, he wanted to study medicine, but his father encouraged him to become an engineer. A high school teacher inspired him to follow his interest in chemistry. Chemical engineering became Wang's course of study, and he earned a B.S. in 1959 from National Taiwan University. Continuing his studies in the field of chemistry, he earned a masters from the University of South Dakota in 1961, and a doctorate from the University of Missouri in 1964.

In the same year that Wang received his doctorate, he became a research fellow at the California Institute of Technology, remaining there until 1966. He then taught at the University of California in Berkeley from 1966 until 1977, when he joined the faculty at Harvard University. He was named the Mallinckrodt Professor of Biochemistry and Molecular Biology at Harvard in 1988.

Wang has served on the editorial boards of several scholarly journals, including *Journal of Molecular Biology, Annual Review of Biochemistry,* and *Quarterly Review of Biophysics.* He was a Guggenheim fellow in 1986, and was elected to the U.S. National Academy of Sciences in 1984.

See also Biotechnology; Double helix; Enzyme therapy; Enzymes, genetic manipulation of; Inborn errors of metabolism; Nucleic acid

WASTE CONTROL AND ENVIRONMENTAL CLEANUP

Scientists have known for many years that microorganisms, particularly **bacteria**, are capable of removing contaminants from water, soil, and sediments. This ability of bacteria has been exploited in the development of biologically based treatment systems. The use of microorganisms to recycle organic contaminants and sequester inorganic ions is referred to as **bioremediation**.

Since 1985, these systems have gained recognition as valid alternatives to conventional treatment technologies. **Biotechnology** as a waste control and environmental cleanup

strategy is, in fact, becoming a preferred treatment method over conventional treatments. Conventional treatments include mechanical removal of soils and sediments, vacuuming, absorption, burning, dispersal of surface contaminants on water and high-pressure steam cleaning. These conventional methods are often prohibitively expensive and do not always achieve the treatment standards for acceptable remediation. As more stringent environmental regulations are enacted around the world, the popularity of biologically based control and cleanup strategies increases.

Using bacteria to digest waste and environmentally harmful compounds has several advantages over conventional treatment systems. The costs of remediation are less than conventional treatments. The diversity of microbes available makes the treatment of a wide variety of waste and environments possible. Bacterial degradation of compounds is usually complete under appropriate conditions; toxic compounds are often completely converted into innocuous compounds. Once the bacteria have become established in the particular niche, they tend to persist with little **operator** involvement and maintenance. Finally, using genetically engineered bacteria, which are tailored to degrade certain target compounds, can provide a buffer zone between the region of waste or contamination and the natural environment. The latter has been achieved in some environments by the installation of a so-called bio-barrier, a dike containing genetically engineered bacteria that separates a waste pool from a natural watercourse.

Pseudomonas aeruginosa is a bacterial **species** found in the many environments, including natural environments. The organism has been successfully exploited to degrade a variety of chemicals. For example, *P. aeruginosa* strains can be selected that effectively degrade hydrocarbons and pentachlorophenol (PCP), both of which are particularly persistent and noxious environmental contaminants. Degradation is usually complete, with the treated soil being chemically indistinguishable from untreated, uncontaminated soil.

Other compounds that can be successfully degraded by bacteria are aromatic hydrocarbons (benzene and naphthalene derivatives), aliphatics (paraffins, cyclohexanes, cyclopentanes), creosote, phenolic compounds, crude oil, chlorinated solvents, pesticides, agricultural chemicals such as nitrogen-based fertilizers and compounds found in asphalt.

A newsworthy application of bacteria in environmental cleanup was the remediation following the oil spill from the Exxon Valdez. This tanker ran aground on March 24, 1989, spilling about 11 million gallons of oil into Prince William Sound and eventually, the Gulf of Alaska. Even in the cold Alaskan environment, the degradative action of bacteria resulted in significantly cleaner sites, relative to sites that were not bioremediated.

Biological treatment falls into two categories, biostimulation and bioaugmentation. Biostimulation refers to the addition of specific nutrients to a waste in an effort to stimulate the target microorganisms to grow and begin to digest the contaminant. This approach assumes that the target microorganism is present in the particular location to begin with. The alternate approach, bioaugmentation, involves the introduction of specific genetically tailored bacteria to an environment. The latter approach permits precise control of the type and amount of bacteria that are added to the contaminated site.

A common type of bacterial waste control is the treatment of wastewater. Bacteria use the organic matter in the wastewater as food. The organic material is converted to carbon dioxide and water and energy to make new bacteria. Ultimately, the pollutants are converted into an insoluble mass, which can be removed mechanically and sent for disposal.

Two of the most common general categories of wastewater treatment systems are the once-through lagoon system and the activated sludge system. In a lagoon, the wastewater enters at one point and exits at another point. In between, bacterial decontamination occurs. In an activated sludge system, the resident bacteria act upon the incoming wastewater. As the organic matter in the water declines, the bacteria settle out. The settled bacteria are recycled back into undigested wastewater, where they can continue degradation.

Genetic engineering of bacteria for waste control and environmental cleanup has traditionally been achieved through selective breeding. Exposure of bacterial populations to the target compound will eventually select for bacteria capable of utilizing the compound. These bacteria can then be isolated and their numbers increased to useable levels.

Now, molecular biology is beginning to be used to design bacteria for remediative tasks. Various species of *Pseudomonas* and *E. coli* have had genes inserted into their genomes in order to instill in them the power to specifically degrade compounds such as toluene, chlorinated benzenes and chlorinated phenolics. Recently, researchers have been exploring the engineering of the merA **gene** of *Escherichia coli* into *Deinococcus radiodurans*. The merA gene encodes a protein that converts a toxic form of mercury to a less toxic and volatile form of mercury. *D. radiodurans* is a bacterium that is resistant to radiation. The bio-engineered *D. radiodurans* is capable of degrading mercury in radioactive sites. Other degradative genes are also engineered into *D. radiodurans*. In the United States, thousands of sites are contaminated with radioactive species and toxic heavy metals like mercury. Because of the complexity of the mixed wastes, conventional treatments are too dangerous and expensive to contemplate. The genetically engineered *D. radiodurans* will be useful in the remediation of such sites.

See also Ecological and environmental genetics

WATSON AND CRICK MODEL · *see* DNA, DEVELOPMENT OF THE WATSON-CRICK MODEL

WATSON, JAMES D. (1928-)
American molecular biologist

James D. Watson won the 1962 Nobel Prize in physiology and medicine along with **Francis Crick** and **Maurice Wilkins** for discovering the structure of **DNA**, or **deoxyribonucleic acid**, which is the **carrier** of genetic information at the molecular level. Watson and Crick had worked as a team since meeting in the early 1950s, and their research ranks as a fundamental advance in molecular biology. More than thirty years later, Watson became the director of the **Human Genome Project**, an enterprise devoted to a difficult goal: the description of every human **gene**, the total of which numbers over thirty thousand. This project would not be possible without Watson's groundbreaking work on DNA.

James Dewey Watson was born in Chicago, Illinois, on April 6, 1928, to James Dewey and Jean (Mitchell) Watson. He was educated in the Chicago public schools, and during his adolescence became one of the original Quiz Kids on the radio show of the same name. Shortly after this experience in 1943, Watson entered the University of Chicago at the age of fifteen.

Watson graduated in 1946, but stayed on at Chicago for a bachelor's degree in zoology, which he attained in 1947. During his undergraduate years Watson studied neither **genetics** nor biochemistry—his primary interest was in the field of ornithology. In 1946, Watson spent a summer working on advanced ornithology at the University of Michigan's summer research station at Douglas Lake. During his undergraduate career at Chicago, Watson had been instructed by the well-known population geneticist **Sewall Wright**, but he did not become interested in the field of genetics until he read Erwin Schrödinger's influential book *What is Life?* It was then, Horace Judson reports in *The Eighth Day of Creation: Makers of the Revolution in Biology,* that Watson became interested in finding out the secret of the gene.

Watson enrolled at Indiana University to perform graduate work in 1947. Indiana had several remarkable geneticists who could have been important to Watson's intellectual development, but he was drawn to the university by the presence of the Nobel laureate **Hermann Joseph Muller**, who had demonstrated twenty years earlier that x rays cause **mutation**. Nonetheless, Watson chose to work under the direction of the Italian biologist **Salvador Edward Luria**, and it was under Luria that he began his doctoral research in 1948.

Watson's thesis was on the effect of x rays on the rate of phage lysis (a phage, or **bacteriophage**, is a bacterial **virus**). The biologist **Max Delbrück** and Luria—as well as a number of others who formed what was to be known as "the phage group"—had demonstrated that phages could exist in a number of mutant forms. A year earlier Luria and Delbruck had published one of the landmark papers in **phage genetics**, in which they established that one of the characteristics of phages is that they can exist in different genetic states so that the lysis (or bursting) of bacterial host cells can take place at different rates. Watson's Ph.D. degree was received in 1950, shortly after his twenty-second birthday.

Watson was next awarded a National Research Council fellowship grant to investigate the molecular structure of proteins in Copenhagen, Denmark. While Watson was studying enzyme structure in Europe, where techniques crucial to the study of macromolecules were being developed, he was also attending conferences and meeting colleagues.

From 1951 to 1953, Watson held a research fellowship under the support of the National Foundation for Infantile Paralysis at the Cavendish Laboratory in Cambridge, England. Those two years are described in detail in Watson's 1965 book, *The Double Helix: A Personal Account of the Discovery of the Structure of DNA.* An autobiographical work, *The Double Helix* describes the events—both personal and professional—that led to the discovery of DNA. Watson was to work at the Cavendish under the direction of Max Perutz, who was engaged in the **x-ray crystallography** of proteins. However, he soon found himself engaged in discussions with Crick on the structure of DNA. Crick was twelve years older than Watson and, at the time, a graduate student studying protein structure.

Intermittently over the next two years, Watson and Crick theorized about DNA and worked on their model of **DNA structure**, eventually arriving at the correct structure by recognizing the importance of x-ray diffraction photographs produced by Rosalind Franklin at King's College, London. Both were certain that the answer lay in model-building, and Watson was particularly impressed by Nobel laureate Linus Pauling's use of model-building in determining the alpha-helix structure of protein. Using data published by Austrian-born American biochemist **Erwin Chargaff** on the symmetry between the four constituent nucleotides (or **bases**) of DNA molecules, they concluded that the building blocks had to be arranged in pairs. After a great deal of experimentation with their models, they found that the **double helix** structure corresponded to the empirical data produced by Wilkins, Franklin, and their colleagues. Watson and Crick published their theoretical paper in the journal *Nature* in 1953 (with Watson's name appearing first due to a coin toss), and their conclusions were supported by the experimental evidence simultaneously published by Wilkins, Franklin, and Raymond Goss. Wilkins shared the Nobel Prize with Watson and Crick in 1962.

After the completion of his research fellowship at Cambridge, Watson spent the summer of 1953 at Cold Spring Harbor, New York, where Delbruck had gathered an active group of investigators working in the new area of molecular biology. Watson then became a research fellow in biology at the California Institute of Technology, working with Delbruck and his colleagues on problems in phage genetics. In 1955, he joined the biology department at Harvard and remained on the faculty until 1976. While at Harvard, Watson wrote *The Molecular Biology of the Gene* (1965), the first widely used university textbook on molecular biology. This text has gone through seven editions, and now exists in two large volumes as a comprehensive treatise of the field. In 1968, Watson became director of Cold Spring Harbor, carrying out his duties there while maintaining his position at Harvard. He gave up his faculty appointment at the university in 1976, however, and assumed full-time leadership of Cold Spring Harbor. With John Tooze and David Kurtz, Watson wrote *The Molecular Biology of the Cell,* originally published in 1983.

In 1989, Watson was appointed the director of the Human **Genome** Project of the National Institutes of Health, but after less than two years he resigned in protest over policy differences in the operation of this massive project. He continues to speak

James D. Watson.

out on various issues concerning scientific research and is a strong presence concerning federal policies in supporting research. In addition to sharing the Nobel Prize, Watson has received numerous honorary degrees from institutions, including one from the University of Chicago, which was awarded in 1961, when Watson was still in his early thirties. He was also awarded the Presidential Medal of Freedom in 1977 by President Jimmy Carter. In 1968, Watson married Elizabeth Lewis. They have two children, Rufus Robert and Duncan James.

Watson, *The Double Helix* confirms, has never avoided controversy. His candor about his colleagues and his combativeness in public forums have been noted by critics. On the other hand, his scientific brilliance is attested to by Crick, Delbruck, Luria, and others. The importance of his role in the DNA discovery has been well supported by Gunther Stent—a member of the Delbruck phage group—in an essay that discounts many of Watson's critics through well-reasoned arguments.

Most of Watson's professional life has been spent as a professor, research administrator, and public policy spokesman for research. More than any other location in Watson's professional life, Cold Spring Harbor (where he is still director) has been the most congenial in developing his abilities as a scientific catalyst for others. Watson's work there has primarily been to facilitate and encourage the research of other scientists.

See also DNA structure; Double helix

WATSON-CRICK MODEL · *see* HISTORY OF GENETICS: DEVELOPMENT OF THE WATSON-CRICK MODEL OF DNA

WEINBERG, ROBERT A. (1942-)

American molecular biologist and biochemist

Robert A. Weinberg has made important discoveries in the field of cancer research. Along with colleagues, he produced tumors in healthy mice by transferring individual cancer-causing genes, called oncogenes, to normal cells. These oncogenes were almost indistinguishable from normal genes—in some cases the difference between a normal **gene** and an oncogene was a single **amino acid** along the chain. Weinberg used new forms of genetic engineering to isolate genes in the cells of human tumors. He demonstrated that these oncogenes, when introduced into normal mouse cells grown in a laboratory environment, modified the normal cells and made them cancerous. This work was of great importance to **cancer** research, as it shifted the focus of biomedical research to investigations at the molecular level. Medical researchers had previously thought cancer was caused in several different ways: by chemical carcinogens, **tumor** viruses, and radiation. Weinberg's work with oncogenes made it apparent that normal cells have genes with malignant potential, and that those previously mentioned causes of cancer must be viewed in terms of their effect in activating genes that exist in a dormant state in normal cells.

Robert Allan Weinberg was born in Pittsburgh, Pennsylvania, on November 11, 1942. He was the son of dentist Fritz E. Weinberg and Lore (Reichhardt) Weinberg, who both had escaped Nazi Germany and emigrated to the United States in 1938. Weinberg studied at the Massachusetts Institute of Technology, receiving a B.S. in 1964, an M.A. in 1965, and a Ph.D. in 1969, all in biology. He also worked as an instructor in biology at Stillman College in Tuscaloosa, Alabama, from 1965 to 1966. Upon graduation, Weinberg went to Israel where he spent two years as a research fellow at the Weizmann Institute in Rehovoth, working with Ernest Winocour. He received a fellowship from the Helen Hay Whitney Foundation in 1970, and worked with **Renato Dulbecco** as a fellow of the Salk Institute in LaJolla, California, from 1970 to 1972. Returning to MIT as a research associate in 1972, he began work with **David Baltimore**.

Before the 1970s, scientists had spent much time searching for viruses as the cause of cancer. But in spite of all efforts, researchers could not establish a connection between viruses and the great majority of human cancers. Then cancer research began to take another direction—this was the beginning of the scientific search for oncogenes.

From 1972 to 1973, Weinberg also worked as an assistant professor in the Department of Biology and at the Center for Cancer Research at MIT. He received the resident scholar award from the American Cancer Society from 1974 to 1977. In 1976, he was promoted to associate professor at MIT; he was also designated as Rita Allen Foundation scholar from 1976 to 1980.

By the late 1970s, although about a dozen oncogenes had been identified, no one had managed to prove that these oncogenes could cause cancer without the presence of a **virus** to activate them. Weinberg finally demonstrated this in 1980. Taking **DNA** from active cancer cells in humans, he transferred it into normal mouse cells. The altered cells became cancerous.

Weinberg's next step was to attempt to identify particular genes as the oncogenes associated with specific cancers in humans. In 1981 he and his colleagues were able to identify genes for human leukemia, as well as colon and bladder cancers. Once the molecular basis of carcinogenesis had been established, and once it was possible to isolate and characterize specific oncogenes, it still remained to analyze those genes in exact structural detail, to find out precisely what changes in the DNA had induced the normal **cell** to behave differently.

Weinberg, through **cloning** bladder and lung cancer oncogenes in 1981, proved that the transforming genes are actually present in normal cells, but are either dormant or active at much lower levels, until they are activated or stimulated by some change. This activation might come about as a response to a seemingly insignificant change—in theory, a carcinogen might affect the gene, causing a slightly different amino acid protein to be manufactured along the chain, and somehow disrupting the cell's normal regulatory mechanism. This implied that there might be a two-step system for developing cancer: first, the creation of an **oncogene**, and second, an exposure to a carcinogen of some sort, perhaps even years later. And this two-step system, if accurate, might offer some explanation for the fact that the incidence of cancer increases with advancing age.

In 1982, Weinberg was promoted to professor of biology at MIT. During that same year, he was made Millard Schult lecturer at Massachusetts General Hospital, became a member of the Whitehead Institute for Biomedical Research, and was named 1982 Scientist of the Year by *Discover* Magazine. In 1983, while teaching at MIT, he was awarded the Warren Triennial Prize and the Robert Koch Foundation Medal (Bonn, Germany). In 1984, Weinberg was the recipient of the National Academy of Science's Armand Hammer Cancer Foundation Award and U.S. Steel Foundation Award, as well as the Howard Taylor Ricketts Award from the University of Chicago Medical Center, the Brown-Hazen Award from the New York State Department of Health, and the Antonio Feltrinelli Prize from Academia Lincei in Rome. He also received the Bristol-Meyer Award for Distinguished Achievement in Cancer Research, and was awarded an honorary doctorate by Northwestern University.

In 1985 Weinberg was made honorary professor of biology by the American Cancer Society. He received the Katherine Berkann Judd Awardfrom the Memorial Sloan-Kettering Cancer Center in 1986, the Sloan Prize from General Motors Cancer Research Foundation in 1987, and was made a member of the MIT Medical Consumers Advisory Committee in 1988, at which time he was also the recipient of an honorary doctorate from State University of New York at Stonybrook.

The next year, 1989, brought Weinberg yet another honorary doctorate, this time from City University, New York. He became a member of the Committee on **Biological Warfare**, Federation of American Scientists, and received the Lucy Wortham James Award from the Society of Surgical Oncologists. In 1990, Weinberg was awarded the Research Recognition Award by the Samuel Roberts Noble Foundation, the distinguished basic scientist award by the Milken Family Medical Foundation, and the Lila Gruber cancer research award by the American Academy of Dermatology.

Weinberg currently serves as the Daniel K. Ludwig Professor for Cancer Research at MIT.

See also Cancer; Cancer genetics; Proto-oncogenes

WEISSMANN, AUGUST (1834-1914)
German biologist

August Weissmann was a German biologist who is best known for his contributions to evolutionary theory, specifically for his germ-plasm theory of **heredity**. Weissmann was born in Frankfurt, Germany. At the age of 18, he began his study of medicine at Göttingen University which he attended until 1856. After completing his education, Weissmann practiced medicine until 1863, at which time he turned his attention to the study of biology and zoology. In 1867, Weissmann became a professor of zoology at the University of Freiburg. He first served at the medical school associated with the university and later at a new Institute of Zoology. Weissmann remained in this capacity until 1912.

During this time Weissmann explored a variety of areas, most notably the **embryology** of insects and crustaceans. He is especially noted for his evaluations of the two-winged flies, the Diptera. He also conducted studies of Hydrozoa (a class of invertebrates which include jellyfish) that led to the development of his famous germ-plasm theory. In this theory, Weissmann proposed that the genetic code for an organism is contained in the germ cells (the egg and sperm), which explains how the information can be transmitted unchanged from generation to generation. In accordance with this theory, Weissmann favored the **natural selection** ideas championed by Charles Darwin, who believed that specific inherited characteristics are passed from generation to generation. An opposing theory, supported by Jean-Baptiste Lamarck and others of the time, postulated that organisms acquired physical characteristics through exposure to their environment, and that they passed on these characteristics to their offspring. Weissmann was the first scientists to conduct experiments to disprove that acquired characteristics can be transmitted from parents to offspring. He did this using an experiment where he cut off the tails of mice for five consecutive generations. He then observed the offspring to determine if this tail-less characteristic was passed on to subsequent generations. These studies confirmed that such environmental stimuli are not passed on in a hereditary fashion. In other studies, Weissmann noted that the process of **meiosis** involved some form of reduction division because the genetic material did not double when the **cell** replicated.

Weissmann published his theories in a series of essays that were translated as *Essays upon Heredity and Kindred Biological Problems* which were published between 1889 and 1892. In 1902, he published *Vorträgeuber Descendenztheorie* which is still considered to be a valuable contribution to evolutionary theory. He was the first to provide a detailed explanation of the neuro-humoral organ that was later named after him, the Weissmann ring. Weissmann died at age 80 in 1914.

See also Darwinism; Evolution; Evolutionary mechanisms; History of genetics: ancient and classical views of heredity

WESTERN BLOTTING ANALYSIS · *see*
BLOTTING ANALYSIS

WEXLER, NANCY SABIN (1945-)
American neuropsychologist

Nancy Wexler's research on Huntington's disease has led to the development of a presymptomatic test for the condition, as well as the identification of the genes responsible for the disease. The symptoms of this fatal, genetically based disorder (for which Wexler herself is at risk) usually appear around middle age, and the disease leads to the degeneration of mental, psychological, and physical functioning. For her pivotal role in these achievements, Wexler was granted the Albert Lasker Public Service Award in 1993.

Nancy Sabin Wexler was born to Milton Wexler, a Los Angeles psychoanalyst, and Leonore Sabin Wexler. She studied social relations and English at Radcliffe and graduated in 1967. Wexler subsequently traveled to Jamaica on a Fulbright scholarship and studied at the Hampstead Clinic Child Psychoanalytic Training Center in London.

In 1968, Wexler learned that her mother had developed the symptoms of Huntington's disease, a condition to which Wexler's maternal grandfather and three uncles had already succumbed. Efforts to fight the disease became a primary mission for Wexler and her family: Her father founded the Hereditary Disease Foundation in 1968, and Wexler herself, who was then entering the doctoral program in clinical psychology at the University of Michigan, eventually wrote her doctoral thesis on the "Perceptual-motor, Cognitive, and Emotional Characteristics of Persons at Risk for Huntington's Disease," and received her Ph.D. in 1974.

After graduating from University of Michigan, Wexler taught psychology at the New School for Social Research in New York City and worked as a researcher on Huntington's disease for the National Institutes of Health (NIH). In 1976, she was appointed by congress to head the NIH's Commission for the Control of Huntington's Disease and it's Consequences. In 1985, Wexler joined the College of Physicians and Surgeons at Columbia University.

Nancy Wexler.

In 1979, Wexler's research led her to Lake Maracaibo in Venezuela, where she studied a community with a high incidence of Huntington's disease. Wexler kept medical records, took blood and skin samples, and charted the transmission of the disease within families. Wexler sent the samples she collected to geneticist **James Gusella** at Massachusetts General Hospital, who used the blood samples to conduct a study to locate the gene—the first such **genetic mapping** of a disease. Gusella eventually discovered a deoxyribonucleic acid(DNA)marker close to the Huntington's **gene**. Based on this study, Gusella introduced a test that was ninety-six percent accurate in detecting whether an individual bears the Huntington's gene. Because there was still no cure for the Huntington's disease, the test proved to be controversial, raising many issues involving patient rights, childbearing decisions, and discrimination by employers and insurance companies. In her interviews and writings, Wexler has stressed the importance of keeping such genetic information confidential.

In 1993, the Huntington's gene was identified through research based on the Venezuelan blood samples and the work of the Huntington's Disease Collaborative Research Group. In October, 1993, Wexler received an Albert Lasker Public Service Award for her role in this effort. In addition, she has served as an advisor on social and medical ethics issues to the Human Genome Project—a massive international effort to map and identify the approximately 30,000–40,000 genes in the human body. Wexler also has assumed directorship of the Hereditary Disease Foundation founded by her father, to

which she donated the honorarium that accompanied the Lasker Award.

See also Genetic mapping; Genetic marker; Neurological and opthalmological genetics

WIESCHAUS, ERIC FRANCIS (1947-)

American molecular biologist

Eric Wieschaus, by studying the fruit fly *Drosophila melanogaster,* made important discoveries concerning genetic mechanisms of control of early embryonic development, For this research, Wieschaus, along with colleagues Christiane Nüsslein-Volhard and **Edward B. Lewis** received the 1995 Nobel Prize in physiology or medicine.

Wieschaus was born in South Bend, Indiana, but grew up in Alabama. He received his bachelor's degree in biology from the University of Notre Dame in 1969 and his doctorate from Yale in 1974. His doctoral dissertation involved using genetic methods to label the progeny (offspring) of single cells in fly embryos. He showed that even at the earliest cellular stages, cells were already determined to form specific regions of the body called segments.

Wieschaus began his Nobel-winning work in the latter part of the 1970s. The Alabama native spent three years with Christiane Nüsslein-Volhard in the European Molecular Biology Lab at the University of Heidelberg, Germany, tackling the question of why individual cells in a fertilized egg develop into various specific tissues. They elected to study *Drosophila,* or fruit flies, because of their extremely fast embryonic development. New generations of fruit flies can be bred in a week. In addition, fruit flies have only one set of genes controlling development compared to the four sets humans possess. This means that testing each fruit fly **gene** individually takes a quarter of the time it would involve to test human genes.

To begin their experiment, Nüsslein-Volhard and Wieschaus damaged male fruit fly **deoxyribonucleic acid (DNA)** by applying ultraviolet light to the genes or by feeding the flies sugar water laced with chemicals. Then the team "knocked out" one gene from the fly, breeding generations of fruit flies without that particular piece of code. In this way, Nüsslein-Volhard and Wieschaus were able to isolate all the genes crucial to the early stages of embryonic development. When the flies were bred, the females produced dead embryos. These lifeless embryos resulted from only 150 different mutations of the 40,000 mutations applied. These 150 genes proved to be essential to the proper development of the fly embryo because, when damaged, the genes caused extraordinary deformities that killed the embryo. By viewing the fly embryos with a two-person microscope, Wieschaus and Nüsslein-Volhard were able to simultaneously view and classify a large quantity of malformations caused by gene mutations. Next, they identified 15 different genes, that, when mutated, eliminate specific body segments in the fly embryos. Wieschaus also established that systematic categorizing of genes that control the various stages of development could be accomplished.

Their first research results reported that the number of genes controlling early development was not only limited, but could also be classified into specific functional groups. They also identified genes that cause severe congenital defects in flies. After additional experimentation, the principles involved with the fruit fly genes were found to apply to higher animals and humans. This led to the realization that many similar genes control human development, and this finding could have a tremendous impact on the medical world. The applications of their research extend to **in vitro fertilization**, identifying **congenital birth defects**, and increased knowledge of substances that can endanger early stages of pregnancy.

It wasn't until 1995, however, that Wieschaus won the Nobel Prize in Physiology or Medicine, along with Edward B. Lewis and Christiane Nüsslein-Volhard, for his work on identifying key genes that make a fertilized fruit fly egg develop into a segmented embryo. His research could help improve knowledge of how genes control embryonic development in higher organisms, including identifying genes that cause human **birth defects**.

See also Developmental genetics

WIGLER, MICHAEL (1947-   )
American geneticist

Michael Wigler made a series of important discoveries that had major impact on the progress of modern molecular biology and **genetics** during the last three decades. He participated in the early experiments that allowed the transfer and stable integration of **exogenous DNA** into the **genome** of eukaryotic cells as well the preparation of enucleated cells. These early experiments provided a useful model system for studies on the transfer of genes to mammalian cells in culture.

In 1981, Wigler discovered, together with geneticists Robert Weinberg and Geoffrey Cooper, the first human **cancer gene** or **oncogene**, (i.e., the *ras* gene). The function of this gene was studied in great detail in a **yeast** model system that was developed in Wigler's laboratory a few years later. *Ras* was shown to be a guanine trinucleotide (GTP) dependant molecule with important function in various biological signaling processes in the **cell**. The role of *ras* in signal **transduction** as well the interaction of oncogenes with a class of cancer related genes called **tumor suppressor genes** or anti-oncogenes were also investigated in Wigler's laboratory.

Together with Clark Still of Columbia University, Wigler developed the powerful method of encoded combinatorial synthesis. This method, which is commercially available, helped accelerate the design, testing and analysis of novel drug candidates and small synthetic peptides that can be used as inhibitors/regulators of biologically active biomolecules like *ras*.

Wigler, together with Nikolai Lisitsyn, developed genomic representational difference analysis (RDA), a powerful genetic technology that allows the identification in the genomes of different cells and tissues. Wigler and colleagues identified several oncogenes and **tumor** suppressor genes using RDA. Several other groups who identified novel genes and even whole viruses have also used this technique. The method has been modified such that, besides the identification of genomic **DNA** variation, it now allows the comparison of differences in the **mRNA** expression profile of different cells and tissues. The research in Wigler laboratory is now focused on molecular genetics of cancer. Applying the RDA technology, the laboratory identified several cancer related genetic lesion and many oncogenes and tumor suppressor genes.

Born in New York, Michael Wigler got a Bachelor of Arts degree from Princeton in 1970, a Masters degree at Rutgers in 1972, and a PhD in microbiology from Columbia University in 1978. He held several positions at Columbia University and was awarded the Pfizer Biomed Award, the Lifetime Research Professor Award of the American Chemical Society and the National Institute of Health (NIH) Outstanding Investigator Award. In 1989, Wigler became a member of the National Academy of Sciences, at the age of 42. He is now professor at the Cold Spring Harbor Laboratory in New York.

See also Cancer genetics; Oncogene; Proto-oncogenes; Tumor suppressor genes

WILD TYPE

In descriptions of a given population of organisms, the alleles most commonly present (genotype), or the most commonly expressed set of characteristics (phenotype) is referred to as the wild type. Because the wild type (also often printed in a hyphenated form as "wild-type") represents the most common genotype, alleles that are not a part of that genotype are often considered mutant alleles. Accordingly, the designation of wild type is based upon a quantitative (numerical) representation or estimation of the norm (normal) or standard in a population. Accordingly, a wild-type allele the standard allele (form of a **gene**) for a particular organism.

For example, one of the first descriptions of a wild type gene was made with reference to the *Drosophila* fruit fly. In early studies of genetic traits of *Drosophila,* the American geneticist **Thomas Hunt Morgan** noted a white-eyed fly in an isolated breeding population of red-eyed *Drosophila* flies (the flies were isolated in a bottle). Because the vast majority of *Drosophila* have red eyes, Morgan considered the white-eyed fly a mutant and termed the gene for red eyes in *Drosophila* the wild type gene.

Initial forms of **gene therapy** were essentially gene replacement therapies that sought to introduce complete copies of the relevant wild type gene into the organism suffering from a genetic disease. The theory was that a wild type gene, introduced via an appropriate vector, would allow for wild type (normal) **gene expression**. In the case of an enzyme deficiency, for example, such an introduction of the wild type gene for the enzyme would allow the **cell** to produce the needed enzyme.

Outside of strict reference to genotype or phenotype, the term wild type is also used to denote the natural state of an organism, or the natural life cycle of an organism. When wild type is used to describe an entire organism the sub-population of most prevalent phenotypes with the population is often referred to as the wild type strain.

Geneticists use a variety of symbols and type scripts (capitals, italics, etc.) to denote wild-type alleles of a gene. One method commonly used indicates a wild type gene by the presence of a plus sign (+). Most often, this symbol is used as a superscript next to the notation for the allele. For example, the notation Pax1+ denotes the wild type allele of a Pax1 gene in mice that is the most prevalent allele for the gene. In contrast, when an organism undergoes a **mutation** that reverts the gene back to the wild type the plus sign is associated with a superscripted allele symbol. Addition reversions are usually identified by numbers preceding the allele in question. Geneticists also often use the letter "w" to denote the wild type gene. In the case of *Drosophila* the allele for red eyes is often designated by the letter "w" or the plus sign

A revertant is a mutation that restores the phenotype to the wild type (most prevalent form). In a true revertant, the original mutation itself is mutated back to the original wild type. With pseudo-revertants, or with pseudo reversions, the original mutation remains while another mutation that takes place within the same gene restores the wild type phenotype. In the case of *Drosophilae* a revertant would restore red eyes to the fly regardless whether it was a true revertant or a pseudo revertant.

See also Adaptation and fitness; Allele frequency; Chromosome mapping and sequencing; Gene dose; Gene frequency; Hardy-Weinberg equilibrium; Immunogenetics; Population genetics

WILKINS, MAURICE HUGH FREDERICK
(1916-)
New Zealand English biophysicist

Maurice Hugh Frederick Wilkins is best known for the assistance he provided to molecular biologists **James D. Watson** and **Francis Crick** in their quest to uncover the structure of **deoxyribonucleic acid** (**DNA**), the genetic blueprint of **heredity** in humans and many other organisms. Specifically, Wilkins' contribution to their discovery involved discerning the structure of DNA through the use of X-ray diffraction techniques. For his efforts, Wilkins shared the 1962 Nobel Prize in physiology or medicine with Watson and Crick.

Wilkins was born in Pongaroa, New Zealand to Irish immigrants Edgar Henry, a physician, and Eveline Constance Jane (Whittaker) Wilkins. Euperior education began at an early age for Wilkins, who began attending King Edward's School in Birmingham, England, at age six. He later received his B.A. in physics from Cambridge University in 1938. After

graduation, he joined the Ministry of Home Security and Aircraft Production and was assigned to conduct graduate research on radar at the University of Birmingham. Wilkins' research centered on improving the accuracy of radar screens.

Soon after earning his Ph.D. in 1940, Wilkins, still with the Ministry of Home Security, was relocated to a new team of British scientists researching the application of uranium isotopes to atomic bombs. A short time later Wilkins became part of another team sent to the United States to work on the Manhattan Project —the military effort to develop the atomic bomb—with other scientists at the University of California at Berkeley. He spent two years there researching the separation of uranium isotopes.

Wilkins' interest in the intersection of physics and biology emerged soon after his arrival to the United States. He was significantly influenced by a book by Erwin Schrödinger, a fellow physicist, entitled *What is Life? The Physical Aspects of the Living Cell*. The book centers on the possibility that the science of quantum physics could lead to the understanding of the essence of life itself, including the process of biological growth. In addition to Schrödinger's book, the undeniable and undesirable ramifications of his work on the atomic bomb also played a role in Wilkins' declining interest in the field of nuclear physics and emerging interest in biology.

After the war, the opportunity arose for Wilkins to begin a career in biophysics. In 1945, Wilkins' former graduate school professor, Scottish physicist John T. Randall, invited him to become a physics lecturer at St. Andrews University, Scotland, in that school's new biophysics research unit. Later, in 1946, Wilkins and Randall moved on to a new research pursuit combining the sciences of physics, chemistry and biology to the study of living cells. Together they established the Medical Research Council Biophysics Unit at King's College in London. Wilkins was, for a time, informally the second in command. He officially became deputy director of the unit in 1955 and was promoted to director in 1970, a position he held until 1972.

It was at this biophysics unit, in 1946, that Wilkins soon concentrated his research on DNA, shortly after scientists at the Rockefeller Institute (now Rockefeller University) in New York announced that DNA is the constituent of genes. Realizing the enormous importance of the DNA molecule, Wilkins became excited about uncovering its precise structure. He was prepared to attack this project by a number of different methods. However, he fortuitously discovered that the particular makeup of DNA, specifically the uniformity of its fibers, made it an excellent specimen for x-ray diffraction studies. X-ray diffraction is an extremely useful method for photographing atom arrangements in molecules. The regularly-spaced atoms of the molecule actually diffract the x rays, creating a picture from which the sizing and spacing of the atoms within the molecule can be deduced. This was the tool used by Wilkins to help unravel the structure of DNA.

Physical chemist Rosalind Franklin joined Wilkins in 1951. Franklin, who had been conducting research in Paris, was adept in X-ray diffraction. Although their personal relationship was not ideal, (Franklin was more outgoing whereas

Wilkins was a quiet, non-confrontational person), together they were able to retrieve some very high quality DNA patterns. One initial and important outcome of their research was that phosphate groups were located outside of the structure, which overturned **Linus Pauling** 's theory that they were on the inside. In another important finding, Wilkins thought the photographs suggested a helical structure, although Franklin hesitated to draw that conclusion. Subsequently, Wilkins, some say unbeknownst to Franklin, passed on to Watson one of the best X-ray pictures Franklin had taken of DNA. These DNA images provided clues to Watson and Crick, who used the pictures to solve the last piece of the **DNA structure** puzzle.

Consequently, in 1953, Watson and Crick were able to reconstruct the famous double-helix structure of DNA. Their model shows that DNA is composed of two strands of alternating units of sugar and phosphate on the outside, with pairs of bases—including the molecular compounds adenine, thymine, guanine, and cytosine—inside, bonded by hydrogen. It is important to note that while Wilkins' contribution to the discernment DNA's structure is undeniable, controversy surrounds the fact that Franklin was not recognized for this scientific breakthrough, particularly in terms of the Nobel Prize. Some feel that Franklin, who died of cancer in 1958 (the Nobel Prize is only awarded to living scientists), did not receive due recognition, whereas others maintain that it was solely Watson's ability to discern the structure in Franklin's photograph that made possible the discovery of the DNA structure.

The knowledge of the DNA structure, which has been described as resembling a spiral staircase, has provided the impetus for advanced research in the field of **genetics**. For example, scientists can now determine predispositions for certain diseases based on the presence of certain genes. Also, the exciting but sometimes controversial area of genetic engineerng has developed.

Wilkins, Watson, and Crick were awarded the 1962 Nobel Prize for physiology or medicine for their work which uncovered the structure of hereditary material DNA. After winning the Nobel Prize, Wilkins focused next on elucidating the structure of ribonucleic acids (**RNA**)—a compound like DNA associated with the control of cellular chemical activities—and, later, nerve **cell** membranes. In 1962 he was able to show that RNA also had a helical structure somewhat similar to that of DNA. Besides his directorship appointments at the Medical Research Council's Biophysics Unit, Wilkins was also appointed director of the Council's Neurobiology Unit, a post he held from 1974 to 1980. Additionally, he was a professor at King's College, teaching molecular biology from 1963 to 1970 and then biophysics as the department head from 1970 to 1982. In 1981, he was named professor emeritus at King's College. Utilizing some of his professional expertise for social causes, Wilkins has maintained membership in the British Society for Social Responsibility in Science (of which he is president), the Russell Committee against Chemical Weapons, and Food and Disarmament International.

Wilkins is an honorary member of the American Society of Biological Chemists and the American Academy of Arts and Sciences. He was also honored with the 1960 Albert Lasker Award of the American Public Health Association (given jointly to Wilkins, Watson, and Crick), and was named Fellow of the Royal Society of King's College in 1959.

Wilkins, known to be a quiet and polite man, married Patricia Ann Chidgey in 1959. The couple have four children, two sons and two daughters.

See also Chromosome structure and morphology; DNA (deoxyribonucleic acid); DNA structure; DNA replication; Double helix

WILLIAMS SYNDROME

Williams syndrome is a genetic disorder caused by a deletion of a series of genes on **chromosome** 7q11. Individuals with Williams syndrome have distinctive facial features, mild mental retardation, heart and blood vessel problems, short stature, unique personality traits and distinct learning abilities and deficits.

Williams syndrome, also known as Williams Beuren syndrome, was first described in 1961 by Dr. J.C.P. Williams of New Zealand. At that time it was noted that individuals with Williams syndrome had an unusual constellation of physical and mental findings. The physical features include a characteristic facial appearance, heart and cardiovascular problems, high blood calcium levels, low birth weight, short stature and other connective tissue abnormalities. The intellectual problems associated with Williams include a mild mental retardation and a specific cognitive profile. That is, individuals with Williams syndrome often have the same pattern of learning abilities and disabilities, as well as many similar personality traits. The findings in Williams syndrome are variable—that is, not all individuals with Williams syndrome will have all of the described findings. In addition to being variable, the physical and mental findings associated with Williams syndrome are progressive and change over time.

Williams syndrome is a genetic disorder due to a deletion of chromosome material on the long arm of chromosome 7. A series of genes are located in this region. Individuals with Williams syndrome may have some or all of these genes deleted. Because of this, Williams syndrome is referred to as a contiguous **gene** deletion syndrome. Contiguous refers to the fact that these genes are arranged next to each other. The size of the deletion can be large or small which may explain why some individual with Williams syndrome are more severely affected than others. If you think of these genes as the letters of the alphabet, some individuals with Williams syndrome are missing A to M, some are missing G to Q and others are missing A to R. While there are differences in the amount of genetic material that can be deleted, there is a region of overlap. Everyone in the above example was missing G to M. It is thought that the missing genes in this region are important causes of the physical and mental findings of Williams syndrome.

Two genes in particular, ELN and LIMK1, have been shown to be important in causing some of the characteristic symptoms of Williams syndrome. The ELN gene codes for a protein called elastin. The job of elastin in the human body is to provide elasticity to the connective tissues such as those in the arteries, joints and tendons. The exact role of the LIMK1 gene is not known. The gene codes for a substance known as lim kinase 1 that is active in the brain. It is thought that the deletion of the LIMK1 gene may be responsible for the visuospatial learning difficulties of individuals with Williams syndrome. Many other genes are known to be in the deleted region of chromosome 7q11 responsible for Williams syndrome and much work is being done to determine the role of these genes in Williams syndrome.

Williams syndrome is an autosomal dominant disorder. Genes always come in pairs and in an autosomal dominant disorder, only one gene need be missing or altered for an individual to have the disorder. Although Williams syndrome is an autosomal disorder, most individuals with Williams syndrome are the only people in their family with this disorder. When this is the case, the chromosome deletion that causes Williams syndrome is called *de novo*. A *de novo* deletion is one that occurs for the first time in the affected individual. The cause of *de novo* chromosome **deletions** is unknown. Parents of an individual with Williams syndrome due to a *de novo* deletion are very unlikely to have a second child with William syndrome. However, once an individual has a chromosome deletion, there is a fifty percent chance that they will pass it on to their offspring. Thus individuals with Williams syndrome have a 50% chance of passing this deletion (and Williams syndrome) to their children.

Williams syndrome occurs in 1/20,000 births. Because Williams syndrome is an autosomal dominant disorder, it affects an equal number of males and females. It is thought that Williams syndrome occurs in people of all ethnic backgrounds equally.

Williams syndrome is a multi-system disorder. In addition, to distinct facial features individuals with Williams syndrome can have cardiovascular, growth, joint, and other physical problems. They also share unique personality traits and have intellectual differences. Infants with Williams syndrome are often born small for their family and 70% are diagnosed as failure to thrive during infancy. These growth problems continue throughout the life of a person with Williams syndrome and most individuals with Williams syndrome have short stature (height below the third percentile). Infants with William syndrome can also be extremely irritable and have "colic-like" behavior. This behavior is thought to be due to excess calcium in the blood (hypercalcemia). Other problems that can occur in the first years include strabismus (crossed eyes), ear infections, chronic constipation and eating problems.

Individuals with Williams syndrome can have distinct facial features sometimes described as "elfin" or "pixie-like." While none of these individual facial features are abnormal, the combination of the different features is common for Williams syndrome. Individuals with Williams syndrome have a small upturned nose, a small chin, long upper lip with a wide mouth, small widely spaced teeth and puffiness around the eyes. As an individual gets older, these facial features become more pronounced.

People with Williams syndrome often have problems with narrowing of their heart and blood vessels. This is thought to be due to the deletion of the elastin gene and is called elastin arteriopathy. Any artery in the body can be affected but the most common narrowing is seen in the aorta of the heart. This condition is called supravalvar aortic stenosis (SVAS) and occurs in approximately 75% of individuals with Williams syndrome. The degree of narrowing is variable. If left untreated, it can lead to high blood pressure, heart disease and heart failure. The blood vessels that lead to the kidney and to other organs can also be affected.

Deletions of the elastin gene are also thought to be responsible for the loose joints of some children with Williams syndrome. As individuals with Williams syndrome age, their heel cords and hamstrings tend to get tight which can lead to a stiff awkward gait and curving of the spine.

Approximately 75% of individuals with Williams syndrome have mild mental retardation. They also have a unique cognitive profile (unique learning abilities and disabilities). This cognitive profile is independent of their I.Q. Individuals with Williams syndrome generally have excellent language and memorization skills. They can have extensive vocabularies and can develop a thorough knowledge of a topic of interest. Many individuals are also gifted musicians. Individuals with Williams syndrome have trouble with concepts that rely on visuospatial ability. Because of this, many people with Williams syndrome have trouble with math, writing and drawing.

The diagnosis of Williams syndrome is usually made by a physician familiar with Williams syndrome and based upon a physical examination of the individual and a review of their medical history. It is often made in infants after a heart problem (usually SVAS) is diagnosed. In children without significant heart problems, the diagnosis may be made after enrollment in school when they are noted to be "slow-learners." While a tentative diagnosis can be made based upon physical examination and medical history, the diagnosis is confirmed by a **DNA** test.

Because Williams syndrome is caused by a deletion of genetic material from the long arm of chromosome 7, a specific technique called fluorescent *in situ* hybridization testing, or **FISH** testing, can determine whether there is genetic material missing. A FISH test will be positive (detect a deletion) in over 99% of individuals with Williams syndrome. A negative FISH test for Williams syndrome means that no genetic material is missing from the critical region on chromosome 7q11.

Prenatal testing (testing during pregnancy) for Williams syndrome is possible using the FISH test on DNA sample obtained by chorionic villi sampling (CVS) or by **amniocentesis**. Chorionic villi sampling is a prenatal test that is usually done between 9 to 11 weeks of pregnancy and involved removing a small amount of tissue from the placenta. Amniocentesis is a prenatal test that is usually performed between 16-19 weeks of pregnancy and involves removing a small amount of the amniotic fluid that surrounds the fetus. DNA is obtained from these samples and tested to see if the deletion responsible for Williams syndrome is present. While prenatal testing is possible, it is not routinely performed. Typically, the test is only done if there is a family history of Williams syndrome.

See also Chromosomal mutations and abnormalities; Genetic disorders and diseases; Hereditary diseases; Inherited cancers; Mutations and mutagenesis

WILMUT, IAN (1944-)
English embryologist

Ian Wilmut and his associates were the first to clone a mammal from the fully differentiated cells of an adult animal, and thus, thrust the practical concepts and ethics of **cloning** into scientific and public debate.

Wilmut was born in Hampton Lucey, Warwick, England. He attended the University of Nottingham, where he became fascinated with **embryology** after meeting G. Eric Lamming, a world-renowned expert in reproduction. The meeting became a turning point for Wilmut, who set out on a singular quest—to understand the genetic engineering of animals. Wilmut graduated from Nottingham in 1967, with a degree in agricultural science.

Wilmut continued his studies at Darwin College at Cambridge University in England. There he received his doctoral degree in 1973, awarded after he completed his thesis on the techniques for freezing boar semen. Wilmut immediately took a position at the Animal Breeding Research Station, an animal research institute supported by government and private funds. The research station eventually became the Roslin Institute. It is headquartered in Roslin, near Edinburgh, Scotland.

In 1973, after receiving his doctorate, Wilmut produced the first calf ("Frosty") born from a frozen embryo that had been implanted into a surrogate mother. The motivation for such an experiment was to harvest cows that provide the best meat and milk by implanting their embryos into other females. The average cow can birth five to ten calves during their lifespan. With the ability to transfer embryos, cattle breeders could increase the quality of their animal stock.

Wilmut continued his research during the 1980s, despite other scientists' growing discouragement in the possibility of cloning. In 1996, Wilmut overheard a story in an Irish bar, while attending a scientific meeting, that solidified his interest in cloning. The story concerned a researcher in Texas who had cloned a lamb using a differentiated **cell** from an already developing embryo.

Like a fertilized egg that contains enough **deoxyribonucleic acid (DNA)** to build an entire organism, a differentiated cell carries a full complement of the genetic material for DNA, which forms a blueprint for an animal's characteristics. To clone an animal, an adult animal cell would have to be harvested, and the **nucleus** placed in an embryo cell, thereby replacing the nucleus of the embryo cell. The problem Wilmut

Ian Wilmut. *AFP/CORBIS. Reproduced by permission.*

considered was how to get the new nucleus to spawn growth in the embryo cell.

Keith Campbell, a biologist at the Roslin Institute, had an insight that proved to be crucial. Campbell deduced that an egg probably will not use genetic material from a transplanted adult cell because the cycles of each cell are not synchronized. Cells go through specific cycles, growing and dividing and making an entirely new package of chromosomes each time. In order to synchronize the cells, Campbell slowed down adult mammal cells—in fact, nearly stopping them—so they would actually exist in synchrony with the embryos. Then each embryo could be joined with an adult cell, and in turn, they could join together and grow. To slow an already developing or adult cell down, Campbell forced it into a hibernating state by depriving it of nutrients. With this method, Campbell and Wilmut were able to clone two sheep from developing embryo cells. They named the sheep Megan and Morag.

To clone an adult sheep, Wilmut and Campbell harvested udder, or mammary, cells from a six-year-old ewe. The cells were preserved in test tubes and starved by reducing their serum concentration for five days. Out of 277 attempts, 29 embryos survived to the stage where implantation was possible. The two scientists implanted an embryo when it was six days old into a surrogate mother, and on July 5, 1996, Dolly (named after country singer Dolly Parton) was born. In late November 1997, Dolly was successfully mated with a four-year-old Welsh Mountain ram (David) and her lamb (Bonnie) was born on the morning of April 13, 1998. The birth of the lamb confirmed that while Dolly's embryonic origins were unique, she was able to breed normally and produce a healthy offspring.

Wilmut remains passionate about his work, asserting the necessity of animal cloning research, and its eventual applications for medicine.

See also Clones and cloning

WILSON, ALLAN C. (1934-1991)
New Zealander biochemist

Allan Wilson, an evolutionary biochemist was born in Ngaruawahia, New Zealand in 1934. Wilson earned a BS at Otago University and in 1955 began work on his PhD at the University of California, Berkeley. At Berkeley, Wilson set up a biochemistry lab where he worked to explain the origins of the human **species**. Wilson devised a "molecular clock" that was determined by genetic mutations rather than the traditional use of fossils to model human origins. The molecular clock started at the time the human species diverged from a common ancestor and accounted for all of the genetic mutations that had taken place since the time of divergence.

Together with doctoral student Vince Sarich, Wilson discovered that the **DNA** of humans is 99 percent identical with that of chimpanzee DNA. Utilizing this information, Wilson and Sarich concluded that proto-hominids evolved five million years ago. Anthropologists of that time believed that proto-hominids had evolved 20 million years ago. In conjunction with Wilson's findings of human and chimpanzee DNA, Wilson and his colleagues were among the first to utilize the **Polymerase chain reaction** (PCR) and the Relative Rate Test to study the DNA of extinct species. For the first time researchers were able to analyze DNA from fossils as well as DNA from a human brain that was 7000 years old. Although these findings erupted controversy among contemporary anthropologists and creationists, the data provided evidence to link molecular biology and **evolution**.

During the 1980s, Allan Wilson began studying the **mitochondrial DNA** (mDNA) of humans. Mitochondrial DNA is not located in the cell's **nucleus**, but rather in the **mitochondria** of the **cell** and is inherited only from the mother to all of her offspring. **Mutation** rates of mitochondria are much greater than that of nuclear DNA allowing mitochondria to evolve more rapidly than nuclear genes. Wilson determined that mitochondria were ideal for studying evolutionary change over a short time period and used this information to hypothesize when and where humans first originated. By comparing restriction maps of mDNA from different races and geographic regions, Wilson and his colleagues theorized that humans originated 200,000 years ago in Africa. A single ancestor is hypothesized to be the mother to all modern humans and has been named *Mitochondrial Eve*. This "Out of Africa" theory initially proved to be even more controversial

that Wilson's first theory of modern proto-humans. However, *Mitochondrial Eve* is now one of the accepted theories for the origin of the human species.

Allan Wilson died in 1981 at the age of 55 from leukemia.

See also Evolution

WILSON, EDMUND BEECHER (1856-1939)
American biologist

Edmund Beecher Wilson emphasized careful experimentation and analysis in biology at a time when the field was rife with theories based on little more than speculation. Indeed, Wilson's work was instrumental in transforming biology into a rigorous, scientific discipline. Although known for his meticulous approach to the study of the structure and function of the **cell**, he never lost sight of biology as a unified field that included **embryology**, **evolution**, and **genetics**. His influence in biology was felt through his position as a professor first at Bryn Mawr College and then at Columbia University, and through his highly influential textbook, *The Cell in Development and Inheritance*. Wilson's study of chromosomes, and especially his discovery of the sex chromosomes, helped lay the foundation for the study of genetics and evolution in the early-twentieth century. Many of the problems that Wilson tackled, including the details of cell development, remain unsolved today.

Edmund Wilson was born in Geneva, Illinois, the second of four surviving children of Isaac Grant Wilson, a lawyer and judge, and Caroline Louisa Clark, both of whom were originally from New England. When Edmund was two years old, his father was appointed a circuit court judge in Chicago. Rather than separate him from her childless sister and brother-in-law in Geneva, Edmund's mother left him to live with them while the rest of the family moved to Chicago. In this manner, he was "adopted" by Mr. and Mrs. Charles Patten and grew up counting himself very lucky to have two homes and four parents.

Shortly before he turned 16, Wilson taught school for one year from 1872 to 1873. As his older brother, Charles, had done the previous year, Wilson taught everything, including reading and arithmetic, to twenty-five pupils aged six to eighteen in a one-room schoolhouse. The following year he attended Antioch College (Yellow Springs, Ohio), following in the footsteps of an older cousin, Samuel Clarke. At Antioch, Wilson decided to devote himself to the study of biology, which, at that time, largely meant natural history.

In the fall of 1874, Wilson did not return to Antioch because he wished to prepare for studying at the Sheffield Scientific School of Yale University, which had been highly recommended to him by his cousin. To ready himself for Yale, Wilson moved to Chicago where he lived with his parents and took courses at the old University of Chicago from 1874 to 1875. He entered Yale in 1875 and received his bachelor's degree in 1878.

Although Wilson's particular focus of research changed many times in his long career, his work was always concerned with gaining a better understanding of how the single fertilized egg gave rise to a complete individual, whether that individual be an earthworm, jellyfish, or human. This interest in the development of the organism led Wilson to study cell structure and function, **heredity**, and evolution.

During his years of graduate and postgraduate work, Wilson studied the embryology and morphology of earthworms, sea spiders, the colonial jellyfish (*Renilla*), and other invertebrates. After Yale, he again followed Sam Clarke's educational path, this time to Johns Hopkins University. A close friend, William T. Sedgwick, entered Johns Hopkins along with him. From 1878 to 1881, Wilson worked closely with William Keith Brooks, obtained his Ph.D. in 1881, and remained at Johns Hopkins for an additional year of postdoctoral work. In 1882, Wilson studied in Europe with the help of a loan from his older brother, Charles. He studied in Cambridge, and, with Thomas H. Huxley's recommendation, gave a paper on *Renilla* before the Royal Society in London. From England, he went to Leipzig, Germany, and then to the Zoological Station at Naples. Wilson worked for almost a year there and formed strong friendships with director Anton Dohrn and zoologist Theodor Boveri. (For Wilson, the embryos of marine invertebrates were more easily studied than those of terrestrial animals, and for almost 50 years, Wilson spent his summers working at the Marine Biological Laboratory in Woods Hole, Massachusetts.)

To visit Naples, Wilson had worked out an arrangement with Clarke, who was then teaching at Williams College (Massachusetts). The college would pay for a laboratory bench at Naples for two years as part of a professorship at Williams. Wilson would work at Naples the first year while Clarke taught at Williams, then the two would switch places. Wilson's stint at Williams College lasted between 1883 and 1884.

From Williams, Wilson moved to the Massachusetts Institute of Technology as an instructor from 1884 to 1885. There, he collaborated with his friend, William T. Sedgwick, in the creation of a textbook titled *General Biology* (1886). Wilson's next teaching appointment, unlike his previous two, offered him the time and opportunity to continue his research. M. Carey Thomas, the first dean of Bryn Mawr College (Bryn Mawr, PA), invited Wilson to become the first professor of biology at the new women's college. he taught there between 1885 and 1891. While at Bryn Mawr, the scientist tackled the problem of cell **differentiation**—the way in which the fertilized egg gives rise to many kinds of specialized cells. To do this, he studied the cell-by-cell development of the earthworm and *Nereis,* a marine worm. This work, known as "cell lineage," established Wilson's reputation as a biologist of considerable skill. His 1890 and 1892 papers on *Nereis* demonstrated the value of cell lineage and inspired other scientists to pursue this fruitful avenue of research.

In 1891, Wilson accepted an appointment to become an adjunct professor of zoology in the new zoology department at Columbia University being organized by Henry Fairfield Osborn. He spent the rest of his career at Columbia, eventually becoming chair of the department, and retiring as DaCosta Professor in 1928. Before settling on campus, however, Wilson spent another fruitful year in Munich and Naples from 1891 to 1892. A series of lectures on the study of the cell that he gave during his first teaching year at Columbia formed the basis of his textbook *The Cell in Development and Inheritance,* published in 1896. Written before the fundamentals of heredity were understood, the book added a balanced, careful voice to the fierce debates over modes of **inheritance** and cell development that were occurring in biology at that time. The book, which illuminated Wilson's penchant for observation and experimentation, was hugely influential and further cemented his already substantial reputation. The book was dedicated to Boveri, the Italian zoologist.

On September 27, 1904, Wilson married Anne Maynard Kidder. Kidder and her family lived in Washington, D.C., but spent their summers at their cottage in Woods Hole, and it was there that the two met. Their only child, Nancy, became a professional cellist. Wilson himself was an avid amateur musician, and his trips to Europe were warmly remembered as much for the music he heard as for the science he learned. A flutist as a young man, he began taking cello lessons while he was living in Baltimore. For the rest of his life, in Bryn Mawr and then New York, he always found himself a quartet of amateur musicians with which to play.

In 1900, the modern era of genetics was born. Three scientists, working independently from each other, stated that inherited characteristics were determined by the combination of two hereditary units, one from each parent. (Today, those two hereditary units are known as genes.) This theory had actually been published 36 years earlier by Gregor Johann Mendel, but had lain dormant until it was "revived" at the turn of the 19th century by **Hugo De Vries**, Karl Erich Correns, and Eric Tschermak von Seysenegg.

Wilson quickly saw the connection between the rediscovery of the laws of heredity and his own work with cells and cell structures. The laws of heredity stated that the fertilized egg received half of the blueprint for its own expression from each parent. Chromosomes, he theorized, were the cell structures responsible for transmitting the units of inheritance. By following instructions from the chromosomes, the fertilized egg gave rise to a complete individual.

In 1905, Wilson and **Nettie Maria Stevens** of Bryn Mawr College independently showed that the X and Y chromosomes carried by the sperm were responsible for determining gender: in many **species**, including humans, females had an XX pair of chromosomes while males had an XY pair. In eight papers published from 1905 to 1912 entitled "Studies on Chromosomes," Wilson brilliantly extended his study of the chromosomal theory of **sex determination**, and it is for this work with chromosomes that he is best remembered. He is also recognized for setting the stage for the zoology department's future excellence in genetics, as personified by **Thomas Hunt Morgan** and **Hermann Joseph Muller**.

In the last years of his career, Wilson continued his study of cell structures. Despite failing health, he also wrote the third edition of *The Cell in Development and Inheritance,* over 1200 pages, which was published in 1925. In most respects, this was actually a completely new book that included the new discoveries in biology of the twentieth century. Wilson retired from Columbia University in 1928. He died in New York, at age 82.

See also Mendelian genetics and Mendel's laws of heredity; Sex chromosomes; Sex determination

WILSON, EDWARD OSBORNE (1929-)

American entomologist

World-renowned entomologist Edward O. Wilson is nicknamed "Dr. Ant," but his achievements impact much of the field of biology. He is co-founder of the modern field of **sociobiology**, believed by some to be one of the great paradigms of science, which has touched off much controversy but also a great deal of research in animal and human social behavior. From his posts as Harvard Univeristy's Frank B. Baird, Jr. Professor of Science and Mellon Professor of Science, Wilson is the recipient of Sweden's Crafoord Prize, a 1979 Pulitzer Prize for literature, and the 1977 National Medal of Science. He has influenced the field of animal taxonomy through his work in speciation theory, conducted research which led to the discovery of pheromones —chemicals which cause behavior in animals—and has been a harbinger of the threat of mass extinction resulting from man's unchecked use of the environment.

Edward Osborne Wilson was born on June 10, 1929 in Birmingham, Alabama. A descendant of farmers and shipowners in subtropical Alabama, Wilson had already decided to become a naturalist explorer by age seven. Fate intervened, however, when on a fishing trip, he vigorously pulled his catch out of the water, and its fin hit and damaged his right eye. He thus developed the habit of examining animals and objects close-up with his keen left eye, and when he subsequently read a National Geographic article entitled "Stalking Ants, Savage and Civilized" at age 10, the entomologist was born. Wilson later studied biology at the University of Alabama, obtaining a B.S. degree in this discipline in 1949 and an M.S. in 1950. In 1955, at age 26, he received his Ph.D. in biology from Harvard. He gained full professorship in 1964, and became Frank B. Baird, Jr. Professor of Science in 1976.

The field of new systematics—the attempt to classify **species** based on the principles of evolutionary theory—occupied Wilson during the early years of his career. With his colleague William L. Brown, Wilson critiqued the utilization of the subspecies category, prompting revised procedures among taxonomists. In 1956, Wilson also co-developed the concept of "character displacement," which occurs when two similar species begin a process of genetic **differentiation** to avoid **competition** and cross-breeding.

During the mid- to late–1950s, Wilson traveled to Australia, the South Pacific islands, and Melanesia to further study and classify ants native to those regions. Because of his fieldwork in the Melanesian archipelagoes, he developed the concept of the taxon cycle, which has since been found among birds and other insects. Wilson described the taxon cycle of Melanesian ants as the process through which a species disperses to a new, harsher habitat and evolves into one or more new "daughter" species, which then adapt to the new habitat.

All the while, Wilson was developing the foundation for what would he would term "sociobiology" two decades later. In 1959, influenced by the rise of molecular biology, he proved his hypothesis that social insects such as ants communicate through chemical releasers. Wilson crushed a venom gland extracted from a fire ant and created a trail of the chemical near a colony of the same species. He had anticipated that a few ants would trace the chemical path. Instead, dozens of fire ants swarmed out of the colony to follow the trail, and were baffled at its end. "That night I couldn't sleep," Wilson notes. "I envisioned accounting for the entire social repertory of ants with a small number of chemicals." Indeed, the chemicals came to be known as pheromones, and this discovery launched an "explosion of research" on the behavior of social insects— research which continues still. Wilson wrote later that pheromones were "not just a guidepost, but the entire message." These chemicals communicate complex instructions for fellow ants—everything from the location of food and how to obtain it to a call for help when in distress.

In the early and middle 1960s, Wilson collaborated with Princeton University mathematician Robert H. Mac Arthur to develop the first quantitative theory of species equilibrium. Prior to their work in this area, it was believed that the regularity of species in a given area was maintained through incomplete colonization. Wilson and his coauthor hypothesized that the number of species on a small island would remain constant, though the variety of species would undergo constant reshuffling from extinction and immigration, the two factors affecting the number of species in an ecosystem.

Wilson's greatest milestone probably was his 1975 book, *Sociobiology: The New Synthesis*. In it, Wilson defines sociobiology as "the systematic study of the biological basis of all social behavior." The term was in use prior to Wilson's landmark book, but he identified the interdisciplinary endeavor as one which was to change the way animal and human behavior is viewed and researched by the scientific community. Arthur Fisher, in *Society* magazine, declares, "Many biologists believe that sociobiology is indeed one of the great scientific paradigms, a powerful new tool for understanding some of the most baffling phenomena in the living world." Fisher compares the new framework to Darwin's theory of **natural selection** and Einstein's revolution of space/time theory. In fact, many of the tenets Wilson put forth in his book have gained widespread acceptance, and have aroused controversy over the ideological implications for human behavior.

The roots of Wilson's journey into this field lay in the beginning of his career. "In the forties and fifties," he says in *Society,* "we were in the midst of a very exciting development, called the new synthesis, which reinvigorated evolutionary biology by applying modern **population genetics** to what had previously been scattered and highly descriptive subjects... It was a period of grand synthesis in which it seemed possible to understand some of the most intrinsically interesting phenomena."

Simultaneously, the field of molecular biology was gaining prominence, and Wilson observed that this threatened to relegate the softer study of animal behavior to a tiny corner of Harvard's biology department. Wilson began focusing on the significance of organisms as carriers of genetic information. Viewing the complex behavior of ants and other social insects in this framework prompted Wilson to describe behavior which served survival not of the individual, but of the population.

Thus Wilson was able to explain, in Darwinian terms, such characteristics as **altruism**, significance of kinship, communication, and specialization of labor—characteristics which had previously confounded scientists. Cooperation among individuals or between species was consistent with early evolutionary theory because it enabled individuals to survive and carry on the **gene pool**. But altruism (behavior in which one individual helps another at possible or certain cost to itself) and spiteful behavior (when an individual harms another and itself) were largely unexplained by biologists.

The answer lies in the broader view of population survival. In a colony of ants, sterile members will work for their family members who share similar genes and who will reproduce on their behalf. Wilson maintains that selflessness is a characteristic of most ant species. He describes their colonies as "superorganisms," in which the welfare of the colony—not the individual—is paramount. On the other end of the behavioral spectrum, a species of Malaysian ants will rupture glands of poison on their own bodies if invaded by enemies—killing themselves and their intruders, while signaling for help from members of their own colony. Other complex and intricate behavior is explained by Wilson's sociobiology. The European red amazon ant, for example, is an aggressive creature which actually invades the nests of more peaceful ant species, killing some individuals and capturing others for use as slaves in their own nests. The slave ants actually do "housework," digging chambers and feeding and nurturing the young Amazons.

It was the twenty-seventh chapter of *Sociobiology* that touched off a controversy that continues today. In "Man—From Sociobiology to Biology," Wilson argued for expanded research on the role of biology in human behavior. "There is a need for a discipline of anthropological genetics," he wrote. "By comparing man with other primate species, it might be possible to identify basic primate traits that lie beneath the surface and help to determine the configuration of man's higher social behavior."

Wilson noted that humans have always been characterized by "aggressive dominance systems, with males generally dominant over females." He also wrote, "a key early step in human social evolution was the use of women in barter." In a separate article, Wilson wrote: "In hunter-gatherer societies, men hunt and women stay at home. This strong bias... appears to have a genetic origin. Even with identical education and equal access to all professions, men are likely to continue to

Edward O. Wilson. *Photos by John Chase/Harvard. Reproduced by permission.*

play a disproportionate role in political life, business, and science."

The anger with which Wilson's words were received led to noisy protests at a 1978 meeting of the American Association for the Advancement of Science. Intruders on the meeting first yelled a diatribe against him and then poured a pitcher of water over him. A letter of protest signed by, among others, two of Wilson's colleagues at Harvard, asserted that theories such as his in the past had led to the "sterilization laws and restrictive immigration laws by the United States between 1910 and 1930 and also for the **eugenics** policies which led to the establishment of gas chambers in Nazi Germany."

Wilson's worst detractors believed that the inevitable conclusion of his theories was "biological determinism." Harvard Professor **Stephen Jay Gould** (who had signed the letter of protest) sought a middle ground, since Wilson's theory did not preclude the possibility that "peacefulness, equality, and kindness are just as biological as violence, sexism and general nastiness." Gould maintained that there is no direct evidence in existence that specific human behaviors are genetically determined. In 1978, Wilson penned a follow-up to *Sociobiology,* the Pulitzer Prize–winning *On Human Nature.* In this volume he attempted to defend his hypotheses forwarded in chapter twenty-seven of *Sociobiology,* as well as to clear up certain areas that had become targets for controversy and prejudice. In particular, Fisher notes, Wilson "aimed for a fuller explanation of his views of the issues of free will, ethics, and development." Wilson's continuing research led to a collaboration with University of Toronto professor Charles Lumsden, with whom he penned 1981's *Genes, Mind and Culture* and 1983's *Promethean Fire*—the latter of which Wilson describes as his "last word on the subject" of human sociobiology.

Whatever the ramifications of Wilson's attempt to apply his entomological expertise to human behavior, his books and life's research represent great forward strides in the field of biology. His fascination with ants, begun in his childhood, culminated with the publication of 1990's *The Ants,* which he co-authored with German entomologist Bert Holldobler. Wilson, the world's leading authority on the creature with 8,800 species, believes they are essential to the world's ecosystems.

Wilson also is forthright in arguing for increased protection of the environment to minimize the mass species extinction now underway. He has warned that the current extinctions due to rainforest destruction will rival those which marked the end of the dinosaur age. Wilson has argued for surveyance of the earth's flora and fauna (the majority of which remain unclassified), the promotion of sustainable development, the wise use of the earth's plant and animal resources for food and medicine, and the restoration of terrains already damaged.

See also Ecological and environmental genetics

WOBBLE THEORY

The wobble theory was proposed in 1966 by **Francis Crick**, one of the co-discoverers of the structure of **deoxyribonucleic acid**. The hypothesis explained the mechanics of part of the **translation** process, where information carried on the **messenger RNA** is recognized and used to manufacture protein. Specifically, the wobble theory addresses how the limited number of **transfer RNA** (tRNA) molecules are able to recognize all the combinations of codons—three base sequences of amino acids—on the messenger **RNA**. Recognition of these **codons** by tRNAs is an absolutely vital step in **protein synthesis**.

The tRNAs read the **mRNA** codons in a specific manner. Specific tRNAs bind specific codons. The rules by which codons are interpreted by the translation system are called the genetic code.

The three base arrangement of a codon means that there are 64 possible amino acids that can be coded. Because there only 20 amino acids, and codons do not overlap (except in some transposable elements and **bacteriophage**), but are read sequentially during translation, more than one three-base sequence must code for each **amino acid**. In other words the genetic code contains some degree of redundancy, with more than one codon coding for an amino acid.

The wobble theory had its genesis in 1961. Francis Crick and colleagues experimented with a bacteriophage called rII. Various **deletions** and insertions of **bases** were constructed in the **DNA** of the bacteriophage, where certain number of amino acid were deleted or inserted in each case. The base change altered the sequence in which the remaining DNA bases were transcribed into messenger RNA, and so would affect the translation of the mRNA into protein. When a single base was deleted, and when two bases were deleted, the ability of the bacteriophage to lyse *Escherichia coli* K12 **bacteria** was lost. However, combinations where three bases were either deleted or inserted restored the ability of the phage to lyse the bacteria. It was this work that established that the mRNA sequence arose from a triplet of bases on the bacteriophage DNA. This pattern has proved to be the case for other microorganisms and other life forms, including humans.

Analysis of the translation of each resulting mRNA into protein enabled them to identify the amino acid encoded by a particular combination of bases. From such experiments it was shown that several codons could encode the same amino acid, revealing the redundancy of the genetic code. Furthermore, the redundancy appeared to reside in the third amino acid in the codon (reading from the same end of the codon each time). This pattern of redundancy of the genetic code suggests that the pairing between the bases of the codon on the mRNA and the bases of the anticodon (the region that recognizes the corresponding codon) on the tRNA is unusual. The pairing between the first two bases of the tRNA with the bases on the mRNA occurs as normal—one base recognizes and pairs only with a certain other base. But, the third base of the tRNA, at the 5' end of the anticodon, seems to be less restricted in its ability to pair with other bases. The wobble theory, published in 1966, proposed that this third base can pair with more than one type of base at the 3' end of the mRNA codon. This was proposed to be possible due to the increased spatial flexibility that could be attained by the 5' base. Subsequent experiments have supported this theory. However, the molecular mechanics of the wobble are not universal; bacteria and **yeast** display differences in their degree of wobble, with less flexibility in yeast. Thus, yeast requires more tRNAs to recognize all the codons.

See also Genetic code; Translation

WOLLMAN, ELIE (1917-)
French geneticist

Elie Wollman is a bacterial geneticist, long affiliated with the Institute Pasteur in Paris. The emergence of bacterial **genetics** as an important research discipline in the 1950s is due in large measure to the work of Wollman. His collaborative research efforts detailing with the nature of the **prophage** and the mechanism of bacterial **chromosome** transfer are of fundamental importance in molecular biology.

Wollman is best known for his fruitful collaborations with Francois Jaçob, beginning in 1954. Together they studied the relationships between the prophage, a **virus** that specifically infects certain **bacteria**, and the genetic material of the target bacterium. A series of experiments led to a definition of the mechanism of bacterial conjugation. In 1954, their experiment with the **bacteriophage** lambda provided evidence that led to the concept of genetic repressor molecules. Another advance occurred in 1957, with their now-classic experiment called the "interrupted mating procedure." Bacteria with wild-type genes for particular traits were mixed with bacteria whose corresponding genes were defective. After increasing lengths of time, the mating between the bacterial types was stopped. The observations that more of the defective traits were cured with increasing time demonstrated that **gene** transfer begins at a certain point on the chromosome (the origin of transfer) and that transfer is linear. They also provided decisive evidence for the circular nature of the chromosome. This remarkable period of collaboration was described in the 1961 book *Sexuality and the Genetics of Bacteria,* co-authored by Wollman and Jaçob.

Among his honors, Wollman was elected to the National Academy of Sciences in 1991.

See also Microbial genetics; Repression

WRIGHT, SEWALL (1889-1988)

American geneticist

During his long and productive life, Sewall Wright achieved international standing in the disciplines of experimental physiological **genetics**, the study of **heredity**, as well as quantitative evolutionary biology. Wright made significant contributions to the fields of genetics, zoology, biometrics (the use of statistics to analyze biological data) and animal breeding, but is best known for his comprehensive theory of **evolution**, known as the shifting-balance theory. The shifting-balance theory accounts for the spread of certain **gene** combinations within a population. The shifting-balance theory took a mathematical and analytical approach to **population genetics**. Wright's work brought serious statistical analysis to the forefront of biological science and touched off a long-running debate about the nature of animal **species** development.

Wright, the oldest of three children, was born to Philip Green Wright, a college professor, and Elizabeth Quincy (Sewall) Wright. When Wright was seven years old, he wrote a small booklet he called "The Wonders of Nature," a hand-sewn volume printed in capital letters. The precocious Wright spent only five years in grade school as his learning was supplemented at home by his intellectual parents. In 1902, he entered Galesburg High School in Illinois, where he excelled at languages, especially Latin and German. The courses that intrigued Wright most, however, were algebra, geometry, and physics. Wright then attended Lombard College in Galesburg, where his father was employed. His original intention was to continue the study of languages. Wright enrolled in several classes taught by his father, including general mathematics and economics, and a course on the fiscal history of the United States. In his senior year, Wright took two biology classes from Wilhemine Key, who introduced him to the relatively new discipline of theoretical biology and to R. C. Punnett's groundbreaking article, "Mendelism," which had just appeared in the eleventh edition of *Encyclopedia Britannica.* Key steered Wright toward graduate study in biology and set

up an internship in zoology for him at Columbia University's Cold Spring Harbor laboratory on Long Island, New York. After graduating from Lombard College in 1911, Wright spent the summer at Cold Spring Harbor, where he studied marine invertebrates. While there, Wright also met a number of influential geneticists and began to take an interest in the field.

With the help of a modest state scholarship, Wright moved to the University of Illinois and received his M.S. in zoology in 1912. The same year, he attended a lecture by the prominent Harvard zoologist, W. E. Castle, who spoke of his **selection** and mammalian genetics experiments in hooded rats. Castle's work centered on the notion that Mendelian factors, such as recessive and dominant traits, were sometimes variable. He later altered this view, but his experiments seemed to indicate that certain genetic combinations could yield unexpected results. Intrigued by these ideas, Wright signed on as Castle's personal assistant and graduate student.

In addition to his doctoral classwork, Wright also worked at Harvard's Bussey Institution, a biological research facility, helping Castle maintain a colony of hooded rats and working with other researchers to develop a guinea pig colony. Wright had learned about the genetics of guinea pigs while at Cold Spring Harbor, and Castle assumed that Wright would eventually use the colony for his own research. Wright thought the guinea pig was a valuable research animal, despite the fact that they are disease prone, relatively large and cumbersome, and have long reproductive cycles. At that time, no one knew exactly how many chromosomes the guinea pig had, and a number of questions remained about their **inheritance** patterns. Wright's work with the guinea pig would continue until 1961 and answer many of these questions. His first major finding occurred in 1914, when he discovered a series of four alleles (a series of two or more genes that can occupy the same position on a **chromosome**) that produced various effects on coat and eye color. Over the next forty years, he would study the inheritance factors of color patterns, hair direction, digit size, and abnormalities in guinea pigs. When Wright received his Sc.D. in zoology from Harvard in 1915, his dissertation was entitled *An Intensive Study of the Inheritance of Color and of Other Coat Characters in Guinea Pigs, with Especial Reference to Graded Variations.*

In 1915, Wright accepted a position as senior animal husbandman at the U.S. Department of Agriculture (USDA) in Washington, D.C. Inheriting an extensive **inbreeding** study of guinea pigs (which the USDA had begun in 1906), Wright was charged with analyzing the mountains of data generated by this on-going experiment. To make his task easier, he developed a mathematical theory of inbreeding in 1920, the methods of which were published in 1921 under the title *Correlation and Causation.* Wright's early work with Castle had led him to the notion that interaction systems between genes had important implications for evolution and that inbreeding in small populations led to variation within a given species. Wright's mathematical theory enabled him to quantify the effects of inbreeding, and he used this theory extensively during the next ten years.

Wright spent the summer of 1920 at Cold Spring Harbor, where he met Louisa Williams, then an instructor at Smith College in Massachusetts. She had earned her master's degree from Denison University while working under Harold Fish, who had been Wright's colleague. She had come to Cold Spring Harbor to help Fish set up a rabbit colony for genetic research and, because of similar interests, became one of Wright's close friends. The couple married on September 10, 1921, and moved to Washington. They had two sons over the next four years, Richard and Robert.

Wright left the USDA in 1925, and accepted a position in the department of zoology at the University of Chicago, where he would remain until 1954. During the late 1920s, Wright refined his ideas on evolution and developed a more comprehensive theory. Wright considered the random drift of genes due to inbreeding and the isolation of small groups within a species were important factors in the evolution of any species. These ideas led to his often spirited, life-long scientific debate with geneticist **Ronald A. Fisher**, who proposed that **natural selection** worked best in large populations in which more mutant genes—genes in which the hereditary material has changed—were available. Fisher believed that each population had a complex gene structure and that many genes affected each characteristic. Fisher postulated the idea of the population as a "gene pool" in which **gene frequency** was determined by natural selection.

When Fisher's *Genetical Theory of Natural Selection* appeared in 1930, Wright reviewed the book, pointing out several errors and stating his overall objections to the theory. The two corresponded and agreed on some common points, but never reached a consensus. In 1931, Wright published a long paper describing in detail his own evolutionary theory, which he called the three-phase shifting-balance theory. The three phases were (1) random gene-frequency drift within subpopulations, (2) increase of the preferred combination of genes or what has become known as mass selection, and (3) the dispersal of the preferred gene combination throughout the population.

During the early 1930s, as Wright's reputation grew, his theory attracted worldwide attention. As a result, he was elected to membership in the American Philosophical Society (1932), the National Academy of Sciences (1934), and the Genetics Society of America (1934). He was asked to serve on a number of scientific boards and was sought out as a guest lecturer and reviewer. Wright received numerous requests from researchers to perform quantitative data analysis on their experimental data. Wright's work with Russian-born geneticist **Theodosius Dobzhansky**, then at the California Institute of Technology, helped further the cause of quantitative genetics. Dobzhansky used Wright's mathematical methods and conclusions to develop an extension of Wright's own evolutionary theory. In 1937, Dobzhansky published *Genetics and the Origin of the Species*, which set the research agenda in population genetics for decades to come.

During the 1940s, Wright divided his research between physiological genetics and theoretical population genetics, becoming one of the most respected scientists in the nation as more and more researchers began to grasp the importance of his theories. He was awarded nine honorary doctorates over the next thirty years and also received the National Academy of Sciences Daniel Giruad Elliot Award and Oxford University's Weldon Memorial Medal. During this period, Wright authored a series of papers which argued that genes were responsible for replicating and coding the **enzymes** (complex proteins which facilitate biochemical-chemical reactions) that determine an organism's physiology. Wright continued his genetic experiments with guinea pigs, but the 1947 discovery of DNA—the molecular components of heredity—moved the field toward the realm of molecular biology. When Wright retired from the University of Chicago in 1954 and moved on to the University of Wisconsin in Madison, he left his guinea pigs behind. It took Wright years to analyze the cumulative data from these experiments and his final papers on inbreeding in guinea pigs did not appear until 1961.

For the next twenty-five years Wright worked on his massive *Evolution and the Genetics of Populations*, a four-volume text that explained the history of genetics in minute scientific and mathematical detail. In recognition of his remarkable achievements in genetics and quantitative evolution, Wright received the National Medal of Science in 1967, the Darwin Medal of the Royal Society of London in 1980, and the Balzan Prize in 1984. During the 1980s, Wright continued his habit of taking long, vigorous walks and retained his interest in the sciences. Although he began to lose his vision in 1980, Wright continued to read vociferously with the aid of a magnifying closed-circuit television device. Wright's last paper appeared in 1988 in *American Naturalist*, one of his favorite outlets, when he was ninety-nine years old.

See also Animal husbandry

X

X CHROMOSOME

The X **chromosome** is one of two sex chromosomes and is present in nearly all animals that undergo **sexual reproduction**. Depending on the organism, it may occur alone, in two copies, or paired with a **Y chromosome**. The sex of an individual is directly related to the presence or absence of specific genes located on these chromosomes.

In humans, at least one X chromosome is required for life to exist. It is usually paired with either another X in a female or a Y in a male. The X chromosome is much larger than the Y, and, with the exception of a small region at one end of the X and Y chromosomes, the pseudoautosomal region, the genes present on the two chromosomes are very different. Genes located uniquely on the X chromosome are said to be X-linked.

Females have two X chromosomes and, thus, two copies of each X-linked **gene**. In order to express an X-linked recessive disease, a female must have two mutant genes. If only a single **mutation** is present, the female will not be affected with the disorder but is a mutation **carrier** and has a 50:50 chance of transmitting that mutation on to her offspring. Female children will also be carriers unless they receive a second mutation from the father. Males, however, have only a single X chromosome, so if a male child inherits an X-linked mutation from his mother, he will be affected with the disease. X-linked dominant disorders are expressed in both males and females, but are often lethal in males.

An interesting feature of the X chromosomes in females is that one of the pair is always turned off, or inactivated, in every **cell**. This mechanism equalizes the amount of protein produced from X-linked genes between males and females.

The fact that X-linked recessive traits are clearly expressed in males, allowed early geneticists to easily map genes to the X chromosome. There are currently 445 known genes on the X chromosome that are involved in a many different functions. The gene for **colorblindness** is found on the distal long arm of the X chromosome. Flanking this locus are a pair of genes, Factors VIII and IX, that are important in blood clotting and are associated with hemophilias A and B. Mutations in the gene retinitis pigmentosum cause a devastating disease of the eyes. Steroid sulfatase mutations produce a disorder characterized by scalyness of the skin that is known as ichthyosis. The most common inherited form of mental retardation, **fragile X syndrome**, is another X-linked gene.

As expected, there are also many genes involved in **sex determination**. Genes on the short arm are critical for female development, and loci on the long arm have been associated with female fertility. The androgen receptor gene functions in male development, and if this is nonfunctional, testicular feminization may occur. This disorder results from the interruption of the male **differentiation** pathway and gives rise to an individual with a female body, but no functional internal genitalia. In the 1950s and 1960s, the Soviet Bloc countries entered these women in Olympic events since they were able to out compete their rivals. It was suggested that the higher levels of testosterone resulted in increased strength. Once the situation was identified, Olympic officials decided to test all participants to ensure that the **sex chromosome** complement matched the physical appearance of each individual.

Imbalance in the number of sex chromosomes gives rise to a diverse set of disorders. The occurrence of three copies of an X chromosome has few adverse consequences, and individuals are apparently normal, fertile females. Apparently normal, fertile males can have one X and two Y chromosomes. However, a single X chromosome gives rise to a female individual with Turner syndrome who is short, usually infertile, with a possible learning deficit and other phenotypic anomalies. A male with two X and one Y chromosomes has **Klinefelter syndrome** and is usually tall, infertile, with some type of learning deficit and development of female secondary sex characteristics.

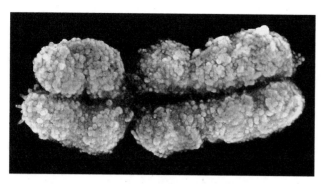

Enhanced scanning electron microscope image of a human X chromosome. *Photo Researchers. Reproduced by permission.*

The X chromosome is critical to normal human development and has proven to be one of the most unusual of the chromosome set due to the unique feature of X inactivation.

See also Sex chromosomes; Sex determination; Y chromosome

X CHROMOSOME INACTIVATION

X chromosome inactivation, also called the Lyon hypothesis, is a phenomenon that occurs in every female mammal that has the double X chromosome configuration. Based on her own work on mice with mottled coats, Mary F. Lyon of the Medical Research Council's Mammalian Genetics Unit in Harwell, England, proposed in 1960 the idea of X inactivation. To ensure that the sexes work with similar doses of X genes, which scientists believe is critical for embryonic development, female mammals evolved the ability to inactivate one of their sex chromosomes. Early in embryonic life, one of the X chromosomes in every cell randomly shuts itself down. Researchers have uncovered evidence that DNA sequences usually dismissed as junk DNA without any function actually may help determine what genes on the X chromosome are suppressed. Some of the X chromosome genes, however, resist inactivation. Huntington F. Willard of Case Western Reserve University School of Medicine in Cleveland and his colleagues recently analyzed the activity of more than 200 genes mapped to the X chromosome, about 10 percent of the chromosome's estimated total. In *Proceedings of the National Academy of Sciences,* Willard's team reported that an unexpectedly large number of X chromosome genes, 34 of 224, are not inactivated. Of the 34 escapees, 31 are located on the short arm of the chromosome. In addition, most of these don't have a partner on the Y chromosome. "Strict dosage compensation of all genes on the chromosome isn't necessary," stated coauthor Laura Carrel. The mechanism of how this inactivation occurs remains controversial. Somehow, the inactivated gene stays tightly coiled so that it becomes unavailable for replication. This inactivated chromosome appears in female cells as a small dark staining spot called a Barr body. In oocytes, the inactivation is reversed so that the females full DNA complement can be replicated.

As one chromosome is inherited from the father and one from the mother, this inactivation means that females are actually mosaics of both parents. This has definite consequences for genetic health, in that females are chimeric for the gene products of the X chromosome. Some mutations that cause genetic disease only occur on the X chromosome. The defective blood clotting disease Hemophilia is one such disease. Males only have one copy of the X chromosome, which they have inherited from their mother. If the mutation occurs on their X chromosome then they will express the disease because they do not have another copy of the gene that is not defective. On the other hand, females may or may not express the disease clinically, depending on how many of their cells contain the active chromosome with the defective gene. Thus, one sees gradations in the severity of symptoms with females, from none at all, to prolonged but non-fatal clotting times. They would be carriers, and could pass the disease on to their children.

See also Mosaic

X LINKAGE • *see* X-CHROMOSOME

XENOGAMY

Xenogamy (also called outbreeding) is a term used to describe the union of genetically unrelated organisms within the same species. Xenogamy promotes genetic variability and vitality within a breeding population by eliminating homozygous individuals. Accordingly, with dominance, lethal or deleterious alleles have a better chance of being repressed by the dominant alleles in heterozygous organisms, thus reducing the impact of lethal or deleterious alleles within a population.

Xenogamy, or outbreeding promotes genetic diversity and thus, also enhances the overall fitness of a species. By producing new and varied combinations, outbreeding is a fundamental part of natural selection and an essential element of evolution. As such, outbreeding is an important tool in the continued survival and evolution of a species.

In plant species, there are several natural mechanisms that can result in xenogamy, including self-incompatibility. With self-incompatibility, there is an inability on the part of sex cells (gametes) from the same species of plants to produce a viable embryo. With such species, it is usually the case that pollen, unable to induce fertilization on its own style, is able to successfully grow down the style of other plants of the same species. The process involving the transfer of pollen to a foreign stigma is termed allogamy. Regardless of the exact mechanism, self-incompatibility mechanisms promote xenogamy (outcrossing) and heterozygosity while acting to prevent inbreeding. Self-incompatibity is often the result of incomplete nuclear or genetic fusion. The failure of the fusion processes are usually traced to a single genetic locus (S locus) that exists as multiple alleles.

In some circumstances, xenogamy and outbreeding are also referred to as cross breeding. Regardless of the exact terminology, the core concept involves an increase in genetic variability. Such increases in genetic variability exert measurable influences on the frequency of genes, types of alleles, and traits within a population.

With crossbreeding, genetically dissimilar or unrelated animals from the same breed can be crossed in a process known as outcrossing. True crossbreeding exists when, for example, differing breeds of cattle are allowed to mate and produce offspring. Extreme xenogamy exists with species crossing (a mating between organisms from two different species).

Xenogamy or crossbreeding is often the attempt by scientists to genetically combine desirable traits to form two differing species or breeds in order to produce offspring with more desirable characteristics. Successful crossbreeding, whether in plants or animals, usually results in increased **hybrid** vigor (heterosis). Hybrid vigor refers to a mating system criteria that measure the better than normal growth rates, immunological tolerance, strength, and other desired characteristics of a hybrid species.

In contrast to xenogamy or allogamy there are the processes of inbreeding and autogamy (self-pollination).

See also Evolutionary mechanisms; Plant genetics; Plant reproduction; Sexual reproduction; Species

X-LINKED AGAMMAGLOBULINEMIA

X-linked agammaglobulinemia (XLA) or Bruton's agammaglobulinemia is present at birth (congenital) and is characterized by low or completely absent levels of immunoglobulins in the bloodstream. Immunoglobulins are protein molecules in blood serum that function like antibodies. Without them, the body lacks a fully functioning immune system. Persons with XLA are vulnerable to repeated, potentially fatal bacterial infections.

XLA occurs in 1 in 50,000 to 1 in 100,000 newborns. Almost all persons with the disorder are males. Although persons with XLA carry the genes to produce immunoglobulins, a genetic defect on the **X chromosome** prevents their formation. This defect is not associated with the immunoglobulins themselves, but rather with the B cells in the bloodstream that ordinarily secrete the immunoglobulins.

B cells are a type of white blood **cell**. They are the sole producers of immunoglobulins in the body. B cells are produced in the bone marrow and carried to the spleen, lymph nodes, and other organs as they mature. The maturation process depends on an enzyme called Bruton's agammaglobulinemia tyrosine kinase (Btk). If Btk is missing or defective, the B cells cannot mature and cannot produce immunoglobulins.

The **gene** for Btk is on the X **chromosome**. Certain changes (mutations) in this gene result in defective Btk. Since the gene is carried on the X chromosome, XLA individuals are almost always male. Females have two X chromosomes, which means they have two copies of the Btk gene, one of which is normal. Males have only one X chromosome.

XLA is caused by a defect in the gene that codes for Btk. This defect leads to blocked maturation of B cells, the cells that produce immunoglobulins. Because other portions of the immune system are functional, people with XLA can fight off some types of infection, such as fungal and most viral infections. Immunoglobulins, however, are vital to combat bacterial infections. Infants with XLA usually do not show symptoms during the first 6 months of life because immunoglobulins from their mothers are circulating in their bloodstreams. As the mother's supply decreases, the baby becomes increasingly vulnerable to bacterial infections.

Common symptoms of immunoglobulin deficiency appear after the infant is six months old. They include frequent ear and sinus infections, pneumonia, and gastroenteritis. Certain viruses, such as hepatitis and polio viruses, can also pose a threat. Children with XLA grow slowly, have small tonsils and lymph nodes, and may develop chronic skin infections. Approximately 20% of these children develop arthritis, possibly as a result of joint infections.

Frequent bacterial infections, a lack of mature B cells, and low-to-nonexistent levels of immunoglobulins point to a diagnosis of XLA. A sample of the infant's blood serum can be analyzed for the presence of immunoglobulins by a technique called immunoelectrophoresis. To make a definitive diagnosis, the child's X chromosome is analyzed for defects in the Btk gene. Similar analysis can be used for prenatal diagnosis or to detect carriers of the defective gene.

Treatment of XLA consists of regular intravenous doses of commercially prepared gamma globulin (sold under the trade names Gamimune or Gammagard) to ward off infections. **Antibiotics** are used to treat infections as they occur. Children with XLA must be treated promptly for even minor cuts and scrapes, and taught to avoid crowds and people with active infections.

Prior to the era of gamma globulin and antibiotic treatment, approximately 90% of XLA individuals died before age eight. Early diagnosis and current therapy allows most individuals with XLA to reach adulthood and lead relatively normal lives.

Parents of a child with XLA often consider genetic counseling if they are planning to have more children.

See also Immunogenetics; Prenatal diagnostic techniques

X-RAY CRYSTALLOGRAPHY

X-ray **Crystallography** is the study and determination of crystalline structures through x-ray diffraction techniques. In 1953, Watson and Crick used x-ray crystallography to discover the double-helix structure of **DNA**. X-ray crystallography is most commonly used in biological and medicinal fields to determine the structures of complex proteins. The protein structures are invaluable in drug and medicine research, resulting in advances in treatment and more efficient production of pharmaceuticals.

The first step in an x-ray diffraction experiment is to grow crystals of the molecule being studied. Several methods are available for growing crystals, the most common being vapor diffusion. Other approaches currently in use are macroseeding, microseeding, batch crystallization, microbatch crystallization, free interface diffusion, and dialysis. To yield crystals of the desired size, several techniques are used in any one diffraction experiment because of the extreme difficulty in controlling the crystal formation.

When electromagnetic waves of an exceedingly small wavelength (comparable to the size of the molecule's bonds) are passed through a molecule, the waves are diffracted by the electron clouds of the constituent atoms. The diffraction is amplified greatly because of the crystal's characteristic repetition of structure. If there were a way to focus x rays, an x-ray "picture" could be produced from these diffraction patterns. However, because x rays are so penetrative, focusing them is impossible. To provide a picture of the individual atoms on such a small scale, the patterns of interference must be recorded and thoroughly analyzed. The diffraction patterns produced can be used by computers to create electron density maps that show the electron probability distribution of the molecule in a topological style. The relative positions of atoms in a molecule can be determined because electrons are most densely packed around the atom's **nucleus**. One problem of this approach is that atoms of hydrogen, with only one electron, are very hard to pinpoint. Using electron density maps, a 2-D picture of the molecule is generated. To extrapolate a 3-D structure from the diffraction patterns, the experiment is repeated through small increments in the angle of the crystal's rotation with respect to the x-rays. Using many electron density maps, and watching how they change, scientists can find a relative 3-D orientation for each atom within the crystal.

See also Biotechnology; History of genetics: The discovery of the Watson-Crick model of DNA

XX MALE SYNDROME

XX male syndrome occurs when the affected individual appears as a normal male, but has a female genotype (two X chromosomes). Two types of XX male syndrome can occur: those with detectable **SRY gene** and those without detectable SRY. SRY is the main genetic switch for determining that a developing embryo will become male.

XX male syndrome is a condition where the sex chromosomes of an individual do not agree with the physical sex of the affected person. Normally, there are 46 chromosomes, or 23 pairs of chromosomes, in each **cell**. The first 22 pairs are the same in men and women. The last pair, the sex chromosomes, is two X chromosomes in females (XX) and an X and a **Y chromosome** in males (XY).

With XX male syndrome, the person has female chromosomes but male physical features. The majority of persons with XX male syndrome have the Y **chromosome gene** SRY attached to one of their X chromosomes. SRY is the main genetic switch for determining that a developing embryo will become male. The rest of the individuals with XX male syndrome do not have SRY detectable in their cells. Hence, others genes on other chromosomes in the pathway for determining sex must be responsible for their male physical features.

In XX male syndrome caused by the gene SRY, a **translocation** between the **X chromosome** and Y chromosome causes the condition. A translocation occurs when part of one chromosome breaks off and switches places with part of another chromosome. In XX male syndrome, the tip of the Y chromosome that includes SRY is translocated to the X chromosome. As a result, an embryo with XX chromosomes with a translocated SRY gene will develop the physical characteristics of a male. Typically, a piece of the Y chromosome in the pseudoautosomal region exchanges with the tip of the X chromosome. In XX male syndrome, this crossover includes the SRY portion of the Y.

In individuals with XX male syndrome that do not have an SRY gene detectable in their cells, the cause of the condition is not known. Scientists assert that one or more genes that are involved in the development of the sex of an embryo are mutated or altered and cause physical male characteristics in a chromosomally female person. These genes could be located on the X chromosome or on one of the 22 pairs of **autosomes** that males and females have in common. As of 2001, no genes have been found to explain the female to male sex reversal in people affected with XX male syndrome that are SRY negative. Approximately 20% of XX males do not have a known cause and are SRY negative. It is thought that SRY is a switch point, and the protein that is made be SRY regulates the activity of one or more genes (likely on an autosomal chromosome) that contribute to sex development. Although there have been some studies which demonstrate autosomal recessive **inheritance** for the XX male.

XX male syndrome occurs in approximately one in 20,000 to one in 25,000 individuals. The vast majority, about 90%, has SRY detectable in their cells. The remaining 10% are SRY negative, although some research indicates that up to 20% can be SRY negative. XX male syndrome can occur in any ethnic background and usually occurs as a sporadic event, not inherited from the person's mother of father. However, some exceptions of more than one affected family member have been reported.

Males with SRY positive XX male syndrome look like and identify as males. They have normal male physical features including normal male body, genitals, and testicles. All males with XX male syndrome are infertile (can not have biological children) because they lack the other genes on the Y chromosome involved in the making of sperm. Men with XX male syndrome are usually shorter than an average male, again because they do not have certain genes on the Y chromosome involved in height. A similar syndrome that affects males with two X chromosomes is Klinefelter's syndrome. Those individuals with 46XX present with a condition similar to Klinefelter's, such as small testes and abnormally long legs.

People with SRY negative XX male syndrome are more likely to be born with physical features that suggest a condition. Many have hypospadias, where the opening of the penis is not at the tip, but further down on the shaft. They can also have undescended testicles, where the testicles remain in the body and do not drop into the scrotal sac. Occasionally, an SRY negative affected male can have some of the female structures such as the uterus and fallopian tubes present. Men with SRY negative XX male syndrome can also have gynecomastia, or breast development during puberty, and puberty can be delayed. As with SRY positive XX male syndrome, these men are infertile and shorter than average because they lack other Y specific genes. The physical features can vary within a family, but most affected people are raised as males.

A small portion of people with SRY negative XX male syndrome are true hermaphrodites. This means they have both testicular and ovarian tissue in their gonads. They are usually born with ambiguous genitalia, where the genitals of the baby have both male and female characteristics. Individuals with XX male syndrome and true hermaphrodites can occur in the same family, suggesting there is a common genetic cause to both. Research indicates that 15% of 46XX true hermaphrodites have the SRY gene.

For people with XX male syndrome who have ambiguous genitalia, hypospadias, and/or undescended testicles, the diagnosis is suspected at birth. For males with XX male syndrome and normal male features, the diagnosis can be suspected during puberty when breast development occurs. Many men do not know they have XX male syndrome until they try to have their own children, are unable to do so, and therefore are evaluated for infertility.

When the condition is suspected in a male, chromosome studies can be done on a small sample of tissue such as blood or skin. The results show normal sex chromosomes, or XX chromosomes. Further genetic testing is available and needed to determine if the SRY gene is present.

Some affected individuals have had SRY found in testicular tissue, but not in their blood cells. This is called mosaicism. Most males only have their blood cells tested for SRY and not their testicular tissue. Hence, some men who think they have SRY negative XX male syndrome may actually be **mosaic** and have SRY in their gonads.

XX male syndrome can be detected before birth. This occurs when a mother has prenatal testing done that shows female chromosomes but on ultrasound male genitals are found. Often the mother has had prenatal testing for a reason other than XX male syndrome, such as for an increased risk of having a baby with **Down syndrome** due to her age. Genetic testing for the presence of the SRY gene can be done by an **amniocentesis**. An amniocentesis is a procedure where a needle is inserted through the mother's abdomen into the sac of fluid surrounding the baby. Some of the fluid is removed and used to test for the presence of the SRY gene. Amniocentesis has a slight risk for miscarriage to occur.

For those with XX male syndrome with normal male genitals and testicles, no treatment is necessary. Affected males with hypospadias or undescended testicles may require one or more surgeries to correct the condition. If gynecomastia is severe enough, breast reduction surgery is possible. The rare person with true hermaphrodism usually requires surgery to remove the gonads, as they can become cancerous. Men with XX male syndrome have normal intelligence and a normal life span. However, all affected men will be infertile.

See also Chromosomal mutations and abnormalities; Genetic disorders and diseases; Hereditary diseases; Inherited cancers; Mutations and mutagenesis

XYY SYNDROME

XYY syndrome is a **chromosome** disorder that affects males. Males with this disorder have an extra **Y chromosome**.

The XYY syndrome was previously considered the "supermale" syndrome where men with this condition were thought to be overly aggressive and more likely to become a criminal. These original stereotypes came about because several researchers in the 1960s found a high number of men with XYY syndrome in prisons and mental institutes. Based on these observations, men with XYY syndrome were labeled as overly aggressive and likely to be criminals.

These original observations did not consider that the majority of males with XYY syndrome were not in prisons or mental institutes. Since then, broader, less biased studies have been done on males with XYY syndrome. Though males with XYY syndrome may be taller than average and have an increased risk for learning difficulties, especially in reading and speech, they are not overly aggressive. Unfortunately, some text books and many people still believe the inaccurate stereotype of the supermale syndrome.

Chromosomes are structures in the cells that contain genes. Genes are responsible for instructing our body how to grow and develop. Usually, an individual carries 46 chromosomes in their cells (i.e., 23 pairs). The first 22 pairs are the same in males and females and the last pair, the sex chromosomes, consist of two X chromosomes in a female, and an **X chromosome** and an Y chromosome in a male.

XYY syndrome occurs when an extra Y chromosome is present in the cells of an affected individual. People with XYY syndrome are always male. The error that causes the extra Y chromosome can occur in the fertilizing sperm or in the developing embryo.

XYY is not considered an inherited condition. An inherited condition usually is one in which the mother and/or father has an alteration in a **gene** or chromosome that can be passed onto their children. Typically, in an inherited condition, there is an increased chance that the condition will reoccur. The risk of the condition reoccurring in another pregnancy is not increased above the general population incidence.

XYY syndrome has an incidence of one in 1,000 newborn males. However, since many males with XYY syndrome

Chromosomal disorders

Disorder	Chormosome affected	Karyotype	Incidence	Symptoms
Turner syndrome	X	45,X (monosomy)	1 in 2,000	Growth retardation; Infertility; Cardiovascular malformations; Learning disabilities
Klinefelter syndrome	X	47,XXY (trisomy)	1 in 500–800	Taller than average; Poor upper body strength (clumsiness); Mild intentional tremor (20–50%); Breast enlargement (33%); Decreased testosterone production; Infertility; Dyslexia (50%)
Triple X	X	47,XXX (trisomy)	1 in 1,000	Mild delays in motor, linguistic, and emotional development; Learning disabilities; Slightly taller than average
XYY syndrome	Y	47,XYY (trisomy)	1 in 1,000	Taller than average; Lack of coordination; Acne; Some infertility; Learning disabilities (50%); Behavior problems, especially impulse control
XX male syndrome	Y	46,X,t(X,Y) (translocation of the SRY gene [90%] or other gene responsible for the male sex determination)	1 in 20,000–25,000	Usually normal male physical features, but may have ambiguous genitalia, hypospadias or undescended testes; Infertility; Shorter than average

look like other males without XYY syndrome, many males are never identified.

There are no physical abnormalities in most males with XYY syndrome. However, some males can have one or more of the following characteristics. Males who have XYY syndrome are usually normal in length at birth, but have rapid growth in childhood, usually averaging in the 75th percentile (taller than 75% of males their same age). Many males with XYY syndrome are not overly muscular, particularly in the chest and shoulders. Individuals with XYY syndrome often have difficulties with their coordination. As a result, they can appear to be awkward or clumsy. During their teenage years, males with XYY syndrome may develop severe acne that may need to be treated by a dermatologist.

Men with XYY syndrome have normal, heterosexual function and most are fertile. However, numerous case reports of men with XYY syndrome presenting with infertility have been reported. Most males with XYY syndrome have normal hormones involved in their sperm production. However, a minority of males with XYY syndrome may have increased amounts of some hormones involved in sperm production. This may result in infertility due to inadequate sperm production. As of 2001, the true incidence of infertility in males with XYY syndrome is unknown.

When XYY men make sperm, the extra Y chromosome is thought to be lost resulting in a normal number of sex chromosomes. As a result, men with XYY syndrome are not at an increased risk for fathering children with chromosome abnormalities. However, some men with XYY syndrome have been found to have more sperm with extra chromosomes than what is found in men without XYY syndrome. Whether these men have an increased risk of fathering a child with a chromosome abnormality is unknown as of 2001.

Men with XYY syndrome usually have normal intelligence, but it can be slightly lower than their brothers and sisters. Approximately 50% of males with XYY syndrome have learning difficulties, usually in language and reading. Speech delay can be noticed in early school years. Males with XYY syndrome may not process information as quickly as their peers and may need additional time for learning.

Males with XYY syndrome have an increased risk of behavior problems. Hyperactivity and temper tantrums can occur more frequently than expected, especially during childhood. As males with XYY syndrome become older, they may have problems with impulse control and appear emotionally immature.

As of 2001, men with XYY syndrome are not thought to be excessively aggressive or psychotic. Most men with XYY syndrome are productive members of society with no criminal behavior.

Most individuals with XYY (also described as 47,XYY) go through their entire life without being diagnosed with this condition. Chromosome studies can be done after birth on a skin or blood sample to confirm the condition. This syndrome can also be diagnosed coincidentally when a pregnant mother undergoes prenatal testing for other reasons. Prenatal tests that can determine whether or not an unborn baby will be affected with 47,XXY are the chorionic villi sampling and **amniocentesis** procedures. Both procedures are associated with potential risks of pregnancy loss and there-

fore are only offered to women who have an increased risk of having a baby born with a chromosome problem or some type of genetic condition.

Most males who have learning disabilities and/or behavior problems due to XYY syndrome have an excellent prognosis. Learning disabilities are mild and most affected males learn how to control their impulsiveness and other behavior problems. XYY syndrome does not shorten life span.

See also Chromosomal mutations and abnormalities; Genetic disorders and diseases; Hereditary diseases; Inherited cancers; Mutations and mutagenesis

Y

Y CHROMOSOME

The Y **chromosome** is one of two sex chromosomes present in many organisms. It never occurs alone but is paired with an **X chromosome**. The Y is usually the smaller of the two, though the size and shape may vary.

In mammals, there is a significant homology between the Y chromosomes of different **species**, but the human Y chromosome is somewhat unique in having a large region of **heterochromatin**, non-coding **DNA**, that makes up most of the long arm. This heterochromatin provides a useful tool in chromosome identification, since it is the brightest region of the human **genome** when the chromosomes are stained with a fluorescent dye such as quinicrine. A quick determination about the presence or absence of a Y chromosome in a newborn with ambiguous genitalia can be made by examining metaphase cells for this bright staining region.

The short arm of the Y chromosome also has a unique structure. The distal end is called the pseudoautosomal region because it shares DNA sequences and genes with the tip of the short arm of the X chromosome, and **recombination** occurs between these two areas in exactly the same way as with recombination on the **autosomes**. The homology also serves as a point of association during chromosome pairing and **cell division**.

A comprehensive physical map of the human Y is available, and the DNA sequence is nearly complete with 34 genes mapped along the length of the chromosome. The **Human Genome Project** has confirmed that the Y chromosome is the smallest of the set, with the fewest number of base pairs, only 21.8 Mb, and less than 30% of the known genes of the second shortest chromosome. But included in its **gene** complement are some of the most critical genes for human development. SRY, sex-determining region of the Y, located on the short arm of the Y is the trigger for male development. For males, the azoospermia genes on the long arm are important in determining fertility. The gonadoblastoma gene is involved in

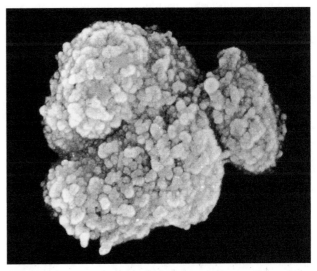

Color enhanced scanning electron microscope image of a human Y chromosome. *Photo Researchers. Reproduced by permission.*

gonadal tumors. Other known genes include such characters as stature, tooth size, blood group determination, and histocompatibility.

Genes located uniquely on the Y chromosome are said to be Y-linked. These genes are passed exclusively from father to son. Early in the history of **medical genetics**, such traits as baldness were thought to be Y-linked because males express the condition fair more commonly than females. Additional study has shown that baldness and other similar characteristics are in fact sex-influenced. This means that rather than a specific Y-linked gene for baldness, it is the higher production of male hormones that is the contributing factor to baldness. There are very few confirmed Y-linked disorders. There is one report of male to male only transmission of a form of retinitis pigmentosa (eye disorder) that has been tentatively assigned Y linkage. Hairy pinnae, or hair growing from the ear lobe, has also been suggested as Y linked, though some reports suggest there may be an autosomal form in some families.

Although there are few genes on the Y chromosome that allow clear analysis of male to male transmission, investigators have identified a number of unique DNA sequences that have contributed to population and family studies. Evaluation of these sequences has produced interesting data on evolutionary patterns of different populations. By charting the frequency, distribution, and different mutations found in these DNA regions, investigators have been able to better understand the dynamics of various populations including migration and invasion. On a smaller scale, DNA analysis of male descendents of the slave Sally Hemmings has suggested that one or more of her children may have been fathered by Thomas Jefferson. DNA patterns in the male Hemmings line bear a close resemblance to the same sequences in male descendents of Jefferson's brother (Jefferson himself has no confirmed direct male descendents).

The Y chromosome, despite its small size, remains a chromosome with a very high profile primarily due to its association with **sex determination**.

See also Sex chromosomes; SRY gene; X chromosome

Y LINKAGE · *see* Y CHROMOSOME

YANOFSKY, CHARLES (1925-)
American geneticist

Charles Yanofsky has made fundamental contributions to the understanding of how the genetic code is read and translated into proteins.

Yanofsky was born in New York City in 1925. He received his education at the City College of New York and obtained a Ph.D. in 1951 from Yale University. Yanofsky's thesis research was concerned with chemistry and microbiology. During his studies at Yale he demonstrated that a suppressor **mutation** results in the reappearance of an enzyme missing in an *Escherichia coli* mutant. Also when at Yale, he participated in research that demonstrated using immunological methods of detection that certain mutated genes are capable of producing proteins but that the proteins are inactive.

Yanofsky moved to Case Western Reserve University Medical School in Cleveland, Ohio from 1954 to 1958. In 1958 he moved to Stanford University in Palo Alto, California, where he has remained. Currently is Morris Herzstein Professor of Biology. Among his noteworthy accomplishments at Stanford is the first demonstration, using the bacterium *Escherichia coli,* that the linear **nucleotide** sequence (the sequence of **bases**) of a **gene** is responsible for a linear sequence of amino acids in its corresponding protein. In other words, a stretch of **DNA** is read in a linear fashion and

used as a template for a corresponding linear manufacture of amino acids. He also discovered the mechanisms of translational suppression and **transcription** regulation by attenuation in the regulation of activity of a target enzyme in mutants of the mold *Neurospora crassa.*

The focus of Yanofsky's research continues to be the control of **gene expression**, in particular the molecular regulatory mechanisms of bacterial transcription. He has received numerous awards and two honorary doctorates and has served as Presidents of the Genetics Society of America and the American Society of Biological Chemists. He has also been an influential mentor. Past graduate students and post doctoral fellows have become faculty members at universities such as Yale, Harvard, Columbia, UCLA and Princeton.

See also Gene regulation; Suppression; Transcription; Translation

YEAST

Yeast is a unicellular fungus. There are many different **species** of yeast, but the majority of them are acomycetes. They all reproduce by a process of budding and they all have the ability to ferment sugars. Yeasts are a rich source of protein and vitamins, particularly of the B complex. There are two main industries that revolve around the use of yeast and their metabolic products. The first of these industries is baking, where carbon dioxide from the breakdown of sugar bubbles through the dough, making the bread rise. The second industry is the alcohol industry where the sugar is turned to alcohol by the action of the yeast.

One of the most common species encountered both in bread and alcohol production is *Saccharomyces cerevisiae.* Historically, this species has also been used in genetic research because it contains 17 linkage groups, which have some 150 known mutants associated with them. Mitochondrial mutants have also been extensively studied. Many other species of yeast exist and some companies have their own species that they use for the production of alcohol. (Different yeasts supposedly impart radically different tastes to alcohol.)

Yeasts are not a formal taxonomic group, but a growth form shown by a widely unrelated group of organisms. Some filamentous forms of fungi will show a yeast form under certain conditions. There are between five and six hundred species of yeast.

See also Fungal genetics; Mitochondrial DNA

YEAST ARTIFICAL CHROMOSOMES (YAC)

Yeast artificial chromosomes (YAC) are **cloning vectors**. Regular **cloning** vectors have size limits for their **DNA** inserts. **Plasmids** can take up to 10 kilobases (kb) of DNA. Bacteriophages take 15 kilobases of DNA. Cosmids can take up to 50 kilobases of DNA. YACs take DNA stretches of 50–2000 kb. YACs possess all the sequence elements for a **chromosome** to replicate in yeast. YACs have a yeast **replication** origin, and have a centromere to allow the proper segregation of the two chromosomes into daughter cells. YACs have two telomeres to seal the ends of the chromosome. YACs also have a cloning site like other vectors, and two arms. On each arm, there is a selectable marker. There is usually a third marker that overlaps with the cloning site. Loss of the third marker distinguishes YACs with foreign DNA from the ones that do not have insert.

YACs have unique restriction sites at the cloning site and at the DNA sequence between the two telomeres. Restriction digestion with these two **enzymes** results in two pieces of linear DNA fragments. Each of the fragments has a **telomere** at one end and a sticky or blunt terminus at the other end. Chromosomal DNA of the targeted organism is digested with the same enzyme as the one at the cloning site. Ligation (attachment) of foreign DNA with the digested YAC allows the joint of the two ends. A perfect chromosome with the foreign DNA insert is then formed. The ligated DNA can then be transformed into yeast cells for further screening and identification. YAC's capacity for large DNA fragments allows researchers to study large intact DNA regions in detail. YACs make it possible to clone one large **gene**, or gene complexes that exceed the limits of other cloning vectors. By restriction digestion of YAC clones, researchers are able to construct physical chromosomal maps. Another major advantage of cloning in with YACs is that it allows the cloning of some sequences that are stable and intact in yeast, but unstable or absent when cloned into prokaryotic systems.

See also Genomic library

Z

ZAMECNIK, PAUL CHARLES (1912-)
American physician and geneticist

In late 1955, and early 1956, Paul Zamecnik proposed the existence of **transfer RNA**, a **ribonucleic acid** like **DNA** that could form a complement to the genetic strand, and carry its information from the **nucleus** of the **cell** out into the cytoplasm, where proteins are made. Zamecnik then announced the discovery of **ribosomes**, small globular bodies in the cell's cytoplasm that appear to read the stretch of transfer **RNA** and bring in the **amino acid** building blocks it specifies to assemble the protein. "It seemed to be a spool on which the reaction took place," Zamecnik said. Then two competing labs were able to put the finishing touches on Zamecnik's idea, decoding the language of transfer RNA and identifying yet another code, **messenger RNA**, which worked between DNA and transfer RNA. For the next 20 years, Zamecnik turned his attention to research that created a whole new field of inquiry for the pharmaceutical industry, fighting diseases of **protein synthesis** by jamming the cell's DNA signals. The mechanism for this was something now called antisense DNA, a complementary short strand of DNA that can be used to bind to a piece of messenger RNA and stop it from working.

Zamecnik targeted a **virus** for his research, the Rous sarcoma virus, which caused **cancer** in chickens. As he struggled to sequence the virus's genes and then make the complementary DNA sequences that would prevent it from copying itself, the tools of **biotechnology** took great strides forward. By 1978, Zamecnik finally had his breakthrough against Rous sarcoma in a strand of man-made DNA that blocked the virus' ability to copy itself. Other scientists previously learned how to sequence genes, chop them into manageable pieces, and make new sequences to order. Using these new tools, Zamecnik was able to make his antisense approach work. "Our results indicated that the small pieces could get in and affect the metabolism of cells. I was astonished that the (man-made sequences) did get into the cell, and blocked the **replication** of the Rouse's sarcoma cells." Zamecnik has been called "the father of antisense," but Zamecnik said he fought the use of the term for years, because it sounds so much like "nonsense." Drug companies have been striving, since his 1978 breakthrough, to devise man-made antisense strands that will effectively block the genetic signals that cause protein-related diseases. While most drugs in use today treat the disorder after the cell has manufactured an errant protein, the antisense approach is believed to have great promise in the treatment of hepatitis, cancers, coronary artery disease and many other disorders, before they occur in the cell.

Born in Cleveland, Ohio, Zamecnik attended Dartmouth College where he obtained his M.D. degree. Zamecnik was a professor of oncologic medicine and director of the Huntington Lab at Harvard from 1956 to 1979. He was also a physician at Massachusetts General Hospital for the same period. Upon retiring from Harvard, he joined the Worcester Foundation for Experimental Biology in Shrewsbury, Massachusetts, as principal scientist. In 1989, he founded a company, Hybridon, to pursue antisense drugs. He left the Worcester Foundation in July 1997,when it was acquired by the University of Massachusetts, and went to work at Hybridon full time, keeping long hours in the lab well into his 80s. "I'm too old to retire now," Zamecnik told colleagues a month before his 85th birthday. "I've muffed it. I'm not very good at gardening or at hammering nails and I don't consider this work. It's interesting. Like watching a horse race." When he received the first ever-awarded Albert Lasker Award for Special Achievement in Medical Science in 1996, Zamecnik was cited for his "brilliant and original science that revolutionized **biochemistry** and created an entirely new field of scientific inquiry." He was also awarded a National Medal of Science in 1991, The National Cancer Society National Award in 1968, and several honorary degrees. Zamecnik has been married to Mary Zamecnik since 1936, and she still assists him in the lab. They have three children.

See also Antisense RNA; Transcription; Translation; Viral genetics

ZINDER, NORTON DAVID (1928-)
American molecular geneticist

Norton Zinder is the John D. Rockefeller Jr. professor of molecular **genetics** at Rockefeller University in New York City. Zinder also serves as the university's dean of graduate and postgraduate studies. Zinder is known primarily for his research during the late 1940s and early 1950s, when he discovered a new mechanism of genetic transfer called bacterial **transduction**. This process refers to the transfer of genetic material between **bacteria** through bacterial viruses. The discovery has shed new light on the location and behavior of bacterial genes. Zinder is the recipient of numerous awards and honors, including the 1962 Eli Lilly Award in Microbiology and Immunology from the American Society of Microbiology.

Norton David Zinder, the older of two boys, was born in New York City to Harry Zinder, a manufacturer, and Jean (Gottesman) Zinder, a homemaker. He attended New York City public schools, graduating from the prestigious Bronx High School of Science, and went on to attend Columbia University, where he received his B.A. in biology in 1947. The following year, at the recommendation of Francis Ryan, a professor of zoology at Columbia and in whose laboratory he had worked, Zinder commenced his graduate career at the University of Wisconsin. There, he studied under American geneticist **Joshua Lederberg**, who had already discovered genetic conjugation (a specialized cellular reproductive process) a few years earlier, and who would win a Nobel Prize in 1958 for his viral and bacterial research. Zinder focused his research on **microbial genetics** (the study of the genetics of microorganisms), at a time when the field was relatively new and when many basic phenomena were as yet undiscovered.

In 1946, Lederberg had researched reproduction of *Escherichia coli*— a bacterium that is found in the intestinal tract of animals and which can cause bacterial dysentery. Zinder wished to continue Lederberg's investigations, and he chose to study the closely related genus of *Salmonella*— bacteria that cause illnesses such as typhoid fever or food poisoning in humans and other warm-blooded animals. For his work, Zinder needed to obtain large numbers of mutant strains, which were, at the time, acquired by randomly testing the survivors among bacteria that had been treated with mutagens, or agents that increase both the chance and extent of **mutation**. Zinder, however, wanted to experiment with a different method of acquisition. He knew that mutant bacteria will not grow in a nutritionally deficient medium and that the antibiotic penicillin will kill only growing bacteria. So, he was able to collect bacteria into an environment that was nutritionally inadequate, then kill any normal bacteria by administering penicillin.

Zinder obtained large numbers of mutant bacteria using this method, and he began his experiments to investigate conjugation in *Salmonella;* however, instead of observing conjugation, he stumbled upon a different method of genetic transfer in bacteria: genetic transduction. As Zinder continued his research, he determined that genetic material is transferred from one bacterial **cell** to another by means of a phage, or a **virus** that invades the bacterial cell, assumes control over the cell's genetic material, reproduces, then eventually destroys the cell. Zinder's discovery of this genetic transfer has led to further studies into the mapping and behavior of genes found in bacteria. For example, Milislav Demerec and other researchers at New York's Cold Spring Harbor Laboratory later found that the bacterial genes that regulate biosynthetic steps are grouped in what have become known as "operons," a term coined in 1960 to describe closely linked genes that function as an integrated whole.

In subsequent investigations, Zinder and his team also discovered the F2 phage, very small in size and the only virus known to contain **RNA** (**ribonucleic acid**) as its genetic substance. The researchers ascertained that the RNA generated by the virus contains codes for specific amino acids—the building blocks of protein molecules—as well as signals to control the termination and initiation of protein chains.

Zinder received his M.S. in genetics in 1949 from the University of Wisconsin and married Marilyn Estreicher in December of that same year; the couple eventually had two sons, Stephen and Michael. In 1952, Zinder completed his Ph.D. in medical microbiology, then accepted the post of assistant professor at Rockefeller University (then Rockefeller Institute for Medical Research). By 1964. he had become a full professor of genetics, and approximately ten years later he was named John D. Rockefeller Jr. Professor of Molecular Genetics; in 1993 he was appointed dean of graduate and postgraduate studies. The primary focus of Zinder's research has been in the molecular genetics of phages.

In addition to his positions at Rockefeller, Zinder also has been associated with other institutions. In the mid–1970s he began lengthy affiliations with the science departments of Harvard University, Yale University, and Princeton University, and, beginning in the same period, he also worked in the viral cancer program at the National Cancer Institute. In 1988, he assumed the position of chair of the program advisory committee for the National Institute of Health (NIH) **human genome project**, and remained in that capacity for three years. He has served in editorial capacities for scientific journals, such as *Virology* and *Intervirology,* and has published numerous articles in professional journals.

Throughout his career Zinder has received several honors, including the United States Steel Award in Molecular Biology from the National Academy of Sciences in 1966, the Medal of Excellence from Columbia University in 1969, and an honorary doctorate of science from the University of Wisconsin in 1990. He was named a fellow of the American Academy of Arts and Sciences, and is associated with such organizations as the National Academy of Sciences, the American Society of Microbiology, Genetics Society of America, the American Society of Virology, and HUGO (Human **Genome** Organization).

See also Bacterial genetics; Bacteriophage; Operon

ZINKERNAGEL, ROLF M (1944-)

Swiss microbiologist

Rolf M. Zinkernagel, a Swiss microbiologist, was born in Basel, Switzerland on January 6, 1944. Zinkernagel attended medical school in 1962 at the University of Basel and graduated in 1968. From 1969 to 1970, he worked as a postdoctoral fellow on electron microscopy at the Institute of Anatomy, University of Basel and from 1971 to 1973 at the Institute of Biochemistry, University of Lausanne, Switzerland.

In 1973, Zinkernagel joined the Department of Microbiology at the Australian National University in Canberra to study immunity of infectious disease. Zinkernagel worked with Peter Doherty who studied inflammatory processes of the brain. Utilizing mice with lymphocytic choriomeningitis **virus** (LCMV), Zinkernagel and Doherty researched the immune responses that led to the discovery of **major histocompatibility complex** (**MHC**) restriction. They determined that T-lymphocytes (white blood cells) must recognize the foreign microorganism, in this case the virus, as well as the self molecules in order to effectively kill virus-infected cells. This discovery came about when one strain of mice with LCMV developed killer T-lymphocytes that were able to protect the mice from the virus. However, when the T-lymphocytes were placed with virus-infected cells from another strain of mice in vitro, the T-lymphocytes did not kill the cells infected with the virus. The self molecules that are necessary in order for T-lymphocytes to recognize the foreign microorganism became known as major histocompatibility antigens. Zinkernagel and Doherty proposed a structural model to explain how the T-lymphocytes recognize both foreign microorganisms and major histocompatibility antigens. A peptide from a foreign microorganism becomes bound with a major histocompatibility antigen that forms a complex recognized by T-cell receptors, recognition molecules of T-lymphocytes.

In 1996, Zinkernagel was awarded the Nobel Prize in Physiology or Medicine in conjunction with Doherty for their discovery on the specificity of the **cell** mediated immune defense. Their discovery has provided further understanding of how the immune system can determine the difference between foreign microorganisms and major histocompatibility antigens and is relevant to certain diseases such as **cancer**, multiple sclerosis, diabetes, and rheumatic conditions. Understanding how the immune system works provides new avenues for the development of vaccines.

Since 1992, Zinkernagel has served as the head of the Institute of Experimental Immunology in Zurich, Switzerland.

See also Antibody and antigen; Immunogenetics; Major histocompatibility complex (MHC)

ZUCKERKANDL, ÉMILE (1922-)

Austrian molecular biologist

Émile Zuckerkandl's fame as a geneticist rests almost entirely upon three papers he co-wrote with **Linus Pauling** (1901-

1994): "Molecular Disease, Evolution and Genetic Heterogeneity," in *Horizons in Biochemistry*, "Evolutionary Divergence and Convergence in Proteins," in *Evolving Genes And Proteins*, and "Molecules as Documents of Evolutionary History," in the *Journal of Theoretical Biology*. In these three articles, Zuckerkandl and Pauling presented their hypothesis of the molecular evolutionary clock. According to this theory, organic macromolecules such as proteins and nucleic acids evolve at constant and discoverable rates.

Zuckerkandl, not to be confused with the anatomist Emil Zuckerkandl (1849-1910), was born as Emil in Vienna, Austria, but later changed his first name to the French form. As a member of a prominent and established Austrian-Jewish clan, he is related not only to the anatomist, but also to the surgeon Otto Zuckerkandl (1861-1921), the novelist and journalist Berta Zuckerkandl-Szeps (1864-1945), the political economist Robert Zuckerkandl (1856-1926), the musician and philosopher Victor Zuckerkandl (1896-1965), the translator Mimi Zuckerkandl (fl. 1920s), and Amalie Zuckerkandl, who sat for an unfinished portrait by Gustav Klimt (1862-1918) and died in the Nazi holocaust. His parents, Friedrich (later Frédéric) Zuckerkandl and Gertrud (later Gertrude) Zuckerkandl-Stekel, took the family out of Austria in 1938 just after the Anschluss, settled first in Paris, but soon came to America to avoid further danger. After World War II,the Zukerkandls returned to France and became French citizens.

Zuckerkandl earned his M.S. at the University of Illinois at Urbana in 1947 and his doctorate in biochemistry at the Sorbonne in 1959. In 1960, an abridged version of his doctoral dissertation appeared as *Hémocyanine et cuivre chez un crustacé décapode, dans leurs rapports avec le cycle d'intermue* (Hemocyanin and Copper in a Decapod Crustacean in Their Relation to the Cycle Between Molts). As a postdoctoral fellow at the California Institute of Technology from 1959 to 1964, Zuckerkandl was Pauling's student. After 1963, he had several professional affiliations with the Stanford University Department of Biological Sciences. From 1967 to 1980, Zuckerkandl was research director of the National Center for Scientific Research in Montpellier, France, affiliated with the Center for Biophysical and Biochemical Research.

In the 1960s and 1970s, Zuckerkandl established a significant scientific correspondence with **Joshua Lederberg**. He founded in 1971, and until 1999, edited the *Journal of Molecular Evolution*. In 1976, he co-authored *Évolution: génétique des populations, évolution moléculaire* (Evolution: **Population Genetics**, Molecular Evolution) with Claudine Petit. From 1980 to 1992, Zuckerkandl was president and director of the Linus Pauling Institute of Science and Medicine, Palo Alto, California. In 1992, Zuckerkandl co-founded with Teni Boulikas the Institute of Molecular Medical Sciences, also in Palo Alto, and served as its first president and director. The International Society of Molecular Evolution, which he helped to found in 1993, honored him in Guanacaste, Costa Rica, in January 1997 with a symposium called "**Junk DNA**: The Role and the

Evolution of Non-Coding Sequences." Zuckerkandl's decorations include the French Order of Merit and the Austrian Cross of Honor.

See also Amino acid; Molecular biology and molecular genetics; Molecular clocks; Mutation; Phylogeny; Sequencing

ZYGOTE

In animals, a zygote is a fertilized egg, formed by the fusion of a male **gamete** (or sperm) and a female gamete (or egg). Male and female gametes (collectively, these are referred to as sex cells) are the unicellular products of **meiosis**, a kind of reduction cellular division that occurs in specialized organs of sexually reproducing animals (the ovary of females, and the testes of males). Meiosis results in the formation of cells having only one of the two complementary (or homologous) sets of chromosomes possessed by animals (that is, gametes are **haploid** cells). However, the zygote formed from their union has two sets of chromosomes (i.e., it is **diploid**). Because one of the sets of chromosomes has been obtained each parent, the genetic information (or **genome**) of the offspring represents a unique combination of **DNA** (**deoxyribonucleic acid**, the genetic biochemical of animals). Zygotes genetically are capable of developing into adult animals.

In plants, the product of meiosis is not gametes. Rather, this reduction division produces multicellular, haploid organisms, which then go on to produce haploid sex cells (or true gametes). The male plant gamete is known as pollen, and the female as an ovule. Fertilized ovules are diploid, and represent a unique genome, having obtained one of its two sets of chro-

Scanning electron micrograph of fertilized human egg (zygote). *Photograph by Dr. Yorgos Nikas/Science Photo Library/Photo Researchers, Inc. Reproduced by permission.*

mosomes from each of its parents. Fertilized ovules develop into seeds, which are capable of germinating and growing in to an adult plant.

See also Cell proliferation; Sexual reproduction

ZYGOTIC SELECTION · *see* SELECTION

Sources Consulted

Books

Adelmann, H. B. *The Embryological Treatises of Hieronymus Fabricius of Aquapendente*. Ithaca, NY: Cornell University Press, 1942.

Alberts, B., et al. *Molecular Biology of the Cell* 3rd ed. NewYork: Garland Publishing, Inc., 1994.

Alberts, B., et al. *Molcular Biology of the Gene*, 2nd ed. New York: Garland Publishing, Inc., 1989.

Allaby, M. *The Concise Oxford Dictionary of Botany*. Oxford: Oxford University Press, 1992.

Allen, Garland E. William E. Castle, Charles C. Gillispie, eds. *Dictionary of Scientific Biography*, Vol. 3, New York: Scribner, 1971.

Allen, Garland Edward. *Thomas Hunt Morgan: The Man and His Science*. Princeton, NJ: Princeton University Press, 1978.

American Council of Learned Societies, Marshall DeBruhl ed. *Dictionary of Scientific Biography*. New York: Scribner, 1980.

Anderson, L. *Charles Bonnet and the Order of the Known*. Boston, MA: D. Reidel, 1984.

Ashburner, Michael, and Edward Novitski, eds. *The Genetics and Biology of Drosophila*. New York: Academic Press, 1976-1986.

Astor, Gerald. *The "Last" Nazi: The Life and Times of Dr. Joseph Mengele*. New York: Fine, 1985.

Attwood, T.K., and D.J. Parry-Smith. *Introduction to Bioinformatics*. New York: Longman Higher Education. 1999.

Audesirk, T., and G. Audesirk. *Biology: Life on Earth*, 4th ed. New Jersey: Prentice Hall Publishing, Inc.,1996.

Axford, R.F.E., ed. *Breeding for Disease Resistance in Farm Animals*. 2nd ed. New York: CABI Pub., 1999.

Bainbridge, B.W. *Genetics of Microbes,* Glasgow and London: Blackie and Son Ltd, 1980.

Baltzer, F. *Theodor Boveri: Life and Work of a Great Biologist*. Berkeley, CA: University of California Press, 1967.

Barch, M. J., T. Knutsen, and J. L. Spurbeck, eds. *The AGT Cytogenetics Laboratory Manual,* 3rd ed. Philadelphia: Lippincott-Raven Publishers. 1997.

Bernard, K., and R. Utz. *Bacillus thuringiensis, an Environmental Biopesticide: Theory and Practice.*, John Wiley & Sons, 1993.

Beurton, Peter, Raphael Falk, Hans-Jörg Rheinberger., eds. *The Concept of the Gene in Development and Evolution*. Cambridge, UK: Cambridge University Press, 2000.

Blobel, G. "Protein targeting (Nobel lecture)." *Chembiochem* 1(2) (18 August 2000): 87–102.

Bock, G. R., and Goode, J. A., eds. *Hyperthermophiles in the History of Life*. New York: John Wiley & Sons, 1996.

Bodmer, W. F., L. L. Cavalli-Sforza. *Genetics, Evolution and Man*. San Francisco: W.D. Freeman, 1976.

Bonner, J. T. *First Signals: The Evolution of Multicellular Development*. Princeton, NJ: Princeton University Press, 2000.

Bonner, J. T. *The Ideas of Biology*. New York: Harper & Row, 1962.

Bouchard, Thomas J., and Peter Propping, eds. *Twins as a Tool of Behavioral Genetics*. New York: Wiley, 1993.

Bowler, P. "Lamarckism" in *Keywords in Evolutionary Biology,* edited by Keller, E. and E. Lloyd, Cambridge: Harvard University Press, 1992.

Bowler, Peter J. *The Mendelian Revolution: The Emergence of Hereditarian Concepts in Modern Science and Society.* Baltimore: Johns Hopkins University Press, 1989.

Boylan, M. *Method and Practice in Aristotle's Biology.* Lanham, MD: University Press of America, 1983.

Briant, Keith Rutherford. *Marie Stopes: A Biography.* London: Hogarth, 1962.

Briggs, D. and S. M. Walters. *Plant Variation and Evolution.* Cambridge: Cambridge Univ. Press, 1984.

Brooker, R. *Genetics Analysis and Principals.* Menlo Park: Benjamin Cummings, 1999.

Buchanan, B. B., W. Gruissem, and R. L. Jones. *Biochemistry and Molecular Biology of Plants.* Rockville, MD: American Society of Plant Physiologists, 2000.

Cairns, J., G. S. Stent, and J. D. Watson, eds. *Phage and the Origins of Molecular Biology,* 2nd ed. New York: Cold Spring Harbor Laboratory of Quantitative Biology, 1992.

Campbell, N., J. Reece, and L. Mitchell. *Biology,* 5th ed. Menlo Park: Benjamin Cummings, Inc. 2000.

Capecchi, Mario, ed. *Molecular Genetics of Early Drosophila & Mouse Development.* Cold Spring Harbor, New York: Cold Spring Harbor Laboratory Press, 1989.

Caplan, Arthur L., ed. *When Medicine Went Mad: Bioethics and the Holocaust* Totowa, N.J.: Humana, 1992.

Carlson, E. A. *The Gene: A Critical History.* Philadelphia, PA: Saunders, 1966

Castle, W. E., et al. *Heredity and Eugenics.* Chicago: University of Chicago Press,

Chesler, Ellen. *Woman of Valor: Margaret Sanger and the Birth Control Movement in America.* New York: Simon and Schuster, 1992

Clark, A.J., ed. *Animal Breeding: Technology for the 21st Century.* Amsterdam: Harwood Academic, 1998.

Clarke, C. A.*Human Genetics and Medicine.* 3rd ed. Baltimore, MD: E. Arnold, 1987.

Clarke, C. A. and R. B. McConnell. *Prevention of Rh-Haemolytic Disease.* Springfield, IL: C. C. Thomas, 1972.

Coigney, Virginia. *Margaret Sanger: Rebel With a Cause.* Garden City, New York: Doubleday, 1969.

Comstock, Ralph E. *Quantitative Genetics with Special Reference to Plant and Animal Breeding.* Ames: Iowa State University Press, 1996.

Cooper, Geoffrey M. *The Cell: A Molecular Approach.* Washington D.C.: ASM Press, 1997.

Creese, I., and Claire M. Fraser, eds. *Dopamine Receptors.* New York: A.R. Liss, 1987.

Daintith, John and D.Gjertsen, eds. *A Dictionary of Scientists.* New York: Oxford University Press, 1999.

Darnell, J., H. Lodish, and D. Baltimore. *Molecular Cell Biology.* New York: Scientific American Books, Inc., 1986.

Darwin, C.R. *The Origin of the Species.* London: John Murray, 1859.

Dawkins, R. *The Selfish Gene.* Oxford: Oxford University Press. 1989.

DeGrood, David H. *Haeckel's Theory of the Unity of Nature: A Monograph in the History of Philosophy.* Boston: Christopher, 1965.

Duboule, D. *Guidebook to the Homeobox Genes.* Oxford: Oxford University Press, 1994.

Elseth, G. D., and K. D. Baumgardner. *Principles of Modern Genetics,* Minnesota: West Publishing Co., 1995.

Emde, Robert N., and John K. Hewitt, eds. *Infancy to Early Childhood: Genetic and Environmental Influences on Developmental Change.* New York: Oxford University Press, 2001.

Emery, A. E. H. *Neuromuscular Disorders: Clinical and Molecular Genetics.* Chicester : John Wiley & Sons, 1998.

Engs, R.C., ed. *Controversies in the Addiction's Field.* Dubuque: Kendal-Hunt, 1990.

Ernst & Young. "The economic contributions of the biotechnology industry to the U.S. economy." *Biotechnology Industry Organization.* 2000.

Fitzgerald, D. *The Business of Breeding: Hybrid Corn in Illinois, 1890-1940.* Ithaca: Cornell University Press, 1990.

Friedman, J., F. Dill, M. Hayden, B. McGillivray. *Genetics.* Maryland: Williams & Wilkins, 1996.

Fruton, J. S. *Molecules and Life. Historical Essays on the Interplay of Chemistry and Biology.* New York: Wiley-Interscience, 1972.

Futuyama, D. J. *Evolutionary Biology.* Sunderland, MA: Sinauer Associates, Inc., 1979.

Gasman, Daniel. *Haeckel's Monism and the Birth of Fascist Ideology.* New York: Peter Lang, 1998.

Gasman, Daniel. *The Scientific Origins of National Socialism: Social Darwinism in Ernst Haeckel and the German Monist League.* London: Macdonald, 1971.

Gehring, W. J. *Master Control Genes in Development and Evolution: The Homeobox Story.* New Haven, CT: Yale University Press, 1998.

Gilbert, S. F., ed. *A Conceptual History of Modern Embryology.* Baltimore, MD: Johns Hopkins Press, 1991.

Glick, B.R. and J. J. Pasternak. *Molecular Biotechnology, Principles and Applications of recombinant DNA, 2nd ed.* Washington: American Society of Microbiology Press, 1998.

Gould, Stephen Jay. *Ever Since Darwin: Reflections in Natural History.* New York: W.W. Norton & Co., 1977.

Grafe, A. *A History of Experimental Virology.* New York: Springer-Verlag, 1991.

Graham, L. *Science in Russia and the Soviet Union.* Cambridge: Cambridge University Press, 1993.

Grant, P. R. *Ecology and Evolution of Darwin's Finches.* New Jersey: Princeton University Press, 1986.

Griffiths, A. et al. *Introduction to Genetic Analysis,* 7th ed. New York: W.H. Freeman and Co., 2000.

Hall, Ruth. *Marie Stopes: A Biography.* London: Andre Deutsch, 1977.

Hamburger, V. *Heritage of Experimental Embryology.* New York: Oxford University Press, 1988.

Hamburger, V. *The Heritage of Experimental Embryology: Hans Spemann and the Organizer.* Oxford: Oxford University Press, 1988.

Haraway, D. J. *Crystals, Fabrics and Fields: Metaphors of Organicism in Twentieth-Century Developmental Biology.* New Haven: Yale University Press, 1976.

Herskowitz, I. H. *Genetics,* 2nd ed. Boston: Little, Brown and Company, 1965.

Hintzsche, Erich. "Rudolf Albert von Koelliker." *Dictionary of Scientific Biography,* v. 7 (1973): 437-440.

Ho, M.W. *Genetic Engineering Dream or Nightmare? The Brave New World of Bad Science and Big Business,* Dublin: Gateway, Gill & Macmillan, 1998.

Hoagland, Mahlon. *Discovery, the Search for DNA's Secrets.* Boston: Houghton-Mifflin Co., 1981.

Hoagland, Mahlon. *The Roots of Life.* Boston: Houghton-Mifflin, 1978.

Hogg, R.V., and E. A. Tanis. *Probability and Statistical Inference,* 6th ed. New Jersey: Prentice Hall, Inc. 2001.

Houdebine, L. M., ed. *Transgenic Animals: Generation and Use.* Amsterdam: Harwood Academic Publishers, 1997.

Horder, T. J., J. A. Witkowski, and C. C. Wylie, eds. *A History of Embryology.* New York: Cambridge University Press, 1986.

Hosmer, David W. Jr., and Stanley Lemeshow. *Applied Logistic Regression.* 2nd ed. John Wiley & Sons, Inc., 1989.

Hubbard, R., and Wald, E. *Exploding the Gene Myth.* Boston: Beacon Books, 1999.

Hughes, Arthur. *A History of Cytology* London: Abelard-Schuman, 1959.

Hughes, S. *The Virus: A History of the Concept.* New York: Science History Publications, 1977.

Ingram, V. M. *Hemoglobins in Genetics & Evolution.* New York: Columbia University Press, 1963.

Jacob, François. *The Logic of Life: A History of Heredity.* New York: Pantheon, 1973.

Jenkins, John B. *Human Genetics,* 2nd ed. New York: Harper & Row, 1990.

Jorde, L. B., J. C. Carey, M. J. Bamshad, and R. L. White. *Medical Genetics,* 2nd ed. Mosby-Year Book, Inc., 2000.

Johnson, George, and Peter Raven. *Biology: Principles & Explorations.* Austin: Holt, Rinehart, and Winston, Inc., 1996.

Joravsky, D. *The Lysenko Affair.* Cambridge, Massachusetts: Harvard University Press, 1970.

Judson, H. F. *The Eighth Day of Creation: Makers of the Revolution in Biology.* New York: Simon & Schuster, 1979.

Juel-Nielsen, Niels. *Individual and Environment: A Psychiatric-Psychological Investigation of Monozygotic Twins Reared Apart.* Copenhagen: Munksgaard, 1965.

Kay, L. E. *Who Wrote the Book of Life? A History of the Genetic Code.* Stanford: Stanford University Press.

Keenan, Katherine. "Lilian Vaughan Morgan (1870–1952)." *Women in the Biological Sciences: A Biobibliographic Sourcebook* Ed. Grinstein, Louise A., Carol A. Biermann, and Rose K. Rose. Westport, CT: Greenwood Press, 1997.

Keller, E. F. *A Feeling for the Organism: The Life and Work of Barbara McClintock.* New York: Freeman, 1983.

Kendrew, J., et al.*The Encyclopedia of Molecular Biology,* Oxford: Blackwell Science Ltd., 1994.

Khoury MJ, et al. *Fundamentals of Genetics Epidemiology.* Oxford: Oxford University Press, 1993.

King, R., J. Rotter, A. Motulsky, G. Arno. *The Genetic Basis of Common Genetic Diseases.* New York: Oxford University Press, 1992.

Klaassen, Curtis D. *Casarett and Doull's Toxicology,* 6th ed. McGraw-Hill, Inc. 2001.

Klug, W. and M. Cummings. *Concepts of Genetics,* 6th ed. Upper Saddle River: Prentice Hall, 2000.

Kmiec, E. B., ed. *Gene Targeting Protocols.* Totowa, N.J.: Humana Press, 2000.

Knapp, Rebecca, et al. *Clinical Epidemiology and Biostatistics.* Baltimore: Williams & Wilkins, 1992.

Kohler, R. E. *Lords of the Fly: Drosophila Genetics and the Experimental Life.*, Chicago, IL: University of Chicago Press.

Kor, Eva Mozes. *Echoes from Auschwitz: Dr. Mengele's Twins: The Story of Eva and Miriam Mozes.* Terre Haute: C.A.N.D.L.E.S., 1995.

•

Kuhn-Schnyder, Emil. *Lorenz Oken, 1779-1851: Erster Rektor der Universitat Zurich.* Zurich: Rohr, 1980.

Lewin, B. *Genes VII.* New York, Oxford University Press Inc., 2000

Lewis, Ricki. *Human Genetics: Concepts and Applications,* 2nd ed. Iowa: William C. Brown, Publishers, 1997.

Lifton, Robert Jay. *The Nazi Doctors: Medical Killing and the Psychology of Genocide.* New York: Basic Books, 1986.

Louro, Iuri D., Juan C. Llerena, Jr., Mario S.Vieira de Melo., Patricia Ashton-Prolla, Gilberto Schwartsmann, Nivea Conforti-Froes, eds. *Genética Molecular do Cancer.* Sao Paulo: MSG Produçao Editorial Ltda., 2000.

Lysenko, T. D. *Soviet Biology: A Report to the Lenin Academy of Agricultural Sciences.* New York: Birch Books, International Publishers, 1948.

Lysenko, T. D. *Agrobiology.* Moscow: Foreign Language Press, 1954.

Magner, L. *A History of the Life Sciences.* New York: Marcel Dekker, Inc., 1994

Maijala, Kalle, ed. *Genetic Resources of Pig, Sheep, and Goat.* Amsterdam and New York: Elsevier Science Publishers, 1991.

Mange, E. and A. Mange. *Basic Human Genetics,* 2nd ed. Massachusetts: Sinauer Associates, Inc., 1999.

Martini, F. H., et al. *Fundamentals of Anatomy and Physiology,* 3rd ed. New Jersey: Prentice Hall, Inc., 1995.

Matalon Lagnado, Lucette, and Sheila Cohn Dekel. *Children of the Flames: Dr. Josef Mengele and the Untold Story of the Twins of Auschwitz.* New York: Morrow, 1991.

Mayr, E. *The Growth of Biological Thought.* Cambridge, MA: Harvard University Press, 1982.

Mayr, E., and P. D. Ashlock. *Principles of Systematic Zoology,* 2nd ed. New York: McGraw-Hill, Inc., 1991.

McClatchey, K. *Clinical Laboratory Medicine.* Baltimore: Williams & Wilkins, 1994.

Mearna, J., and W.C. Koller, eds. *Parkinson's Disease and Parkinsonism in the Elderly.* New York: Cambridge University Press, 2000.

Mertz, L., *Recent Advances and Issues in Biology.* Phoenix, AZ: Oryx Press, 2000.

Meyer, A. W. *An Analysis of William Harvey's Generation of Animals.* Stanford, CA: Stanford University Press.

Micklos, David, A., Greg A. Freyer. *DNA Science, A First Course in Recombinant DNA Technology.* United States: Cold Spring Harbor Laboratory Press and Carolina Biological Supply Company, 1990.

Mischer, Sibille. *Der verschlungene Zug der Seele: Natur, Organismus und Entwicklung bei Schelling, Steffens und Oken.* Würzburg: Königshausen & Neumann, 1997.

Morgan, Thomas Hunt. *The Theory of The Gene.* New Haven: Yale University Press, 1926.

Müller-Hill, Benno. *Murderous Science: Elimination by Scientific Selection of Jews, Gypsies, and Others, Germany 1933–1945.* Oxford: Oxford University Press, 1988.

Needham, J. *The Rise of Embryology..* New York: Cambridge University Press.

Nei, M. *Molecular Evolutionary Genetics.* New York: Columbia University Press, 1987.

Nelkin, D., and M. S. Lindee. *The DNA Mystique.* New York: Freeman, 1995.

Nyiszli, Miklûs. *Auschwitz: A Doctor's Eyewitness Account.* New York: Arcade; Boston: Little, Brown, 1993.

Olby, Robert C. *Origins of Mendelism,* 2nd ed. Chicago: University of Chicago Press, 1985.

Olby, R. *The Path to the Double Helix.* Seattle, WA: University of Washington Press, 1974.

Oppenheimer, J. M. *Essays in the History of Embryology and Biology.* Cambridge, MA: MIT Press, 1967.

Ottaway, J. H., D. K. Apps. *Biochemistry,* 4th ed. Edinburgh: Baillier Tindall, 1986.

Pinto-Correia, C. *The Ovary of Eve: Egg and Sperm and Preformation.* Chicago, IL: University of Chicago Press, 1997.

Portugal, F. H. and J. S. Cohen. *A Century of DNA.* Cambridge, MA: The MIT Press.

Primrose, S. P. *Principles of Genome Analysis.* Oxford: Blackwell, 1995.

Renfrew, C., and K. Boyle, eds. *Archaeogenetics: DNA and the population prehistory of Europe.* McDonald Institute.

Rieger, R., A. Michaelis, and M. M. Green. *Glossary of Genetics and Cytogenetics,* 4th ed. Berlin: Springer Verlag, 1976.

Rifkin, J.*The Biotech Century.* Putnam Publishing Group. 1998.

Ritter, B., et al. *Biology,* B.C. ed. Scarborough: Nelson Canada, 1996.

Roe, S. A. *Matter, Life, and Generation. Eighteenth Century Embryology and the Haller-Wolff Debate.* New York: Cambridge University Press.

Roger, P.L. and Fleet, G.H. *Biotechnology and the Food Industry.* New York: Godon and Breach Science Publishers, 1987.

Rose, June. *Marie Stopes and the Sexual Revolution.* London: Faber and Faber, 1992.

Rothwell, Norman V. *Human Genetics.* New Jersey: Prentice-Hall, 1977.

Rothwell, Norman V. *Understanding Genetics,* 4th ed. New York: Oford University Press, 1988.

Russell, P. J. *Genetics,* 3rd ed. New York: Harper Collins, 1992.

Sachs, J .*History of Botany (1530-1860),* trans. Henry E. F. Garnsey. Oxford: Clarendon Press, 1890.

Sager, R and F. J. Ryan. *Cell Heredity.* New York: John Wiley & Sons, 1961.

Sambrook, J., E. F. Fritsch, and T. Maniatis. *Molecular Cloning: A Laboratory Manual,* 2nd ed. Cold Spring Harbor, N.Y.: Cold Spring Harbor Laboratory Press, 1989.

Sanderson, Michael J., and Larry Hufford, eds. *Homoplasy: The Recurrence of Similarity in Evolution.* San Diego: Academic Press, 1996.

Sapp, J. *Beyond the Gene: Cytoplasmic Inheritance and the Struggle for Authority in Genetics.* New York: Oxford University Press, 1987.

Sapp, J. *Where the Truth Lies: Franz Moewus and the Origins of Molecular Biology.* New York: Cambridge University Press, 1990.

Sayre, A. *Rosalind Franklin & DNA.* New York: Norton, 1975.

Scheinfeld, Amram. *Twins and Supertwins.* London: Chatto & Windus, 1968.

Schottenfeld, D., and J.F. Fraumeni Jr., eds. *Cancer Epidemiology and Prevention* New York: Oxford University Press, 1996.

Scriver, Charles R., et al. *The Metabolic and Molecular Bases of Inherited Disease.* 8th ed. New York: McGraw-Hill Professional Book Group, 2001

Seashore, M. and R. Wappner. *Genetics in Primary Care & Clinical Medicine.* Stamford: Appleton and Lange, 1996 .

Segal, Nancy L. *Entwined Lives: Twins and What They Tell Us About Human Behavior.* New York: Plume, 2000.

Seibel, H. R. "Chaperonin Structure and Conformational Changes." in *The Chaperonins.* New York: Academic Press, 1996.

Shine, Ian, and Sylvia Wrobel. *Thomas Hunt Morgan: Pioneer of Genetics.* Lexington: University Press of Kentucky, 1976.

Simpson, G. G. *Principles of Animal Taxonomy.* New York: Columbia University Press, 1961.

Singer, M. and P. Berg. *Genes and Genomes.* Mill Valley, CA: University Science Books, 1991.

Smith, A.D., et al. *Oxford Dictionary of Biochemistry and Molecular Biology.* New York: Oxford University Press, Inc., 1997.

Snustad, D. Peter, Michael J. Simmons, and John B. Jenkins. *Principles of Genetics.* New York: John Wiley, 1997.

Soyfer, V. *Lysenko and the Tragedy of Soviet Science.* New Jersey: Rutgers University Press, 1994.

Solomon, Eldra Pearl, Linda R. Berg, and Diana W. Martin. *Biology,* 5th ed. New York: Saunders College Publishing, 1999.

Soyfer, V. N. *Lysenko and the Tragedy of Soviet Science,* trans. L. Gruliow and R. Gruliow, New Brunswick: Rutgers University Press, 1994.

Spector, D. L., R. D. Goldman, and L. A. Leinwand. *Cells: A Laboratory Maual,* Plainview, N.Y.: Cold Spring Harbor Laboratory Press, 1998.

Spemann, H. *Embryonic Development and Induction.* New York: Hafner Publishing Company, 1967.

Stent, G. S., ed. *The Double Helix. A Personal Account of the Discovery of the Structure of DNA.* New York: Norton, 1980.

Stern, M.B., and H.I. Hurtig. *The Comprehensive Management of Parkinson's Disease* New York: PMA Publishing Corp., 1988.

Strachan, T., and A. Read. *Human Molecular Genetics.* 1st ed. Oxford: Bios Scientific Publishers Ltd., 1996.

Strachan, T. and A. Read. *Human Molecular Genetics.* New York: Bios Scientific Publishers, 1998.

Stryer, L., *Biochemistry,* 4th ed. New York: W.H: Freeman and Co., 1995.

Sturtevant, A. H. *A History of Genetics.* New York: Harper & Row, 1965.

Syvanen, M., and Kado, C., eds. *Horizontal Gene Transfer.* New York: Chapman and Hall, 1998.

Tanzi, Rudolph, Ann B. Parsons. *Decoding Darkness.* Cambridge, MA: Perseus Publishing, 2000.

Thompson, M., et al. *Genetics in Medicine.* Philadelphia: Saunders, 1991.

Tomei, L. David, F. O. Cape,eds. *Apoptosis: The Molecular Basis of Cell Death.* Cold Spring Harbor, N.Y.: Cold Spring Harbor Laboratory Press, 1991.

Vassiliki, Betty Smocovitis. *Unifying Biology: The Evolutionary Synthesis and Evolutionary Biology.* Princeton, N.J.: Princeton University Press, 1996.

Verma, R.S., and A. Babu. *Human Chromosomes Principles and Techniques,* 2nd ed. McGraw-Hill Inc., Health Professions Division. 1995.

Voet, D., and J. Voet, *Biochemistry,* 2nd ed. New York: John Wiley and Sons, Inc., 1995.

Vogelstein, B & Kinzler, K., eds. *The Genetic Basis of Human Cancer,* 1st ed. New York: McGraw-Hill, 1998.

Waterson, A. and L. Wilkinson. *An Introduction of the History of Virology.* Cambridge: Cambridge University Press, 1978.

Watson, J. D., et al. *Recombinant DNA.* 2nd ed. New York: Scientific American Books, Inc., 1992.

Willier, B. H. and J. M. Oppenheimer, eds. *Foundations of Experimental Embryology.* New York: Hafner Press, 1974.

Wilson, E. O. *The Diversity of Life.* Cambridge, MA: The Belknap Press of Harvard University Press, 1992.

Winsor, M. P. *Starfish, Jellyfish, and the Order of Life. Issues in Nineteenth-Century Science.* New Haven, CT: Yale University Press, 1976.

Periodicals

Acierno, L.J., and T. Worrell. "Profiles in Cardiology: James Bryan Herrick." *Clinical Cardiology* no. 23 (2000): 230–232.

Adams, K.L., et al. *Intracellular Gene Transfer in Action: Dual Transcription and Multiple Silencings of Nuclear and Mitochondrial Cox2 Genes in Legumes.* Proceedings of the National Academy of Sciences of the United States of America 96 (1999): 13863–13868.

Adams, K.L., et al. "Repeated, Recent and Diverse Transfers of a Mitochondrial Gene to the Nucleus in Flowering Plants." *Nature* 408 (2000): 354-357.

Adams, M.D., et al. "Complementary DNA Sequencing: Expressed Sequence Tags and Human Genome Project." *Science* 252 (21 June 1991): 1651–1656.

Alexander, Leo. "Medical Science Under Dictatorship." *New England Journal of Medicine* 241, no. 2 (1949): 39–47.

Alonso, S., and J. A. Armour. "A Highly Variable Segment of Human Subterminal 16p Reveals a History of Population Growth for Modern Humans Outside Africa." *Proceedings of the National Academy of Sciences of the United States of America* 98 (2001): 864–869.

American Judicial Society. "Genes and Justice: The Growing Impact of the New Genetics on the Courts," *Judicature* 83 (November-December 1999): special issue.

Amir, R.E., et al. "Rett Syndrome is Caused by Mutations in X-linked MECP2, Encoding Methyl-CpG-binding Protein 2." *Nature Genetics* 23 (October 1999): 185–188.

Artlett, Carol M. et al. "Identification of Fetal DNA and Cells in Skin Lesions From Women With Systemic Sclerosis." *The New England Journal of Medicine* 338:1186–1191.

Associated Press. "New Cancer Treatment Starves Tumors." *The Detroit News,* 28 November 1997.

Audic, S., and E. Beraud-Colomb. "Ancient DNA is thirteen years old." *Nature Biotechnology,* 15 (1997): 855–860.

Averbeck, D., and S. Averbeck. "DNA Photodamage, Repair, Gene Induction and Genotoxicity Following Exposures to 254 nm UV and 8-methoxypsoralen Plus IVA in an Eukaryotic Cell System." *Photochemistry & Photobiology* 68 (1998): 289–295.

Avery, O. T., C. M. MacLeod, and M. McCarty. "Studies on the Chemical Nature of the Substance Inducing Transformation of Pneumococcal Types." *Journal Exp. Med.* 79 (1944)137–158.

Ayala, Francisco J., Walter M. Fitch, and Michael T. Clegg, eds. "Variation and Evolution in Plants and Microorganisms: Toward a New Synthesis 50 Years after Stebbins." *Proceedings from the National Academy of Sciences of the United States of America* 2000.

Barlowe, C. et al. "COPII: a Membrane Coat Formed by Sec Proteins That Drive Vesicle Budding From the Endoplasmic Reticulum." *Cell* 77 (June 1994): 895–907.

Barbajosa, Cassandra. "DNA Profiling, The New Science of Identity." *National Geographic* (May 1992): 112–124.

Baxevanis, A.D. "The Molecular Biology Database Collection: an Updated Compilation of Biological Database Rresources." *Nucleic Acids Research* 29 (January 2001): 1–10.

Baylor College of Medicine. "Sensorineural Hearing Loss." 2000.

Baylor College of Medicine. "Childhood Hearing Loss." 2000.

Berg, P., et al. "Asilomar Conference on Recombinant DNA Molecules." *Science* no. 188 (6 June 1975): 991–994.

Bergers, Gabriele, et al. "Effects of Angiogenesis Inhibitors on Multistage Carcinogenesis in Mice." *Science* 284 (April 1999): 808.

Betsch, D. F. "DNA Fingerprinting in Agricultural Genetics Programs." *Biotechnology Information Series (Bio-7), North Central Regional Extension Publication.* Iowa State University 1994.

Birmes, A., A. Sattler, K. H. Maurer, D. Riesner. "Analysis of the Conformational Transitions of Proteins by Temperature-gradient Gel Electrophoresis." *Electrophoresis* 11 (1990): 795–801.

Boguski, M. S. "The Turning Point in Genome Research." *Trends in Biochemical Sciences* 20 (August 1995): 295–296.

Bonner, J. T. "The Origins of Multicellularity." *Integrative Biology* 1 (1998): 27–36

Bouchez, D., and Höfte, H. "Functional Genomics in Plants" *Plant Physiology* 118 (November 1998): 725–732.

Brinster, R. "The Effect of Cells Transferred Into Mouse Blastocyst on Subsequent Development." *Experimental Medicine* (1974): 1049–1056.

Britten, R. J., D. B. Stout, and E. H. Davidson. "The Current Source of Human Alu Retroposons is a Conserved Gene Shared with Old World Monkey." *Proceedings of the National Academy of Sciences of the United States of America* 86 (1989): 3718–3722.

Britten, R. J. and D. E. Kohne. "Repeated Sequences in DNA." *Science* 161 (1968): 529–540.

Britten, R. J. and E. H. Davidson. "Repetitive and Non-repetitive DNA Sequences and a Speculation on the Origins of Evolutionary Novelty." *Quarterly Review of Biology* 46 (1971): 111–138.

Brown, W. L., and E. O. Wilson. "Character Displacement." *Systematic Zoology* 5 (June 1956): 49–64.

Bueler, H., A. Aguzzi, A. Sailer, R.-A. Greiner, P. Autenried, M. Aguet, and C. Weissmann. "Mice Devoid of PrP are Resistant to Scrapie. *Cell* 73 (1993): 1339–1347.

Bunick, D., et al. "Mechanism of RNA Polymerase II - Specific Initiation of Transcription *In Vitro*: ATP Requirement and Uncapped Runoff Transcripts." *Cell* 29(3) (July 1982): 877–886.

Caballero A. and M. A. Toro. "Interrelations Between Effective Population Size and Other Pedigree Tools for the Management of Conserved Populations." *Genet Res* 75(3) (June 2000):331–43.

Campisi, Judith. "Replication Senescence: An Old Lives' Tale?" *Cell* 84 (23 February 1996): 497–500.

Capecchi, Mario. R. "Altering the genome by homologous recombination." *Science* no. 244 (1989):1288–1292.

Caplan, A. L."If Gene Therapy is the Cure, What is the Disease?" *Gene mapping* (1992): 128-141.

Chagnon, Y. C., et al. "The Human Obesity Gene Map: the 1999 Update." *Obesity Research* 8 (2000): 89–117.

Chakravarti, A. "To a Future of Genetic Medicine" *Nature* 409 (2001): 822–823.

Chaw, S. M., et al. "Seed Plant Phylogeny Inferred From All Three Plant genomes: Monophyly of Extant Gymnosperms and Origin of Gnetales from Conifers." Proceedings of the National Academy of Sciences of the United States of America 97 (2000): 4086–4091.

Cho, Y., and J. D. Palmer. "Multiple Acquisitions Via Horizontal Transfer of a Group I Intron in the Mitochondrial Cox1 Gene During Evolution of the Araceae Family." *Molecular Biology and Evolution* 16 (1999): 1155–1165.

Cho, Y., et al. "Explosive Invasion of Plant Mitochondria By a Group I Intron." *Proceedings of the National Academy of Sciences of the United States of America* 95 (1998): 14244–14249.

Church, G. M., and S. Kiefer-Higgins. "Multiplex DNA Sequencing." *Science* 240 (April 1998): 185–188.

Collins F. S., and V. A. McKusick. "Implications of the Human Genome Project for Medical Science." *JAMA* 285 (7 February 2001): 540–544.

Conforti-Froes, N., et al. "Predisposing Genes and Increased Chromosome Aberrations in Lung Cancer Cigarette Smokers." *Mutation Research* 379 (1997): 53–59.

Crewdson, John. "Trial Success in the Details, Folkman says." *Chicago Tribune* 17 October 1999.

Crick, F. H. C. "The Origin of the Genetic Code" *Journal of Molecular Biology* 38 (1968): 367–379.

Csonka, E., et al. "Novel Generation of Human Satellite DNA-based Artificial Chromosomes in Mammalian cells." *Journal of Cell Science* 113 (2000): 3207–3216.

Darlington, C. D., T. D. Lysenko (Obituary). *Nature* 266 (1977): 287–288.

Darnell, J. E., Jr. "The Processing of RNA." *Scientific American* 249 (1983): 90-100.

Darnell, J. E., Jr., et. al. "Jak-STAT Pathways and Transcription Activation in Response to IFNs and Other Extracellular Signaling Proteins." *Science* 264 (1994): 1415-1421.

DaSilva, E., "Biological Warfare, Terrorism, and the Biological Toxin Weapons Convention." *Electronic Journal of Biotechnology* no. 3 (December 15, 1999): 1–17.

Dieter Kotzot D. "Abnormal Phenotypes in Uniparentaldisomy (UPD): Fundamental Aspects and a Critical Review With Bibliography of UPD Other Than 15." *American Journal of Medical Genetics* 82 (January 1999): 265–274.

Dingeon, B. "Gender Verification and the Next Olympic Games." To the Editor. *Journal of the American Medical Association* 269 (20 January 1993): 357.

Doi, N., H. Yanagawa,, and D. A. Largaespada. "Insertional Gene Fusion Technology." *FEBS Lett.* 457 (20 August 1999): 1–4.

Donnelly, S., C. R. McCarthy, and R. Singeleton, Jr. "The Brave New World of Animal Biotechnology." *Special Supplement, Hastings Center Report* (1994).

Douglas, J. T., and D. T. Curie." Targeted Gene Therapy." *Tumor Targeting* (1995): 67–84.

Drouin, R., and J-P. Therrien. "UVB-induced Cyclobutane Pyrimidine Dimer Frequency Correlates With Skin Cancer Mutational Hotspots in p53." *Photochemistry and Photobiology* 66 (November 1997): 719–724.

Dutton, Gail. "Biotechnology Counters Bioterrorism." *Genetic Engineering News* no. 21 (December 2000): 1–22ff.

Easton, D. F., D. Ford, D. T. Bishop, et al. "Breast and Ovarian Cancer Incidence in BRCA1 Mutation Carriers." *American Journal of Human Genetics* 56 (1995): 265–271

Editor. "Understanding the Causes of Schizophrenia." *The New England Journal of Medicine* No. 8 (25 February 1999).

Edmondson, D. G. and Roth, S. Y." Interactions of Transcriptional Regulators with Histones." *Methods* 15 (1998): 355–384.

Ewen, S.W., and A. Pusztai. "Effect of Diets Containing Genetically Modified Potatoes Expressing Galanthus Nivalis Lectin on Rat Small Intestine." *Lancet* 354 (16 October 1999): 1353–1354.

Farina A., et al. "A Latent Class Analysis Applied to Patterns of Fetal Sonographic Abnormalities: Definition of Phenotypes Associated With Aneuploidy." *Prenatal Diagnostics*19 (September 1999): 840–845.

Farina A, and Bianchi DW. "Fetal Cells in Maternal Blood as a Second Non-invasive Step for Fetal Down Syndrome Screening." *Prenatal Diagnosis* no. 18 (September 1998): 983–984.

Farina A., et al. "When Are We Allowed to Use a Marker in Down's Syndrome Screening?" *Prenatal Diagnostics* 19 (November 1999): 1084–1085.

Fields, S. "Proteomics in Genomeland." *Science* 291 (16 February 2001): 1221–1224.

Fischer, S. G., and L. S. Lerman. "DNA Fragments Differing by Ssingle Base-pair Substitutions are Separated in Denaturing Gradient Gels: Correspondence With Melting Theory." *Proceedings of the National Academy of Sciences of the United states of America* 80 (1983): 1579–1583.

Fitch, W. M. "Molecular Clocks Are Better Than You Think." *Journal of General Physiology* 102 (1994): 1a.

Fitch, W.M., F. J. Ayala. "The Superoxide-dismutase Molecular Clock Revisited." *Proceedings National the National Academy of Sciences of the United States of America* 91(1994): 6802–6807.

Fitts, R. "Development of DNA-DNA Hybridization Test for the Presence of Salmonella in Foods." *Food Technology* 39 (March, 1985): 95–102.

Folkman, Judah. "Fighting Cancer by Attacking its Blood Supply." *Scientific American*, September 1996.

Foroud, T., and T.K. Li. "Genetics of alcoholism: a review of recent studies in human and animal models." *American Journal of Addiction* 8(4) (1999): 261–278.

Foster, P. and J. Cairns. "Mechanism of directed mutation." *Genetics* 131 (1992): 783–789.

Foster, W. K., and I. D. Louro. "Expressao da Enzima Gliceraldeìdo-3-Fosfato-Desidrogenase em Displasias e Tumores." *Revista da Sociedade Brasileira de Cancerologia* 13 (March 2001): 23–30.

Fraenkel-Conrat, H., and B. Singer. "Virus Reconstitution II. Combination of Protein and Nucleic Acid From Different Strains." *Biochim. Biophys. Acta.* 24 (1957): 540–548.

Franch, T., and K. Gedes. "U-turns and Regulatory RNAs." *Current Opinion in Microbiology* 3 (2000): 159–164.

Fraser, C.M., J. Eisen, R.D. Fleischmann, K.A. Ketchum, and S. Peterson. "Comparative Genomics and Understanding of Microbial Biology." *Emerging Infectious Diseases* 6, no. 5 (September-October 2000).

Frauenfelder, H., and H.C. Berg. "Physics and Biology." *Physics Today* no. 2 (February 1994): 20–21.

Futami, J., Y. Tsushima, H. Tada, M. Seno, and H. Yamada. "Convenient and Efficient *In Vitro* Folding of Disulfide-containing Globular Protein From Crude Bacterial Inclusion Bodies." *Journal of Biochemistry* 127 (2000): 435–441.

Gaudeul M., P. Taberlet, and I. Till-Bottraud. "Genetic Diversity in an Endangered Alpine Plant, Eryngium Alpinum L. (Alpiaceae), Inferred From Amplified Fragment Length Polymorphism Markers." *Molecular Ecology* 9(10) (October 2000): 1625–37.

Gardner, R. C., A. J. Howarth, P. Hahn, M. Brown-Luedi, R. J. Shepherd, J. Messing. "The Complete Nucleotide Sequence of an Infectious Clone of Cauliflower Mosaic Virus by M13mp7 Shotgun Sequencing." *Nucleic Acids Research* 9 (1981): 2871–2888.

Gething, M. J., and J. Sambrook. "Protein Folding in the Cell." *Nature* 355 (1992): 33–45.

Gilbert, W., and D. Dressler. "DNA Replication, the Rolling Circle Model." *Cold Spring Harbor Symposium of Quantitative Biology* 33 (1968): 473.

Gillespie, John. H. "The Molecular Clock May Be an Episodic Clock." *Proceedings of the National Academy of Sciences of the United States of America* 81 (1984): 8009–8013.

Glavac, D., and M. Dean. "Applications of Heteroduplex Analysis for Mutation Detection in Disease Genes." *Human Mutation* 6 (1995): 281–287.

Golden, Frederic. "Mental Illness: Probing the Chemistry of the Brain." *Time* 157 (January 2001).

Grant, P. R. "Convergent and Divergent Character Displacement." *Biological Journal of the Linnean Society* 4 (March 1972): 39–68.

Greb, A. and D. Womble, eds. "Profile: Meet Mark Hughes, M.D., Ph.D." *Advances* (June 1998): 3.

Gregor, J. W. "The Ecotype." *Cambridge Biol. Rev.* 19 (1944): 20–30.

Grundfast, K. M., J. L. Atwood, and D. Chuong. "Genetics and Molecular Biology of Deafness." *Otolaryngologic Clinics of North America* no. 32 (December 1999):1067–1088.

Hamon, P., M. Robert, P. Schamasch, M. Pugeat. "Sex Testing at the Olympics." *Nature* 358 (1992): 447.

Han, D. S., et al. "Keratinocyte Growth Factor-2 (FGF-10) Promotes Healing of Experimental Small Intestinal Ulceration in Rats." *Am Journal Physiol. Gastrointest. Liver Physiol.* 279 no. 5 (November 2000): G1011–22.

Hanis, C. L., E. Boerwinkle, R. Chakraborty, D. L. Ellsworth, et al. "A Genome-wide Search for Human Non-insulin-dependent (Type 2) Diabetes Genes Reveals a Major Susceptibility Locus on Chromosome 2. *Nature Genetics* no. 13 (1996): 161–166.

Hardin, P.E. "From Biological Clock to Biological Rhythms." *Genome Biology* 1 (2000): 1023.1–1023.5.

Hardy, G. H. "Mendelian Proportions in a Mixed Population." *Science* 28 (1908): 49–50.

Hayflick, L. "Be fruitful and immortalize." *Nature Genetics* no. 19 (June 1998): 103–4.

Hayflick, L. "The Future of Aging." *Nature* no.408 (2000): 103–4.

Hayflick, Leonard. "How and Why We Age." *Experimental Gerontology* 33 (1998): 639–653, 1998.

Herrick, J. B. "Peculiar Elongated and Sickle-shaped Red Blood Corpuscles in a Case of Severe Anemia." *Archives of Internal Medicine* no. 6 (1910): 517–521.

Hershey, A. D., and M. Chase. "Independent Functions of Viral Protein and Nucleic Acid in Growth of Bacteriophage." *Journal Gen. Physiol.* 36. (1952): 39–56.

Hofreiter, M., et al. "Ancient DNA," *Nature Reviews Genetics* 2 (2001): 353–359.

Hoskins, K. P., J. E. Stopfer, K. A. Calzone, et cols. "Assessment and Counseling For Women With a Family History of Breast Cancer—A Guide For Clinicians." *JAMA* 273 (1995): 577–585.

The Huntington's Disease Collaborative Research Group. "A Novel Gene Containing a Trinucleotide Repeat that is Expanded and Unstable on Huntington's Disease Chromosomes." *Cell* (26 March 1993): 1–20.

Hutchinson, G. E. "Homage to Santa Rosalia or Why Are There So Many Kinds of Animals?" *The American Naturalist* 93 (1959): 145–159.

Huxley, J. "The Tree Types of Evolutionary Process." *Nature* 180 (7 September 1957): 454–455.

Hyman, S. E. "The Genetics of Mental Illness: Implications for Practice." *Bulletin of the World Health Organization* 78 (April 2000): 455–463.

Ingram, V. M. "A Case of Sickle-cell Anemia: a Commentary by Vernon M. Ingram." *Biochem. Biophys. Acta.* no. 1000 (1989): 147–150.

Ingram, V. M. "A Specific Chemical Difference Between Globins of Normal and Sickle-cell Anemia Humoglobins." *Nature* no. 178 (1956): 792–794.

Ingram, V. M. "Gene Mutations in Human Hemoglobin: the Chemical Difference Between Normal and Sickle Hemoglobin." *Nature* no. 180 (1957): 326–328.

International Genome Sequencing Consortium, "Initial Sequencing and Analysis of the Human Genome." *Nature* 409 (2001): 860–921.

The International Mouse Mutagenesis Consortium. "Functional Annotation of Mouse Genome Sequences." *Science* 291 (16 February 2001): 1251–1255.

The International SNP Map Working Group. "A Map of Human Genome Sequence Variation Containing 1.42 Million Single Nucleotide Polymorphisms." *Nature* 409 (2001): 928–933.

Jeffords, J. M. and Tom Daschle. "Political Issues in the Genome Era," *Science* 291 (16 February 2001): 1249–50.

Jehaes, E., H. Pfeiffer, K. Toprak, R. Decorte, B. Brinkmann, J. J. Cassiman. "Mitochondrial DNA Analysis of the Putative Heart of Louis XVII, Son of Louis XVI and Marie-Antoinette." *European Journal of Human Genetics* 9 (2001): 185–190.

Jimenez PA, and M. A. Rampy. "Keratinocyte Growth Factor-2 Accelerates Wound Healing in Incisional Wounds." *Journal Surg. Res.* 81, no. 2 (Febrary 1999): 238–242.

Johnson, P., and D. A. Hopkinson. "Detection of ABO Blood Group Polymorphism by Denaturing Gradient Gel Electrophoresis." *Human Molecular Genetics* 1 (1992): 341–344.

Karner, M. B., E. F. DeLong,, and D.M. Karl. "Archaeal Dominance in the Mesopelagic Zone of the Pacific Ocean." *Nature* 409 (2001): 507–510.

Ke, S. H., and R. M. Wartell. "Influence of Nearest Neighbor Sequence on the Stability of Base Pair Mismatches in Long DNA; Determination by Temperature-gradient Gel Electrophoresis." *Nucleic Acids Research* 21: 5137–5143.

Keeling, P. J. and J. D. Palmer. "Parabasalia Flagellates are Ancient Eukaryotes." *Nature* 405 (2000): 635–637.

Keenan, Katherine. "Lilian Vaughan Morgan (1870–1952): Her Life and Work." *American Zoologist* 23 (1983): 867–876.

Keller, L., and K.G . Ross. "Selfish genes: Green beard in the red fire ant." *Nature* 394 (06 August 1998): 573–574.

Kerr, J. F. K., A.Wyllie, and A.H. Currie. "Apoptosis, a Basic Biological Phenomenon With Wider Implications in Tissue Kinetics." *British Journal of Cancer* 26 (1972): 239–45.

Khler, S., et al. "A Plastid of Probable Green Algal Origin in Apicomplexan Parasites." *Science* 275 (1997): 1485–1489.

Kostrikis, L. G., S. Tyagi, M. M. Mhlanga, D. D. Ho, and R. F. Kramer. "Spectral Genotyping of Human Alleles." *Science* 279 (1998): 1228–1229.

Krainer, M., S. S. Arrieta, M. G. Fitzgerald, et al. "Differential Contributions of BRCA1 and BRCA2 to Early-onset Breast Cancer." *New England Journal of Medicine* 336 (1997): 1416–1421.

Kyogoku, Y., R. C. Lord, and A. Rich. "Hydrogen Bonding Specificity of Nucleic Acid Purines and Pyrimidines in Solution." *Science* 154 (28 October 1966): 518–520.

Kyrpides, N. C. and C. A. Ouzounis. "Transcription in Archaea." *Proceedings of the National Academy of Sciences of the United States of America (20 July 1999).*

Lacayo, Richard. "For Whom the Bell Curves." *Time,* October 4, 1994.

Landman, O. "The inheritance of acquired characteristics." *Ann. Rev. Genet* 25 (1991): 1-20.

Lasky, T., and E. Silbergeld. "p53 Mutations Associated With Breast, Colorectal, Liver, Lung, and Ovarian Cancers." *Environmental Health Perspectives* 104 (December 1996): 1324–1331.

Lewin, R. "Conflict Over DNA Clock Results." *Science* 241:1598–1600; and continued in "DNA Clock Conflict Continues." *Science* 241(1988):1756–1759.

Lewis, R. "A Survey of Clock Genes." *The Scientist* (Dec. 1995): 18–19.

Lindahl, T. "Facts and Artefacts of Ancient DNA." *Cell,* 90, (1997): 1–3.

Liu, B., and B. Alberts. "Head-on Collision Between DNA Replication Aapparatus and RNA Polymerase Transcription Complex. *Science* 267 (1995): 1131–1137.

Logsdon, J.M., Jr., et al. "Seven Newly Discovered Intron Positions in the Triose-phosphate Isomerase Gene: Evidence for the Introns-late Theory." *Proceedings of the National Academy of Sciences of the United states of America* 92 (1995): 8507–8511.

Losey, J. E., L. S. Rayor, and M. E. Carter. "Transgenic Pollen Harms Monarch Larvae." *Nature* 399: 214.

Maassen, J. A., J. J. Jansen, T. Kadowaki, J. M. van den Ouweland, L.M. 't Hart and H.H. Lemkes. "The Molecular Basis and Clinical Characteristics of Maternally Inherited Diabetes and Deafness (MIDD), a Recently Recognized Diabetic Subtype." *Experimental and Clinical Endocrinology and Diabetes* no.104(3) (1996): 205–11.

Mark A., et al. "The Genome Sequence of Drosophila melanogaster." *Science* 287 (March 2000): 2185–2195.

Marshall, E. "Gene Therapy's Growing Pains." *Science* 269 (23 August 1995): 1050–1055.

Maxam, A., and W. Gilbert. "A New Method of Sequencing DNA." *Proceedings of the National Academy of Sciences of the United States of America* 74 (1998): 560–564.

Merten, Thomas R. "Introducing Students to Population Genetics and the Hardy-Weinberg Principle." *The American Biology Teacher* 54 (1992): pp 103–107.

Millen, R. S., et al. "Many Parallel Losses of InfA From Chloroplast DNA During Angiosperm Evolution With Multiple Independent Transfers to the Nucleus." *The Plant Cell* 13 (2001): 645–658.

Morell, V. "Huntington's Gene Finally Found." *Science* 260 (1993): 28–30.

Morgan, Lilian V. "A Closed X Chromosome in *Drosophila melanogaster.*" *Genetics* 18 (1933): 250–283.

Morgan, Thomas H., Alfred H. Sturtevant, and Lilian V. Morgan. "Maintenance of a *Drosophila* Stock Center, in Connection with Investigations on the Germinal Material in Relation to Heredity." *Carnegie Institution of Washington Yearbook* 44 (1945): 157–160.

Narod, S., D. Ford, P. Devilee, et cols. "An Evaluation of Genetic Heterogeneity and Penetrance Analysis of the BRCA1 and BRCA2 Genes in Breast Cancer Families." *American Journal of Human Genetics* 62 (1998): 676–689.

Nath, J., et al. "A Review of Fluorescence *In Situ* Hybridization (FISH): Current Status and Future Prospects." *Biotech Histochem* 75 (March 2000): 54–78.

Nelson, J. Lee, et al. "Microchimerism and HLA-compatible Relationships of Pregnancy in Scleroderma." *The Lancet* 351 (1998): 559–562.

Newman, T., et al. "Genes Galore: a Summary of Methods for Accessing Results From Large-scale Partial Sequencing of Anonymous *Arabidopsis* cDNA Clones." *Plant Physiology* 106 (December 1994): 1241–1255.

Nickerson, M. L., Weirich, G., Zbar, B., L. S. Schmidt. "Signature-based Analysis of MET Proto-oncogene Mutations Using DHPLC. *Human Mutation* 16 (2000): 68–76.

Nussbaum, R. L., and M. H. Polymeropoulos "Genetics of Parkinson's disease." *Human Molecular Genetics* 6 (1997):1687–1691.

O'Donovan, M., et al. "Blind Analysis of Denaturing High-Performance Liquid Chromatography as a Tool for Mutation Detection." *Genomics* 52 (1998): 44–49.

Ohlsson, G., J. Muller, and M. Schwartz. "Genetic Diagnosis of 21-hydroxylase Deficiency: DGGE-based Mutation Scanning of CYP21." *Human Mutation* 13 (1999): 385–389.

Oldroyd, G. E.D., B. J. Staskawicz. "Genetically Engineered Broad-spectrum Disease Resistance in Tomato."

Proceedings of the National Academy of Sciences of the Unied States of America 95 (18 August 1998): 10,300–5.

Palmer, J. D., and C. F. Delwiche. "The Origin and Evolution of Plastids and their Genomes." *Molecular Systematics of Plants, II.* (1998): 375-409.

Palmer, J. D., et al. "Dynamic Evolution of Plant Mitochondrial Genomes: Mobile Genes and Introns and Highly Variable Mutation Rates." *Proceedings of the National Academy of Sciences of the United States of America* 97 (2000): 6960–6966.

Parkinson, C. L., et al. "Multigene Analyses Identify the Earliest Lineages of Extant Flowering Plants." *Current Biology* 9 (1999): 1481–1485.

Pennisis, E. "Architecture of Hearing." *Science* no. 278 (1997): 1223.

Perera, F. P., and I. B. Weinstein. "Molecular Epidemiology: Recent Advances and Future Directions." *Carcinogenesis* 21 (2000): 517–524.

Perna, N. T., et al. "Genome Sequence of Enterohaemorrhagic *Escherichia coli* O157:H7." *Nature* 409 (25 January 2001): 529–533.

Pertl, B.,and D. W. Bianchi. "First Trimester Prenatal Diagnosis: Fetal Cells in the Maternal Circulation." *Seminar Perinatology* no. 23 (October 1999): 393–402.

Prusiner, S. B. "Molecular Biology of Prion Diseases." *Science* 252 (14 June 1991): 1515–1522.

Prusiner, S. B. "The Prion Diseases." *Scientific American* 272 (January 1995): 48–57.

Qiu, Y.-L., et al. "The Gain of Three Mitochondrial Introns Iidentifies Liverworts as the Earliest Land Plants." *Nature* 394 (1998): 671–674.

Ranford, J. C., A. R. M. Coates, and B. Henderson. "Chaperonins Are Cell-signaling Proteins: the Unfolding Biology of Molecular Chaperones." *Expert Reviews in Molecular Medicine* (September 2000): 1–17.

Renier, M., et al. "Use of a Membrane Potential-sensitive Probe to Assess Biological Expression of the Cystic Fibrosis Transmembrane Conductance Regulator." *Human Gene Therapy* 6 (1994): 1275–1283.

Roman H. "Boris Ephrussi." *Annual Review of Genetics* 14 (1980): 447–450.

Russo, E. "1998 Lasker Award Recipients Honored For Their Groundbreaking Achievements." *The Scientist* 12 (12 October 1998): 1.

Russo, Eugene. "Proteomic Factories," *The Scientist* 14 (7 February, 2000).

Sanger, F., S. Nicklen, and A.R. Coulson. "DNA Sequencing With Chain-terminating Inhibitors." *Proceedings of the National Academy of Sciences of the United States of America* 74: 5463–5467.

Sasadeusz, J. J., et al. "Homopolymer Mutational Hot Spots Mediate Herpes Simplex Virus Resistance to Acyclovir." *Journal of Virology* 71 (May 1997): 3872–3878.

Schwartz, D. C. and C. R. Cantor. "Separation of Yeast Chromosomal-sized DNAs by Pulsed Field Gel Electrophoresis." *Cell* 37 (1984): 67–75.

Scott, J. D. and T. Pawson. "Cell Communication: The Inside Story" *Scientific American* 282 (June 2000): 54–61.

Selwa, R. "Researcher Talks About Ethics of Genetic Therapy" *Macomb Daily* (25 October 2000): 1A, 8A.

Shay, Jerry W., and Woodring E. Wright. " Hayflick, His Limit, and Cellular Aging", *Nature reviews/ Molecular Cell Biology* (1 October 2000): 72–76.

Shields, R., "The Emperor's New Clothes." *Trends in Genetics* 17 (2001) 189.

Shuckit, M. A.. "New Findings on the Genetics of Alcoholism." *Journal of the American Medical Association* 281(20) (1999): 1875–1876.

Siepel, A., et al. "ISYS: a Decentralized, Component-based Approach to the Integration of Heterogeneous Bioinformatics Resources." *Bioinformatics* 16 (2000): 1–12.

Silberg, J., et al. "The Influence of Genetic Factors and Life Stress on Depression Among Adolescent Girls." *Journal of the American Medical Association* 56 (March 1999): 225–232.

Singer, M. F. "News and Comment. 1968 Nobel Laureate in Medicine or Physiology." *Science* 162 (1968): 433–36.

Sirover, M.A. "Role of the Glycolitic Protein, Glyceraldehyde-3-phosphate Dehydrogenase, in Normal Cell Function and in Cell Pathology." *Journal of Cellular Biochemistry* 66 (1997): 133–140.

Smaglik, P. "Gene Therapy—The Next Generation." *The Scientist* 12 (1998): 4.

Soudek, D., "Small Supernumerary Inactive Marker Chromosomes." *Karyogram* 10 (1984): 13–15.

Southern, E.M. "Detection of Specific Sequences Among DNA Fragments Separated by Gel Electrophoresis." *Journal of Molecular Biology* no. 98 (1975): 503–517.

Spillman, T. "Genetic Diseases of Hearing." *Current Opinion in Neurology* no.7(1) (February 1994): 81–87.

Stebbins, G.L. "A Brief Summary of My Ideas on Evolution." *American Journal of Botany* 86 (1999): 1207–1208.

Stunkard, A., et al. "The Body-mass Index of Twins Who Have Been Reared Apart." *New Engand Journal of Medicine* 322 (1990): 1483–1487.

Suter, U., et al. "Trembler Mouse Carries a Point Mutation in a Myelin Gene." *Nature* 356 (19 March 1992): 241–244.

Tang, Y.-P. et al. "Genetic enhancement of learning and memory in mice." *Nature* 401 (1999):63–69.

Thomson, J., et al. "Embryonic Stem Cell Lines Derived From Human Blastocysts." *Science* 282 (6 November 1998): 1145–1147.

Thorpe, J. P. "The Molecular Clock Hypothesis: Biochemical Evolution, Genetic Differentiation, and Systematics." *Annual Review of Ecology and Systematics* 13 (1982): 139–168.

Van G., J. Romero-Severson, M. Walton, D. D. Chadee, and D.W. Severson. "Population Genetics of the Yellow Fever Mosquito in Trinidad: Comparison of Amplified Fragment Length Polymorphism (AFLP) and Restriction Fragment Length Polymorphism (RFLP) mLarkers." *Molecular Ecology* 8(6) (June 1999): 951–63.

Van Oosterhout C., et al. "Inbreeding Depression and Genetic Load in Laboratory Metapopulations of the Butterfly Bicyclus anynana." *Evolution Int J Org Evolution* 54, no.1, 218–25.

Van Slyke, D. D., and W. A. Jacobs. "Phoebus Aaron Levene." *Biographical Memoirs of the National Academy of Sciences of the United States of America* 23 (1945): 1–12.

Venter, J. C., et al. "The Sequence of the Human Genome." *Science* 291 (2001): 1304–1351.

Veuille, E. "Genetics and the evolutionary process." *C. R. Acad. Sc.i III* 323, no.12 (December 2000):1155–65.

Via, S. and R. Lande. "Genotype-environment Interaction and the Evolution of Phenotypic Plasticity." *Evolution* 39 (1985): 505–522.

Voet, D. and Rich, A. "The Crystal Structures of Purines, Pyrimidines and their Intermolecular Structures." *Prog. Nucleic Acid Res. Mol. Biol.,* 10 (1970):183–265.

Wang, A.H., Quigley, G. J., Kolpak, F. J., Crawford, J. L., van Boom, J. H., van der Marel, G., and Rich, A. "Molecular Structure of a Left-handed Double Helical DNA Fragment at Atomic Resolution." *Nature,* 282 (13 December 1979): 680–686.

Watson, J. D., and F. H. C. Crick. "Genetical Implications of the Structure of Deoxyribonucleic Acid." *Nature* 171 (1953): 964–969.

Watson, J. D., and F. H. C. Crick. "Molecular structure of nucleic acids." Nature, 171 (1953): 737–738.

Wenrich, D.H. "Clarence Erwin McClung." *Journal of Morphology* 66 (1940): 635–688.

Westphal, C.H., and P. Leder. "Transposon-generated 'Knock-out' and 'Knock-in' Gene Targeting Constructs for Use in Mice." *Current Biology* 7 (1994): 530–533.

White, J.A., et al. "A New Set of *Arabidopsis* Expressed Sequence Tags From Developing Seeds. The Metabolic Pathway From Carbohydrates to Seed Oil." *Plant Physiology* 124 (December 2000): 1582–1594.

Wiese, U., M. Wulfert, S. B. Prusiner, D. Riesner. "Scanning For Mutations in the Human Prion Protein Open Reading Frame By Temporal Temperature Gradient Gel Electrophoresis." *Electrophoresis* 16 (1995): 1851–1860.

Wilkinson, D. "When You Wish Upon a Star: Molecular Beacons: Real Time in a Twinkle." *The Scientist* 13 (March 1999).

Wilmut, Ian, and Keith Kasnot. "Cloning for Medicine." *Scientific American* 279 (1 December 1998): 58–63.

Wilmut, I., A. E. Schnieke, J. McWhir, A. J. Kind, and K. H. S. Campbell. "Viable Offspring Derived From Fetal and Adult Mammalian Cells. *Nature* 385 (1997): 810–813.

Wilson, J. F., et al. "Genetic Evidence for Different Male and Female Roles During Cultural Transitions in the British Isles." *Proceedings of the National Academy of Sciences of the United States of America* 98 (2001): 5078–5083.

Woese, C. R., O. Kandler, and M. L. Wheelis. "Towards a Natural System of Organisms: Proposal for the Domains Archae, Bacteria, and Eukarya." *Proceedings of the National Academy of Sciences of the United States of America* 87 (1990): 4576–4579.

Wolfe, K. H., et al. "Function and Evolution of a Minimal Plastid Genome from a Non-photosynthetic Parasitic Plant." *Proceedings of the National Academy of Sciences of the United States of America* 89 (1992): 10648–10652.

Xu, D., et al. "Downregulation of Telomerase Reverse Transcriptase mRNA Expression by Wild Type p53 in Human Tumor Cells." *Oncogene* 19 (26 October 2000).

Yang, T., et al. "A Mouse model for Prader-Willi Syndrome." *Abstract: Prader-Willi Syndrome Association 12th Annual Scientific Day* (16 July 1997).

Yanofsky, C. "Attenuation in the Control of Expression of Bacterial Operons." *Nature* 289 (1981): 751–758.

Yanofsky, C., et al. "On the Colinearity of Gene Structure and Protein Structure." *Proceedings of the National Academy of Science of the United States of America* 51 (1964): 266–272.

Yaspo, M.-L. et al. "The DNA Sequence of Human Chromosome 21." *Nature* 6784 (May 2000): 311–319.

Yuspa, S. H. "Overview of Carcinogenesis: Past, Present and Future." *Carcinogenesis* 21 (2000): 341–344.

Zinkernagel R. M., and P. C. Doherty. "MHC Restricted Cytotoxic T Cells: Studies on the Bbiological Role of Polymorphic Major Transplantation Antigens Determining T Cell Restriction Specificity." *Advances in Immunology* 27 (1979): 151–177.

Web Sites

(Editor's note: As the World Wide Web is constantly expanding, the URLs listed below may be altered or nonexistent as of June 15, 2001.)

Access Excellence. "The collaboration of proteins during replication." (02 April 2000) <http://www.accessexcellence.com/AB/GG/collaboration.html>

Access Excellence. "DNA Vaccine Outlook." (1999) <http//www.accessexcellence.com/WN/SUA11/dnavax1297.html>

Access Excellence. "Polymerase Chain Reaction—Xeroxing DNA." (1999) <http//www.accessexcellence.com/AB/IE/PCR_Xeroxing_DNA.html>

Access Excellence Classic Collection. "Restriction Enzymes Background Paper." (2001). <http//www.accessexcellence.com/AE/AEC/CC/restriction.html>

American Heart Association, " Stroke Risk Factors" (2001). <http://www.americanheart.org/Heart and Stroke A Z Guide/strokeri.html>

American Psychological Association. "Concern follows psychological testing." APA Monitor (December 1999). <http://www.apa.org/monitor/dec99/ss4.html>

American Psychological Association. "APA Task Force Examines the Knowns and Unknowns of Intelligence." (1996). <http://www.apa.org/releases/intell.html>

American Society of Human Genetics. "Solving the Puzzle: Careers in Genetics". (2001). <http://www.faseb.org/genetics/gsa/careers/bro-menu.htm>

American Society for Microbiology. "ASM Issues Statement on Genetically Modified Organisms." (06 March 2001) <http://unisci.com/stories/20003/0719001.htm>

Apoptosis Interest Group. "About Apoptosis." (15 June 2001). <http//www.nih.gov/sigs/aig/Aboutapo.html>

Bartlett, J. "AAV Vectors in Gene Therapy for Cystic Fibrosis." (2000). <http://www.newswise.com/articles/2000/5/CFGENE.SPR.html>

Bayer News. "High-throughput Screening: the Fast Way to the Right Compound." (29 April 2001). <http://www.news.bayer.com/News/news.nsf/ID/NT0000792E>

Biologisk Institutt. "RAPD" (01 April 2001). <http//www.biologi.uio.no/FellesAvdelinger/DNA_KAFFE/Kaffe_Resources/RAPD.html>

Biology Pages. "Antisense RNA." (2001). <http//www.ultranet.com/~jkimball/BiologyPages/A/AntisenseRNA.html>

Biology Pages. "Apoptosis." (1999). <http//www.ultranet.com/~jkimball/BiologyPages/A/Apoptosis.html>

BioOnline. "Bioinformatics." (2001). <http//www.bio.com/articles/bioinformatics.html>

Brescia University. "The Wobble Hypothesis." (29 April 2001). <http://www.med.unibs.it/~marchesi/protsyn.html>

The British Council. "Alec Jeffreys—genetic evidence." (1 March 2001). <http://www.britishcouncil.org/science/science/personalities/text/ukperson/jeffreys.htm>

Brown University. "VSLI Models of Neural Systems." (02 May 2001). <http://landow.stg.brown.edu/cpace/sciencedgneuro/present/systems.html>

Cancer Genetics. "Mutation and variation." (31 March 2001). <http//www.cancergenetics.org/mutate.htm>

C.A.N.D.L.E.S. Holocaust Museum. (14 May 2001). <http://www.candles-museum.com>

City of Hope Medical Center. "Genetics of schizophrenia and other mental illness." (05 April 2001). <http://www.cityofhope.org/molgen/mental.asp>

City of Hope National Medical Center. "Renowned Genetics and Evolution Researcher, Susumo Ohno, Dies." (2001). <www.cityofhope.org/cohpress/Medical_Science/Ohno.obit.htm>

Cold Spring Harbor Laboratory. "James E. Darnell, Jr."(9 February 2001). <http://vector.cshl.orl/dnaftb/concept_35/con35bio.html>

Colorado State University. "Transgenic crops: An Introduction and Resource Guide." (2000). <http//www.colostate.edu/programs/lifesciences/TransgenicCrops/>

Colorado State University. "What are transgenic plants?." (02 April 2001). <http//www.colostate.edu/programs/lifesciences/TransgenicCrops/what_page.html>

Conklin Lab Technical Information. "Conditional expression of Activated G Protein Alpha Subunits in Transgenic Mice." (2001). <http://gladstone.ucsf.edu/labs/conklin/Technical/Tetinmice.html>

Costa, A. B. "Levene, Pheobus Aaron Theodor." American National Biography Online, (3 April 2001). <http://www.anb.org/articles/13/13-00988.html>

Darnell, J., H. Lodish, and D. Baltimore. Molecular Cell Biology, New York: Scientific American (2001).<http://www.americanheart.org/Heart and Stroke A Z Guide/riskfact.html>

The Field Museum. "Computational Biology." (2001). <http//cb.fmnh.org/>

Government of Australia. "Genetically modified organisms." (06 March 2001). <http://www.agric.nsw.gov.au/news/hotopics/gmo.htm>

Harvard University. "Arabidopsis Genomic Sequencing." (1999). <http://weeds.mgh.harvard.edu/goodman/8261/8261description.shtml>

Harvard University. "Biological clock genes identified." (2000.) <http//www.news.harvard.edu/science/current_stories/25.May.00/bioclock.html>

Harvard University. "Peering at a machine that pries apart DNA." (02 April 2001). <http//www.hms.harvard.edu/news/releases/1099helicase.html>

Homeobox Genes Data Base "RAREs: Kinetics of Activation of Hox Genes by RA." (2001). <http://www.mssn.edu/molbio/hoxpro/new/ra2hox.html>

The Hospital for Sick Children. "Researchers Pinpoint Region for Cystic Fibrosis Modifier Gene." (04 March 2001). <http://www.sickkids.on.ca/releases/cfres.asp>

Howard Hughes Medical Institute. "Cancer: the role of genetic collaboration." (01 April 2001). <http//www.hhmi.org/research/investigators/leder.html>

Indiana University. "Genetic factors weigh-in as clues to obesity." (02 April 2001). <http//www.medicine.indiana.edu/mini_med/1999/mms6.htm>

International Food Information Council Foundation. "Backgrounder—Food Biotechnology." (06 March 2001) <http://ificinfo.health.org/backgrond/BKGR14.htm>

Journal of Science. "In Defense of Evolution: that Persistent Homeobox." (06 March 2001). <http://journalof-science.wlu.edu/winter1999/articles/misc/homeobox.html>

Kher, Unmesh. "The Next Frontier:Proteomics," *Time* 156(1) (3July 2000). <http://www.time.com/time/magazine/articles/0,3266,48107,00.html>

Li, Q-X., et al. "Ribozyme Gene Vector Libraries Identify Putative Tumor Supressor Genes." (2001). <http://www.oral.gov/hgmis/publicat/00santa/function.html>

Massachusetts Institute of Technology, Department of Biology. Faculty, Labs & Research. (April 9, 2001). <http://web.mit.edu/biology/www/Ar/ingram.html>

Massachusetts Institute of Technology. "The Lac Operon." (30 March 2001). <http//esg-www.mit.edu:8001/esgbio/pge/lac.html>

Massachusetts Institute of Technology. "Prokaryotic Gene Regulation." (05 March 2001). <http//esg-www.mit.edu:8001/esgbio/pge/intro.html>

Mesdcape CME. "COX-2 Inhibitors and Cancer Cemoprevention (CME)" (2001). <http://www.medscape.com/Medscape/Oncology/TreatmentUpdate/2000/tu10/pat-tu10.html>

Michigan State University. "Interpreting the genetic code II." (29 April 2001). <htpp://www.arbor.edu/~michaelb/ingencod2.htm>

Microbiology Institute at the Federal Institute of Technology at Zurich. "Of Archea and Evolution." (2001). <http//www.micro.biol.ethz.ch/new/overview4.htm>

Murdoch University Bioinformatics Research Group. "Amplified Fragment Length Polymorphisms." (2001). <http//www.lupindb.org/aflp.shtml>

Mukai, Y., E. Matsuo, S. Y. Roth, and S.Harashima, "Conservation of Histone Biding and Transcriptional Repressor Functions in an Schizosaccharomyces pombe Tup1 Homolog." (Abstract) (1999). <http://www3.mdanderson.org/DEPARTMENTS/biochem/roth.htm>

National Animal Disease Center (10 February 2001). <http://www.nadc.ars.usda.gov/>

National Bioethics Advisory Commission. "Executive Summary of Cloning Human Beings." (2001). <http://bioethics.gov/pubs/cloning1/executive.htm>

National Center for Chronic Disease Prevention and Health Promotion, "Physical Activity and Good Nutrition: Essential Elements for Good Health" (2001). <http://www.cdc.gov/nccdphp/dnpa/dnpaaag.html>

National Human Genome Research Institute (2001). <http://www.nhgri.nih.gov/>

National Human Genome Research Institute. "Ethical, Legal and Social Implications of Human Genetic Research." (October 2000). <http://www.nhgri.nih.gov/ELSI/>

National Human Genome Research Institute. "Twenty Questions About DNA Sequencing (and the Answers)." (28 April 2001). <http://www.nhgri.nih.gov/NEWS/Finish_sequencing_early/twenty_questions_about_DNA.html>

National Institute on Alcohol Abuse and Alcoholism. "The Genetics of Alcoholism (10 February 2001). <http://silk.nih.gov/silk/niaaa1/publication/aa18.htm>

National Institutes of Health, Online Mendelian Inheritance in Man. "Friedreich Ataxia." (2001). <http://www.ncbi.nlm.nih.gov/omim/>

National Institutes of Health. "Precise Gene Insertion Using Introns." (04 March 2001). <http://www.nigms.nih.gov/news/releases/intons.html>

National Institutes of Health. "Stem Cells: A Primer." (2000). <http://www.nih.gov/news/stemcell/primer.htm>

Nobel Prize Internet Archive. "Gunter Blobel." (2000). <http//nobelprizes.com/nobel/medicine/1999a.html>

OER. "Multiplexing approach to DNA sequencing." (31 March 2001). <http//www.spie.org/web/oer/july/dna_sequencing.html>

Oak Ridge National Laboratory. "Careers in Genetics and the Biosciences". 6 (March 2001). <http://www.ornl.gov/hgmis/education/careers.html>

Oak Ridge National Laboratory. "Genetics Privacy and Legislation." (2001). <http://www.ornl.gov/hgmis/elsi/legislat.html>

Oak Ridge National Laboratory. "Potential Benefits of Human Genome Project Research." (21 January 2001). <http//www.ornl.gov/hgmis/project/benefits.html>

Oklahoma State University. "Somatic hybridization." (07 March 2001). <htpp://opbs.okstate.edu/~melcher/MG/MGW1/MG1213.html>

Physicians and Scientists for Responsible Application of Science and Technology. "How Are Genes Engineered?" (04 March 2001).<http//www.psrast.org/whisge.htm>

Princeton University. "Physical maps and positional cloning." (31 March 2001). <http//www.princeton.edu/~lsilver/book/MG10.html>

Rockefeller University News. "Rockefeller University Cell Biologist, Günter Blobel, Wins 1999 Nobel Prize in Physiology and Medicine." (2001). <http//www.rockefeller.edu/pubinfo/blobel.nr.html>

Rutgers University. "Structure-based Functional Genomics Workshop Concept." (03 March 2001). <http//www.cabm.rutgers.edu/bioinformatics_meeting/concept.html>

Samford University. "Fluorescence-activated cell sorting as a form of flow cytometry." (2001). <http//faculty.samford.edu/~gekeller/rousseau.html>

Scleroderma Research Foundation (2001). <http://www.srfcure.org/>

Stanford University. "Center for Computational Genetics and Biological Modeling." (04 May 2001). <http://corporate.stanford.edu/research/programs/compgen.html>

Stanford University. "Charles Yanofsky." <http://www.stanford.edu/group/biosci/faculty/yanofsky.html> (29 April 2001).

Stubbs, L. "Of Mice and Men." *Science and Technology Review* <http://www.llnl.gov/str/Stubbs.html> (December 1999).

Sunday Observer. "Orchids in bloom." (07 March 2001). <http://www.lanka.net/lakehouse/1999/11/14/fea12.html>

Swedish University of Agricultural Sciences. "Antisense RNA Control." (2001). <http//www.mikrob.slu.se/genexp/general/antisens.htm>

Tatum, O. L., et al. "Contamination of BAC Clones by E. coli IS186 Insertion Elements." <http://www.oral.gov/hgmis/publicat/00santa/function.html> (2000).

Taylor, R. "Bioterrorism Special Report." *New Scientist.* (2001). <http://www.newscientist.com/nsplus/insight/bioterrorism/allfall.html>

Texas A & M University. "Amino Acid Composition and Protein Sequencing." (28 April 2001). <http://www.ntri.tamuk.edu/graduate/sequence.html>

Texas A & M University. "Use of a Protoplast Regeneration System for African Violet Improvement." (07 March 2001). <htpp://aggie-horticulture.tamu.edu/tisscult/proto/wink/wink.html>

United States Environmental Protection Agency, Office of Pesticide Programs. "What are Biopesticides?" (2000). <http//www.epa.gov/pesticides/biopesticides/what_are_biopesticides.htm>

United States Geological Society. "Bioremediation: Nature's Way to a Cleaner Environment." (05 May 2001). <http://water.usgs.gov/wid/html/bioremed.html>

University of Arizona. "Bacteriophage lambda: a complex viral operon." (02 April 2001). <http//www.blc.arizona.edu/marty/411/Modules/lambda.html>

University of Arizona. "Restriction Enzyme Analysis of DNA." (1998). <http//biotech.biology.arizona.edu/labs/DNA_analysis_RE_student.html>

University of Basel. "The Homeobox Page." (06 March 2001). <http//copan.bioz.unibas.ch/homeo.html>

University of California at Berkeley. Drosophila Genome Project. (December 2000). <http://www.fruitfly.org>

University of California at Los Angeles. "Introduction to Autoradiography." (1999). <http//laxmi.nuc.ucla.edu:8248/M248_98/autorad/auto_index.html>

University of California at Los Angeles. "M267 Protein Targeting and Secretory Pathway." (04 March 2001). <http//www.medsch.ucla.edu/som/biochem/cm267/lectures/001ec_vanderblick.htm>

University of Edinburgh. "Microorganisms and the Tree of Life." (2001). <http/helios.bto.ed.ac.uk/bto/microbes/microbes.htm>

University of Minnesota Department of Psychology: Bouchard. (14 May 2001). <http://www.psych.umn.edu/psyfac/core/Diff_BG/bouchard.htm>

University of Minnesota Institute of Human Genetics: Faculty Biographies: Bouchard. (14 May 2001). <http://www.ihg.med.umn.edu/Fac.biographies/Bouchard.html>

University of Nottingham "Giemsa for Helicobacter pylori." (06 March 2001). <http://www.nottingham.ac.uk/~mpzjlowe/protocols/giemshel.html>

University of Pennsylvania. "Flow Cytometry Primer." (02 March 2001). <http//biochem.dental.upenn.edu/~sdmfacs/flowprim.htm>

University of Rochester. "Virus vectors: phage basics." (2000). <http://www.urmc.rochester.edu/smd/mbi/grad2/phage01.html>

University of Texas. "*Escherichia coli.*" (01 March 2001). <http//www.medic.med.uth.tmc.edu/path/00001497.htm>

University of Texas. "Episomes and Plasmids: Gene Transport Vehicles Between Cells." (1999). <http//www.botany. utexas.edu/facstaff/facpages/ksata/ecpf96/12b/trans.htm>

University of Texas Medical Branch. "Genetics." (29 April 2001). <http://gsbs.utmb.edu/microbook/ch005.htm>

University of Utah, Department of Human Genetics.(April 9, 2001). <http://www.biology.utah.edu/people/regfaculty/ ~capecchi/capecchi.html>

University of Utah School of Medicine, The Eccles Institute of Human Genetics, and The Utah Museum of Natural History."The Natural History of Genes." (8 March 2001). <http//raven.umnh.utah.edu>

University of Virginia. "Human HOX Affects Limbs and Genitals and May Explain Why Humans Have 5 Digits on Each Appendage." (06 March 2001). <http://www. people.virginia.edu/~rjh9u/humanhox.html>

University of Wisconsin. "Crime Gene Investigation, Handiest Tool." (29 February 2001). <http://whyfiles.org/ 126dna_forensic/2.html>

University of Wisconsin. "McArdle Laboratory for Cancer Research." (02 April 2001). <http//www.mcardle. oncoogy.wisc.edu/donations.html>

University of Wisconsin. "Mutagenesis in vitro." (2000). <http//www.bact.wisc.edu/microtextbook/BactGenetics/ mutainvitro.html>

University of Wisconsin. "Transduction." (29 April 2001). <http://www.bact.wisc.edu/MicrotextBook/ BactGenetics/transduct1.html>

Wellcome Trust Tour of the Human Genome (2001). <http://www.wellcome.ac.uk/en/genome/>

Whitehead Center for Genome Research at WIMR (2001). <http://www.wi.mit.edu/news/genome/gc.html>

Williams Inference Center. "Gene chips." (04 March 2001). <http://www.williamsinference.com/2410gene.htm>

WorldMedicus. "Insertional mutagenesis." (04 March 2001). <http://www.worldmedicus.com/servlet/Controller/ $700a00273f7e0000.sj_viewr/index.html>

World Medicus. "Operator regions." (31 March 2001). <http//www.worldmedicus.com/servet/Controller/ $700e001b26940000.sj_viewd/>

Worthington Biochem. "DNA and RNA Polymerase." (31 March 2001). <http//www.worthington-biochem.com/ manual/D/dnap.html>

Zinn, A., M. Watson, N. Inoue, K. Hess, R. Moreadith, "Newly Discovered Gene Family May Regulate Important Biological Processes." (1999). <http://www.eurekalert. org/releases/utsmc-ndg060899.html>

c. 50,000 B.C.   *Homo sapiens sapiens* emerged as a conscious observer of nature.

c. 10,000 B.C.   Neolithic Revolution: transition from a hunting and gathering mode of food production to farming and animal husbandry, that is, the domestication of plants and animals.

c. 3500 B.C.   Sumerians described methods of managing the date harvest.

c. 1800 B.C.   Male flower of the date plant mentioned in a Babylonian document. Even though scientists did not experimentally demonstrate plant sexuality until the seventeenth century, in ancient Mesopotamia, the date palm was cross-pollinated and the ratio of male to female trees was regulated in order to secure good crops.

c. 600 B.C.   Thales, the founder of the Ionian school of Greek philosophy, identified water as the fundamental element of nature. Other Ionian philosophers constructed different theories about the nature of the universe and living beings.

c. 500 B.C.   Alcmaeon, Pythagorean philosopher and naturalist, pursued anatomical research, concluded that humans are fundamentally different from animals, and established the foundations of comparative anatomy.

c. 450 B.C.   Empedocles, Greek philosopher, asserted that the universe and all living things are composed of four fundamental elements: earth, air, fire, and water.

c. 400 B.C.   Democritus, Greek philosopher, argued that atoms are the building blocks of the universe and all living things. Democritus was an early advocate of the preformation theory of generation (embryology).

c. 400 B.C.   Hippocrates, Greek physician, founded a school of medicine on the Aegean island of Cos. According to Hippocratic medical tradition, the four humors that make up the human body correspond to the four elements that make up the universe. Hippocrates suggested using the developing chick egg as a model for embryology. Hippocrates noted that offspring inherit traits from both parents.

c. 350 B.C.   Aristotle, the Greek philosopher who has been called the founder of biology, attempted to classify animals and described various theories of generation, including sexual, asexual, and spontaneous generation. Aristotle argued that the male parent contributes "form" to the offspring and the female parent contributes "matter." Aristotle discussed preformation and epigenesis as possible theories of embryological development, but believed that development occurs by epigenesis.

c. 300 B.C.   Theophrastus, Aristotle's disciple and the founder of botany, attempted to establish a classification system for plants.

c. 50 B.C.   Lucretius proposed a materialistic, atomistic theory of nature in his poem *On the Nature of Things*. He favored the preformation theory of embryological development.

c. 200   Galen, the preeminent medical authority of late antiquity and the Middle Ages, created a philosophy of medicine, anatomy, and physiology that was virtually unchallenged until the sixteenth and seventeenth centuries. He believed that embryological development was epigenetic, although he disagreed with Aristotle about which organs were formed first and which were most important.

1543   Andreas Vesalius published his epoch-making treatise *The Fabric of the Human Body.* He generally accepted Galenic physiological doctrines and ideas about embryology.

1583   Andreas Cesalpino (Cesalpinus) wrote *On Plants,* a landmark in the development of botanical taxonomy.

1561   Gabriello Fallopio (Fallopius) published *Observationes anatomicae,* which increased knowledge of the sexual organs. The Fallopian tubes are named after him.

1600   Girolamo Fabrizzi (Fabricus ab Aquapendente) published *De formato foetu (On the Formation of the Fetus).* The illustrations in the embryological works of Fabricus were a great advance on previous work.

1621   Girolamo Fabrizzi (Fabricus ab Aquapendente) published *De formatione ovi et pulli (On the Formation of the Egg and the Chick).*

1651   William Harvey published a landmark treatise on embryology entitled *On the Generation of Animals,* stating that all living things come from eggs. Harvey demonstrated that oviparous and viviparous generation are analogous to each other. Although Harvey discovered many errors in Aristotle's ideas, he supported the Aristotelian doctrine that generation occurs by epigenesis.

1665   Robert Hooke published *Micrographia,* an account of observations made with the new instrument known as the microscope. He presented his drawings of the tiny box-like structures found in cork and called these tiny structures "cells." Although the cells he observed were not living, the name was retained. He also described the streaming juices of live plant cells.

1667   Nicolaus Steno discovered the "female testicles" of the shark and introduced the term "ovary." He argued that the female testicles contain ova (eggs).

1668   Regnier de Graaf published his treatise on the human sex organs, and described the structures that are now known as the Graafian follicles. He also confirmed Harvey's analogy between oviparous and viviparous reproduction.

1668   Francesco Redi published *Experiments on the Generation of Insects,* which described the results of experiments that disproved the theory that maggots are spontaneously generated by rotting meats. He demonstrated that maggots develop from eggs laid by flies.

1669   Jan Swammerdam began his pioneering work on the metamorphosis of insects and the anatomy of the mayfly. Swammerdam suggested that new individuals were embedded, or preformed, in their predecessors. Nicolas de Malebranche reformulated Swammerdam's preformationist ideas into a more sophisticated philosophical doctrine that involved a series of embryos preexisting within each other like a nest of boxes.

1677   Antoni van Leeuwenhoek discovered "little animals" (spermatozoa) in semen. His observations were published in *The Philosophical Transactions of the Royal Society* in 1679.

1681   Nehemiah Grew introduced the term "comparative anatomy."

1682   John Ray published his book *Methodus plantarum* in which he divided flowering plants into monocotyledons and dicotyledons.

1683   Antoni van Leeuwenhoek discovered different types of infusoria (minute organisms found in decomposing matter and stagnant water). He also described protozoa and bacteria.

1686   John Ray published the first volume of his three-volume *General History of Plants.* This work introduced the idea of species, that is, groups of individual plants with similar seeds, into botany and provided much of the inspiration for further developments in taxonomy.

1694   Rudolph Jakob Camerarius provided experimental evidence that plants reproduce sexually.

1700   Joseph Pitton de Tournefort developed an early version of the binomial method of classification, which was later developed by Carl Linnaeus.

1735   Carl Linnaeus published his *Systema Naturae, or The Three Kingdoms of Nature Systematically Proposed in Classes, Orders, Genera, and Species,* a methodical and hierarchical classification of all living beings. He developed the binomial nomenclature for the classification of plants and animals. In this system, each type of living being is classified in terms of genus (denoting the group to which it belongs) and species (its particular, individual name). His classification of plants was primarily based on the characteristics of their reproductive organs.

1737   *The Bible of Nature,* by Jan Swammerdam, was published by Hermann Boerhaave, Dutch naturalist, who helped to establish Swammerdam's reputation as the founder of modern entomology.

1740 Abraham Trembley discovered that the fresh water hydra, or "polyp," appeared to be an animal rather than a plant. When the hydra was cut into pieces, each part could regenerate a complete new organism. These experiments raised many philosophical questions about the "organizing principle" in animals and the nature of development.

1745 Charles Bonnet published *Insectology,* in which he described his experiments on parthenogenesis in aphids (the production of offspring by female aphids in the absence of males).

1746 Pierre-Louis Moreau de Maupertuis published *Venus Physique.* Maupertuis criticized preformationist theories because offspring inherit characteristics of both parents. He proposed an adaptationist account of organic design. His theories suggested the existence of a mechanism for transmitting adaptations.

1754 Pierre-Louis Moreau de Maupertuis suggested that species change over time, rather than remaining fixed.

1757 Albrecht von Haller published the first volume of his eight-volume *Elements of Physiology of the Human Body* (1757–1766), a landmark in the history of modern physiology.

1759 Kaspar Friedrich Wolff published *Theory of Generation,* which argues that generation occurs by epigenesis, that is, the gradual addition of parts. This book marks the beginning of modern embryology.

1760 Carl Linnaeus described hybrids and their possible role in the generation of new varieties of plants in his book on plant sexuality.

1762 Marcus Anton von Plenciz, Sr., suggested that all infectious diseases are caused by living organisms and that there is a particular organism for each disease.

1765 Lazzaro Spallanzani published his *Microscopical Observations.* Spallanzani's experiments refuted the theory of the spontaneous generation of infusoria.

1765 Abraham Trembley observed and published drawings of cell division in protozoans and algae.

1780 Lazzaro Spallanzani carried out experiments on fertilization in frogs and attempted to determine the role of semen in the development of amphibian eggs.

1789 Antoine-Laurent de Jussieu published his *Plant Genera,* a widely acclaimed book which incorporated the Linnaean system of binomial nomenclature. This book is generally regarded as the foundation of the natural system of botanical classification.

Following Ray, Jussieu classified plants on the basis of cotyledons, dividing all plants into acotyledons, monocotyledons, and dicotyledons.

1793 Christian Konrad Sprengel clearly described the role of insects in plant fertilization in his *Nature's Revealed Secret in the Structure and Pollination of Flowers.*

1796 Erasmus Darwin, grandfather of Charles Darwin and Francis Galton, published his *Zoonomia.* In this work, Darwin argues that evolutionary changes are brought about by the mechanism primarily associated with Jean-Baptiste Lamarck, that is, the direct influence of the environment on the organism.

1797 Georges-Léopold-Chrétien-Frédéric Dagobert Cuvier established modern comparative zoology with the publication of his first book, *Basic Outline for a Natural History of Animals.* Cuvier studied the ways in which an animal's function and habits determine its form. He argued that form always followed function and that the reverse relationship did not occur.

1800 Marie-François-Xavier Bichat published his first major work, *Treatise on Tissues,* which established histology as a new scientific discipline. Bichat distinguished 21 kinds of tissue and related particular diseases to particular tissues.

1802 Jean-Baptiste-Pierre-Antoine de Monet de Lamarck and Gottfried Reinhold Treviranus propose the term "biology" to denote a new general science of living beings that would supercede studies in natural history.

1809 Jean-Baptiste-Pierre-Antoine de Monet de Lamarck introduced the term "invertebrate" in his *Zoological Philosophy,* which contains the first influential scientific theory of evolution. He attempted to classify organisms by function rather than by structure and was the first to use genealogical trees to show relationships among species.

1812 Georges-Léopold-Chrétien-Frédéric Dagobert Cuvier founded vertebrate paleontology with his *Investigations of the Fossil Bones of Quadrupeds.*

1817 Georges-Léopold-Chrétien-Frédéric Dagobert Cuvier publishes his major work, *The Animal Kingdom,* which expands and improves Linnaeus's classification system by grouping related classes into broader groups called phyla. He is also the first to extend this system of classification to fossils.

1818 William Charles Wells suggested the theory of natural selection in an essay dealing with human color variations. Wells noted that dark skinned people were more resistant to tropical diseases than lighter skinned people. Wells also called attention to selection carried out by animal breeders. Jerome Lawrence, James Cowles Prichard, and others made similar suggestions. However, these individuals did not develop their ideas into a coherent and convincing theory of evolution.

1821 Jean-Louis Prévost and Jean-Baptiste-André Dumas jointly published a paper that demonstrates that spermatozoa originate in tissues of the male sex glands. Three years later they published the first detailed account of the segmentation of a frog's egg.

1824 René-Joachim-Henri Dutrochet suggested that tissues are composed of living cells.

1825 Jan Evangelista Purkinje described the "germinal vesicle," or nucleus, in a hen's egg.

1826 James Cowles Prichard presented his views on evolution in the second edition of his book *Researches into the Physical History of Man* (first edition 1813). These ideas about evolution were suppressed in later editions.

1827 Karl Ernst von Baer published *On the Mammalian Egg,* documenting his 1826 discovery of the mammalian ovum. In his book *On the Developmental History of Animals* (two volumes, 1828-1837), von Baer demonstrated that embryological development follows essentially the same pattern in a wide variety of mammals. Early mammalian embryos are very similar, but they diverge at later stages of gestation. His work established the modern field of comparative embryology.

1828 Friedrich Wöhler synthesized urea. This is generally regarded as the first organic chemical produced in the laboratory and an important step in disproving the idea that only living organisms can produce organic compounds. Work by Wöhler and others established the foundations of organic chemistry and biochemistry.

1828 Robert Brown observed a small body within the cells of plant tissue and called it the "nucleus." He also discovered what became known as "Brownian movement."

1831 Robert Brown recognized the cell nucleus as a regular feature of all plant cells.

1831 Patrick Mathew included a discussion of evolution and natural selection in his book *On Naval Timber and Arboriculture.* Matthew later claimed priority in the discovery of evolution by means of natural selection in an article published in 1860 in the journal *Gardeners' Chronicle.*

1831 Charles Robert Darwin began his historic voyage on the H.M.S. *Beagle* (1831–1836). His observations during the voyage lead to his theory of evolution by means of natural selection.

1832 Anselme Payen isolated "diastase" from barley. This substance stimulates the conversion of starch into sugar. Diastase was eventually characterized as one of the organic catalysts (now called enzymes) found in cells.

1836 Félix Dujardin described the "living jell" of the cytoplasm. He called it "sarcode."

1836 Theodor Schwann carried out experiments that refuted the theory of the spontaneous generation of infusoria. He also demonstrated that alcoholic fermentation depended on the action of living yeast cells. The same conclusion was reached independently by Charles Caignard de la Tour.

1837 Karl Theodor von Siebold and Michael Sars described the division of invertebrate eggs.

1837 Robert Remak described the relationship between nerve cells and nerve fibers. Remak described and named the neurolemma (the myelin sheath around many nerve fibers). He later described and named the three germinal layers in the developing embryo: ectoderm (outer skin), mesoderm (middle skin), and endoderm (inner skin).

1838 Matthias Jakob Schleiden noted that the nucleus, which had been described by Robert Brown, was a characteristic of all plant cells. Schleiden described plants as a community of cells and cell products. He helped establish cell theory and stimulated Theodor Schwann's recognition that animals are also composed of cells and cell products.

1839 Theodore Schwann extended the theory of cells to include animals and helped establish the basic unity of the two great kingdoms of life. In *Microscopical Researches into the Accordance in the Structure and Growth of Animals and Plants,* Schwann asserted that all living things are made up of cells, each of which contains certain essential components. He also coined the term "metabolism" to describe the overall chemical changes that take place in living tissues.

1839 Jan Evangelista Purkinje used the term "protoplasm" to describe the substance within living cells.

1840 Rudolf Albert von Kölliker established that spermatozoa and eggs are derived from tissue cells. He attempted to extend the cell theory to embryology and histology.

1840 Karl Bogislaus Reichert introduced the cell theory into embryology. He proved that the segments observed in fertilized eggs develop into individual cells, and that organs develop from cells.

1840 Friedrich Gustav Jacob Henle published the first histology textbook, *General Anatomy*. This work includes the first modern discussion of the germ theory of communicable diseases.

1842 Charles Robert Darwin wrote out an abstract of his theory of evolution, but he did not plan to have this theory published until after his death.

1842 Theodor Ludwig Wilhelm Bischoff published the first textbook of comparative embryology, *Developmental History of Mammals and Man.*

1843 Martin Berry observed the union of sperm and egg of a rabbit.

1844 Robert Chambers anonymously published *Vestiges of the Natural History of Creation,* which advocated the theory of evolution. This controversial book became a best seller and introduced the general reading public to the theory of evolution.

1845 Karl Theodor Ernst von Siebold realized that protozoa are single-celled organisms. He was the first scientist to define protozoa as organisms.

1846 Giovanni Battista Amici established sexuality in flowering plants.

1848 Karl Theodor Ernst von Siebold published his *Textbook of Invertebrate Comparative Anatomy,* one of the first major textbooks on invertebrate anatomy.

1849 Rudolf Wagner and Karl Georg Friedrich Rudolf Leuckart reported that spermatozoa are a definite and essential part of the semen, and that the liquid merely keeps them in suspension. They also reject the old hypothesis that spermatozoa are parasites, and argue that spermatozoa are essential for fertilization.

1849 Karl Friedrich Gärtner published the first comprehensive treatment of plant hybridization. Charles Darwin and Gregor Mendel closely studied his work.

1850 Wilhelm Friedrich Hofmeister proved the fundamental similarity between flowering and non flowering plants and demonstrated the alternation of sexual and non-sexual generations.

1851 Hugo von Mohl published his *Basic Outline of the Anatomy and Physiology of the Plant Cell,* in which he proposed that new cells are created by cell division.

1854 George Newport performed the first experiments on animal embryos. He suggested that the point of sperm entry determines the planes of the segmentation of the egg.

1854 Gregor Mendel began studying 34 different strains of peas. He selected 22 kinds for further experiments. From 1856 to 1863, Mendel grew and tested over 28,000 plants and analyzed seven pairs of traits.

1855 Alfred Russell Wallace wrote an essay entitled *On the Law Which has Regulated the Introduction of New Species* and sent it to Charles Darwin. Wallace's essay and one by Darwin were published in the 1858 *Proceedings of the Linnaean Society.*

1856 Nathanael Pringsheim observed the sperm of a freshwater algae plant enter the egg.

1857 Louis Pasteur demonstrated that lactic acid fermentation is caused by a living organism. Between 1857 and 1880, he performed a series of experiments that refuted the doctrine of spontaneous generation. He also introduced vaccines for fowl cholera, anthrax, and rabies, based on attenuated strains of viruses and bacteria.

1858 Rudolf Ludwig Carl Virchow published his landmark paper "Cellular Pathology" and established the field of cellular pathology. Virchow asserted that all cells arise from preexisting cells (*Omnis cellula e cellula*). He argued that the cell is the ultimate locus of all disease.

1858 Charles Darwin and Alfred Russell Wallace agree to a joint presentation of their theory of evolution by natural selection.

1859 Charles Darwin published his landmark book *On the Origin of Species by Means of Natural Selection.*

1860 Ernst Heinrich Haeckel described the essential elements of modern zoological classification.

1860 Louis Pasteur carried out experiments that disproved the doctrine of spontaneous generation.

1860 Max Johann Sigismund Schultze described the nature of protoplasm and showed that it is fundamentally the same for all life forms.

1861 Carl Gegenbaur confirmed Theodor Schwann's suggestion that all vertebrate eggs are single cells.

1861 Louis Pasteur concluded a series of simple, but rigorous experiments that disproved the theory of spontaneous generation.

1861 Rudolf Albert von Kölliker published *Developmental History of Man and Higher Animals,* the first treatise on comparative embryology.

1863 Thomas Henry Huxley extended Darwin's theory of evolution to include humans in his book *Evidence As to Man's Place in Nature.* He became the champion and defender of Darwinism in England.

1865 Gregor Mendel presented his work on hybridization of peas to the Natural History Society of Brno, Czechoslovakia. The paper was published in the 1866 issue of the Society's *Proceedings.* In a series of papers on "Experiments on Plant Hybridization" published between 1866 and 1869, Mendel presented statistical evidence that hereditary factors were inherited from both parents. His experiments provided evidence of dominance, and the laws of segregation and independent assortment. His work was generally ignored until 1900.

1865 Franz Schweiger-Seidel proved that spermatozoa consist of a nucleus and cytoplasm.

1866 Ernst Heinrich Haeckel published his book *A General Morphology of Organisms*. Haeckel summarized his ideas about evolution and embryology in his famous dictum "ontogeny recapitulates phylogeny." Haeckel suggested that the nucleus of a cell transmits hereditary information. He introduced the use of the term "ecology" to describe the study of living organisms and their interactions with other organisms and with their environment.

1868 Charles Darwin published *The Variation of Animals and Plants under Domestication* (2 volumes).

1868 Francis Galton published his book *Hereditary Genius*. Galton argued that the study of human pedigrees proved that intelligence was a hereditary trait.

1868 Thomas Henry Huxley introduced the term "protoplasm" to the general public in a lecture entitled "The Physical Basis of Life."

1869 Johann Friedrich Miescher discovered nuclein, a new chemical isolated from the nuclei of pus cells. Two years later he isolated nuclein from salmon sperm. This material is now known as nucleic acid.

1870 Lambert Adolphe Jacques Quetelet showed the importance of statistical analysis for biologists and provided the foundations of biometry.

1871 Charles Robert Darwin published *The Descent of Man, and Selection in Relation to Sex*. This work introduces the concept of sexual selection and expands his theory of evolution to include humans.

1871 Ferdinand Julius Cohn coined the term bacterium.

1872 Ferdinand Julius Cohn published the first of four papers entitled "Research on Bacteria," which established the foundation of bacteriology as a distinct field. He systematically divided bacteria into genera and species.

1872 Franz Anton Schneider observed and described the behavior of nuclear filaments (chromosomes) during cell division in his study of the platyhelminth Mesostoma. His account was the first accurate description of the process of mitosis in animal cells.

1873 Walther Flemming discovered chromosomes, observed mitosis, and suggested the modern interpretation of nuclear division.

1873 Franz Anton Schneider described cell division in detail. His drawings included both the nucleus and chromosomal strands.

1873 Camilo Golgi discovered that tissue samples could be stained with an inorganic dye (silver salts). Golgi used this method to analyze the nervous system and characterized the cells known as Golgi Type I and Golgi Type II cells and the "Golgi Apparatus." Golgi won a Nobel Prize in 1906 for his studies of the nervous system.

1874 Francis Galton demonstrated the usefulness of twin studies for elucidating the relative influence of nature (heredity) and nurture (environment).

1874 Wilhelm August Oscar Hertwig concluded that fertilization in both animals and plants consists of the physical union of the two nuclei contributed by the male and female parents. Hertwig carried out pioneering studies of reproduction of the sea urchin.

1874 Eduard Strasburger, German embryologist, accurately described the processes of mitotic cell division in plants.

1875 Eduard Adolf Strasburger published *Cell-Formation and Cell-Division*, in which he described nuclear division in plants. Strasburger accurately described the process of mitosis and argued that new nuclei can only rise from the division of preexisting nuclei. His treatise helped establish cytology as a distinct branch of histology.

1876 Edouard G. Balbiani observed the formation of chromosomes.

1876 Wilhelm August Oscar Hertwig observed the fertilization of a sea urchin egg and established that both parents contribute genetic material to the offspring. He also proved that fertilization is due to a fusion of the sperm nucleus with the nucleus of the egg.

1876 Robert Koch described new techniques for fixing, staining, and photographing bacteria.

1877 Wilhelm Friedrich Kühne proposed the term enzyme (meaning "in yeast"). Kühne established the critical distinction between enzymes, or "ferments," and the microorganisms that produce them.

1878 Charles-Emanuel Sedillot introduced the term "microbe." The term became widely used as a term for a pathogenic bacterium.

1879 Hermann Fol observed the penetration of the egg of a sea urchin by a sperm. He demonstrated that only one spermatozoon is needed for fertilization and suggested that the nucleus of the sperm was transferred into the egg.

1879 Walther Flemming described and named chromatin, mitosis, and the chromosome threads. Fleming's drawings of the longitudinal splitting of chromosomes provided the first accurate counts of chromosome numbers.

1880 The first issue of the journal *Science* was published by the American Association for the Advancement of Science.

1880 The basic outlines of cell division and the distribution of chromosomes to the daughter cells were established by Walther Flemming, Eduard Strasburger, Edouard van Beneden, and others.

1881 Eduard Strasburger coined the terms cytoplasm and nucleoplasm.

1881 Walther Flemming discovered the lampbrush chromosomes.

1881 Wilhelm Roux the founder of experimental embryology, published *The Struggle of the Parts in the Organism: A Contribution to the Completion of a Mechanical Theory of Teleology.* Roux argued that his experimental approach to embryonic development, based on mechanistic principles, provided evidence that development proceeded by means of self-differentiation.

1882 Pierre Émile Duclaux suggested that enzymes should be named by adding the suffix "-ase" to the name of their substrate.

1882 Edouard van Beneden outlined the principles of genetic continuity of chromosomes and reported the occurrence of chromosome reduction during the formation of the germ cells.

1882 Wilhelm Roux offered a possible explanation for the function of mitosis.

1882 Walther Flemming published *Cell Substance, Nucleus, and Cell Division,* in which he described his observations of the longitudinal division of chromosomes in animal cells. Flemming observed chromosome threads in the dividing cells of salamander larvae.

1883 Walther Flemming, Eduard Strasburger and Edouard Van Beneden demonstrated that chromosome doubling occurs by a process of longitudinal splitting. Strasburger described and named the prophase, metaphase, and anaphase stages of mitosis.

1883 Wilhelm Roux suggested that chromosomes in the nucleus carry the hereditary factors.

1883 August F. Weismann began work on his germplasm theory of inheritance. Between 1884 and 1888, Weismann formulated the germplasm theory which held that the germplasm was separate and distinct from the somatoplasm. He argued that the germplasm was continuous from generation to generation and that only changes in the germplasm were transmitted to further generations. Weismann proposed a theory of chromosome behavior during cell division and fertilization and predicted the occurrence of a reduction division (meiosis) in all sexual organisms.

1883 Francis Galton coined the word "eugenics," which he defined as the science of improving the human race by means of selective breeding.

1884 Louis Pasteur and coworkers published a paper entitled "A New Communication on Rabies." Pasteur proved that the causal agent of rabies could be attenuated and the weakened virus could be used as a vaccine to prevent the disease. This work served as the basis of future work on virus attenuation, vaccine development, and the concept that variation is an inherent characteristic of viruses.

1884 Elie Metchnikoff discovered the antibacterial activity of white blood cells, which he called "phagocytes," and formulated the theory of phagocytosis.

1884 Oscar Hertwig, Eduard Strasburger, Albrecht von Kölliker, and August Weismann independently reported that the cell nucleus served as the basis for inheritance.

1884 Karl Rabl suggested the concept of the individuality of the chromosomes. He argued that each chromosome originated from a preexisting chromosome in the mother cell that was like it in form and size.

1884 Walther Flemming observed sister chromatids passing to opposite poles of the cell during mitosis.

1885 Francis Galton devised a new statistical tool, which is known as the correlation table.

1886 Edouard van Beneden proved that chromosomes persist between cell divisions. He discovered that the cell nucleus of each species has a fixed and specific number of chromosomes. He demonstrated chromosome reduction in gamete maturation, thereby confirming August Weismann's predictions. That is, he showed that the division of chromosomes during one of the cell divisions that produced the sex cells was not preceded by the doubling of chromosomes.

1886 Adolf Mayer published the landmark article "Concerning the Mosaic Disease of Tobacco." This paper is considered the beginning of modern experimental work on plant viruses. Mayer assumed that the causal agent was a bacterium, although he was unable to isolate it.

1887 Theodor Boveri observed the reduction division during meiosis in *Ascaris* and confirmed August

Weismann's predictions of chromosome reduction during the formation of the sex cells.

1888 Francis Galton published *Natural Inheritance,* considered a landmark in the establishment of biometry and statistical studies of variation. Galton also proposed the Law of Ancestral Inheritance, a statistical description of the relative contributions to heredity made by previous generations.

1888 Heinrich Wilhelm Gottfried Waldeyer coined the term "chromosome." Waldeyer introduced the use of hematoxylin as a histological stain.

1888 Theodor Heinrich Boveri discovered and named the centrosome (the mitotic spindle that appears during cell division).

1888 Woods Hole Marine Biological Station, which later became the headquarters of the Woods Hole Oceanographic Institution and the Marine Biological Laboratory, was established in Massachusetts.

1889 Theodor Boveri and Jean-Louis-Léon Guignard established the numerical equality of the paternal and maternal chromosomes at fertilization.

1889 Richard Altmann developed a method of preparing nuclein that was apparently free of protein. He called his protein-free nucleins "nucleic acids."

1891 Charles-Edouard Brown-Sequard suggested the concept of internal secretions (hormones).

1891 Hermann Henking distinguished between the sex chromosomes and the autosomes.

1892 August Weismann published his landmark treatise *The Germ Plasm: A Theory of Heredity,* which emphasized the role of meiosis in the distribution of chromosomes during the formation of gamates.

1892 George M. Sternberg published his *Practical Results of Bacteriological Researches.* Sternberg's realization that a specific antibody was produced after infection with vaccinia virus and that immune serum could neutralize the virus became the basis of virus serology. The neutralization test provided a technique for diagnosing viral infections, measuring the immune response, distinguishing antigenic similarities and differences among viruses, and conducting retrospective epidemiological surveys.

1893 Hans Adolf Eduard Driesch discovered that he could separate sea urchin embryos into individual cells and that the separated cells continued to develop. Driesch concluded that all the cells of the early embryo were capable of developing into whole organisms. Therefore, embryonic cells were equipotent, but differentiated and developed in response to their position within the embryo.

1893 William Bateson published *Materials for the Study of Variation,* which emphasized the importance of discontinuous variations (the kinds of variation studied by Mendel).

1893 Karl Pearson published the first in a long series of contributions to the mathematical theory of evolution. Pearson proposed useful statistical methods for analyzing the distribution of traits within populations.

1894 Wilhelm Konrad Roentgen discovered x rays.

1896 Edmund Beecher Wilson, American zoologist, published the first edition of his highly influential treatise *The Cell in Development and Heredity.* Wilson called attention to the relationship between chromosomes and sex determination.

1898 The First International Congress of Genetics was held in London.

1898 William Bateson published a paper on the importance of hybridization and crossbreeding experiments.

1898 Carl Benda discovered and named mitochondria, the subcellular entities previously seen by Richard Altmann.

1898 Martin Wilhelm Beijerinck published his landmark paper "Concerning a Contagium Vivum Fluidum as Cause of the Spot Disease of Tobacco Leaves." Beijerinck thought that the etiological agent, which could pass through a porcelain filter that removed known bacteria, might be a new type of invisible organism that reproduced within the cells of diseased plants. He realized that a very small amount of the virus could infect many leaves and that the diseased leaves could infect others.

1898 Friedrich Loeffler and Paul Frosch published their *Report on Foot-and-Mouth Disease.* They proved that this animal disease was caused by a filterable virus and suggested that similar agents might cause other diseases.

1899 Jacques Loeb proved that it was possible to induce parthenogenesis in unfertilized sea urchin eggs by means of specific environmental changes.

1900 Carl Correns, Hugo de Vries, and Erich von Tschermak independently rediscovered Mendel's laws of inheritance. Their publications mark the beginning of modern genetics. Using several plant species, de Vries and Correns had performed breeding experiments that paralleled Mendel's earlier studies and had independently arrived at similar interpretations of their results. Therefore, upon reading Mendel's publication they immediately recognized its significance. William Bateson immediately described the importance of Mendel's contribution in an address to the Royal Society of London.

1900 Hugo Marie de Vries described the concept of genetic mutations in his book *Mutation Theory*. He used the term mutation to describe sudden, spontaneous, drastic alterations in the hereditary material.

1900 Karl Pearson developed the chi-square test.

1900 Karl Landsteiner discovered the blood-agglutination phenomenon and the four major blood types in humans.

1900 Thomas H. Montgomery studied spermatogenesis in various species of Hemiptera. He concluded that maternal chromosomes only pair with corresponding paternal chromosomes during meiosis.

1901 Clarence E. McClung argued that particular chromosomes determine the sex of the individual carrying them. Although his work was done with insects, he suggested that this might be true for human beings and other animals.

1901 Theodor Boveri discovered that in order for sea urchin embryos to develop normally, they must have a full set of chromosomes. He concluded that this meant that the individual chromosomes must carry different hereditary determinants.

1901 Archibald Edward Garrod reported that a human disease, alkaptonuria, seemed to be inherited as a Mendelian recessive trait.

1901 William Bateson coined the terms genetics, F1 and F2 generations, allelomorph (later shortened to allele), homozygote, heterozygote, and epistasis.

1902 Walter Sutton presented evidence that chromosomes have individuality, that chromosomes occur in pairs (with one member of each pair contributed by each parent), and that the paired chromosomes separate from each other during meiosis. Sutton concluded that the concept of the individuality of the chromosomes provided the link between cytology and Mendelian heredity.

1902 Wilhelm Ludwig Johannsen introduced and defined the terms phenotype, genotype, and selection in terms of the new science of genetics.

1902 Carl Neuberg introduced the term biochemistry.

1903 Walter S. Sutton published the paper in which he presented the chromosome theory of inheritance. This important theory, which states that the hereditary factors are located in the chromosomes, was independently proposed by Theodor Boveri and was generally referred to as the Sutton-Boveri hypothesis.

1903 Archibald Edward Garrod provided evidence that errors in genes caused several hereditary disorders in human beings. His book *The Inborn Errors of Metabolism* (1909) was the first treatise in biochemical genetics.

1904 Lucien Claude Cuénot's crossbreeding experiments on mice that provided proof that Mendel's newly rediscovered laws applied to animals as well as to plants. His research on the crossbreeding of yellow mice also provided the first example of genes that were lethal when homozygous. In 1910 other scientists proved that yellow homozygotes always die *in utero*.

1905 Nettie Maria Stevens, American geneticist, discovered the connection between chromosomes and sex determination. She determined that there are two basic types of sex chromosomes, which are now called X and Y. Stevens proved that females are XX and males are XY. Stevens and Edmund B. Wilson independently described the relationship between the so-called accessory or X chromosomes and sex determination in insects.

1905 William Bateson and Reginald C. Punnett reported the discovery of two new genetic principles: linkage and gene interaction.

1907 William Bateson urged biologists to adopt the term "genetics" to indicate the importance of the new science of heredity.

1907 Godfrey Harold Hardy proposed that Mendelian mechanisms acting alone would have no effect on allele frequencies. This suggestion forms the mathematical basis for population genetics and what became known as the Hardy-Weinberg Law.

1908 Thomas H. Morgan published a paper expressing doubts about Mendelian explanations for inherited traits.

1908 George H. Shull proposed using self-fertilized lines in the production of commercial seed corn. This resulted in highly successful hybrid corn programs.

1908 Nilsson H. Ehle proposed the multiple-factor hypothesis to explain the quantitative inheritance of seed-coat color in wheat.

1908 Godfrey Harold Hardy and Wilhelm Weinberg independently published similar papers describing a mathematical system that accounts for the stability of gene frequencies in succeeding generations of population. Their resulting Hardy-Weinberg law links the Mendelian hypothesis with population studies.

1908 Margaret A. Lewis successfully cultured mammalian cells *in vitro*.

1909   Archibald E. Garrod published *Inborn Errors of Metabolism*, in addition to Garrod's papers, one of the earliest discussions of biochemical genetics.

1909   Wilhelm Ludwig Johannsen realized the necessity of distinguishing between the appearance of an organism and its genetic constitution. He invented the terms "gene" (carrier of heredity), "genotype" (an organism's genetic constitution), and "phenotype" (the appearance of the actual organism).

1909   Thomas Hunt Morgan selected the fruit fly *Drosophila* as a model system for the study of genetics. Morgan and his coworkers confirmed the chromosome theory of heredity and realized the significance of the fact that certain genes tend to be transmitted together. Morgan postulated the mechanism of "crossing over." His associate, Alfred Henry Sturtevant demonstrated relationship between crossing over and the rearrangement of genes in 1913.

1910   Thomas Hunt Morgan discovered the *Drosophila* mutant called white eye. Research on this mutant led to the discovery of sex-linked traits. Morgan suggested that the genes for white eyes, yellow body, and miniature wings in *Drosophila* are linked together on the X chromosome.

1911   Francis Peyton Rous published the landmark paper "Transmission of a Malignant New Growth by Means of a Cell-Free Filtrate." His work provided the first rigorous proof of the experimental transmission of a solid tumor and suggested that a filterable virus was the causal agent.

1912   Alfred H. Sturtevant working with Thomas Hunt Morgan at Columbia, provided experimental evidence of the linkage of genes and began constructing the first genetic map for *Drosophila* chromosomes.

1913   Calvin Blackman Bridges discovered evidence of nondisjunction of the sex chromosomes in Drosophila. This evidence helped support Thomas Hunt Morgan's new chromosome theory of heredity.

1913   Alfred H. Sturtevant published the landmark paper that contained the first chromosome map. "The Linear Arrangement of Six Sex-Linked Factors in *Drosophila,* as Shown by Their Mode of Association" was published in the *Journal of Experimental Zoology*. Within two years Thomas Hunt Morgan's group had described four groups of linked factors; these groups corresponded to the four pairs of *Drosophila* chromosomes. The chromosome theory replaced "beanbag" genetics with the image of genes as beads on a string.

1914   Frederick William Twort and Felix H. D'Herelle independently discovered bacteriophages, viruses that infect and destroy bacteria.

1914   Thomas Hunt Morgan, Alfred Henry Sturtevant, Calvin Blackman Bridges, and Hermann Joseph Muller published the classic treatise of modern genetics, *The Mechanism of Mendelian Heredity.*

1915   Katherine K. Sanford isolated a single mammalian cell *in vitro* and allowed it to propagate to form identical descendants. Her clone of mouse fibroblasts was called L929, because it took 929 attempts before a successful propagation was achieved. The picture of this clone has often been reproduced. It was an important step in establishing pure cell lines for biomedical research.

1915   Frederick William Twort published the landmark paper "An Investigation of the Nature of Ultra-Microscopic Viruses." Twort noted the degeneration of bacterial colonies and suggested that the causative agent was an ultra-microscopic-filterable virus that multiplied true to type.

1916   Felix Hubert D'Herelle carried out further studies of the agent that destroyed bacterial colonies and gave it the name "bacteriophage" (bacteria eating agent). D'Herelle and others unsuccessfully attempted to use bacteriophages as bactericidal therapeutic agents.

1917   D'Arcy Wentworth Thompson published *On Growth and Form,* which suggested that the evolution of one species into another occurs as a series of transformations involving the entire organism, rather than a succession of minor changes in parts of the body.

1917   Calvin Blackman Bridges discovered the first chromosome deficiency in *Drosophila.*

1918   Thomas Hunt Morgan and coworkers published *The Physical Basis of Heredity,* a survey of the remarkable development of the new science of genetics.

1918   Thomas Hunt Morgan demonstrated further evidence of the correlation between the number of linkage groups in *Drosophila melanogaster* and the haploid number of chromosomes in that species.

1918   Calvin B. Bridges discovered chromosomal duplications in *Drosophila.*

1921   Lillian V. Morgan discovered attached-X chromosomes in *Drosophila.*

1922   Calvin B. Bridges discovered chromosomal translocations in *Drosophila.*

1923   A. E. Boycott and C. Diver described a classic example of "delayed Mendelian inheritance." The direction of the coiling of the shell in the snail *Limnea peregra* is under genetic control, but the gene acts on the egg prior to fertilization. Thus, the direction of coiling is determined by the egg cytoplasm, which is controlled by the mother's genotype.

1924   Alfred H. Sturtevant analyzed the Barr-eye phenomenon in *Drosophila* and discovered position effect.

1925   Alfred H. Sturtevant discovered the first example of the chromosome inversion phenomenon (change in the location of a block of genes within a chromosome by rotation through 180°) in *Drosophila*.

1926   J.B.S. Haldane suggested that the genes known to control certain coat colors in various rodents and carnivores may be evolutionarily homologous.

1926   James B Sumner published a report on the isolation of the enzyme urease and his proof that the enzyme was a protein. This idea was controversial until 1930 when John Howard Northrop confirmed Sumner's ideas by crystallizing pepsin. Sumner, Northrop, and Wendell Meredith Stanley shared the Nobel Prize in chemistry in 1946.

1926   Bernard O. Dodge initiated genetic studies on *Neurospora.*

1926   Thomas C. Vanterpool published a paper that clarified the problem of "mixed infections" of plant viruses. His study of the condition known as "streak" or "winter blight" of tomatoes showed that it was the result of simultaneous infection of tomato plants by tomato mosaic virus and a potato mosaic virus.

1927   Hermann Joseph Muller induced artificial mutations in fruit flies by exposing them to x rays. His work proved that mutations result from some type of physical-chemical change. Muller wrote extensively about the danger of excessive x rays and the burden of deleterious mutations in human populations.

1927   Lewis J. Stadler produced artificial mutations in barley and maize, and demonstrated that the dose-frequency response is linear.

1928   Fred Griffith discovered that certain strains of pneumococci could undergo some kind of transmutation of type. After injecting mice with living R type pneumococci and heat-killed S type, Griffith was able to isolate living virulent bacteria from the infected mice. Griffith suggested that some unknown "principle" had transformed the harmless R strain of the pneumococcus to the virulent S strain.

1929   Frank M. Burnet and Margot McKie reported critical insights into the phenomenon known as lysogeny (the inherited ability of bacteria to produce bacteriophage in the absence of infection). Burnet and McKie postulated the presence of a "lytic unit" as a normal hereditary component of lysogenic bacteria. The "lytic unit" was capable of liberating bacteriophage when it was activated by certain conditions. This concept was confirmed in the 1950s.

1929   Francis O. Holmes introduced a valuable technique, known as the "local lesion" as a means of measuring the concentration of tobacco mosaic virus. The method became extremely important in virus purification.

1930   Max Theiler demonstrated the advantages of using mice as experimental animals for research on animal viruses. Theiler used mice in his studies of the yellow fever virus.

1930   Ronald A. Fisher published *Genetical Theory of Natural Selection,* a formal analysis of the mathematics of selection.

1930   Curt Stern, and, independently, Harriet B. Creighton and Barbara McClintock, demonstrated cytological evidence of crossing over.

1931   Phoebus A. Levene published a book that summarized his work on the chemical nature of the nucleic acids. His analyses of nucleic acids seemed to support the hypothesis known as the tetranucleotide interpretation, which suggested that the four bases were present in equal amounts in DNAs from all sources. This indicated that DNA was a highly repetitious polymer that was incapable of generating the diversity that would be an essential characteristic of the genetic material.

1931   Joseph Needham published his landmark work *Chemical Embryology,* which emphasized the relationship between biochemistry and embryology.

1931   Alice Miles Woodruff and Ernest W. Goodpasture demonstrated the advantages of using the membranes of the chick embryo to study the mechanism of viral infections.

1932   William J. Elford and Christopher H. Andrewes developed methods of estimating the sizes of viruses by using a series of graded collodion membranes as filters. Later studies proved that the viral sizes obtained by this method were comparable to those obtained by electron microscopy.

1932   Theophilus S. Painter demonstrated the value of the giant salivary gland chromosomes of *Drosophila* for cytogenetic studies.

1932   Barbara McClintock demonstrated that a single exchange within the inversion loop of a paracentric inversion heterozygote in maize generates an acentric and a dicentric chromatid.

1932   Thomas Hunt Morgan received the Nobel Prize in medicine or physiology for his development of the theory of the gene. He was the first geneticist to receive a Nobel Prize.

1932   Aldous Huxley published the novel *Brave New World,* which presented a dystopian view of genetic manipulations of human beings.

1934 J.B.S. Haldane presented the first calculations of the spontaneous mutation frequency of a human gene.

1934 George W. Beadle, working with Boris Ephrussi, in collaboration with A. Kuhn and A. Butenandt, worked out the biochemical genetics of eye-pigment synthesis in *Drosophila* and *Ephestia,* respectively.

1934 Calvin B. Bridges published salivary gland chromosome maps for *Drosophila melanogaster.*

1935 Alfred H. Sturtevant and Theodosius Dobzhansky published the first account of the use of inversions in constructing a chromosomal phylogenetic tree.

1935 Wendell Meredith Stanley reported that he had purified and crystallized a protein having the infectious properties of tobacco mosaic virus. Stanley concluded that tobacco mosaic virus was an autocatalytic protein that could only reproduce within living cells. Further work proved that the virus was a combination of protein and nucleic acid. Stanley won the Nobel Prize in chemistry in 1946.

1936 Theodosius Dobzhansky published *Genetics and the Origin of Species,* a text considered a classic in evolutionary genetics.

1936 Max Schlesinger reported that partially purified bacteriophage consist of DNA and protein. Other researchers showed that the nucleic acid in some plant viruses was RNA.

1936 George P. Berry and Helen M. Dedrick reported that the Shope virus could be "transformed" into myxomatosis/Sanarelli virus. This virological curiosity was referred to as "transformation" or "recombination" or "multiplicity of reactivation." Later research suggested that it was the first example of genetic interaction between animal viruses, but some scientists warned that the phenomenon might indicate the danger of reactivation of virus particles in vaccines and in cancer research.

1937 Richard Benedict Goldschmidt postulated that the gene is a chemical entity rather than a discrete physical structure.

1938 Emory L. Ellis and Max Delbrück performed studies on phage replication that marked the beginning of modern phage work. They introduced the "one-step growth" experiment, which demonstrates that after bacteriophages attack bacteria, they replicate within the bacterial host during the "latent period," after which viral progeny are released in a "burst."

1939 Richard E. Shope reported that the swine influenza virus survived between epidemics in an intermediate host. This discovery was an important step in revealing the role of intermediate hosts in perpetuating specific diseases.

1939 Moses Kunitz reported the purification and crystallization of ribonuclease from beef pancreas.

1940 Kenneth Mather coined the term "polygenes" and described polygenic traits in various organisms.

1941 George W. Beadle and Edward L. Tatum published their classic study on the biochemical genetics of *Neurospora,* "Genetic Control of Biochemical Reactions in *Neurospora.*" Beadle and Tatum irradiated the red bread mold, *Neurospora,* and proved that genes produce their effects by regulating particular enzymes. This work led to the one gene-one enzyme theory.

1942 Salvador E. Luria and Thomas F. Anderson published the first electron micrographs of bacterial viruses. The phage appeared to have a round, or polyhedral head and a thin tail.

1942 Salvador E. Luria and Max Delbrück proved that bacteria undergo spontaneous mutation. This made the study of bacterial genetics possible.

1944 Oswald T. Avery, Colin M. MacLeod, and Maclyn McCarty published a landmark paper on the pneumococcus transforming principle ("Studies on the chemical nature of the substance inducing transformation of pneumococcal types"). Avery suggested that the transforming principle seemed to be DNA, but contemporary ideas about the structure of nucleic acids suggested that DNA did not possess the biological specificity of the hypothetical genetic material.

1944 Salvador E. Luria proved that mutations occur in bacterial viruses.

1944-1945 New techniques and instruments, such as partition chromatography on paper strips and the photoelectric ultraviolet spectrophotometer, stimulated the development of biochemistry after World War II. New methodologies made it possible to isolate, purify, and identify many important biochemical substances, including the purines, pyrimidines, nucleosides, and nucleotides derived from nucleic acids.

1945 Joshua Lederberg and Edward L. Tatum demonstrated genetic recombination in bacteria.

1945 Max Delbrück organized the first session of the phage course at Cold Spring Harbor Laboratory. The widely influential phage course, which was taught for 26 consecutive years, served as the training center for the first two generations of molecular biologists

1946 Hermann J. Muller was awarded the Nobel Prize in medicine or physiology for his contributions to radiation genetics.

1946   James B. Sumner, John H. Northrop, and Wendell M. Stanley were awarded the Nobel Prize in chemistry for their independent work on the purification and crystallization of enzymes and viral proteins.

1946   Joshua Lederberg and Edward L. Tatum demonstrated that genetic recombination occurs in bacteria as the result of sexual matings. Lederberg and Tatum announced their discovery at the 1946 Cold Spring Harbor Symposium on Microbial Genetics, which is often described as the landmark event in the development of molecular biology.

1946   Max Delbrück and W. T. Bailey, Jr. published a paper entitled "Induced Mutations in Bacterial Viruses." Despite some confusion about the nature of the phenomenon in question, this paper established the fact that genetic recombinations occurred during mixed infections with bacterial viruses. Alfred Hershey and R. Rotman made the discovery of genetic recombination in bacteriophage simultaneously and independently. Hershey and his colleagues proved that this phenomenon could be used for genetic analyses. They constructed a genetic map of phage particles and showed that phage genes could be arranged in a linear fashion.

1947   Hermann J. Muller coined the term "dosage compensation."

1947   Joshua Lederberg and Norton Zinder, and, independently, Bernard D. Davis developed the penicillin selection technique for isolating biochemically deficient bacterial mutants.

1948   James V. Neel reported evidence that the sickle-cell disease is inherited as a simple Mendelian autosomal recessive.

1948   Barbara McClintock published her research on transposable regulatory elements in maize. McClintock had begun the work on a class of unstable mutant strains that led to this concept about six years before she published her results. Her work was not appreciated until similar phenomena were discovered in the 1960s and 1970s in bacteria and fruit flies. These elements are sometimes referred to as "jumping genes." McClintock won the Nobel Prize in medicine or physiology in 1983.

1949   John F. Ender, Thomas H. Weller, and Frederick C. Robbins published "Cultivation of Polio Viruses in Cultures of Human Embryonic Tissues." The report by Enders and coworkers was a landmark in establishing techniques for the cultivation of poliovirus in cultures on non-neural tissue and for further virus research. The technique led to the polio vaccine and other advances in virology.

1950   Erwin Chargaff demonstrated that the Tetranucleotide Theory was incorrect and that DNA was more complex than the model that had been developed by Phoebus A. Levene. Chargaff proved that the nucleic acids are not monotonous polymers. Chargaff also discovered interesting regularities in the base composition of DNA; these findings were later known as "Chargaff's rules." Chargaff discovered a one-to-one ratio of adenine to thymine and guanine to cytosine in DNA samples from a variety of organisms.

1950s   Ruth Sager's work on the algae *Chlamydomonas* proved that cytoplasmic genes existed and that they could undergo mutation. She showed that such genes could be mapped on a "cytoplasmic chromosome." Confirmation was provided when other researches reported similar findings in yeast and *Neurospora,* and the DNA was located to cytoplasmic organelles.

1950   Douglas Bevis, British physician, demonstrated that amniocentesis could be used to test fetuses for Rh-factor incompatibility.

1951   Esther M. Lederberg discovered a lysogenic strain of *Escherichia coli* K12 and isolated a new bacteriophage, which was called lambda.

1951   Joshua Lederberg and Norton Zinder reported the discovery of a phenomenon they called "transduction." Lederberg and Zinder proved that transduction in Salmonella was caused by phage particles that occasionally carried assorted host genes into new hosts. That is, bacteriophage particles served as the vectors of genetic exchange. The discovery of transduction was announced at Cold Spring Harbor in 1951. The next year Zinder and Lederberg published their results in a paper entitled "Genetic Exchange in Salmonella. New mechanism for the heritable transfer of genetic traits from one bacterial strain to another."

1951   Rosalind Franklin obtained sharp x-ray diffraction photographs of DNA.

1952   William Hayes isolated a strain of *E. coli* that produced recombinants thousands of times more frequently than previously observed. The new strain of K12 was named Hfr Hayes, to signify high-frequency recombination.

1952   Alfred Hershey and Martha Chase published their landmark paper "Independent Functions of Viral Protein and Nucleic Acid in Growth of Bacteriophage." The famous "blender experiment" suggested that DNA was the genetic material. When bacteria were infected by a virus, at least 80% of the viral DNA entered the cell and at least 80% of the viral protein remained outside.

1952   Rosalind Franklin gave a seminar on her x-ray crystallography studies of two forms of DNA.

1952   Renato Dulbecco developed a practical method for studying animal viruses in cell cultures. His plaque method was comparable to that used in studies of bacterial viruses and it was important in genetic studies of viruses. These methods were described in his paper "Production of Plaques in Monolayer Tissue Cultures by Single Particles of an Animal Virus."

1952   Karl Maramorosch demonstrated that some viruses could multiply in both plants and insects. This work led to new questions about the origins of viruses.

1953   James D. Watson and Francis H. C. Crick published two landmark papers in the journal *Nature*: "Molecular structure of nucleic acids: a structure for deoxyribose nucleic acid" and "Genetical implications of the structure of deoxyribonucleic acid." Watson and Crick proposed a double helical model for DNA and called attention to the genetic implications of their model. Their model seemed to explain how the genetic material was transmitted.

1954   Seymour Benzer worked out the fine structure of the rII region of the bacteriophage T4 of *E. coli,* and coined the terms cistron, recon, and muton.

1954   Frederick Sanger determined the entire sequence of the amino acids in insulin.

1955   François Jacob and Elie L. Wollman determined the mechanism of the transmission of genetic information during bacterial mating. Jacob and Wollman used a blender to interrupt the mating process and then determined the sequence of genetic transfer between bacterial cells.

1955   Fred L. Schaffer and Carlton E. Schwerdt reported the crystallization of the poliovirus, the first successful demonstration of the crystallization of an animal virus.

1955   Heinz Fraenkel-Conrat and Robley C. Williams proved that tobacco mosaic virus could be reconstituted from its nucleic acid and protein subunits. The reconstituted particles exhibited normal morphology and infectivity.

1956   Arthur Kornberg demonstrated the existence of DNA polymerase in *E. coli.*

1956   Alfred Gierer and Gerhard Schramm demonstrated that naked RNA from tobacco mosaic virus was infectious. Subsequently, infectious RNA preparations were obtained for certain animal viruses.

1956   Vernon M. Ingram reported that normal and sickle cell hemoglobin differ by a single amino acid substitution.

1956   Joe Hin Tijo and Albert Levan proved that the number of chromosomes in a human cell is 46, and not 48, as believed since the early 1920s.

1956   Mary F. Lyon proposed that one of the X chromosomes of normal females is inactivated. This concept became known as the Lyon hypothesis and helped explain some confusing aspects of sex-linked diseases. Females are usually "carriers" of genetic diseases on the X chromosome because the normal gene on the other chromosome protects them, but some X-linked disorders are partially expressed in female carriers. Based on studies of mouse coat color genes, Lyon proposed that one X chromosome is randomly inactivated in the cells of female embryos.

1957   François Jacob and Elie L. Wollman demonstrated that the single linkage group of *E. coli* is circular and suggested that the different linkage groups found in different Hfr strains result from the insertion at different points of a factor in the circular linkage group that determines the rupture of the circle.

1957   Francis Crick proposed that during protein formation each amino acid was carried to the template by an adapter molecule containing nucleotides and that the adapter was the part that actually fits on the RNA template. Later research demonstrated the existence of transfer RNA.

1957   Alick Isaacs and Jean Lindenmann published their pioneering report on interferon, a protein produced by interaction between a virus and an infected cell that can interfere with the multiplication of viruses.

1958   Frederick Sanger was awarded the Nobel Prize in chemistry for his work on the structure of proteins, especially for determining the primary sequence of insulin.

1958   Matthew Meselson and Frank W. Stahl published their landmark paper "The replication of DNA in *Escherichia coli,*" which demonstrated that the replication of DNA follow the semiconservative model.

1958   George W. Beadle, Edward L. Tatum, and Joshua Lederberg were awarded the Nobel Prize in medicine or physiology. Beadle and Tatum were honored for the work in *Neurospora* that led to the one gene-one enzyme theory. Lederberg was honored for discoveries concerning genetic recombination and the organization of the genetic material of bacteria.

1959   Severo Ochoa and Arthur Kornberg were awarded the Nobel Prize in medicine or physiology for their discovery of the mechanisms in the biological synthesis of ribonucleic acid and deoxyribonucleic acid.

1959   Jerome Lejeune, Marthe Gautier and Raymond A. Turpin reported that Down syndrome is a chromosomal aberration involving trisomy of a small telocentric chromosome. That is, patients with Down syndrome have 47 chromosomes instead of the normal 46, because they have three copies of chromosome 21.

1959   Arthur Kornberg and Severo Ochoa were awarded the Nobel Prize in medicine or physiology for their discovery of enzymes that produce artificial DNA and RNA.

1959   Robert L. Sinsheimer reported that bacteriophage ÿX174, which infects *E. coli,* contains a single-stranded DNA molecule, rather than the expected double stranded DNA. This provided the first example of a single-stranded DNA genome.

1959   Sydney Brenner and Robert W. Horne published a paper entitled "A Negative Staining Method for High Resolution Electron Microscopy of Viruses." Brenner and Horne developed a method for studying the architecture of viruses at the molecular level, using the electron microscope.

1961   François Jacob and Jacques Monod published "Genetic regulatory mechanisms in the synthesis of proteins," a paper which described the role of messenger RNA and proposed the operon theory as the mechanism of genetic control of protein synthesis.

1961   Marshall Warren Nirenberg synthesized a polypeptide by using an artificial messenger RNA, a synthetic RNA containing only the base uracil, in a cell-free protein-synthesizing system. The resulting polypeptide contained only the amino acid phenylalanine, indicating that UUU was the codon for phenylalanine. This important step in deciphering the genetic code was described in the landmark paper by Marshall Nirenberg and J. Heinrich Matthaei, "The Dependence of Cell-Free Synthesis in *E. coli* upon Naturally Occurring or Synthetic Polyribonucleotides." This work established the messenger concept and a system that could be used to work out the relationship between the sequence of nucleotides in the genetic material and amino acids in the gene product.

1962   James D. Watson, Francis Crick, and Maurice Wilkins were awarded the Nobel Prize in medicine or physiology for their work in elucidating the structure of DNA.

1963   Ruth Sager discovered DNA in chloroplasts. Boris Ephrussi found DNA in mitochondria.

1964   Barbara Bain published a classic account of her work on the mixed leukocyte culture (MLC) system that was critical in determining donor-recipient matches for organ or bone marrow transplantation. Her work was published in *Blood* in 1964 and became a citation classic in *Current Contents.* Bain showed that the MLC phenomenon was caused by complex genetic differences between individuals.

1965   François Jacob, André Lwoff, and Jacques Monod were awarded the Nobel Prize in medicine or physi-

ology for their discoveries concerning genetic control of enzymes and virus synthesis.

1966   Marshall Nirenberg and Har Gobind Khorana led teams that deciphered the genetic code. All of the 64 possible triplet combinations of the four bases (the codons) and their associated amino acids were described.

1967   Charles T. Caskey, Richard E. Marshall, and Marshall Warren Nirenberg suggested that there is a universal genetic code shared by all life forms.

1967   Charles Yanofsky demonstrated that the sequence of codons in a gene determines the sequence of amino acids in a protein.

1967   The first volume of *Annual Review of Genetics* was published, adding ARG to the *Annual Review* series. Volume 1 was dedicated to H. J. Muller.

1968   Robert W. Holley, Har Gobind Khorana, and Marshall W. Nirenberg were awarded the Nobel Prize in Medicine or Physiology for their interpretation of the genetic code and its function in protein synthesis.

1968   Werner Arber discovered that bacteria defend themselves against viruses by producing DNA-cutting enzymes. These enzymes quickly become important tools for molecular biologists.

1968   Mark Steven Ptashne and Walter Gilbert independently identified repressor genes.

1969   Max Delbrück, Alfred D. Hershey, and Salvador E. Luria were awarded the Nobel Prize in Medicine or Physiology for their discoveries concerning the replication mechanism and the genetic structure of viruses.

1969   Stanford Moore and William H. Stein determine the sequence of the 124-amino acid chain of the enzyme ribonuclease.

1969   Jonathan R. Beckwith isolated the lac Z gene.

1970   Har Gobind Khorana announced the synthesis of the first wholly artificial gene. Khorana and his coworkers synthesized the gene that coded for alanine transfer RNA in yeast.

1970   Howard Martin Temin and David Baltimore independently discovered reverse transcriptase in viruses. This is an enzyme that catalyzes the transcription of RNA into DNA.

1970   Hamilton Smith and Kent Wilcox isolated the first restriction enzyme, HindII, an enzyme that cuts DNA molecules at specific recognition sites.

1971   Christian B. Anfinsen, Stanford Moore, and William H. Stein were awarded the Nobel Prize in chemistry.

Anfinsen was cited for his work on ribonuclease, especially concerning the connection between the amino acid sequence and the biologically active conformation, and Moore and Stein were cited for their contribution to the understanding of the connection between chemical structure and catalytic activity of the active center of the ribonuclease molecule.

1972 Paul Berg and Herbert Boyer produced the first recombinant DNA molecules.

1972-73 Recombinant technology emerged as one of the most powerful techniques of molecular biology. Scientists were able to splice together pieces of DNA to form recombinant genes. As the potential uses, therapeutic and industrial, became increasingly clear, scientists and venture capitalists established biotechnology companies.

1973 Joseph Sambrook and coworkers refined DNA electrophoresis by using agarose gel and staining with ethidium bromide.

1973 Annie Chang and Stanley Cohen showed that a recombinant DNA molecule could be maintained and replicated in *E. coli*.

1973 Herbert Wayne Boyer and Stanley H. Cohen created recombinant genes by cutting DNA molecules with restriction enzymes, joining the molecules together with ligases, and introducing the recombinant genes into *E. coli*, where these novel genes could be expressed and replicated. These experiments are generally referred to as the beginning of genetic engineering.

1973 Concerns about the possible hazards posed by recombinant DNA technologies, especially work with tumor viruses, led to the establishment of a meeting at Asilomar, California. The proceedings of this meeting were published by the Cold Spring Harbor Laboratory as a book entitled *Biohazards in Biological Research*.

1975 David Baltimore, Renato Dulbecco, and Howard Temin shared the Nobel Prize in medicine or physiology for their discoveries concerning the interaction between tumor viruses and the genetic material of the cell and the discovery of reverse transcriptase.

1975 Scientists at an international meeting in Asilomar, California, called for the adoption of guidelines regulating recombinant DNA experimentation.

1975 César Milstein discovered monoclonal antibodies.

1976 Har Gobind Khorana constructed a functional synthetic gene containing a complete system of regulatory mechanisms.

1976 Michael J. Bishop, Harold Elliot Varmus, and coworkers established definitive evidence of the oncogene hypothesis. They discovered that normal genes could malfunction and cause cells to become cancerous.

1977 Philip Allen Sharp and Richard John Roberts independently discovered that the DNA making up a particular gene could be present in the genome as several separate segments. Although both Roberts and Sharp used a common cold-causing virus, called adenovirus, as their model system, researchers later found "split genes" in higher organisms, including humans. Sharp and Roberts won the Nobel Prize in medicine or physiology in 1993 for the discovery of split genes.

1977 Carl R. Woese and George E. Fox published an account of the discovery of a third major branch of living beings, the Archaea. Woese suggested that an rRNA database could be used to generate phylogenetic trees.

1977 Fred Sanger developed the chain termination (dideoxy) method for sequencing DNA.

1977 Genentech, the first genetic engineering company was founded, in order to use recombinant DNA methods to make medically important drugs.

1978 Walter Gilbert and others established Biogen, a pioneering biotechnology company.

1978 Werner Arber, Daniel Nathans, and Hamilton O. Smith were awarded the Nobel Prize in medicine or physiology for the discovery of restriction enzymes and their application to problems of molecular genetics.

1977 Frederick Sanger reported the sequencing of the complete genetic information of a microorganism.

1978 Somatostatin became the first human hormone produced by using recombinant DNA technology.

1978 Scientists from Genentech cloned the gene for human insulin.

1980 Researchers successfully introduced a human gene, which codes for the protein interferon, into a bacterium.

1980 Martin Cline constructed a transgenic mouse.

1980 Leroy Hood developed the first automated DNA sequencing machine.

1980 Paul Berg, Walter Gilbert, and Frederick Sanger shared a Nobel Prize in chemistry. Berg was honored for his fundamental studies of the biochemistry of nucleic acids, with particular regard to recombinant-DNA. Gilbert and Sanger were honored for their contributions to the sequencing of nucleic acids. This was Sanger's second Nobel Prize.

1980   The United States Supreme Court ruled that a living organism developed by General Electric (a microbe used to clean up an oil spill) can be patented.

1981   Karl Illmensee cloned baby mice.

1981   Researchers in China cloned a fish.

1982   The United States Food and Drug Administration approved the first genetically engineered drug, a form of human insulin produced by bacteria.

1982   Thomas Cech discovered ribozymes. Within a year Sidney Altman also reported the discovery of RNAs with catalytic properties.

1983   James Gusella and coworkers used blood samples collected by Nancy Wexler to establish a genetic marker on chromosome 4 in order to provide a screening test for Huntington's disease (an incurable genetic disease). The gene for Huntington's disease was established in 1993.

1983   Andrew W. Murray and Jack William Szostak created the first artificial chromosome.

1983   Barbara McClintock received the Nobel Prize in medicine or physiology for her discovery of mobile genetic elements, which are also known as transposons, or "jumping genes." McClintock discovered transposons in maize in the 1940s, but this work was not fully appreciated until the 1960s when similar phenomena were discovered in bacteria. In the 1970s transposable elements were discovered in fruit flies.

1983   Los Alamos National Laboratory, a Department of Energy Laboratory (LANL), and Lawrence Livermore National Laboratory, a Department of Energy Laboratory (LLNL), began production of DNA clone (cosmid) libraries representing single chromosomes.

1984   Steen A. Willadsen successfully cloned sheep.

1984   Allan Charles Wilson and Russell Higuchi cloned genes from an extinct species by removing a gene from the preserved skin of a quagga, a type of zebra.

1984   The United States Department of Energy (DOE), Office of Health and Environmental Research, U.S. Department of Energy (OHER, now Office of Biological and Environmental Research), and the International Commission for Protection Against Environmental Mutagens and Carcinogens (ICPEMC) cosponsored the Alta, Utah, conference highlighting the growing role of recombinant DNA technologies. OTA incorporates the proceedings of the meeting into a report acknowledging the value of deciphering the human genome.

1985   Kary Mullis, who was working at Cetus Corporation, developed the polymerase chain reaction (PCR), a new method of amplifying DNA. This technique quickly became one of the most powerful tools of molecular biology. Cetus patented PCR and sold the patent to Hoffman-LaRoche, Inc. in 1991.

1985   Alec Jeffreys developed "genetic fingerprinting," a method of using DNA polymorphisms (unique sequences of DNA) to identify individuals. The method, which has been used in paternity, immigration, and murder cases, is generally referred to as "DNA fingerprinting."

1985   Robert Sinsheimer organized a meeting at University of California, Santa Cruz to discuss the feasibility of launching the Human Genome Initiative. The meeting was commissioned by officials at OHER.

1985   Elizabeth Blackburn and Carol Greider discovered the enzyme telomerase, an unusual RNA-containing DNA polymerase that can add to the telomeres (specialized structures found at the ends of chromosomal DNA). Telomeres seem to protect the integrity of the chromosome. Most normal somatic cells lack telomerase, but cancer cells have telomerase activity, which might explain their ability to multiply indefinitely.

1986   Robert A. Weinberg and coworkers isolated a gene that inhibits growth and appears to suppress retinoblastoma (a cancer of the retina).

1986   The United States FDA approved the first genetically engineered vaccine for humans, for hepatitis B.

1986   Following the Santa Fe conference, the United States Department of Energy officially initiated the Human Genome Initiative. Pilot projects at DOE national laboratories were given a budget of $5.3 million to develop critical resources and technologies.

1987   The United States Congress chartered Department of Energy advisory committee, Health and Environmental Research Advisory Committee (HERAC), recommended a 15-year, multidisciplinary, scientific, and technological undertaking to map and sequence the human genome. DOE designated multidisciplinary human genome centers. National Institute of General Medical Sciences at the National Institutes of Health (NIH NIGMS) began funding genome projects.

1987   The United States Department of Energy began funding the Human Genome Project.

1987   Maynard Olson created and named yeast artificial chromosomes (YACs), which provided a technique to clone long segments of DNA.

1987   David C. Page and colleagues discovered the gene responsible for maleness in mammals. It is a single gene on the Y chromosome that causes the development of testes instead of ovaries.

1988 The Human Genome Project officially adopted the goal of determining the entire sequence of DNA comprising the human chromosomes.

1988 Reports by the congressional Office of Technology Assessment (OTA) and committees established by the National Research Council (NRC) and the National Academy of Sciences (NAS) recommended a concerted effort to carry out genome research programs.

1988 The Human Genome Organization (HUGO) was established by scientists in order to coordinate international efforts to sequence the human genome.

1988 Philip Leder and Timothy Stewart, at Harvard University, were granted the first patent for a genetically altered animal. Leder and Stewart constructed a mouse that is highly susceptible to breast cancer.

1989 Sidney Altman and Thomas R. Cech were awarded the Nobel Prize in chemistry for their discovery of ribozymes (RNA molecules with catalytic activity). Cech proved that RNA could function as a biocatalyst as well as an information carrier.

1989 The journal *Science* selected polymerase chain reaction as the "Molecule of the Year."

1989 James D. Watson was appointed head of the National Center for Human Genome Research. The agency was created to oversee the $3 billion budgeted for the American plan to map and sequence the entire human DNA by 2005.

1989 Francis Collins and Lap-Chee Tsui identified the gene coding for the cystic fibrosis transmembrane conductance regulator protein (CFTR) on chromosome 7. Mutations in this gene cause cystic fibrosis.

1989 DNA sequence tagged sites (STSs) were recommended as a method for correlating diverse types of DNA clones.

1989 The United States Department of Energy and National Institutes of Health establish the Joint Ethical, Legal, and Social Issues (ELSI) Working Group.

1989 Cells from one embryo were used to produce seven cloned calves.

1990 The United States Department of Energy and National Institutes of Health presented their joint 5-year HGP plan to Congress, thus formally launching the U.S. Human Genome Project. Because techniques for sequencing genes were slow and expensive in the early 1990s, the Genome Project adopted a "map-first, sequence-later" strategy. The initial

strategy of the Project was to construct a higher-resolution map that could then be used to sequence and assemble the human genome. By the late 1990s rapid sequencing was made possible by the ABI PRISM 3700 DNA Analyzer, developed by Michael Hunkapiller of PE Biosystems.

1990 The International Human Genome Project was officially launched.

1990 Michael R. Blaese and French W. Anderson conducted the first gene replacement therapy experiment on a four-year-old girl with adenosine deaminase (ADA) deficiency, an immune-system disorder. T cells from the patient were isolated and exposed to retroviruses containing an RNA copy of a normal ADA gene. The treated cells were returned to her body where they helped restore some degree of function to her immune system.

1990 Researchers at various laboratories began projects to mark gene sites on chromosome maps as sites of mRNA expression.

1990 Research and development began for efficient production of more stable, large-insert bacterial artificial chromosomes (BACs).

1990 Michael Crichton, inspired by advances in molecular biology, published the novel *Jurassic Park,* in which bioengineered dinosaurs roam a paleontological theme park.

1991 The human chromosome mapping data repository, known as the Genome Database (GDB), was established.

1991 Mary-Claire King concluded, based on her studies of the chromosomes of women in cancer-prone families, that a gene on chromosome 17 causes the inherited form of breast cancer and also increases the risk of ovarian cancer.

1992 The United States Army began collecting blood and tissue samples from all new recruits as part of a "genetic dog tag" program aimed at better identification of soldiers killed in combat.

1992 American and British scientists described a technique for testing embryos *in vitro* for genetic abnormalities such as cystic fibrosis and hemophilia.

1992 Francis Collins replaced James Watson as head of the National Center for Human Genome Research at the National Institutes of Health. Watson had clashed with Craig Venter, then at NIH, over the patenting of DNA fragments known as "expressed sequence tags."

1992 Craig Venter established The Institute for Genomic Research (TIGR) in Rockville, Maryland. TIGR later sequenced *Haemophilus influenzae* and many other bacterial genomes.

1992 Harry F. Noller demonstrated that RNA plays a greater role in protein synthesis than had been assumed.

1992 Low-resolution genetic linkage maps of the entire human genome were published.

1992 Guidelines for data release and resource sharing related to the Human Genome Project were announced by the United States Department of Energy and National Institutes of Health.

1993 Richard J. Roberts, and Phillip A. Sharp were awarded the Nobel Prize in medicine or physiology for the discovery of split genes.

1993 Kary Mullis and Michael Smith were awarded the Nobel Prize in chemistry. Mullis was honored for the discovery of polymerase chain reaction. Smith was cited for his fundamental contributions to the establishment of oligonucleiotide-based, site-directed mutagenesis.

1993 FlavrSavr™ tomatoes, genetically engineered for longer shelf life, were marketed.

1993 The international Integrated Molecular Analysis of Gene Expression (IMAGE) Consortium was established to coordinate efficient mapping and sequencing of gene-representing cDNAs.

1993 The United States ELSI Working Group's Task Force on Genetic and Insurance Information released recommendations for dealing with genetic information.

1993 Then United States Department of Energy and National Institutes of Health revised their 5–year goals for the HGP.

1993 French Gépnéthon made its mega-YACs available to the genome community.

1993 IOM released the U.S. HGP-funded report "Assessing Genetic Risks."

1993 The Lawrence Berkeley National Laboratory, a Department of Energy Laboratory (LBNL) implemented a novel transposon-mediated chromosome-sequencing system.

1993 The Gene Recognition and Analysis Internet Link (GRAIL) sequence-interpretation service provided Internet access at Oak Ridge National Laboratory, a Department of Energy Laboratory (ORNL).

1993 Scientists identified p53, a tumor suppressor gene, as a critical factor preventing uncontrolled cell growth. The gene also performs a variety of functions ensuring cell health.

1993 After analyzing the family trees of gay men and the DNA of pairs of homosexual brothers, biochemists at the United States National Cancer Institute reported that at least one gene related to homosexuality resides on the X chromosome, which is inherited from the mother.

1993 George Washington University researchers cloned human embryos and nurtured them in a Petri dish for several days. The project provoked protests from ethicists, politicians and critics of genetic engineering.

1993 An international research team, led by Daniel Cohen, of the Center for the Study of Human Polymorphisms in Paris, produced a rough map of all 23 pairs of human chromosomes.

1994 Biologists discovered that both vertebrates and invertebrates share certain developmental genes.

1994 Researchers identified a metastasis-suppressor gene and determined that Tamoxifen, an anti-cancer drug, blocked the blood supply that supported the growth of malignant tumors.

1994 Geneticists determined that DNA repair enzymes perform several vital functions, including preserving genetic information and protecting the cell from cancer.

1994 The five-year goal for genetic-mapping was achieved one year ahead of schedule.

1994 LLNL and LBNL announced that they had completed the second-generation DNA clone libraries representing each human chromosome.

1994 The Genetic Privacy Act, the first United States Human Genome Project legislative product, proposed regulation of the collection, analysis, storage, and use of DNA samples and genetic information obtained from them. These rules were endorsed by the ELSI Working Group.

1994 DOE announced the establishment of the Microbial Genome Project as a spin-off of the Human Genome Project.

1994 Sequencing by hybridization (SBH) technologies from Argonne National Laboratory (ANL), a Department of Energy Laboratory, was commercialized.

1994 The Human Genome Project Information Web site was made available to researchers and the public.

1995 LANL and LLNL announced the completion of high-resolution physical maps of chromosome 16 and chromosome 19.

1995 Moderate-resolution maps of chromosomes 3, 11, 12, and 22 maps were published.

1995 A physical genome map with more than 15,000 STS markers published.

1995 Edward B. Lewis, Christiane Nüsslein-Volhard, and Eric F. Wieschaus, developmental biologists, shared the Nobel Prize in Medicine or Physiology for work that led to the discovery of a family of genes later named "homeotic selector genes." These genes play a critical role in the organization of temporal and spatial aspects of genome expression patterns during development.

1995 The whole genome of the bacterium *Haemophilus influenzae* was sequenced.

1995 The sequence of *Mycoplasma genitalium,* which is generally regarded as the smallest known bacterium, was completed. *Mycoplasma genitalium* is considered a model of the minimum number of genes needed for independent existence.

1995 Equal Employment Opportunity Commission (EEOC) guidelines extended Americans with Disabilities Act(ADA) employment protection to cover discrimination based on genetic information related to illness, disease, or other conditions.

1995 Peter Funch and Reinhardt Moberg Kristensen created a new phylum, Cycliophora, for a novel invertebrate called *Symbion pandora* found living in the mouths of Norwegian lobsters.

1995 Researchers at Duke University Medical Center reported that they had transplanted hearts from genetically altered pigs into baboons. All three transgenic pig hearts survived at least a few hours, suggesting that xenotransplants (cross-species organ transplantation) might be possible.

1995 Former football player O. J. Simpson was found not guilty in a high-profile double-murder trial in which prosecutors unsuccessfully attempted to win a conviction based, at least in part, on evidence from PCR and DNA fingerprinting techniques.

1996 International participants in the genome project met in Bermuda and agreed to formalize the conditions of data access. The agreement, known as the "Bermuda Principles," called for the release of sequence data into public databases within 24 hours.

1996 The sequence of the *Methanococcus jannaschii* genome provided further evidence of the existence of third major branch of life on earth.

1996 The United States Department of Energy initiated six pilot projects on BAC end sequencing.

1996 The Health Care Portability and Accountability Act incorporated provisions to prohibit the use of genetic information in certain health-insurance eligibility decisions. The Department of Health and Human Services was charged with the enforcement of health-information privacy provisions.

1996 The United States Department of Energy and National Center for Human Genome Research (NCHGR) at the National Institutes of Health issued guidelines on the use of human subjects for large-scale sequencing projects.

1996 The Wellcome Trust sponsored a large-scale sequencing strategy meeting for international coordination of human genome sequencing.

1996 Gerard Schellenberg and colleagues identified the gene that causes Werner's syndrome, a condition that leads to premature aging.

1996 William R. Bishai and co-workers reported that SigF, a gene in the tuberculosis bacterium, enables the bacterium to enter a dormant stage.

1996 Researchers reported important progress in understanding BRCA1 and BRCA2, the genes responsible for some types of breast cancer.

1996 Scientists reported further evidence that individuals with two mutant copies of the CC-CLR-5 gene are generally resistant to HIV infection.

1996 Chris Paszty and co-workers successfully employed genetic engineering techniques to create mice with sickle-cell anemia, a serious human blood disorder.

1997 Ian Wilmut of the Roslin Institute in Edinburgh, Scotland, announced the birth of a lamb called "Dolly," the first mammal cloned from an adult cell (a cell in a pregnant ewe's mammary gland).

1997 Donald Wolf and co-workers announced that they had cloned rhesus monkeys from early stage embryos, using nuclear transfer methods.

1997 William Jacobs and Barry Bloom created a biological entity that combines the characteristics of a bacterial virus and a plasmid (a DNA structure that functions and replicates independently of the chromosomes). This entity is capable of triggering mutations in *Mycobacterium tuberculosis.*

1997 Christof Niehrs identified a protein responsible for the creation of the head in a frog embryo.

1997 Researchers identified a gene that plays a crucial role in establishing normal left-right configuration during organ development.

1997 Researchers reported progress in using the study of genetic mutations in humans and mice to decipher the molecular signals that lead undeveloped neurons from inside the brain to their final position in the cerebral cortex.

1997   The National Center for Human Genome Research (NCHGR) at the National Institutes of Health became the National Human Genome Research Institute (NHGRI).

1997   The DNA sequence of *Escherichia coli* was completed.

1997   The DNA sequence of the yeast *Saccharomyces cerevisiae* was completed.

1997   The second large-scale sequencing strategy meeting was held in Bermuda.

1997   High-resolution physical maps of chromosomes X and 7 were completed.

1997   The United States Department of Energy and National Institutes of Health Task Force on Genetic Testing released its final report and recommendations.

1997   The United States Department of Energy and National Institutes of Health formed the Joint Genome Institute for implementing high-throughput activities at DOE human genome centers, initially in sequencing and functional genomics.

1997   United Nations Educational, Scientific, and Cultural Organization (UNESCO) adopted the Universal Declaration on the Human Genome and Human Rights

1997   Scientists discovered p123, a protein component of telomerase. Mutations in this gene seem to be associated with chromosomal shrinkage.

1998   Ian Wilmut announced the birth of "Polly", a transgenic lamb containing human genes.

1998   "Dolly," the first cloned sheep, gave birth to a lamb that had been conceived by a natural mating with a Welsh Mountain ram. Researches said the birth of "Bonnie" proved that "Dolly" was a fully normal and healthy animal.

1998   Scientists at Japan's Kinki University announced the cloning of eight identical calves using cells from an adult cow.

1998   A telomere sequence having implications for aging and cancer research was identified at LANL.

1998   British and American scientists completed the genetic map of the nematode, *Caenorabditis elegans*. The genetic map, showing the 97 million genetic letters, in correct sequence, derived from the worm's 19,900 genes, was the first completed genome of an animal.

1998   University of Hawaii scientists, improved on Ian Wilmut's technique, cloning a mouse from cumulus cells. The process was repeated for three generations, yielding over fifty cloned mice by the end of July,

1998. The "Honolulu Technique's" success rate of 50:1 was almost six times better than that of the Roslin Institute's success rate of 277:1.

1998   Two research teams succeeded in growing embryonic stem cells.

1998   The Unites States Department of Energy and National Institutes of Health announced a new five-year plan for the Human Genome Project, which predicted that the project would be completed by 2003.

1998   The United States Department of Energy's Joint Genome Institute (JGI) in Walnut Creek, California, which houses the DOE's production sequencing facility, announced that it had exceeded its sequencing goal and had completed 20 Mb for FY 1998.

1998   GeneMap'98 containing 30,000 markers was released.

1998   Incyte Pharmaceuticals announced plans to sequence the human genome in 2 years.

1998   *Mycobacterium tuberculosis* bacterium sequenced.

1998   DOE funded production BAC end sequencing projects

1998   An ELSI meeting, sponsored by DOE; Whitehead Institute; and the American Society of Law, Medicine, and Ethics, was attended by over 800 scientists and scholars from diverse disciplines.

1998   Cold Spring Harbor Laboratory held its first meeting on human genome mapping and sequencing.

1998   The United States Department of Energy and National Institutes of Health signed a memorandum of understanding that outlined plans for cooperation on genome research.

1998   Scientists in Korea claimed to have cloned human cells.

1998   Craig Venter formed a company, later named Celera, and predicted that the company would decode the entire human genome within three years. Celera planned to use a "whole genome shotgun" method, which would assemble the genome without using maps. Venter said that his company would not follow the Bermuda principles concerning data release.

1999   The public genome project responded to Craig Venter's challenge with plans to produce a draft genome sequence by 2000. Most of the sequencing was done in five centers, known as the "G5": the Whitehead Institute for Biomedical Research in Cambridge, Massachusetts; the Sanger Centre near Cambridge, United Kingdom; Baylor College of Medicine in Houston, Texas; Washington University in St. Louis, Missouri; the DOE's Joint Genome Institute (JGI) in Walnut Creek, California.

1999 Scientists announced the complete sequencing of the DNA making up human chromosome 22. The first complete human chromosome sequence was published in December 1999.

1999 HGP advanced its goal for obtaining a draft sequence of the entire human genome from 2001 to 2000.

1999 Researchers reported that transgenic mice known as "Doogie" mice, performed better than normal counterparts on learning and memory tests. Two years later other researchers found that the Doogie mice were also more sensitive to chronic pain.

2000 The sequence of Chromosome 21 was published in May 2000. The sequence was a collaborative effort led by German and Japanese groups under the direction of André Rosenthal and Yoshiyki Sakaki, respectively.

2000 The complete sequence of the fruit fly *Drosophila melanogaster* was announced in March.

2000 The complete sequence of the plant *Arabidopsis thaliana* was announced in December.

2000 On June 26, 2000, leaders of the public genome project and Celera announced the completion of a working draft of the entire human genome sequence. Ari Patrinos of the DOE helped mediate disputes between the two groups so that a fairly amicable joint announcement could be presented at the White House in Washington, D.C.

2000 President Clinton signed an executive order prohibiting federal departments and agencies from using genetic information in hiring or promoting workers.

2000 The first volume of *Annual Review of Genomics and Human Genetics* was published. Genomics was defined as the new science dealing with the identifi-cation and characterization of genes and their arrangement in chromosomes and human genetics as the science devoted to understanding the origin and expression of human individual uniqueness.

2001 In February 2001, the complete draft sequence of the human genome was published. The public sequence data was published in the British journal *Nature* and the Celera sequence was published in the American journal *Science*.

2001 The International Rice Genome Sequencing Project announced that it would complete the sequence of the *Oryza japonica* rice genome by the end of the year in order to ensure that the sequence data would be accurate and freely available. The project was publicly funded by a ten nation consortium, led by a Japanese research team and planned to complete the sequence in 2004. The target date was revised when a private company announced that it had completed the sequence.

2001 The European Patent Office's Appeals Board confirmed the validity of two controversial patents on transgenic plants. Protesters argued that the European Patent Convention prevented plant varieties from being patented. The Board overruled objections filed by environmentalist groups against patents on a herbicide-resistant plant developed by AgrEvo and Monsanto's FlavrSavr™ tomato. In 2000 the patent office lifted a four-year moratorium on applications for patents on plants.

2001 Scientists, officials, executives, and members of the Human Proteome Organization (HUPO), held a conference in McLean, Virginia, titled "Human Proteome Project: Genes Were Easy." The goal of the participants was to discuss the advisability of a proteome project. HUPO is dedicated to organizing an international effort to study the output of all human genes.

A

A-DNA, 202

A-T (Ataxia-telangiectasia), **48**

A2M (Alpha-2 macroglobulin) gene, 668

Abbe, Ernst, 161

ABCR gene, 505

ABO blood group
  adaptation, 4
  antigens, 37
  genetics, 90

Acetylation, 353

Achondroplasia, 86

Achromatopsia, 156

Acquired character, **2**

Acquired immune deficiency syndrome (AIDS), **7–10**
  Brown's research, 105–106
  protease inhibitors, 569
  provirus hypothesis, 574
  retroviruses, 596–597
  slow viruses, 634–635
  Temin's research, 672–673

Acrasin, 94

Acrodermatitis enteropathica (AE), 324–325

Active sites, **3**
  allosteric regulation, 13
  enzyme research, 229
  multimeric enzymes, 490–491
  protease, 569

ADA deficiency. *See* Adenosine deaminase deficiency

Adaptation
  character displacement, 128
  fitness, **3–5**
  phenotypic variation, 544
  social Darwinism, 639–640
  survival of the fittest, 663

Adaptive immune system, 382

Adenosine deaminase deficiency, **1–2**
  enzyme therapy, 228
  gene therapy, 29
  severe combined immunodeficiency, 620

Adenosine diphosphate (ADP), 5

Adenosine triphosphate (ATP), **5**
  biochemistry, 76
  chaperones, 127
  enzyme research, 229
  Miller-Urey experiment, 470
  mitochondria, 473
  Ochoa's research, 518
  Oparin's research, 523
  Ponnaperuma's research, 558

Adenoviruses, 604–606, 706

Admixture linkage analysis, 190

Adoption research, 12

ADP (Adenosine diphosphate), 5

AFLP (Amplified fragment length polymorphism), **26–27**, 585

AFP (Alpha-fetoprotein), 563–564, 667

Aggressive mimicry, 472

Aging
  colorblindness, 156
  Hayflick's research, 342
  life expectancy, **5–6**
  mutagens, 493

Agricultural genetics, **7**
  animal husbandry, 33–34
  Borlaug's research, 95–97
  Carver's research, 115–117
  ecotypes, 220
  Emerson's research, 227
  genetically modified organisms, 311–312
  Lysenkoism, 439–440
  Mayer's research, 449–450
  plant breeding, 548–549
  plant hybridization, 382
  seeds and seed saving, 613–614
  Stebbins's research, 652–653
  Tijo's research, 677
  tobacco mosaic virus, 677–678
  von Tschermak-Seysenegg's research, 709
  See also Plant genetics

Agrobacterium method, 7

Ahlquist, Jon, 628

AIDS. *See* Acquired immune deficiency syndrome
Albinism, **10–11,** 302
Alcoholism, genetic factors, **11–12**
Alkalinity, 493–494
Allele specific polymerase chain reaction, 632
Alleles, **12–13**
    dominance relations, 207
    fixed, **256–257**
    heterozygotes, 351–352
    homologous chromosomes, 366
    loci, 434
    polygenic inheritance, 555–556
    population genetics, 559–560
    Punnett square, 577–579
    viability, 702–703
    wild type, 719–720
Allergens, 35–36
Allergic reactions, 35
Allograft, 685
Allopatric speciation, 642
Allosteric regulation, **13,** 533
Allotype, **12–13**
Alloway, James, 54
Alpha-2 macroglobulin (A2M) gene, 668
Alpha-amino acids, 23–24
Alpha-fetoprotein (AFP), 563–564, 667
Alpha-thalassemia, 675
ALS (Amyotrophic lateral sclerosis), 369, 534
Alteration of generation, **14**
Alternate genetic codes, **14–15**
Altman, Sidney, **15–16,** 118–119, 599–600
Altruism, **16–17**
    Hamilton's research, 337
    sociobiology, 640
    Wilson's research, 727–728
Alzheimer, Alois, 17–18, *19*
Alzheimer's disease, **17–19,** *18*
    Down syndrome, 210
    genetic testing, 307
    Parkinson's disease research, 534
    Selkoe's research, 617
    Tanzi's research, 668
Amaurosis, Leber's congenital, 504
Amber mutation, **20**
American Phage Group, 436–437
Ames, Bruce N., **20–22**
Ames test, 20–22, 493
Amino acids, **22–23**
    allotypes, 13
    bases, 64–66
    behavioral genetics, 69–70
    Chargaff's research, 130
    chemistry, **23–25**
    disorder screening, **25**
    enzyme structure, 30–31
    Miller-Urey experiment, 469–470
    Miller's research, 468–469
    molecular clock research, 479–480
    Moore's research, 484
    mutagenesis, 492–493
    Nirenberg's research, 508–509
    one gene-one enzyme theory, 521
    origin of life, 527–528
    Pauling's research, 538–539
    phenylketonuria, 544–545
    Ponnaperuma's research, 558
    protein electrophoresis, 569–571
    protein synthesis, 572
    proteins and enzymes, 572–573
    Sanger's research, 610–611
    sequencing, 101–102, 618
    sickle cell anemia, 628–629
    Stein's research, 654
    translation, 684–685
    tRNA, 604, 682
Amniocentesis, **25–26**
    albinism, 11
    birth defects diagnosis, 85
    chromosomal abnormalities, 138–139
    congenital birth defects, 161
    karyotype analysis, 404–405
    Prader-Willi syndrome, 562
    prenatal diagnosis, 563–564
Amplified fragment length polymorphism (AFLP), **26–27,** 585
Amyloid beta-protein, 17–19, 668
Amyloid plaques, 18–19
Amyotrophic lateral sclerosis (ALS), 369, 534
Anagenesis vs. cladogenesis, **27**
Ancestral relations, **293**
Ancient DNA, **27–28,** 531, 696–697
Anderson, W. French, 1, **28–29,** 291–292
Andreasen, Nancy C., **29–30**
Anemia, sickle cell. *See* Sickle cell anemia
Anencephaly, 502–503
Aneuploidy, 137–138, 552–553
Anfinsen, Christian Boehmer, **30–31,** 484
Angelman syndrome (AS), **31–33,** 561
Animal husbandry, **33–34**
    cloning, 149–150
    inbreeding, 390
    Wright's research, 729–730
Aniridia, 504
Antibiotics, **34**
    bioterrorism, 83
    rare genotype advantage, 585
    Twort's research, 694
    X-linked agammaglobulinemia, 733
Antibodies, **34–37,** *36*
    autoimmune disorders, 49–50
    Köhler's research, 413–414
    Milstein's research, 470–471
    monoclonal, 37, **38,** 413–414
    polyclonal, 386–387
    transplantation genetics, 685–686
    Zinkernagel's research, 745
Anticancer drugs, **38–39,** 691
Anticipation, 509–510, 699–700
Antigen-presenting cells (APCs), 443
Antigens, **34–37**
    major histocompatibility complex, 442–443
    transplantation genetics, 685–686
    Zinkernagel's research, 745
Antiphospholipid antibody (APL), 48–50
Antisense RNA, **39,** 743
Antithrombin III, 89
AP-2 transcription factor, 612–613
Apaf-1, 39–40
APCs (Antigen-presenting cells), 443

Apert syndrome, 584
ApoB genes, 299
ApoE (Apolipoprotein E), 19, 668
Apolipoprotein E (ApoE), 19, 668
Apomixis, 381
Apoptosis, **39–40**
    aging, 6
    cancer genetics, 111
    eukaryotic cell cycle, genetic regulation, 122
    Hayflick's research, 342
    Horvitz's research, 369
    oncogene research, 520
    trinucleotide repeats, 687–688
Arabidopsis thaliana, **40–42,** *41,* 244
Arber, Werner, **42–43,** 497, 635
Archaea, **43**
    Miller-Urey experiment, 469–470
    prokaryotic genomes, 567–568
    Urey's research, 696–697
Archaeogenetics, **43–45,** *44,* 531
Aristotle, 226–227, 354
Arrestin, 505
AS (Angelman syndrome), **31–33,** 561
Asexual reproduction, **45–46**
    cloning, 148–150
    hybrids, 381
    parthenogenesis, 534–535
    plants, 551
Asilomar Conference, **46–47**
Asparagine, 23
Assisted reproductive techniques
    GIFT therapy, 250, 388
    in vitro fertilization, 388–389
    zygote intrafallopian tube transfer, 388–389
Ataxia
    Friedrich, **267–268**
    telangiectasia, **48**
Ataxia-telangiectasia (A-T), **48**
ATM gene, 48
ATP. *See* Adenosine triphosphate
ATRX gene
    Smith-Fineman-Myers syndrome, 637–638
Autograft, 685
Autoimmune disorders, **48–50,** 443
Automated sorting devices, **50**
    flow cytometry, 257–258
    fluorescent dyes, 260
Autoradiography, **50–51**
    blotting analysis, 93
    DNA probes, 198
Autosomal inheritance, **51–52**
    dominant, 51, 540–541
    recessive, 51
Autosomes, **52**
    human population genetics, 378
    inheritance, 397
    mitochondrial DNA, 473–474
    multifactorial transmission, 490
    reciprocal crossing, 587
    sex-linked traits, 622–623
    structure and morphology, 144–145
    XX male syndrome, 734–735

Avery, Oswald Theodore, **52–54,** *53,* 131
    history of genetics, 355–357
    MacCleod's association, 441
    McCarty's collaboration, 452
    microbial genetics research, 466
    transformation, 682
Ayala, Francisco J., **54–55**
AZT, 10

B

B cells
    ADA deficiency, 1
    HIV antibodies, 9
    major histocompatibility complex, 443
    X-linked agammaglobulinemia, 733
B-chromosomes. *See* Supernumerary chromosomes
B-DNA, 201–202
BAC Resource Consortium, 374
Bachrach, Howard L., **57–58**
Bacillus subtilis, 2
BACs. *See* Bacterial artificial chromosomes
Bacteria, **58–60,** *59*
    antibiotics, 34
    bacteriophage therapy, 542–543
    biological warfare, 80–81
    bioremediation, 81–82
    biotechnology, 82–83
    bioterrorism, 83–84
    cell division, 123–124
    chaperones, 127–128
    Chase's research, 131–132
    chromosome structure, 144–145
    cloning research, 151
    cloning vectors, 151–152
    coordinate genetic regulation, 163
    Gaia hypothesis, 271–272
    lysogenic, 438
    MacLeod's research, 441–442
    McCarty's research, 452
    microbial genetics, 466
    Monod's research, 482–483
    mRNA, 463–464
    origin of life, 527–528
    origin recognition complex, 592
    Pasteur's research, 535–536
    polymerase chain reaction, 556
    prokaryotes, 567
    prokaryotic genomes, 567–568
    rare genotype advantage, 585
    repeated sequences, 590
    restriction enzymes, 593
    ribonuclease, 600
    sexual reproduction, 623
    spontaneous generation, 647
    structural genes, 661
    transposition, 596
    Twort's research, 694
    viruses, 706
    waste control/environmental cleanup, 713–714
    See also specific bacteria
Bacterial artificial chromosomes (BACs), **60**
    chromosome mapping, 141–142

contig genes and mapping, 162–163
genetic mapping, 305
human genome analysis, 374
human genome programs, 375–376
Human Genome Project, 376–377
Bacterial genetics
    antibiotic resistance, 34
    fertility factor, 249–250
    Lwoff's research, 437–439
    model organism research, 476
    Smith's research, 635–636
    transduction, 681
    Wollman's research, 728–729
Bacteriophages, **60–62,** *63*
    Arber's research, 42
    Benzer's T4 research, 71
    Brenner's research, 101–102
    Chase's research, 132
    Delbrück's research, 183–185
    D'Hérelle's research, 189–190
    Dulbecco's research, 213
    ecological and environmental genetics, 218–219
    Escherichia coli, 234, *235*
    Hershey's research, 349–350
    history of genetics, 356–358
    Luria's research, 435–437
    lytic infection, 705–706
    phage genetics, 542
    prokaryotic genomes, 568
    prophage, 568–569
    RNA replication, 15
    Sinsheimer's research, 633
    Stahl's research, 649
    structure, 705
    therapy, 542–543
    transduction, 681
    Twort's research, 694
    viral genetics, 703–704
    Watson's research, 714
    wobble theory, 728
    Zinder's research, 744
Baer, Karl Ernst von, **62–63,** 78, 227
Bakh, Aleksei N., 522
Baltimore, David, **63–64,** 574, 598–599
    Dulbecco's collaboration, 212–213
    Temin's collaboration, 87, 672
    Weinberg's collaboration, 716
Barr bodies
    Lyon's research, 439
    Morgan's research, 485
    Ohno's research, 518
    X chromosome inactivation, 732
    See also X chromosomes
Bart test, 298, 652
Basal body temperature, 393
Base pairing, **64**
    amber, ocher, and opal mutations, 20
    Chargaff's rules, 131
    DNA repair, 199
    history of genetics, 357–358
    palindromes, 532
    protein synthesis, 572
    replication, 591

resolution, 593
RNA, 603–604
synapsis, 665
transversions, 686–687
Bases, **64–66,** *65*
    biochemical analysis, 75
    cancer genetics, 111
    Chargaff's research, 130
    chromosome structure and morphology, 144–145
    cutting sites, 170
    degenerate code, 183
    DNA, 194
    DNA fingerprinting, 196–197
    DNA structure, 201–202
    exons and introns, 242–243
    hot spots, 369–370
    mathematical theories, 448–449
    Miller-Urey experiment, 470
    mRNA, 464
    Ponnaperuma's research, 558–559
    ribonuclease, 600
    sequencing, 618
    shotgun method, 627
    Stern's research, 658
    wobble theory, 728
    See also Nucleotide bases
Bates, Henry Walter, **66–67**
    mimicry research, 471, 488
    Wallace's association, 711
"Batesian mimicry," 67
Bateson, William, **67,** 166
    agricultural research, 548
    gene linkage research, 283
    genetics research, 313
    history of genetics, 354
    Punnett's collaboration, 577
    Vavilov's association, 701
Bayes' theorem, *297,* 298, 318
Bayesian statistics
    chi-square test, 134–135
    probability, 566–567
    risk calculations, 318–319
Bcl-2 gene, 39–40
Beadle, George Wells, **67–69,** *68,* 476–477
    Ephrussi's collaboration, 230–231
    one gene-one enzyme theory, 520–521
    protein synthesis research, 571
    Tatum's collaboration, 669
Becker muscular dystrophy, 215
Beckwith-Wiedemann syndrome, 398
Behavioral genetics, **69–70**
    Benzer's research, 71–72
    legal and public policy, 429
    mental illness, 462
    Prader-Willi syndrome, 561–562
    risk factors, 602
    selection, 615
Bell, Graham, 585
Benacerraf, Baruj, **70–71,** 179, 639
Benign tumors, 690
Benzer, Seymour, **71–72,** 101–102
"Berg letter," 74

Berg, Paul, **72–74, 73,** 323, 611
     recombinant DNA research, 588
     Singer's association, 630–631
Bernal, J. D., 405
Bernard-Soulier syndrome, 89
Beta-thalassemia, 675
Binary fission, 60, 124–125
Bioadhesion molecules, 84
Bioaugmentation, 713–714
Bio-barriers, 713–714
Biochemical analysis techniques, **75–76**
     Fischer's research, 253–254
     phylogeny, 546–547
     prenatal diagnosis, 563–564
     serial analysis of gene expression, 619–620
Biochemistry, **76–77**
     alternate genetic codes, 15
     Borst's research, 97–98
     Chargaff's research, 129–131
     Feulgen's research, 252
     Fischer's research, 253–254
     Garrod's research, 276–277
     genetics, 313
     history of genetics, 357
     Hopkins' research, 367–368
     Hoppe-Selyer's research, 368–369
     Karle's research, 402
     Luria's research, 435–437
     Oparin's research, 522
     phylogenetics, 546
     RNA, 603–604
     Roberts's research, 605
     Sanger's research, 609–611
     Selkoe's research, 617
     Stanley's research, 649–651
     Stein's research, 654
     Todd's research, 678–679
     Twort's research, 694
     Urey's research, 696–697
     Wilson's research, 724
Biodiversity, **77–78**
     Brown's research, 106
     character displacement, 128
     conservation genetics, 162
Biogenetic law, **78–79,** 521
Bioinformatics, **79**
     careers, 312
     gene chips and microarrays, 279
     proteomics, 573–574
     very large scale integration, 702
Biolistics, 549
"Biological determinism," 727–728
Biological risk factors, 602
Biological warfare, **79–81,** 80
     bioterrorism, 83–84
     Lederberg's research, 427–428
     Todd's research, 678–679
Biology
     computational, **79,** 574, 702
     Wilson's research, 724–726
     See also Molecular biology; Sociobiology
Biomolecules, 75–76, 539
Biopesticides, 7

Bioremediation, **81–82,** 713–714
Biorobots, 83–84
Biosensors, 84
Biostimulation, 713–714
Biotechnology, **82–83**
     animal husbandry, 34
     Baltimore's research, 63
     Benzer's research, 71
     Chargaff's research, 130
     chromosome sorters, 144
     cloning, 150
     Cohen's research, 153
     density gradient centrifugation, 187
     economics, 219–220
     enzymes, genetic manipulation, 229–230
     ethical issues, 235
     genetically modified organisms, 311–312
     Gilbert's research, 322–324
     human genome programs, 374–376
     industrial genetics, 391
     Mullis's research, 489–490
     mutagenesis, 496
     plant genetics, 550–551
     seeds and seed saving, 613–614
     Smith's research, 636–637
     Stebbins's research, 653
     Tatum's research, 668–670
     transgenics, 683–684
     Wang's research, 712–713
     waste control/environmental cleanup, 713–714
Bioterrorism, 79–81, **83–84**
Birth control, 611–612, 659–660
Birth defects, **84–87,** 85
     congenital birth defects, 161, 673–674, 719
     embryology, 225–226
     intergenic regions, 398
     multifactorial transmission, 490
     Nüsslein-Volhard's research, 515
     Optiz's research, 524–525
     Patau syndrome, 537
     pleiotropy, 552
     pluripotent stem cells, 554
     random genetic syndrome, 584
     stem cell research, 655–656
     synapsis, 665
     teratology, 673–674
     viability, 702–703
     Wiechaus's research, 719
     See also specific defects
Bishop, J. Michael, **87–88,** 700
Biston betularia, 191–192
Blackburn, Elizabeth H., **88**
Blaese, Michael, 1, 29
Bleeding disorders, **89**
Blindness, 504
     See also Colorblindness
Blobel, Günter, **89–90**
Blood coagulation defects, **89,** 345–346
     See also Hemophilia
Blood group genetics, **90,** 91
     Clarke's research, 146–147
     Dausset's research, 178–179
     Landsteiner's research, 423–425

Prévost's research, 565
Rous's research, 606
thalassemia, 674–676
transplantation, 685–686
Blood tests, 25
"Blood theory," 545
Blotting analysis, **92–93,** 507–508
Bone marrow transplantation, 1, 656
Bonner, John Tyler, **93–94**
Bonnet, Charles, **94–95**
Borlaug, Norman Ernest, **95–97**
Borst, Piet, **97–98**
Bouchard, Thomas Joseph, 692
Boveri, Theodore Heinrich, **98,** 725
Boyer, Herbert Wayne, **98–99,** 153, 588
Braconnot, Henri, 23
Bragg, William Henry, 168
Bragg, William Lawrence, 168
BRCA 1 / BRCA 2
    breast/ovarian cancer, hereditary, 99–101
    cancer, 109
    genetic testing, 308
    penetrance, 286–287
    tumor suppressor genes, 690
BrdUrd (Bromodoxyuridine), 633–634
Breast cancer, *100,* **99–101,** 408–409
Breast/ovarian cancer syndrome, hereditary, *100,* **99–101**
Breeding. *See* Cross breeding; Plant breeding
Brenner, Sydney, 15, 71, **101–102,** 165–166, 463
Bridges, Calvin Blackman, **102–104,** *103,* 487
Britten, Roy John, **104**
Brockway, L. O., 539
Bromodoxyuridine (BrdUrd), 633–634
Brown, Michael S., **104–105**
Brown, Patrick O., **105–106**
Brown, Robert, **106**
Brown, William L., 726
Brown, William S., 128
Brownian motion, 106
Bruton's agammaglobulinemia tyrosine kinase (Btk), 733

C

C-value paradox, 604
Caenorhabditis elegans, 102, 243–244
Cairns, John, 2, **107**
Calcium-phosphate transfection, 682
Calmet, Jeanne, 5
Calmodulin, 13
Camerarius, Rudolf Jacob, **107–108**
cAMP (Cyclic adenosine monophosphate), 13, **170–171**
Campbell, Keith, 723
Cancer, **108–109**
    Ames test, 20
    aneuploidy, 554
    antigen function, 36–37
    apoptosis, 39–40
    Bishop's research, 87–88
    carcinogenic agents, 113
    cell proliferation, 125
    Hayflick's research, 342
    Holley's research, 362
    hox genes, 371

mutations, 495
Petermann's research, 541–542
pluripotent stem cells, 554
repifermin, 591
retroviruses, 597
Rous's research, 606–607
stem cell research, 656
telomerase, 671
tumor formation, 690
two-hit hypothesis, 411–412
von Waldeyer-Hartz's research, 710
Weinberg's research, 716–717
See also Carcinogens; Oncogenes; specific types of cancer
Cancer genetics, **109–112,** *111*
    epidemiology, 231–232
    familial polyposis, 248
    Lwoff's research, 438–439
    mutagenesis, 492–493
    oncogene research, 520
    proto-oncogenes, 574
    retinoblastoma, 594–596
    Rowley's research, 608
    Singer's research, 631
    Smith's research, 637
    Snell's research, 639
    somatic cells, 641
    Temin's research, 672–673
    Varmus's research, 700
    Weinberg's research, 716–717
    Wigler's research, 719
CAP (Catabolite activator protein), 288
Capecchi, Mario Renato, **112**
Capillary transfer method, 92–93
Capsids, 705
Carcinoembryonic antigen test, **112–113**
Carcinogenesis
    Bishop's research, 87–88
    genetic toxicology, 309
    tumors, 689–690
    Weinberg's research, 716
Carcinogenic agents, **113**
Carlsson, Arvid, **113–114**
Carrier molecules, 58
Carriers, **114–115**
    birth defects, 86
    cloning, 150
    Cohen's research, 154
    colorblindness, 155
    DNA, 194
    genetic testing, 307
    hemophilia, 343–344
    non-Mendelian inheritance, 510
    pedigree analysis, 540–541
    phenylketonuria, 545
    prenatal diagnosis, 563–564
    sickle cell anemia, 628–629
    Tay-Sachs disease, 670–671
    thalassemia, 675–676
    viruses, 706
Carver, George Washington, **115–117**
Casein, 23
Caspersson, T. O., 140–141
Castle, William Ernest, **117–118**

Catabolite activator protein (CAP), 288

Catalytic RNA, 119

Catastrophism vs. gradualism, **330**

Catecholamines, 24

CD4 receptors, 597

CD4-T lymphocytes, 10

CD45 protein, 285

cDNA. *See* Complementary DNA

CEA test, **112–113**

Cech, Thomas R., 15–16, **118–119**

ced-9 gene, 344

Celera, 374–375, 702

Cell cycle
  eukaryotic, genetic regulation, **121–122**
  housekeeping genes, 370
  prokaryotic, genetic regulation, **122–123**
  sister chromatid exchange, 633–634

Cell death. *See* Apoptosis

Cell division, **123–124**
  cancer, 108–109
  cancer genetics, 111
  chromosome segregation and rearrangement, 142–143
  chromosome structure and morphology, 139–140, 145
  eukaryotic cell cycle, 121–122
  Flemming's research, 257
  Murray's research, 491
  nucleosomes, 513
  ovum, 529
  ploidy, 553–554
  prokaryotic cell cycle, 122–123
  prophage, 569
  replication, 591–592
  Roux's research, 607–608
  sexual reproduction, 623–625
  stem cells, 654–656
  tumor suppressor genes, 691
  von Waldeyer-Hartz's research, 710
  See also Meiosis; Mitosis

Cell proliferation, **124–125**
  breast/ovarian cancer, hereditary, 99–100
  cancer, 109
  euchromatin, 237
  eukaryotic cell cycle, 121–122
  human papilloma viruses, 707
  ribosomal RNA, 600
  tumor suppressor genes, 690–691
  zygotes, 746

Cell therapy, **126**
  cloning, 151
  stem cell research, 655

Cells
  fusion, 300
  history of genetics, 354
  nucleus (*See* Nucleus)
  Schleiden's research, 613
  structure and properties, **119–121,** *120,* 132, 137–139
  Von Mohl's research, 708
  wall structure, bacteria, 59–60
  Wilson's research, 724–725
  See also specific types of cells

Cellular respiration, 60

CentiMorgan unit, 141–142

Centrifugation, density gradient. *See* Density gradient centrifugation

Cervical abnormalities, 393

CFTR gene, 171–173, 688–689

CGH (Comparative genomic hybridization), 374

Chambers, Robert, **126–127**

Chang, Annie C. Y., 153

Chaperones, **127–128**

Character displacement, **128,** 726

Characteristics and traits, **128–129**
  complementation test, 158–159
  gene pool, 287–288
  genetic load, 302
  genotype and phenotype, 321–322
  homozygotes, 367
  intergenic regions, 398
  linkage, 432
  multifactorial transmission, 490
  mutations, 495
  one gene-one enzyme theory, 521
  polygenic inheritance, 555–556
  polymorphism, 557
  recessive, 586–587
  Riddle's research, 601
  sex-linked, 622–623
  sickle cell anemia, 629
  twins, 693
  variation, **699–700**
  wild type, 719–720

Charcot-Marie-Tooth disease, 265

Chargaff, Erwin, 65–66, **129–131**
  DNA structure, 165, 715
  history of genetics, 355–358

Chargaff's rules, 65–66, **131**

Chase, Martha Cowles, **131–132,** 349–350, 356

Chemical evolution, 557–559

Chemical mutagenesis, **132,** 492–493

Chemistry. *See* Biochemistry; Immunochemistry

Chemurgy, 116–117

Chiasma, **132–133**

Child, Charles Manning, **133–134**

Chi-square test, **134–135**

Chlamydomonas, 476–477

Chloroplast genomes, 525, 549

Cholesterol, 327–328

Chorionic villus sampling, **135–136**
  albinism, 11
  congenital birth defects, 161
  genetic testing, 308
  Prader-Willi syndrome, 562
  prenatal diagnostic techniques, 564
  sickle cell anemia, 629

Chromatin, **136–137**
  chromosome banding, 140–141
  euchromatin, 236–237
  eukaryotic cell cycle, genetic regulation, 122
  Flemming's research, 257
  heterochromatin, 236–237
  methylation, 464–465
  mitosis, 475
  nucleosomes, 513
  nucleus, 514
  replication, 592
  Strasburger's research, 660

Chromatography, 187

Chromosome 13
    Patau syndrome, 536–537
    retinoblastoma, 110, 594–596
Chromosome 15
    Angelman syndrome, 32
    Prader-Willi syndrome, 560–562
Chromosome 21
    Down syndrome, 19, 86, 208, 210–211
    Tanzi's research, 668
Chromosomes, **139–140**
    abnormalities, **137–139,** 307–309
    amniocentesis, 25–26
    autosomes, 52
    carcinogenic agents, 113
    Castle's research, 117
    Cech's research, 118–119
    cell division, 123–124
    cell proliferation, 125
    characters and traits, 129
    chiasma, 133
    chorionic villus sampling, 135–136
    chromatin, 136–137
    congenital birth defects, 161
    double helix structure, 208
    duplication, 213–214
    history of genetics, 355
    identification and banding, **140–141,** 322
    inherited variation, 699–700
    inversion, 398–399
    jumping, 149
    karyotypes, 402–405
    law of independent assortment, 425
    mapping and sequencing, **141–142**
    McClintock's research, 453–454
    mitochondrial DNA, 473–474
    mitosis, 475
    morphology, **144–145**
    mutations, **137–139,** 307–309, 494–495
    ploidy, 552–554
    prokaryotic cell cycle, 123
    Réaumur's research, 585–586
    replication, 592
    Rowley's research, 608
    segregation and rearrangement, **142–143**
    sorter, **143–144**
    Strasburger's research, 660
    structure, **144–145**
    supernumerary, 663
    Sutton's research, 663–664
    tetrad analysis, 674
    Tijo's research, 677
    transgenics, 683–684
    translocation, 685
    walking, 149
    See also specific chromosomes and chromosome types
Chromsome 7q11, 721–723
Chronic myelogenous leukemia (CML), 520, 574
Circular RNA/DNA, **567–568**
Cladistics, **145–146**
    homoplasy, 366
    phylogeny, 547
Cladogenesis, **27**
Clarke, Cyril Astley, **146–147**

Clarke, Frieda Hart, 146
Cleft lip and palate, 85, **146–147**
Clock genes, **148**
Clonal selection hypothesis, 382, 386–387
Clones and cloning, **148–150**
    Asilomar Conference, 47
    bacterial artificial chromosomes, 60
    biological applicatoins, **151**
    biotechnology, 82–83
    Brown's research, 105
    Collins's research, 155
    contig genes and mapping, 162–163
    crop improvement, 548–549
    Driesch's research, 211
    embryo transfer, 224–225
    gene splicing, 99
    genetic engineering technology, 300
    human cloning, **373**
    lambda phage, 542
    Monaco's research, 481–482
    Mullis's research, 489–490
    neo-Darwinism, 546
    nucleus structure, 514
    plant breeding, 548–549
    polymerase chain reaction, 556
    provirus hypothesis, 574
    Prusiner's research, 575
    recombinant DNA, 588
    restriction enzymes, 593
    shotgun method, 627
    Sinsheimer's research, 633
    social impact, 317–318
    transfection, 681–682
    trinucleotide repeats, 687–688
    Tsui's research, 688–689
    in vitro research, 389
    Wilmut's research, 723–724
Cloning vectors, **151–152**
    chromosome mapping, 141–142
    DNA probes, 198
    Human Genome Project, 376–377
    Murray's research, 491
    plasmids, 551
    Szostak's research, 665
    yeast artificial chromosomes, 740–741
Clubfoot, 85
Cluster analysis, 652
CML (Chronic myelogenous leukemia), 520, 574
Coalescence, **152**
Coat proteins (COP), 285
Co-dominance, 207
Codons, **153–154**
    amber, ocher, and opal mutations, 20
    amino acid, 22–23
    Brenner's research, 101–102
    chromosome structure and morphology, 144–145
    degenerate code, 183
    Matthei's research, 449
    missense mutations, 473
    mRNA, 464
    nonsense mutation, 510
    point mutations, 554
    protein synthesis, 572

proteins and enzymes, 573
    same-sense mutation, 609
    silent mutation, 630
    translation, 685
    translocation, 685
    transversions, 687
    tRNA, 682
Coenzymes, 415–417
Cohen, Stanley N., 99, **153–155,** 588
Cohn, Ferdinand Julius, 59
Cole, Rufus, 53
Collins, Francis Sellers, **155**
Colorblindness, **155–156,** 504
    McKusick's research, 455
    X chromosome, 731
Combinatorics, 440
Comparative genomics, **156–157**
    bioinformatics and computational biology, 79
    hybridization, 374
    very large scale integration, 702
Competition, **157–158**
    biodiversity, 78
    Borlaug's research, 96
    character displacement, 128
    selection, 615–616
    social Darwinism, 639–640
Competitive displacement, 158
Competitive release, 157
Complement system, 158–159, 442–443
Complementary DNA (cDNA), **159**
    Human Genome Project, 376–377
    reverse transcriptase, 599
    Sargent's research, 612–613
    sequencing, biochemical analysis, 76
    serial analysis of gene expression, 618–620
Complementary RNA (cRNA), **159**
Complementation tests, **158–159**
    multimeric enzymes, 491
    replication, 592
    viral genetics, 704
Compound microscopes, **160–161,** 223–224
Computational biology, **79**
    proteomics, 574
    very large scale integration, 702
Cone-rod dystrophy, 504
Congenital amaurosis, 504
Congenital birth defects, **161**
    teratology, 673–674
    Wiechaus's research, 719
Congenital heart defects, 85–86, 721–722
Conjugation, 568, 623–624
Conservation genetics, **162,** 353
Conservative transposition, 686
Constitutive heterochromatin, 351
Contiguous genes and mapping, **162–163**
    Down syndrome, 210–211
    Williams syndrome, 721
Convention on the Prohibition of the Development, Production and Stockpiling of Bacteriological (Biological) and Toxin Weapons and on Their Destruction, 80–81
Coordinate genetic regulation, **163**
COP (Coat proteins), 285
Cordocentesis, 564

Correns, Carl Erich, **163–164,** 709
Cortical inheritance, 2
Coryell, Charles, 539
Cosmids, 152
Covalently closed circular molecules, 551
COX-2 inhibitors, 552
Creationism, 223
Creighton, Harriet, 453
Crick, Francis, 15, 28–29, **164–166**
    Brenner's collaboration, 101–102
    Chargaff's association, 130
    Delbrück's research, 185
    DNA, 194–195, 201, 202
    Franklin's association, 265–267
    neutral mutation research, 506
    replication research, 591
    Watson-Crick model, 355–358, *356*
    Watson's collaboration, 714–715
    Wilkins's association, 720–721
    wobble theory, 728
cRNA (complementary RNA), **159**
Crops, 7
    genetically modified organisms, 311–312
    improvement of, **548–549,** 627
Cross and crossing, **166–167,** 303–304, 587
Cross breeding, 390
    animal husbandry, 33–34
    Punnett square, 577–579
    xenogamy, 733
Crossing over, **167–168**
    autosomes, 52
    chiasma, 132–133
    Darlington's research, 175
    heredity, 347
    linkage, 432
    meiosis, 457
    Müller's research, 487
    sexual reproduction, 623
    Stahl's research, 649
    statistical analysis, 652
    Sturtevant's research, 661–662
    synapsis, 664–665
Crow, James F., **168,** 407
Crystallographic electron microscopy, 411
Crystallography, **168–169,** 402, 605–606
    See also X-ray crystallography
CTR1 gene, 324–325
Cuénot, Lucien Claude, **169–170**
Culver, Kenneth, 1, 29
Curran, James, 9
Cutting sites, **170**
CVS (Chorionic villus sampling), 11, **135–136**
Cyclic AMP, 13, **170–171**
Cyclooxygenase-2 inhibitors, 552
CYP2D6 gene, 543–544
Cystic fibrosis, **171–173,** *172*
    autosomal inheritance, 51–52
    cloning applications, 151
    cloning vectors, 152
    Collins's research, 155
    enzyme therapy, 228
    frame shifts, 264
    genetic testing, 307

molecular diagnostics, 481
non-Mendelian inheritance, 510
Tsui's research, 688–689
Cystine, 23
Cytochrome 280–281, 450, 543–544
Cytogenetics
chromosome mapping, 141–142
human genome analysis, 374
medical genetics, 456
molecular, **480**
prenatal diagnostic techniques, 563–564
Stevens's research, 658–659
Cytokines, 552
Cytology, 357
Cytoplasmic inheritance, **173–174,** 509–510
Ephrussi's research, 231
organelle evolution, 525
Sonneborn's research, 641–642

D

Dam methylases, 592
Darlington, Cyril Dean, **175**
Darnell, James Edwin Jr., **175–176**
Darwin, Charles Robert, 78, 126, **176–177**
character displacement, 128
heredity research, 346
history of genetics, 354
Kölliker's research, 414
Malthus theories, 445
natural selection research, 498–499
Naudin's research, 500
Oparin's research, 523
pollination research, 554–555
Pouchet's association, 560
punctuated equilibrium theory, 576
survival of the fittest theory, 663
theories of, 177–178
Wallace's association, 711
Darwinism, **177–178**
Bates's research, 66–67
Dobzhansky's research, 205–206
Eldridge's research, 222–223
genealogy, 293
Gould's research, 328–330
history of genetics, 354
Huxley's research, 380–381
neo-Darwinism, 545–546
Pouchet's research, 560
race and ethnicity concepts, 295–296
See also Social Darwinism
Dausset, Jean, **178–179,** 639
Davenport, Charles Benedict, **179–180**
Dawid, Igor Bert, **180**
Dawkins, Richard, 285, 615
De Buffon, Georges-Louis Leclerc, 177, **181**
De novo imprinting
methylation, 465
viability, 703
Williams syndrome, 722
De Vries, Hugo, 163, **181–182**
gene mutation research, 284–285
heredity research, 347

history of genetics, 354
Morgan's research, 486
von Tschermak-Seysenegg's research, 709
DEAE-dextran transfection, 682
Deafness, **182–183,** 408–409
Defense Against Weapons of Mass Destruction Act, 81
Degenerate code, **183**
homozygotes, 367
mathematical theories, 448–449
molecular biology, 477–479
same-sense mutation, 609
silent mutation, 630
Deinococcus radiodurans, 714
Delbrück, Max, 71, 131, **183–185,** *184*
Dulbecco's association, 213
history of genetics, 358
Luria's collaboration, 61–62, 435–437
spontaneous mutation research, 647
Stahl's association, 649
Watson's association, 714–715
Deletions, **185–186**
cancer, 108–109
carcinogenic agents, 113
chromosomal abnormalities, 138
chromosome segregation, 142–143
chromosome structure, 140
mutagenesis, 496
oncogene research, 520
Prader-Willi syndrome, 562
radiation mutagenesis, 583–584
SRY gene, 647–648
translocation, 685
transposition, 596
viral genetics, 704
Williams syndrome, 721–723
wobble theory, 728
Denaturing gradient gel electrophoresis (DGGE), **186–187,**
570–571, 676
heteroduplex analysis, 351
single nucleotide polymorphisms, 632
Denaturing high performance liquid chromatography (DHPLC), **187**
Density gradient centrifugation, **187,** 649, 667
Deoxyribonucleases, 484, 507–508, 511–512
Deoxyribonucleic acid (DNA), **193–195,** *196*
alternate genetic codes, 14
ancient, **27–28,** 531, 696–697
bacteriophage structure, 60–62
base pairing, 64
bases, 64–66
biochemistry, 76–77
cancer genetics, 110
cell structure, 119
Chargaff's research, 129–131
Chase's research, 131–132
chromatin, 136–137
chromosomal abnormalities, 137–138
chromosome structure, 139–140
cloning, 150
Crick's research, 164–166
Delbrück's research, 183–185
double helix structure, 208, *209*
genetic code, 293–295
genetic information, 15

Gilbert's research, 322–324
history of genetics, 355–357
hydrogen bonds, 382–383
Levene's research, 430
MacLeod's research, 441–442
McCarty's research, 452
Mieschner's research, 467–468
mitosis, 474–475
molecular biology, 477–479
Nirenberg's research, 508–509
operons, 524
orthology and paralogy, 528
primers, 565
replisomes, 592
Sanger's research, 610–611
Sharp's research, 625–626
shotgun method, 627
Sibley's research, 627–628
single-cell gene disorders, 632–633
Smith's research, 635–636, 636–637
transcription, 681
transgenics, 683–684
transversions, 686–687
viral genetics, 703–704
viruses, 704–708
Watson's research, 714–715
Wilkins's research, 720–721
See also Nucleic acids; specfic types of DNA
DeSilva, Ashanthi, 1
Developmental and generational theory, **226–227**, 644–645
Developmental genetics, **188–189**
Hall's research, 335
Lewis's research, 430–431
Mangold's research, 446
random genetic syndrome, 584
stem cells, 656
Turner's syndrome, 691
Wiechaus's research, 719
DGGE. *See* Denaturing gradient gel electrophoresis
D'Hérelle, Félix H., 61, **189–190**, 694
DHPLC (Denaturing high performance liquid chromatography), **187**
Diabetes, 182–183, **190**
Dictyostelium discoideum, 243–244
Diet and nutrition
amino acids, 23, 25
birth defects, 84
neural tube defects, 503
risk factors, 602
Differentiation, **190–191**
biochemical analysis, 76
Child's research, 133–134
cloning, 151
developmental genetics, 188–189
eukaryotic cells, 121
homeobox research, 364
Lyon's research, 439
Nirenberg's research, 509
retroviruses, 597
selfish DNA, 617
sexual reproduction, 624–625
somatic cell genetics, 641
Spemann's research, 644–645
stem cells, 656

tumor formation, 690
Wilson's research, 725
X chromosome, 731–732
Dimerization, 199
Diploid cells, **191**
alleles, 12
alteration of generation, 14
autosomes, 52
carriers, 115
chromosomal abnormalities, 137–138
chromosome structure, 140, 144–145
conservation genetics, 162
cytoplasmic inheritance, 174
genome, 319
hemizygosity, 342–343
Levan's research, 430
ovum, 529
parthenogenesis, 535
plant breeding, 548–549
ploidy, 552
population genetics, 559
quantum speciation, 581
sex determination, 621–622
sexual reproduction, 623–625
somatic cell genetics, 641
Directional selection, **191–192**, 614–616, 648
evolutionary mechanisms, 241
neutral selection, 507
Discontinuous polymorphism, 557
Disruptive selection, **191–192**, 614–616, 648
DNA. *See* Deoxyribonucleic acid
DNA amplification
cancer genetics, 111
polymerase chain reaction, 556
randomly amplified polymorphic DNA, 585
DNA binding proteins, **192–193**
bases, 66
replisomes, 592
DNA fingerprinting
forensics, 195–197, *196*
restriction fragment length polymorphisms, 594
social impact, 314–315
DNA hybridization, **197**
Britten's research, 104
genomic library, 320–321
Sibley's research, 628
species research, 642
DNA ligase, 648
DNA methylation
Angelman syndrome, 33
bases, 66
genomic imprinting, 319–320
DNA microarrays, **197–198**
DNA polymerases, 556–557
chain reaction, 556
DNA replication, 199–200
Kornberg's research, 415–417
mitochondrial inheritance, 474
organelles, 525
replisomes, 592
See also Reverse transcriptase
DNA probes, **198**
biochemistry, 77

chromosome mapping, 141–142
fluorescence in situ hybridization, 259
Fodor's research, 260
nick translation reaction, 508
oncogene research, 520
parentage testing, 538
polymerase chain reaction, 556
DNA repair, **199**
bases, 65–66
cancer, 109
deletions, 186
missense mutations, 472–473
mutant sensitivity, 493–494
nucleases, 511–512
radiation mutagenesis, 583–584
spontaneous mutation, 647
trinucleotide repeats, 687–688
DNA replication, **199–201,** 591–592
bases, 66
Blackburn's research, 88
Cohen's research, 154
deletions, 186
enzymes, 572–573
history of genetics, 357
Meselson's research, 462–463
mutagenesis, 492
polymerase chain reaction, 556
polymerases, 556–557
spontaneous mutation, 647
Stahl's collaboration, 648–649
synapsis, 664–665
telomeres, 671
DNA sequencing, 618
alternate genetic codes, 14
chromosomes, 142
coalescence, 152
molecular diagnostics, 481
multiplexing, 491
phage genetics, 542
plant breeding, 549
population genetics, 559–560
protein electrophoresis, 571
tandem repeats, 667–668
DNA structure, **201–202**
bases, 64–66
Cech's research, 118–119
chemical mutagenesis, 132
Crick's research, 164–166
histones, 353
history of genetics, 356–358
human genome programs, 375–376
Stahl's research, 649
telomeres, 671
Wang's research, 712–713
Watson-Crick model, 164–166, 715
Wilkins's research, 720–721
wobble theory, 728
DNA synthesis, **202,** 203
AIDS therapy, 10
ATP, 5
complementary DNA, 159
enzymes, 572–573
nucleases, 511–512

prokaryotic cell cycle, 123
Temin's research, 672
DNA topoisomerases, 105–106, 712–713
DNA vaccines, **202, 204**
Dobzhansky, Theodosius, **204–206,** 205, 546
Stebbins's association, 653
Sturtevant's collaboration, 662
Wright's collaboration, 730
Dochez, A. Raymond, 53
Doherty, Peter C., **206–207**
Dominance
law of, 460, 586–587
relations, 128–129, **207**
Dopamine, 114, 535
Double helix, **208,** 209
bases, 64–66
cell properties, 119
Chargaff's research, 130
Chargaff's rules, 131
chromosome structure, 144–145
complementary RNA, 159
Crick's research, 164–166
DNA structure, 201–202
Franklin's research, 266–267
heterochromatin, 350–351
history of genetics, 356–358
hydrogen bonds, 382–383
Sibley's research, 628
transversions, 687
Wang's research, 712–713
Watson's research, 715
Wilkins's research, 721
X-ray crystallography, 733–734
Double stranded DNA, 551
Down, J. Langdon, 210
Down syndrome, **208–211**
aging, 6
Alzheimer's disease, 19
amniocentesis, 26
behavioral genetics, 69
birth defects, 86
chromosomal abnormalities, 137–139
chromosome segregation and rearrangement, 142–143
chromosome structure and morphology, 145
genetic testing, 308
meiosis, 458
neurological effects, 503
ploidy, 553
Tanzi's research, 668
XX male syndrome, 735
Dreschel, Edmund, 23
Driesch, Hans Adolf, **211**
Drosophila melanogaster, **211–212**
Benzer's research, 72
clock genes, 148
coordinate genetic regulation, 163
Dobzhansky's research, 204–206
Ephrussi's research, 230–231
experimental applications, 243–244
homeobox research, 364
hox genes, 370–371
locus control region, 433–434
McClintock's research, 453–454

Morgan's research, 485, 486
Müller's research, 487
Nüsslein-Volhard's research, 515
reciprocal crossing, 587
Stern's research, 658
Sturtevant's research, 661–662
Tatum's research, 669
Wiechaus's research, 718–719
wild type traits, 719–720
dsDNA (Double stranded DNA), 551
Duchenne muscular dystrophy, 215, 419
Dulbecco, Renato, 63, **212–213,** 574, 672, 716
Duplication, **213–214**
    cell proliferation, 125
    chromosomal abnormalities, 138
    chromosome structure and morphology, 145
    Ohno's research, 518
    single nucleotide polymorphisms, 632
    translocation, 685
    uniparental disomy, 695
Dysbetalipoproteinemia, 299
Dysgenics, **214–215**
Dyslexia, 70
Dyslipidemias, **298–299**
Dystrophinopathies, **215,** 419, 687–688

E

E-cadherin gene, 398
East, Edward Murray, 627
Ebola virus, **217–218,** 708
Ecological and environmental genetics, **218–219**
    bioremediation, 81–82
    birth defects, 84
    Borlaug's research, 96–97
    Cairns's research, 107
    competitive, 158
    Parkinson's disease, 535
    risk factors, 602
    speciation, 642–644
    twin studies, 692
    waste control/environmental cleanup, 713–714
    Wilson's research, 726–728
Economics, **219–220**
Ecotypes, **220**
Edelman, Gerald M., 38
Edwards, Robert G., 657
Edward's syndrome, **220–222,** *221*
Egg cells, 528–529
    heredity, 346–347
    mitosis, 475
    Prévost's research, 565
    sexual reproduction, 624–625
    See also Ovum
Eldridge, Niles, **222–223,** 576
Electron microscope, 160, **223–224**
    crystallographic electron microscopy, 411
    ribosome research, 601
    sperm cell research, 645
Electrophoresis, 508, **569–571,** 570, 593
    See also Gel electrophoresis
Electrophoretic transfer, 92–93, 508
Electroporation, 549, 682

ELISA (Enzyme-linked-immunosorbent assay), 9, 387
ELN gene, 721–722
Elongation, 592, 681
Embryo transfer, **224–225**
Embryology, **225–227,** *226*
    biogenetic law, 78
    Bonner's research, 94
    developmental and generational theory, **226–227**
    differentiation, 191
    Driesch's research, 211
    Fabrici's research, 247
    Hamburger's research, 336–337
    Harrison's research, 340–341
    Holtfreter's research, 363–364
    Kölliker's research, 414
    Mangold's research, 445–446
    Nüsslein-Volhard's research, 514–515
    pluripotent stem cells, 554
    prenatal diagnostic techniques, 563–564
    Prévost's research, 565
    Roux's research, 607–608
    sex determination, 622
    sexual reproduction, 624–625
    Spemann's research, 644–645
    Steptoe's research, 657
    Stevens's research, 658–659
    twins, 693
    Von Baer's research, 62–63
    Weissmann's research, 717
    Wiechaus's research, 718–719
    Wilmut's research, 723–724
Emerson, Rollins Adams, **227**
Encephaloceles, 502–503
Endometriosis, 392
Endonucleases, 42, 512, 550, 593
Endosymbiosis, 525
Environmental genetics. *See* Ecological and environmental genetics
Enzyme-linked immunosorbent assay (ELISA), 9, 387
Enzyme therapy, **228**
Enzymes, **228–229**
    active site, 3
    allosteric regulation, 13
    amino acid chemistry, 24–25
    Brown's research, 105
    cAMP, 171
    cell properties, 120
    Chargaff's research, 130
    chromosome mapping, 141–142
    chromosome structure and morphology, 144–145
    eukaryotic cell cycle, 121–122
    genetic manipulation, **229–230**
    inborn errors of metabolism, 390
    Kornberg's research, 415–417
    methylation, 464–465
    mitochondria, 473
    Moore's research, 483–484
    multimeric, 159, 490–491
    Oparin's research, 522
    operons, 524
    origin of life, 527–528
    palindromes, 532
    Pauling's research, 538–539
    plasmids, 551

prokaryotic cell cycle, 123
proteins, 572–573
recombinant DNA, 587–588
replication, 591
RNA, 15–16, 603
Stein's research, 653–654
structure-function relationships, 30–31
Wang's research, 712–713
See also specific enzymes
Ephrussi, Boris, 173–174, **230–231,** 482
Epidemics, **531**
Epidemiology
    genetics, **231–232**
    prions, 565–566
    viral genetics, 704
Epigenetics, 226–227, **232–233,** 510
Episomes, **233**
Epistasis, 232–233
Equilibrium, 13
    See also Hardy-Weinberg equilibrium; Punctuated equilibrium
Escherichia coli, **233–234**
    experimental applications, 243–244
    polymerase research, 556–557
    prokaryotic genomes, 567–568
    Prusiner's research, 575
    recombinant DNA, 587–588
    Tatum's research, 670
    waste control/environmental cleanup, 714
    Yanofsky's research, 740
ESTs. *See* Expressed sequence tags
Ethical issues
    Berg letter, 74
    dysgenics theory, 214–215
    gene therapy, 292
    genetically modified organisms, 311–312
    genetics, **234–236**
    human cloning, 373
    in vitro fertilization, 657
    intelligence testing, 315–316
    legal and public policy, 429
    newborn genetic screening, 507
    predictive genetic testing, 562–563
    Shockley's research, 626
    Singer's research, 630–631
    Sinsheimer's research, 633
    stem cell research, 655–656
    transgenics, 684
    twin studies, 692
    Wilmut's research, 723–724
    Wilson's research, 726–728
    See also Eugenics; Society and genetics
Ethnicity
    genetic concepts, **295–296**
    HLA variation, 443
    Tay-Sachs disease, 670–671
    thalassemias, 675–676
Euchromatin, 137, **236–237,** 433–434
Eugenics, **237**
    Castle's research, 117
    Davenport's research, 179–180
    ethical issues, 316
    Fisher's research, 255–256
    Galton's research, 272–273

human cloning, 373
    intelligence testing, 315–316
    Koltsov's research, 415
    Müller's research, 488
    Sanger's research, 611–612
    Shockley's research, 626
    Shull's research, 627
    Stopes's research, 660
    twin studies, 692
    Wilson's research, 727–728
Eukaryotes, **237–238,** 567
    bacteria, 58
    cell cycle, genetic regulation, 121–122
    cell properties, 119–121
    Drosophila melanogaster research, 211–212
    evolutionary mechanisms, 241
    genetics, **238–239**
    Margulis's research, 446–447
    organelles, 525
    polymerase, 556
    regulatory and sequencing regions, 590
    replication, 591
    retroposons, 596
    ribosomal RNA, 600
    ribosomes, 601
    RNA splicing, 604
    sexual reproduction, 623
    structural genes, 661
Euploidy, 552
Evolution, **239**
    acquired character, 2
    altruism, 16–17
    anagenesis vs. cladogenesis, 27
    ancient DNA, 28
    Ayala's research, 54
    Bates's research, 66–67
    Bateson's research, 67
    biogenetic law, 78
    Bonner's research, 94
    Bonnet's research, 94–95
    Britten's research, 104
    cancer genetics, 110
    Chambers's research, 126–127
    character displacement, 128
    chemical evolution, 557–559
    cladistics, 146
    clock genes, 148
    comparative genomics, 156–157
    competition, 157
    Cuénot's research, 170
    Darwinism, 177–178
    Darwin's research, 176–177
    de Buffon's research, 181
    developmental genetics, 188–189
    DNA hybridization, 197
    Dobzhansky's research, 204–206
    evidence of, **240**
    fitness, 256
    Gaia hypothesis, 271–272
    genetic variation, 310–311
    Goldschmidt's research, 326
    Gould's research, 328–330
    Haeckel's research, 333–334

history of genetics, 354
human population genetics, 377–378
Huxley's defense, 381
Kimura's research, 407–408
Kropotkin's research, 418–419
Lamarckism, 333–334, 423
Margulis's research, 446–447
maternal inheritance, 447–448
Mayr's research, 450–451
molecular clocks, 479–480
Müller's research, 488–489
neo-Darwinism, 545–546
neutral mutation, 506
neutral selection, 506–507
ontogeny and phylogeny, 521–522
Oparin's research, 522
origin of life, 527–528
phenotypic variation, 544
phylogenetics, 546
pollination, 555
Pouchet's research, 560
race and ethnicity concepts, 295–296
rare genotype advantage, 585
seeds and seed saving, 614
social Darwinism, 639–640
speciation, 642–644
spontaneous generation, 647
stasigenesis, 651
Stebbins's research, 652–653
survival of the fittest, 663
transposition, 43, 596
Wallace's research, 711
Wilson's research, 724
Wright's research, 729–730
Evolutionary mechanisms, **240–241**
Darwin's research, 177–178
Fox's research, 261
Gould's research, 328–330
gradualism vs. catastrophism, 330
Malthus's research, 445
mathematical theories, 448–449
molecular clocks, 479–480
phenotypic variation, 544
phylogeny, 547
Ponnaperuma's research, 557–559
punctuated equilibrium theory, 576
quantum speciation, 581
social Darwinism, 639–640
sociobiology, 640
speciation, 642–644
spontaneous generation, 647
stasigenesis, 651
viability, 702–703
Exogenous DNA, **242**
gene insertion, 282–283
transformation, 682–683
Wigler's research, 719
Exons, **242–243,** 604
Exonucleases, 511–512
Experimental organisms, **243–244,** 586–587
Expressed sequence tags (ESTs), **244–245**
economics, 219–220
human genome analysis, 374

serial analysis of gene expression, 619–620
Venter's research, 701–702
Extra Y hypothesis, **245**
Extranuclear inheritance, **245–246,** 509–510
acquired character, 2
Lwoff's research, 437–438
organelle evolution, 525
Extremophiles, 494

F

F factor. *See* Fertility factor
Fabrici, Girolamo, 226–227, **247**
FACS (Fluorescence activated cell sorter)
chromosome sorters, 143–144
flow karyotyping, 258
Factor VIII, 89, 588
Factor IX deficiency, 89, 588
Factor XI deficiency, 89
Facultative heterochromatin, 351
Familial Alzheimer's disease, 19
Familial combined hyperlipidemia (FCHL), 299
Familial hypercholesterolemia (FH)
Brown's research, 104–105
genetic dyslipidemias, 299
single-cell gene disorders, 632–633
Familial hypoalphalipoproteinemia, 299–300
Familial polyposis, **248,** 308
Family
Sanger's research, 611–612
socio-ethno-genetic concepts, **248–249**
See also Heredity
FAS (Fetal alcohol syndrome), 84
FCHL (Familial combined hyperlipidemia), 299
Feline leukemia virus (FELV), 597
FELV (Feline leukemia virus), 597
Female infertility, **391–395,** *394*
Fertility factor, **249–250**
Fertilization, **250**
bioremediation, 81–82
Brown's research, 106
cell proliferation, 124–125
chromosomal abnormalities, 138
chromosome segregation and rearrangement, 142
chromosome structure, 140
cross and crossing, 166–167
Cuénot's research, 170
diploid cells, 191
embryo transfer, 224–225
hermaphrodite, 347–348
hybrids, 381
in vitro fertilization, 388–389
maternal inheritance, 447–448
Mendelian genetics, 459–461
ovum, 529
parthenogenesis, 535
plant reproduction, 551
pollination, 554–555
Prévost's research, 565
Roux's research, 607–608
self, 390
self-incompatibility, 616
sex chromosomes, 621

sex determination, 621–622
sperm cells, 645
stem cell research, 654–655
Steptoe's research, 657
twins, 693
uniparental disomy, 695
xenogamy, 732–733
Fetal alcohol syndrome (FAS), 84
Fetal cells, **250–251**
    cloning research, 149
    congenital birth defects, 161
    microchimerism, 467
    mutations, 495
    prenatal diagnostic techniques, 563–564
Feulgen, Robert Joachim, **251–252**
FGFR2 gene, 584
FH (Familial hypercholesterolemia)
    Brown's research, 104–105
    genetic dyslipidemias, 299
    single-cell gene disorders, 632–633
Filo viruses, 708
Finger, Ernest, 424
Fischer, Emil Hermann, 3, 23, **253–254**
FISH. *See* Fluorescence *in situ* hybridization
Fisher, Ronald A., **254–256**, 559, 730
Fitness, **256**
    adaptation, 3–5
    altruism, 17
    Brown's research, 105
    character displacement, 128
    Crow's research, 168
    eugenics, 237
    natural selection, 498–499
    selection, 615–616
    stabilizing selection, 648
    viability, 702–703
    xenogamy, 732–733
    See also Survival of the fittest
Fixed alleles, **256–257**
Flavin adenine dinucleotide, 679
Flaviviruses, 708
Flemming, Walther, **257**, 347, 660
Flow cytometry, **257–258**
    automated sorting devices, 50
    chromosome mapping, 141–142
    chromosome sorters, 143–144
    immunological techniques, 387
Flow karyotyping, **258**
Fluctuation test, 436
Fluorescence activated cell sorter (FACS), 143–144, 258
Fluorescence in situ hybridization (FISH), **258–259**
    Angelman syndrome, 33
    chromosome mapping, 141–142
    fetal cells, 250–251
    human genome analysis, 374
    molecular cytogenetics, 480
    oncogene research, 520
    Prader-Willi syndrome, 562
    Williams syndrome, 722
Fluorescent dyes, **260**
    biochemical analysis, 76
    flow cytometry, 257–258
    karyotype analysis, 403–404

throughput screening, 677
FMR-1 gene, 262–263, 687–688
Fodor, Stephen Philip Alan, **260**
Folic acid, 502–503, 645–646
Food, genetically modified, **311–312**, 548–550, 631
Food and Drug Administration, 126
Forensics
    DNA fingerprinting, 195–197, *196,* 314–315
    parentage testing, 538
    Wallace's research, 712
Founder effect, **261**
    conservation genetics, 162
    quantum speciation, 581
    stasigenesis, 651
Fox, Sidney Walter, **261**
Fox, William Darwin, 176
Fraenkel-Conrat, Heinz, **261–262,** 651
Fragile sites, **262–263**
Fragile X syndrome, **262–264**
    genetic testing, 308
    methylation, 465
    neurological effects, 503
    trinucleotide repeats, 687–688
    X chromosome, 731
Frame shifts, 241, **264**
Francke, Uta, **265**
Franklin, Rosalind Elsie, 165, **265–267,** *266*
    DNA structure, 195, 715
    history of genetics, 357
    Klug's association, 410–411
    Wilkins's collaboration, 720–721
Fraser, John, 210
Frataxin gene, 267–268
Fraternal twins, 693
Friedewald, W. F., 607
Friedrich ataxia, **267–268**
Functional genomics, **268,** 574
Fungal genetics, **268–269,** 674, 740

G

G-banding, 140–141
G-protein receptors, 504–505
Gabor, Dennis, 224
Gaia hypothesis, **271–272,** 448
Gall, Joseph, 88
Gallo, Robert, 8–9
Galton, Francis, 70, 180, 182, **272–273**
    heredity research, 346
    twin studies, 691–692
Gamete intra-fallopian transfer (GIFT), 250, 388, 393
Gametes, **273**
    cell proliferation, 124–125
    chromosome segregation and rearrangement, 142
    fertilization, 250
    inherited variation, 700
    parthenogenesis, 535
    Sutton's research, 664
    uniparental disomy, 695
    zygotes, 746
Gametogenesis, **273–274**
    history of genetics, 354
    homologous chromosomes, 366

meiosis, 457–458
Mendelian genetics, 460
Gametophytic self-incompatibility, 616
Gamma globulins, 35
Gamow, George, **274–276**, *275*, 295
Gamow-Teller Selection Rule, 275–276
GAPDH (Glyceraldehyde-3-phosphate dehydrogenase), 552
Garrod, Archibald, 228–229, **276–277**, 390, 571, 669
Gel electrophoresis, 570–571
    denaturing gradient gel electrophoresis, 186–187
    pulsed field gel electrophoresis, 435
    thermal gradient, 676
    thermal gradient gel electrophoresis, 676
Gene amplification, **278–279**
    developmental genetics, 188–189
    hox genes, 371
    proto-oncogenes, 574
Gene banks, 614
Gene chips and microarrays, **279**, 370
Gene expression, **280**
    agricultural genetics, 7
    allosteric regulation, 13
    Angelman syndrome, 32
    antisense RNA, 39
    bioinformatics and computational biology, 79
    Britten's research, 104
    cancer genetics, 112
    Capecchi's research, 112
    Cech's research, 118–119
    coordinate genetic regulation, 163
    Dawid's research, 180
    developmental genetics, 188–189
    DNA binding proteins, 193
    DNA microarrays, 197–198
    ecological and environmental genetics, 218–219
    epigenetics, 232–233
    locus control regions, 433–434
    mutant sensitivity, 493
    nucleosomes, 513–514
    phages, 542
    proteins and enzymes, 572–573
    regulation genes, 589
    regulatory and sequencing regions, 589–590
    restriction fragment length polymorphisms, 593–594
    serial analysis, 618–620, *619*
    transfection, 681–682
    wild type alleles, 719–720
    Yanofsky's research, 740
Gene families, **280–281**
Gene flow, **281**, 284–285, 532–533
Gene frequency, **281–282**
    fixed alleles, 256–257
    Kimura's research, 408
    population genetics, 559–560
    selection, 615–616
    selfish DNA, 616–617
    Wright's research, 730
Gene gun method, 7, 202, 204
Gene induction, **282**
Gene insertion, **282–283**
Gene linkage, 67, **283**
Gene pool, **287–288**
    biodiversity, 78

conservation genetics, 162
    founder effect, 261
    neo-Darwinism, 546
    race and ethnicity concepts, 296
    species, 642–644
    Wilson's research, 727–728
    Wright's research, 730
Gene regulation, **288–289**
    Arabidopsis thaliana, 41
    Britten's research, 104
    cAMP, 170–171
    complementation test, 159
    coordinate genetic regulation, 163
    Darnell's research, 175–176
    history of genetics, 358
    Nathans's research, 498
    nucleosomes, 513
    Ptashne's research, 576
    Stern's research, 658
Gene splicing, 29, **289**
    Arber's research, 42
    Boyer's research, 98–99
    carrier molecules, 58
Gene targeting, **290**
    Capecchi's research, 112
    cloning research, 149, 151
    Kmiec's research, 411
    therapeutic strategies, **290–291**
Gene therapy, **291–292**
    ADA deficiency, 1, 29
    Anderson's research, 28–29
    antisense RNA, 39
    Berg's research, 72–74
    biotechnology, 82–83
    Chargaff's research, 130
    cystic fibrosis, 171–173
    exogenous DNA, 242
    gene targeting, 290–291
    history of genetics, 358
    human artificial chromosomes, 372–373
    Kmiec's research, 411
    medical genetics, 456–457
    Parkinson's disease, 535
    phages, 543
    phenylketonuria, 545
    retroviruses, 596–597
    Rowley's research, 608
    severe combined immunodeficiency, 620
    single-cell gene disorders, 632–633
    Smith's research, 635–636
    Tay-Sachs disease, 671
    transplantation genetics, 686
    viral genetics, 704
    wild type alleles, 719–720
Genealogy, **293**
Generalized transduction, 681
Genes, **277–278**
    characters and traits, 128–129
    Chargaff's research, 130
    chromosome structure, 140, 144–145
    mutations and genetic change, **284–285**
    names and functions, **285–286**
    penetrance, **286–287**

pseudogenes, 575–576
regulation, 589
RNA function, 603
sex-linked traits, 622–623
suppression, **289–290**
Wilson's research, 725
See also specific genes and gene types
Genetic code, 101–102, **293–295**, *294*
Genetic counseling, **296–297**
chi-square test, 134–135
pedigree analysis, 540–541
risk calculations, 318–319
statistical analysis, 652
Tay-Sachs disease, 670–671
testing and screening procedures, 306–309
Genetic disorders and diseases, **297–298**
Francke's research, 265
Hall's research, 335
Optiz's research, 524–525
Rimoin's research, 601–602
sex-linked traits, 623
Singer's research, 630–631
single-cell gene disorders, 632
Smith-Fineman-Myers syndrome, 637–638
statistical analysis, 651–652
viability, 702–703
See also Hereditary diseases; specific disorders
Genetic drift
evolutionary mechanisms, 241
mutation and genetic change, 284–285
neutral mutation, 506
neutral selection, 506–507
Genetic dyslipidemias, **298–299**
Genetic engineering technology, **300**
cloning, 149
Cohen's research, 154
plasmids, 551
recombinant DNA, 588
reverse transcriptase, 599
Singer's research, 631
Temin's research, 672–673
waste control/environmental cleanup, 714
Genetic equilibrium, 13
Genetic fingerprinting. *See* DNA fingerprinting
Genetic load, 168, **302**
Genetic mapping, **302–305**, *303–304*
chromosome mapping, 141–142
cloning research, 149
deletions, 185–186
Fodor's research, 260
Huntington's disease, 718
linkage, 302–303
physical maps, 303–305
resolution, 592–593
restriction enzymes, 593
sequence tagged sites, 617
shotgun method, 627
Sturtevant's research, 661–662
Tsui's research, 688–689
Genetic markers, **305**
chromosome sorters, 144
restriction enzymes, 593
Wexler's research, 717–718

Genetic technology transfer, **219–220**
Genetic testing and screening, **305–309**, *308*
ethical issues, 235–236, 316–317
Hughes's research, 371–372
newborn genetic screening, 507
parentage testing, 537–538
Prader-Willi syndrome, 562
predictive, **562–563**
prenatal, 563–564
Rett syndrome, 598
sickle cell anemia, 629
Tay-Sachs disease, 670–671
thalassemia, 675–676
thermal gradient gel electrophoresis, 676
throughput screening, 590–591, 677
Williams syndrome, 722
Genetic toxicology, 21, **309**, 634
Genetic transmission patterns, **309–310**, 692
Genetic variation, **310–311**
Ayala's research, 54–55
behavioral genetics, 70
cell division, 124
evolutionary mechanisms, 240–241
gene flow, 281
human population genetics, 378
inherited characteristics, 699–700
intergenic regions, 398
molecular diagnostics, 480–481
mutations, 494–496
neo-Darwinism, 546
neutral mutation research, 506
plant breeding, 548–549
population genetics, 559–560
rare genotype advantage, 585
Stern's research, 658
very large scale integration, 702
Genetically modified foods and organisms, **311–312**, 548–550, 631
Geneticists, **312**
Genetics, **312–314**
archaeogenetics, **43–45**, *44*, 531
careers in, **312**
Chase's research, 131–132
chiasma, 133
cladistics, 145–146
Clarke's research, 146–147
DNA, 193–195
Dobzhansky's research, 204–206
Drosophila melanogaster, 211–212
epidemiology, 231–232
Fisher's research, 255–256
industrial applications, **391**
Koltsov's research, 414–415
Kropotkin's research, 418–419
laboratory mice, 421, *422*
legal and public policy, 429
mathematical theories, 448–449
McKusick's research, 454–455
Mendel's research, 458–459
mental illness, 462
models, **448–449**
Moewus's research, 476–477
natural selection research, 499
obesity, 517–518

Opitz's research, 524–525
opthalmalogical, 504–505
paleopathology, 531
pedigree analysis, 539–541
philosophy, **545–546**
Punnett's research, 576–577
Shockley's research, 626
Shull's research, 627
Singer's research, 630–631
social issues, **314–318**
somatic cells, 640–641
subcellular, 525
transplantation, 685–686
twin studies, 692
See also Ethical issues; History of genetics; specific branches of genetics
Geneva Protocol for the Prohibition of the Use in War of Asphyxiating, Poisonous or Other Gases and of Bacteriological Methods of Warfare, 80
Genomes, **319**
biochemical analysis, 75–76
bioinformatics and computational biology, 79
carrier, 114–115
cell proliferation, 125
chloroplast, 525, 549
chromosome banding, 140–141
chromosome mapping and sequencing, 141–142
chromosome structure, 144–145
comparative genomics, 156–157
differentiation, 191
Drosophila melanogaster, 212
eukaryotic cell cycle, 121–122
human population genetics, 378
mitochondrial inheritance, 474
organelles, 525
prokaryotic, 122–123, 567–568
proto-oncogenes, 519–520, 574
pseudogenes, 575–576
repeated sequences, 590
restriction fragment length polymorphisms, 593–594
selection, 615–616
sequence tagged sites, 617
sequencing, 618
shotgun method, 627
somatic cell genetics, 641
trinucleotide repeats, 687–688
viral genetics, 703–704
Wallace's research, 712
Genomic imprinting, **319–320**
epigenetics, 232–233
inherited variation, 700
uniparental disomy, 695
Genomic libraries, **320–321**
cloning research, 149
functional genomics, 268
Human Genome Project, 376–377
very large scale integration, 702
Genomic sequencing
bacterial artificial chromosomes, 60
careers, 312
functional genomics, 268
multiplexing, 491
neurological/ophthalmological genetics, 505

Venter's research, 702
very large scale integration, 702
Genotype, **321–322**
polygenic inheritance, 555–556
Punnett square, 577–579
rare advantage, **585**
transduction, 681
Germ cells and cell line, **322**
Germ plasm theory, 717
Germline mutations, 492–493
Giemsa stain, **322**
chromosome banding, 140–141
heterochromatin, 351
karyotype analysis, 403–405
GIFT. *See* Gamete intra-fallopian transfer
Gilbert, Walter, **322–324**, 576, 611
Gitscher, Jane, **324–325**
Glanzmanns thrombasthenia, 89
Globin genes, 628–629, 674–676
Glucose-6-phosphate dehydrogenase gene, 543–544
Glyceraldehyde-3-phosphate dehydrogenase (GAPDH), 552
Glycine, 23
Gold, Lois Swirsky, 21
Goldschmidt, Richard B., **325–327**
Goldstein, Joseph L., 104–105, **327–328**
Golgi, Camillo, 710
Gorer, Peter, 638–639, 685
Gould, Stephen Jay, 222, 326, **328–330**, *331,* 576, 727–728
Graaf, Regnier de, 62
Gradualism, **330,** 576
Graft-versus-host disease (GVHD)
HLA variation, 443
microchimerism, 467
transplantation genetics, 686
Grant, Peter, 128
Green beard genes, 17, 285
Green Revolution, 95–97
Greengard, Paul, 113–114
Griffith, Frederick, 54, 131, **331,** 441, 466
Growth factors, 121–122
See also specific growth factors
Growth inhibitors, 362–363
Grunberg-Manago, Marianne, 518
Guerrier-Takada, Cecilia, 16
Gusella, James, **331–332,** 668, 718
GVHD. *See* Graft-versus-host disease

H

HAC (Human artificial chromosomes), **372–373**
Haeckel, Ernst Heinrich Phillip August, 78, **333–334,** 521, 660
Haldane, J. B. S., **334–335,** 528, 559
Hall, Judith Goslin, **335**
Hamburger, Viktor, **335–337**
Hamilton, William D., 16–17, **337**
Hanahan, Douglas, **337–338**
Haploid cells, **338**
alleles, 12
alteration of generation, 14
Arabidopsis thaliana, 41
chromosomal abnormalities, 137–138
chromosome structure, 140, 144–145
diploid cells, 191

division, 124
fungal genetics, 269
gametogenesis, 273–274
gene amplification, 279
genome, 319
parthenogenesis, 535
plant breeding, 548–549
ploidy, 552
proliferation, 124–125
repeated sequences, 590
self-incompatibility, 616
sexual reproduction, 623–625
tetrad analysis, 674
Hardy, Godfrey Harold, **338–339,** 408
Hardy-Weinberg equilibrium, **339–340**
allelic frequency, 13
ecological and environmental genetics, 218–219
founder effect, 261
gene frequency, 281–282
Komura's research, 408
marriage and mating customs, 300–302
population genetics, 559–560
random mating, 584
Harrison, Ross Granville, **340–341**
Hartwell, Leland H., **341**
Hayflick, Leonard, 6, **342**
HD. *See* Huntington disease (HD)
HDN (Hemolytic disease of the newborn), 90
Hearing loss, **182–183,** 408–409
Heart defects, congenital, 85–86, 721–722
Heat shock proteins, 493
Hedin, Sven, 23
Heidelberger, Michael, 53
Helicases, 592
Helling, Robert B., 153
Hemizygous organisms, **342–343,** 586–587
Hemoglobin, 628–629, 674–676
Hemolytic disease of the newborn (HDN), 90
Hemophilia, 89, **343–344**
cloning research, 149–150
enzyme therapy, 228
genetic testing, 307
Gitscher's research, 325
history of genetics, 358
marriage and mating customs, 302–303
X chromosome inactivation, 732
Hengartner, Michael Otmar, **344**
Hepadnaviruses, 707
Hepatitis B virus, 707–708
Hereditary cancers. *See* Inherited cancers
Hereditary diseases, **344–345**
ethical issues, 235
gene targeting, 290–291
gene therapy, 291–292
genetic mapping, 304–305
genomic imprinting, 320
Gitscher's research, 324
history of genetics, 358–359
industrial genetics, 391
neurological disorders, 503–504
newborn genetic screening, 507
Nirenberg's research, 509
non-Mendelian inheritance, 509–510

ophthalmological disorders, 504–505
pedigree analysis, 540–541
polygenic inheritance, 555–556
predictive genetic testing, 563
prenatal diagnosis, 563–564
restriction fragment length polymorphisms, 594
risk factors, 602
Sharp's research, 625–626
Smith's research, 636–637
trinucleotide repeats, 687–688
viability, 702–703
Wexler's research, 717–718
See also Genetic disorders and diseases; specific diseases
Hereditary hearing loss, **182–183,** 408–409
Hereditary hemorrhagic telangiectasia (HHT), **345–346**
Hereditary radicular neuropathy, 504
Heredity, **346–347**
alcoholism genetics, 12
Bateson's research, 67
birth defects, 84–87
Bridges's research, 103
Castle's research, 117
Chargaff's research, 129–131
chromosome theory, 485
Correns's research, 163
Cuénot's research, 170
Darwinism, 178, 545
Davenport's research, 179–180
De Vries's research, 181–182
DNA, 193–195
DNA hybridization, 197
extranuclear inheritance, 245–246
Garrod's research, 277
genealogy, 293
genes, 277–278
Haldane's research, 334–335
history of genetics, 354–355
intelligence, 315–316
Margulis's research, 446–447
Mendelian laws of, 459–461
Mendel's research, 458–459
Morgan's research, 485–486
Naudin's research, 499–500
polymorphism, 557
probability and statistics, 566–567
Punnett's research, 577
race and ethnicity concepts, 295–296
Riddle's research, 601
Roberts's research, 604–606
sex-linked traits, 622–623
Stern's research, 658
tetrad analysis, 674
twin studies, 691–692
von Tschermak-Seysenegg's research, 709–710
Weissmann's research, 717
Wilson's research, 725–726
Heritability, 490
Hermaphrodites, 117, **347–348,** 734–735
HERP (Daily Human Exposure dose/Rodent Potency dose), 21
Herpes viruses, 706
Herrick, James Bryan, **348**
Hershey, Alfred Day, 71, **349–350**
Chase's collaboration with, 131–132

history of genetics, 356
    Luria's association, 183–184, 436–437
Hertwig, Oscar, 710
Heterochromatin, 137, 236–237, **350–351**
    chromosome banding, 141
    inherited variation, 699–700
    locus control region, 433–434
    Y chromosomes, 739–740
Heteroduplex analysis, **351**
Heterogeneous nuclear RNA (hnRNA)
    function, 602–604
    mRNA synthesis, 463–464
    transcription, 681
Heterogony, 535
Heterozygosity, **352–353**
Heterozygotes, **351–352**, 367
    adaptation, 4
    advantage, **352**
    carriers, 115
    genetic load, 302
    inversion, 399
    stabilizing selection, 648
    viability, 703
HEXA gene, 670
Hexosaminidase A, 670
HGP. *See* Human Genome Project
HHT (Hereditary hemorrhagic telangiectasia), **345–346**
High mobility group proteins, 193
High performance liquid chromatography (HPLC), 187
Hillier, James, 223
Histidines, 417–418
Histones and histone conservation, 278–279, 280–281, **353**
History of genetics
    ancient and classical views, **353–355**
    genetic disease imprinting, 359–360
    modern genetics, **357–359**
    Needham's research, 501
    Oparin's research, 523
    paleopathology, 531
    Vavilov's research, 701
    von Nägeli's research, 709
    Watson-Crick DNA model, 355–357, *356*
    X-ray crystallography, 733–734
Hitchhiking genes, **360**
HIV. *See* Human immune deficiency virus
HLA. *See* Human leukocyte antigens
HMG proteins, 193
hnRNA. *See* Heterogeneous nuclear RNA
Hoagland, Mahlon Bush, **360–361**
Hogness box, 590
Holley, Robert William, **361–363**, 406, 508–509
Holoprosencephaly, 503
Holtfreter, Johannes, **363–364**
Homeobox genes, **364**
    coordinate genetic regulation, 163
    regulatory and sequencing regions, 590
    Sargent's research, 613
Homologous chromosomes, **364–366**, *365*
    alleles, 12
    autosomal inheritance, 51
    chiasma, 132–133
    chromosome segregation and rearrangement, 142
    crossing over, 167–168

fixed alleles, 256–257
heterozygotes, 352–353
homozygotes, 352–353, 367
inversion, 399
McClintock's research, 453–454
meiosis, 457
structure and morphology, 145
synapsis, 664–665
transposition, 596
Homologous recombination, 589
Homology, 366
Homoplasmy, 246
Homoplasy, **366**
Homozygosity, **352–353**
Homozygotes, **366–367**
    adaptation, 4
    genetic load, 302
    viability, 702–703
Hooke, Robert, 161, 601
Hopkins, Frederick Gowland, 23, **367–368**
Hoppe-Selyer, Ernst Felix, **368–369**, 418, 468
Hormones, 362
Horowitz, Norman, 528
Horvitz, H. Robert, **369**
Hot spots, **369–370**
Housekeeping genes, **370**
Hox genes, 364, **370–371**, 646
HPG (Human population genetics), **377–378**
HPLC (High-performance liquid chromatography), 187
HTLV (Human T-cell leukemia virus), 597, 705, 708
Huang, Alice Shih, 63
Huebner, Robert, 87
Hughes, Mark, **371–372**
Human artificial chromosomes (HAC), **372–373**
Human cloning, **373**
Human genome analysis, **374**
Human genome programs, **374–376**
Human Genome Project (HGP), **376–377**
    Arber's research, 43
    bacterial artificial chromosomes, 60
    Baltimore's influence, 63
    Berg's contributions, 74
    bioinformatics and computational biology, 79
    Brenner's contributions, 102
    chromosome sequencing, 142
    Collins's research, 155
    Down syndrome, 210–211
    economics, 219–220
    ethical issues, 236
    familial socio-ethno-genetic concepts, 248–249
    gene chips and microarrays, 279
    gene pool, 288
    gene therapy, 291–292
    genetic counseling, 296–297
    genetic mapping, 305
    genome research, 319
    genomic library, 320–321
    Gilbert's research, 322–324
    Gusella's research, 332
    history of genetics, 358
    McKusick's research, 455
    medical genetics, 457
    Monaco's research, 481–482

pharmacogenetics, 543–544
ploidy research, 553–554
polymerase chain reaction, 489–490, 556
predictive genetic testing, 563
programs, 374–376
proteomics, 573–574
repifermin, 590–591
resolution, 593
sequence tagged sites, 617
sequencing, 618
Sinsheimer's research, 633
Sturtevant's research, 661–662
transgenics, 684
Venter's research, 701–702
very large scale integration, 702
Watson's research, 714–715
Y chromosomes, 739–740
Human immune deficiency virus (HIV), 596–597, *707*, 708
AIDS, 7–10
protease inhibitors, 569
slow viruses, 634
Temin's research, 672–673
Human leukocyte antigens (HLA)
disease associations, 443–444
immunogenetics, 383
major histocompatibility complex, 442–444
structure and function, 37
Human papoviruses, 706–707
Human population genetics (HPG), **377–378**
Human T-cell leukemia virus (HTLV), 597, 705, 708
Huntington disease (HD), **378–380,** *379,* 504
autosomal inheritance, 52
behavioral genetics, 69
biotechnology, 82
Collins's research, 155
founder effect, 261
Gusella's research, 332
predictive genetic testing, 562–563
trinucleotide repeats, 687–688
Wexler's research, 717–718
Hutchinson, G. Evelyn, 128
Huxley, Julian, 27
Huxley, Thomas Henry, 126, 346, **380–381**
Hybrid resistance, 639
Hybridization
biotechnology, 83
double helices, 197
Naudin's research, 499–500
plants, **381–382**
See also DNA hybridization
Hybridomas, 38, 470
Hybrids, **381**
Cuénot's research, 170
Karpetchenko's research, 402
Koelreuter's research, 412–413
plant breeding, 548–549
recessive genes and traits, 586–587
Sharp's research, 625–626
Shull's research, 627
Hydrocephaly, 502–503
Hydrogen bonds, 100, **382–383**
Hypercholesterolemia
Brown's research, 104–105

genetic dyslipidemias, 299
single-cell gene disorders, 632–633
Hyperlipidemia, 299
Hyperthermia, malignant, 543
Hypoalphalipoproteinemia, 299–300

I

ICSI (Intracytoplasmic sperm injection), 388, 395–396
Identical twins, 693
Idioplasm theory, 660, 709
IEF (Isoelectric focusing), 570–571
Immune system, 382, 385, 442–443
Immunochemistry
Avery's research, 52–54
Benacerraf's research, 70–71
major histocompatibility complex, 442–444
Pasteur's research, 536
Immunogenetics, **385–386**
Doherty's research, 206–207
Landsteiner's research, 423–425
Pasteur's research, 536
Snell's research, 638–639
Tonegawa's research, 679–680
transplantation, 685–686
viruses, 706
X-linked agammaglobulinemia, 733
Zinkernagel's research, 745
Immunoglobulin E (IgE), 35
Immunoglobulin G (IgG), 35, 49
Immunoglobulin M (IgM), 35
Immunoglobulins
antibody, 34–35
immunogenetics, 382–383
X-linked agammaglobulinemia, 733
Immunological memory, 382
Immunological techniques, **386–387,** 513–514, 538
In vitro fertilization (IVF), **388–389,** *388–389*
female infertility, 393, *394*
male infertility, 395–396
Nüsslein-Volhard's research, 515
pluripotent stem cells, 554
prenatal diagnostic techniques, 563–564
Steptoe's research, 657
In vitro research, **389–390**
In vivo research, **389–390**
Inborn errors of metabolism, **390**
anticancer drugs, 38
Garrod's research, 276–277
genetic transmission patterns, 309–310
medical genetics, 455–456
phenylketonuria, 544–545
Inbreeding, **390**
animal husbandry, 33
Crow's research, 168
Hardy-Weinberg equilibrium, 340
marriage and mating customs, 302
population genetics, 559–560
self-incompatibility, 616
Stopes's research, 659–660
Wright's research, 729–730
Independent assortment
meiosis, 457–458

Mendel's research, 459–460, *461*

Punnett square, 579

See also Law of independent assortment

Induction, 483

Industrial applications of genetics, **391**

Infertility

female, **391–395**, *394*

male, **395–396**

prenatal diagnostic testing, 563–564

Steptoe's research, 657

in vitro fertilization, 388–389

XYY syndrome, 736–737

Ingram, Vernon, 165, **396**

Inheritance, **397**, 447–448

acquired character, 2

alleles, 12

autosomal inheritance, 51–52

Batson's research, 67

Castle's research, 117

characters and traits, 128–129

Chase's research, 131

chromosomal theory, 347

chromosome mapping, 141–142

chromosome segregation and rearrangement, 142–143

Clarke's research, 146–147

congenital birth defects, 161

Correns's research, 163–164

Cuénot's research, 170

cytoplasmic inheritance, 173–174

Davenport's research, 180

Down syndrome, 210

epidemiology and genetics, 231–232

extranuclear, 245–246

gene frequency, 281–282

genes, 277–278

genetic disorders, 297–298

genetic variation, 310–311

Hardy's research, 339

immunogenetics, 382

Kölliker's research, 414

Lederberg's research, 428

Lyon's research, 439

Mendel's research, 459

mitochondrial, 474, 505

Morgan's research, 486

multifactorial, 490

natural selection research, 499

neo-Darwinism, 545–546

non-Mendelian, 447, 509–510

pedigree analysis, 540–541

polygenic, 555–556

recessive genes and traits, 586–587

reciprocal crossing, 587

severe combined immunodeficiency, 620

Shull's research, 627

sickle cell anemia, 629

slow viruses, 634–635

Stevens's research, 658–659

Tatum's research, 669

uniparental disomy, 695

variation, 699–700

von Nägeli's research, 709

Wallace's research, 712

See also Maternal inheritance

Inherited cancers, **397–398**

breast/ovarian, **99–101**

gene penetrance, 286–287

genetic testing, 307–308

genomic imprinting, 320

predictive genetic testing, 563

retinoblastoma, 594–596

Smith-Fineman-Myers syndrome, 637–638

See also specific cancers

Innate immune system, 385

Insertions

sequences, **233**

viral genetics, 704

The Institute for Genomic Research (TIGR), 375, 702

Insulin, 610–611

Insulin-dependent diabetes mellitus (IDDM), 190

Intelligence and intelligence testing, **315–316**, 626

Intergenic regions, **398**, 434

International System for Human Cytogenetic Nomenclature (ISCN), 404–405

Interphase, 122

Interspersed repeats, 590

Intracytoplasmic sperm injection (ICSI), 388, 395–396

Introns, **242–243**, 532–533, 604

Inversion, **398–399**, *399*

Angelman syndrome, 32

chromosomal abnormalities, 138

chromosome segregation and rearrangement, 143

chromosome structure and morphology, 145

palindromes, 532

transposition, 596

Ion channels, 505

ISCN (International System for Human Cytogenetic Nomenclature), 404–405

Isochromosomes, 142–143, 210

Isodisomy, 695

Isoelectric focusing (IEF), 570–571

Isograft, 685

Ivanovsky, Dimitri, 678

IVF. *See In vitro* fertilization

J

J receptor, 97

Jacob, François, 166, 176

Brenner's collaboration, 102

gene regulation research, 288

history of genetics, 358

Meselson's association, 463

Monod's collaboration, 483

operon research, 422

Pardee's collaboration, 533

regulation gene research, 589

Wollman's collaboration, 728–729

Janssen, Zacharias, 161

Janssens, Frans Alfons, 132–133

Jeffreys, Alec, 314–315

Jenner, Edward, 37, 706

Jerne, Niels K., 470

Johannsen, Wilhelm, 321, 347, 354

Jumping gene theory, 104, 686

"Junk" DNA, 590, **616–617**, 745–746

K

Kandel, Eric, 113–114

Kandiyohi morph, 557

Kappa factor, **641–642**

Karle, Isabella, **401–402**

Karpetchenko, Georgiy Dmitrievich, **402**

Karyotype and karyotype analysis, 258, **402–405**, *403–404*, 520

Kendrew, John Cowdery, **405–406**, 410–411

Khorana, Har Gobind, 361, **406–407**, 508–509, 636–637

Kimura, Moto, 168, **407–408**
    neutral selection research, 479, 506–507, 546
    population genetics, 559

Kin selection, 16–17, 256, 337

King, Mary-Claire, **408–409**

Klinefelter syndrome, **409–410**
    genetic testing, 308
    meiosis, 458
    X chromosome, 731–732

Klug, Aaron, 267, **410–411**

Kmiec, Eric B., **411**

"Knocked-out" mice, 112

Knockout genes, 612–613
    cloning, 148–149
    Stahl's research, 649
    Wiechaus's research, 718–719

Knoll, Max, 223

Knot theory, 448–449

Knudson Jr., Alfred G., 397, **411–412**

Koch, Robert, 59

Koelreuter, Joseph Gottlieb, **412–413**

Kohler, Georges, 37, 38, 470–471

Köhler, Georges, **413–414**

Kölliker, Albert von, **414**, 431, 519, 709

Koltsov, Nikolai Konstantinovich, **414–415**

Kornberg, Arthur, 73, 201, 406–407, **415–417**, *416*
    Ochoa's collaboration, 518
    polymerase research, 556–557

Kornberg, Roger D., **417**

Kossel, Albrecht, 23, 355, **417–418**
    DNA research, 194
    Hoppe-Selyer's association, 369
    nucleic acid research, 511

Kropotkin, Petr Alekseevich, **418–419**

Kühnlein, Urs, 42

Kunkel, Louis, **419**

L

L-bodies, 2

Laboratory mice
    genetics, 421
    Leder's research, 427
    model organism research, 476

Lac operons, 218–219, 288, **422–423**, 524

lac repressor, 193

Lamarck, Jean Baptiste, 2, 499

Lamarckism, **423**
    Haeckel's research, 333–334
    Lysenkoism, 439–440
    neo-Lamarckism, 545–546

Lambda phage, 542
    lysogenic cycle, 706

    Sanger's research, 611
    Stahl's research, 649
    viral genetics, 703–704
    Wollman's research, 728–729

Landsteiner, Karl, 38, 90, 146, **423–425**, *424*

Latent class analysis, 652

Latham, Harriet, 175

Law of dominance, 460

Law of independent assortment, **425**
    inheritance, 397
    meiosis, 457–458
    Mendel's research, 459–460, *461*
    Punnett square, 579

Law of segregation, 460
    inheritance, 397
    Naudin's research, 500
    non-Mendelian inheritance, 509–510

LCR (Locus control region), **433–434**

LDL receptors, 104–105

Leaky genes, 286

Leber's congenital amaurosis, 504

Leber's hereditary optic neuropathy (LHON), 505

Leder, Philip, **427**

Lederberg, Joshua, 67–68, 356–358, **427–429**
    microbial genetics research, 466
    spontaneous mutation research, 648
    Tatum's collaboration, 668–670
    Zinder's association, 744
    Zuckerkandl's association, 745–746

Leeuwenhoek, Antoni van, 58, 160–161

Lejeune, Jerome, 210

LEP gene, 517

LEPR, 517

Leptin, 517

Leucine, 23

Leukemia
    chronic myelogenous, 520, 574
    feline leukemia virus, 597
    human T-cell leukemia virus, 597, 705, 708

Leukoencephalopathy, progressive multifocal, 634, 707

Levan, Johan Albert, **429–430**, 677

Levene, Phoebus Aaron Theodore, 357–358, 419, **430**

Levine, Philip, 146

Lewis, Edward B., **430–431**, 515, 718

Leydig, Franz, **431**

LHON (Leber's hereditary optic neuropathy), 505

Li-Fraumeni syndrome, 111, 397

Licensing factors, 592

Life expectancy, 5–6

Likelihood ratio, 652

LIMK1 gene, 721–722

Linderstrom-Lang, Kai, 31

Linkages, 24, 302–303, **432**, 652

Linnaeus, Carl, 108, **432–433**, *433*, 585

Liposome-mediated transfection, 682

Liquid chromatography, 187

Lisitsyn, Nikolai, 719

Littlewood, J. E., 339

Loci, **434**

Locus control region (LCR), **433–434**

Logistic regression, 652

Long-range restriction mapping, **435**

Long terminal repeats (LTR), 590, 596

Louis-Bar syndrome. *See* Ataxia-telangiectasia
Low birth weight, 84
LPL deficiency, 299
LTR (Long terminal repeats), 590, 596
Luria, Salvador Edward, 164, 212, **435–437**, *436*
    Delbrück association, 61–62
    Hershey's association, 183–184
    spontaneous mutation research, 647
    Watson's association, 714
Luxury genes, 286
Lwoff, André, 358, **437–439**, *438,* 482
Lyell, Charles, 498
Lyon, Mary Francis, **439**
Lyon hypothesis, 439, 732
Lysenko, Trofim Denisovich, 439–440, 523
Lysenkoism, **439–440**
    Karpetchenko's research, 402
    Oparin's research, 523
    Sonneborn's research, 642
    Vavilov's research, 701
Lysine, 23
Lysogenic bacteria, 438
Lysogenic cycle, 542, 568–569, 706
Lytic cycle, 542–543, 568–569, 705–706

M

M1 RNA, 16
MacLeod, Colin Munro, 54, 131, **441–442**, 452, 682
Magendie, François, 23
Major histocompatibility complex (MHC), **442–444**
    autoimmune disorders, 49–50
    Benacerraf's research, 70–71
    Dausset's research, 179
    Doherty's research, 206–207
    immunogenetics, 382–383
    Snell's research, 639
    transplantation genetics, 685–686
    Zinkernagel's research, 745
Male infertility, **395–396**
Malignancies, 690
Malignant hyperthermia, 543
Malpighi, Marcello, 227, **444–445**
Malthus, Thomas Robert, 272–273, **445**, 498–499
Mangold, Hilde Proescholdt, **445–446**
Marburg virus, 708
Margulis, Lynn, **446–447**
Markov, Georgi, 81
Marriage and mating customs
    genetic implications, **300–302**, *301*
    Sanger's research, 611–612
    Stopes's research, 660
    twin studies, 692
    See also Random mating
Martin, G. Steven, 87
Martin-Bell syndrome, 262–264
Master genes, 286, 590
Maternal inheritance, **447–448**, 509–510
    cytoplasmic inheritance, 174
    extranuclear inheritance, 245–246
    inborn errors of metabolism, 390
    Prader-Willi syndrome, 560–561
Maternally Inherited Diabetes and Deafness (MIDD), 182–183

Mathematical theories
    chi-square test, 134–135
    genetic models, **448–449**
    population genetics, 559–560
    probability and statistics, 566–567
    statistical analysis, 651–652
Mating. *See* Marriage and mating customs
Matthaei, Johann Heinrich, **449**
Maturation promoting factor (MPF), 457–458
Maxam-Gilbert technique, 618
Maximum tolerated dosage (MTD), 21
Mayer, Adolf, **449–450**
Mayr, Ernst, **450–451**
MC4R, 517
McCarty, Maclyn, 54, 131, 441–442, **452**, 682
McClintock, Barbara, 104, **452–454**, *453,* 686
McClung, Clarence Erwin, **454**, 664
McKusick, Victor A., **454–455**
MDR (Multidrug resistance), 97, 291
MECP2 gene, 598
Medical genetics, **455–457**, *456*
    biotechnology, 82–83
    Chargaff's research, 130
    Clarke's research, 146–147
    Collins's research, 155
    Crow's research, 168
    Galton's research, 272–273
    regenerative medicine, 6
    Rimoin's research, 601–602
    very large scale integration, 702
    Y chromosomes, 739–740
Megasatellite DNA, 667
Meiosis, **457–458**
    aging, 6
    alleles, 12
    autosomal inheritance, 52
    carriers, 115
    cell division, 123–124
    cell proliferation, 124–125
    chiasma, 132–133
    chromosomal abnormalities, 137–138
    chromosome segregation and rearrangement, 142–143
    chromosome structure, 140
    crossing over, 167–168
    Darlington's research, 175
    diploid cells, 191
    eukaryotic cell cycle, 121–122
    linkage, 432
    Murray's research, 491
    parthenogenesis, 535
    plant reproduction, 551
    selfish DNA, 617
    sex chromosomes, 621
    sex determination, 621–622
    sexual reproduction, 623–625
    Sutton's research, 663–664
    synapsis, 664–665
    tetrad analysis, 674
    Weissmann's research, 717
    zygotes, 746
Mendel, Johann Gregor, 163, **458–459**
    cross and crossing, 166–167
    DNA research, 193–195

genetic research, 277–278, 312–313
heredity research, 347
history of genetics, 354
inheritance theory, 397, 545
law of independent assortment, 397, 425, *426*
plant hybridization, 381–382
pollination research, 554–555
recessive genes and traits, 586–587
rediscovery of, 67
von Nägeli's research, 709
Mendel, Lafayette B., 23
Mendelian genetics, **459–461,** *460*
probability and statistics, 566–567
Punnett's research, 577
random genetic syndrome, 584
Menkes disease, 324–325
Mental illness, 30, 462
MerA gene, 714
Meselson, Matthew Stanley, 102, **462–463,** 648–649
Messenger RNA (mRNA), **463–464**
alternate genetic codes, 14
amber, ocher, and opal mutations, 20
amino acid, 22–23
Arber's research, 42–43
Berg's research, 73
blotting analysis, 93
Brenner's research, 101–102
Cech's research, 118–119
cell properties, 119–120
codons, 152–153
complementary RNA, 159
DNA microarrays, 197–198
function, 602–603
history of genetics, 358
molecular biology, 477–479
Monod's research, 482–483
nonsense mutation, 510
nucleic acid research, 511
operators, 523–524
operons, 524
polymerase, 556–557
prokaryotic genomes, 567–568
protein synthesis, 571–572
regulatory and sequencing regions, 589–590
ribosomal RNA, 600
ribosomes, 601
RNA splicing, 604
sequencing, 618
serial analysis of gene expression, 619–620
Sharp's research, 625–626
transcription, 681
translation, 684–685
tRNA, 682
wobble theory, 728
Metabolic disorders. *See* Inborn errors of metabolism
Metaphase spread, 677
Methylation, **464–465**
bases, 66
cancer genetics, 112
DNA, 33, 66, 319–320
epigenetics, 232–233
histones, 353
non-Mendelian inheritance, 510

origin recognition complex, 592
Prader-Willi syndrome, 562
restriction enzymes, 593
MHC. *See* Major histocompatibility complex
Mice, laboratory. *See* Laboratory mice
Michurin, I. V., 439–440
Microbial genetics, **465–466**
bioremediation, 81–82
Moewus's research, 476–477
Sonneborn's research, 641–642
Twort's research, 694
Wollman's research, 728–729
Zinder's research, 744
Microchimerism, **466–467**
Microglia, 18–19
Microsatellite DNA, 360, 667
Microscopes
compound, 160–161, 223–224
Malpighi's research, 444–445
See also Electron microscopes
MIDD (Maternally Inherited Diabetes and Deafness), 182–183
Miescher, Johann Friedreich, 355, **467–468,** 511
Miller, A. Dusty, 1
Miller, Stanley L., **468–469,** 523
Miller-Urey experiment, **469–470,** 528, 558–559, 665, 696–697
Milstein, César, 37, 38, 413, **470–471**
Mimicry, **471–472**
Clarke's research, 146–147
HLA variation, 443–444
Müller's research, 488–489
Minisatellite DNA, 667
Mismatch repair genes, 111
Missense mutations, **472–473,** 496
Mitchell, Arthur, 210
Mitochondria, **473**
aging, 6
apoptosis, 39
cell structure, 121
cloning, 150
cytoplasmic inheritance, 173–174
maternal inheritance, 447–448
molecular diagnostics, 481
Palmer's research, 532–533
prokaryotes, 567
pseudogenes, 575–576
Wallace's research, 712
Mitochondrial DNA, **473–474**
ancient DNA, 28
archaeogenetics, 45
DNA fingerprinting, 315
Friedrich ataxia, 268
genetic disorders, 298
homoplasmy, 246
human population genetics, 378
organelles, 525
parentage testing, 538
Wallace's research, 712
Wilson's research, 724
yeast, 740
Mitochondrial inheritance, **474,** 505
Mitosis, **474–476,** *475*
asexual reproduction, 45–46
cell division, 123–124, *124*

cell proliferation, 124–125
chromosomal abnormalities, 137–138
chromosome segregation and rearrangement, 142–143
chromosome structure, 139–140
crossing over, 167–168
eukaryotic cell cycle, 121–122
Flemming's research, 257
mosaicism, 487
ovum, 529
prenatal diagnostic techniques, 563–564
prokaryotic cell cycle, 122–123
replication, 591
Roux's research, 608
Mitzutani, Satoshi, 674
Model organism research, **476**
Modifier genes, 285
Moewus, Franz, **476–477**
Molecular biology, **477–479,** *478*
nucleases, 512
Sanger's research, 611
Urey's research, 696–697
Venter's research, 701–702
waste control/environmental cleanup, 714
Wigler's research, 719
Molecular clocks, 6, **479–480,** 506–507, 546
Molecular cytogenetics, **480**
Molecular diagnostics, **480–481**
medical genetics, 456
predictive genetic testing, 562–563
Wallace's research, 712
Molecular genetics, 313, **477–479,** *478*
Delbrück's research, 183
Hershey's research, 349–350
Kimura's research, 408
model organism research, 476
orthologous molecules, 528
Sargent's research, 612–613
Tatum's research, 668–670
Molecular neurogenetics, 71–72
Monaco, Anthony, **481–482**
Monoclonal antibodies, 37, **38,** 413–414
Monod, Jacques Lucien, 164, 166, 176, **482–483** *483*
history of genetics, 358
operon research, 422
Pardee's collaboration, 533
regulation gene research, 589
Monophyly, 533
Monosomy, 552–553
Monozygotic twins, 693
Montagnier, Luc, 8
Montagu, Mary Wortley, 37
Moore, Stanford, 30–31, **483–484,** 653–654
Morgan, Lilian Vaughan Sampson, **485**
Morgan, Thomas Hunt, 67, 164, **485–486,** *486,* 659
Bridges's association, 103
chiasma research, 133
DNA research, 193
Dobzhansky's collaboration, 204–206
Drosophila melanogaster, 211–212
gene linkage, 283, 302–303
Monod's association, 482
Müller's association, 487
Sturtevant's association, 661–662

wild type trait, 719–720
Mosaic cell line, 143, 695, 735
Mosaic viruses, 7, 262, 627
Mosaics, **486–487**
Hall's research, 335
Mayer's research, 449–450
microchimerism, 466–467
ploidy, 553
Motor neurons, 503–504
MPF (Maturation promoting factor), 457–458
mRNA. *See* Messenger RNA
MRPs (Multidrug resistance proteins), 97
Mulder, Gerardus, 23
Müller, Hermann Joseph, 68, **487–488,** 714
Bridges's collaboration, 103
history of genetics, 355
Snell's association, 638–639
Sturtevant's association, 662
Müller, Johann Friedrich Theodor (Fritz), 471, **488–489**
Müller, Johannes, 333
Müllerian inhibitory factor, 647
Mullis, Kary Banks, **489–490,** 556, 637
Multidrug resistance (MDR), 97, 291
Multidrug resistance proteins (MRPs), 97
Multifactorial traits, 509–510, 534, 693
Multifactorial transmission, **490**
Multifocal leukoencephalopathy, progressive, 634, 707
Multimeric enzymes, 159, **490–491**
Multiplexing, **491,** 538
Murray, Andrew, **491**
Muscular dystrophies, 215, 419, 481–482
Mutagenesis, **495–496**
chemical, **132,** 492–493
point mutations, 554
radiation, **583–584**
spontaneous mutation, 647
transposition, 596
Mutagens, **492–493**
cancer genetics, 111
genetic toxicology, 309
teratology, 673–674
Mutants, **493–494**
Mutations, **494–496,** *495*
active site, 3
amber, **20**
Beadle's research, 68
cancer, 109
chemical mutagenesis, 132
chromosomal abnormalities, 137–139
coalescence, 152
complementation test, 158–159
De Vries's theory, 182
deletions, 185–186
frame shifts, 241, 264
genetic change, 284–285
genetic toxicology, 309
germline, 492–493
Luria's research, 436–437
McClintock's research, 453–454
missense, 472–473, 496
molecular biology, 478–479
Müller's research, 487–488
mutagenesis, 495–496

neutral, **506**
nonsense, 20, 496, **510**
ocher, **20**
opal, **20**
phenotypic variation, 544
phylogeny, 547
polygenic inheritance, 555–556
polymorphism, 557
Prader-Willi syndrome, 560–562
proto-oncogenes, 574
random genetic syndrome, 584
random mating, 584
ribosomal RNA, 600
same-sense, **609**
sex-linked traits, 623
Sharp's research, 625–626
silent, 506, 609, **629–630**
spontaneous, **647**
stasigenesis, 651
Sturtevant's research, 662
teratology, 673–674
thermal gradient gel electrophoresis, 676
transversions, 687
tumor formation, 689–690
viability, 702–703
viral genetics, 703–704
wild type alleles, 719–720
Yanofsky's research, 740
See also Point mutations
Mycoplasma pneumoniae, 342
Myoglobin, 405–406

N

Nathans, Daniel, 42–43, **497–498,** 635
Natural immune response, 442–443
Natural killer cells, 383, 442–443
Natural selection, **498–499**
altruism, 16–17
Ayala's research, 54
Bates's research, 67
biogenetic law, 78
Castle's research, 117–118
character displacement, 128
comparative genomics, 156–157
competition, 157
Crow's research, 168
Darwinism, 177–178
Darwin's research, 176–177
De Vries's theory, 182
developmental genetics, 188–189
Dobzhansky's research, 205–206
ecotypes, 220
evolution, 239–240
fitness, 3
gene pool, 287–288
genetic variation, 310–311
Huxley's defense, 381
Lederberg's research, 428
Lysenkoism, 439–440
Malthus's research, 445
Mayr's research, 450–451
mimicry, 471–472

Morgan's research, 486
Müller's research, 488
Neo-Darwinism, 545–546
Oparin's research, 523
phylogenetics, 546
population genetics, 559–560
Pouchet's research, 560
punctuated equilibrium theory, 576
social Darwinism, 639–640
sociobiology, 640
Stebbins's research, 652–653
Wallace's research, 711
Weissmann's research, 717
Wright's research, 729–730
See also Survival of the fittest; specific types of selection
Naudin, Charles, **499–500**
Needham, John Turberville, **500**
Needham, Joseph, **501**
Neel, James V., **501–502**
Neo-Darwinism, 545–546
Neo-Lamarckism, 545–546
Neonatal genetic screening, **507**
Neoplasms. *See* Tumors
Neuberger, Albert, 610
Neural tube defects, 231–232, **502–503,** 563–564
See also specific defects
Neurocutaneous syndromes, 503
Neurofibrillary tangles, 18
Neurofibromatosis, 503
Neurofibromin genes, 398
Neurogenetics, molecular, 71–72
Neurological disorders
inherited, **503–504**
Monaco's research, 482
Selkoe's research, 617
tandem repeats, 667
Wexler's research, 718
See also specific disorders
Neurological genetics, **504–506**
Hamburger's research, 335–337
molecular, 71–72
Prusiner's research, 574–575
von Waldeyer-Hartz's research, 710
Neuronal dystrophies, 505
Neurons, 503–504
Neurospora crassa, 669–670
See also Yeast cells
Neurotransmitters, 24
Neutral mutation, **506**
Neutral selection, **506–507,** 546, 559–560
Newborn genetic screening, 507
Nick translation reaction, **507–508,** 589
Niehans, Paul, 126
Night blindness, 504–505
Nirenberg, Marshall Warren, 29, 166, 406, **508–509**
Goldstein's association, 327–328
Matthei's association, 449
protein synthesis, 572
tRNA research, 361
Noninsulin-dependent diabetes mellitus (NIDDM), 190
Non-Mendelian inheritance, 447, **509–510**
Nonsense mutations, 20, 496, **510**
Northern blotting, 93

Nuclear transfer, 148–149
Nucleases, 434, **511–512**, *512,* 648
Nucleic acids, **511**
    Hoppe-Selyer's research, 369
    nucleotides, 514
    origin of life, 527–528
    See also Deoxyribonucleic acid; Ribonucleic acid
Nuclein, 417–418, 468
Nucleocytoplasmic transport, 90
Nucleolus, 514
Nucleosomes, **512–514**
    chromatin, 136–137
    heterochromatin, 350–351
    Kornberg's research, 417
    mRNA, 464
    nucleases, 512
Nucleotide bases
    Brenner's research, 101–102
    Chargaff's rules, 131
    chromosome structure and morphology, 144–145
    degenerate code, 183
    deletions, 185–186
    DNA structure, 201–202
    DNA synthesis, 202
    double helix, 208
    history of genetics, 357
    mutations, 494–496
    replication, 591
    Todd's research, 678–679
    transcription, 681
    transversions, 686–687
    tRNA, 682
Nucleotide sequences
    codons, 153–154
    enzymes, 573
    gene splicing, 99
    genetic mapping, 305
    Human Genome Project, 377
    Huntington disease, 378–380
    mRNA, 463–464
    phenotypic variation, 544
    proteins, 573
    reverse transcription, 599
    Sanger's research, 611
    silent mutation, 630
    thermal gradient gel electrophoresis, 676
    translation, 684–685
    transversions, 687
    trinucleotide repeats, 687–688
    Yanofsky's research, 740
Nucleotides, **514**, 518
Nucleus, **514**
    Boveri's research, 98
    Brownian motion, 106
    cell proliferation, 124–125
    cell structure, 119
    Chargaff's research, 130
    chromatin, 136–137
    chromosome structure, 139–140, 144–145
    eukaryotic cell cycle, 121–122
    Gamow's research, 274–276
    radiation mutagenesis, 583–584
    RNA synthesis, 603–604

Roux's research, 608
Schleiden's research, 613
Strasburger's research, 660
transcription, 680–681
Von Mohl's research, 708
von Waldeyer-Hartz's research, 710
Number theories, 449–450
Nüsslein-Volhard, Christiane, **514–515**, 718–719
Nutrition. *See* Diet and nutrition

O

Ob gene, 517
Obesity
    genetic factors, *517,* 517–518
    Prader-Willi syndrome, 561–562
    Turner's syndrome, 691
Ocher mutation, **20**
Ochoa, Severo, 130, 406–407, 415–416, **518**
Ocular albinism, 10
Oculocutaneous albinism, 10
Ohno, Susumo, **518–519**
Oken, Lorenz, 414, **519,** 708
Olovnikov, Aleksei M., 671
Oncogenes, **519–520**
    Bishop's research, 87
    cancer genetics, 110–111
    human papillloma viruses, 707
    recombinant DNA, 588
    research, **520**
    Varmus's research, 700
    Weinberg's research, 716–717
    Wigler's research, 719
    See also Carcinogenesis; Proto-oncogenes
Oncoproteins, *99–100,* 99–101
One gene-one enzyme hypothesis, 67–69, 228–229, **520–521,** 668–670
    history of genetics, 358
    protein synthesis, 571–572
Ontogeny, 78, 188–189, **521–522**
Opal mutation, **20**
Oparin, Aleksandr Ivanovich, **522–523,** 528
Operation Whitecoat, 80–81
Operators, **523–524,** 590
Operons, **524**
    coordinate genetic regulation, 163
    history of genetics, 358
    lac operon, 218–219, 288, **422–423,** 524
    microbial genetics, 465–466
    Monod's research, 483
    Pardee's research, 533
    plasmids, 551
    punctuated equilibrium theory, 576
    regulatory and sequencing regions, 590
    structural genes, 661
Ophthalmological genetics, **504–506**
Opitz, John Marius, **524–525**
Optic neuropathy, 505
ORC (Origin recognition complex), 591–592
Organelles, 447–448, **525**
Organizer effect, 644–645
Origin of life, *527,* **527–528**
    Ponnamperuma's research, 557–559
    Szostak's research, 665

Urey's research, 697
See also Evolution
Origin recognition complex (ORC), 591–592
Oro, Juan, 470
Orthologous molecules, 528
Orthology, **528**
Orthomyxoviruses, 708
Osborne, Thomas B., 23
Outbreeding, 300–302, 390, 732–733
Ovarian cancer, hereditary, **99–101**
Ovulatory disorders, 393
Ovum, **528–529,** 645
Oxygen thermometer, 697

P

p16 gene, 690
p21 genes, 285
p53 gene, 40, 111–112, 690–691
Packman, Seymour, 324
Paleopathology, **531**
Palindromes, **531–532,** 593
Palmer, Jeffrey Donald, **532–533**
Pancreatic ribonuclease, 600
Pappenheimer, Alwin M., 541
Paralogous genes, 528
Paralogy, **528**
Paramecium, 2
Pardée, Arthur Beck, 166, **533**
Parkinson's disease, 114, **533–534**
Parthenogenesis, 46, 94–95, **534–535**
Parvoviruses, 708
Pasteur, Louis, 58–59, **535–536**
        enzymes, 228–229
        Pouchet's association, 560
        spontaneous generation, 522, 647
Patau syndrome, 137–138, **536–537**
Paternity and parentage testing, **537–538**
Pathogens
        biological warfare, 80–81
        cancer, *109*
        classification, 60
        rare genotype advantage, 585
Pauling, Linus, 165, 266–267, **538–539**
        DNA structure, 721
        history of genetics, 357–358
        molecular clock, 479
        Zuckerkandl's collaboration, 745–746
PAX gene, 646
PCR. *See* Polymerase chain reaction
PCSK1 gene, 517
Pearson, Karl, 182
Pedigree analysis, **539–541**
        Huntington disease, *379*
        symbol guide, *303*
Pelvic inflammatory disease, 391–392
Peptide linkages, 24
Perutz, Max, 715
Petermann, Mary Locke, **541–542**
Pfefferkorn, Elmer, 87
PFGE (Pulsed field gel electrophoresis), 435
pH levels, 493–494
Phage Treaty of 1944, 184–185

Phages
        genetics, **542,** 714–715
        therapy, **542–543**
Phagocytes, 442–443
Pharmacogenetics, **543–544**
        anticancer drugs, 38–39
        medical genetics, 456
        model organism research, 476
        throughput screening, 677
Phenotype, **321–322**
        inherited variation, 700
        Lamarckism, 423
        polygenic inheritance, 555–556
        Punnett square, 577–579
        transduction, 681
        variation, **544**
Phenylketonuria (PKU), 25, 69, **544–545**
Pheromones, 726
Phi X, 174, 416
Philosophy of genetics, 501, **545–546**
Phocomelia, 85
Phosphate ions, 13
Phosphorylation, 13
Phototransduction, 504–505
Phylogenetics, **546**
        cladistics, 145–146
        comparative genomics, 156–157
        homoplasy, 366
        organelles, 525
        Palmer's research, 533
        species research, 642
        trinucleotide repeats, 687–688
Phylogeny, **521–522, 546–548**
        biogenetic law, 78
        developmental genetics, 188–189
        Sibley's research, 628
Pigment pattern polymorphisms, 557
PKU. *See* Phenylketonuria (PKU)
Plagues, **531**
Plant breeding, **548–549,** 627
Plant genetics, **550–551**
        Carver's research, 115–117
        Darlington's research, 175
        ecotypes, 220
        hybridization, 381–382, 499–500
        Koelreuter's research, 412–413
        Oparin's research, 522–523
        Palmer's research, 532–533
        pollination, 554–555
        reproduction, **551**
        seeds and seed saving, 613–614
        self-incompatibility, 616
        Stebbins's research, 652–653
        Strasburger's research, 660
        tobacco mosaic virus, 677–678
        Vavilov's research, 701
        von Tschermak-Seysenegg's research, 709–710
        See also Agricultural genetics
Plasma proteins, 541–542
Plasmids, **233, 551**
        archae, 43
        bacteria, 58
        cell properties, 120

cloning vectors, 151–152
Cohen's research, 153
gene splicing, 99
mutagenesis, 496
plant genetics, 550
prokaryotic cell cycle, 123
prokaryotic genomes, 568
recombinant DNA, 588
repeated sequences, 590
replication termini, 592
transfection, 681–682
yeast artificial chromosomes, 740–741
Platelet disorders, 89
Pleiotropy, 524–525, **552**
Ploidy, **552–554**, 663
Pluripotent stem cells, **554**, 655, 656
PML (Progressive multifocal leukoencephalopathy), 634, 707
Point mutations, **554**
amber, ocher, and opal, 20
deletions, 185–186
DNA repair, 199
enzymes, genetic manipulation, 229–230
fragile X syndrome, 263–264
Ingram's research, 396
prions, 566
proto-oncogenes, 574
same-sense mutation, 609
sickle cell anemia, 628–629
viral genetics, 704
Polio, 424–425
Pollination, **554–555**
plant reproduction, 551
self-incompatibility, 616
sexual reproduction, 624–625
Polyclonal antibodies, 386–387
Polydactyly, 86, 699–700
Polygenic inheritance, 555–556, 628
Polymerase chain reaction (PCR), **556**
amplified fragment length polymorphism, 26–27
ancient DNA, 27–28
bases, 66
complementary DNA, 159
denaturing gradient gel electrophoresis, 186–187
denaturing high performance liquid chromatography, 187
DNA synthesis, 202
genetic markers, 305
heteroduplex analysis, 351
microchimerism, 467
molecular diagnostics, 481
Mullis's research, 489–490
oncogene research, 520
parentage testing, 538
primers, 565
radiation mutagenesis, 583–584
randomly amplified polymorphic DNA, 585
recombinant DNA, 588
resolution, 593
sequence tagged sites, 617
single nucleotide polymorphisms, 632
Wilson's research, 724
Polymerases, 508, **556–557**, 591
See also DNA polymerases; RNA polymerases

Polymorphisms, **557**
adaptation, 4
Ayala's research, 54
chemical mutagenesis, 132
genetic testing, 306–309
hitchhiking, 360
human population genetics, 377–378
neutral selection, 507
parentage testing, 538
parthenogenesis, 535
population genetics, 559–560
single nucleotide polymorphisms, 632
See also Restriction fragment length polymorphisms (RFLPs)
Polyomaviruses, 707
Polypeptides, 23
Polyploidy, 381, 552
Polyposis, familial, **248,** 308
Ponnamperuma, Cyril, 523, **557–559**
Population genetics, 313, **559–560**
adaptation, 3
Arabidopsis thaliana, 41
coalescence, 152
conservation genetics, 162
Crow's research, 168
DNA fingerprinting, 315
Dobzhansky's research, 204–206
dysgenics theory, 214–215
gene flow, 281
gene pool, 287–288
Hardy's research, 338–339
Hardy-Weinberg equilibrium, 339–340
human, **377–378**
Kimura's research, 407–408
Malthus's research, 445
marriage and mating customs, 300–302
Neel's research, 501–502
paleopathology, 531
Parkinson's disease, 534
race and ethnicity concepts, 296
random mating, 584
very large scale integration, 702
Wilson's research, 726–728
Wright's research, 729–730
Zuckerkandl's research, 745–746
Porter, Rodney R., 38
Positive predictive value (PVP), 318–319
Pouchet, Felix-Archimede, **560**
Pox viruses, 705–706
Prader-Willi Syndrome (PWS), 265, **560–562**
pRB gene, 594–596
Pre-implantation diagnosis, 371–372
Predictive genetic testing, **562–563**
Preformationist theory, 354
Prenatal diagnostic techniques, **563–564**
molecular diagnostics, 480–481
Patau syndrome, 537
statistical analysis, 652
thalassemia, 675–676
Williams syndrome, 722
X-linked agammaglobulinemia, 733
XYY syndrome, 737
See also Amniocentesis; Chorionic villus sampling
Presenilin genes, 617, 668

Prévost, Jean-Louis, **565**
Primary interaction immunoassays, 387
Primase, 592
Primers, **565**
    chromosome mapping and sequencing, 141–142
    polymerase chain reaction, 556
    polymerases, 556–557
    replication, 591
Prions, **565–566**
    chaperones, 127
    Prusiner's research, 574–575
    slow viruses, 634–635
    thermal gradient gel electrophoresis, 676
Privacy issues, 429
Probability and statistics, **566–567,** 577–579
    See also Bayesian statistics
Progressive multifocal leukoencephalopathy (PML), 634, 707
Prokaryotes, **567**
    bacteria, 60
    cell cycle, genetic regulation, 122–123
    cell properties, 119–120
    genomes, 122–123, **567–568**
    Margulis's research, 446–447
    regulatory and sequencing regions, 590
    ribosomes, 601
Prolactin, 601
Promoter sequence, 7, 680–681
Prophage, **568–569**
    bacteriophage structure, 61
    lysogenic cycle, 706
    Wollman's research, 728–729
Proteases/protease inhibitors, **569**
Protein C deficiency, 89
Protein electrophoresis, 508, **569–571,** 593
Protein folding, 127–128
Protein S deficiency, 89
Protein synthesis, **571–572**
    alternate genetic codes, 14
    amber, ocher, and opal mutations, 20
    amino-acid chemistry, 23
    bacteria, 58
    Capecchi's research, 112
    Cech's research, 118–119
    cell properties, 119–120
    circular RNA/DNA, 567–568
    Crick's research, 166
    degenerate code, 183
    DNA synthesis, 202
    eukaryotic cell cycle, 122
    history of genetics, 358
    Khorana's research, 406–407
    Matthei's research, 449
    molecular biology, 477–479
    Monod's research, 483
    nonsense mutation, 510
    nucleic acids, 511
    Pardee's research, 533
    prokaryotes, 567
    ribosomal RNA, 600
    ribosomes, 600–601
    RNA function, 603
    silent mutation, 630
    translocation, 685

Proteins, **572–573**
    amino acids, 23–24
    Kossel's research, 417–418
    proteomics, 122, 573–574
    Sanger's research, 610–611
    Stein's research, 654
    See also Enzymes; specific proteins
Proteomics, 122, **573–574**
Proto-oncogenes, **574**
    Bishop's research, 87
    cancer genetics, 110
    Hengartner's research, 344
    Weinberg's research, 716–717
    See also Oncogenes
Protoplasts, 2
Prout, William, 23
Provirus hypothesis, **574**
Prp gene, 566, 575, 634
Prusiner, Stanley, 565–566, **574–575**
Pseuacreaea eurytus, 192
Pseudogenes, **575–576**
Pseudohermaphrodism, 347
Pseudomonas aeruginosa, 713
Psychiatric disorders, 30, 462
Ptashne, Mark Steven, **576**
Pulsed field gel electrophoresis (PFGE), 435
Punctuated equilibrium, 222–223, **576**
    evolution, 239
    Gould's research, 329–330
    quantum speciation, 581
    stasigenesis, 651
Punnett, Reginald Crundall, 67, 283, **576–577**
Punnett square, **577–579**
PWS (Prader-Willi Syndrome), 265, **560–562**

Q

Quantum speciation, **581**
Q-banding, 140–141

R

R-banding, 140–141
Race, **295–296**
    See also Ethnicity
Rachischisis, 646
Radiation
    Karle's research, 401–402
    mutagenesis, 492–493, **583–584**
    Neel's research, 501–502
Radicular neuropathy, 504
Radioactive elements, 50–51
Radioimmunoassay (RIA), 387
Randall, John T., 720
Random genetic syndrome, **584**
Random mating, **584–585**
    gene flow, 281
    Hardy-Weinberg equilibrium, 340
    marriage and mating customs, 303
    quantum speciation, 581
Randomly amplified polymorphic DNA (RAPD), **585**
Rapkine, Louis, 482
Rare genotype advantage, **585**

ras oncogene, 719
RB-1 gene, 594, 690
RDA (Representational difference analysis), 719
RDS protein, 504
Reactive oxygen species, 583–584
Réaumur, Réné Antoine Ferchault de, 94, **585–586**
Recapitulation, 521
     See also Phylogeny
Receptor genes, 286
Receptor molecule, 327–328
Recessive genes and traits, 128–129, 540–541, **586–587**
Reciprocal crossing, **587**
Recombinant DNA, **587–588**
    Asilomar Conference, 46–47
    Bachrach's research, 57–58
    Berg's research, 72–74
    Cohen's research, 153
    laboratory mouse genetics, 421
    nucleases, 511–512
    Singer's research, 630–631
    technology and engineering, **588**
    transgenics, 683–684
Recombination, **589**
    archaeogenetics, 45
    behavioral genetics, 70
    Benzer's research, 71
    chiasma, 132–133
    chromosome mapping and sequencing, 141–142
    fungal genetics, 269
    Hershey's research, 350
    Lederberg's research, 427–429
    phages, 543
    plant genetics, 550
    prokaryotic cell cycle, 123
    restriction fragment length polymorphisms, 593–594
    sexual reproduction, 623
    Stahl's research, 649
    Sturtevant's research, 662
    synapsis, 665
    Tatum's research, 670
    Tonegawa's research, 679–680
    trinucleotide repeats, 687–688
    viral genetics, 703–704
Red Queen hypothesis, 585
Regenerative medicine, 6
Regulation genes, **589**, 630
Regulatory regions, **589–590**
Rendu-Osler Weber disease. *See* Hereditary hemorrhagic telangiectasia
    (HHT)
Rensch, Bernhard, 27
Repeated sequences, **590**
    deletions, 186
    eukaryotic genetics, 238–239
    intergenic regions, 398
    pseudogenes, 575–576
    telomeres, 671
Repifermin, **590–591**
Replication, **591**
    autoimmune disorders, 50
    bacteriophage, 62
    Baltimore's research, 63
    base pairing, 64
    Blackburn's research, 88

    Brown's research, 105–106
    cancer genetics, 112
    cell division, 123–124
    cell properties, 119
    chromosomal abnormalities, 137–138
    chromosome structure, 140
    Cohen's research, 154
    complementary DNA, 159
    crossing over, 167–168
    DNA polymerase, 556
    eukaryotic genetics, 122, 238–239
    genetics, 313–314
    hydrogen bonds, 382–383
    Meselson's research, 462–463
    molecular biology, 478–479
    mutagenesis, 496
    origin of life, 528
    process manipulation, **591–592**
    prokaryotic cell cycle, 123
    retroviruses, 597
    RNA, 15
    sequencing, 618
    sister chromatid exchange, 633–634
    Szostak's research, 665
    trinucleotide repeats, 687–688
    viral, 568–569, 705
    See also DNA replication; RNA replication
Replicative transposition, 686
Replisomes, **592**
Reporter genes, 285–286
Representational difference analysis (RDA), 719
Repressor proteins, 524, 590
Resolution, 141–142, 196–197, **592–593**
Restriction endonucleases, 550
Restriction enzymes, **593**
    amplified fragment length polymorphism, 26–27
    bacteriophages, 62
    bases, 66
    Berg's research, 73–74
    chromosome mapping and sequencing, 141–142
    cloning, 149
    cutting sites, 170
    DNA fingerprinting, 195–197
    gene splicing, 289
    genetic mapping, 303–304
    genetic testing, 306–309
    Gilbert's research, 324
    long-range mapping, 435
    Meselson's research, 463
    mutagenesis, 496
    Nathans's research, 497–498
    nucleases, 512
    primers, 565
    protein electrophoresis, 569–571
    randomly amplified polymorphic DNA, 585
    recombinant DNA, 587–588
    resolution, 593
    Roberts's research, 606
    sequencing, 618
    serial analysis of gene expression, 619–620
    Smith's research, 635–636
Restriction fragment length polymorphisms (RFLPs), **593–594**
    chromosome mapping and sequencing, 141–142

DNA fingerprinting, 314–315
    Gusella's research, 332
    parentage testing, 538
    predictive genetic testing, 563
    randomly amplified polymorphic DNA, 585
Restriction-modification system, 42
Retinitis pigmentosa, 504–505
Retinoblastoma, 397, **594–596,** *595*
    cancer genetics, 110
    genetic testing, 308
    Knudson's research, 411–412
    tumor suppressor genes, 690
Retroposons, **596**
Retroviral-like elements, 590
Retroviruses, **596–597,** 708
    acquired character, 2
    ADA deficiency, 1
    Baltimore's research, 63
    Brown's research, 105–106
    gene therapy, 29
    history of genetics, 358
    oncogenes, 519–520
    reverse transcriptase, 599
    Rous sarcoma virus, 87
    Rous's research, 606–607
    schematic, *9*
Rett syndrome, 503, **598**
Reverse transcriptase, **598–599**
    acquired character, 2
    Baltimore's research, 63
    Brown's research, 105–106
    cancer research, 87
    history of genetics, 358
    HIV retrovirus, 8
    provirus hypothesis, 574
    retroviruses, 596–597
    Temin's research, 672–673
    transposition, 596
    viral genetics, 703–704
Reverse transcription, **599**
Reversions, spontaneous, **647**
RFLPs. *See* Restriction fragment length polymorphisms
Rh blood group, 90, 146–147
Rhabdoviruses, 708
Rhesus disease, 37, 146–147
Rhodopsin, 504
Rhogam, 146–147
RIA (Radioimmunoassay), 387
Ribonucleases, 511–512, **599–600**
    Chargaff's research, 130
    DNA synthesis, 572–573
    Moore's research, 483–484
    prokaryotic cell cycle, 123
    RNase P, 16
    Stein's research, 653–654
    structure-function relationship, 31
Ribonucleic acid (RNA), **603–604**
    alternate genetic codes, 14
    antisense, **39,** 743
    antisense RNA, 39
    base pairing, 64
    bases, 64–66
    biochemistry, 76–77

    catalytic, 119
    Cech's research, 118–119
    cell properties, 119–120
    Chargaff's research, 130
    circular, **567–568**
    complementary, **159**
    Crick's research, 166
    discovery, 15
    DNA research, 195
    eukaryotic cell cycle, 121–122
    function, **602–603**
    heterogeneous nuclear, 463–464, 602–604, 681
    history of genetics, 355–357
    Khorana's research, 406–407
    Leven's research, 430
    M1, 16
    Miller's research, 469
    molecular biology, 477–479
    Ochoa's research, 518
    retroviruses, 596–597
    reverse transcriptase, 598–599
    ribonuclease structure, 31, 599–600
    Szostak's research, 665
    transversions, 686–687
    viral genetics, 651, 703–704
    viruses, 704–708
    Wilkins's research, 721
    See also Nucleic acids; specific types of RNA
Ribosomal RNA (rRNA), **600**
    alternate genetic codes, 14
    function, 602–604
    gene amplification, 279
    nucleolus structure, 514
    origin of life, 528
    polymerase, 557
    protein synthesis, 572
    repeated sequences, 590
    ribosome function, 601
    translation, 684–685
Ribosomes, **600–601**
    Blobel's research, 89–90
    Cech's research, 118–119
    cell properties, 119–120
    enzymes, 573
    Petermann's research, 541–542
    proteins, 571, 573
    ribosomal RNA, 600
    RNA function, 603
    translation, 684–685
    Zamecnik's research, 743
Riddle, Oscar, **601**
Rimoin, David, **601–602**
Risk calculations
    Bayesian statistics, 318–319
    chi-square test, 134–135
    genetic disorders, 297–298
    genetic testing, 306–309
    probability and statistics, 566–567
    statistical analysis, 652
Risk factors, **602**
    Alzheimer's disease, 19
    epidemiology and genetics, 231–232
    genetic dyslipidemias, 298–299

genetic markers, 305
Parkinson's disease, 534
predictive genetic testing, 562–563
teratology, 673–674
RNA. *See* Ribonucleic acid
RNA phage, 97
RNA polymerases, 556–557
Ochoa's research, 518
operators, 523–524
palindromes, 532
phage transcription, 542
polymerase II, 417, 464
protein synthesis, 571–572
ribosomal RNA, 600
transcription, 680–681
RNA primers, 123
RNA replication, 572, 591
RNA splicing, **604,** *605*
Cech's research, 118–119
nucleases, 511–512
transcription, 681
RNA synthesis, 603–604
RNA viruses, 217–218
RNase P (Ribonuclease P), 16
Roberts, Richard J., **604–606,** 625–626
Robertson, O. H., 606
Robertsonian translocation, 210
Rod photoreceptors, 504
ROM1 gene, 504
Rose, William, 23
Rous, Peyton, 87, **606–607**
cancer genetics, 108–110
Temin's collaboration, 672
viral research, 708
Rous sarcoma virus (RSV), 87, 597, 606–607
provirus hypothesis, 574
structure, 708
Temin's research, 672
Zamecnik's research, 743
Roux, Wilhelm, 211, 227, **607–608**
Rowley, Janet Davison, **608**
RSV. *See* Rous sarcoma virus
Rubin, Henry, 672
Ruska, Ernst, 161, 223
Ryan, Francis, 476

S

Saccharomyces cerevisiae, 243–244, 740
See also Yeast cells
SAGE (Serial analysis of gene expression), 245, **618–620**
Salmonella, 198
Same-sense mutation, **609**
Sanger-Coulson procedure, 618
Sanger, Frederick, 31, 38, 323, 470, **609–611**
bacteriophage research, 62
Moore's collaboration, 484
one gene-one enzyme theory, 521
tRNA research, 606
Sanger, Margaret Louisa Higgins, **611–612,** 659
Sargent, Thomas Dean, **612–613**
Satellite DNA, 667
Scherrer, Klaus, 175

Schizophrenia, 30, 462
Schleiden, Matthias Jacob, **613,** 708
SCID. *See* Severe combined immunodeficiency
Screening, genetic. *See* Genetic testing and screening
SDS polyacrylamide gel electrophoresis, 570
Sedgwick, Adam, 127
Seeds and seed saving, **613–614**
Segregation, law of, 397, 460, 500, 509–510
Selection, **614–616**
cancer genetics, 110
competitive, 158
Darwin's research, 177–178
Davenport's research, 180
Gamow's research, 275–276
microbial genetics, 466
mutant sensitivity, 494
mutation and genetic change, 284–285
phenotypic variation, 544
population genetics, 559–560
random mating, 584
recessive genes and traits, 586–587
transgenics, 684
See also specific types of selection
Self-fertilization, 390
Self-incompatibility, **616,** 732–733
Selfish DNA, **616–617,** 625–626
Selfish genes, 285
Selkoe, Dennis James, **617**
Semen sampling, 395–396
Senility, 17–19
Sensory neurons, 504
Sequenced tagged sites (STSs), **617**
Sequencing, **618**
bioinformatics and computational biology, 79
Brenner's research, 101–102
Brown's research, 105
chromosome, 141–142
cloning vectors, 152
Collins's research, 155
comparative genomics, 156–157
economics, 219–220
expressed sequence tags, 245
gene splicing, 289
Gilbert's research, 323–324
human genome, 374–376
Human Genome Project, 376–377
mathematical theories, 448–449
McKusick's research, 455
multiplexing, 491
neo-Darwinism, 546
nucleases, 512
organelles, 525
orthology and paralogy, 528
palindromes, 531–532
regulatory regions, 589–590
repeated sequences, 590
repifermin, 590–591
resolution, 593
Roberts's research, 606
same-sense mutation, 609
Sanger's research, 611
shotgun method, 627
Venter's research, 701–702

Serial analysis of gene expression (SAGE), 245, **618–620**
Severe combined immunodeficiency (SCID), 1, 421, **620**
Sex chromosomes, **620–621**, *621*
    chromosomal abnormalities, 138
    germ cells and cell line, 322
    hemizygosity, 342–343
    heterozygotes, 352
    karyotype analysis, 404
    McClung's research, 454
    ploidy, 553
    Smith-Fineman-Myers syndrome, 637–638
    Wilson's research, 724–726
    X chromosome, 731–732
    Y chromosome, 739–740
Sex determination, **621–622**, *622*
    autosomes, 52
    Brenner's research, 102
    chromosome structure and morphology, 144–145
    Cuénot's research, 169–170
    genotype, 321–322
    McClung's research, 454
    Ohno's research, 518
    SRY gene, 647–648
    Stevens's research, 658–659
    Wilson's research, 725–726
    X chromosome, 731–732
    Y chromosome, 740
Sex-linked traits, **622–623**
    inherited variation, 699–700
    Stevens's research, 659
    Sturtevant's research, 661–662
Sexual reproduction, **623–625**
    cloning, 150
    female infertility, 392
    Gaia hypothesis, 271–272
    gametes, 273
    gametogenesis, 273–274
    germ cells and cell line, 322
    hermaphrodite, 347–348
    Leydig's research, 431
    Moewus's research, 476–477
    parthenogenesis, 534–535
    plants, 107–108, 412–413, 432–433, 551
    random mating, 584
    Riddle's research, 601
    Sanger's research, 611–612
    self-incompatibility, 616
    Sonneborn's research, 641–642
    sperm cells, 645
    Sutton's research, 663–664
    Tatum's research, 668–670
    X chromosome, 731–732
    zygotes, 746
Sexually transmitted diseases, 9
SFMS (Smith-Fineman-Myers syndrome), 637–638
Sharp, Phillip A., 118, 605, **625–626**
Shifting-balance theory, 729–730
Shockley, William, **626**
Short tandem repeats (STRs), 314–315
Shotgun method, 613–614, **627**
Shull, George Harrison, **627**
Sibley, Charles G., **627–628**
Sickle cell anemia, **628–629**
    adaptation, 4
    amino acid chemistry, 24
    birth defects, 84, 86
    Herrick's research, 348
    heterozygous advantage, 352
    history of genetics, 358
    Ingram's research, 396
    missense mutations, 473
    molecular diagnostics, 481
    Neel's research, 501–502
    one gene-one enzyme theory, 521
    Pauling's research, 538–539
    recessive genes and traits, 586–587
    restriction fragment length polymorphisms (RFLPs), 594
Signal transduction pathways, 504–505, 573
Silent mutation, 506, 609, **629–630**
Simian immunodeficiency virus (SIV), 597
Simple sequence length polymorphism (SSLP), 141–142
Simpson, George Gaylord, 27
Singer, Maxine, **630–631**, *631*
Single-cell analysis, 371–372
Single-cell gene disorders, **632–633**
Single-gene disorders, 286–287, 509–510
Single nucleotide polymorphisms (SNPs), 198, 538, **632**
Sinsheimer, Robert Louis, **633**
Sister chromatid exchange, 309, **633–634**
Sitoserolemia, 299
SIV (Simian immunodeficiency virus), 597
Slow viruses, **634–635**
Small nuclear RNA (snRNA), 603
Smallpox, 83
Smith-Fineman-Myers syndrome, **637–638**
Smith, Hamilton O., 42–43, 497, **635–636**
Smith, Michael, **636–637**
Snell, George David, **638–639**
SNP (Single nucleotide polymorphisms), 198, **632**
SNPRN gene, 561–562
Social Darwinism, **639–640**
    Darwin's research, 177–178
    Haeckel's research, 334
    race and ethnicity concepts, 296
    Stopes's research, 659–660
    twin studies, 692
    Wallace's research, 711
    See also Darwinism
Society and genetics
    DNA fingerprinting, 314–315
    intelligence testing, 315–316
    modern research, 317–318
    technology and, 316–317
Sociobiology, **640**
    familial socio-ethno-genetic concepts, 248–249
    Malthus's research, 445
    Sanger's research, 611–612
    Stopes's research, 659–660
    Wilson's research, 726–728
SOD (Superoxide dismutase), 6, 369
Sodium dodecyl sulfate polyacrylamide gel electrophoresis, 570
Somatic cells
    genetics, **640–641**
    mutagenesis, 492–493
    nuclear transfer, 655
    proliferation, 124–125

Somatic fusion, 549
Sonneborn, Trace Morton, **641–642**
Sorby's fundus dystrophy, 505
Southern blotting, 93, 520, 593–594
Southern, E. M., 92
Specialized transduction, 681
Species, **642–644,** *643*
    cancer genetics, 110
    Chambers's research, 127
    character displacement, 128
    Chargaff's research, 129–131
    cladistics, 145–146
    Clarke's research, 146
    competition, 157
    Cuénot's research, 169–170
    Darwinism, 178
    Darwin's research, 176–177
    de Buffon's research, 181
    De Vries's research, 181–182
    evolution, 239
    Linnaeus's classification, 432–433
    Mayr's research, 450–451
    mimicry, 471–472
    natural selection, 498–499
    phenotypic variation, 544
    phylogeny, 547
    plants, 108
    polymorphism, 557
    prokaryotic genomes, 567–568
    punctuated equilibrium theory, 576
    quantum speciation, 581
    selection, 614–616
    Sibley's research, 627–628
    sociobiology, 640
    stasigenesis, 651
    Sturtevant's research, 662
    systematics theory, 726
    xenogamy, 732–733
Species Survival Plane (SSP), 162
Specific recombination, 589
Specificity, 382
Spector, Deborah, 87
Speech impairment, 32–33
Spemann, Hans, 336, 363–364, 445–446, **644–645**
Sperm cells, **645**
    fertilization, 250
    heredity, 346–347
    Kossel's research, 418
    mitosis, 475
    Prévost's research, 565
    sexual reproduction, 624–625
    von Nägeli's research, 709
Sperm count, 395–396
Spermatogenesis, 273–274
Spina bifida, 85, 502–503, **645–646**
    congenital birth defects, 161
    prenatal diagnosis, 563–564
Spliceosomes, 604
Split genes, 604–605
Spongiform encephalopathies, 565–566
Spontaneous generation, **646**
    Needham's research, 500
    Oparin's research, 522

Pasteur's research, 536
Pouchet's research, 560
Sporophytic self-incompatibility, 616
SRY gene, **647–648,** 734–735
SSLP (Simple sequence length polymorphism), 141–142
St. Victor, Niepce de, 50
Stabilizing selection, 191–192, 614–616, **648**
Stagard's macular dystrophy, 505
Stahl, Franklin W., 201, 462–463, **648–649**
Stanley, Wendel Meredith, 58, **649–651,** 678
Stasigenesis, **651**
Stationary night blindness, 504–505
Statistical analysis, **651–652**
    See also Bayesian statistics; Probability and statistics
Stebbins, George Ledyard Jr., **652–653**
Stehelin, Dominique, 87
Stein, William Howard, 30–31, 483–484, **653–654**
Stem cells, **656**
    ADA deficiency, 2
    cloning, 151
    ethical issues, 236, 315–318
    gene targeting, 291
    human cloning, 373
    ovum, 529
    pluripotent, **554,** 655, 656
    research, **654–655**
    severe combined immunodeficiency, 620
    sperm cells, 645
    structure, 120–121
Steptoe, Patrick, **657**
Stern, Curt, 501, **658**
Stetson, R. E., 146
Steudel, Hermann, 252
Stevens, Nettie M., 98, **658–659,** 725
STI-571 (Gleevec), 520
Stopes, Marie Charlotte Carmichael, **659–660**
Strasburger, Eduard Adolf, **660**
Streptococcus pneumoniae, 441–442, 682–683
STRs (Short tandem repeats), 314–315
Structural genes, **660–661**
    Monod's research, 483
    mutant sensitivity, 494
    neurological disorders, 503–504
    repeated sequences, 590
    silent mutation, 629–630
STSs (Sequenced tagged sites), 617
Sturtevant, Alfred H., 103, 193, 487, **661–662**
Subcellular genetics, **525**
Substrates, 3
Suicide genes, 286, 290–291
Supergenes, 146–147
Supernumerary chromosomes, **663,** 695
Superoxide dismutase (SOD), 6, 369
Survival of the fittest, **663**
    cancer genetics, 110
    Malthus's research, 445
    selection, 615–616
    sociobiology, 640
    speciation, 642–644
Sutton, Walter Stanborough, 98, 347, 454, **663–664**
SV40 virus, 73–74
Swammerdam, Jan, 227
Sympatric speciation, 642, 644

Synapsis, 399, 457, **664–665**
Systematics, 726
Szostak, Jack William, **665**

T

T cells
    ADA deficiency, 1
    immunogenetics, 385–386
    immunological techniques, 386–387
    major histocompatibility complex, 443
    retroviruses, 597
    transplantation genetics, 685–686
T4 bacteriophage, 649
Tabula rasa hypothesis, 237
Tandem repeats, **667–668**
    chromosome segregation and rearrangement, 143
    DNA fingerprinting, 314–315
    parentage testing, 538
    repeated sequences, 590
    trinucleotide, 687–688
Tangier disease, 300
Tanzi, Rudolph Emile, **668**
Tatum, Edward Lawrie, 67–68, 476–477, **668–670**
    Lederberg's collaboration, 427–429
    one gene-one enzyme theory, 520–521
    protein synthesis research, 571
Taxol, 547
Taxonomy, 27, 432–433
Tay-Sachs disease, 86, 307, **670–671**
Telomerase
    Blackburn's research, 88
    Hayflick's research, 342
    tandem repeats, 667
    telomere synthesis, 671
Telomeres, **671**
    aging, 6
    euchromatin, 237
    human artificial chromosomes, 372–373
    karyotype analysis, 403–404
    Ptashne's collaboration, 576
    yeast artificial chromosomes, 741
Temin, Howard, 63, 87, 212–213, 574, 598–599, **672–673**
Temperate phages, 542
Temperature sensitivity
    inherited variation, 699–700
    molecular biology, 696–697
    mutants, 493–494
    Réaumur's research, 586
Teratology, **673–674**
Testing, genetic. *See* Genetic testing and screening
Tetrad analysis, **674**
Tetrahymena thermophila, 118
TGGE (Thermal gradient gel electrophoresis), **676**
Thalassemia, 29, 510, **674–676**
Thalidomide, 84–85, 218–219
Theory of recapitulation, 521
Thermal gradient gel electrophoresis (TGGE), **676**
Thermus aquaticus, 556
Threonine, 23
Throughput screening, 590–591, **677**
TIGR (The Institute for Genomic Research), 374–375, 702
Tijo, Joe-Hin, 430, **677**

TIMP3 (Tissue inhibitor of metalloproteinase 3), 505
Tiselius, Arne, 542, 570–571
Tissue cultures, 340–341
Tissue inhibitor of metalloproteinase 3 (TIMP3), 505
TMV. *See* Tobacco mosaic virus
TNF (Tumor necrosis factor), 282
Tobacco mosaic virus (TMV), **677–678,** 704–705
    Fraenkel-Conrat's research, 262
    Mayer's research, 449–450
    Stanley's research, 650
Todaro, George, 87
Todd, Alexander Roberts, **678–679**
Tolerance, 493–494
Tonegawa, Susumu, **679–680**
Traits. *See* Characteristics and traits
Transcription, **680–681**
    bases, 66
    behavioral genetics, 69
    Cech's research, 118
    cell proliferation, 125
    chromatin, 137
    codons, 153
    history of genetics, 358
    homeobox research, 364
    hot spots, 369–370
    Kornberg's research, 417
    methylation, 464–465
    mRNA, 463
    operators, 523–524
    palindromes, 532
    phages, 542
    protease inhibitors, 569
    protein synthesis, 571–572
    pseudogenes, 575–576
    Ptashne's research, 576
    regulation genes, 589
    regulatory and sequencing regions, 589–590
    restriction fragments, 43
    reverse, **599**
    ribosomal RNA, 600
    RNA polymerase, 556–557
    same-sense mutation, 609
    serial analysis of gene expression, 620
    structural genes, 661
Transcription factors, 193, 371
Transduction, **681**
    Capecchi's research, 112
    Carlsson's research, 113–114
    eukaryotic cell cycle, genetic regulation, 122
    gene targeting, 291
    mutant sensitivity, 493–494
    prokaryotic genomes, 568
    Wigler's research, 719
    Zinder's research, 744
Transfection, **681–682**
    exogenous DNA, 242
    gene targeting, 290–291
    transgenics, 684
Transfer RNA (tRNA), **682**
    alternate genetic codes, 14
    amber, ocher, and opal mutations, 20
    Benzer's research, 71–72
    Berg's research, 73

codons, 153
function, 16, 602–604
gene suppression, 289–290
history of genetics, 358
Hoagland's research, 360–361
Holley's research, 361–362
Khorana's research, 407
missense mutations, 473
nonsense mutation, 510
nucleic acid research, 511
polymerase, 557
protein synthesis, 572
repeated sequences, 590
ribonuclease, 599–600
ribosomes, 601
Roberts's research, 606
Sanger's research, 611
translation, 684–685
wobble theory, 728
Zamecnik's research, 743
Transformation, **682–683**
cancer genetics, 110–111
cancer research, 87
cloning vectors, 151–152
exogenous DNA, 242
Griffith's research, 331
Hayflick's research, 342
MacLeod's research, 441–442
McCarty's research, 452
Oparin's research, 523
plant genetics, 550
prokaryotic genomes, 568
somatic cell genetics, 641
transgenics, 684
Transgenics, **683–684**
agricultural genetics, 7
cloning, 148–149, 151
Hanahan's research, 337–338
hybridization, 382
in vivo research, 389–390
laboratory mice, 421
Leder's research, 427
plants, 550–551
recombinant DNA, 588
Translation, **684–685**
amino acid, 22–23
bases, 66
behavioral genetics, 69
cell properties, 120
codons, 152–153
Crick's research, 166
enzymes, 573
history of genetics, 358
missense mutations, 472–473
nick translation reaction, 507–508
nonsense mutation, 510
prokaryotic genomes, 567–568
proteins, 571–572, 573
repeated sequences, 590
ribosomal RNA, 600
same-sense mutation, 609
silent mutation, 630
transfer RNA, 682

Translocase enzymes, 685
Translocation, **685**
Angelman syndrome, 32
cancer genetics, 111–112
chaperones, 127
chromosomal abnormalities, 138
chromosome segregation and rearrangement, 143
chromosome structure and morphology, 145
Down syndrome, 210
meiosis, 458
mutations, 494–495
oncogene research, 520
Prader-Willi syndrome, 561
proto-oncogenes, 574
Rowley's research, 608
SRY gene, 648
XX male syndrome, 734–735
Transmissible link encephalopathy, 566
Transplantation
bone marrow transplantation, 1, 656
genetics, **685–686**
Snell's research, 638–639
Transposase, 596
Transposition, **686**
evolution, 43
McClintock's research, 454
prokaryotic genomes, 568
recombination, 589
retroposons, 596
Transposons, **233,** 568, 686
Transversions, **686–687**
Trinucleotide repeats, 465, 667, **687–688**
Triplet DNA coding
Brenner's research, 101–102
Friedrich ataxia, 267–268
transversions, 687
Tris phosphate tris, 21
Trisomy
autosomes, 52
chromosomal abnormalities, 137–138
ploidy, 553
Trisomy 13, 137–138, 536–537
Trisomy 18, 220–222, *221*
Trisomy 21. *See* Down syndrome
tRNA. *See* Transfer RNA
Tryptophan, 23
Tsui, Lap-Chee, 155, *688,* **688–689**
Tumor necrosis factor (TNF), 282
Tumor suppressor genes, **690–691**
breast/ovarian cancer, hereditary, 99–100
cancer genetics, 110–112
eukaryotic cell cycle, genetic regulation, 122
inherited cancers, 397–398
Knudson's research, 412
methylation, 465
proto-oncogenes, 574
testing and screening, 307–308
tumor formation, 689–690
Wigler's research, 719
Tumors, **689–690**
anticancer drugs, 38
eukaryotic cell cycle, 121–122
monoclonal antibodies, 38

retinoblastoma, 594
See also Cancer
Tunneling, 274–276
Turner, J. R., 606
Turner's syndrome, 138, **691**
Twin studies, **691–692**
    aging, 5–6
    alcoholism, 12
    Andreasen, 30
    diabetes, 190
    mental illness, 462
    obesity, 517
    Parkinson's disease, 534
    polygenic inheritance, 556
Twins, **693**
    cloning, 148–150
    microchimerism, 467
    pedigree analysis, 540–541
    Smith-Fineman-Myers syndrome, 638
Two-dimensional electrophoresis, 570
Two-hit cancer hypothesis, 411–412
Twort-d'Hérelle phenomenon, 694
Twort, Frederick William, 61, **694**
Tyron, Robert, 70

U

UBE3A gene, 32–33
UCP2 gene, 517
Ultrasound, 563–564
Ultraviolet light, 528
Uniparental disomy, 32, 560–562, **695**
Uniparental inheritance, 510
UPD. *See* Uniparental disomy
Urey, Harold, 468–469, 523, 528, **696–697**
Urinalysis, 25
U.S. Food and Drug Administration, 126
Usher's syndrome, 182
Uterine disorders, 392–393

V

Vaccination
    antigens, 37
    DNA vaccine, 202, 204
    Pasteur's research, 536
    viruses, 706
Van Beneden, Edouard, 98
Variability, 699–700
Variant Surface Glycoprotein (VSG), 97
Varmus, Harold E., 87–88, **700**
Vauquelin, Louis-Nicolas, 23
Vavilov, Nikolay Ivanovich, 415, 440, **701**
Venter, John Craig, 375, **701–702**
Very large scale integration (VLSI), **702**
Viability, **702–703**
    biological warfare, 80
    chaperones, 127
    conservation genetics, 162
    mutant sensitivity, 493–494
    plant hybridization, 382
    ploidy, 553
Viral genetics, **703–704**

Dulbecco's research, 213
Fraenkel-Conrat's research, 261–262
provirus hypothesis, 574
Prusiner's research, 575
reverse transcription, 599
Rous's research, 606–607
slow viruses, 634–635
Temin's research, 672–673
tobacco mosaic virus, 677–678
Twort's research, 694
Varmus's research, 700
Virchow, Rudolf, 710
Virulence, 615
Viruses, **704–708**, *707*
    acquired character, 2
    cloning vectors, 152
    Lwoff's research, 437–439
    replication, 568–569
    RNA replication, 15
    Rous's research, 606–607
    slow viruses, **634–635**
    somatic cell genetics, 641
    Stanley's research, 649–651
    See also specific viruses by name
Vitamins, 539, 672, 678–679
VLSI (Very large scale integration), **702**
Vogel, Orville A., 96
Von Baer. *See* Baer, Karl Ernst von
Von Mohl, Hugo, **708**
von Nägeli, Carl Wilhelm, 163, 519, 660, 708, **709**
von Tschermak-Seysenegg, Eric, **709–710**
von Waldeyer-Hartz, Heinrich Wilhelm Gottfried, **710**

W

Waardenberg, P. J., 210
Waardenburg's syndrome, 182
Waldeyer's throat ring, 710
Wallace, Alfred Russell, 66, 126, 498, 639, **711**
Wallace, Douglas C., **712**
Wang, James C., **712–713**
Waste control, **713–714**
Wastewater treatment, 714
Watson, James D., 15, 28–29, **714–715**
    Brenner's collaboration, 101–102
    Crick's collaboration, 164–166
    Delbrück's research, 185
    DNA, 194–195, 201, 202
    Franklin's association, 265–267
    Gilbert's association, 323–324
    Human Genome Project, 375
    replication research, 591
    Watson-Crick model, 355–358, *356*
    Wilkins's association, 720–721
Watson-Crick model, **355–358**, 733–734
Weinberg, Robert A., **716–717**
Weinberg, Wilhelm, 408
Weiner, A. S., 146
Weissmann, August, 347, **717**
Wexler, Nancy Sabin, **717–718**
Wiechaus, Eric Francis, 515, **718–719**
Wigler, Michael, **719**

Wild-type traits, **719–720**
    reciprocal crossing, 587
    reversions, 648
    viral genetics, 704
Wilkins, Maurice Hugh Frederick, 165, 195, **720–721**
    Crick's collaboration, 164–166
    Franklin's collaboration, 265–267
    history of genetics, 357
    Watson's collaboration, 714–715
Willard, Huntington F., 732
Williams, J. C. P., 721
Williams, R. C., 651
Williams syndrome, **721–723**
Wilmut, Ian, 149–150, 211, **723–724**
Wilson, Allan C., **724**
Wilson, Edmund Beecher, 98, 355, 640, **724–726**
    McClung's association, 454
    Stevens's association, 659
    Sutton's association, 664
Wilson, Edward Osborne, 128, **726–728**
Witschi, Emil, 524
Wobble theory, **728**
Woese, Carl, 43
Wolff, Gustav, 644
Wollaston, William Hyde, 23
Wollman, Elie, **728–729**
Wollman, Eugene, 438
Wright, Sewall, 559, 714, **729–730**

X

X chromosomes, 621, **731–732**
    abnormalities, 138
    albinism, 10
    chromosome segregation and rearrangement, 142–143
    colorblindness, 155
    dystrophinopathies, 215
    fragile X sites, 262–263
    fragile X syndrome, 263–264
    genetic transmission patterns, 310
    Goldschmidt's research, 325–327
    Haldane's research, 334–335
    hemizygosity, 342–343
    hemophilia, 343–344
    hereditary diseases, 345
    heterochromatin, 351
    inactivation, **732**
    Klinefelter syndrome, 409–410
    Lyon hypothesis, 439
    methylation, 465
    Morgan's research, 485
    mosaicism, 487
    Ohno's research, 518
    recessive genes and traits, 586–587
    reciprocal crossing, 587
    Rett syndrome, 598
    severe combined immunodeficiency, 620
    sex determination, 621–622
    sex-linked traits, 622–623
    single-cell gene disorders, 632–633
    Smith-Fineman-Myers syndrome, 637–638
    SRY gene, 647–648
    Turner's syndrome, 691

    X-linked agammaglobulinemia, 733
    XX male syndrome, 734–735
    XYY syndrome, 735–737
Xenogamy, **732–733**
Xenograft, 685
XLA (X-linked agammaglobulinemia), **733**
X-linked agammaglobulinemia (XLA), **733**
X-ray crystallography, 168–169, **733–734**
    Cowdery's research, 405–406
    DNA research, 195
    Klug's research, 410–411
    mutant sensitivity, 494
    proteomics, 573–574
    replisomes, 592
    ribonuclease, 599–600
X-ray diffraction
    Crick's research, 164
    Franklin's research, 266–267
    Karle's research, 401–402
    Pauling's research, 539
    Watson's research, 715
    Wilkins's research, 720–721
XX male syndrome, **734–735**
XYY karyotype, 245
XYY syndrome, **735–737**, *736*

Y

Y chromosomes, 621, *739*, **739–740**
    archaeogenetics, 45
    chromosomal abnormalities, 138
    extra Y hypothesis, 245
    fragile X syndrome, 263–264
    genetic transmission patterns, 310
    hemizygosity, 342–343
    human genome analysis, 374
    human population genetics, 378
    karyotype analysis, 405
    microchimerism, 467
    mosaicism, 487
    parentage testing, 538
    reciprocal crossing, 587
    sex determination, 621–622
    sex-linked traits, 622–623
    single nucleotide polymorphisms, 632
    SRY gene, 647–648
    structure and morphology, 144–145
    Wallace's research, 712
    X chromosome, 731
    XX male syndrome, 734–735
    XYY syndrome, 735–737
YACs. *See* Yeast artificial chromosomes
Yanofsky, Charles, **740**
Yeast artificial chromosomes (YACs), **740–741**
    cloning vectors, 151–152
    contig genes and mapping, 163
    human genome programs, 375–376
    Human Genome Project, 376–377
    Murray's research, 491
    shotgun method, 627
    Szostak's research, 665
Yeast cells, **740**
    chaperones, 127–128

chromosome mapping and sequencing, 141–142
cloning research, 151
cytoplasmic inheritance, 173–174
enzymes, 228–229
experimental applications, 243–244
Hartwell's research on, 341
Hopkins' research, 368
Hoppe-Selyer's research, 369
locus control region, 434
Mieschner's research, 468
origin recognition complex, 592
Pasteur's research, 535–536
prokaryotic genomes, 567–568
ribonuclease, 599–600
Stanley's research, 650
structure and properties, 119–121
tetrad analysis, 674
Yuan, Robert, 463

Z

Zamecnik, Paul Charles, **743**
Zaug, Arthur, 118

Z-DNA, 195, 202
Zeiss, Carl, 161
Zinder, Norton David, **744**
Zinkernagel, Rolf M., 206–207, **745**
Zuckerkandl, Emile, 479
Zuckerkandl, Émile, **745–746**
Zworykin, Vladimir Kosma, 223–224
Zygote intrafallopian tube transfer, 388–389, 393
Zygotes, **746**
    Brenner's research, 102
    chromosomal abnormalities, 137–139
    chromosome segregation and rearrangement, 142
    chromosome structure, 140
    fertilization, 250
    ploidy, 553
    Roux's research, 607–608
    sex chromosomes, 621
    twins, 693
    viability, 702–703
    Wallace's research, 712
    See also Heterozygotes; Homozygotes